THE RACHEL LAMBERT MELLON COLLECTION

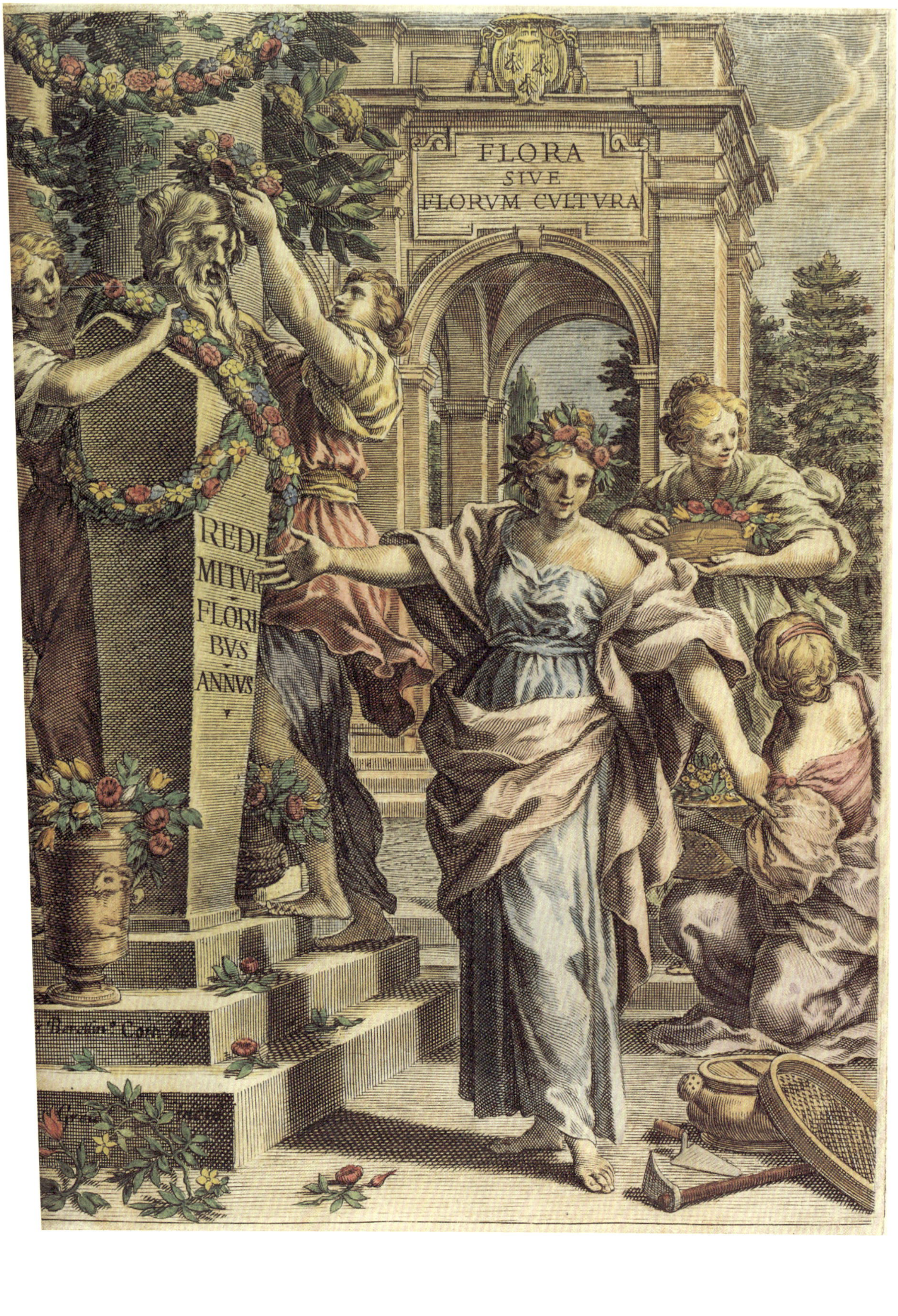
FLORA
SIVE
FLORVM CVLTVRA
REDI
MITVR
FLORI
BVS
ANNVS

AN OAK SPRING
FLORA

FLOWER ILLUSTRATION FROM THE FIFTEENTH CENTURY TO THE PRESENT TIME

A Selection of the Rare Books, Manuscripts and Works of Art in the Collection of Rachel Lambert Mellon

By LUCIA TONGIORGI TOMASI

UPPERVILLE VIRGINIA
OAK SPRING GARDEN LIBRARY
1997

DISTRIBUTED BY YALE UNIVERSITY PRESS, NEW HAVEN & LONDON

PRINTED IN THE UNITED STATES OF AMERICA

LIBRARY OF CONGRESS CATALOG CARD NUMBER 96–71751

ISBN 0–9654508–0–5

Translated
from the Italian by Lisa Chien
with bibliographical descriptions by
Julia Dupuis Blakely

FRONTISPIECE:

GIOVANNI BATTISTA FERRARI, *Flora*, ROME, 1638

FRONTISPIECE PORTRAIT OF FLORA

For my dear and loyal friend

JACQUELINE BOUVIER KENNEDY ONASSIS

(1929–1994)

CONTENTS

I · FLOWERS AND DEVOTION

II · FLORILEGIA: FLOWER BOOKS FOR COLLECTORS AND CONNOISSEURS

III · THE FLORAL STILL-LIFE IN THE FLEMISH AND DUTCH TRADITION

IV · THE FLOWER-GARDEN AND THE RISE OF THE 'JARDINIER FLEURISTE'

V · FLOWERS FOR MONARCHS, VIRTUOSI AND MEN OF SCIENCE

VI · FLOWERS IN THE DECORATIVE ARTS AND MANUFACTURING

VII · TULIPOMANIA AND OTHER FLOWER MANIAS

VIII · EXTRAORDINARY WOMEN PAINTERS

IX · A GENTEEL PASTIME: FLOWER PAINTING FOR DILETTANTES

X · THE ROMANTIC PERIOD: SENTIMENT AND EXTRAVAGANCE

XI · FLOWERS FROM DISTANT LANDS

XII · FLOWER PAINTERS OF OUR TIME

LIST OF ILLUSTRATIONS

FOREWORD

FLORA, the third volume in the Oak Spring Garden Library series of catalogues, begins with my first flower book, a small volume given to me by my grandfather when I was just eleven. It was part of a series written for children as a guide to natural history. Bound in brown cloth, it travelled everywhere with me, a bible in its importance.

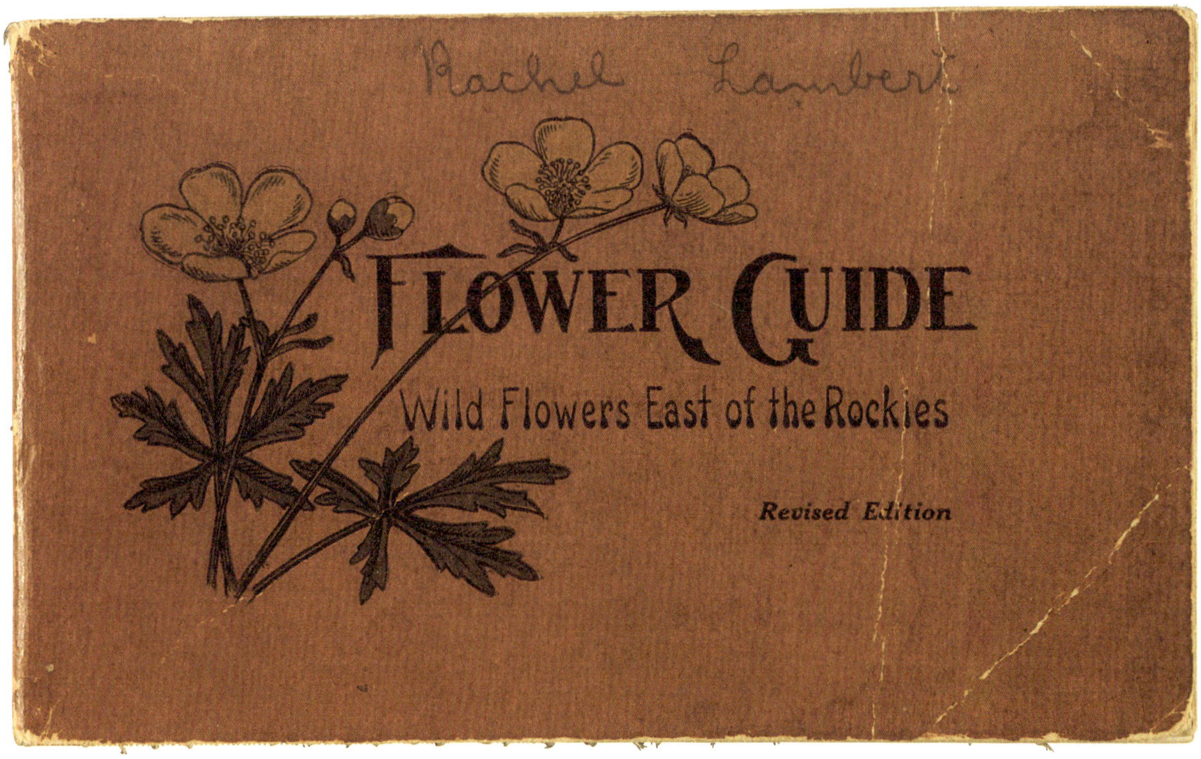

CHESTER A. REED, *Flower Guide: Wild Flowers East of the Rockies*, New York, 1920. Cover with buttercups

Our family house in Princeton was surrounded by open fields that led me into the pleasure of discovery. As a child, wild flowers were part of my feeling of freedom—hidden under larger plants or creating fields of lavender thistles that coloured the landscape like a sea in the wind. The intense bright yellow of buttercups made me think that if ever I had to live alone in a cellar room, I would paint it yellow and never miss the sun.

My curiosity went beyond the family's garden flowers. Flowers planted in broad strokes of single colours were remembered from seeing acres of yellow mustard and blue flax throughout Europe. Flowers are the paintbox of garden design, and they can create a sense of peace and simplicity.

Countless imaginative creations have found their expression in flowers, and the cycle of their life

ROBERT FURBER,
The Flower-Garden Display'd,
London, 1734. Title-page.
The first rare book purchased by Rachel
Lambert Mellon

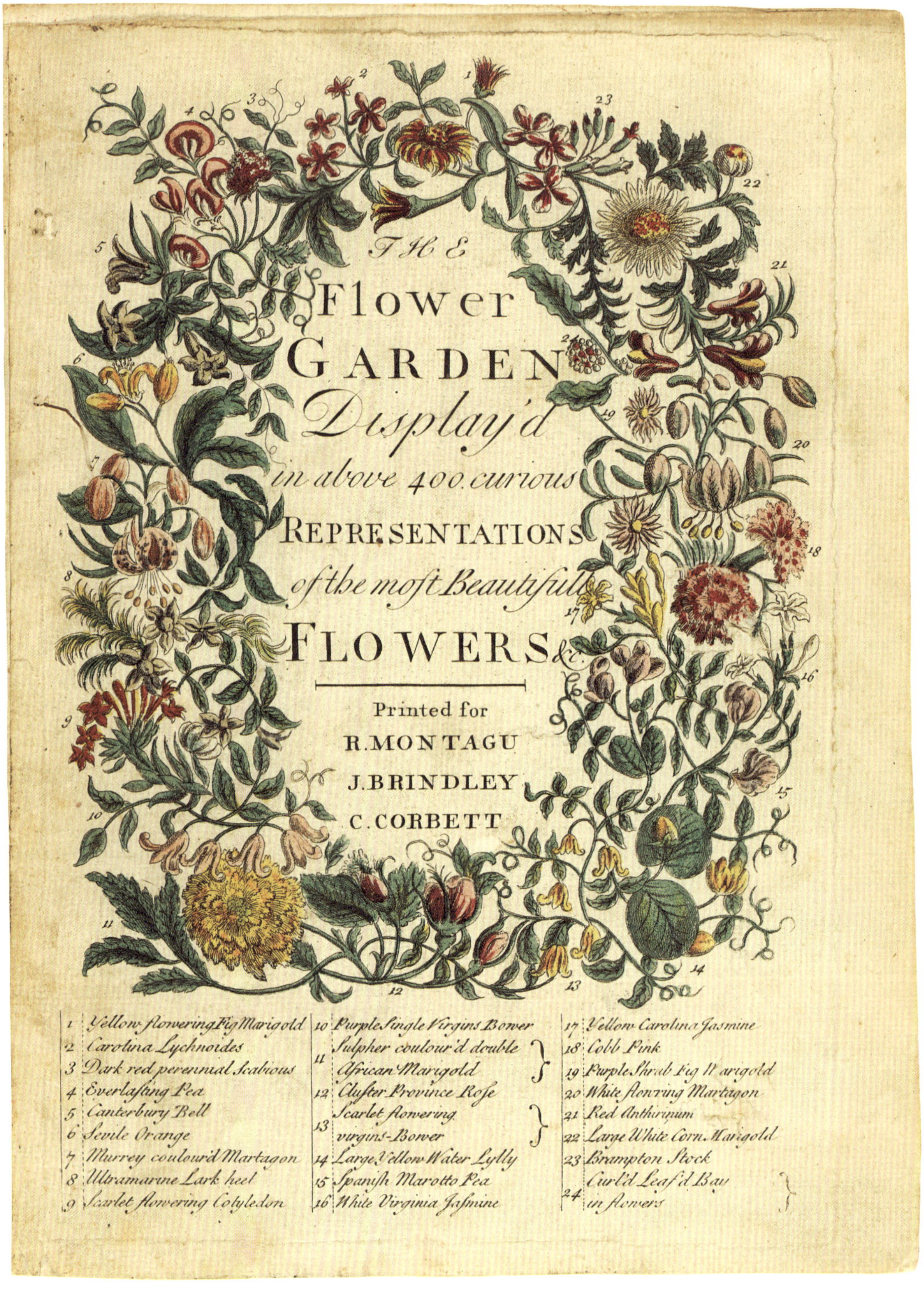
THE
Flower
GARDEN
Display'd
in above 400 curious
REPRESENTATIONS
of the most Beautifull
FLOWERS &c.

Printed for
R. MONTAGU
J. BRINDLEY
C. CORBETT

1 Yellow flowering Fig Marigold
2 Carolina Lychnoides
3 Dark red perennial Scabious
4 Everlasting Pea
5 Canterbury Bell
6 Sevile Orange
7 Murrey coulour'd Martagon
8 Ultramarine Lark heel
9 Scarlet flowering Cotyledon
10 Purple Single Virgins Bower
11 Sulpher coulour'd double African Marigold
12 Cluster Province Rose
13 Scarlet flowering virgins-Bower
14 Large Yellow Water Lylly
15 Spanish Marotto Pea
16 White Virginia Jasmine
17 Yellow Carolina Jasmine
18 Cobb Pink
19 Purple Shrub Fig Marigold
20 White flowring Martagon
21 Red Anthirinum
22 Large White Corn Marigold
23 Brampton Stock
24 Curl'd Leaf'd Bay in flowers

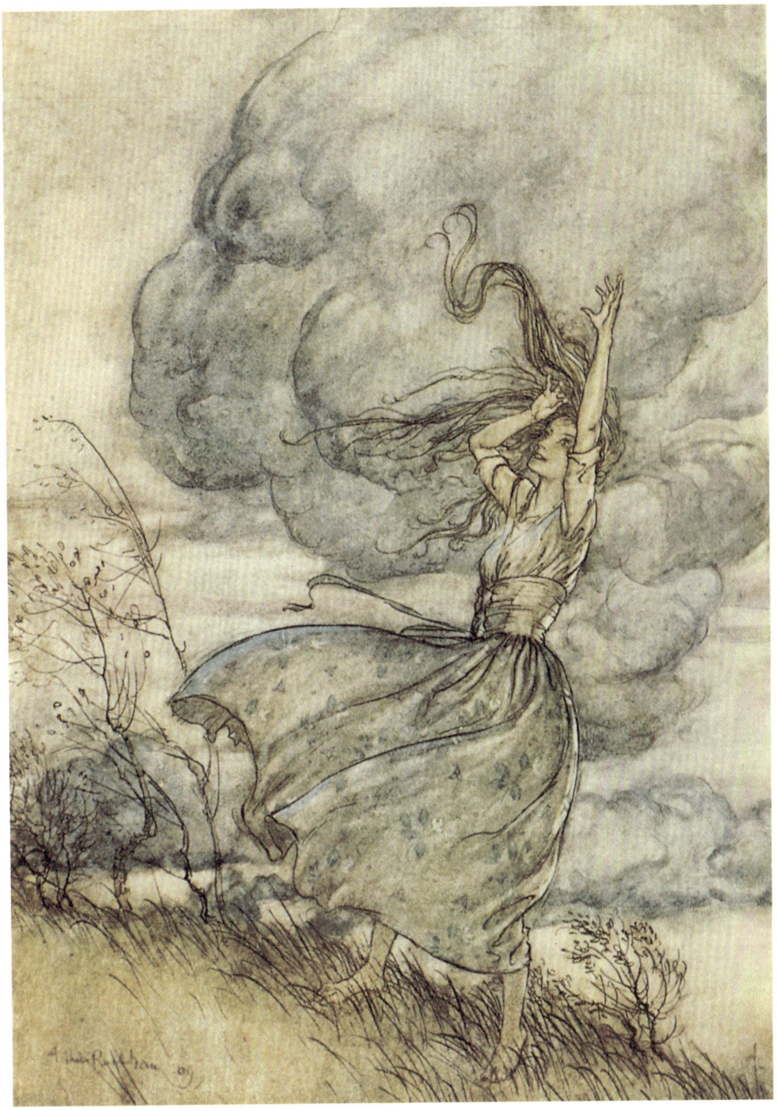

FRIEDRICH DE LA MOTTE FOUQUÉ, *Undine*, adapted from the German by W. L. Courtney, illustrations by Arthur Rackham, London and New York, 1909. Undine during a thunderstorm

has the strength of sensual pleasure with their scent, fruits and seeds. Their presence inspires our tired spirits with their fragile being, and allow our minds to go beyond its earthly limits. Poets and lovers wander into their secret realms, hoping for permission to share a part of their mystery. My collection of books gathered over the years since my childhood was picked one by one for its unique quality and interest, not as part of a long list of well-known flower books. Time and space did not allow for clutter or random acquisition.

Years of collecting brought with it new ideas and new friends that have built a life for me beyond the Garden Library's walls, and opened unexpected doors. A lovely example is my friendship with Lucia Tongiorgi Tomasi, an Italian professor who came to the Library from the University of Pisa. As the author of this volume of flora, with her extraordinary knowledge of flowers and botanical

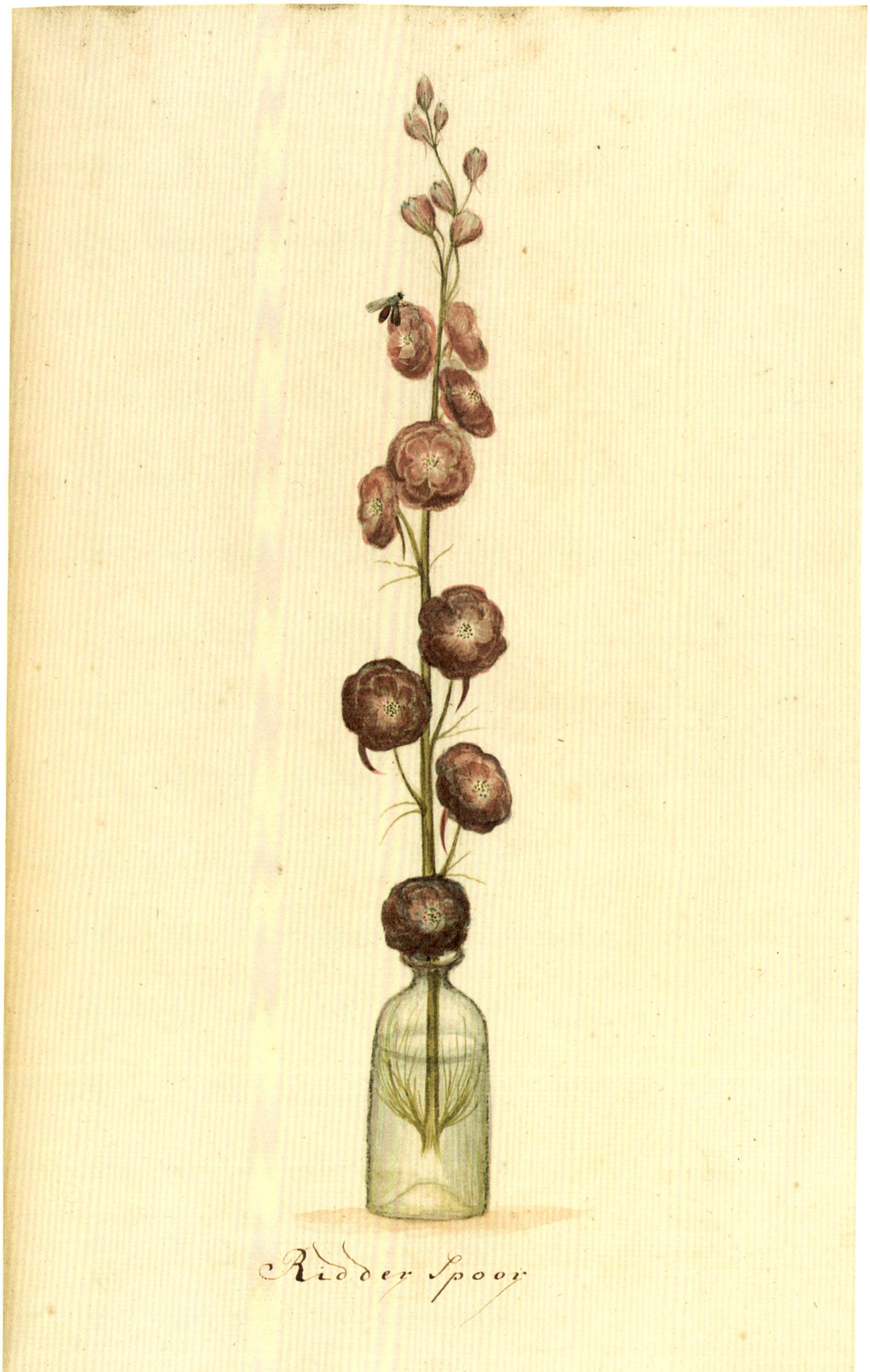

CORNELIUS MARKEE, *Naauwkeurige Verzameling van Europasche Insekten alles naauwkeurig gelchetst en gelchildert*, 1756, volume II. A stalk of flowers in a glass vase, 'Ridder Spoor'

illustration she has helped me not only to describe these books, but as a gentle and understanding friend she has shared many discoveries and insights which bring them alive. I hope this catalogue will bring every reader and gardener the same interest and enthusiasm that Lucia has brought to me.

It is a great pleasure to work again with Mark Argetsinger. His exceptional taste and experience in designing books is combined with the enthusiasm we share working together. Greg Heins, a talented photographer and friend, has again photographed the works of art, books and manuscripts with flawless reality. I thank him for his patience and caring over many years. I am sincerely grateful to Julia Blakely, a rare-books librarian who has brought years of experience with her and is now a permanent part of the Garden Library. Her bibliographical descriptions cover each book included in *Flora*. At the suggestion of William Robertson of The Andrew Mellon Foundation, the Library was fortunate in securing the help and advice of Susanna Tadlock as production manager. She had previously been in charge of The North Point Press in California, and has brought confidence and direction to all our efforts. I am especially pleased and thankful as I find organization the most difficult part of creativity. We appreciate very much Robert Williams, who as copyeditor corrects each manuscript as it arrives on his desk in London. And last, for my two dear collaborators from the beginning—when our books were housed not in a building but in cardboard boxes piled in cellars—Dita Amory and Tony Willis. They have been present in rain or shine with the encouragement and strength to carry the Library from a dream to reality. My arms go around them both.

RACHEL LAMBERT MELLON
Oak Spring
January 1997

RACHEL LAMBERT MELLON, *Nemesia suttonii*, watercolour, 1937

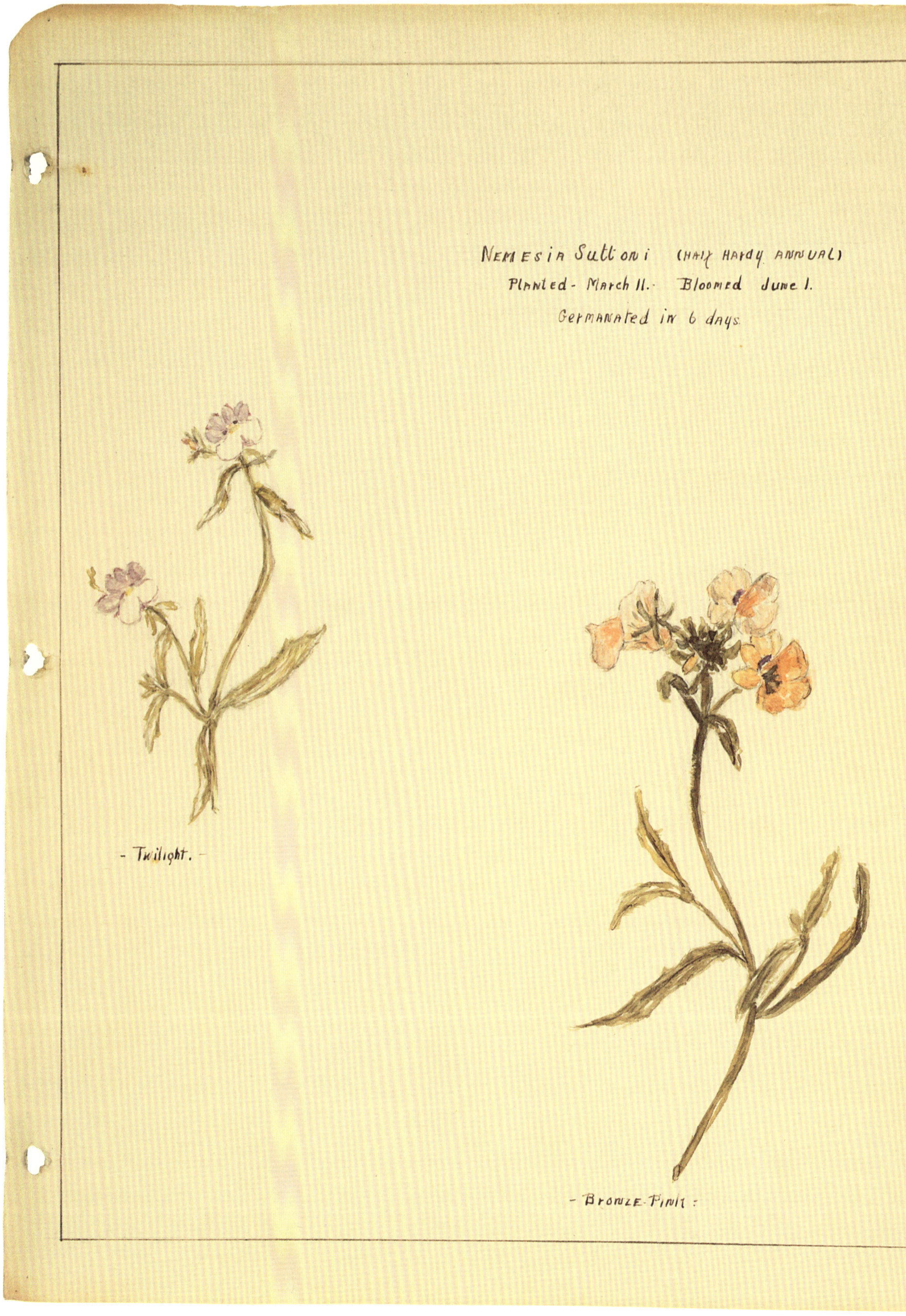

ACKNOWLEDGEMENTS

ONE DAY, during the course of my explorations of the precious material conserved at the Oak Spring Garden Library, I came upon an album notebook whose well-worn covers intimated its loving use over a period of many years. It was the property of the young Rachel Lambert Lloyd, later to become Mrs Paul Mellon, who had kept in it a careful record of her gardening activities, noting the characteristics and prices of the plants she acquired, and recording her observations as she planted and tended them in her garden at Millwood in Virginia. There are hand-drawn garden plans, designs for flowerbeds, lists of gardening tools and instructions for the cultivation of specific flowers, while tucked between its pages are letters, old bills and photographs of gardens and greenhouses—countless details testifying to an interest in gardening that was no mere passing whim, but rather a genuine passion that would deepen with experience and the passage of time.

The notebook opens with two watercolours painted by Rachel Lloyd herself, the first depicting 'a twilight and a bronze pink' and the second 'a shirley poppy and a delphinium'. Above the bronze pink the budding gardener punctiliously noted: 'Nemesia Suttonii (Half Hardy Annual) Planted March 11. Bloomed June 1. Germinated in 6 days'.

This little book helps us to understand the origins of Mrs Mellon's lifelong commitment to gardening and horticulture, and it also, in a certain sense, explains how, many years later, she conceived the idea of the Garden Library at Oak Spring.

The two watercolour sketches hint that even in her youth Mrs Mellon was already aware of the importance of keeping a visual record of the plants under her care, as if she had intuitively grasped one of the main principles laid down in the mid-sixteenth century by the fathers of modern botany. Later, as her knowledge of the subject grew and deepened, she came to realize that for botanists and gardeners in centuries past, as for herself in the present, sound theory and sound practice must always go hand in hand. Thus Mrs Mellon began to search for manuscripts and rare books on gardening and horticultural subjects documenting the activities of botanists and virtuosi down through the centuries. Gradually she amassed a unique collection of manuscripts, printed books, botanical prints and paintings. At this point an irresistible comparison comes to mind, for we may recall that one of the favourite metaphors of naturalists in the sixteenth and seventeenth centuries was that of the 'great book of Nature', expressing the utopian idea that the total sum of man's knowledge of the natural world could be contained within the pages of a single volume, a vast encyclopaedia describing every aspect of the wondrous universe.

To house her valuable books and manuscripts, so eagerly sought out and lovingly preserved in a period when such works were of interest to but a few specialists, Mrs Mellon conceived the idea of a library—but a somewhat anomalous one. In what way anomalous? Fortified by her conviction that books on gardens and botany could be fully appreciated only if studied in a setting that permitted a close and direct contact with nature itself, Mrs Mellon conceived of a library immersed in its own vast landscape. The great windows running along the walls of the austere white building that resulted allow a free interchange between the outside and the inside, between nature as described and depicted within the pages of Mrs Mellon's books, and nature itself, spreading beyond the windows as far as the eye can see. During the many happy hours that I spent examining illuminated manuscripts, prints and paintings at my table in the Oak Spring Garden Library, I would often look out beyond these windows to study the landscape in which I too was immersed. Returning, gaze refreshed, to the works of art lying before me I would find that they had acquired a new significance, almost new life, in my eyes.

It was certainly not an easy task to choose from the vast collection of manuscripts, books, prints and paintings conserved at Oak Spring a single series of works that could best illustrate the development of the floral image in art over the centuries. In the first place, because every attempt at selection required the sacrifice of many other works of equal beauty and importance. But second, because the floral image itself often poses a perplexing dilemma, having throughout its history borne a double significance, almost a divided identity. As Martyn Rix has observed in his catalogue on Rory McEwen: 'The distinctions between flower painting and botanical illustration are of purpose and intent, rather than of execution. At times the two have been separate, while at other times the same painter has been master of both botanical illustration and the art of painting flowers'. Floral images have, in fact, often been studied in a wholly dichotomous manner, with art historians analysing them as 'works of art' and botanists, floriculturists and scholars of the history of science scrutinizing them as scientific documents whose authors' sole purpose had been to capture as accurately as possible the botanical characteristics of a species rather than the unique beauty of the individual specimen. Only recently has this divided conception of the botanical illustration undergone a welcome revision, finally allowing us to appreciate the genre of flower painting on its own merits. We also realize now that during the course of history, the 'illustration' and the 'image'—in themselves terms difficult to pin down and define—have often influenced one another, co-mingling, interweaving, and sometimes fusing in a vital and ever-changing symbiosis.

In *An Oak Spring Flora* I can only touch on some of the complex and fascinating problems posed by the genre of flower painting; nevertheless, I have attempted to tease out a single bright 'iconographic thread' to follow in my meanderings through the brilliant forest of floral images that has been produced by artists down through the centuries. Furthermore, rather than presenting the works in *Flora* in simple chronological fashion, I have sought to focus on certain important themes that reflect key moments in the evolution of the floral image. I have also included a few works normally placed in other categories altogether, such as the 'herbal' or the 'flora', because their images seem to me to embody the very essence of flower painting.

An Oak Spring Flora would never have been realized without the generous help of many persons. Of great importance to me were my preliminary discussions with John Dixon Hunt, author of *An Oak Spring Hortus*, who helped me to clarify my own ideas regarding certain conceptual issues, and to decide on the best approach to *Flora*.

I owe a profound debt to numerous friends, colleagues and distinguished scholars for their valuable suggestions, advice, and in many cases their practical assistance. I would particularly like to thank Lina Bolzoni, Shelby Caison, Beverly Carter, Luigi A. Cassuolo, Roberto Paolo Ciardi, Lee Kimball Clark, Carolyn Colbert, Linda Cook, David Corse, Gigetta Dalli Regoli, Carlo Del Prete, Ray Desmond, Donata Devoti, Anita Engel, Giancarlo Fasano, Francesca Fedi, Sylvia FitzGerald, John Flanagan, Carla C. de Glopper, Eugene Howard, Jay Keys, Theo H. Lunsingh Scheurleer, Oliver Murray, Patrick Noon, Giuseppe Olmi, Michon Ornellas, Jerry Pauley, Duncan Robinson, Franco Sabatelli, Francesco Solinas, Alfredo Stussi, Susanna Tadlock, Dorothy Tines, Grazia Tomasi, Michelle Tompkins, Alessandro Tosi, Liesbeth J. van Soest, Bruce Whiton, Krystyna Wasserman, Kees Watermann, and Patrizia Zambrano.

Paolo Tongiorgi, my husband, identified the many insects that flit or scurry through the paintings and engravings reproduced in *Flora*. And as always, he read with great patience many sections of my work, and was generous with his advice. My thanks also to Dita Amory and John Baskett, both of whom read portions of the text.

I would also like to extend my sincere thanks to Joachim Wolschke-Bulmahn, former Director of Studies in Landscape Architecture at Dumbarton Oaks, in whose library I spent many hours, helped by all the librarians, in particular Annie Thacher.

My most affectionate thanks go to Else Terwen Dionisius, who was of constant and invaluable help to me as I investigated various problems relating to the history of art in the Low Countries.

Fabio Garbari was of inestimable help to me in resolving many of the botanical questions that came up during the course of my research, tracking down plant names and identifying species on the basis of the slender clues I sent him with the same insatiable curiosity and enthusiasm for the pursuit of knowledge that characterize his own research.

I am also grateful to Lisa Chien, who skilfully translated the Italian text of *Flora*, and always managed to capture the tenor of my thoughts; to Robert Williams who took on the task of editing this work with great sensitivity and attention to detail; to Greg Heins, responsible for the splendid photographs, which do full justice to the originals; and to Mark Argetsinger, who completed the design and layout of *Flora* with the refinement and intelligence that distinguishes all his work.

I would also like to express my heartfelt gratitude and thanks to my friends Julia Blakely and Tony Willis, respectively librarian and assistant librarian at Oak Spring. Julia, with her great expertise of antique books and manuscripts, and Tony, with his intimate knowledge of the many secrets and hidden delights of the Library, helped me with unfailing generosity, efficiency and good humour during my extended visits to Oak Spring over three long years. *An Oak Spring Flora* owes much to their untiring efforts.

To Mr Paul Mellon, who allowed me to study, and granted permission to reproduce in *Flora*, several works in his extraordinary collection, in particular one flower painting by John Constable and one by Henri Fantin-Latour, I extend my deepest thanks.

In closing, I would like to express my most profound gratitude to Rachel Lambert Mellon. Our long and stimulating conversations on the iconographic problems relating to floral painting, and above all the delightful work that we shared when choosing the images to be included in *An Oak Spring Flora*, with the lively satisfaction that we derived from a particularly significant discovery (often owing to her sensitive and attentive eye rather than to my own) have constituted important moments in my life as a scholar. To Mrs Mellon my sincere thanks, and friendship, which I dare to hope is returned.

LUCIA TONGIORGI TOMASI

INTRODUCTION
The Iconography of Flowers

THE term *flora*, which today is used to refer to the entire plant kingdom, was originally the name of an ancient Italic goddess of fecundity who had the power to bring all green things into bloom. Not only in the Western world was this phenomenon associated with divinity; the Aztecs, for example, had their own fertility goddess, Xochiquetzal, who could be recognized by the floral garlands she wore in her hair and carried in her hands. In early Roman times, from late April into the first days of May a festival of unbridled licentiousness—the *Floralia*—was held in honour of Flora, during which courtesans, scantily dressed and crowned with garlands, ran through the streets of Rome scattering flowers as they passed. The figure of *Flora meretrix* (Flora the Prostitute) lingered on in the imagination of men, and references to her can be found in the satires of Juvenal, in the writings of some of the early Church Fathers (Lactantius and Minucius Felix), and in the works of numerous poets, for example Giovanni Boccaccio's *De claris mulieribus* of 1362 or François Villon's *Ballade des dames du temps jadis* of 1462 (see Held in Further Reading).

Gradually, however, the figure of Flora underwent a process of improving moralization and her original attribute as prostitute was transformed. In his *Fasti* (v.193–222), the Roman poet Ovid linked Flora to a figure from Greek mythology, the innocent earth nymph Chloris (the 'Verdant'), who was kidnapped by Zephyr, the West wind and bringer of spring flowers. Following their marriage, Flora was designated to reign henceforth over all the flowers of the garden and the cultivated fields. Generally she was portrayed as a maiden clothed in diaphanous, fluttering robes woven with a pattern of flowers, carrying flowers in her arms which she scattered wherever she went, while sweetly perfumed blooms fell from her hair at every movement. Since each season was ruled by a specific wind and Zephyr was the wind of spring, Flora herself gradually became identified with Spring and with one of the allegorical figures in the cycle of the Five Senses—that of the sense of smell (Tongiorgi Tomasi and Tosi, p. 69). On occasion, erotic connotations—echoes from her fallen past—returned, lending her the glorious face of Venus (Dempsey, p. 260; *Il Giardino di Flora*, p. 18).

The story of the kidnapping of Chloris and of her metamorphosis from maiden into woman ('Chloris eram quae Flora vocor': I once was Chloris who am now called Flora) is presented, in a Neoplatonic vein, by Sandro Botticelli in his celebrated painting *La Primavera* in the Uffizi Gallery, Florence (Warburg; Wind; Levi D'Ancona 1983, pp. 25–67). Renaissance artists were fascinated by the ambiguous connotations of the goddess *ab antiquo*, and she was painted in many guises. A lost prototype by Leonardo da Vinci is known by the many copies his followers made (these include a *Flora* at Hampton Court Palace by Bernardino Luini, one at Basildon Park, Berkshire, attributed to Giampietrino, and the *Colombina* in the Hermitage, St Petersburg, thought to be by Francesco Melzi); each one is a half-length portrait of a beautiful young woman, usually with one breast exposed, offering a bouquet of flowers. Jacomo Negretti Palma Vecchio (National Gallery, London) and Titian (Uffizi, Florence) depicted Flora in a similar pose, while an artist from the school of Tintoretto used the figure as a model for his *Venetian Courtesan* (Prado, Madrid; see Held). The sixteenth-century Flemish painter Jan Massys executed two pendant pictures, one the portrait of a reclining nude, *Flora–Venus*, with the city of Genoa in the background (Nationalmuseum, Stockholm), and another, *Flora*, with uncovered breast and the city of Antwerp in the background (Kunsthalle, Hamburg). Each of these figures, shown holding a nosegay of flowers, is charged with erotic symbolism.

Recalling the gentler, more innocent, aspect of Flora, the Huguenot artist Jacques Le Moyne de Morgues depicted *A Young Daughter of the Picts* (the Picts were the ancient inhabitants of northern Britain, afterwards Scotland) as a beautiful virgin warrior clad in a gossamer veil decorated with flowers, reminiscent of Botticelli's *Primavera* (*English Drawings and Watercolours*, pl. 1). Le Moyne's painting in bodycolour on vellum is preserved at the Yale Center for British Art, New Haven (Paul Mellon Collection). This work was engraved by Theodor de Bry and used in *Admiranda narratio, fida tamen, de commodis et incolarum ritibus Virginiae* (Frankfurt, 1590) to illustrate the supposed similarity in physiognomy between the ancient Britons and the natives of Virginia.

The disastrous phenomenon of 'tulipomania'—the uncontrolled speculation in tulip bulbs that swept over the Low Countries in 1634–7, bringing with it financial ruin for thousands of investors great and small—resulted in a curious but distinct revival of the negative persona of Flora. In many admonitory pamphlets from this period she was described as a seductive whore who led men astray. She is portrayed thus in a well-known print by Crispijn van de Passe the younger, *Floraes Mallewaagen alias het valete der Bloemisten* (The Fools' Chariot, or the decline of the florists), while in another print, *Floraes Gecks-Kap*, she is shown fleeing the wrath of the townspeople, ignominiously mounted on a mule.

Flora in a more beneficent, often classicized, guise appeared in paintings, prints and book illustrations throughout the seventeenth century, when an interest in flowers and their allegorical associations pervaded every aspect of European culture. The eccentric artist Arcimboldo painted a remarkable portrait of Flora (*c.* 1591; private collection) composed entirely of flowers; in an ingenious synthesis of human and floral elements her cheeks are formed of full-blown roses, her eyelids of translucent petals, and her lips from two flower buds. Another version of Flora was engraved by Crispijn van de Passe the younger for his *Hortus Floridus*, the most popular florilegium of the seventeenth century. Here she is shown as a munificent figure holding a large cornucopia filled with flowers—an attribute long associated with Zephyr. In another superb florilegium, *Theatrum florae*

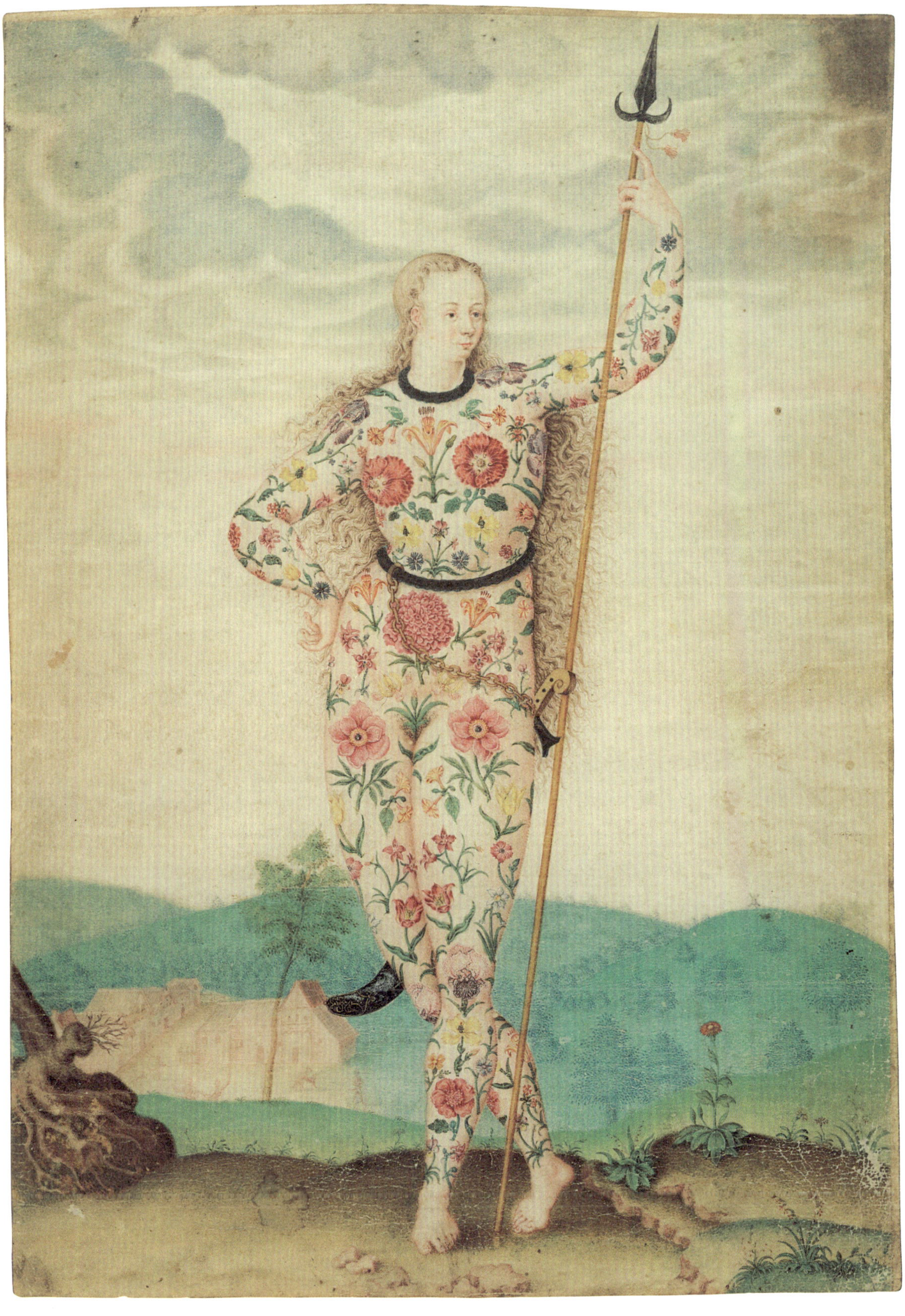

JACQUES LE MOYNE DE MORGUES, *A Young Daughter of the Picts*, watercolour and body-colour on vellum. Gift from Paul Mellon to the Yale Center for British Art, New Haven, Connecticut by the French artist Daniel Rabel—which presents the reader with the most beautiful flowers 'ab ipsus Deae sinu proferuntur' (That are produced from the breast of this goddess)—Flora appears in an elegant frontispiece opposite Vertumnus, while between them extends the vast prospect of a formal garden. The goddess is shown here in a more informal attitude, seated on a low wall holding a rake in one hand, but the traditional attributes of the floral crown and the mantle decorated with flowers have not been forgotten.

A monumental figure of Flora, shown seated on a wall while below her Ceres and Pomona offer tributes, dominates the title-page of John Rea's *Flora, seu, de florum cultura* (London, 1665). Shortly after, in 1668, the Jesuit naturalist Filippo Arena portrayed Flora, in a reversal of roles, offering gifts to Pomona and the newly invented allegorical figures of Botany and Physics. In the frontispiece to Jean Paul de Rome d'Ardène's *Traité des renoncules* (1746), an elegantly attired Flora can be seen wandering through a garden carrying a basket of ranunculi (one of the most popular flowers of the seventeenth century). Thus in Italy, particularly at the beginning of the seventeenth century, when a cultivated circle of intellectuals—many of whom were passionately interested in flowers and horticulture—became attached to the papal court of the Barberini, the figure of Flora was endowed with an ever more vast aureole of learned associations. In this period wealthy collectors vied with one another to obtain exotic flowers for their gardens, and to commission flower paintings for their art collections. As a result numerous *fioranti*, artists specializing in floral themes, established studios in Rome. The most skilled of these earned appropriate soubriquets, for example Mario Nuzzi was 'Mario dei Fiori', while Nicolas Guillaume Delafleur of Lorraine was universally known as 'Monsù Fiore'. In 1638 Monsù Fiore prepared the plates for a rare collection of floral engravings that were printed in Paris in the same year by Pierre Mariette.

The French 'painter–philosopher' Nicolas Poussin executed two imposing works on the theme of Flora during his long sojourn in Rome. Taking Ovid's *Metamorphoses* as his inspiration, in *The Triumph of Flora* (*c.* 1627; Louvre, Paris) he shows the carriage of Flora surrounded by a throng of ill-fated lovers from classical mythology—Narcissus, Hyacinth, Adonis, Ajax, Clytie and Smilax, all of whom were transformed into flowers by the pitying gods. In *The Realm of Flora* (1631; Gemäldegalerie, Dresden) Poussin depicts these same figures, together with Smilax's lover, Crocus, at the very moment of their metamorphosis, setting the scene in a garden presided over by Flora and (in the form of a herm) the fertility god Priapus. This painting was inspired by a verse from *Rime* (1602) by the Neopolitan poet Giovanni Battista Marino, a friend and patron of the artist.

The figure of Flora also dominates the frontispiece of one of the most important books published in Rome in the seventeenth century, *De florum cultura* by the Jesuit Giovanni Battista Ferrari, a work so successful that it was soon translated from the original Latin into Italian as *Flora overo cultura di fiori*. Ferrari was a close friend of Poussin, and his lavishly illustrated *Flora* was published in Rome just two years after the French artist had completed *The Realm of Flora*. The frontispiece to Ferrari's work was designed by Pietro Berrettini (Pietro da Cortona), another important artist associated with the papal court, and etched by the German artist Matthäus Greuter. The process of moralization that had already altered the goddess almost beyond recognition was here carried a step further. In his introduction, addressed to the reader and friend of flowers ('Al lettore amico dei fiori'), Ferrari alludes to 'a finally rediscovered chaste Flora, who does not contaminate customs, but rather sows her flowers in the souls of men with even greater skill than in the earth'. Surrounded by a retinue of four graceful nymphs symbolizing the seasons of the year rather than by Ovid's unhappy lovers, Flora indicates with one hand the gardening tools that 'invite one to a pleasing labour' and with the other a herm of Janus, inventor of garlands, symbol of the new year and of the passage of time. This herm was introduced to replace the more erotic one of Priapus, herms of whom were traditionally set up in gardens to serve as a kind of scarecrow. The herm of Janus is here decorated with flowers and the inscription 'Redimitur Floribus Annus' (The year returns with its flowers), an image that reappears in the frontispiece to *Hortus Cliffortianus* by Linnaeus in 1737. The background of Ferrari's frontispiece prominently displays the coat of arms, with its three bees, of the Barberini family, and beyond it one glimpses the vista of an elegant garden.

In 1634 Rembrandt portrayed his young wife Saskia as Flora (Hermitage, St Petersburg), but it was only towards the end of the seventeenth century that many aristocratic ladies—particularly in France, where *manières galantes* were much cultivated—chose to have themselves portrayed in the guise of that goddess. In the Musée Granet in Aix-en-Provence is a charming portrait of Mme Gueidan as Flora by Nicolas de Largillière, while in the Musée de Versailles can be found a similar portrait of Mme Henriette, the beloved daughter of Louis XV, painted by Jean-Marc Nattier. The image of Flora herself, beautiful and infinitely mutable, continued to appear throughout the eighteenth century in various symbolic trappings. In the title-page designed by the Romantic artist Johann Heinrich Füssli for Part II of Erasmus Darwin's poem *The Botanic Garden* (1789), the goddess appears in a languid pose holding Cupid's bow and arrow, this being a learned reference to the Linnaean concept of 'the loves of the plants'. In *The Temple of Flora*, a remarkable work by the physician Robert John Thornton, there are two engravings of Flora. In the first she is shown flying over the earth scattering her gifts, while in the second she appears in the company of Asclepius, Ceres and Cupid, who have gathered together to honour the bust of the great Swedish botanist Linnaeus.

It was apparently only after the remarkable success of Ferrari's *Flora overo cultura di fiori* that the word 'flora' came to be used as a term to refer to the entire kingdom of plants and flowers (Migliorini, p. 49). The German botanist Johannes Loeselius (active in the first half of the seventeenth century) adopted the term to designate those plants native to a specific geographical region. Loeselius's *Flora Prussica* was published posthumously in Königsberg in 1703, while Linnaeus published his *Flora Lapponica* in 1737.

The flower has always been esteemed as one of the most precious gifts of a bountiful nature. With its infinite variety of form, colour and perfume and because of its presage of a sweet and abundant harvest to come, in every society it has been used as a votive offering and as a love token. In almost all languages the word for flower is also used to denote the quality of youth, or the precious essence of a substance (Goody, p. 5), and while its symbolism may vary widely from culture to culture, its associations have always been positive, linked to such concepts as the vital principle, the joy of living, the victory of life over death, the end of winter. In the Hindu mythology of India, Brahma is believed to have emerged from a lotus-flower. Analogously, in the cosmology of the ancient Egyptians the world was thought to have been born when the omnipotent sun-god emerged from this same flower. In the Bible, Paradise is described as a garden filled with trees and flowering plants, while flowers themselves appear frequently as a sign of divine goodness, as in the miraculous flowering staffs of Joseph and Aaron. In a more melancholy vein, the flowers of the field are used in the Psalms as a symbol of the ephemerality of life, a symbol that would be borrowed by poets and artists in later ages, providing an inexhaustible source of metaphors and poetic inspiration (Borchardt, p. 23).

The flower has naturally been a favourite motif for artists since earliest times. The celebrated Kamáres-style pottery of Crete (1900–1700 B.C.) was often decorated with plant motifs, at first highly stylized, then later of a remarkable realism. Indeed, through the ages the image of the flower has oscillated between realism and stylization, modulated by the culture, taste and personality of the artists who have been called on to depict them. And although *An Oak Spring Flora* chiefly focuses on examples drawn from Western art, it must not be forgotten that the floral motif has also played a central role from earliest times in Eastern art from Persia to India, China to Japan (Hulton and Smith, pp. 52–112).

In the classical period flowers were widely used in the fashioning of wreaths and in the manufacture of perfumed essences, but were not very often portrayed in art. Nevertheless, Pliny the elder recounts the legend—later the subject of a famous work painted by Peter Paul Rubens in collaboration with Jan Brueghel the elder (Ringling Museum of Art, Sarasota)—of the Greek artist Pausias, who depicted a splendid bouquet of flowers for his beloved Glycera, the Garland-maker. Later, the fresco decorations of Pompeii would include verdant landscapes spangled with flowering plants, while a gifted artist in the sixth century A.D. would paint images of many flowering plants in the earliest illuminated manuscript, the *Codex Vindobonensis*, now in Vienna (Nationalbibliothek).

During the Middle Ages the floral motif gained new dignity as specific symbolic meanings came to be associated with particular flowers, based on metaphors drawn from the Bible and other sacred texts and on a rigorously Christian reinterpretation of the classical authors. The primarily symbolic reality of these plants was reflected in the manuscript herbals of the period, whose texts might contain careful descriptions of their supposed medicinal properties but whose schematic illustrations made little effort at accurate representation. The garden meanwhile, once the symbol of Eden, now flourished behind the high walls of the convent and was dedicated to the cultivation of simples (medicinal plants), aromatic herbs and flowering plants whose essences could be used in the preparation of soothing syrups and distilled spirits. A handful of flowers ennobled by Christian symbolism were also cultivated, such as the white rose, emblem of purity and divine love, and the red rose of Christ's Passion. Both of these flowers were also linked to the Virgin Mary and the Christ-Child, while the lily and the iris were symbols of virginity, and thus of the Annunciation. In addition, in this period of devout and simple piety, many plants and flowers were named after saints, and they were punctiliously listed by the botanist Johann Bauhin in his *De plantis a divis sanctisve nomen habentibus*, published in Basel in 1591. Gradually, common wild flowers also began to steal into the illuminated manuscripts, altarpieces and frescoes of the Middle Ages. In Hubert and Jan van Eyck's magisterial altarpiece (completed 1432) at St Bavo, Ghent, the numerous plants that fill the sunny meadow of the glorious scene to be found in the central lower panel (the *Adoration of the Mystic Lamb*) present us with a veritable medieval herbal (*Fleurs et jardins*, p. 13).

From the late Middle Ages lay persons too began to plant gardens and to take an interest—both scientific and aesthetic—in plants and flowers. It is therefore not surprising that the beautifully illuminated Breviaries and Books of Hours produced for the wealthy laity contained in their border decorations not only the Christian rose but also garden flowers such as the columbine, violet, daisy and strawberry plant. The artists of the Ghent-Bruges School in particular depicted these flowers in their manuscripts in a highly naturalistic manner, as if the flowers had just been plucked from the ground and captured *à trompe l'œil*. This intimate rapport with the natural world was also reflected in the famous *millefleurs* designs of tapestries made in France and the Low Countries. Plants and flowers were no longer considered mere decorative elements, but as subjects in themselves that were worthy of the artist's full attention.

The art of botanical painting was transformed during the Renaissance by Leonardo da Vinci and Albrecht Dürer, whose perception of the natural world combined the aesthetic sensibility of the artist with the analytical eye of the naturalist. Although many of their greatest works, such as Leonardo's *Virgin of the Rocks* and Dürer's *Madonna with an Iris*, contain floral motifs, both artists also painted masterly studies of single flowers. Among the 'molti fiori ritratti al naturale' (Many flowers portrayed from nature) by the young Leonardo is a celebrated pen and ink drawing of a *Lilium candidum* (Royal Library, Windsor Castle) whose leaves, curving bracts, and flowers have a precise, almost palpable reality. Dürer's life-size watercolour of an *Iris* (Kunsthalle Kupferstichkabinett, Bremen) and his *Large Piece of Turf* (Albertina, Vienna)—a veritable microcosm, in which every stone and blade of grass has been carefully depicted—represented precocious examples of the art of botanical painting (as well as the mixed media of watercolour and bodycolour) that many artists were inspired to emulate or even to copy down to the last detail (Koreny, pp. 176–226).

During the Renaissance floral motifs with impeccable classical associations—the acanthus, the floral wreath, garlands laden with fruit and flowers—were also resurrected and inserted, in a modernized and more naturalistic form, as arabesques in paintings and sculpture on classical and modern historical themes. The figures in Renaissance portraiture were sometimes depicted with a flower or

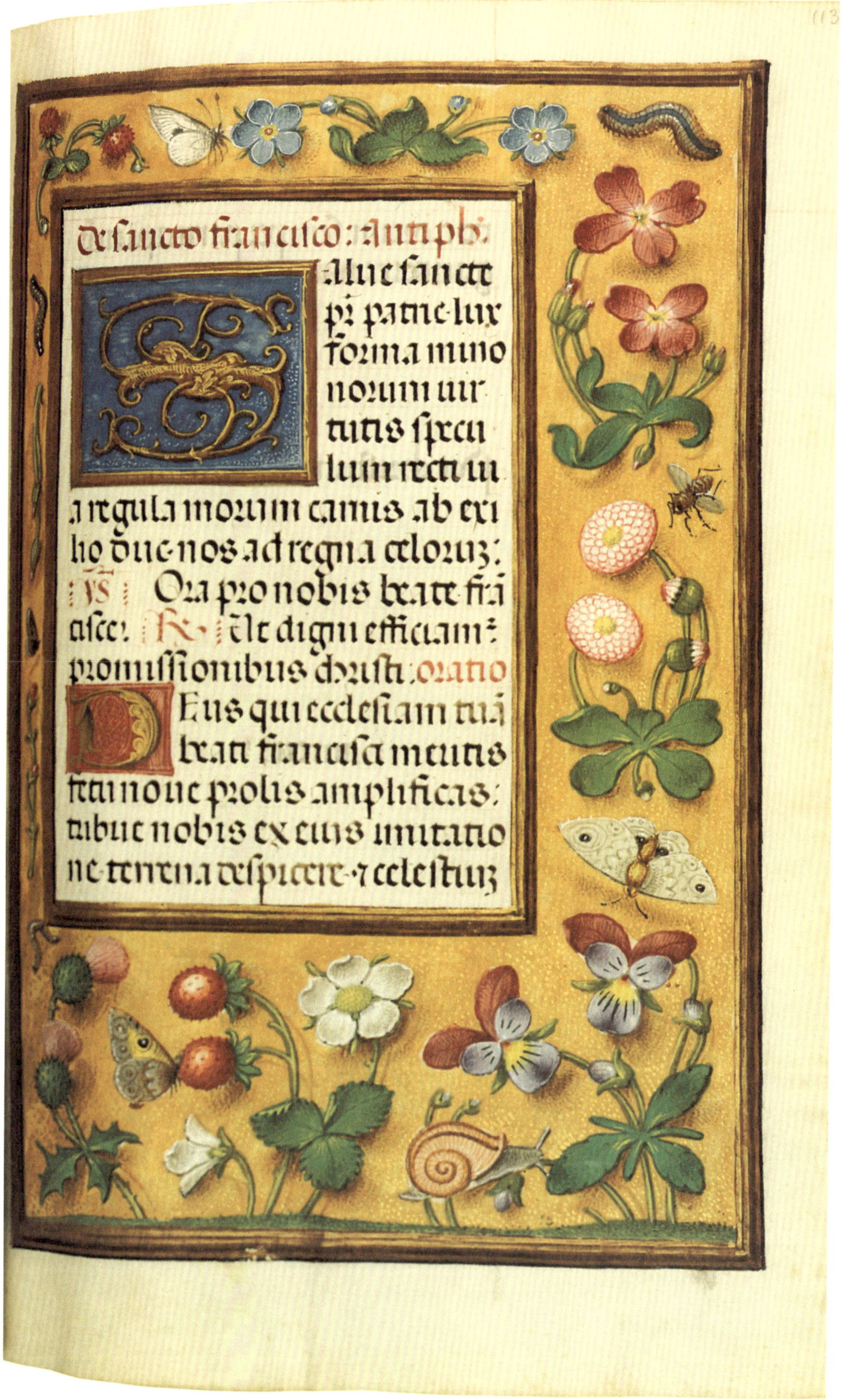

SIMON BENING, *Horae beatae Mariae Virginis ad usum Romanum.* Folio 113r, floral border with insects a small nosegay in one hand, as in the *Benois Madonna* (Hermitage, St Petersburg) by Leonardo or in Joos de Cleve's *Emperor Maximilian holding Three Carnations* (Musée Jacquemart-André, Paris). These flowers often carried an emblematic meaning: the carnation, for example, as a symbol of love appeared frequently in the portraits of betrothed or young wedded couples. In the sixteenth century it became common practice to include a portrait of the author on the title-page of a botanical text, showing him absorbed in the task of identification, with a plant in one hand and a book in the other (Tosi).

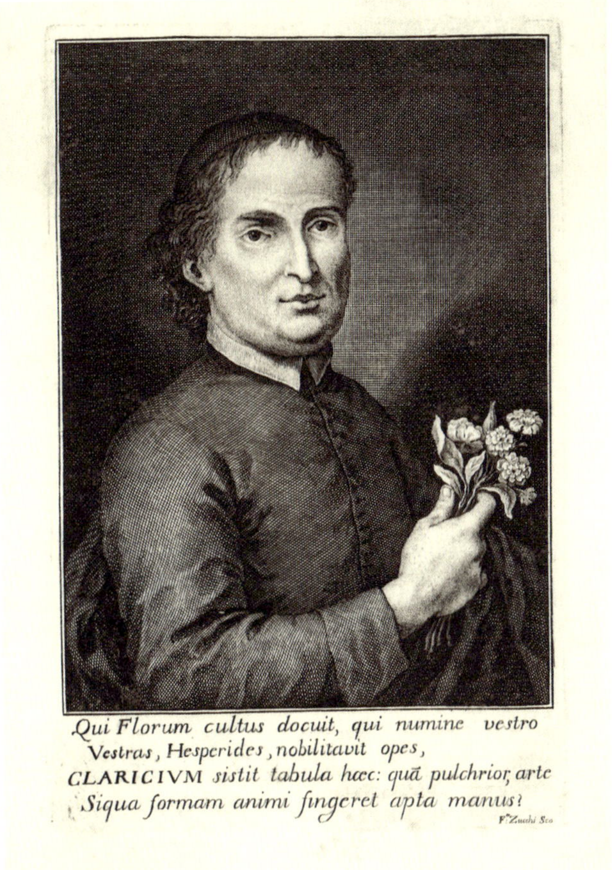

PAOLO BARTOLOMEO CLARICI, *Istoria e coltura delle piante*, Venice, 1726. Frontispiece portrait

In the first decades of the sixteenth century the study of botany underwent a process of renovation that, in a remarkably short space of time, transformed it into a modern science. In the Middle Ages plants were studied in a disparate and empirical manner, being valued almost exclusively for their medicinal properties. During the Renaissance, however, scholars began to undertake a systematic study of the problems of morphology, classification and even the economic aspects of botany, as the pursuit of knowledge received an added impetus from commercial pressures.

This revolution began, as it did in many other areas of knowledge, with a critical revision of the classics. There was a clear need for such a revision, since a flood of new species unknown to the authors of antiquity (such as Dioscorides, Theophrastus and Pliny the elder) was reaching Europe from the Near and Far East and from the New World. The German botanists Otto Brunfels and Leonhart Fuchs, and the Italian Pietro Andrea Mattioli, were among the first to attempt a more rigorous and systematic study of the plant world, and they soon realized that the simple herbals and *horti sanitatis* of the medieval period were quite inadequate for the purposes of documenting their researches. Works provided with detailed descriptions and realistic illustrations were necessary for the study and transmission of this new knowledge. Thus, the modern botanical illustration was born, the product of a close collaboration between the artist and the scientist, who together studied nature's wonders and decided how best to capture them in permanent form.

The surprisingly delicate woodcuts to be seen in the botanical texts of Brunfels (1530–36), Fuchs (1545) and Mattioli (1554) therefore generally show not just the flowers but the roots too, as if these plants had just been gently drawn from the soil. Some of the woodcuts, such as that of the turgid and ductile water-lily in Brunfels (p. 37) or the elegant columbine in Fuchs (chapter XXXV) may be counted among the most beautiful botanical images produced in Europe during the first half of the sixteenth century. The *Commentarii* of Mattioli opens with an illustration of the iris, one of the most admired flowers of the period; in the text the author explains that the native and cultivated species of this flower most familiar to gardeners could all be traced back to a single wild species. These works, with their innovative visual component, not only served the botanist in his studies, they also stimulated the imaginations of artists and craftspeople, offering an inexhaustible source of motifs for paintings, tapestry work, embroidery and ceramics.

The first (albeit unillustrated) monograph dedicated to a flower—and not by accident the rose—was *De rosa et partibus*, published in Antwerp in 1551 together with some short treatises on medical subjects by the physician and botanist Nicolas Monardes of Seville. The very first comprehensive study of flowers, *Florum, et coronariarum odoratarumque nonnullarum herbarum historia*, by the Flemish botanist Rembert Dodoens, was published in Antwerp in 1568. In his epistolary dedication, which is decorated with charmingly realistic woodcuts by Petrus van der Borcht, the author observes that flowers, once appreciated solely for their medicinal properties, were now considered subjects worthy of scientific study and, in addition, as a source of 'honest pleasure' and refreshment for the soul. This theme was destined to form a recurrent motif in books on gardening and horticulture ever after. A century later, John Rea was to define gardening as 'a most pleasing diversion', while in the nineteenth century Berthe Hoola Van Nooten found comfort in nature's 'soft and balmy influences' during her widowed exile on the island of Java.

Since the science of botany has always striven to be firmly based on the direct observation of natural phenomena, in the Renaissance and Baroque eras increasing emphasis was placed on the accurate visualization of reality and hence on the image itself. With the invention of the microscope at the beginning of the seventeenth century, the curiosity of the scientist could extend itself in the study of ever more minute phenomena invisible to the naked eye. At first the microscope repre-

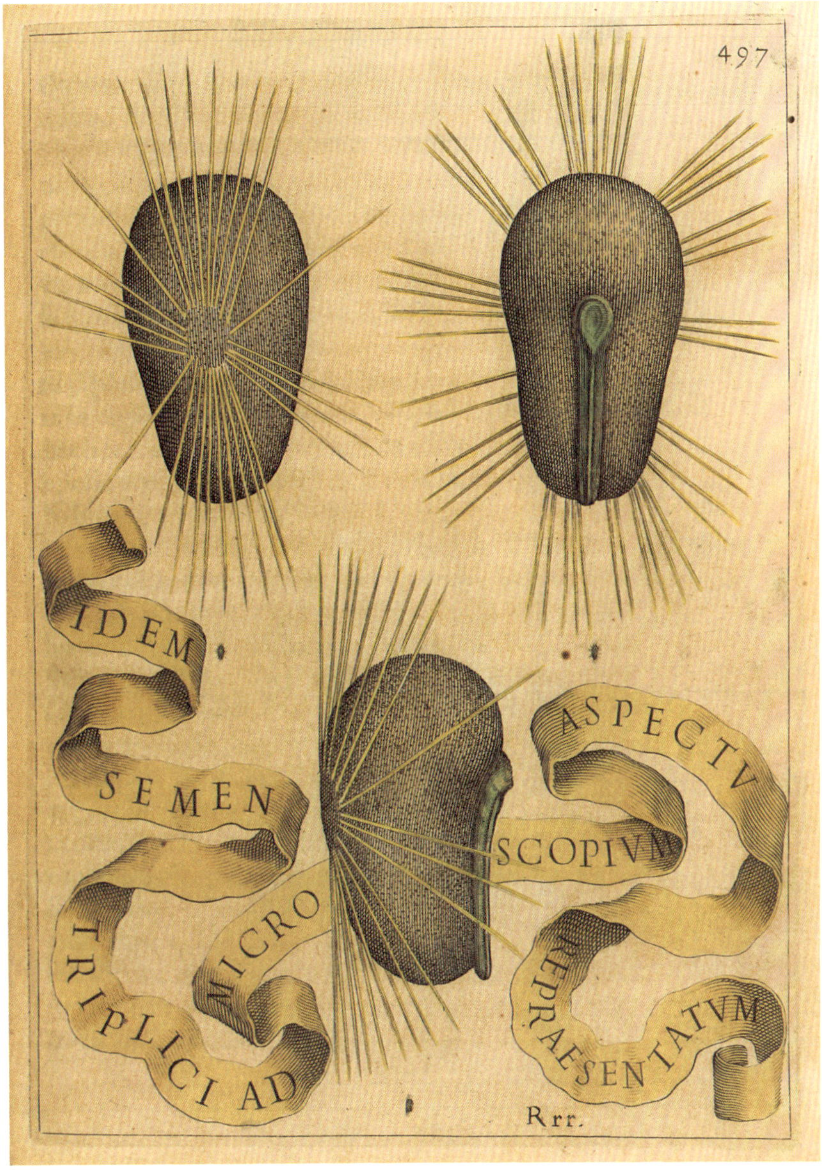

GIOVANNI BATTISTA FERRARI, *Flora*, Rome, 1638. Seeds of a Chinese rose (*Hibiscus rosa-sinensis plenus*) observed through a microscope, page 497

sented a curiosity rather than a serious scientific tool, for the magic 'little eyeglass', as Galileo described it, seemed to open on a mysterious world filled with symbolic meaning rather than on a mere prosaic reality greatly magnified. One celebrated plate in Ferrari's *Flora* shows three enlarged seeds of the exotic 'Chinese rose', or hibiscus, whose form was reminiscent, according to Ferrari, of the 'kidneys of a baby goat'; the seeds have been deliberately arranged to echo the pattern of tiny bees in the coat of arms of the Barberini family, in graceful homage to Ferrari's powerful patrons. Although the microscope at first appeared to be little more than a fascinating toy for wealthy col lectors, scientists soon grasped its potential as a tool for research. The enlightened Prince Federico Cesi of Rome, founder of the first scientific academy in Europe, the prestigious Accademia dei

Lincei, utilized the microscope frequently in his studies, and his five-volume *Plantae et flores* (Paris, Institut de France, MSS. 974–8) includes many plates with microscopic details illustrated for the first time.

During the course of the sixteenth century the flower had shed many of its symbolic associations and came to be appreciated on its own merits, in part as a result of the scientific studies being carried out by naturalists and in part due to the very worldly pleasure that aristocratic amateurs, dilettantes and connoisseurs drew from the cultivation of unusual species in their gardens. The enthusiasm for flowers that marked the Baroque period had far-reaching cultural and economic consequences. Increasing knowledge led to a greater sensibility to the beauty of different species, while the resulting spiral in demand set into motion forces that had important economic repercussions.

These changes were so rapid and so marked that observers in the England of James I could write of a veritable 'Gardening Revolution' (Thomas, p. 224). The flower became an article of luxury—longed for, sought after, and dearly paid for. One result was the emergence of two new professions—those of the nurseryman and the floriculturist—dedicated to experimentation with new methods of cultivation and hybridization. The profession of botany itself acquired new lustre, and scientists were summoned to the most important courts of Europe by princes and emperors keen to add to their own prestige by establishing opulent gardens and important scientific collections. Other botanists were generously subsidized in their researches and in the preparation of magnificently illustrated scientific works.

Scientific expeditions to distant lands were organized for the purposes of discovering and bringing back new species of flora and fauna. Regular cargo ships were often equipped with special compartments in their holds or on deck for the transportation of plants and animals, and many sailors learned how to gather and preserve specimens properly since they knew they could earn considerable sums by selling rarities to the collectors who gathered at the quay to meet the ships arriving home after a long voyage to distant lands. The botanist and gardener John Parkinson noted in his *Paradisi in sole* of 1629 (p. 133) that he had obtained a specimen of the 'Scilla alba' in just such a manner.

Flower shows and flower markets sprang up everywhere during the seventeenth century, becoming a regular event in many cities; the annual Frankfurt fair was particularly celebrated, for in addition to books one could procure there almost any species of rare plant or flower. In this period a newly prosperous bourgeoisie began to take up the pastime of gardening too, certainly on a more modest scale than that of the great princes, but with analogous enthusiasm, particularly in England, France and the Low Countries. An extremely dynamic market developed, for the flower—like all beautiful and superfluous things—was subject to the whims of fashion, and one species after another enjoyed a spell of immense popularity, the vogue for the tulip being followed in its turn by a vogue for the carnation, the hyacinth, the rose and, nearer our own time, the orchid.

With this widespread interest in horticulture, it is not surprising that paintings which had flowers as their main subject also began to appear. Furthermore, since the purpose of these works was to delight the eye rather than to document the studies of scientists, the artist could interpret his subject-matter in a highly personal manner. This marked a crucial turning-point in the history of botanical painting, when the artist began to take precedence over the illustrator. Flower paintings became increasingly popular, and the ownership of a work by a celebrated painter became one among the many status symbols enjoyed by the wealthy. Publishers were quick to note and exploit this phenomenon, producing sumptuous flower books illustrated by draughtsmen and engravers whom they encouraged to specialize in botanical subjects. In this way two new genres of painting were born: the florilegium or *hortus floridus* (books of flower paintings or engravings), and the floral still-life or *nature morte* (an engaging contradiction in terms, since such works were supposedly painted 'from life'). In both, flowers were presented either in elegant single studies or in sophisticated compositions, arranged in vases, baskets or other containers, or heaped on tables or ledges. Such works reflected an entirely new approach to flower painting, one inspired by the beauty of the subject-matter yet informed by the disciplined technique of botanical illustration that by then had reached full maturity. Nevertheless, the accuracy and clarity that were the primary goals of the illustrator were now replaced by aesthetic and painterly concerns, and the artist could concentrate on the purely decorative qualities of his subject.

Various flowers enjoyed moments of exorbitant popularity that were directly reflected in the flower books and still-life paintings of the period. The tulip, for example, generated a veritable collectors' mania that resulted in the production of the many sumptuous *tulpenboeken* or tulip catalogues during the first half of the seventeenth century. The passion-flower (also known as the 'Maracoc', the 'Granadilla' or the 'Clematis virginiana')—a climbing plant from Peru that was first described by the Spanish botanist Monardes—fascinated artists and flower lovers, who associated the bizarre form of its stamens and pistils with the crown of thorns of Christ's crucifixion. The English—and stoutly Protestant—botanist Parkinson dismissed this as an absurd notion propagated by 'superstitious' Jesuits, and in his *Paradisi in sole* included two plates, one showing a real passion-flower and the other a distorted image, 'The Jesuites Figure of the Maracoc' (see Pozzi, p. 335).

The simple religious symbolism that had grown up around the flower in the Middle Ages was replaced in the seventeenth century by new and more complex moral associations. The first book that dealt with this subject, mainly on the basis of the Aristotelian tradition concerning colour symbolism, is *Il mostruosissimo mostro, diviso in due trattati. Nel primo dei quali si ragiona del significato dei colore. Nel secondo si tratta dell'herbe e fiori,* published in Ferrara in 1584. A little later the English botanist John Gerard wrote in his *Herball* (first published in London in 1597) that the exquisite beauty, colour and form of flowers such as the violet naturally evoked contemplation of 'honesty, dignity and every other human virtue' (p. 850). At the same time, the inherently fleeting nature of its beauty invited one to meditate on the passage of time, the vanity of earthly things, and one's own mortality. In this univocal reading, the withered flower frequently appeared—together with the

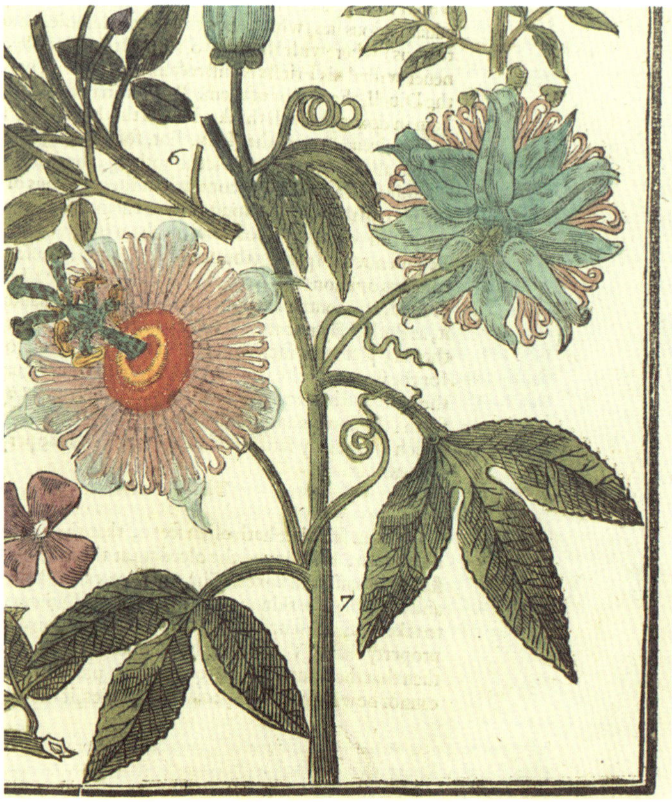

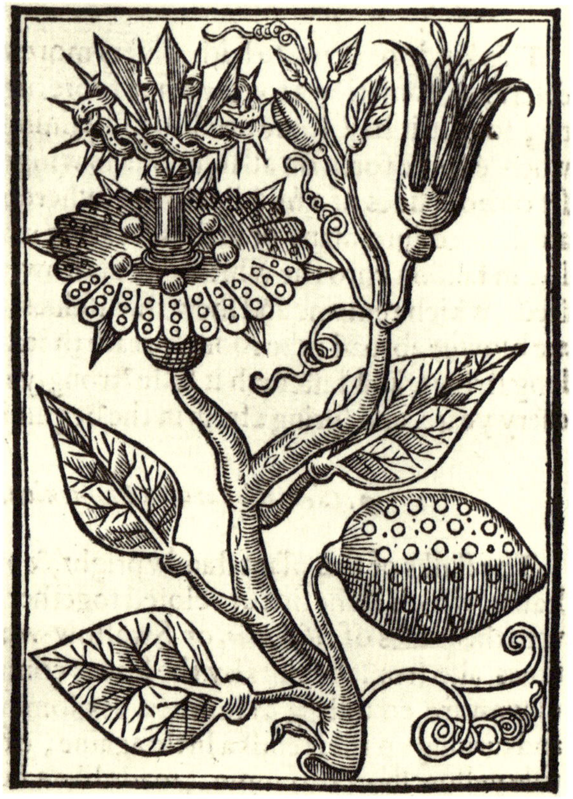

JOHN PARKINSON, *Paradisi in sole*, London, 1629. Page 395, detail of a *Passiflora* contrasted with 'The Jesuites Figure of the Maracoc', page 394

hour-glass, the snuffed candle and the human skull—in still-life paintings on the *memento mori* theme. Even so, the floral still-life lent itself to a variety of more positive interpretations, for the beauty of the flower, however brief, was also eternal, returning every spring and summer in all its glory. It thus became a symbol of rebirth and immortality, a much more appealing notion to those wealthy art collectors who by now had tired of the grandiose subjects associated with religious and historical painting and greatly preferred the charm of simpler, more intimate genres.

The flower in its role as 'moral symbol' was also frequently co-opted as a visual element in the *corpo* (body) or figurative portion of the elaborate symbolic emblems that intellectuals delighted in devising in the sixteenth and seventeenth centuries. The motto or verbal component (also referred to as the *anima*) that explained the moral significance of the device was generally drawn from familiar sacred or classical texts, or from the works of contemporary poets. Appropriately enough, the author of the first book entirely dedicated to botanical and floral emblems, *Symbolorum et emblematum ex re herbaria desumtorum centuria* (Nuremberg, 1590) was a physician and naturalist, Joachim Camerarius the younger. Within the pages of his work we find the tulip (p. 98) linked to the motto 'Languesco sole latente' (I languish for lack of sun), while the fritillary is presented as the symbol both of the modesty of youth—because the stem with its flower bud bends modestly earthward—and of virtuous old age, because as the flower wilts the stem begins to straighten. As the motto declares, 'Modesta iuventus, honesta senectus' (p. 81).

Religious symbolism was also used in the devising of emblems, and the abbot Filippo Piccinelli of Milan compiled an encyclopaedia of such devices, *Il Mondo simbolico* (Milan, 1653). In the eleventh book, which is entirely given over to floral symbols, the author describes in twenty chapters the emblems and mottoes, together with their textual sources, of twenty flowers ranging in alphabetical sequence from the amaranth to the violet. The many editions of Jacob Cats's *Silenus Alcibiadis, sive Proteus*, first published in 1618, also contain a variety of floral emblems (Segal 1982, p. 80).

Cut flowers have been used since time immemorial to create floral decorations, and the ways in which they have been represented by artists tells us much about their cultural and symbolic significance in different periods. Many sacred pictures from the Middle Ages, especially scenes of the Annunciation, include a handful of flowers arranged in a simple glass or vase (the formal flower bouquet would make its appearance much later). Hugo van der Goes placed in the centre foreground of his dazzling *Portinari Altarpiece* (*c.* 1475; Uffizi, Florence) a majolica apothecary's jar containing two irises and a lily, and a drinking-glass with some carnations and columbines. In the marginal decorations of Books of Hours from the Ghent–Bruges School one might find two or three flowers (rather than actual bouquets) arranged in receptacles ranging from bottles and jugs to double-handled majolica vases, and on occasion such homely details as an earthenware pot filled with carnations whose stems are supported by a wickerwork frame—a device actually used by gardeners for many centuries and often depicted in paintings.

During the Renaissance, classical garlands as well as more imposing bouquets arranged in ornately wrought vases began to appear, although paintings on religious themes continued to feature the rose, the lily and a few other flowers. Botticelli depicted his Primavera with wreaths of cornflowers, daisies, lilies of the valley, forget-me-nots, periwinkles and roses woven into her hair and draped around her neck (Moggi). Classical garlands embellished with 'modern' species of fruit and flowers can be found in the works of Mantegna and in the Roman frescoes of Raphael and his assistant Giovanni da Udine (Dacos and Furlan; Caneva). Dürer softened the severity of his portrait *Erasmus of Rotterdam*, engraved in 1526, by placing a small vase of flowers next to the scholar's tomes.

Beginning in the mid-sixteenth century, dense bouquets of horticultural varieties, among them several exotic flowers, began to appear ever more frequently in the backgrounds of paintings and prints and in the title-pages of botanical works until, finally, early in the seventeenth century artists in northern Europe began to concentrate on the flower bouquet as an autonomous subject. The first such paintings had generally consisted of a handful of flowers arranged in carefully balanced compositions, as in a pair of celebrated floral panels (private collection) by Ludger Tom Ring the younger (Veca 1982, p. 128), some paintings by Georg Hoefnagel and certain early works by Ambrosius Bosschaert the elder. As time passed these restrained compositions were replaced by increasingly sophisticated arrangements in which artifice took precedence over realism. Many works by the renowned flower painter Jan 'Velvet' Brueghel, for example, show large bouquets—veritable an-

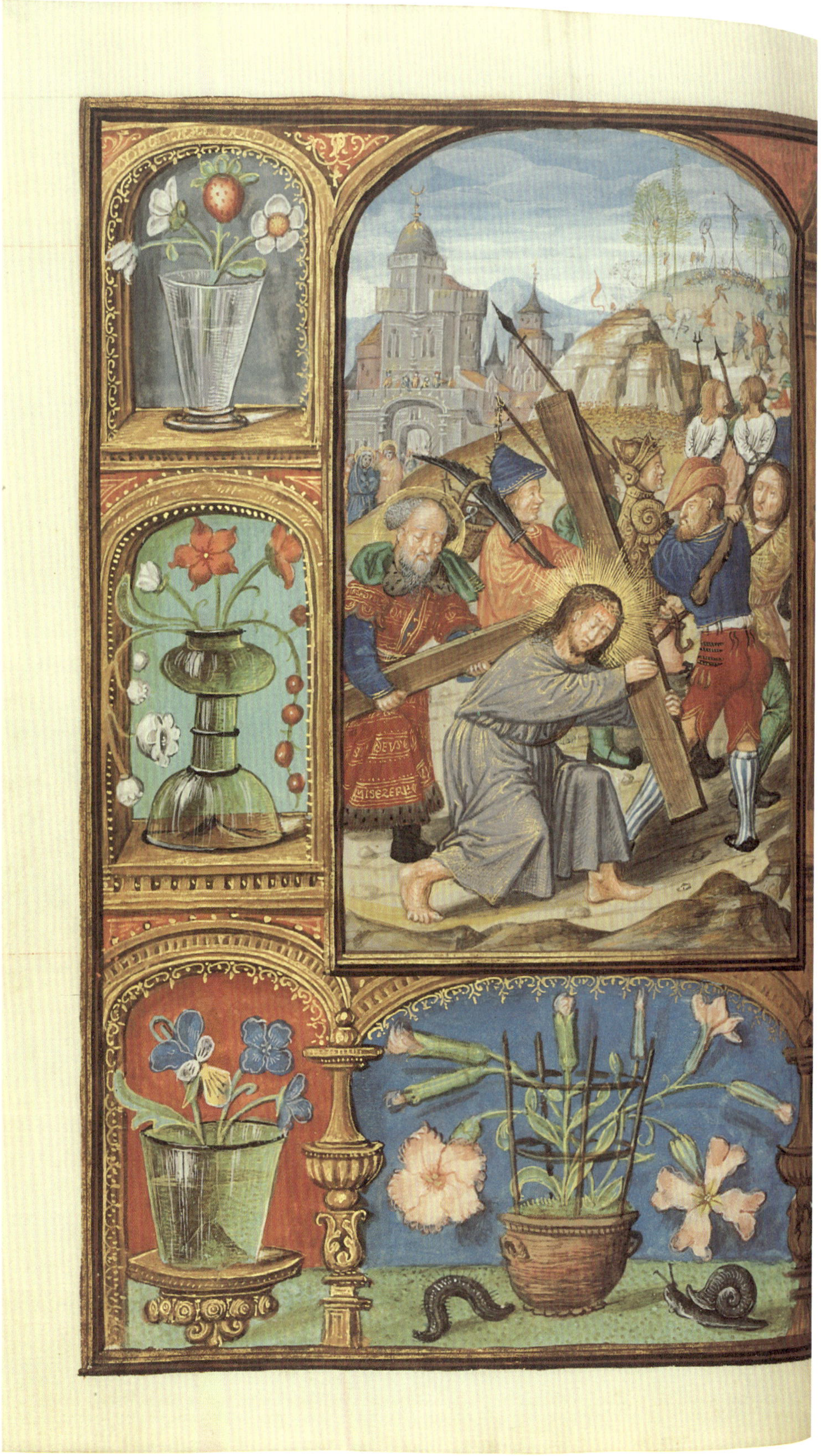

SIMON BENING, *Horae beatae Mariae Virginis ad usum Romanum*. Folio 147v, Christ Carrying the Cross, within a border of flowers arranged in various receptacles thologies of flowers containing dozens of species that, in real life, would never be found simultaneously in bloom—precariously balanced in very small vases.

The apparently spontaneous arrangement of the elements in these paintings was in reality governed by rigid aesthetic canons. According to these rules the apex of the composition should comprise tall flowers such as irises or fritillaries, while shorter-stemmed flowers might be placed (without overlapping) to either side. Examples suitable for copying could be found in many of the engraved florilegia of the period, while an artist who had developed a particularly successful formula might repeat it in more than one painting, with only slight variations in his choice of vase or in the type and dimension of the flowers used. These strict rules of composition underlie the deceptive simplicity of Brueghel's *Basket of Flowers and a Vase of Flowers* of 1615 and Balthasar van der Ast's *Basket of Flowers* of *c.* 1625 (both gifts from Mrs Paul Mellon to the National Gallery of Art, Washington, D.C.). These rules were adopted by many of the Dutch, Flemish, German and French artists who specialized in the genre of flower painting at the beginning of the seventeenth century, among them Roelandt Savery, Jacob Marrel, Daniel Seghers, Georg Flegel and Jacques Linard (see Hairs).

G. B. Ferrari devoted several pages of his highly influential *Flora* (pp. 396–421) to a description of how to compose not only bouquets and baskets of flowers but also elaborate trophies and centrepieces which he defined as 'macchine fiorite' (flowering mechanisms). His account of these quintessentially Baroque inventions is accompanied by six plates illustrating different vases, containers and perforated columns that could be used to construct them. The author even notes that if the flowers in one's composition failed completely to mask the container, the latter might be 'painted with foliage, which could well vie in loveliness with the flowers themselves' (p. 418). Artisans began to produce vases expressly designed for such compositions, utilizing a wide variety of materials—glass, majolica, bronze, precious metals and, later, porcelain—and amusing themselves by creating *trompe l'œil* effects in which natural and artificial elements were confounded (Veca 1982, p. 406).

As the vogue for flower painting spread throughout Europe, distinct 'schools' formed in various countries, although the taste for magnificent and opulent compositions was almost universal. During the reign of Louis XIV a highly ornate decorative style emerged in France that culminated in the appointment of Jean Baptiste Monnoyer as the first official 'peintre des fleurs' at the Académie Royale de Peinture in Paris. His consummate virtuosity can be appreciated in a series of flower paintings executed for his royal patron (Berrall, pp. 49–51), elaborate compositions in which form and colour are perfectly counterbalanced. Indeed, although they are depicted in meticulously accurate detail, Monnoyer's flowers are often shown from unusual perspectives against a theatrical backdrop of richly coloured and textured draperies, carpets and hangings in a vision that is more fantastical than it is 'natural'.

The popularity of this style, which was reflected in all the decorative arts but particularly in the famous tapestry work of Flanders, was to continue for much of the eighteenth century side by side with the more restrained elegance of the Rococo. Flowers were a constant motif, for example, in the effervescent *scènes galantes* painted by artists such as Jean-Honoré Fragonard. In fulfilling this decorative function, however, the flower lost its botanical significance, becoming transformed into pure ornament. This led to such unusual works as the catalogue *Twelve Months of Flowers* by Robert Furber, which, in order to appeal to connoisseurs 'of taste', presented the flowering plants that could be purchased from Furber's nurseries in highly artificial and magniloquent bouquets. Gerard van Spaendonck, the officially appointed 'professeur de peinture des fleurs' at the Jardin des Plantes in Paris, was also an heir to this aesthetic tradition and dedicated himself to floral painting in the *grande manière* rather than according to the rigid canons of scientific illustration.

In the seventeenth century the morphology of the flower played a key role in the vigorously debated problem of botanical description and classification. As explorers, traders and missionaries brought back more and more new species of plants from distant lands that had to be studied, described and named, the need for a rational system of classification and nomenclature became imperative. It was only between the end of the seventeenth century and the first decades of the eighteenth that this problem was resolved, however, thanks to the efforts of botanists, greatly aided by the art of botanical illustration.

In 1694 Joseph Pitton de Tournefort, who just one year earlier had been named professor of botany and director of the Jardin des Plantes, published *Elémens de botanique*, in which he presented a systematic key to the genera of plants based primarily on the characteristics of the corolla. This work was illustrated in an exemplary manner by a close collaborator of Tournefort, the botanical artist Claude Aubriet (Greene, II.963). Other interesting observations were proposed in 1720 by Giulio Pontedera, professor of botany at the University of Padua, in his *Anthologia sive De Floris Natura*. The real breakthrough that laid the groundwork for the modern system of botanical classification was made by Carl Linnaeus, who proposed what he defined as a 'sexual system' that reflected the role played by the flower as the seat of fructification. Linnaeus divided the entire plant world into twenty-four classes based on the number and disposition of the stamens and pistils of the flowers, that is, the plant's sexual organs. This system, together with Linnaeus's thorough revision of binomial nomenclature, made it possible even for female amateur botanists, who were attracted in ever larger numbers to this gentle science, to study and correctly classify botanical specimens. The very first 'Table' illustrating Linnaeus's revolutionary system was drawn, coloured and engraved in Leiden in 1736 by Georg Dionysius Ehret, a German artist whom Linnaeus himself had taught 'to correctly depict the stamens, pistils and other small parts' of the flower. Ehret, perhaps more than any figure in the history of botanical illustration, succeeded in uniting art and science in his remarkable works, for in addition to his talent as an artist he had a thorough knowledge of botany. This unique combination is manifest in the plate on the 'characters of flowers' etched and engraved by Ehret in 1748 for his *Plantae et papiliones rariores*. The unusual beauty, combined with scientific accuracy, of his works were to inspire countless generations of botanical artists.

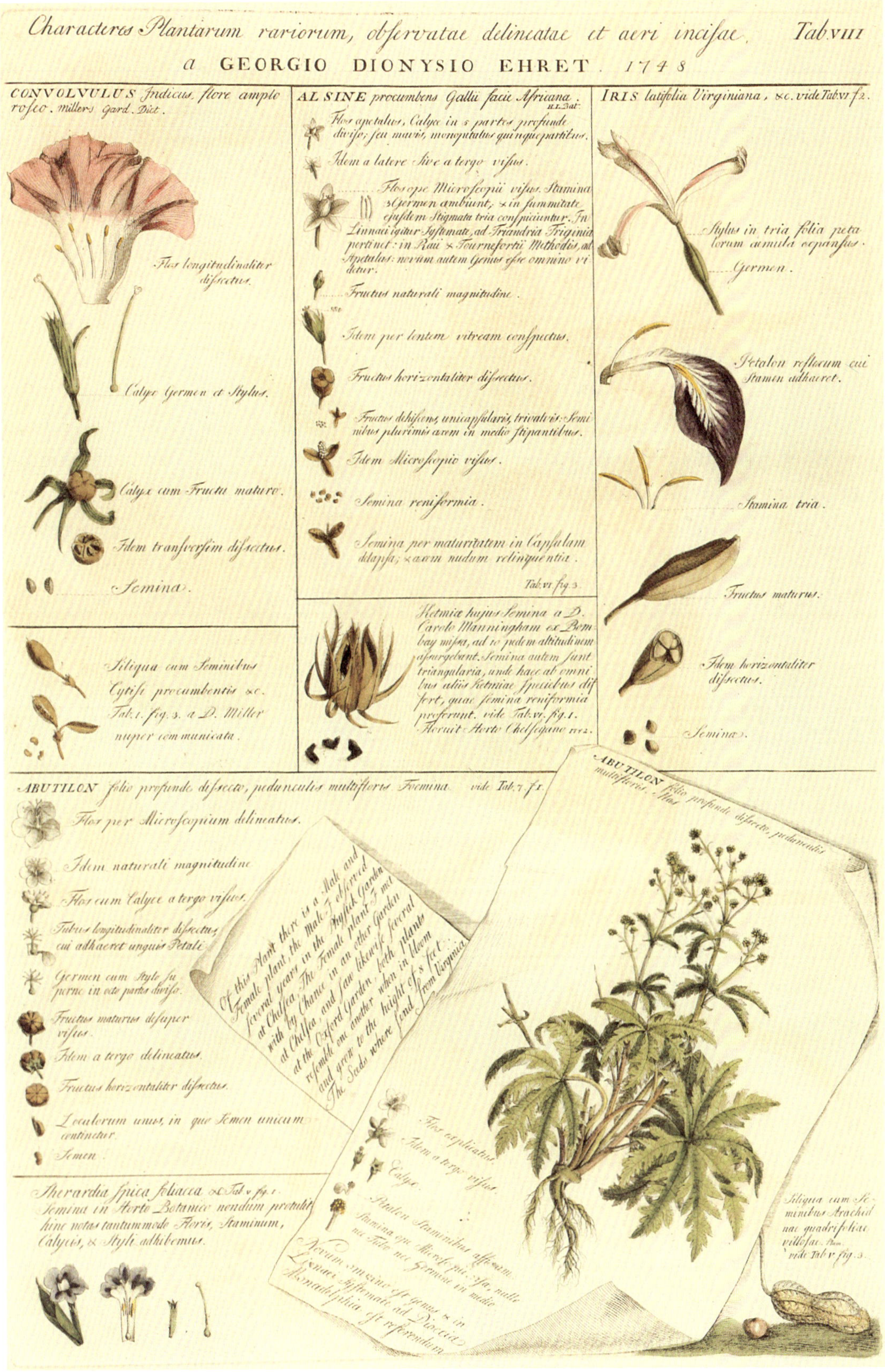

GEORG DIONYSIUS EHRET, *Plantae et papiliones rariores*. Characters of plants, plate VIII

The different media and techniques utilized by artists in various periods also played a fundamental role in the evolution of the genre of botanical painting, for there were profound differences between the images executed in opaque watercolour or bodycolour and those produced by means of engraving or etching. While a painting could show a plant in its true colours, a print was limited to black and white, although a print could, of course, at least be run off in a large number of impressions. The woodcut, which produced a clear, if not very subtle line, was used in the sixteenth century by the authors of the first botanical texts, beginning with Otto Brunfels, and because of its relatively low cost botanists and publishers often continued to rely on it throughout the seventeenth century. Even so, the invention of the techniques of engraving and etching on copper plates had revolutionized the art of printing, for with the burin a skilled engraver could create a remarkably sensitive line and *chiaroscuro* effects on the plate that, once inked, might range from an almost imperceptible grey to the deepest black, while with the etching needle striking effects of light and shadow could be achieved. These techniques were incomparably superior to the woodcut in the rendition of plastic values and the impression of three-dimensionality. In the seventeenth century many scientists would take advantage of engraving and etching, using the former for elegant calligraphic effects that reproduced fine morphological detail, and the latter to capture their specimens' most delicate nuances in hue. At the same time the practice sprang up of enhancing the beauty and realism of these images by hand tinting them in watercolour or, more frequently, in bodycolour. This was done to embellish the de-luxe editions intended for wealthy collectors and the dedication copies presented to august patrons.

The custom of hand colouring illustrations in botanical and zoological texts became widespread from the second half of the sixteenth century; thus we find that in his *Hortus Floridus* Crispijn van de Passe the younger provided a list of the exact colours of the flowers in each plate, so that—as he states on the title-page—they might be correctly painted. Many printers and scientists employed specially trained artists to colour the plates in their books, for this apparently ancillary step contributed information that was of crucial importance to botanists and zoologists. Demanding painstaking accuracy as well as an artist's eye for form and colour, the task was often entrusted to women; indeed, the celebrated printing house of Plantin in Antwerp, which also specialized in the publication of scientific texts, at one point had no less than three women in its employ solely for the purpose of colouring illustrations. We also know that the botanical artist Maria Sibylla Merian asked that only she and her two daughters be allowed to colour the plates of her major work, *Dissertatio de generatione et metamorphosibus insectorum surinamensium* (1705), which describes the insects and flowering plants she had studied *in situ* at the Dutch colony of Surinam. The early experience of colouring the plates for botanical texts also proved to be of fundamental importance for G. D. Ehret later in his career.

Many seventeenth-century texts on natural history were truly impressive works, and the possession of an unusually fine edition, magnificently printed and coloured by hand, conferred great prestige on its owner. The Prince-Bishop of Eichstätt, Johann Conrad von Gemmingen, who had extensive gardens planted around his castle, Wilibaldsburg, near Nuremberg, and who commissioned the naturalist Basilius Besler to document his botanical collection in a lavishly illustrated work, *Hortus Eystettensis*, possessed a copy of *Archetypa studiaque patris Georgii Hoefnagelii* illustrated with plates engraved in 1592 by Jacob Hoefnagel (Vignau-Wilberg, p. 45) and coloured by Daniel Herzog, an artist whom he retained at court for this purpose. This represents one of the rare cases in which the name of an artist responsible for work of this nature has been recorded for posterity, for in general they remained modestly in the background. Even more unusual is the example of a certain 'Guillelmus Théodorus', who actually signed and dated a copy of Daniel Rabel's *Theatrum florae*, which he had coloured by hand, in 1624. It is also interesting to note that the Italian botanical artist Gherardo Cibo not only coloured but also added charming landscape backgrounds to the plates in some copies of the herbals of Fuchs and Mattioli (Tongiorgi Tomasi 1989). The process of hand colouring carried its own risks, however, for without careful supervision by a qualified botanist and without live specimens from which to study, the artist could all too easily fall into the temptation to colour each flower as he or she pleased, creating illustrations that might be beautiful, but that were also inaccurate and misleading.

During the eighteenth century artists continued to experiment with various printmaking techniques, refining them to obtain ever more subtle *chiaroscuro* effects (see *Printmaking in the Service of Botany*). Colour-printed mezzotints became available in this period, but although the process proved to be ideal for both portraiture and the reproduction of paintings, it was rather less well adapted to the scientific illustration, as may be seen from the somewhat smudged plates in John Martyn's *Historia plantarum rariorum* (1728–37) or those in Johann Wilhelm Weinmann's *Phytanthoza iconographia* of 1737–45 (Watson, p. 12). Aquatint was found to be much more suitable, especially in the hands of such brilliant engravers as Richard Havell. Illustrations printed by any one of these techniques still had to be further coloured by hand afterwards, however, in what remained a costly and time-consuming process. Many attempts were therefore made to develop a reliable method of printing in colours using copper plates. The first important experiments were carried out at the beginning of the eighteenth century by Jacob Christoph Le Blon, who, basing himself on Newton's theory of colour, sought to create all the hues of the spectrum by means of a three-step process, printing the three primary colours—red, blue and yellow—in succession. Around 1750, Le Blon's former assistant, Jacques-Fabien Gautier d'Agoty, added black to the colours used by Le Blon to achieve fairly realistic effects in a series of anatomical plates. The botanist Pierre Bulliard also used a sequence of plates, one for each colour, to print the illustrations in his *Flora Parisiensis* (1776–83), although with less than satisfactory results.

Towards the end of the eighteenth century two great botanical artists, Pierre Joseph Redouté and Gerard van Spaendonck, began to experiment separately with a new technique of colour printing from stipple-engraved plates, which proved to be uniquely adapted to the genre of botanical

DANIEL RABEL,
Theatrum florae, Paris,
1622. Title-page coloured
by Guillelmus
Théodorus,
1624

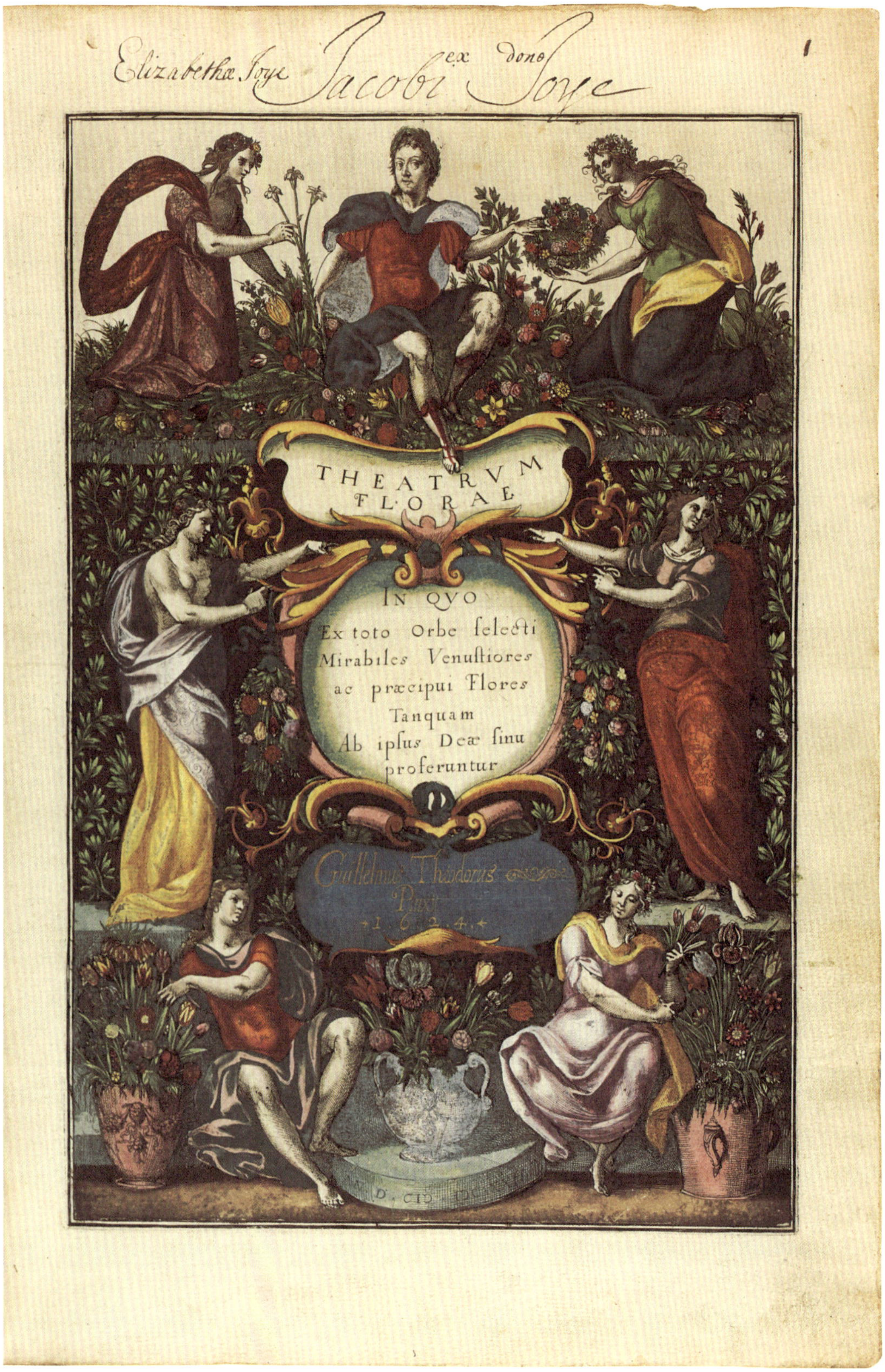

illustration. The young Redouté discovered this technique during a sojourn in England, where stipple-engraving had become extremely popular due to the superb portraits and landscapes produced by the Italian Francesco Bartolozzi. Since it was based on the 'stippling' of a mass of minute dots or strokes on an etching ground rather than on the drawing out of continuous lines, this technique was able to reproduce the delicate shades, textures and details of flowers with remarkable fidelity. Prior to printing in colours, the plate was carefully inked with various pigments before it was passed through the press, although even here each printed impression then had to be touched by hand in order to obtain 'tout le moelleux et tout le brillant de l'acquarelle'. Redouté was the first botanical artist to abandon the traditional medium of opaque bodycolour for the more subtle technique of watercolour in which he excelled, recreating the hues of each flower with great accuracy and extraordinary effects of transparency and luminosity.

The nineteenth century was the age of the lithograph and the chromolithograph, techniques that would further revolutionize the art of book illustration, although traditional media, often combined to achieve highly realistic effects, continued to be used in botanical illustration; indeed, the oldest technique of all, that which depended on blocks of wood, enjoyed a surprising revival in the form of wood-engravings, among the best of which are those by Thomas Bewick. Lithography, however, which was based on the incompatibility of water and the greasy colours that were in turn applied to the surface of the printing 'stone' (porous limestone was originally used for this) was a much more versatile technique, capable of both great precision and great expressiveness. Lithographs were first produced using black crayon, but soon other colours were introduced (a separate 'stone' was needed for each additional colour) and the limitations of engraving and etching were overcome once and for all. The development of the chromolithograph made it possible to print a large number of colour illustrations quickly at a moderate cost, and had an immediate impact on the publishing industry. There was a flourishing market for scientific works in the nineteenth century, and these new techniques were quickly pressed into service on a large scale for the preparation of books, as well as for the printing of scientific journals, the very first of which were founded in this period.

The nineteenth century was also the time of the Romantic movement in the arts, when the image of the flower—particularly in Britain and France—underwent yet another profound transformation. The sophisticated Rococo style of flower painting that had been so greatly admired in the eighteenth century was replaced by a more spontaneous and direct approach to nature, and the lingering prejudice that had dismissed floral painting and the still-life as minor genres depicting 'petty peculiarities', as Sir Joshua Reynolds expressed it, was overcome by a series of artists who were also 'patient pupils of nature'. One such artist was John Constable, whose floral pictures, although not numerous, reflect the same acute sensitivity to the beauties of nature that inspired his airy landscapes.

The *Journal* and correspondence of one of the greatest artists of the Romantic period, Eugène

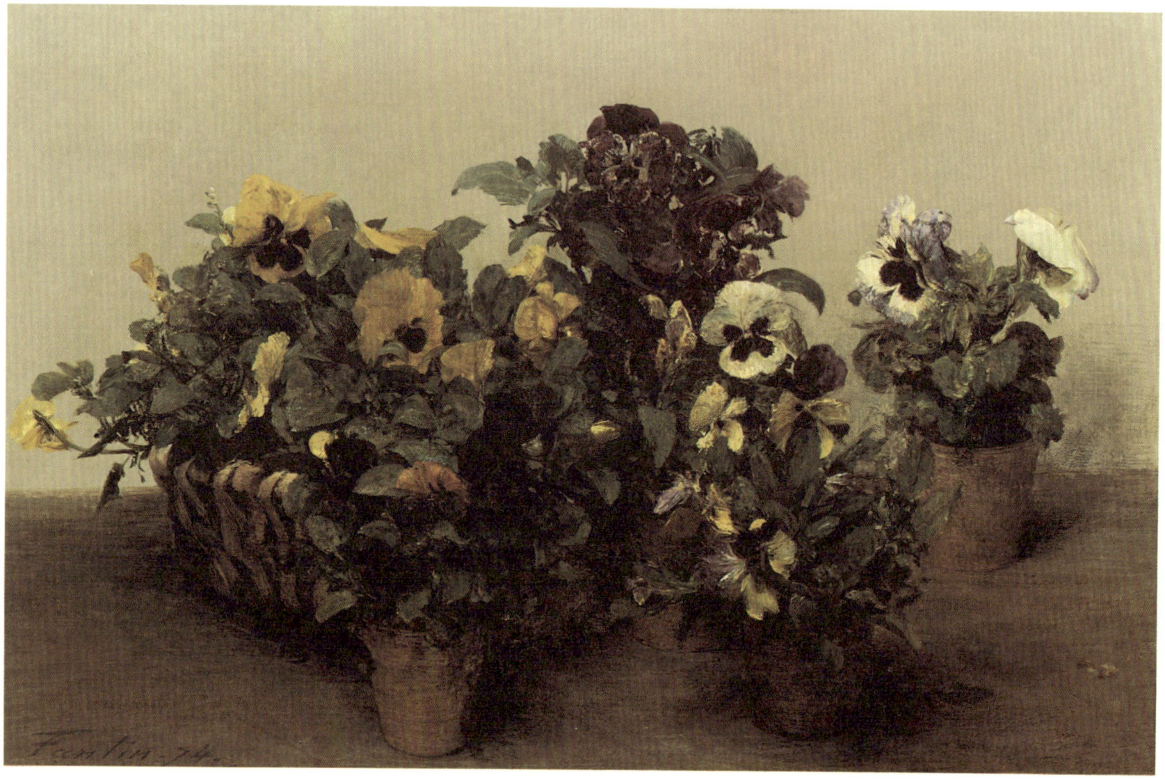

IGNACE HENRI JEAN THEODORE FANTIN-LATOUR, Pansies (*Viola tricolor hortensis*), 1874, oil on canvas. Collection of Mr and Mrs Paul Mellon

Delacroix, document the profound changes in aesthetic sensibility that were taking place and which would eventually transform the art of flower painting. For example, while conceding the undeniable talent of Jean Baptiste Monnoyer, then reckoned to be the greatest French flower painter that had ever lived, Delacroix did not hesitate to write of his work: 'L'étude des détails, poussée à un très haut point, nuit un peu a l'ensemble'. Delacroix himself advocated an entirely different approach: 'J'ai voulu sortir un peu de l'espèce de poncif qui semble condamner tous les peintres de fleurs à faire le même vase avec les mêmes colonnes ou les mêmes draperies'; indeed, he continued: 'J'ai essayé de faire des morceaux de nature comme ils se présentent dans les jardins. . .'. The painting of flowers was thus transformed from what was too often an academic exercise into an unmediated experience and a direct rendition of nature (cited by M. and F. Faré in *Peintres des fleurs*).

In a *Journal* entry dated 14 February 1848, Delacroix noted that the botanist Adrien de Jussieu had warmly invited him to visit the Jardin des Plantes in the spring, promising to take him on a tour of the greenhouses and to obtain permission for him to visit whenever he wished in order to paint there. Many flower studies by Delacroix's hand have survived; mostly watercolours on paper executed *in situ*, they vary from representations of plants in their native habitat to studies of single flowers and small bouquets. Although many of these were actually studies for the great oil paintings that the artist would send to the Salon, they constitute veritable gems in themselves, the 'morceaux de nature' of which Delacroix wrote in his notes. The artist's vividly expressive portrayal of nature

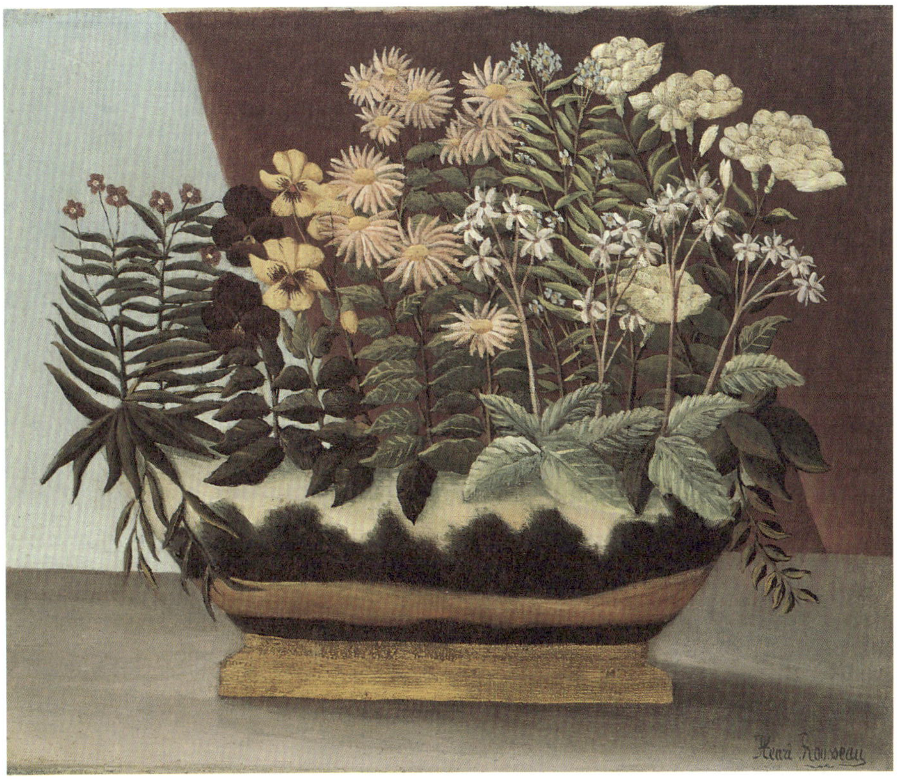

HENRI JULIEN FELIX ROUSSEAU, LE DOUANIER, *Flowers of Poetry*, 1890–95, oil on canvas

can be seen in two small watercolours from the collection of Mr and Mrs Paul Mellon, *Fleurs* and *Narcissus*. One of Delacroix's most fervent admirers, Henri Fantin-Latour, a sensitive and idiosyncratic artist and an intimate friend of both James Whistler and Edouard Manet, also executed many flower paintings, including the remarkable *Pansies* of 1874.

The first group that really grasped and applied the lessons of Constable and Delacroix was that of the French Impressionists. Most of the leading artists in this movement seized on the flower as an ideal subject on which to test their theories, and each produced flower paintings couched in his or her own unique idiom. Thus the genre underwent a radical transformation, as the experiments the Impressionists undertook with light, colour and atmospheric effects swept aside conventional models and opened the way for the creation of entirely new aesthetic forms. Manet, for example, executed various flower studies, including the celebrated *Peonies* (Musée d'Orsay, Paris), although perhaps his most significant contribution to the genre was a series of small paintings of flowers in vases that he painted for his friends in the winter of 1882–3, during the very last months of his life (see A. Forge and R. Gordon, *The Last Flowers of Manet*, New York, 1986). Claude Monet's light-drenched paintings of the roses and irises at Giverny and the visionary 'formes overtes' of his water-lily studies would serve as an inspiration to several twentieth-century abstract painters.

Let us close our consideration of the floral image with a small painting that falls quite outside the classical canons of botanical painting, the *Flowers of Poetry* (1890–95) by Henri Rousseau, le

Douanier, which hangs behind my desk at Oak Spring. While claiming to portray nature in an uncompromisingly realistic manner, Rousseau's intimately lyrical style and apparently 'naïve' perception of the material world allowed him to create still-lifes that were completely transfigured by his imagination, suspended somewhere between fantasy and reality, seemingly tangible yet with the vaporousness of a dream. Thus the shifting, iridescent world of flowers and their *vie silencieuse*, as captured and presented to us by artists and botanical illustrators through the course of the centuries, will never cease to lift our hearts with their fragile, timeless beauty.

DESCRIPTIVE METHOD

EACH ENTRY begins with a description of the item (or items) concerned. Manuscripts and works of art are given a brief physical description. Transcriptions of the titles of printed works are supplied in full, following the typography of the original text, while the use of *sic* following oddities in spelling is severely restricted. Square brackets are employed when the title or date of a work has a source other than the work itself; to indicate inserted material and variants, and supplied letters when special marks of contraction have been employed by the printer; and to describe vignettes, ornaments and devices. Earlier forms of letters and diacritical marks are converted to modern form. The colophon is given only when it contains information not provided on the title-page.

The collation section states the format, measurements in centimetres, gatherings by signatures (for books of the hand-press period), pages or foliation, and the number of leaves of plates. Italics indicate inferred pagination and signatures; errors in numbering or signings are noted. 'Binding' describes briefly material and decoration. The 'Plates' paragraph provides the illustration processes, whether printed separately from the text or not; artists where known; and the presence of hand colouring or hand finishing. 'Provenance' gives information, where available, on previous owners and on their bookplates, inscriptions and marginalia. Also noted are enclosures, for example personal correspondence. Finally, 'References' supplies details of relevant published material; for those given in an abbreviated form, full details will be found in the 'Further Reading'.

FURTHER READING

This guide to Further Reading includes all abbreviated titles referred to in the 'References' and in descriptions of items.

Age of the Marvelous:
The Age of the Marvelous, exhibition catalogue edited by Joy Kenseth, Hood Museum of Art, Dartmouth College, Hanover, N.H., 1991.

ALLAN:
Mea Allan, *The Hookers of Kew, 1785–1911*, London, 1967.

ANDERSON 1977:
F. J. Anderson, *An Illustrated History of the Herbals*, New York, 1977.

ANDERSON 1980:
D. R. Anderson, *American Flower Painting*, New York, 1980.

ARBER:
Agnes Arber, 'The Colouring of Sixteenth-century Herbals', *Nature*, CXLV, 1940, p. 803; reprinted in *Herbals*, Cambridge, 1986, pp. 315–18.

ARNOLD:
E. M. Tucker, *Catalogue of the Library of the Arnold Arboretum of Harvard University*, 3 vols., Cambridge, Mass., 1914–33.

Art of the Printed Book:
Art of the Printed Book, 1455–1955: Masterpieces of Typography through Five Centuries from the Collections of the Pierpont Morgan Library, exhibition catalogue with an Essay by Joseph Blumenthal, Pierpont Morgan Library [1973], reprinted New York, 1978.

AYMONIN and SCHNAPPER:
Gerard G. Aymonin (introduction) and Antoine Schnapper, *Daniel Rabel: cent fleurs et insectes*, Arcueil, 1991.

AZZI VISENTINI:
Margherita Azzi Visentini, *L'Orto botanico di Padova e il giardino del Rinascimento*, Milan, 1984.

BAER:
Winfried Baer, 'Botanical Decors on Porcelain', in *Flowers into Art*, exhibition catalogue edited by Vibeke Woldbye, The Hague, 1991, pp. 82–92.

BALTRUSAITIS:
Jurgis Baltrusaitis, *Le Moyen Age fantastique: Antiquités et exotismes dans l'art gothique*, Paris, 1981.

BARKER:
Nicolas Barker, *Hortus Eystettensis: The Bishop's Garden and Besler's Magnificent Book*, London, 1994.

BAUMANN:
F. A. Baumann, *Das Erbario Carrarese und die Bildtradition des Tractatus de Herbis*, Bern, 1974.

BAZIN 1960:
Germain Bazin, *A Gallery of Flowers*, London, 1960.

BAZIN 1984:
Germain Bazin, *Les Fleurs vues par les peintres*, Lausanne, 1984.

BEHLING:
Lottlisa Behling, *Die Pflanzen in der mittelalterlichen Tafelmalerei*, Weimar, 1957.

BELLI BARSALI:
Isa Belli Barsali, 'Une Fonte per i giardini del Seicento: Il Trattato di Giovan Battista Ferrari', in *Il Giardino storico Italiano*, edited by Giovanna Ragionieri, Florence, 1981, pp. 221–34.

BÉNÉZIT:
Emmanuel Bénézit, *Dictionnaire critique et documentaire des peintres, sculpteurs, dessinateurs et graveurs*, 8 vols., Paris, 1960.

BERGSTRÖM 1955:
Ingvar Bergström, 'Disguised Symbolism in "Madonna" Pictures and Still Life', *The Burlington Magazine*, XCVII, 1955, pp. 303–08 and pp. 342–9.

BERGSTRÖM 1956:
Ingvar Bergström, *Dutch Still-life Painting in the Seventeenth Century*, translated by C. Hedström and G. Taylor, London, 1956.

BERGSTRÖM 1984:
Ingvar Bergström, 'Jacob Marrel's Earliest Tulip Book—Hitherto Unknown', *Tableau*, VII/2, 1984, pp. 33–49.

BERRALL:
J. S. Berrall, *A History of Flower Arrangement*, New York, 1953.

BLEILER:
E. F. Bleiler, editor, *Early Floral Engravings: All 110 Plates from the 1612 Florilegium by Emanuel Sweerts*, New York, 1976.

BLUNT 1950:
Wilfrid Blunt, *Tulipomania*, Harmondsworth, 1950.

BLUNT 1957:
Flowers Drawn from Nature: Reproduced from the 1800 folio 'Fleurs dessinées d'après nature' [by Gerard van Spaendonck], edited by Wilfrid Blunt, Sharpthorne, Sussex, 1957.

BLUNT and RAPHAEL:
Wilfrid Blunt and Sandra Raphael, *The Illustrated Herbal*, London, 1979.

BLUNT and STEARN:
Wilfrid Blunt, with the assistance of William T. Stearn, *The Art of Botanical Illustration*, London, 1950, revised by W. T. Stearn, Woodbridge, Suffolk, 1994.

BOGAERT-DAMIN and PIRON:
Anne-Marie Bogaert-Damin and Jacques Piron, editors, *Livres de fleurs du XVIe au XXe siècle dans les collections de la Bibliothèque Universitaire Moretus Plantin*, Namur, 1984.

BOL:
Laurens J. Bol, *The Bosschaert Dynasty: Painters of Flowers and Fruit*, Leigh-on-Sea, 1960.

Book of Flowers:
Anne Scott-James, Ray Desmond and Frances Wood, *The British Museum Book of Flowers*, London, 1989.

BORCHARDT:
Rudolf Borchardt, *Il Giardiniere appassionato*, Milan, 1992.

Botany in the Low Countries:
Botany in the Low Countries, End of the 15th Century—ca. 1650, exhibition catalogue, Plantin-Moretus Museum and the Stedelijk Prentenkabinet, Antwerp, 1993.

BOUCHOT:
Henri Bouchot, 'Le Portrait-miniature en France', *Gazette des Beaux-Arts*, II, 1893, pp. 392–415.

BOVEN and SEGAL:
Margriet van Boven and Sam Segal, *Gerard & Cornelis van Spaendonck: Twee Brabantse Bloemenschilders in Parijs*, Maarsen, 1980.

BREDEKAMP:
Horst Bredekamp, *Botticelli: 'La Primavera'*, Modena, 1996.

BREDIUS:
Abraham Bredius, *Künstler-Inventare. Urkunden zur Geschichte der holländischen Kunst des XVIten, XVIIten und XVIIIten Jahrhunderts*, 8 parts, The Hague, 1915–22.

BRIQUET:
Charles-Moïse Briquet, *Les Filigranes: Dictionnaire historique des marques du papier des leur apparition vers 1282 jusq'en 1600*, Amsterdam, 1968.

Bruegel: Une dynastie de peintres:
Bruegel: Une dynastie de peintres, exhibition catalogue, Palais des Beaux-Arts, Brussels, 1980.

BRUNET:
J. C. Brunet, *Manuel du libraire et de l'amateur de livres*, 6 vols., Paris, 1860–65.

BULTINGAIRE:
Léon Bultingaire, 'Les Origines de la collection des vélins du Muséum et ses premiers peintres', *Archives du Museum d'Histoire Naturelle*, 6th series, I, 1926, pp. 129–49.

CALMANN:
Gerta Calmann, *Ehret: Flower Painter Extraordinary*, Oxford, 1977.

CANEVA:
Giulia Caneva, *Il Mondo di Cerere nella Loggia di Psiche: Villa La Farnesina, sede dell'Accademia Nazionale dei Lincei*, Rome, 1992.

CARTER:
Thomas M. Carter, *The Victorian Garden*, London, 1984.

A Catalogue of Redoutéana:
A Catalogue of Redoutéana, exhibition catalogue edited by George H. M. Lawrence, Hunt Botanical Library, Pittsburgh, Pa., 1963.

CHASTEL:
André Chastel, *Musca depicta*, Milan, 1984.

CIARDI and TONGIORGI TOMASI:
Immagini anatomiche e naturalistiche nei disegni degli Uffizi: Secc. XVI e XVII, exhibition catalogue edited by R. P. Ciardi and Lucia Tongiorgi Tomasi, Gabinetto Disegni e Stampe degli Uffizi, Florence, 1984.

CLEVELAND:
The Cleveland Herbal, Botanical, and Horticultural Collections. A Descriptive Bibliography of pre-1830 Works from the Libraries of the Holden Arboretum, the Cleveland Medical Library Association, and the Garden Center of Greater Cleveland, compiled by Stanley H. Johnston Jr., Kent, Ohio, 1992.

COATS:
Alice M. Coats, *The Book of Flowers: Four Centuries of Flower Illustration*, New York, 1973.

COFFIN 1972:
The Italian Garden, edited by David R. Coffin, Dumbarton Oaks Colloquium on the History of Landscape Architecture I, Washington, D.C., 1972.

COFFIN 1991:
David R. Coffin, *Gardens and Gardening in Papal Rome*, Princeton, N.J., 1991.

COOKE and MCNICOL:
Dorothy Cooke and Pamela McNicol, editors, *A History of Flower Arranging*, Oxford, 1989.

DACOS and FURLAN:
Nicole Dacos and Caterina Furlan, *Giovanni da Udine, 1487–1561*, Udine, 1987.

DACOSTA KAUFMANN and KAUFMANN:
Thomas DaCosta Kaufmann and Virginia Roehrig Kaufmann, 'The Sanctification of Nature: Observations on the Origins of "trompe l'œil" in Netherlandish Book Painting of the Fifteenth and Sixteenth Centuries', in *The Mastery of Nature: Aspects of Art, Science, and Humanism in the Renaissance*, Princeton, N.J., 1993, pp. 11–48.

DANIELS:
G. S. Daniels, *Artists from the Royal Botanic Gardens, Kew*, Pittsburgh, Pa., 1974.

DBI:
Dizionario Biografico degli Italiani, Rome, 1960–.

DE BRAY:
Lys De Bray, *The Art of Botanical Illustration*, Bromley, Kent, 1989.

DECKERT:
Helmut Deckert, *Maria Sibylla Merians 'Neues Blumenbuch' (Nurnberg 1680)*, 2 vols., Leipzig, 1966.

DELAISSÉ:
L.M.J. Delaissé, 'The Importance of the Books of Hours for the History of the Medieval Book', in *Gatherings in Honor of Dorothy E. Miner*, Baltimore, Md., 1974, pp. 203–25.

DEMPSEY:
Charles Dempsey, '"Mercurius Ver": The Sources of Botticelli's "Primavera"', *Journal of the Warburg and Courtauld Institutes*, XXXI, 1968, pp. 251–73.

DESMOND 1986:
Ray Desmond, *Wonders of Creation: Natural History Drawings in the British Museum*, London, 1986.

DESMOND 1987:
Ray Desmond, *A Celebration of Flowers: Two Hundred Years of 'Curtis's Botanical Magazine'*, London, 1987.

DESMOND 1994:
Ray Desmond, *Dictionary of British and Irish Botanists and Horticulturalists*, 2nd edition, London, 1994.

DE TONI:
Giovanni Battista de Toni, *Le Piante e gli animali in Leonardo da Vinci*, Bologna, 1922.

DUNTHORNE:
Gordon Dunthorne, *Flower and Fruit Prints of the 18th and Early 19th Centuries*, London, 1938.

ELLIOTT:
Brent Elliott, *Treasures of The Royal Horticultural Society*, London, 1994.

EMBODEN:
William A. Emboden, *Leonardo da Vinci on Plants and Gardens*, London, 1987.

English Drawings and Watercolours:
English Drawings and Watercolours, 1550–1850, in the Collection of Mr and Mrs Paul Mellon, exhibition catalogue by J. Baskett and D. Snelgrove, with a Foreword by C. Ryskamp and Introduction by G. Reynolds; Pierpont Morgan Library, New York; Royal Academy of Arts, London; 1972.

EWAN:
Joseph Ewan, *William Bartram: Botanical and Zoological Drawings, 1756–1788*, Philadelphia, Pa., 1968.

FARÉ:
Michel and Fabrice Faré, *La Vie silencieuse en France: La Nature morte au XVIIIe siècle*, Fribourg, 1976.

FIGUIER:
Luigi Figuier, *Vita e costumi degli animali: gl'insetti*, Milan, 1893.

FISHER:
John Fisher, *The Origins of Garden Plants*, London, 1982.

Fleurs et jardins:
Fleurs et jardins dans l'art Flamand, exhibition catalogue by Paul Eeckhout, Brussels, 1960.

Flora Portrayed:
John V. Brindle and James J. White, *Flora Portrayed: Classics of Botanical Art from the Hunt Institute Collection*, Pittsburgh, Pa., 1985.

Floralia:
Floralia: Florilegio dalle collezioni Fiorentine del Sei—Settecento, exhibition catalogue edited by Marilena Mosco and Milena Rizzotto, Palazzo Pitti, Florence, 1988.

Flowers and Nature:
Flowers and Nature: Netherlandish Flower Painting of Four

Flowers and Nature: (cont.)
Centuries, exhibition catalogue by Sam Segal; Nabio Museum of Art, Osaka; Station Gallery, Tokyo; Art Gallery of New South Wales, Sydney; The Hague, 1990.

Flowers in Books and Drawings:
Flowers in Books and Drawings, c. 940–1840, exhibition catalogue with a preface by Charles Ryskamp, introduction by Helen B. Mules, Pierpont Morgan Library, New York, 1980.

Flowers into Art:
Flowers into Art: Floral Motifs in European Painting and Decorative Arts, exhibition catalogue edited by Vibeke Woldbye, Kunstindustrimuseet, Copenhagen; The Hague, 1991.

Flowers of Ten Centuries:
Flowers of Ten Centuries, exhibition catalogue, preface by Virginia V. White, Pierpont Morgan Library, New York, 1947.

Flowers of Three Centuries:
Flowers of Three Centuries: One Hundred Drawings and Watercolours from the Broughton Collection [Fitzwilliam Museum], exhibition catalogue by David Scrase, Washington, D.C., 1983–84.

FRANS HALSMUSEUM *Catalogus*:
Catalogus van Teekeningen, Schilderijen, Boeken, Pamfletten . . . , Haarlem, 1935.

FRATI:
Luigi Frati, *Catalogo dei manoscritti di Ulisse Aldrovandi*, Bologna, 1907.

FREEDBERG 1981:
David Freedberg, 'The Origins and Rise of the Flemish Madonnas in Flower Garlands: Decoration and Devotion', *Münchner Jahrbuch der bildenden Kunst*, XXXII, 1981, pp. 115–50.

FREEDBERG 1989:
David Freedberg, 'From Hebrew and Gardens to Oranges and Lemons: Giovanni Battista Ferrari and Cassiano dal Pozzo', in *Cassiano dal Pozzo: Atti del Seminario Internazionale di Studi*, edited by Francesco Solinas, Rome, 1989, pp. 37–72.

French Master Drawings:
French Master Drawings from the Pierpont Morgan Library, exhibition catalogue edited by Cara D. Denison, Pierpont Morgan Library, New York, 1993.

FRIEDMAN:
Joan Friedman, 'Every Lady her own Drawing Master', *Apollo*, CV, April 1977, pp. 262–7.

GAMMELBO:
P. Gammelbo, 'Some Flower Still Lifes by Jean Baptiste Van Fornenburgh', *Artes*, October 1985, pp. 5–16.

GASCAR and AYMONIN:
Pierre Gascar and Gerard G. Aymonin, *L'Herbier des Quatre Saisons ou le jardin d'Eichstatt, Basilius Besler*, Paris, 1987.

GEISSLER:
H. Geissler, 'Die Stichvorbereitung für Giovan Battista Ferrari Botanische Werke', *Jahrbuch der Staatlichen Kunstsammlungen in Baden-Württemberg*, XV, 1978, pp. 28–35.

GERARD:
Robert A. Gerard, 'Woutneel, de Passe and the Anglo-Netherlandish Print Trade', *Print Quarterly*, XIII, 1996, pp. 363–76.

Il Giardino di Flora:
Il Giardino di Flora: Natura e simbolo nell'immagine dei fiori, exhibition catalogue edited by Marzia Cataldi Gallo and Farida Simonetti, Loggia della Mercanzia, Genoa, 1986.

GINORI CONTI:
Piero Ginori Conti, *Lettere inedite di Charles de L'Escluse (Carolus Clusius) a Matteo Caccini, floricultore Fiorentino*, Florence, 1939.

GOLDSMITH:
V. F. Goldsmith, *A Short Title Catalogue of French Books, 1601–1700, in the Library of the British Museum*, 7 parts, Folkestone and London, 1969–73.

GONCOURT:
Edmond and Jules de Goncourt, *L'Art du dix-huitième siècle*, 3rd edition, 2 vols., Paris, 1880–82.

GONZALES-PALACIOS:
Alvar Gonzáles-Palacios, 'Limiti e contesto di Leonardo van der Vinne', *Paragone*, XXVIII/333, 1977, pp. 37–68.

GOODCHILD:
Peter Goodchild, 'John Rea's Gardens of Delight: Introduction and the Construction of the Flower Garden', *Garden History*, Journal of the Garden History Society, IX/2, 1981, pp. 99–109.

GOODY:
Jack Goody, *The Culture of Flowers*, Cambridge, 1993.

Grandville:
Grandville: Dessins originaux, exhibition catalogue by Clive F. Getty and S. Guillaume, Musée des Beaux-Arts, Nancy, 1986.

GRANT:
M. H. Grant, *Flower Paintings through Four Centuries: A*

Descriptive Catalogue of the Collection formed by Major the Hon. Henry Rogers Broughton, Leigh-on-Sea, 1932.

GREENE:
Edward Lee Greene, *Landmarks of Botanical History*, revised edition by Frank N. Egerton, 2 vols., Stanford, Calif., 1983.

GREGORI 1986:
Mina Gregori, 'Linee della natura morta Fiorentina', in *Il Seicento Fiorentino*, I, exhibition catalogue, Palazzo Strozzi, Florence, 1986, pp. 31–51.

GREGORI 1991:
Mina Gregori, 'La Natura morta in Toscana', *La Natura morta in Italia*, II, Milan, 1991, pp. 513–23.

GREGORIO and CELLI:
Maria Gregorio and Giorgio Celli, editors, *Maria Sibylla Merian: la meravigliosa metamorfosi dei bruchi*, Turin, 1993.

GRIGSON and BUCHANAN:
Thornton's 'Temple of Flora', introduction by Geoffrey Grigson and Handasyde Buchanan, London, 1951.

GUILLOT:
Gerard Guillot, 'Les Solitaires de Lyon', *Apollo*, LXXX, October 1964, pp. 301–05.

HAIRS:
Marie-Louise Hairs, *Les Peintres Flamands de fleurs au XVIIe siècle*, Paris, 1955.

HALLER:
Albrecht von Haller, *Bibliotheca Botanica qua Scripta ad Rem Herbariam facientia a rerum initiis recensentur*, 2 vols., Tiguri [Zurich], 1771–72.

HARDOUIN-FUGIER:
Elisabeth Hardouin-Fugier, *The Pupils of Redouté*, Leigh-on-Sea, 1981.

HARDOUIN-FUGIER and GRAFE 1989:
Elisabeth Hardouin-Fugier and Etienne Grafe, *French Flower Painters of the Nineteenth Century: A Dictionary*, edited by Peter Mitchell, London, 1989.

HARDOUIN-FUGIER and GRAFE 1992:
Elisabeth Hardouin-Fugier and Etienne Grafe, *Les Peintres des fleurs en France de Redouté à Redon*, Paris, 1992.

HARRIS:
John Harris, *Gardens of Delight: The Rococo English Landscapes of Thomas Robins the Elder*, with Botanical Notes by Martyn Rix, 2 vols., London, 1978.

HATTON:
Richard G. Hatton, *Handbook of Plant and Floral Ornament; Selected from the Herbals of the Sixteenth Century*, New York, 1960 (first published as *The Craftsman's Plant-book*, London, 1909).

HAYDEN:
Ruth Hayden, *Mrs Delany: Her Life and her Flowers*, London, 1980.

HEDRICK:
U. P. Hedrick, *A History of Horticulture in America to 1860*, New York, 1950.

HEIKAMP:
Detlef Heikamp, 'Agostino del Riccio: Il Giardino di un re', in *Il Giardino storico Italiano*, edited by Giovanna Ragionieri, Florence, 1981, pp. 65–123.

HELD:
Julius S. Held, 'Flora: Goddess and Courtesan', in *De artibus opuscula XL: Essays in Honor of Erwin Panofsky*, edited by Millard Meiss, 2 vols., New York, 1961, I, pp. 201–25.

HENDRIX and VIGNAU-WILBERG:
Mira Calligraphiae Monumenta: A Sixteenth-century Calligraphic Manuscript Inscribed by Georg Bocksay and Illuminated by Joris Hoefnagel, introduction by Lee Hendrix and Thea Vignau-Wilberg, Malibu, Calif., 1992.

HENREY:
Blanche Henrey, *British Botanical and Horticultural Literature before 1800*, 3 vols., London, 1975.

History of Orchids:
The History of Orchids, exhibition catalogue by Joachim Wolschke-Bulmahn, Dumbarton Oaks, Washington, D.C., 1994.

HOBHOUSE:
Penelope Hobhouse, *Gardening Through the Ages*, London, 1992.

HOFER:
A Catalogue of an Exhibition of the Philip Hofer Bequest in the Department of Printing and Graphic Arts (*Harvard College Library*), Cambridge, Mass., 1988.

HOLLSTEIN 1949:
F.W.H. Hollstein, *Dutch and Flemish Etchings, Engravings, and Woodcuts, c. 1450–1700*, Amsterdam, 1949.

HOLLSTEIN 1954:
F.W.H. Hollstein, *German Engravings, Etchings, and Woodcuts, c. 1400–1700*, Amsterdam, 1954.

HOOZEE:
Robert Hoozee, *L'Opera completa di Constable*, Milan, 1979.

HULTON:
Paul Hulton, *The Work of Jacques Le Moyne de Morgues: A*

HULTON: (*cont.*)
Huguenot Artist in France, Florida and England, 2 vols., London, 1977.

HULTON and SMITH:
Flowers in Art from East and West, exhibition catalogue by Paul Hulton and Lawrence Smith, British Museum, London, 1979.

HUNT:
Catalogue of Botanical Books in the Collection of Rachel McMasters Miller Hunt, compiled by Jane Quinby and Allan Stevenson, 2 vols. (the second in 2 parts), Pittsburgh, Pa., 1958–61.

JALABERT:
Denise Jalabert, 'Memoir de la Flore dans l'Art Français', *Médicine de France*, XXXIII, 1952, pp. 17–28.

JARDINE, SECORD and SPARY:
N. Jardine, J. A. Secord and E. C. Spary, editors, *Cultures of Natural History*, Cambridge, 1996.

JOUANIN:
Christian Jouanin, 'Joséphine and the Natural Sciences', *Apollo*, CVI, July 1977, pp. 50–59.

JRHS:
Journal of the Royal Horticultural Society.

KEMP:
Martin Kemp, '"Implanted in our Natures": Humans, Plants, and the Stories of Art', in *Voyages, Botany, and Representations of Nature*, edited by David Philip Miller and Peter Hanns Reill, Cambridge, 1996, pp. 197–229.

KEUNECKE:
Hortus Eistettensis: zur Geschichte eines Gartens und eines Buches; Schriften der Universitätsbibliothek Erlangen-Nürnberg, edited by Hans Otto-Keunecke, Munich, 1989.

KING:
'The Temple of Flora' by Robert Thornton, introduction by Ronald King, New York, 1981.

Kleuren Raffinement:
Kleuren Raffinement: Tekeningen uit de Unicorno collectie, exhibition catalogue by Charles Dumas and Robert Jan te Rijdt, Museum het Rembrandthuis, Zwolle, 1994.

KORENY:
Fritz Koreny, *Albrecht Dürer and the Animal and Plant Studies of the Renaissance*, Boston, Mass., 1988.

KRELAGE 1942a:
E. H. Krelage, *Bloemenspeculatie in Nederland. De tulpomanie van 1636–'37 en de hyacintenhandel 1720–'36*, Amsterdam, 1942.

KRELAGE 1942b:
E. H. Krelage, *De Pamfletten van den tulpenwindhandel, 1636–1637*, The Hague, 1942.

KREMPEL:
Jan van Kessel d. A. 1629–1679: Die Vier Erdteile, exhibition catalogue by Ulla Krempel, Munich, 1973.

KRÜSSMANN:
Gerd Krüssmann, *Roses*, London, 1981.

LACOMBE:
Paul Lacombe, *Livres d'Heures imprimés au XVe et au XVIe siècle*, Nieuwkoop, 1963.

LAISSUS 1969:
Yves Laissus, 'Les Plantes du Roi: Note sur un grand ouvrage de botanique preparé au XVIIe siècle par l'Academie Royale des Sciences', *Revue d'Histoire des Sciences*, XXII, 1969, pp. 193–236.

LAISSUS 1980:
Yves Laissus, *Nicolas Robert et les Vélins du Muséum National d'Histoire Naturelle*, Paris, 1980.

LANDWEHR:
John H. Landwehr, *Studies in Dutch Books with Coloured Plates published 1662–1875*, The Hague, 1976.

LAUTERBACH:
Iris Lauterbach, *Der französische Garten am Ende des Ancien Régime. 'Schöne Ordnung' und 'geschmackvolles Ebenmass'*, Worms, 1987.

LÉGER:
Charles Léger, *Redouté et son temps*, Paris, 1945.

LEITH-ROSS:
Prudence Leith-Ross, 'A Seventeenth-century Paris Garden', *Garden History*, Journal of the Garden History Society, XXI/2, 1993, pp. 150–57.

LEMMON:
Kenneth Lemmon, *The Golden Age of Plant Hunters*, London, 1968.

LENDORFF:
Gertrud Lendorff, *Maria Sibylla Merian, 1647–1717; ihr Leben und ihr Werk*, Basel, 1955.

LE ROUGETEL:
Hazel Le Rougetel, *The Chelsea Gardener: Philip Miller, 1691–1771*, London, 1990.

LEVI D'ANCONA 1977:
Mirella Levi D'Ancona, *The Garden of the Renaissance: Botanical Symbolism in Italian Painting*, Florence, 1977.

LEVI D'ANCONA 1983:
Mirella Levi D'Ancona, *Botticelli's 'Primaveri': A Botanical Interpretation, including Astrology, Alchemy and the Medici*, Florence, 1983.

LEWIS:
Jan Lewis, *Walter Hood Fitch: A Celebration*, London, 1992.

LUGT:
Frits Lugt, *Répertoire des catalogues de ventes publiques*, 3 vols., The Hague, 1938–64.

MACDOUGALL:
Elisabeth Blair MacDougall, *Fountains, Statues and Flowers: Studies in Italian Gardens of the Sixteenth and Seventeenth Centuries*, Dumbarton Oaks, Washington, D.C., 1994.

MCMILLAN:
Nora McMillan, 'Mrs Edward Bury (née Priscilla Susan Faulkner), Botanical Artist', *Journal of the Society for the Bibliography of Natural History*, V, 1968, pp. 71–5.

A Magnificent Collection:
A Magnificent Collection of Botanical Books: Being the Finest Colour-plate Books from the Celebrated Library formed by Robert de Belder, sale catalogue, Sotheby's, London, 1987.

MAIORINO *et al.*:
Antonella Maiorino, Marcella Minelli, Anna Letizia Monti, Barbara Negroni and Ada Segre, 'L'Uso dei bulbi da fiore nei giardini del Rinascimento', in *Boboli '90: Atti del Convegno Internazionale di Studi*, edited by Cristina Acidini Luchinat and Elvira Garbero Zorzi, Florence, 1991, pp. 277–89.

Martin Johnson Heade:
Martin Johnson Heade, exhibition catalogue by Theodore E. Stebbins, Museum of Fine Arts, Boston, Mass., 1969.

MARTIN and CHARTIER:
Histoire de l'edition française, edited by Henri-Jean Martin and Roger Chartier, 4 vols., Paris, 1983–87.

MASSON 1961:
Georgina Masson, *Italian Gardens*, New York, 1961.

MASSON 1968:
Georgina Masson, 'Italian Flower Connoisseurs', *Apollo*, LXXXVIII, 1968, pp. 164–71.

MASSON 1970:
Georgina Masson, 'Fiori quali pezzi da collezione nell'Italia del secolo XVII', *Arte Illustrata*, III/30–33, 1970, pp. 100–09.

MASSON 1972:
Georgina Masson, 'Italian Flower Collectors' Gardens in Seventeenth-century Italy', in *The Italian Garden*, edited by David R. Coffin, Washington, D.C., 1972, pp. 63–80.

MEYER:
Veronique Meyer, 'L'Oeuvre gravé de Daniel Rabel', *Nouvelles de l'Estampe*, LXVII, 1983, pp. 6–15.

MIGLIORINI:
Bruno Migliorini, *Parole d'autore* (*Onomaturgia*), Florence, 1975.

MITCHELL 1973:
Peter Mitchell, *European Flower Painters*, London, 1973.

MITCHELL 1992:
Peter Mitchell, *Flowers*, London, 1992.

MOGGI:
Guido Moggi, 'Piante e fiori nella "Primavera"', in *Metodo e scienza: Operatività e ricerca nel restauro*, edited by Umberto Baldini, Florence, 1982, pp. 217–24.

MORETON:
Charles Oscar Moreton, *The Auricula: Its History and Character, with Seventeen Coloured Plates reproduced from Paintings by Rory McEwen*, London, 1964.

MORTIMER:
Ruth Mortimer, *Harvard College Library. Department of Printing and Graphic Arts, Catalogue of Books and Manuscripts*, Part I: *French 16th-century Books*, 2 vols.; Part II: *Italian 16th-century Books*, 2 vols., Cambridge, Mass., 1964–74.

MORTON:
Theresa-Mary Morton, *A Souvenir Album of Flowers from the Royal Collection, Windsor Castle*, 1990.

MURRAY:
W. S. Murray, 'The Introduction of the Tulip, and Tulipomania', *JRHS*, XXXV, 1909–10, pp. 8–30.

Natura Viva in Casa Medici:
Natura Viva in Casa Medici, exhibition catalogue edited by Marilena Mosco, Palazzo Pitti, Florence, 1985–86.

NEBEL:
Gerhard Nebel, editor, *Die schönsten Tafeln aus dem grossen Buch der Schmetterlinge und Pflanzen, Metamorphosis insectorum Surinamensium, Maria Sybilla Merian*, Hamburg, 1964.

Nederlandse en Vlaamse:
Nederlandse en Vlaamse tekeningen uit de zeventiende en achttiende eeuw, exhibition catalogue by Jap Bolten, Museum voor Stad en Lande, Groningen, 1965.

NISSEN:
Claus Nissen, *Die botanische Buchillustration. Ihre Geschichte und Bibliographie*, 2 vols., Stuttgart, 1951.

NOCCHI and PELLEGRINI:
Paolo B. Nocchi and Ezio Pellegrini, 'La Collezione Botani-

Nocchi and Pellegrini (*cont.*)
ca del Cardinal Odoardo', in *Gli Orti Farnesiani sul Palatino*, Rome, 1990, pp. 413–29.

An Oak Spring Garland:
An Oak Spring Garland: Illustrated Books, Prints, and Drawings from the Oak Spring Garden Library, Upperville, described by Sandra Raphael (the catalogue for an exhibition in Princeton University Library), Upperville, Va., 1989.

An Oak Spring Hortus:
An Oak Spring Hortus: Garden Design in Italy, France, and the North and South Netherlands, A Selection of the Rare Books, Manuscripts, and Works of Art in the Collection of Rachel Lambert Mellon, by John Dixon Hunt, Upperville, Va., in preparation.

An Oak Spring Pomona:
An Oak Spring Pomona: A Selection of the Rare Books on Fruit in the Oak Spring Garden Library, Upperville, described by Sandra Raphael, Upperville, Va., 1990.

An Oak Spring Sylva:
An Oak Spring Sylva: A Selection of the Rare Books on Trees in the Oak Spring Garden Library, Upperville, described by Sandra Raphael, Upperville, Va., 1989.

OCG:
The Oxford Companion to Gardens, edited by Geoffrey and Susan Jellicoe, Patrick Goode and Michael Lancaster, Oxford, 1986.

Olmi 1989:
Giuseppi Olmi, 'Natura morta e illustrazione scientifica', in *La Natura morta in Italia*, edited by F. Zeri, Milan, 1989, I, pp. 69–91.

Olmi 1992:
Giuseppi Olmi, *L'Inventario del mondo: catalogazione della natura e luoghi del sapere nella prima età moderna*, Bologna, 1992.

Pächt 1947:
Otto Pächt, *The Master of Mary of Burgundy*, London, 1948.

Pächt 1950:
Otto Pächt, 'Early Nature Studies and the Calendar Landscape', *Journal of the Warburg and Courtauld Institutes*, XIII, 1950, pp. 13–47.

Painting in England:
Painting in England, 1700–1850: Collection of Mr and Mrs Paul Mellon, exhibition catalogue by Basil Taylor, Virginia Museum of Fine Arts, Richmond, Va., 1963.

Pallottino:
Paola Pallottino, 'Riflessi dall'ultimo specchio. Da "Les Fleurs animées" di Grandville all'universo delle figurine: un esempio di persistenza, evoluzione e migrazione di schemi iconografici "forti"', in *Figurine! Pubblicità, arte, collezionismo e industria, 1867–1985*, exhibition catalogue, Palazzo Reale, Milan, 1989, pp. 48–61.

Panofsky:
Erwin Panofsky, *Early Netherlandish Painting: Its Origins and Character*, 2 vols., Cambridge, Mass., 1953.

Pavière:
S. H. Pavière, *Floral Art: Great Masters of Flower Painting*, Leigh-on-Sea, 1965.

Peintres des fleurs:
Peintres des fleurs en France du XVIIe au XIXe siècles, exhibition catalogue edited by A. Cacan de Bissy, with an essay by Michel and Fabrice Faré, Musée du Petit Palais, Paris, 1979.

Pfeiffer:
Max Adolf Pfeiffer, *Die Werke der Maria Sibylle Merian*, Meissen, 1931.

Pinault:
Madeleine Pinault, *The Painter as Naturalist from Dürer to Redouté*, Paris, 1991.

Pischel:
Bouquets, introduction by Emy Pischel, with texts by Jean Louis Prévost, London, 1960.

Pizzetti and Cocker:
Ippolito Pizzetti and Henry Cocker, *Il Libro dei fiori*, Milan, 1976.

Plesch:
Mille et un livres botaniques: Répertoire bibliographique de la Bibliothèque Arpad Plesch, by Henri Pierre Gourry, Brussels, 1973.

Pozzi:
Giovanni Pozzi, *Sull'Orlo del visibile parlare*, Milan, 1993.

Printmaking in the Service of Botany:
Gavin D. R. Bridson, Donald E. Wendel and James A. White, editors, *Printmaking in the Service of Botany*, Pittsburgh, Pa., 1985.

Pritzel:
Thesaurus Literaturae Botanicae, omnium gentium [1851], edited by G. A. Pritzel, 2nd edition, Leipzig, 1872–7 (reprinted Milan, 1950).

Puppi:
Lionello Puppi, 'The Villa Garden of the Veneto from the Fifteenth to the Eighteenth Century', in *The Italian Garden*, edited by David R. Coffin, Washington, D.C., 1972, pp. 89–114.

RENONCIAT:
Annie Renonciat, *La Vie et l'œuvre de J. J. Grandville*, Paris, 1985.

REYNOLDS:
Graham Reynolds, *The Later Paintings and Drawings of John Constable*, 2 vols., New Haven and London, 1984.

RHS:
Exhibition of Manuscripts, Books, Drawings, Portraits, Medals and Congratulatory Addresses on the Occasion of the Society's 150th Anniversary Celebrations, exhibition catalogue, Royal Horticultural Society, Vincent Square, London, 1954.

RITTERBUSH:
Philip C. Ritterbush, 'Art and Science as Influences on the Early Development of Natural History Collections', *Proceedings of the Biological Society of Washington*, LXXXII, 1969, pp. 561–78.

RIX:
Martyn Rix, *The Art of the Botanist*, Guildford and London, 1981.

RIZZOTTO:
Milena Rizzotto, 'Le Amene mostruosità: Coltura, selezione e manipolazione delle piante ornamentali nel XVII secolo', in *Le Monde végétal (XIIe—XVIIe siècles): Savoirs et usages sociaux*, edited by Allen J. Grieco, Odile Redon and Lucia Tongiorgi Tomasi, Saint-Denis, 1993, pp. 141–56.

ROHDE:
E. S. Rohde, *The Old English Herbals*, London, 1922.

RORY MCEWEN:
Rory McEwen, 1932–1982: The Botanical Paintings, exhibition catalogue edited by Caroline Cuthbert; Royal Botanic Garden, Edinburgh; Serpentine Gallery, London; 1988.

Rosenwald Collection:
The Lessing J. Rosenwald Collection: A Catalog of the Gifts of Lessing J. Rosenwald to the Library of Congress, 1943 to 1975, Washington, D.C., 1977.

RUSSELL:
Daniel S. Russell, *The Emblem and Device in France*, Lexington, Ky., 1985.

SALMI:
Breviario Grimani, introduction by Mario Salmi, 2nd edition, Milan, 1979.

SAUNDERS:
Gill Saunders, *Picturing Plants: An Analytical History of Botanical Illustration*, London, Berkeley and Los Angeles, 1995.

SAVAGE 1922:
Spencer Savage, 'The Discovery of Some of Jacques Le Moyne's Botanical Drawings', *Gardeners' Chronicle*, 3rd series, LXXI, 1922, p. 44.

SAVAGE 1923:
Spencer Savage, 'The "Hortus Floridus" of Crispijn vande Pas the Younger', *Transactions of the Bibliographical Society*, 2nd series, IV, 1923, pp. 181–206.

SAVAGE 1925:
Spencer Savage, *Catalogue of the Printed Books and Pamphlets in the Library of the Linnean Society of London*, 2nd edition, London, 1925.

SCHAAP:
Ella B. Schaap, *Bloemen op tegels in de Gouden Eeuw Van prent tot tegel / Dutch Floral Tiles in the Golden Age and their Botanical Prints*, Haarlem, 1994.

SCHAMA:
Simon Schama, *The Embarrassment of Riches: An Interpretation of Dutch Culture in the Golden Age*, New York, 1987.

Schlumberger:
Jean Schlumberger, II: *Objets*, introduction by Jean d'Ormesson, Milan, 1976.

SCHNAPPER 1983:
Antoine Schnapper, '"Curieux fleuristes": Collectionneurs de fleurs dans la France du XVIIe siècle', *Commentaire*, XXI, 1983, pp. 171–80.

SCHNAPPER 1985:
Antoine Schnapper, 'Fleurs, fruits, vélins', in *L'Age d'Or du Mécénat (1598–1661)*, CNRS International Colloquium, March 1983, Paris, 1985, pp. 353–62.

SCHNAPPER 1988:
Antoine Schnapper, *Le Géant, la licorne et la tulipe: Collections et collectionneurs dans la France du XVIIe siècle*, Paris, 1988.

SCHNEIDER:
Norbert Schneider. *The Art of the Still Life*, Cologne, 1990.

SCOURSE:
Nicolette Scourse, *The Victorians and their Flowers*, London, 1983.

SEATON:
Beverly Seaton, 'French Flower Books', *Nineteenth-century French Studies*, XI, 1980, pp. 60–72.

SEGAL 1982:
Sam Segal, *A Flowery Past: A Survey of Dutch and Flemish Flower Painting from 1600 until the Present*, Amsterdam, 1982.

SEGAL 1993:
Sam Segal, 'Tulips Portrayed: The Tulip Trade in Holland in the 17th Century', in *The Tulip: A Symbol of Two Nations*, exhibition catalogue edited by Michiel Roding and Hans Theunissen, Utrecht and Istanbul, 1993, pp. 9–24.

SEGRE:
Ada Segre, 'Le Metamorfosi e il giardino Italiano nel Seicento', in *Il Giardino delle Muse: Arti e artifici nel Barocco Europeo*, edited by Maria A. Giusti and Alessandro Tagliolini, Florence, 1995, pp. 97–126.

SÉGUIER:
Jean-François Séguier, *Bibliotheca botanica*, Hagae-Comitum [The Hague], 1740.

SHEPHERD:
Roy E. Shepherd, *History of the Rose*, New York, 1978.

SHERWOOD COLLECTION:
Shirley Sherwood, *Contemporary Botanical Artists: The Shirley Sherwood Collection*, edited by Victoria Matthews, London, 1996.

SHTEIR:
Ann B. Shteir, *Cultivating Women, Cultivating Science: Flora's Daughters and Botany in England, 1760–1860*, Baltimore and London, 1996.

SITWELL and BLUNT:
Sacheverell Sitwell and Wilfrid Blunt, *Great Flower Books, 1700–1900: A Bibliographical Record*, London, 1956.

SOLINAS:
Cassiano dal Pozzo: Atti del Seminario Internazionale di Studi, edited by F. Solinas, Rome, 1989.

SPRENGEL:
Kurt Sprengel, *Geschichte der Botanik*, 2 vols., Altenburg and Leipzig, 1817–18.

STAFLEU and COWAN:
Frans A. Stafleu and Richard S. Cowan, *Taxonomic Literature*, 7 vols., 2nd edition, Utrecht, 1976–88.

STEARN 1965:
William T. Stearn, 'The Origin and Development of Garden Plants', *JRHS*, XC, 1965, pp. 279–91, 322–40, 350.

STEARN 1978:
William T. Stearn, *The Wondrous Transformation of Caterpillars: Fifty Engravings Selected from 'Erucarum Ortus' (1718)*, London, 1978.

STEARN 1990:
William T. Stearn, *Flower Artists of Kew: Botanical Paintings by Contemporary Artists*, London, 1990.

STEARN and RIX:
William T. Stearn and Martyn Rix, *Redouté's Fairest Flowers*, London, 1987.

STECHOW:
Wolfgang Stechow, 'Ambrosius Bosschaert and the Appearance of Independent Flower Painting', *Bulletin of the Cleveland Museum of Art*, LIII, 1965, pp. 59–65.

STEIN:
H. Stein, 'La Collection des Vélins du Muséum d'Histoire Naturelle', *Gazette des Beaux-Arts*, April 1890, pp. 290–306.

Stiftung für Botanik, Vaduz:
The Magnificent Botanical Library of the Stiftung für Botanik, Vaduz, Liechtenstein, Collected by the late Arpad Plesch, 3 vols., sale catalogue, Sotheby's, London, 1975–76.

Stilleben in Europa:
Stilleben in Europa, exhibition catalogue, Westfälisches Landesmuseum für Kunst und Kulturgeschichte, Munster; Staatliche Kunsthalle, Baden-Baden; 1979–80.

Symbolique et botanique:
Symbolique et botanique: le sens caché des fleurs dans la peinture au VIIe siècle, exhibition catalogue, Musée des Beaux-Arts, Caen, 1987.

THIEME–BECKER:
U. Thieme and F. Becker, editors, *Allgemeines Lexikon der bildenden Künstler von der Antike bis zur Gegenwart*, 37 vols., Leipzig, 1907–50.

THOMAS:
Keith Thomas, *Man and the Natural World: Changing Attitudes in England, 1500–1800*, London, 1983.

TOESCA:
Ilaria Toesca, 'A Florilegium by Franciscus de Geest', *Oud-Holland*, LXXII, 1957, pp. 121–5.

TONGIORGI TOMASI 1980:
Lucia Tongiorgi Tomasi, 'Il Giardino dei Semplici dello Studio Pisano', in *Livorno e Pisa: due città e un territorio nella politica dei Medici*, Pisa, 1980, pp. 514–98.

TONGIORGI TOMASI 1983:
Lucia Tongiorgi Tomasi, 'Projects for Botanical and other Gardens: A 16th-century Manual', *Journal of Garden History*, III/1, 1983, pp. 1–34.

TONGIORGI TOMASI 1989:
Lucia Tongiorgi Tomasi, 'Gherardo Cibo: Visions of Landscape and the Botanical Sciences in a Sixteenth-century Artist', *Journal of Garden History*, IX/4, 1989, pp. 199–216.

TONGIORGI TOMASI 1991:
Lucia Tongiorgi Tomasi, 'Arte e natura: Il Giardino dei Semplici dalle origini alla fine dell'età medicea', in *Giardino dei Semplici*, edited by Fabio Garbari, Lucia Tongiorgi Tomasi and Alessandro Tosi, Pisa, 1991, pp. 115–212.

TONGIORGI TOMASI 1993:
Lucia Tongiorgi Tomasi, *I Ritratti di piante di Iacopo Ligozzi*, Pisa, 1993.

TONGIORGI TOMASI 1995:
Lucia Tongiorgi Tomasi, 'Tulipomania Addenda', in *Il Giardino delle Muse: Arti e artifici nel Barocco Europeo*, edited by Maria A. Giusti and Alessandro Tagliolini, Florence, 1995, pp. 79–96.

TONGIORGI TOMASI and FERRARI:
Lucia Tongiorgi Tomasi and Angela Ferrari, 'Botanica Barocca', *Gazzetta del Bibliofilo*, XVII, 1986, supplement in *FMR*, no. 46, 1986, pp. 3–17.

TONGIORGI TOMASI and TOSI:
Flora e Pomona: L'Orticoltura nei disegni e nelle incisioni dei secoli XVI–XIX, exhibition catalogue by Lucia Tongiorgi Tomasi and Alessandro Tosi, Gabinetto dei Disegni e Stampe, Uffizi, Florence, 1990.

TOSI:
Alessandro Tosi, 'Il Ritratto del naturalista tra XVI e XVII secolo', in *Luca Ghini: Cinquecento anni di scienze botaniche (1490–1990): Atti del Convegno Internazionale*, Imola, 1992, pp. 297–326.

ULLMANN:
Ernst and Helga Ullmann, *Les Aquarelles de Leningrad* [*de*] *Maria Sibylla Merian*, Paris, 1974.

VANUXEM:
Jacques Vanuxem, 'Fleurs sur vélin: Nicolas Robert et Jean Joubert', *L'Oeil*, CXVII, 1969, pp. 1–7.

VECA 1981:
Vanitas: Il Simbolismo del tempo, exhibition catalogue by Alberto Veca, Galleria Lorenzelli, Bergamo, 1981.

VECA 1982:
Paradeisos—Dall'Universo del fiore, exhibition catalogue by Alberto Veca, Galleria Lorenzelli, Bergamo, 1982.

VERWEY:
Herman de la Fontaine Verwey, 'The Binder Albertus Magnus and the Collectors of his Age', *Quærendo*, I/3, 1971, pp. 158–78.

VEZIN:
Luc Vezin, *Les Artistes au Jardin des Plantes*, Paris, 1990.

VIGNAU-WILBERG:
Thea Vignau-Wilberg-Schuurman, *Archetypa Studiaque Patris Georgii Hoefnagelii 1592: Natur, Dichtung und Wissenschaft in der Kunst um 1600*, Munich, 1994.

WALKER:
John Walker, *John Constable*, New York, 1978.

WARBURG:
Aby Warburg, *Sandro Botticellis Geburt des Venus und Frühling*, Hamburg, 1893.

WATSON:
William P. Watson, 'Redouté—Artist and Technician', in *Lilies, and Related Flowers by P. J. Redouté*, text by Patrick Mathew, with contributions by W. T. Stearn and W. P. Watson, Woodstock, N.Y., 1982, pp. 8–15.

WHITEHEAD and EDWARDS:
P.J.P. Whitehead and Phyllis Edwards, *Chinese Natural History Drawings Selected from the Reeves Collection in the British Museum* (*Natural History*), London, 1974.

WICK:
Peter A. Wick, *The Court of Flora / Les Fleurs animées: The Engraved Illustrations of J. J. Grandville*, New York, 1981.

WIJNANDS:
D. Onno Wijnands, *The Botany of the Commelins*, Rotterdam, 1983.

WIND:
Edgar Wind, 'Botticelli's "Primavera"', in *Pagan Mysteries in the Renaissance*, New Haven, Conn., 1958, pp. 100–110.

WOLSCHKE-BULMAHN:
Joachim Wolschke-Bulmahn, '"Aus der neuen Welt": Zur Einführung exotischer Pfanzen in die europaïsche Gartenkultur', *Die Gartenkunst*, IV/2, 1992, pp. 317–34.

A World of Flowers:
A World of Flowers: Paintings and Prints, exhibition catalogue edited by Henry Clifford, Philadelphia Museum of Art, 1963.

AN OAK SPRING FLORA

I • FLOWERS AND DEVOTION

BEFORE Johann Gutenberg invented printing by means of moveable type, which introduced both printed text and the means of its mass production, all texts, whether liturgical or lay, were composed, copied and decorated by hand—in the ancient world, in the *scriptoria* of monasteries, and later by specialist artists and artisans. The manuscript, always greatly prized as a vehicle of learning, was also concomitantly appreciated for its value as a work of art created for wealthy patrons. Throughout the Middle Ages and Renaissance, ornate bindings and pages made of ever finer, softer and whiter vellum, elegant calligraphy and, above all, the initials and images magnificently illuminated in bright colours with fine gold all contributed to enhance the value of these works in the eyes of their beholders.

Many botanical and floral images can be found in one of the oldest surviving illuminated manscripts dealing with naturalistic subjects, the Codex Vindobonensis after Dioscorides (Vienna, Nationalbibliothek), which dates from the sixth century. From the Middle Ages on, however, phytomorphic and zoomorphic elements came to be used with increasing frequency by artists in the initials, marginal decorations and full-folio illustrations that embellish medieval manuscripts.

Among the liturgical manuscripts produced in the medieval period, those known as Books of Hours often constituted works of art of surpassing beauty, being executed by highly accomplished artists. Unlike the more ordinary Psalter and Breviary, which were intended for daily use by the clergy, the Book of Hours was designed to assist noble lay persons in the execution of their private devotions, marking off the hours of the day in accordance with the dictates of the Catholic faith and thus allowing the secular to participate indirectly in the liturgy of the Church. In this way the laity could reconcile its religious duties with the many other pressing tasks of daily life, this duty being rendered all the less onerous by the pleasure of owning and using an object of exceptional beauty.

In the period between the end of the fourteenth century and the first decades of the sixteenth, the use of Books of Hours among the wealthy became widespread; indeed, Delaissé (p. 203) could without exaggeration describe them as the 'best sellers of the late medieval period'. In addition to the Little Office of the Sacred Virgin, a Book of Hours often included the Church's Saints' and Feast days, the Seven Penitential Psalms, the Litany, the Office of the Dead and other devotional texts, conceived and laid out in such a way as to leave ample space for the illustrations, initials and decorative marginalia (often executed by different specialists). Indeed, the secular audience for which these works were destined made the continuous interweaving of the text (generally Latin) and *sacrae imagines* (sacred illustrations) a more than happy expedient, for, as one of the early Church Fathers—

Pope Gregory the Great—observed: 'Pictura est laicorum scriptura' (The picture is the text of the lay person).

Since Books of Hours were intended for use by private individuals, their production did not come under the rigid control of the ecclesiastical authorities. As a result they frequently contained errors or omissions committed by the *scriptor* during his transcription of the text, or variations in the ordering of their sections. The calendar was the only fixed point in a Book of Hours, always appearing at the beginning of the manuscript. However, although in those manuscripts that were produced in the diocese of Paris, for example, the calendar was usually quite complete, those produced in Rome were often quite selective.

During the course of the fifteenth century, the pictorial component of the Book of Hours became more or less codified. The calendar was typically embellished with scenes depicting the cycle of the months, the signs of the zodiac or, sometimes, with the images of favourite saints whose commemorative days were celebrated in the course of the year. The Little Office of the Sacred Virgin constituted the most important section, and was generally illustrated with significant episodes from the Virgin's life. The Psalms, another regular component, were usually accompanied by scenes from the life of King David, whose authorship of the Psalms was then unquestioned. The Office of the Dead was illustrated with solemn funeral scenes, while images of different saints appeared next to the Mass for the Souls of the Dead.

The illustrations in Books of Hours were not limited to sacred scenes, however, for the creative impulses of the artists engaged in their production quickly flowered in innumerable ways, and more and more space came to be dedicated to purely ornamental decoration. In this way, during the course of the fifteenth century, naturalistic (and, in particular, botanical and floral) images came to be used ever more widely as decorative elements that sometimes matched, and at times even exceeded, in visual impact the scenes they decorated. A magnificent French Book of Hours, perhaps from the School of Bourges, now at Oak Spring but which originally had belonged to the Italian noblewoman Vittoria della Rovere, contains one of the earliest and most characteristic botanical motifs to be adopted by medieval artists—a flowering vine that forms a graceful frame around the text or painting on each folio (see no. 1).

The Books of Hours decorated by artists of the Ghent–Bruges School between the end of the fifteenth century and the beginning of the sixteenth were among the most admired and sought after in Europe. In them both the text and the full-folio illustrations, which open like windows on marvellously detailed and realistic scenes, were framed by wide borders (*marginalia*) decorated *à trompe l'œil* with flowers, fruits, birds, insects, small animals, shells, jewels and small *objets d'art*. The artists reveal their superb virtuosity in the observation and reproduction of these tiny fragments of reality. Such border decorations represented an important innovation, which tradition attributed to the Master of Mary of Burgundy, later identified as Alexander Bening, a Ghent painter who died in 1519. Bening's sons, Simon and Paul, continued to incorporate this decorative conceit, which in time was taken up in Germany, for example by Nikolaus Glockendon (d. 1534), a student and friend of Albrecht Dürer, and in France by Jean Bourdichon (*c.* 1457–1521) and others. Alexander and Simon Bening were together also responsible for the decoration of one of the greatest masterpieces of the Flemish school of miniature painting, the Grimani Breviary (1514), now in the Biblioteca Nazionale Marciana in Venice. Its borders are lavishly embellished with naturalistic elements, vases of flowers, classicized candelabra and other objects, both humble and precious, some of them inserted in niches (see Salmi). An exceptionally fine Book of Hours (see no. 2) at Oak Spring, dated 1524 and based on the Roman liturgy can also—in the light of its many similarities to the Grimani Breviary—be attributed to Simon Bening.

By the end of the fifteenth century the *printed* Book of Hours was becoming available, and gradually it replaced the extremely costly illuminated manuscript. The first such Book of Hours was published in Paris in 1486 by the miniaturist and calligrapher Antoine Vérard. These works have aptly been described as *manuscrits-imprimés* (Martin and Chartier, II.185), for they closely followed the format of the handwritten, hand-illustrated manuscript. Indeed, the printer usually sought to achieve exactly the same visual effects as the calligrapher and the miniature artist, and to organize the text, images and decorative elements on the page in precisely the same manner. An especially fine example of a printed Book of Hours, published in Paris in 1527 by Geoffroy Tory (see no. 3), is also at Oak Spring. With the passage of time, however, the visual elements of the printed work—including the plant and floral motifs—became simplified and schematized as illustrators adapted themselves to the nature and limits of the printing process.

The hand-illuminated Books of Hours, particularly those produced by the Ghent–Bruges School, offer some of the most striking and significant examples of floral imagery in Western art. Here flowers are painted, as if *d'après nature*, with loving attention to detail, the artist seeking to reproduce as faithfully as possible the effects of volume, and the play of light and shadow on the petals and leaves. Next to them appear *trompe l'œil* fruits, birds and insects, sometimes depicted life-size or even larger than life in order to enhance the illusion of depth (Bazin 1960, pp. 50–51). As Otto Pächt perceptively remarked: 'The border decoration of the Ghent–Bruges school . . . seems to have had the function of keeping alive energies which could find no outlet in the contemporary monumental art, but which later, in the flower-pieces and other still-life paintings of the seventeenth century, were once more to become creative' (Pächt 1950, p. 32). Pächt omitted from his consideration, however, a distinct genre that evolved, beginning in the second half of the sixteenth century, in parallel with the more celebrated genre of the floral still-life—that of the botanical painting on paper or vellum. These were works commissioned by collectors and scientists for the purposes of documenting their collections or their research. One of the greatest masters of this genre was certainly the painter Joris Hoefnagel of Antwerp (1542–1601), although many other artists worthy of comparison to him emerged, including Jacopo Ligozzi (1547–1626), Jacques Le Moyne de Morgues (see nos. 4, 5, 6), and François de Geest (no. 20).

Recently, the felicitous suggestion has been put forward that the flowers, *naturalia* and other objects depicted in the margins of these manuscripts might be interpreted as painted versions of the flowers, insects and devotional objects collected by pilgrims during the course of their journeys and carefully preserved between the folios of their Breviaries (DaCosta Kaufmann and Kaufmann, p. 45).

Of the various elements that abounded in the margins of liturgical texts, flowers were the most frequent, above all those drawn from the traditional repertory of Christian symbolism—the rose, lily, columbine, iris, pink, marguerite, eglantine (sweetbriar), bluebell, daisy, pansy, violet and the strawberry plant. Although the realism with which they are depicted presaged a wholly new form of representation—the botanical illustration—it is well to recall that these motifs were inextricably linked to the iconography of the Church. Many flowers, for example, were associated with specific qualities or attributes of the Madonna or Christ, or both (Levi D'Ancona 1977).

The symbolism of some of these flowers can be traced back to classical times, the rose being an example *par excellence*, for this symbol of Venus, the goddess of Love, was quickly transformed by the Church into a symbol of universal Christian love. The *Lilium candidum*, the emblem of virginity, was naturally incorporated into the iconography of the Annunciation, as was the columbine, since the columbine's flowers are marked by five prominent spurs, evoking the five feathers in the tail of the dove, or Holy Spirit. The iris, emblem of royalty, symbolized the Virgin Mary as Queen of Heaven. The carnation stood for divine love, while the violet—traditional token of humility—was appropriated by the Church as the symbol of Christ's supreme sacrifice and act of humility, his Crucifixion. The pansy was another symbol of humility, although the form of its petals meant it was utilized as an emblem of the Trinity. The strawberry plant, complete with fruit and flowers, was considered to represent Paradise, while also serving as a Marian symbol.

Nevertheless, the artists who decorated these Books of Hours, especially those from the Ghent–Bruges School, often seemed to forget the higher, symbolic aspect of the flowers they depicted. Instead, absorbed in their task, they, like many other artists working in Europe toward the end of the fifteenth century (one need only recall figures such as Leonardo da Vinci and Dürer), created paintings that are at once highly decorative and extremely naturalistic. Their work clearly belies the thesis that painting in miniature constituted a 'minor art' next to the more impressive, larger scale genres of the period. Indeed, the interest of these painters in their subject-matter at times seems to have extended beyond their mere decorative qualities and touched on the scientific, for on occasions they dedicated the marginal decoration of an entire folio to a single flower, showing it from various angles and in various stages of its life-cycle. At times they also added insects and butterflies, elements that would eventually become an integral part of the formal, highly sophisticated genre of flower painting.

1. Book of Hours, of the use of rome [*c.* 1435]

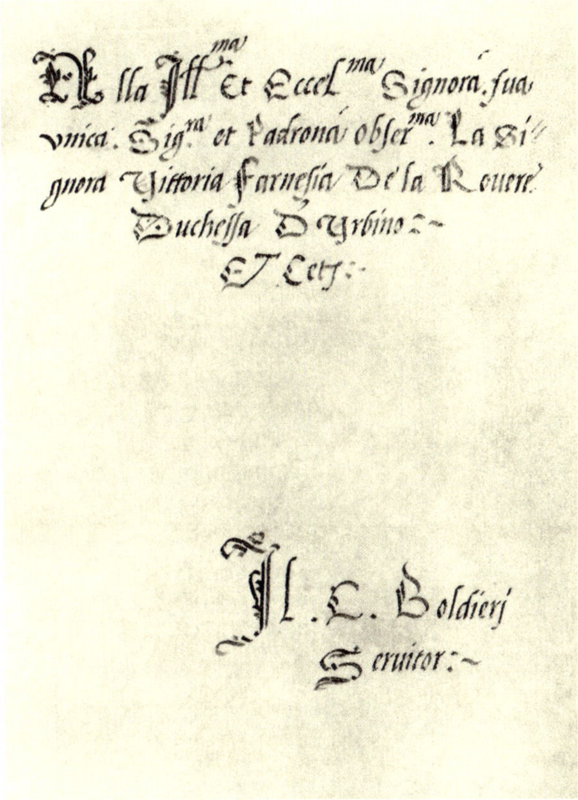

Book of Hours, of the Use of Rome. Manuscript dedication by Boldieri to Vittoria della Rovere

Manuscript on vellum.

21 × 16 cm. A vellum paste-down, 3 blank leaves, a leaf with a notation regarding ownership, leaves numbered by hand 1–182, and finally 3 unnumbered leaves and a vellum paste-down. Folios 91v and 92r–v are blank. A handwritten notation in red, 'N.4', appears on the recto of the first leaf.

BINDING: 17th-century Shagreen leather, showing signs of now missing clasps.

PLATES: 25 miniatures (7.8 × 6.8 cm) depicting St John, St Luke, St Matthew, St Mark, the Archangel Michael, St Peter, the symbol of St Paul, St Francis, St Gregory the Great, the Annunciation, the Visitation, the Nativity, the Annunciation to the Shepherds, the Adoration of the Magi, the Presentation in the Temple, the Flight into Egypt, the Coronation of the Virgin, David in Prayer, the Crucifixion, the Pentecost, St Catherine, a burial scene, the Holy Trinity, the Virgin Mary, the Christ Child with two angels in a garden, and a Mass for the Dead. Every miniature is surrounded by a wide border in which vines with gilt leaves (ivy and holly can be recognized) are interwoven with a variety of flowering plants, birds, insects and little scenes. Similar borders surround the 12 leaves of the calendar and the 15 leaves containing penitential psalms and the Litany.

PROVENANCE: On the fourth preliminary leaf (recto) is written: 'Della Ill.ma et Eccel.ma Signora sua unica Sig.ra et Padrona observ.ma La Signora Vittoria Farnesia De la Rovere Duchessa d'Urbino et Cet. Il C. Boldieri Servitor'. On the first unnumbered leaf at the end is written, in the same hand: 'Confectio pro salute animae' (see Appendix). Acquired in 1894 by the Fontaine Collection.

This magnificent Book of Hours belonged to Vittoria, daughter of Pierluigi Farnese, who married Guidobaldo II della Rovere, Duke of Urbino (1548–1602). Nothing is known of Boldieri, the Duchess's 'servitor' and author of the 'Confectio pro salute animae' (Recipe for the health of the soul) provided at the end of the manuscript (see Appendix), which explains the symbolic meaning of the various parts of the plants following the complex precepts of the art of memory. The manuscript, decorated with miniatures and gilt initial letters and ornamentation, was produced in France in 1430–40, perhaps at Bourges, cited in the calendar and famous for its painters in miniature.

As with most manuscript *horae*, this one owned by the Duchess of Urbino contains a calendar decorated with the cycle of the months and the signs of the zodiac (see fols. 13r, 14v, 16r, 18r,

Book of Hours, of the Use of Rome. Folio 14v, St Luke

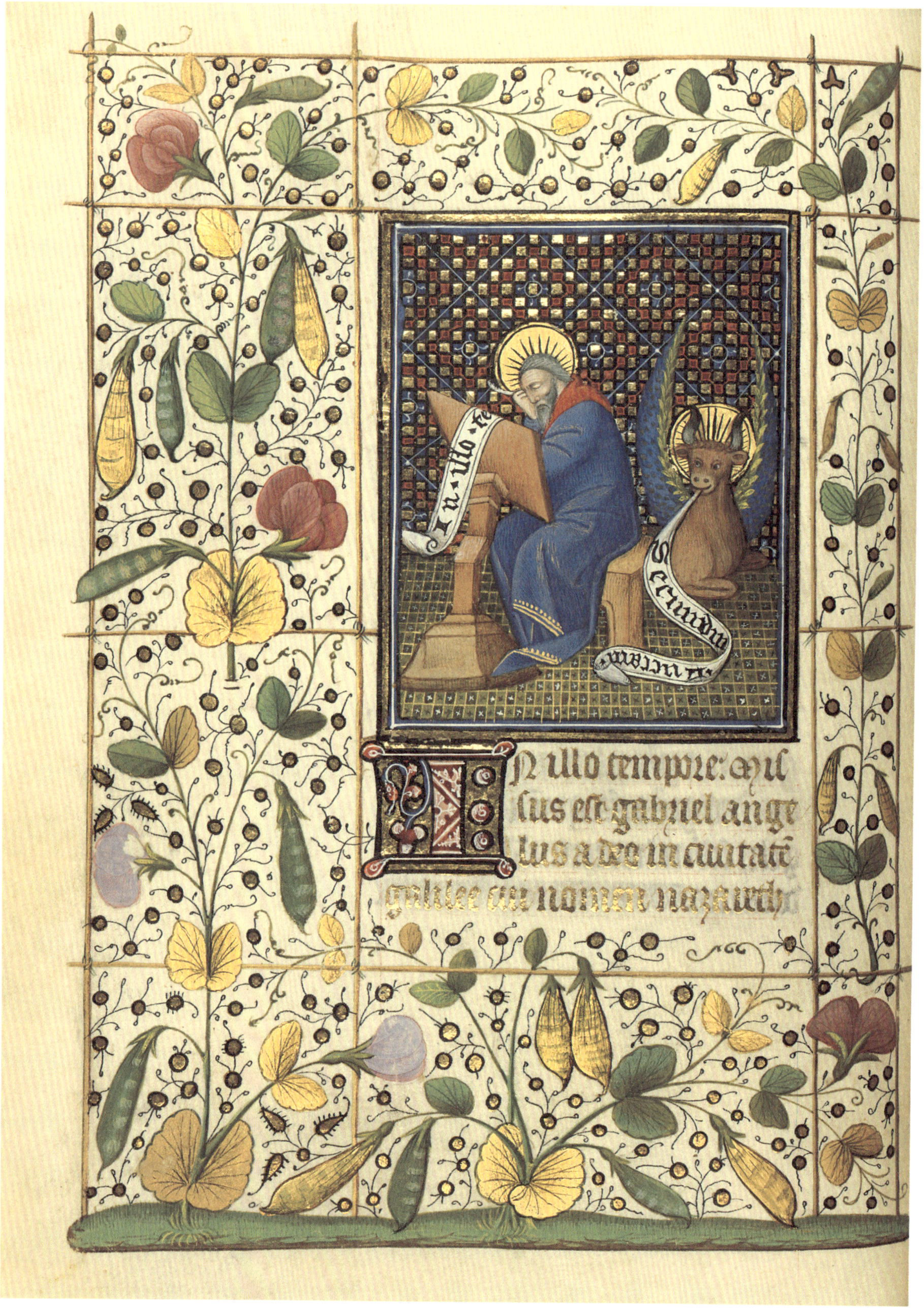

19v, 21r, 22r, 23v, 24v, 25v). The Little Office of the Sacred Virgin is illustrated with scenes from the life of the Virgin (see fols. 27r, 44r, 55r, 59v, 64r, 68r, 72r, 79r). The Psalms are decorated with scenes from the life of David (see fol. 93r). A funeral scene accompanies the Office of the Dead (fol. 131r). In the Suffrages is portrayed St Catherine of Alexandria with the sword and the palm branch of her martyrdom (fol. 125v). A miniature showing the Virgin and Child seated in a garden illustrates the Fifteen Joys of the Virgin (fol. 177v), while the Holy Trinity is depicted next to the Seven Supplications to our Lord (fol. 144v). These two miniatures, as well as the calendar of Saints' days, have elaborate decorated borders.

Two artists contributed to the decoration of this Book of Hours. The full-folio miniatures were executed by a painter whose style is reminiscent of that of the Master of the Boucicaut Hours (Paris, Musée Jacquemart-André, MS 2); the brocaded effect of the painting and the chequered background are particularly characteristic of this Master's work. Another artist was responsible for the elaborate and imaginative border decorations. A charming example is the border ornamenting the miniature of St Francis, along which meanders a rope of the type used for belts by the friars of this order (fol. 24v). The border of the scene depicting the Annunciation to the Shepherds is made up of silver and azure droplets falling from clouds to a landscape containing a lake with swans and geese (fol. 60v). The border to the Adoration of the Magi consists of a delicate trellis of plants with blue and pink flowers against a chequered gilt background (fol. 64r). The naturalistic motifs, the illusionistic effects, and the alternating use of green and gold for the leaves in this border recall the style of the Master of the Hours of Margaret of Orléans, a manuscript dating from *c.* 1426 (Paris, Bibliothèque Nationale, MS lat. 1156B).

A fine example of the collaborative effort of these two artists is folio 14v. The miniature depicts the evangelist St Luke seated at his writing-desk with his attribute, a winged ox, at his feet. The background and pavement are chequered in bright colours alternating with gold. On a cartouche between the two figures is written 'In illo te[mpore] Secundum Luccam'. However, the eye is equally drawn to the border decoration that frames the painting, an elegant lacework of pea flowers supported by a trellis. Here, although the artist depicted the various parts of the pea plant—the leaves, bracts, pods, vine tendrils, and pale mauve and violet flowers—quite realistically, he also added some little round fruit for purely ornamental purposes.

APPENDIX

CONFECTIO PRO SALUTE ANIMAE		RECIPE FOR THE HEALTH OF THE SOUL	
	Rect[a]e fidei		Upright faith
Radicum	Sincerissimi cordis	Roots	Sincere heart
	Largissim[a]e pietatis		Profound piety

Foliorum	Certissim[a]e spei Amplissim[a]e misericordi[a]e Abundatissim[a]e clementi[a]e	Leaves	Certain hope Greatest mercy Abundant clemency
Florum	Violarum humilitatis Rosarum charitatis Liliorum puritatis	Flowers	The humility of the violet The charity of the rose The purity of the lily
Seminis	Castitatis Deiuniorum Elemosinarum Orationum	Seeds	Chastity Fasting Alms-giving Prayer
Absintij Aloes Agarici Mirr[a]e	Contricionis Confessionis Satisfationis Penitenti[a]e	Absinth Aloe Agaric Myrrh	Contrition Confession Satisfaction Penitence
Turis	Contemptus mundi	Incense	Disdain for wordly things
Confectionis matris Christi cum margarittis		Recipe of the mother of Christ with pearls	
Specierum Trociscorum	Di[ct]a Apostolorum Di[ct]a Martirum Di[ct]a Sanctorum Omnium	Spices Pills	The sayings of the Apostles The sayings of the Martyrs The sayings of all the Saints

Disceantur h[a]ec omnia in mortario conscienti[a]e. A terrantur Pistillo doloris. Agitentur baculo iustici[a]e. Cribrentur memoria passionis Jesu Christi et cum saccaro divini amoris: dissoluto in aqua lachrimarum, ad ignem tribulationis, amaritudinis, et patienti[a]e, fiat confectio cordialis in morsellis pura et sincera mente, quotidie mastigandis, gustandis, et diglutiendis.

Place all these things together in the mortar of the conscience. Crush with the pestle of suffering. Stir with the rod of justice. Sift with the remembrance of the passion of Christ and with the sugar of divine love. Dissolve in the water of tears over the fire of tribulation, bitterness and patience, and there will come forth the recipe for the cordial to be taken in small amounts by daily chewing, tasting and swallowing, and with a pure and sincere spirit.

2. Simon Bening (1483/84–1561) [*attributed*]

[Horae Beatae Mariae Virginis ad usum Romanum.]

21.5 × 15.5 cm. Manuscript, 1524. One blank leaf followed by leaves 1–303 numbered in manuscript, then another blank leaf.

BINDING: Modern binding in brown leather with stamped gilt decoration and a gilt brass clasp. On the front cover is a round plaque in gilt brass, decorated around the rim with a frieze of circles stamped in relief, and bearing an impaled heraldic shield charged with three scallop- or cockle-shells (sinister chief) and, on the dexter, six annulets and a raised palm. Gilt and gauffered edges.

PLATES: The leaves of this Book of Hours are devoted to a calendar and the months of the year (fols. 1v–13r), each decorated with a landscape scene surrounded by an illuminated border and, at the top of the page, the corresponding sign of the zodiac (fols. 2r, 3r, 4r, 5r, 6r, 7r, 8r, 9r, 10r, 11r, 13r). There are many other full-page miniatures with decorated borders (fols. 13v, 15v, 17v, 19v, 21v, 24v, 38v, 59v, 62v, 81v, 88v, 103v, 106v, 112v, 114v, 116v, 130v, 147v, 153v, 164v, 187v, 214v, 219v, 228v, 234v, 242v, 265v), as well as leaves of text with similarly decorated borders (fols. 14r, 16r, 18r, 20r, 22r, 25r, 39r, 51r, 63r, 73r, 74r, 82r, 89r, 95r, 100r, 104r, 107r, 108r,

109r, 110r, 111r, 113r, 115r, 117r, 118r, 119r, 125r, 131r, 142r, 148r, 154r, 160r, 165r, 169r, 188r, 200r, 205r, 210r, 215r, 220r, 229r, 235r, 243r, 266r). Initial letters and other decorations illuminated in gold are scattered throughout.

PROVENANCE: Ex-libris 'Charles William Dyson Perrins' (1864–1958), and Paul Mellon.

REFERENCE: *Descriptive Catalogue of Illuminated Manuscripts in the Library of C. W. Dyson Perrins*, Oxford, 1920, 106.

THE SUPERB miniatures and border decorations in this Book of Hours were almost certainly executed by artists from the Ghent–Bruges School, renowned throughout Europe for their skills in the art of illumination. The chief members were Alexander Bening and his son Simon. Both Benings collaborated on the manuscript generally regarded as one of the greatest masterpieces to emerge from the Flemish school of manuscript illumination, the Grimani Breviary, now in the Biblioteca Nazionale Marciana, Venice. Its folios contain borders sumptuously decorated with a delightful variety of flora and fauna, vases, classicized candelabra and other objects set in niches.

The style and quality of the Oak Spring *horae* strongly suggest that this work also may be attributed to Simon Bening, the last of the great Flemish masters of the art of miniature. Born in Ghent, he lived and worked from 1517 in Bruges, Antwerp and London. His patrons included monarchs and aristocrats from across Europe, including Charles V of Spain and the Infante Ferdinando of Portugal.

The miniatures in this Book of Hours include a variety of skilfully composed scenes: interiors carefully drawn in perspective, and outdoor scenes with figures standing in the foreground and with landscapes beyond. Particularly charming are the illuminations that decorate the months of the year and the calendar. They depict traditional rural scenes, such as the pruning of vines in February, or a garden being prepared for planting in March. In a decorative feature in the left-hand margin of the month of December (folio 12v), the artist has inserted the date '1524'.

The borders, reminiscent of those in the Grimani Breviary, are embellished with a variety of elements, ranging from classicizing architectural elements to grotesques, festoons, *trompe l'œil* lettering, complex still-lifes, and elaborate jewelled objects seemingly floating against backgrounds of gold, red, blue, pink or brown. Certain of the illuminations bear an unmistakable resemblance to paintings in the Grimani Breviary, such as the 'Funeral' scene from folio 265v, which corresponds perfectly with the miniature from folio 450r of the Breviary, or 'St Francis receiving the stigmata' on folio 122v, which is almost identical to the Breviary's folio 756v.

As in many other Books of Hours from the Ghent–Bruges School, the Oak Spring manuscript contains a number of borders decorated with plants and flowers that are prominent in Christian iconography: roses, violets, columbines, daisies, carnations, and lilies, shown sometimes in full flower and sometimes with their buds still hidden in green bracts. Although clearly symbolic in meaning, these items are depicted as realistically as the many zoological elements interspersed among them, including butterflies, caterpillars and birds, of which all form part of a living, almost palpable microcosm.

Simon Bening's mastery of the elements of colour and composition can be fully appreciated in

SIMON BENING,
Horae beatae Mariae Virginis ad usum Romanum. Folio 17v, St Luke

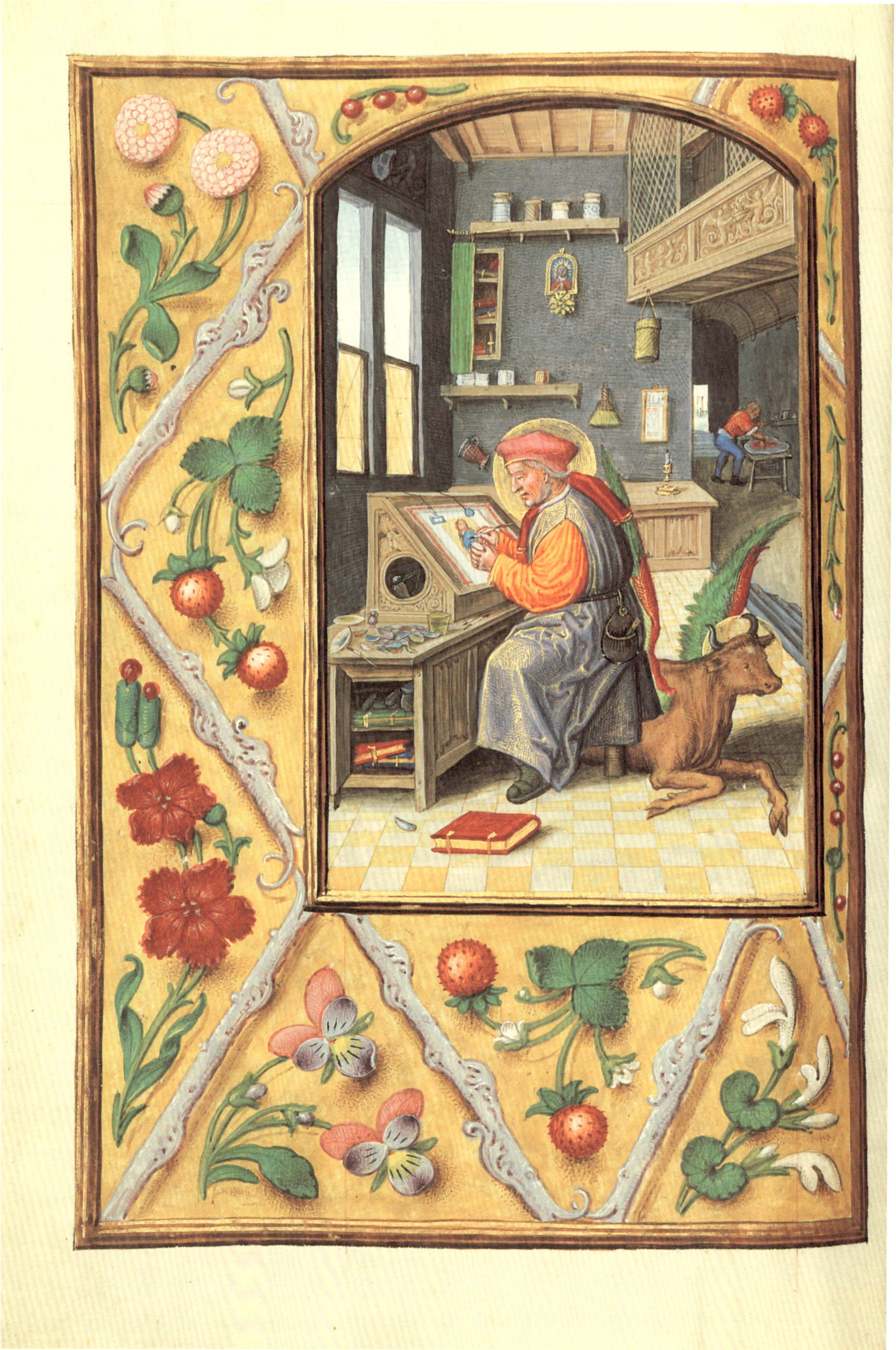

his portrait of Luke the Evangelist, shown here (in accordance with the pleasing, but erroneous tradition that held this Saint to have been an artist) in the act of painting a miniature of the Virgin. Presented in sharp perspective is a well-furnished interior in which the Saint can be seen working at his bench, surrounded by all the paraphernalia of his art, while at his feet lies a winged bull. The illusion of depth is reinforced by a wide gilt border that narrows abruptly along the inside margin of the page. The border is divided by rods decorated with stylized acanthus leaves into a pattern of triangular fields, each containing a simple arrangement of flowers—delicate pink English daisies, crimson carnations, snow-white honeysuckle blossoms, and strawberry plants complete with flowers and ruby red fruit.

3. Book of Hours [1527]

[*a woodcut border decorated with birds, flowers, insects and, at the base, a crowned coat of arms with the lilies of France. In black and red lettering*] Hore in laudem beatissime Virginis Marie: secundum consuetudinem Ecclesie Parisiensis. [*in the centre of the page is a woodcut printer's device, consisting of an ansate vase pierced by a spear, standing on a book. The vase contains various flowers, including an iris and a columbine; at its foot are three chains, two with padlocks and one attached to an unidentifiable object. Sunbeams radiate from the upper-left portion of the page, while at the right appears a cartouche* (7.5 × 4.3 *cm.*) *with the motto* 'Non Plus'. *Beneath, in black and red lettering*] Venales habentur Parrhisiis apud Magistrum Gotofredum Torinu[m] Biturigicum: sub insigni Vasis effracti: gallico sermone Au Pot Casse.

4º 20 × 13.2 cm. a–r^8 s^4 *1–280* p. Gothic lettering in red and black. Privilège du Roy, 5 September 1526 (a1r and a2v); colophon: (s4v) the name of the printer: 'Simon du Bois'; the date: 22 October 1527; and a repetition of the printer's device.

BINDING: Modern brown leather with gilt fillet border and panel design, in each corner gilt fleurons, centrepiece of strap and arabesque work, all edges gilt. On the spine: 'Horae B. Virginis Mariae. Paris. G. Tory 1527'.

PLATES: 12 woodcuts (9.3 × 6.5 cm.) printed from 13 blocks (two were used for the illustration of the Annunciation, which takes up two pages). Every woodcut is surrounded by a decorated border.

PROVENANCE: Leather bookplate: 'Ex Musaeo Huthii'. Ex-libris Paul Mellon.

REFERENCES: Lacombe, p. 205, no. 364; Mortimer, 1.395' no. 304; *Rosenwald Collection*, no. 985.

THIS BOOK OF HOURS was one in a series that were the first to be printed, from 1486, by the Parisian miniaturist and calligrapher Antoine Vérard, in order to provide a less costly alternative to the illuminated manuscript. As was customary in printed Books of Hours, the copy at Oak Spring contains a title-page in which the publisher's name figures prominently. This was Geoffroy Tory (*c.* 1480–*c.* 1533), a humanist, librarian, decorator and engraver, celebrated for his *Champfleury* (1529), a work in which he sought to explain the form and allegorical significance of the letters of the alphabet. Tory's Book of Hours was printed at the sign of the 'Pot Casse' in the rue Saint-Jacques, where a similar *horae* by him had already been published in 1525 (*Art of the Printed Book*, pp. 14–15). The printer, as emerges from the colophon, was Simon du Bois, well known in Paris for the editions of the Gospels he had published.

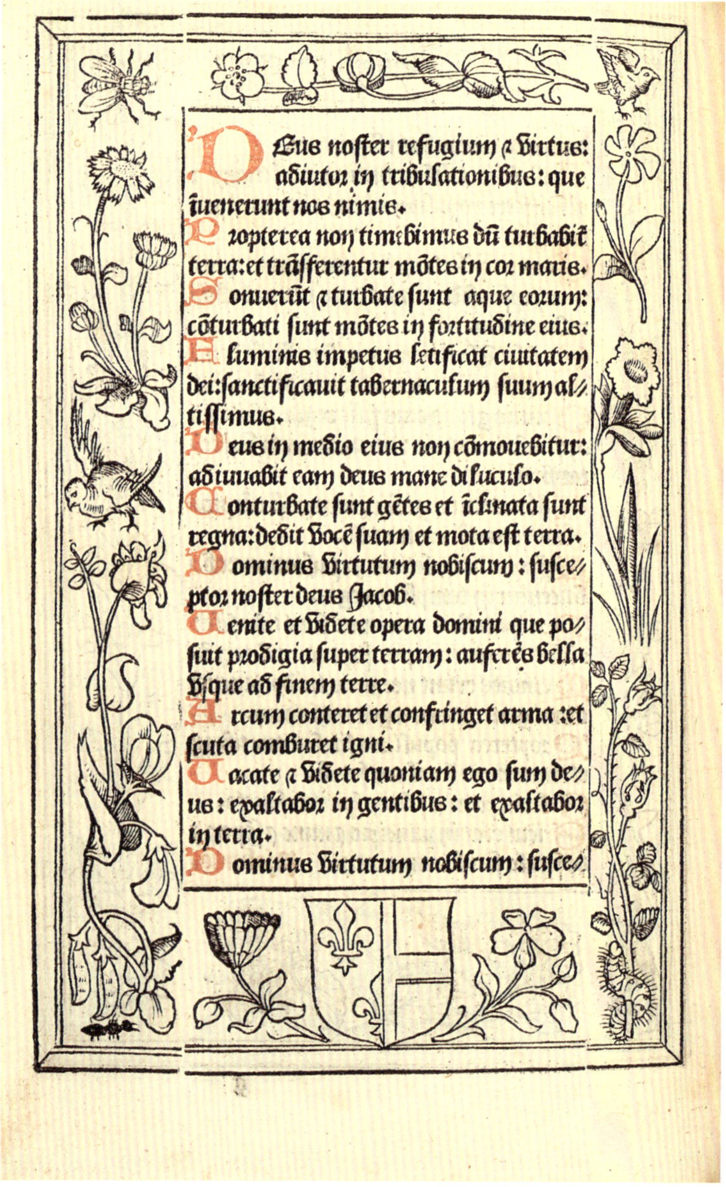

Deus noster refugium ⁊ virtus: adiutor in tribulationibus: que inuenerunt nos nimis.
Propterea non timebimus dū turbabit terra: et trāsferentur mōtes in cor maris.
Sonuerūt ⁊ turbate sunt aque eorum: cōturbati sunt mōtes in fortitudine eius.
Fluminis impetus letificat ciuitatem dei: sanctificauit tabernaculum suum al⸗tissimus.
Deus in medio eius non cōmouebitur: adiuuabit eam deus mane diluculo.
Conturbate sunt gētes et inclinata sunt regna: dedit vocē suam et mota est terra.
Dominus virtutum nobiscum: susce⸗ptor noster deus Jacob.
Venite et videte opera domini que po⸗suit prodigia super terram: auferēs bella vsque ad finem terre.
Arcum conteret et confringet arma: et scuta comburet igni.
Vacate ⁊ videte quoniam ego sum de⸗us: exaltabor in gentibus: et exaltabor in terra.
Dominus virtutum nobiscum: susce⸗

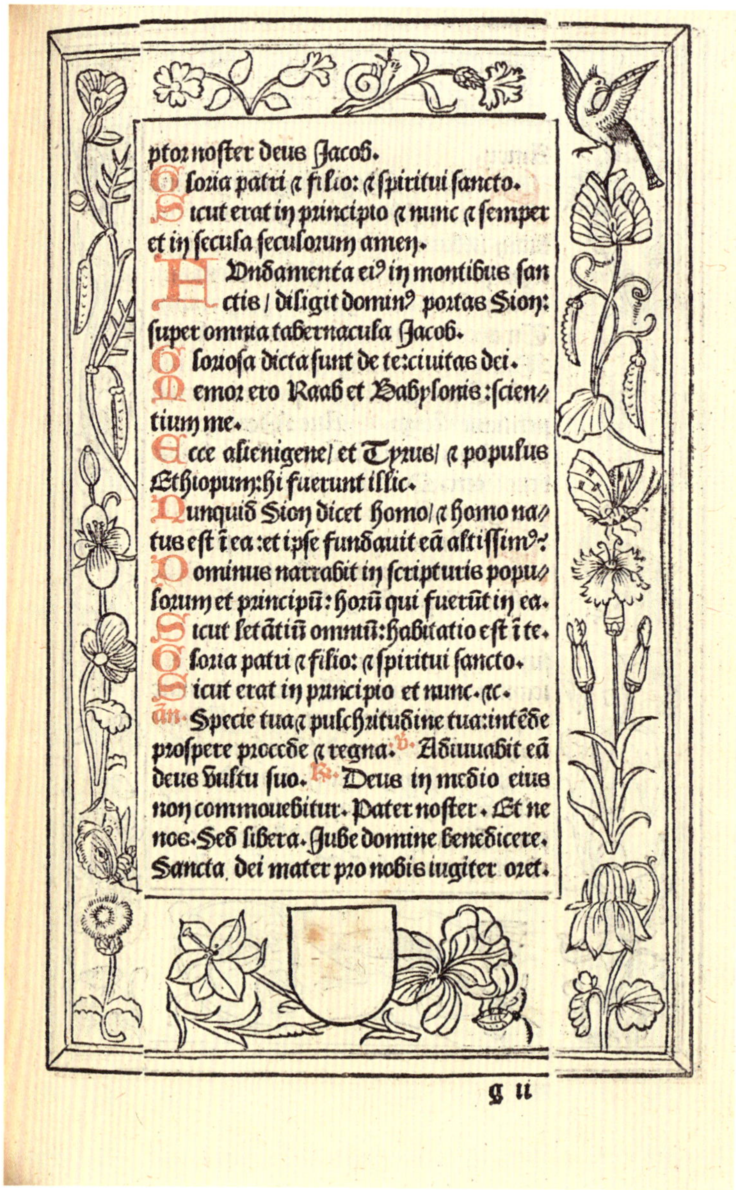

ptor noster deus Jacob.
Gloria patri ⁊ filio: ⁊ spiritui sancto.
Sicut erat in principio ⁊ nunc ⁊ semper et in secula seculorum amen.
Fundamenta ei⁹ in montibus san⸗ctis / diligit domin⁹ portas Sion: super omnia tabernacula Jacob.
Gloriosa dicta sunt de te: ciuitas dei.
Memor ero Raab et Babylonis: scien⸗tium me.
Ecce alienigene/ et Tyrus/ ⁊ populus Ethiopum: hi fuerunt illic.
Nunquid Sion dicet homo/ ⁊ homo na⸗tus est ī ea: et ipse fundauit eā altissim⁹:
Dominus narrabit in scripturis popu⸗lorum et principū: horū qui fuerūt in ea.
Sicut letātiū omniū: habitatio est ī te.
Gloria patri ⁊ filio: ⁊ spiritui sancto.
Sicut erat in principio et nunc. ⁊c.
An. Specie tua ⁊ pulchritudine tua: intēde prospere procede ⁊ regna. V. Adiuuabit eā deus vultu suo. R. Deus in medio eius non commouebitur. Pater noster. Et ne nos. Sed libera. Jube domine benedicere. Sancta dei mater pro nobis iugiter oret.

g ii

Book of Hours, *Hore in laudem beatissime Virginis Marie*. Leaf g1^{v} and leaf g2^{r}

In addition to the sophisticated typographer's emblem and a framed initial letter containing a floral motif, the pictorial component of this Book of Hours consists of thirteen woodcuts, among which the Annunciation, printed on two facing pages, the Circumcision and the Coronation of the Virgin are particularly fine. They betray a marked Italian influence, and indeed it is known that Tory lived for an extended period in Italy (Mortimer, pp. 395–7).

The floral borders conceived *à la moderne*, are characterized by a decorative motif produced by a set of small woodblocks that is repeated, with slight variations, on every page. In a reflection of the celebrated style of miniature painting of the Ghent–Bruges School (see no. 2), the borders of this *horae* are also decorated with a variety of flowers, small birds, coats of arms and other emblems. It seems likely that the figures, engraved in outline without any shading, may have been coloured by hand in some copies. Notwithstanding the simple, somewhat ingenuous, style of the illustrations, which do not take into account the relative sizes of the objects depicted (an insect may appear as large as a bird), the animals and the plants (mostly wild species) have been very realistically depicted, and in almost every case are easily recognized.

In the decorations on one page (g 1v), for example, we can distinguish a marigold, a corn cockle, a pea flower, two periwinkles, a trumpet narcissus and a rose, which share the border with a bird, an ant, a caterpillar and a fly. At the base of the illustration is a coat of arms divided into two fields, one bearing the lilies of France and the other the cross of Savoy, a tribute to François I, the royal patron of this work, and his mother, Louise de Savoy. The following page (g2r) is decorated with two leguminous plants, a corn cockle, a dianthus, a columbine, a centaury, an iris, a thistle, and what appears to be a silene. Among the animals may be discerned a snail, a coleoptera, and a bird with a sweeping tail. Other figurative elements are symbolic in nature, such as the Christological image of the pelican, or the crowned salamander, one of François's emblems. On many pages also appear the crowned initials 'F' and 'L', once again references to this work's royal patrons.

II · FLORILEGIA: FLOWER BOOKS FOR COLLECTORS AND CONNOISSEURS

THE FIRST printed herbals date from the middle decades of the sixteenth century, while the first florilegia began to appear around 1600. The herbals compiled by the great botanists of the sixteenth century, such as Otto Brunfels (1530–36), Leonhart Fuchs (1542) and Pietro Andrea Mattioli (1554), were conceived as scientific texts in the modern sense, for they endeavoured to present as large a number of plants as possible (mostly indigenous species), in a carefully predetermined order, and with detailed descriptions, including observations on the natural habitat and pharmacological properties of each. These texts are always accompanied by numerous illustrations, drawn whenever possible from life and *in situ*, the artists having sought to make each rendition as realistic and three-dimensional as possible. Botanical illustrations of this kind represented a significant advance over the images that had been copied and re-copied from one manuscript to another in earlier herbals—schematic, stereotypal images that eventually were to pass into the first printed herbals too.

The florilegia were intended for a quite different audience, less specialized but more refined in its taste, composed of collectors and *amateurs fleuristes*. Flowers that had been valued only for their medicinal properties or symbolic associations finally came to be appreciated for their own sake, that is, appreciated on purely aesthetic grounds.

During the first half of the seventeenth century, many wealthy aristocrats began to collect rare natural curiosities—not only minerals, bones, stuffed fish and birds, which they proudly displayed in their *Wunderkammern*, but also unusual flowers, which they considered 'miracles' of nature and which they cultivated in order to impress visitors to their vast pleasure gardens. This mania for collecting was fed by the arrival (beginning in the last decades of the sixteenth century) of large numbers of new species from the Balkan peninsula, the Near and Far East and the New World, a phenomenon that was destined to transform the floral panorama of central and Mediterranean Europe. These hitherto unknown plants quickly became fashionable items, avidly sought after by every collector. The bulbous plants in particular, such as the iris, the narcissus, the scarlet lily, the fritillary (called the 'imperial crown') and, above all, the tulip, were in great demand. Collectors were also fascinated by 'flores *singulares*', flowers with peculiar anomalies, such as double blooms, teratologic forms or unusual colours. Such anomalies were actually the fruit of somatic or genetic mutations that occur naturally under certain growing conditions, but sixteenth-century floriculturists were convinced that they were due to the esoteric processes they had devised and jealously sought to keep secret (see Rizzotto, and chapter Four below).

It was only with the inception of the florilegium, however, that the beauty of these new, rare or strange flowers could be captured for posterity. As the notary J. Laurent of Laon remarked in his treatise on plants, fruits and flowers published in 1675, the 'Grands Fleuristes' were those 'qui font peindre leurs fleurs avec remarque de leur noms . . . et en jouissent ainsi toute l'année' (Schnapper 1988, p. 55). Unlike the herbal, the florilegium was accompanied by only the most summary of texts, and sometimes had no text at all; instead, it featured a large number of plates whose object was to reveal the sheer beauty of each flower's form and colour rather than document its botany. Generally only the cut flower was depicted, although sometimes the root or bulb was also shown. As in the herbals that preceded them, the illustrations in these florilegia were meticulously drawn from life, although the artist took pride in arranging the various specimens and parts of the flower in elegant and effective compositions on the page. His task frequently was not to depict 'typical' specimens as clearly as possible for the benefit of the botanist, but to concentrate on unusual flowers bearing features quite unlike those of the ordinary indigenous or cultivated species.

During the fifteenth and sixteenth centuries the woodcut was the least expensive, most widely used technique for the illustration of printed books. Most herbals and early botanical texts relied on this technique, which could produce an admirably clear and precise outline, although shading could only be supplied by means of parallel or cross-hatching. The authors of the florilegia were instead able to take advantage of the more sophisticated techniques of engraving and etching, which replaced the woodcut and which lent themselves admirably to the creation of subtle *chiaroscuro* effects. Since these techniques were, however, limited to black and white, authors whenever possible had a number of copies of their works, particularly those dedicated to important personages, touched by hand in water- or bodycolour. The plates of these florilegia, prepared by highly specialized artists, are often of rare beauty and remarkable realism.

The first florilegia were commissioned by amateur botanists and floriculturists, or by aristocrats who wished to have a permanent record made of the rare flowers being cultivated in their gardens. But soon, artists specializing in flower painting, as well as embroiderers, weavers and other artisans, began to use these works as a source of motifs for their canvases or looms.

A highly singular and eclectic artist, Jacques Le Moyne de Morgues, who divided his active years between France, England and the Americas, was one of the originators of the genre of the florilegium. His early training as a cartographer no doubt contributed to the virtuosity he later displayed as a naturalistic painter. In addition to some extraordinary paintings of fruit and flowers now in London in the Victoria and Albert Museum and the British Museum, a number of manuscripts of flower paintings, all characterized by Le Moyne's sober style and remarkable sensitivity to his subject-matter, have recently come to light.

At Oak Spring are three works that can be attributed, either directly or indirectly, to Le Moyne. One is a manuscript containing sixty watercolour paintings of flowers and fruits (no. 4), many of which bear close affinities to a series of bodycolour paintings by Le Moyne in the Victoria and Albert Museum. Another volume containing thirty-one bodycolour paintings of flowering plants (no. 5) shown with their roots or bulbs—an innovation that many later artists would adopt—was probably executed by an artist working closely within Le Moyne's circle. The extremely rare *La Clef des champs* (no. 6), the last known work by Le Moyne, is illustrated with woodcuts of flowers, animals and fruit that were intended to serve as models for other artists, thus anticipating one of the florilegium's most important functions.

Other works in the Oak Spring collection—by Pierre Vallet (no. 8), Emanuel Sweerts (no. 9), Johann Theodor de Bry (no. 10), Basilius Besler (no. 11), Crispijn van de Passe (no. 12), François Langlois (no. 14), Daniel Rabel (no. 15), Johann Theodor de Bry and Matthaeus Merian (no. 16), and Jeremias Falck (no. 19)—may be counted among the most beautiful florilegia to be published in Europe during the course of the seventeenth century. Many of these works were dedicated to sovereigns or aristocrats who were themselves passionate flower lovers, such as Marie de' Medici, Rudolf II and Prince-Bishop Johann Conrad von Gemmingen of Eichstätt. In them we can see artists such as Langlois and Falck experimenting with a new, more decorative typology.

Some of these florilegia are embellished with exquisite, hand-coloured plates, often the work of extremely talented artists. The copy of Rabel's *Theatrum florae*, for example, at Oak Spring was hand coloured (a full two years after it was printed) by Guillelmus Théodorus; his contribution transformed this printed book into a unique work quite comparable to that of an illuminated manuscript.

Given the enormous popularity of flowers and floriculture in the seventeenth century, it is not surprising that some of these florilegia were even used as catalogues for the sale of bulbs and plants. Sweerts had a magnificent florilegium prepared with 110 plates illustrating more than 560 different flowers that could be purchased from him either in Amsterdam or in Frankfurt during the latter's famous annual fair.

Another printed work, Crispijn van de Passe's *Hortus Floridus*, was perhaps the most popular florilegium ever compiled, for both scientists and artists found its meticulous plates invaluable in their work. It was reissued in numerous editions and translated into various languages, while many of its images reappeared in botanical texts by other authors, or were adopted by artists working in a new genre, the still-life (Olmi 1989). Many other printed florilegia became quite well known too, since they could be produced in large numbers and widely circulated.

During the seventeenth century, florilegia painted entirely by hand on paper or vellum also continued to be produced. Some of them were the work of amateur flower lovers or floriculturists

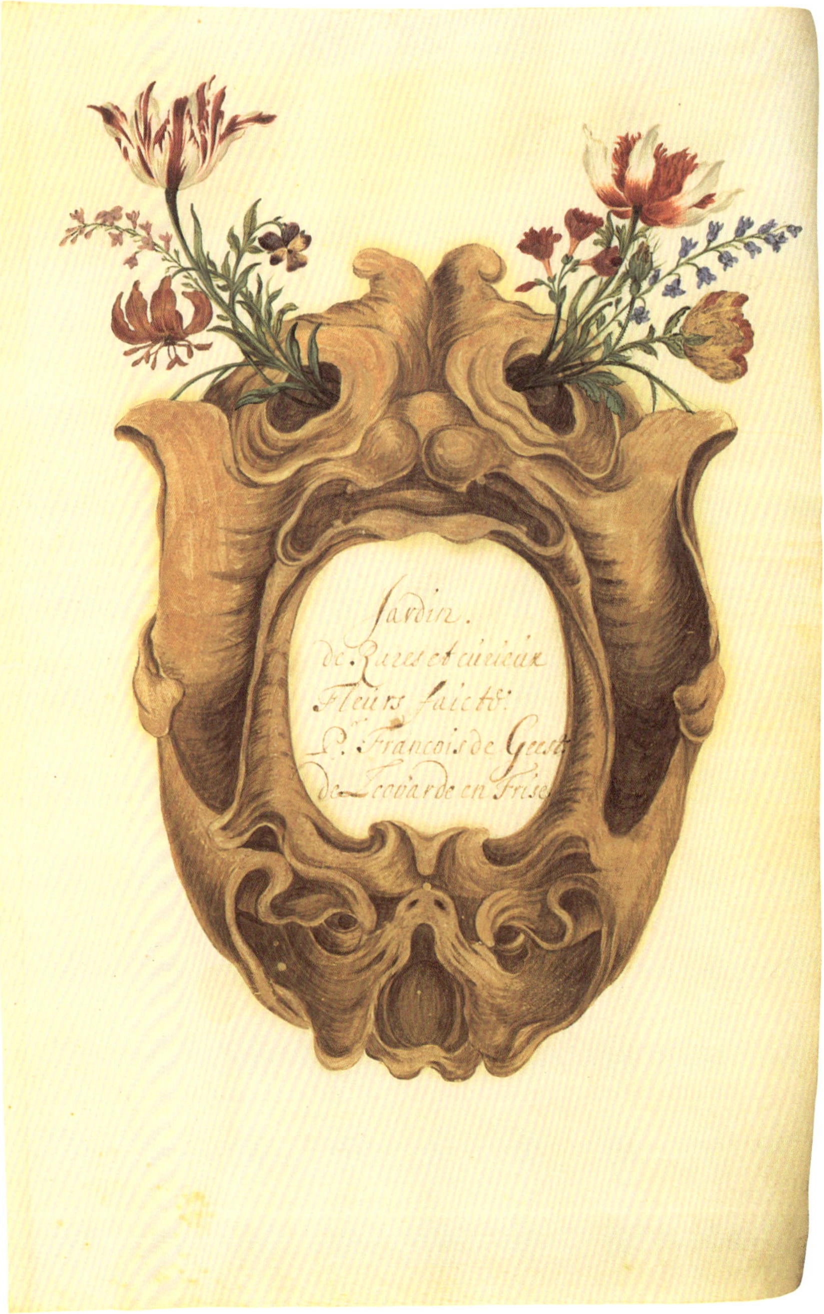

JULIUS FRANÇOIS DE GEEST, *Jardin de rares et curieux fleurs.* Title-page

rather than professional artists, such as the florilegium by an unknown Dutch artist (no. 17), or the charming manuscript on bulbous plants by a certain Domenico Buonvicini, of whom virtually nothing is known (no. 7). Among the many other hand-painted florilegia in the collection at Oak Spring, one by François de Geest (no. 20) and one by Jan Withoos (no. 21) show that these two artists, best known for their work in other genres, were also very talented botanical illustrators. The

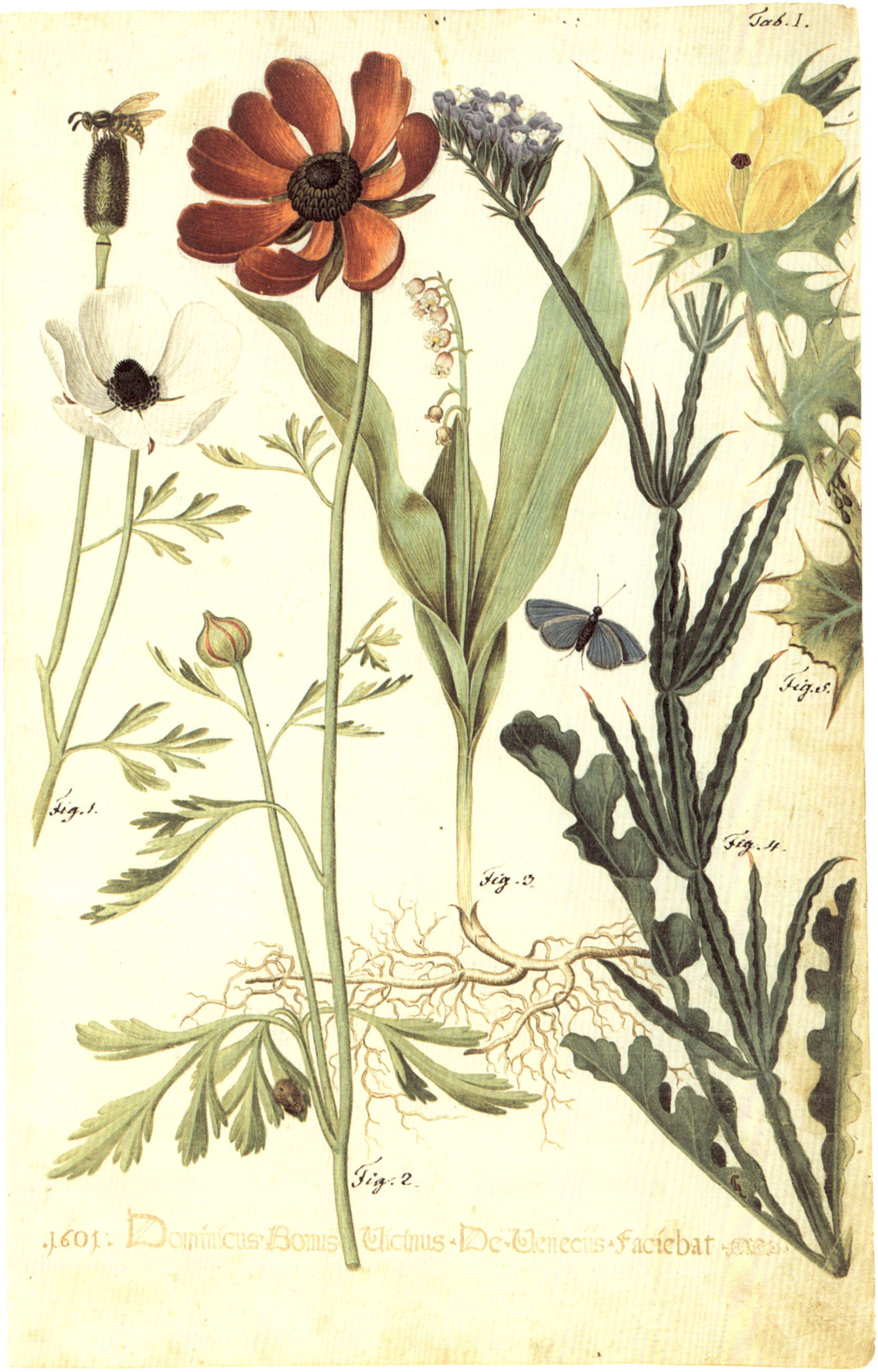

DOMENICO BUONVICINI, Venetian manuscript florilegium. Title-page with two Persian buttercups (*Ranunculus asiaticus*), lily-of-the-valley (*Convallaria majalis*), statice (*Limonium* sp.), Mexican poppy (*Argemone mexicana*) and insects

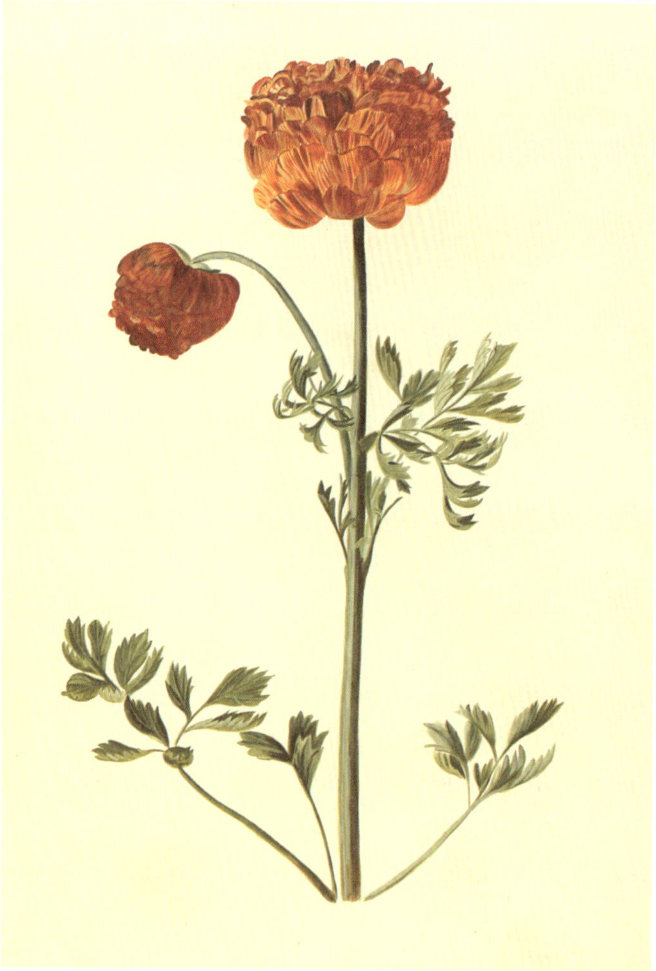

JAN WITHOOS,
A collection of flowers painted on vellum. A type of *Ranunculus*, volume I, folio 13

beautiful paintings on paper by Pieter Holsteyn the younger illustrating a variety of carnations (no. 18) also originally formed part of a florilegium, similar to one preserved intact in the Lindley Library, London.

Finally, two paintings at Oak Spring by the mysterious artist Girolamo Pini (autograph replicas of works in the Musée des Arts Décoratifs, Paris) were apparently directly inspired by the typology of the florilegium (no. 13). They in fact constitute large 'painted catalogues', in which dozens of flowers are carefully presented, with their names listed in an elegant *trompe l'œil* cartouche. A similar display of flowering plants (with each plant punctiliously numbered) can be found in an engraving in *Le Jardin et le cabinet poetique* by the pharmacist Paul Constant of Poitiers, which was published in 1609.

4. Jacques Le Moyne de Morgues (*c.* 1533–1588) [*attributed*]

Manuscript of flowers and fruits.

20 × 29 cm. 60 paintings in water- and bodycolour. None of the leaves bear any inscription except fol. 21, where 'Violete de Mars' is written in the upper margin. There are some traces of writing, subsequently erased, on fol. 24. Many leaves bear a watermark similar to Briquet 12660 (Bruxelles 1534; Néaumur 1535) or 12664 (Mézières 1544; Néaumur 1545–1549).

BINDING: Contemporary limp vellum with gilt fillet border and panel with fleur-de-lis at angles and, in the centre, a floral motif inserted into a rectangular frame with a lily at each corner. The spine is divided by double gilt lines into eight compartments, each containing a cinquefoil. This binding is quite similar, apart from a few variations in the decorative elements, to that of another manuscript of bodycolour paintings (England, *c.* 1600) attributed to Le Moyne and today in the Pierpont Morgan Library.

PLATES: 42 paintings of flowers and 18 of fruits. Each image is presented within a double frame ruled in golden ink. The leaf number appears in the upper-right corner of every folio, although some of the numbers were trimmed away during the binding of the manuscript. Folios 38 and 41 were removed from the manuscript before it entered the collection, and folio 45 is blank.

The artist responsible for the sixty magnificent paintings in water- and bodycolour in this manuscript is here identified for the first time. Many of the illustrations are almost identical to a series of bodycolour paintings on paper contained in a small folio volume (27.4 × 18.8 cm.) acquired in 1856 by what is now the Victoria and Albert Museum's National Art Library in London. It was Spencer Savage who in 1922 first attributed the London paintings to the French Huguenot artist Jacques Le Moyne de Morgues, whose signature appears in the first work, a depiction of narcissus blooms and a butterfly.

Only recently have details regarding the life and cultural background of Le Moyne begun to emerge, and many lacunae remain to be filled before we can document with any certainty his singular *œuvre*, which ranges from botanical illustrations of indigenous European species to works depicting the life of Native Americans. He was born in Dieppe *c.* 1533, where he most probably completed his education as an artist and mapmaker, since by 1564 he had reached a sufficient level of competence to be invited by Lieutenant René Goulaine de Laudonnière to join an expedition bound for Florida. Since the mid-sixteenth century it was customary to recruit artists and draughtsmen for these trips, whose role was to document the exotic flora and bizarre fauna that were being discovered with every new voyage. In Europe these strange species were not only studied by naturalists, they were also sought after by collectors and by authors wishing to depict them in the first printed books devoted to distant lands.

During the course of the expedition led by Laudonnière, Le Moyne produced many drawings that faithfully portrayed the daily life of Florida's natives. Two very fine paintings that are also of great documentary interest, *The Indian Chief Authore Shows Laudonnière the Marker Column set up by Ribault* (New York Public Library) and the *Young Daughter of the Picts* (Yale Center for British Art,

New Haven), were completed by the artist either during the trip itself or immediately on his return home on the basis of his notes and drawings.

Unfortunately, all Le Moyne's sketches from this expedition have been lost, but some idea of their artistic quality may be gathered from a series of engravings that were made from them to illustrate a brief record of the expedition, *Brevis narratio eorum quae in Florida Americae provincia Gallis acciderunt*. This account was included by the printer and engraver Theodor de Bry in the second volume (published in 1591) of his monumental work *America*, a collection of narratives of various voyages to the New World. Although no record exists of any strictly botanical or zoological illustrations completed by Le Moyne during his trip, de Bry's engravings testify to his skill as a naturalistic illustrator, for his scenes of native life contain in the background many finely rendered details of the local flora and fauna. Some of Le Moyne's bodycolour paintings were also copied by the young English artist John White, who two decades after Le Moyne's trip accompanied Sir Walter Ralegh on a voyage to America.

Following Savage's identification in 1922, Le Moyne's *œuvre* was further amplified when fifty-nine botanical paintings in bodycolour in the Victoria and Albert Museum, followed by an album of fifty bodycolour paintings of fruits and flowers acquired in 1962 by the British Museum, were attributed to him. A sonnet in the latter work cites Le Moyne's name and the date 1585, indicating that this album was produced late in the artist's career. One year later Le Moyne printed an elegant and rare little volume containing a vivacious series of woodcuts of various plants, animals and fruit, *La Clef des champs* (see no. 6). This was published at Blackfriars in London, where the Huguenot artist had taken refuge in order to escape religious persecution after his return to France from America.

Two other manuscripts have been tentatively added to his *œuvre*. One, a volume containing twenty-nine leaves of miniature paintings, is now in the Pierpont Morgan Library (see *Flowers in Books and Drawings*, no. 71), a work originally attributed to Thomas Moffet due to the large number of insects that had been incorporated into its floral compositions. The second is a series of eight miniature paintings of floral subjects on vellum now in the Garden Library at Dumbarton Oaks, Washington, D.C.

Much closer in style to the works securely attributed to Le Moyne is this remarkable manuscript of fruit and flowers at Oak Spring. The sixty paintings have been executed in watercolours heightened with bodycolour, a technique favoured by Le Moyne and one well adapted to conveying subtle variations in shading and the changing contours of the different fruits and flowers. Most of the flowers (all common indigenous species) and all of the fruits are to be found also in the manuscript in the Victoria and Albert Museum (see Appendix). Apart from slight variations in the arrangement of the specimens on the page, the only significant difference is that here the stems have been cut, whereas many of the plants in the London manuscript are shown complete with their bulbs or roots.

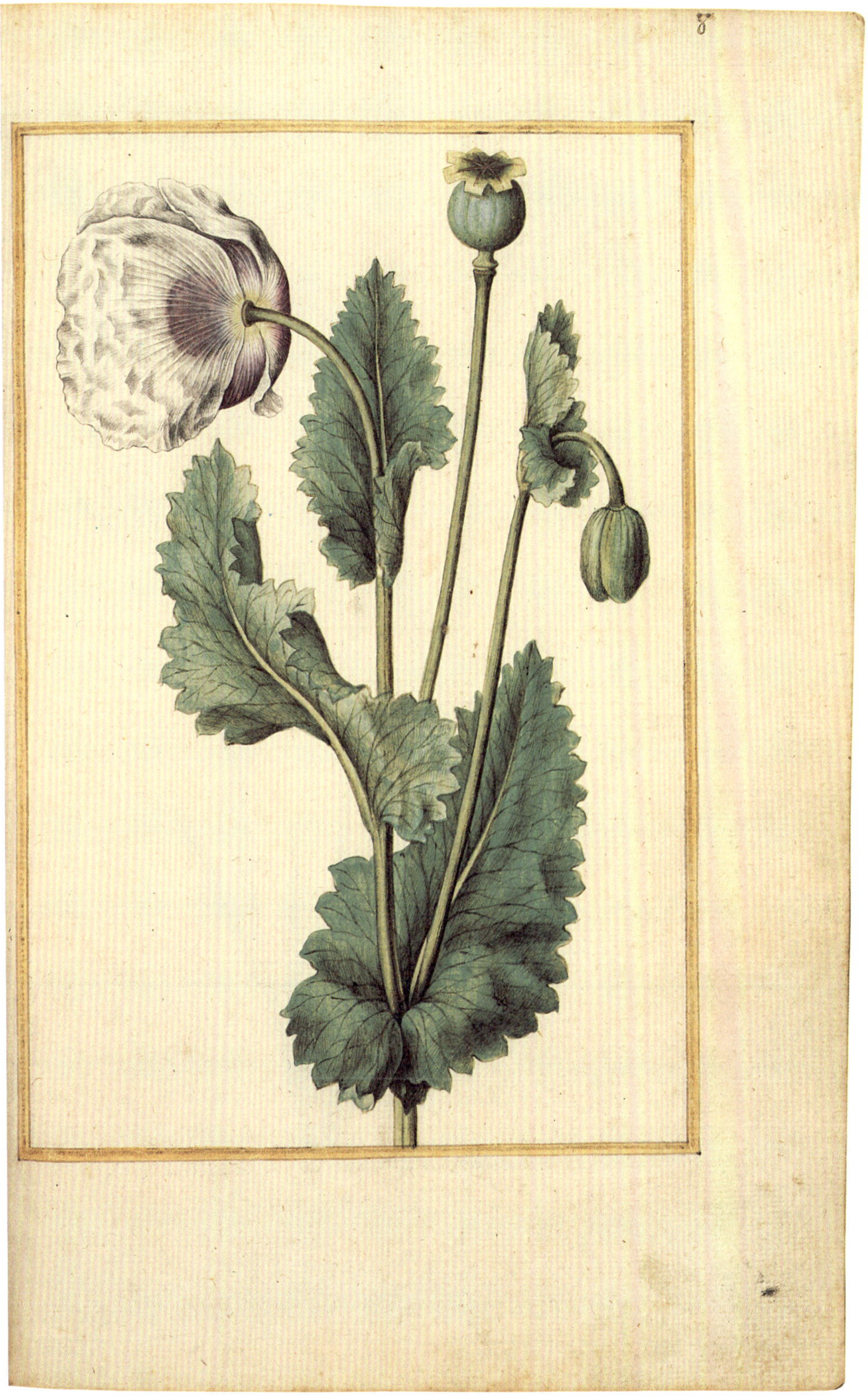

JACQUES LE MOYNE DE MORGUES, attributed, Manuscript of flowers and fruits. Opium poppy (*Papaver somniferum*), folio 8

In addition to the similarities in iconography and technique, there are other reasons to believe that these sixty paintings are the work of Le Moyne—the watermarked paper, for example, which was fabricated in France around the mid-sixteenth century, when Le Moyne was just beginning his career as an artist.

Then again, it is not unusual to find in the sixteenth and seventeenth centuries an artist producing more than one version of his own botanical paintings, all the while affirming that every work was 'painted from the life'. In fact, the Renaissance notion of the artist not as a passive imitator of nature, but rather as one seeking to emulate nature itself in his unceasing creation of new forms, was still widespread. Therefore, reproducing an image was not a simple act of copying, nor was the second version considered in any way inferior to the original. More than one copy of a particularly successful painting or manuscript might be produced, since such works were greatly coveted, not only by the naturalist and the collector but by artists and artisans, who used them as patterns for embroidery, textiles and other decorative arts. Le Moyne himself, in the dedicatory 'epistre' addressed to Mme 'de Sidney' that precedes the woodcuts of *La Clef des champs*, affirms that his work may serve 'tant chez les Nobles que parmi les Artisans, les uns pour leur preparer en l'Art de Peinture, ou Graveure, les autres pour estre Orfeures ou Sculteurs, & aucuns pour la Broderie ou Tapisserie, & mesme pour toute sorte d'ouvrage à l'éguille. . .'. Le Moyne's floral paintings generally were much admired by his contemporaries; indeed Crispijn van de Passe the younger directly copied a number of his images in the 'Altera pars' section of his vastly popular *Hortus Floridus* (no. 12) of 1614 (Savage 1923).

If one compares the paintings at Oak Spring with those in the Victoria and Albert Museum, it becomes quite clear that the former are not merely passively executed copies of the latter. The flowers, fruits and insects are often arranged differently on the page, with the addition or elimination of various details. In some cases the specimens of a fruit (such as those of the citrus species) may appear together on one leaf in the London manuscript, while in the Oak Spring copy the artist has prepared separate illustrations. The two French marigolds in the London manuscript (folio 6r) have been depicted on folio 4 in the Oak Spring manuscript in a much more elaborate form. During this process of creative re-elaboration, Le Moyne added butterflies and other insects to some of his compositions. There are also no less than seventeen new paintings in the Oak Spring manuscript.

There is thus ample evidence to support the attribution of the Oak Spring manuscript either to Le Moyne himself or to a talented pupil working within his immediate circle. If this attribution is confirmed by further archival documentation, an important new work will have been added to the *œuvre* of this singular artist.

The paintings in this manuscript are clearly the work of an artist of unusual talent and sensibility. The fruits and flowers have all been rendered with great naturalism, the artist managing to convey the tactile qualities of each fruit (such as the soft, velvety curve of the peach on folio 46) and the subtle details of every flower, such as the finely lined petals of the crocus in folio 19 or the delicate

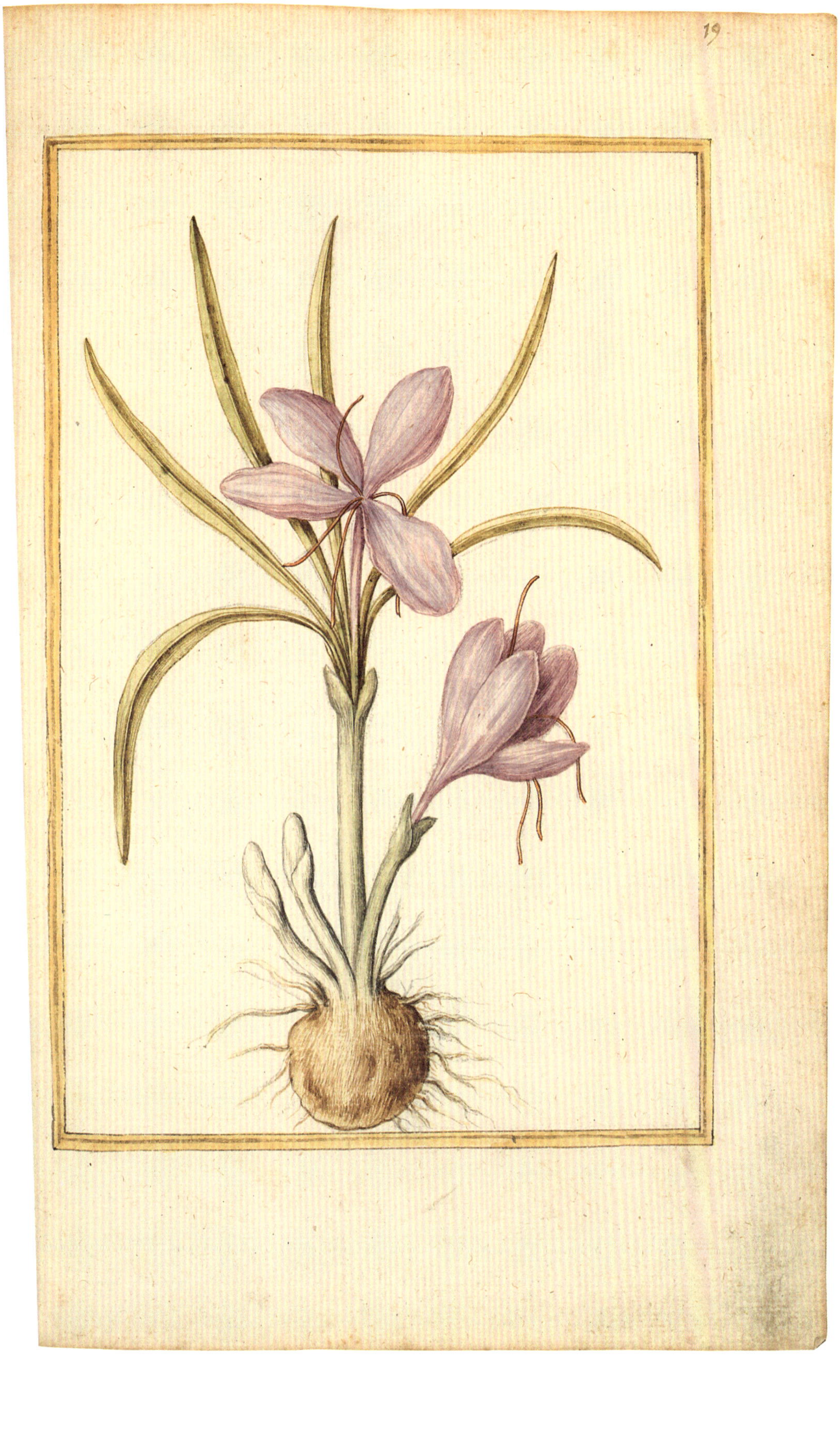

JACQUES LE MOYNE DE MORGUES, attributed, Manuscript of flowers and fruits. Saffron crocus (*Crocus sativus*) with bulb, folio 19

bend of the opium poppy stem under the weight of its bud and flowers. This manuscript is a fine example of the work of an important sixteenth-century artist, whose acute sensitivity and powers of observation served him well in the depiction of botanical and, in particular, floral subjects.

APPENDIX

A comparison of the folios in the Oak Spring Garden Library manuscript and those in the Victoria and Albert Museum's National Art Library manuscript reveals the following parallels:

1: Common mallow and damselfly (cf. V&A, fol. 84).

2: Lily of the valley (cf. V&A, fol. 5r).

3: Common borage (cf. V&A, fol. 11r). The Oak Spring version is considerably simplified, but the artist has added a hairy caterpillar, shown climbing up the stem.

4: The two French marigolds depicted separately in the lower part of fol. 6r in the V&A manuscript are here shown united on a single stem, with the addition of another bud.

5: A rose (cf. V&A, fol. 2v).

6: The corn cockle has been adapted from fol. 13v of the V&A manuscript, but without the view of the flower from behind.

7: Larkspur (cf. V&A, fol. 15v).

8: Opium poppy (cf. V&A, fol. 10r).

9: This is a copy of the cornflower which appears on fol. 13r of the V&A manuscript, but another specimen has been added on the left-hand side of the page.

10: The two gillyflowers that appear here were drawn from fol. 12v of the V&A manuscript (specifically, one from the upper right and one from the bottom of the sheet).

11: Marigold with dragonfly (not in the V&A manuscript).

12: Anemone with dragonfly (not in the V&A manuscript).

13: Wild carnation (not in the V&A manuscript).

14: The dog-rose here is a simplified version of the specimen on fol. 7 of the V&A manuscript.

15: Plantain (not in the V&A manuscript).

16: Chicory plant (not in the V&A manuscript).

17: White clover (not in the V&A manuscript).

18: Unidentified plant with butterfly (not in the V&A manuscript).

19: Crocus (not in the V&A manuscript).

20: German iris, reproduced from fol. 4r of the V&A manuscript.

21: Sweet violets reproduced from fol. 1v, but without the butterfly. On this page has been written 'Violete de Mars'.

22: The corn poppy is a somewhat simplified version of the image on fol. 15r in the V&A manuscript.

23: Dragon arum (cf. V&A, fol. 14v).

24: Peony (cf. V&A, fol. 3v).

25: Lesser periwinkle (cf. V&A, fol. 3r).

26: *Nuphar lutea* with butterfly (not in the V&A manuscript).

27: Yellow clover with dragonfly (not in the V&A manuscript).

28: Red clover with butterfly (not in the V&A manuscript).

29: Light blue bellflower (not in the V&A manuscript).

30: Arum with a snail (not in the V&A manuscript).

31: Clove pinks similar to those in the upper part of fol. 6r in the V&A manuscript.

32: Lily with a butterfly (not in the V&A manuscript).

33: *Phyllitis scolopendrium* and butterfly (not in the V&A manuscript).

34: Delphinium and butterfly.

35: White lily and snail.

36: Sweet violet (cf. V&A, fol. 9v).

37: Pea. This image of the common vetch also appears on fol. 10v of the V&A manuscript, but with the butterfly on the left side of the stem rather than on the right.

38: *lacking*

39: The cyclamen has been copied from fol. 11r in the V&A manuscript, although without its bulb and roots.

40: Pot marigold (cf. V&A, fol. 6v).

41: *lacking*

42: White rose (not in the V&A manuscript).

43: This double daisy appears on fol. 2r of the V&A manuscript, but the roots are less conspicuous and the butterfly has been moved to the left.

44: The columbine is reproduced from the image on the right side of fol. 12r in the V&A manuscript, with the addition of a flying stag-beetle.

45: *blank*

46: Peach (cf. V&A, fol. 25v).
47: Wild cherry (cf. V&A, fol 19r).
48: Cucumber (cf. V&A, fol. 26r).
49: Common fig. (cf. V&A, fol. 21v).
50: Pear reproduced from fol. 23r of the V&A manuscript, but without the leaves attached to the stem.
51: Melon (cf. V&A, fol. 26r).
52: This picture of a Seville orange is a copy in simplified form of the fruit shown in the lower part of fol. 22r in the V&A manuscript.
53: This image of a walnut branch is similar to that on fol. 28r of the V&A manuscript, with slight modifications in the arrangement of the nuts.
54: Wild strawberry and an insect (not in the V&A manuscript).
55: Bullace (cf. V&A, fol. 19v).
56: Quince (cf. V&A, fol. 24r).
57: Lemon reproduced from the upper part of fol. 22r.
58: Apple from fol. 24v, with the omission of the leafy branch.
59: Grape vine (cf. V&A, fol. 23v).
60: Medlar (cf. V&A, fol. 21r).
61: Mulberry (cf. V&A, fol. 22v).
62: Nectarine (cf. V&A, fol. 18r).
63: Pomegranate (cf. V&A, fol. 25r).

5. Jacques Le Moyne de Morgues (*c.* 1533–1588) [*circle of*]

Manuscript of flower paintings.

38.8 × 25 cm. 31 leaves.

BINDING: Late 16th-century limp vellum with double gilt fillets and blocked arabesque and strapwork centrepiece, produced in northern France, perhaps by Huguenot artisans. The front and back endpapers are identical and bear the watermark of the Troyes papermaker Jean Nivelle, whose paper was utilized in France and the Low Countries in the period around 1580 (see Briquet 3640). The leaves on which the bodycolour paintings were executed bear an unidentified watermark. Each leaf is numbered by hand in the upper-right corner. Many leaves contain relatively recent pencilled notes in French with the names of the plants.

PLATES: 31 plates, consisting of paintings of various plants in water- and bodycolour.

These large-scale bodycolour paintings of plants, most of them shown in flower, can be attributed to artists working within the circle of the Huguenot artist Jacques Le Moyne de Morgues. This link can be drawn on the basis of a number of indications. To begin with, the plants depicted in this manuscript are for the most part well-known indigenous species, and the works of Le Moyne show a similar preference for the charms of the familiar. Although his paintings reflect the new approach to depicting plants that emerged in the sixteenth century with the teachings of Brunfels, Fuchs and Mattioli (who stressed the importance of drawing subjects from life), in Le Moyne's work we find none of those exotic and horticultural species that so fascinated the collectors and botanists of the period. It was, therefore, the humble wild flower—the field poppy, cornflower and cowslip—and the simple blooms that could be found in every kitchen-garden (the white lily, pea flower and peony) that attracted the artist's eye.

Almost all the plants in this manuscript are shown complete with their bulbs and roots, in accordance with prevailing botanical iconography, and not cut, as are most of the flowers in Le

JACQUES LE MOYNE DE MORGUES, circle of, Manuscript of flower paintings. Common pea plant (*Pisum sativum*), folio 16

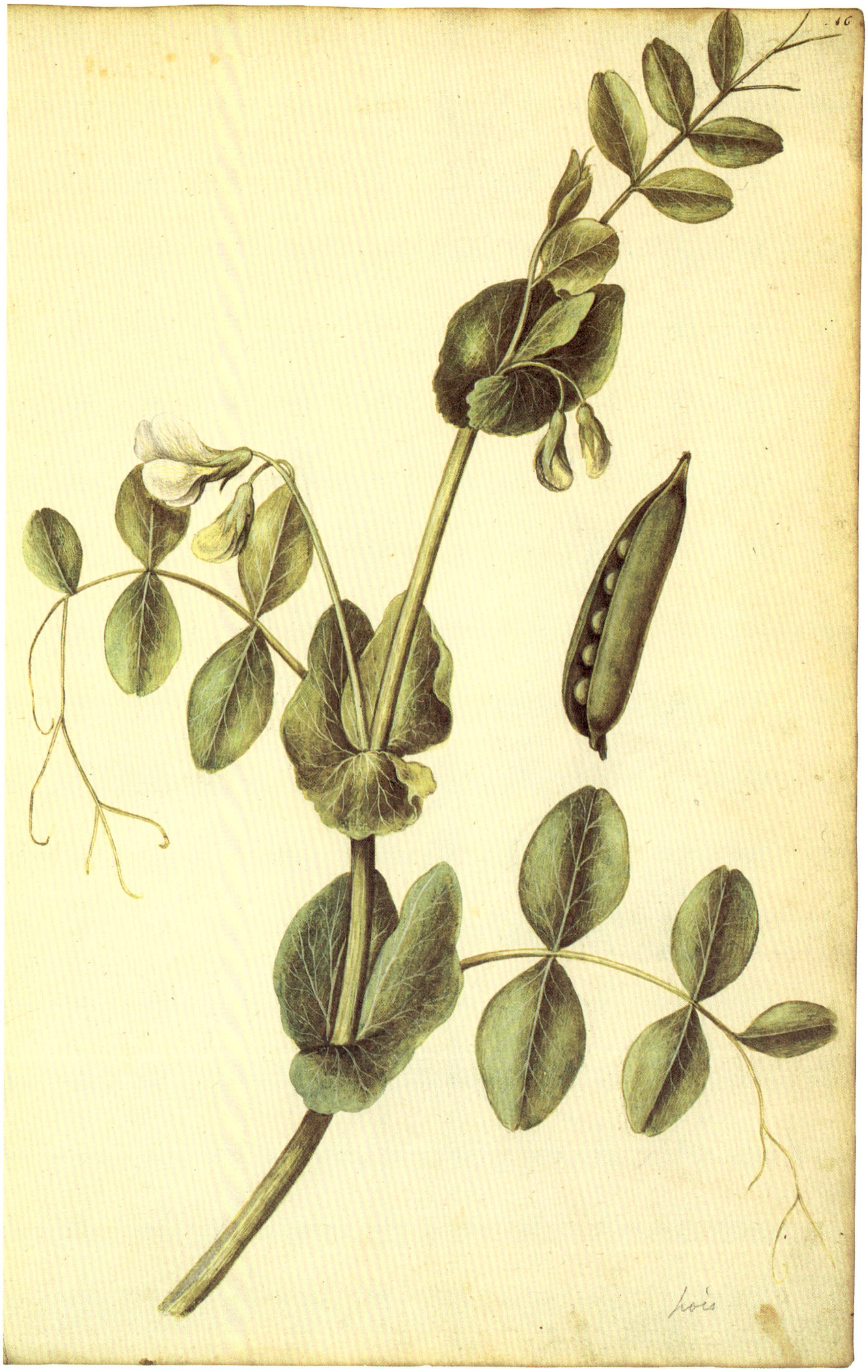

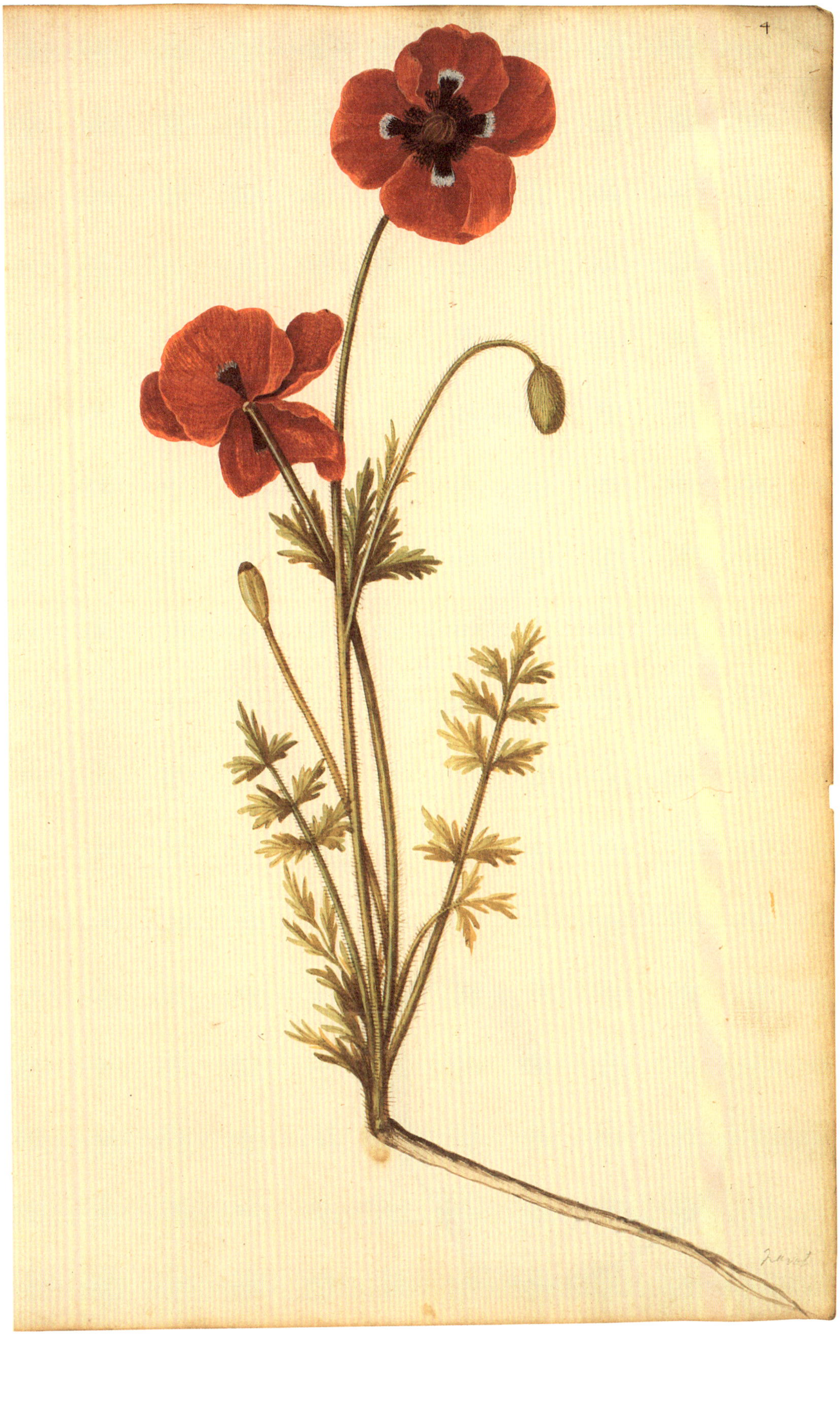

JACQUES LE MOYNE DE MORGUES, circle of, Manuscript of flower paintings. A type of *Papaver*, folio 4

Moyne's works. However, the unusual painting technique, which involved both bodycolour and watercolour, and the mode in which some of the plants are depicted, bring to mind similar works by Le Moyne, including one manuscript in the Victoria and Albert Museum and another at Oak Spring (no. 4). Close stylistic ties exist between paintings of the same plant that can be found in all three manuscripts, such as those of the 'Field poppies', the 'Sweet violet' or the 'Corn cockle'. Other works, for example 'Garden pea' and 'Pasque flower', have such an immediacy that they were obviously drawn directly from life, the artist capturing to perfection the elegantly flexible stem of the one and the thick, downy leaves of the other. Curiously enough, the plate of the mandrake portrays quite realistically the leaves and flowers of the plant, but not its anthropomorphic roots, which had constituted one of the most characteristic images in herbals since the early medieval period.

6. Jacques Le Moyne de Morgues (*c.* 1533–1588)

La Clef des Champs, pour trouver plusieurs Animaux, tant Bestes qu'Oyseaux, avec plusieurs Fleurs & Fruitz. Anno. 1586. Imprimé aux Blackefriers, pour Jaques le Moyne, dit de Morgues Paintre.

2° 16 × 23 cm. A^{6} B–C^{4} E^{4} D^{4} F^{4} $^{2}F^{4}$ $^{2}D^{4}$ $^{2}\mathit{E}^{4}$ ^{2}A–C^{4} (B4 signed Bij). Modern leaf with annotation tipped-in (see Appendix).

BINDING: Modern full leather, all edges gilt. Leaves numbered by a modern hand in pencil in the upper right-hand corner.

PLATES: 50 plates. 48 woodcuts coloured in water- and bodycolour, with two images on each leaf. Folios 3–14 illustrate various animals; fol. 15 birds; fols. 17–26 birds; fols. 27–38 flowers; fols. 38–41 fruits and vegetables; fols. 43–51 fruits. Folios 16v and 42r are bodycolour paintings, one of a peacock and the other of a cherry bough. On fol. 50 'Quercus pedunculata' has been written in pencil in a modern hand. The sequence of the pages varies in each of the 3 known copies.

PROVENANCE: Ex-libris Vernon James Watney (1860–1928), and Paul Mellon.

REFERENCES: Arber, p. 127; Hatton, p. 9; Coats, pp. 22–3; Hulton; Blunt and Stearn, pp. 96–7.

THERE ARE THREE known copies extant of *La Clef des champs*, the only printed work by Jacques Le Moyne de Morgues. Two are in the British Museum, of which the first contains only the plates (the title-page and the three pages of text are missing), while the second, which once belonged to Sir Joseph Banks, lacks the first plate. The latter copy has been described and illustrated in detail by Paul Hulton, who has also written an exhaustive study of the artist. The copy at Oak Spring is unique because it is complete. It also contains two bodycolour paintings (one depicting a peacock, the other a cherry bough), which, though not attributable to Le Moyne himself, bear unmistakable signs of his influence and were probably executed by someone working within his immediate circle.

La Clef des champs was the final work to be completed by Le Moyne, who died just two years after it was published in 1586. It provides important clues to the personality of this fascinating artist, about whom all too little is known even though a substantial body of work, consisting mostly of bodycolour paintings on botanical subjects (see nos. 4 and 5), has been attributed to him. The title of

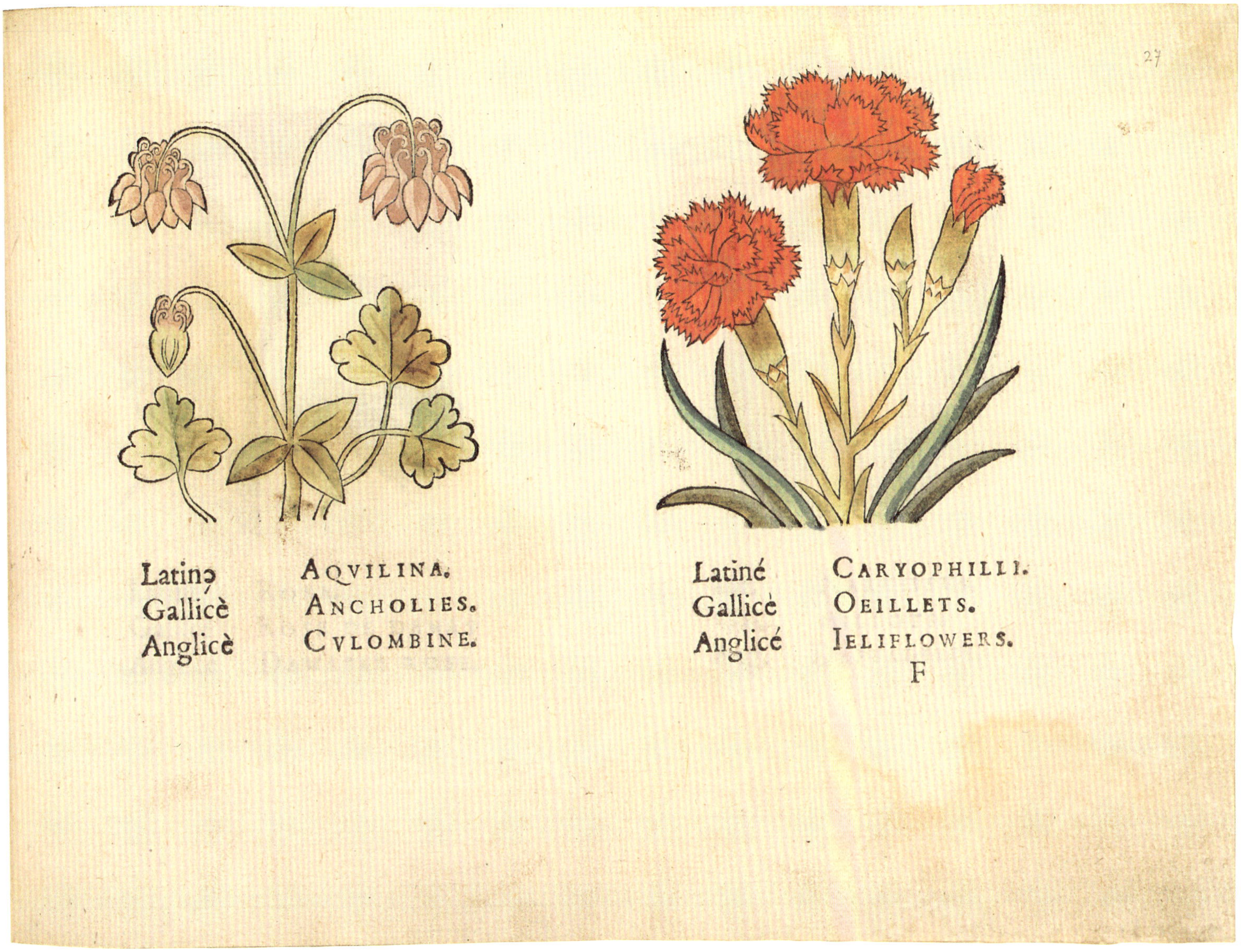

JACQUES LE MOYNE DE MORGUES, *La Clef des champs. Aquilegia* and *Dianthus*

this work alludes to the author's intention of providing the reader with a 'key' to the interpretation and appreciation of the beauties of nature, as exemplified by the various animals, plants, flowers and fruits illustrated in the volume.

La Clef des champs opens with an 'epistre' and a 'sonnet', not devoid of literary merit, dedicating this work to Lady Mary Dudley, the wife of Sir Henry Sidney of Penshurst, under whose patronage the Huguenot artist lived after fleeing French religious persecution. To his august patroness, therefore, the artist offered 'un petit Livret ayant choysi d'entre les Animaux, quelque nombre de Bestes & d'Oyseaux des plus remarquables, non seulement de ceux qui se blasonnent aux Armoiries des Seigneurs: mais aussi qui sont plus plaisans à l'œil, et que l'Admirable Ouvrier de Nature à mieux peins & bigarrez, ayant accompagné les dits Animaux d'autant des plus belles Fleurs & Fruitz, que j'ai pensé les plus propres, le tout apres le vif . . .'.

Thus the reader was to find not only the noble beasts that graced the coats of arms of the aristocracy, but other animals, plants and flowers created by 'the marvellous artifice of Nature'. Le Moyne states that his illustrations were also intended for those who wished to learn 'things both good and honest', for so well were they executed and so faithfully did they adhere to nature that, in keeping with the artistic practices of the period, they could serve as exemplary figures—ideal copies of the original. He therefore recommended *La Clef des champs* as a pattern-book for use by painters, engravers, jewellers, embroiderers, weavers and other artisans.

The illustrations to *La Clef des champs* depict birds and animals, each one carefully labelled with its name in Latin, French, German and English, and flowers and fruits with their names in Latin, French and English. The blocks were most probably cut by Le Moyne himself, who perhaps also printed and coloured the plates. The most interesting illustrations are those showing various plants in flower, which in fact represent skilful reinterpretations in the form of woodcuts of Le Moyne's works in bodycolour. Thus similar images may be found in a manuscript of paintings in the Victoria and Albert Museum (Savage 1922) and in another manuscript at Oak Spring (see no. 4).

APPENDIX

The following annotation is tipped-in the Oak Spring Garden Library's copy:
Jacques Le Moyne at Blackfriars 1586. The Author went to Florida. See Brunet. The book is not mentioned by Lowndes – [. . .] Copy Brunet could record was the Copy of Sir Joseph Banks in the British Museum—which is imperfect. Wanting Plate 1.

The book is dedicated to

Sir Henry Sidney of Pen[s]hurst = Lady Mary Dudley sister to Robert, Earl of Leicester

Sir Philip Sidney

Sir Robert

Elizabeth = Earl of Rutland

Mary = Earl of Pembroke

24 figures of Beastes
25 figures of Birds
49 figures of Flowers
Fruits
——
98

There are 2 drawings
The Peacock
& a bunch of
Cherries

7. Domenico Buonvicini (*dates unknown*)

Venetian manuscript florilegium.

[*plate 1 of the manuscript makes up the title-page of this florilegium. At the foot of the first leaf is written*] 1601 Dominicus Bonus Vicinius. De Veneciis. Faciebat.

34.3 × 23 cm. 95 leaves, all numbered by hand. Each plate is accompanied by a guard sheet.

PLATES: 94 plates of flowers, painted in water- and bodycolour, and one of a mole cricket, followed by 41 blank leaves. The Linnaean name of the plant has been added, in manuscript at a later date, to some of the illustrations. Each specimen has been numbered for an index which does not, however, appear in this manuscript.

BINDING: 18th-century red half calf, labelled 'Dom. vicini Opus'. Gilt and gauffered edges.

Nothing is known about the artist Domenico Buonvicini who in 1601 signed this painted florilegium. The work may have been commissioned by an 'amateur' of natural history, or by a herbalist or apothecary of Venice; it is even possible that Buonvicini himself was a naturalist with sufficient artistic skill to prepare his own herbal.

During the second half of the sixteenth century the Republic of Venice was one of the most important centres in Italy for botanical studies, and many gardens were constructed throughout the city-state in which rare and curious species could be cultivated (Masson 1961, pp. 223–4). Although the city itself did not offer sufficient space, on the islands of Murano and Giudecca, noble families—the Navagero, the Gritti, the Barbaro, the Mocenigo, the Vendramin—succeeded in establishing truly splendid gardens. The great painter Titian designed one such for an aristocratic patron, while the humanist scholar Pietro Aretino described the 'fragrant flowers' blooming in the garden of the Marcolini family (*OCG*, p. 583).

In this period Venice was often the first port of call of ships returning from the Near and Far East, and therefore the first to receive many of the exotic specimens that gradually were to change the botanical landscape of Europe. The neighbouring city of Padua boasted one of the oldest botanical gardens in the world, and naturalists from all over Europe went there to pursue their studies (Azzi Visentini). This scientific activity had a profound influence on the art of botanical illustration. Many talented scientific illustrators flocked to Venice to work with collectors and scientists. For example, around the mid-sixteenth century a Venetian nobleman, Pietro Antonio Michiel, commissioned Domenico dalle Greche to paint hundreds of the plants growing in his garden, including such prized specimens as two precocious tulips, *Tulipa sylvestris* and *T. praecox*, for *I Cinque Libri di piante* (today in Venice's Biblioteca Nazionale Marciana). And, as his correspondence reveals, the eminent naturalist Ulisse Aldrovandi of Bologna often received paintings of plants and animals from his colleagues in Venice (Olmi 1992, pp. 74 75, 77, 78).

This manuscript by Buonvicini is an important example of the art of botanical illustration in Venice at the turn of the century. On each of its ninety-three leaves appear one to four different

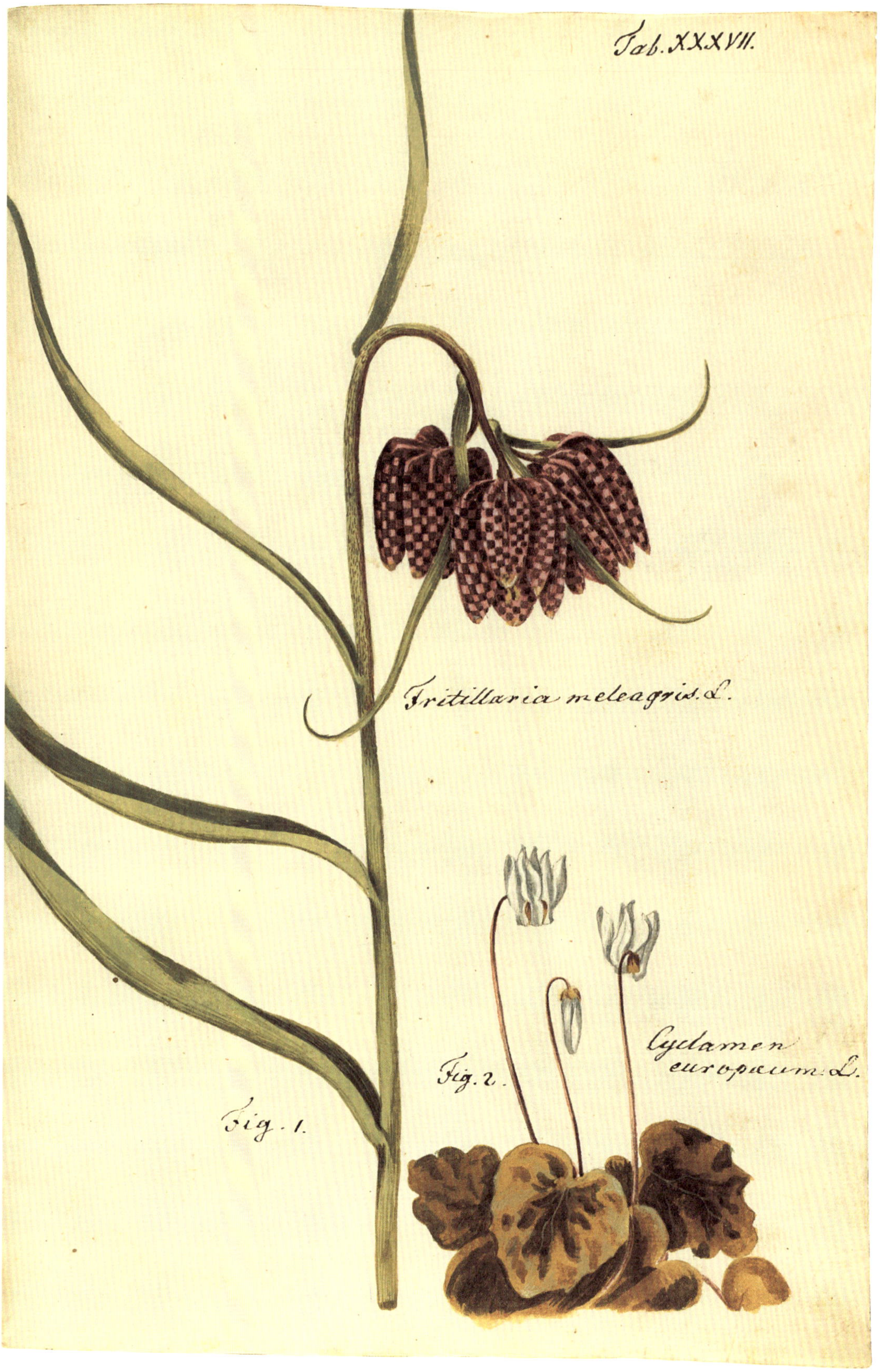

DOMENICO BUONVICINI, Venetian manuscript florilegium. Checkered lily (*Fritillaria meleagris*) and sowbread (*Cyclamen europaeum*), folio XXXVII

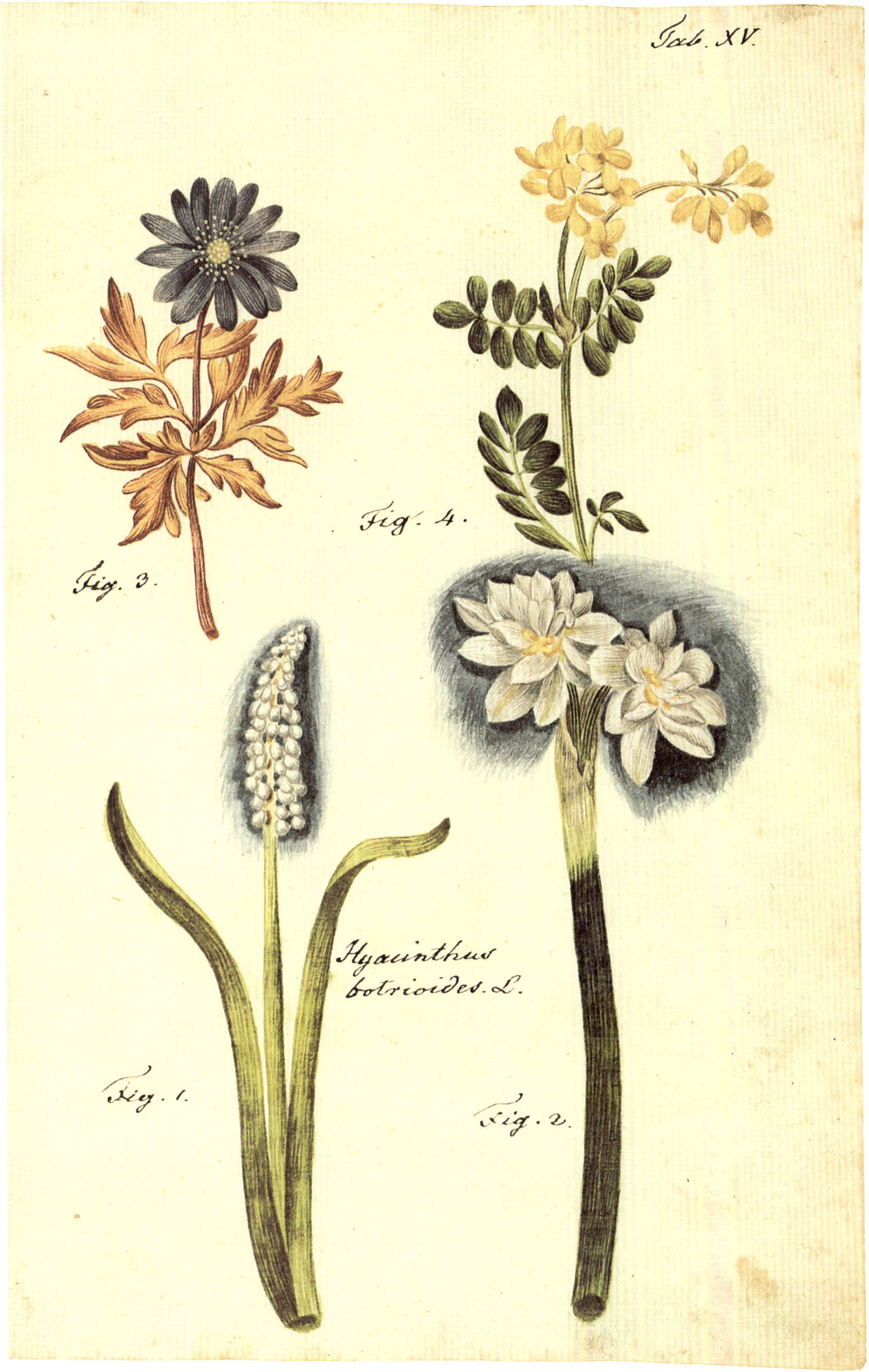

DOMENICO BUONVICINI, Venetian manuscript florilegium. Windflower (*Anemone apennina*), crown vetch (*Coronilla* sp.), grape-hyacinth (*Muscari botryoides*), and a double *Narcissus* sp., folio XV flowering plants, comprising a total of no less than 259 species. It gives an accurate idea of the varieties that were generally to be found growing in a well-stocked Italian garden of the time. In addition to the many species of narcissi, irises, tulips and fritillaries that are illustrated, there are common flowers, such as the violet, primrose and daffodil. It is interesting to note a series of small orchids (pls. 17, 49, 64, 65, 67 and 72)—species not at all easily cultivated in a garden, although they could be found growing wild in northern Europe and perhaps even in the forests of the Veneto.

This manuscript provides ample testimony as to Buonvicini's talent as a botanical illustrator. The title-page in particular could stand on its own as a floral still-life, artfully composed of two anemones, one white and one red, a lily of the valley with long, fibrous roots, some sea lavender and a yellow argemon; a bee is poised on the receptacle containing the first anemone, while a light-blue butterfly hovers to the right. The other plates are less elaborate, but present their specimens in pleasing compositions: plate 15, for example, shows an *Anemone apennina*, a *Coronilla*, a *Muscari botryoides* and a narcissus with a double bloom; this narcissus is surrounded by touches of intense blue that contrast effectively with the white of the flowers. In plate 37 a *Fritillaria meleagris* with long, undulating leaves shares the sheet with a delicate *Cyclamen europaeum*.

8. Pierre Vallet (*c.* 1575–1657)

[*etched title-page with a classical arch framing the vista of a garden. On either side of the arch appear statues of the botanists Carolus Clusius and Matthias de L'Obel; their names are inscribed on the balustrade at the base of the arch, into which is also set a plaque bearing the epigram*] Ver hic perpetuum perennis hic flos Crebescens vario colore pictus, Flora hic condidit hortulos beatos, Et quos suspiciunt Dei Deæque [*Here it is perpetually spring, here grow flowers in a multitude of colours, here Flora created beautiful gardens, which are much admired by the Goddess and the Gods*]. [*inscribed on a cartouche set in the tympanum of the arch*] Le Jardin Du Roy Tres Chrestien Henry. IV. Roy De France Et De Navare Dedie A La Royne [*at the base of the arch*] par Pierre Vallet, brodeur ordinaire du Roy. 1608. [*at the foot of the page*] Avec previle. du Roy. P.V.I.F.L. [*Pierre Vallet Inventor fecit Lutetiae*]. [*at the top of the page, written in an antique hand*] 'N 46'.

F° 33.5 × 21.5 cm. ã² e² *1–8* p., 76 leaves of plates.

BINDING: Modern vellum, with a red label on the spine that reads 'Pierre Vallet—Le Jardin du Roy'.

PLATES: 76 unnumbered etched plates including the title-page. One is a portrait of Pierre Vallet, in an oval inscribed within a rectangle with various artist's tools at the four corners (18 × 11 cm.). 'Pierre Vallet 1608' is inscribed on the oval frame. On a plaque beneath the portrait: 'Anagrammatismus Petri Valettij is vult, et superat. Is vult, et superat Cunctos ad Palladis artes, Cui se se accingunt, arte, Manu Ingenio' (This being an anagram of the name Petrus Valletius: He aspires and surpasses. He aspires, and surpasses all of those whom with great art, and a skilful hand, dedicate themselves to the arts of Pallas). A second plate (16 × 9 cm.) contains a portrait of Jean Robin; at the top of the page has been written 'Aet.LVII AN. 1608' (at the age of 67 years—1608) and on a plaque at the base of the page: 'Joannes Robinus Omnes Herbas novi Quot tulit Hesperidum Mundi quot Fertilis hortus Herbarum species novit hic onus eas' (Jean Robin brought all of these plants from the New World of the Hesperides and he alone knew all of the species in that fertile garden). Followed by 73 plates (29.5 × 17 cm.) of flowering plants, with the name of each given in Latin.

REFERENCES: Nissen 2039; Hunt 187; Coats, p. 25; Tongiorgi-Tomasi 1980, p. 581; Schnapper 1988, pp. 41–2; Blunt and Stearn, pp. 101–02; Saunders p. 55.

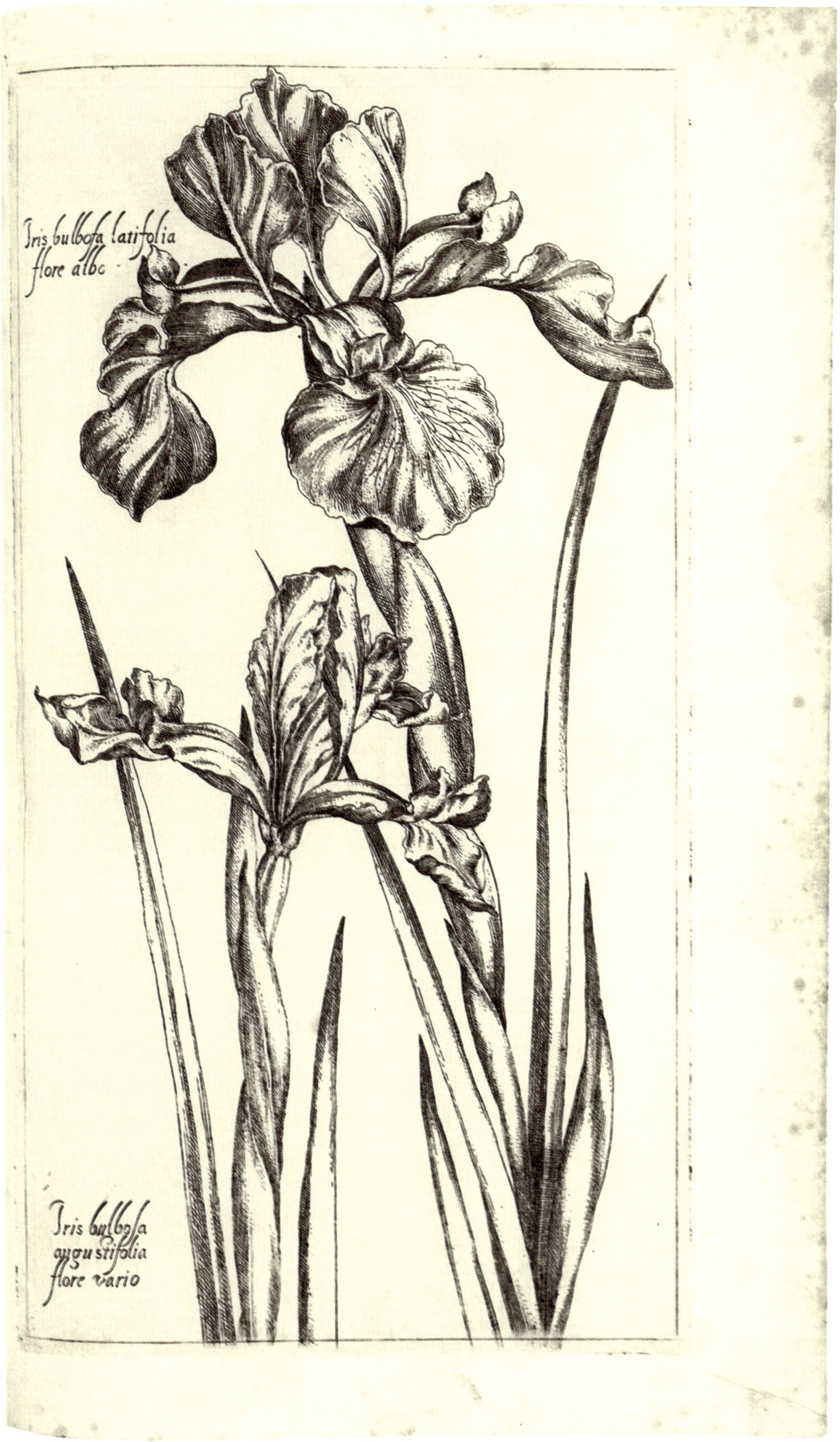

PIERRE VALLET,
Le Jardin du Roy tres chrestien Henry IV. Two varieties of *Iridaceae*

THE FIRST PRINTED FLORILEGIUM in Europe, that by Adrian Collaert, was published *c.* 1590, while the highly influential *Le Jardin du Roy* was published at Paris in 1608. The author of the latter was Pierre Vallet, an eclectic artist who became 'brodeur ordinaire' to the court of Henri IV. Born in Orléans *c.* 1575, Vallet soon transferred to the capital, where he found employment first as an engraver of illustrations and then as a designer of embroidery patterns. Jean Robin the younger (1550–1629), whose etched portrait follows that of Vallet in *Le Jardin du Roy*, was the creator of the royal gardens of the Louvre for Henri IV. As director of the garden, it was no doubt with his encouragement that Vallet began to focus his talent on the painting of floral motifs.

Le Jardin du Roy was immediately recognized as a masterpiece, and was soon copied by other authors. It contained many departures from the conventional typology of the florilegium, for Vallet not only modified the way in which the specimens were selected and presented on the page, he invented an unusually successful visual conceit for the title-page—that of a classical arch, flanked by statues of famous botanists ancient or modern, which opens onto a garden. A further element introduced by Vallet that would henceforth be a regular feature of the printed florilegium is the prominent portrait of the author featured on one of the opening pages. Reflecting the seventeenth-century taste for wordplay, beneath the portrait of Vallet in *Le Jardin du Roy* appears the phrase 'Is vult, et superat' (He aspires and surpasses), an anagram of the Latin form of the artist's name (Petrus Vallettius). The inscription continues, and translated reads 'He aspires, and surpasses all of those whom with great art, and a skilful hand, dedicate themselves to the arts of Pallas [Athena]'.

At the beginning of the seventeenth century, the floral theme was quite the rage at the French court. This was in large part due to the influence of the monarch, Marie de' Medici, who surrounded herself with flowers, no doubt because they reminded her of the luxuriant gardens of her beloved Tuscany. Not only blossoms of every kind, but also paintings and fine embroidery with floral motifs were certain to appeal to her. The States-General of Holland frequently sent her gifts of paintings by Jacob de Gheyn of Antwerp, who was celebrated for his still-lifes and flower bouquets (Blunt and Stearn, p. 101). Thus Vallet felt encouraged to prepare his florilegium, in part to document the exotic species brought back from Spain and Guinea by Jean Robin (*Catalogus stirpium tam indigenarum quam exoticarum quae lutetiae coluntur*, Paris, *c.* 1601), many of which appear in a long list in the text, and in part to provide new floral motifs for the court embroiderers. Vallet dedicated his work to the Queen: 'Je vous offre les fleurs, si vous les regardez / Vous y verrez les lys de France / & de Florence / Que vous avez unis, & bien contregardez' (I offer you flowers, and if you regard them well, you will discern the lily of France and that of Florence, which you have united and so well safeguarded).

Vallet himself prepared both the original drawings and the plates, using a technique that combined etching with engraved highlights, and a stipple-like treatment of the metal plate to enhance the effect of shadowing. The illustrations combine great artistry with impressive scientific accuracy. Vallet opens his work with the fritillary, the 'imperial crown', while among the other flowers

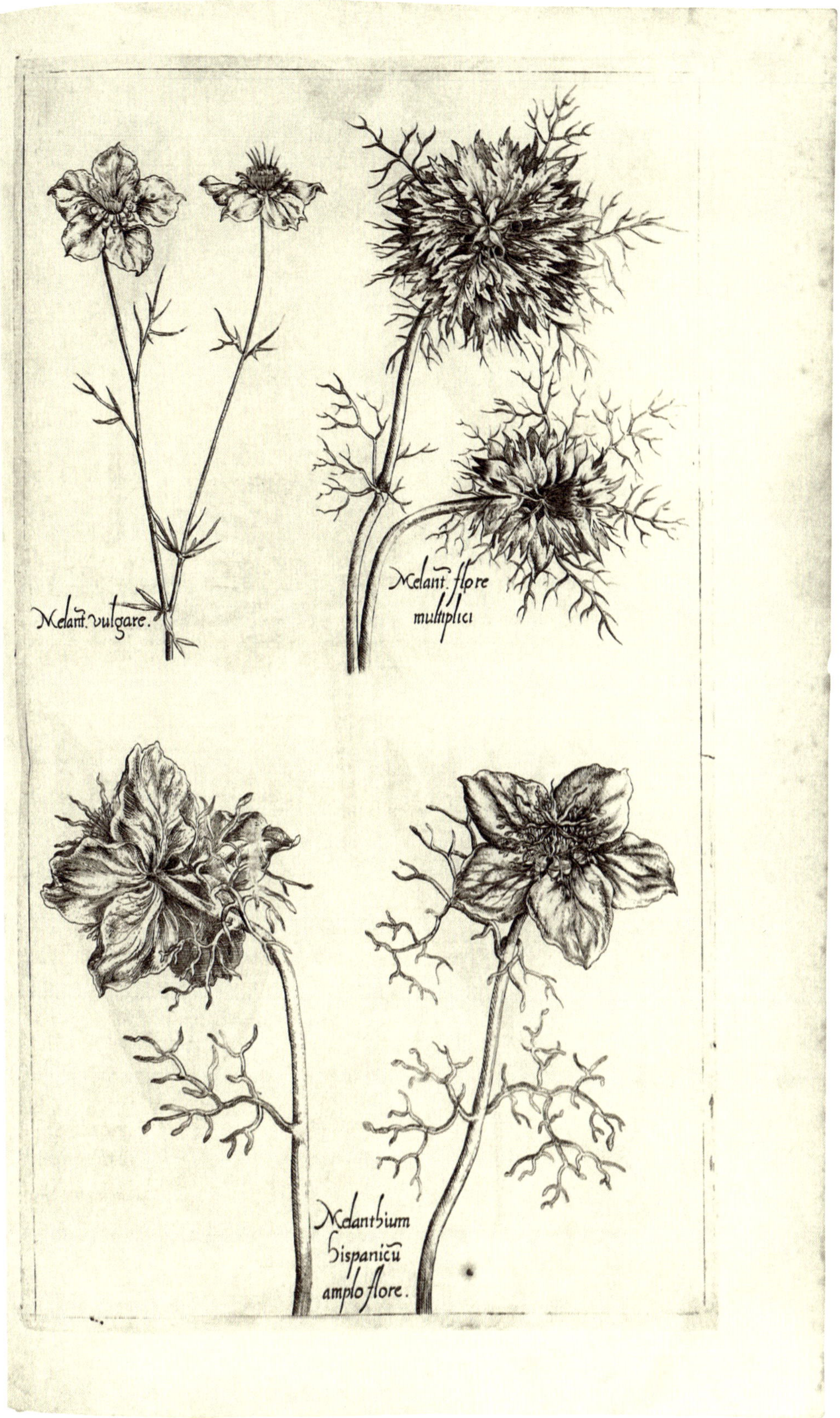

PIERRE VALLET,
Le Jardin du Roy tres chrestien Henry IV. Three varieties of love-in-a-mist (*Nigella*)

depicted are the tulip and the peony, both very popular embroidery motifs. Some of the plates were subsequently copied by other authors, such as Johann Theodor de Bry in his *Florilegium novum* (see no. 10) and Friderico Barbette in his *Florilegium novum* published in 1641. A second edition of Vallet's work, titled *Le Jardin du Roy tres Chrestien Loys XIII*, containing ninety-one plates, was published at Paris in 1623.

9. Emanuel Sweerts (1552–1612)

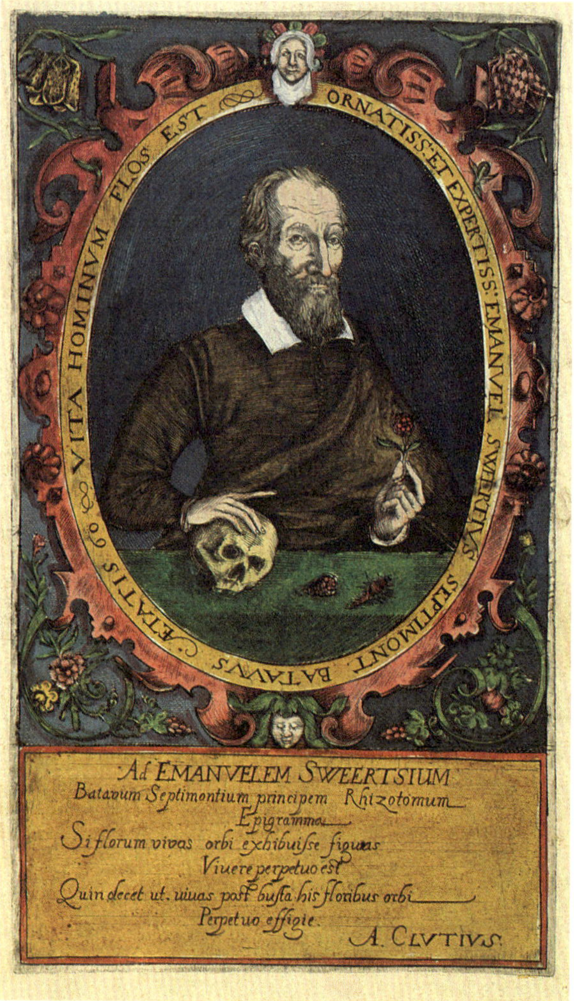

EMANUEL SWEERTS, *Florilegium*. Frontispiece portrait of the author

[*engraved title-page showing a formal garden with flowerbeds and a fountain; overhead floats the Tetragrammaton symbolizing the omnipresent God. In the foreground appear three figures; in the centre Flora is seated holding a vase of flowers and the key to the garden, while on either side of her stand Apollo, the sun god, and Diana, goddess of the moon. At the top of the page is a cartouche, decorated with the heads of putti, on which is written*] Florilegium Emanuelis Sweerti Septimonti Batavi Amsteledami Comorantis, Tractans De Variis Florib[us], Et Aliis Indicis Pla[n]tis ad Vivum delineatum in duabus, Partib[us], Et quatuor Linguis Concinnatum. [*at the foot of the page*] Cum Gratia et Privilegio S. Caesareae Maiestatis. [*beneath this, in a cartouche flanked by masses of fruits and flowers and by two portraits, one of Carolus Clusius and one of Rembert Dodoens*] Prostat Venale Una Cum, Florib[us], et Plantis Ipsis, Apud Ipsum Autorem, Eman; Sweertium Cuius Officina Ante Curiam. Frankofur MDCXII. [*below the engraving*] Impressum Francofurti ad Moenum Apud Anthonium Kempner Sumptib[us]. Autoris .1612. De novo Correctum et emendatum.

[*printed title-page for Pars Secunda*] Florilegii Pars Secunda, In Qua Agitur de Praecipuis Plantis Et Floribus Fibrosas Radices Habentibus: Nec Non Arboribus Speciosis Et Odoriferis, Quibus Horti In Utraque Germania Decorantur: Opera & Impensis Emanuelis Swertii [*sic*] Septimontani Batavi, In Gratiam Rei Herbariae Studiosorum Edita. [*printer's device with the statue of a knight* 8 × 7 *cm.*] Francofurti Ex Officina Typographica Erasmi Kempfferi, Anno M.DC.XIV.

F^{o} 39 × 25 cm. π^{2} $*_{*}*$–$2*_{*}*^{6}$ $3*_{*}*$–$4*_{*}*^{4}$. (Library's copy has $*_{*}*1$ bound following $*_{*}*6$).

BINDING: Contemporary mottled calf binding, full gilt spine and edges of the boards. On the spine is a handwritten label 'E Sweertii Florilegium'.

PLATES: The first part: a portrait of the author and 67 numbered engravings coloured by hand. The second part: 43 numbered engravings coloured by hand.

PROVENANCE: Armorial bookplate (with engraved coat of

arms, surmounted by a crown and supported by a pair of fantastic beasts, each with the head and forefeet of an eagle and the body and hind legs of a lion. The field contains three crescents surrounding a star).

REFERENCES: Pritzel 9073; Nissen 1920; Hunt 196; Bleiler; Blunt and Stearn, pp. 102–04.

THIS FLORILEGIUM by Emanuel Sweerts (or Sweert or Schwertz), first published in 1612, was so immensely successful that it was reprinted in Frankfurt in 1614 and then in Amsterdam in 1620, 1631, 1641, 1647 and 1655. The first two editions were essentially catalogues for the selling of plants and bulbs, while the later editions were true florilegia intended for the connoisseur and the scientist, even if their plates were engraved with somewhat less care.

Sweerts was born in Zevenbergen, Holland, in 1552, but spent the greater part of his life in Amsterdam, where he worked as an artist and as a merchant in *objets d'art* and rare or curious natural specimens. Each year he went to Frankfurt for the celebrated annual fair, setting up shop just in front of the Römer. His inventory included a wide variety of cultivated and bulbous plant species. His coupling of professions was an auspicious one, for rare flowers were considered prized items by collectors, but unlike the other objects in the connoisseur's collection, their beauty was transient and could only be captured on paper or canvas.

Sweerts's fame as a floriculturist spread far and wide, and a white iris was even named after him (the *Iris sweertii*). His name often appears in the correspondence of naturalists, botanists and floriculturists of the period, and he himself corresponded with many of them, including the floriculturist Matteo Caccini (1573–1640) of Florence (Ginori Conti, p. 25). The Habsburg emperor Rudolf II, possessor of many magnificent collections, tried unsuccessfully to entice Sweerts to join his court in Prague as director of the royal gardens.

The dedication 'Ad candidum lectorem', supplied in Latin, Dutch, German and French, furnishes some interesting details about the life and work of the author. Sweerts states that it was Rudolf II, 'le plus grand Admirateur et Amateur du monde de toutes les raritez et artifices mondaines', who encouraged him to depict the more rare and unusual flowers in his collection and to have them etched in copper, in order that they might be published in a book under his patronage. Unfortunately, Rudolf died in 1612 and the work, by then already in press, was dedicated to his brother, Matthias. In compiling this work, Sweerts goes on to declare, he wished also to glorify the infinite power and wisdom of God: 'un miroir auquel l'hom[m]e se peust mirer et en se mirant se souvenir de la briefcté et fragilité de sa vie' (and, in fact, Sweerts himself was to die at the end of this same year). The theme of *vanitas*, central to the Protestant ethic of the United Provinces in the seventeenth century, appears clearly in the portrait of the author, which precedes the plates: in an ornate oval frame Sweerts, already in his sixties, can be seen holding a flower in his left hand and a skull in his right, while two *Fritillaria meleagris* gracefully surronud the upper part of the frame. The inscription 'Vita hominum flos est' (The life of man is like that of a flower) could not be a more clear allu-

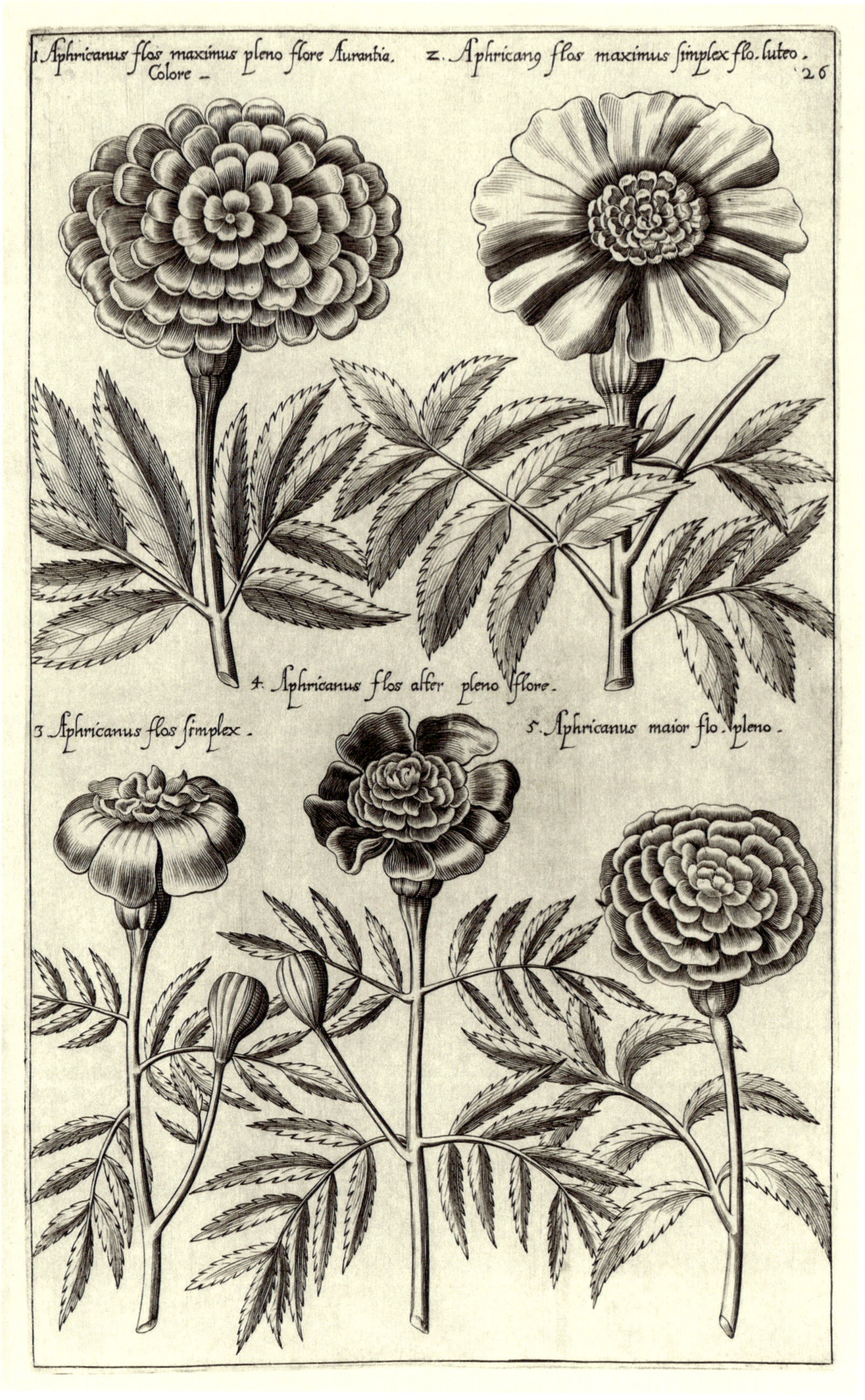

EMANUEL SWEERTS,
Florilegium. Five varieties of French marigolds (*Tagetes patula*), plate 26

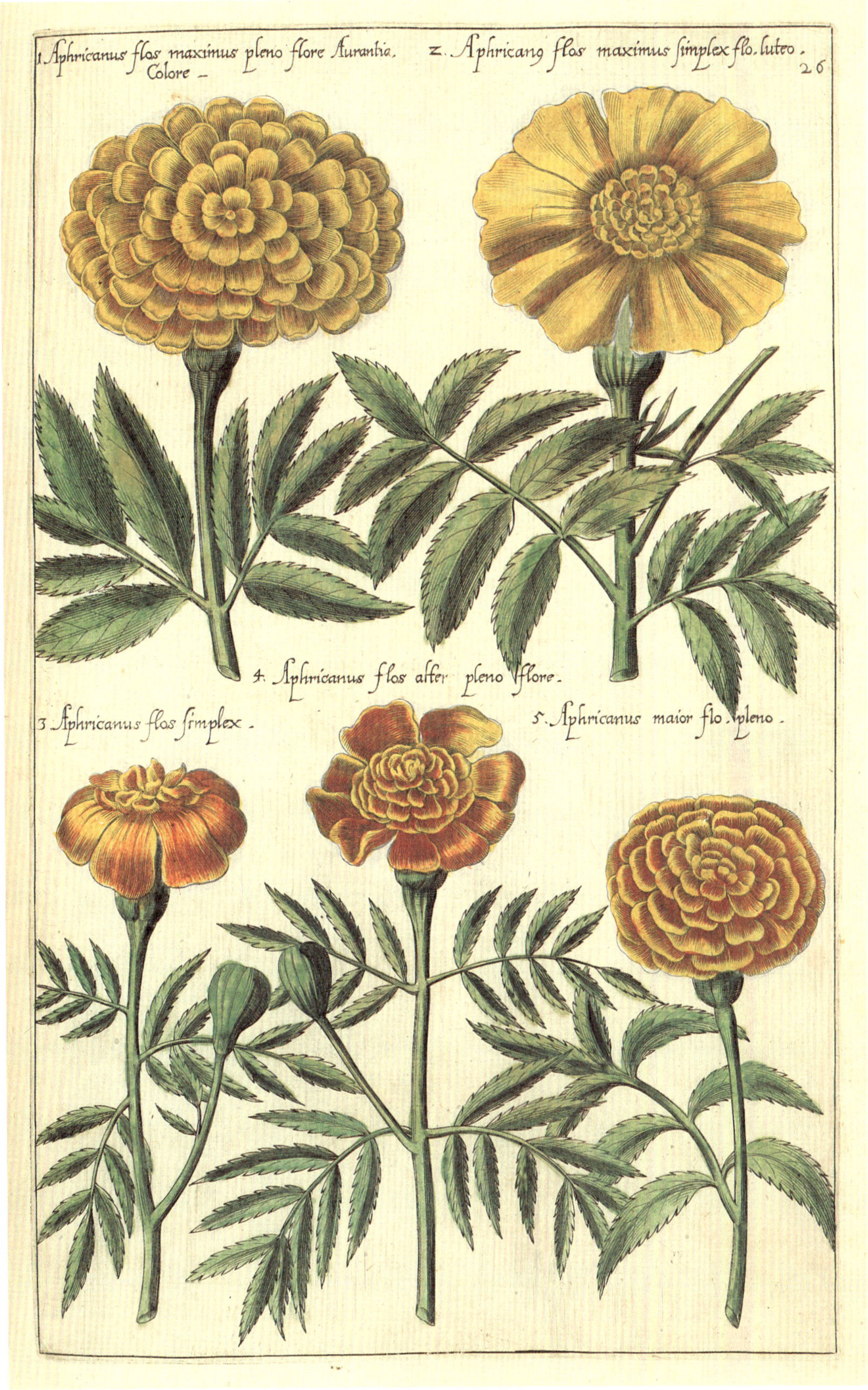

EMANUEL SWEERTS, *Florilegium*. Five varieties of French marigolds (*Tagetes patula*), hand-coloured engraving, plate 26

sion to the transient nature of man's existence. In the same vein the conch shell—permanent and immutable—could represent a symbol of immortality.

From the dedication we also learn that the work has been divided by the author into two parts, the first dealing with the bulbous species, and the second with those species having 'fibrous' roots, plus a number of 'beautiful' and 'fragrant' trees. Each illustration is accompanied by the name of the plant, certain descriptive details regarding colour and other details, and useful information, such as planting instructions.

We do not know the name of the artist who engraved these plates, although some of them (the bulbous species in particular) were directly copied, with slight variations, from the trial proof copy of de Bry's *Florilegium novum* printed just one year earlier (see no. 10), a work that in turn owes much to Vallet's *Jardin du Roy tres Chrestien Henry IV* (see no. 8). The plates in the first part of Sweerts's work are much finer than those in the second, and thus it is likely that more than one engraver was involved in the enterprise. Over 560 different flowers are depicted. They are grouped by species, and in some cases several appear together on a single page. In the case of the tulips and anemones, only the flower is shown. For other species, such as the *Pancratium illyricum* (Part 1, pl. 27), the bulb occupies a large portion of the page.

The copy of Sweerts's florilegium at Oak Spring is particularly fine due to the quality of the printed engravings and their skilful colouring. The Library also possesses another edition of the same work, printed in Amsterdam in 1631, the plates of which are not coloured.

10. Johann Theodor de Bry (1561–?1623)

[*in the upper part of an arch decorated with garlands and urns of flowers, which opens onto a garden with formal flowerbeds and a fountain, is written*] Florilegium Novum, hoc est: variorum maximeque Rariorum Florum ac Plantarum Singularium unacum suis radicibus & cepis, Eicones diligenter aere sculptae & ad vivum ut- plurimum expressae. New Blumbuch darinnen allerhand schöne Blumen und frembde Gewächs, mit ihren Wurtzeln und Zwiebeln, mehrer theils dem Leben nach in Kupffer fleissig gestochen, zu sehen seind: exhibitum nupérq[ue], auctum | A Johanne Theodoro de Bry Cive Oppenheimense M.D.CXII.

F° 32 × 20 cm. The illustrated title-page is missing from this copy and has been replaced by a facsimile, laid-in. In addition, wanting $^2\pi$, the *Brevis Description*. $\pi 1$)(^{2}, 82 engraved plates.

BINDING: Contemporary binding in olive morocco with gilt decoration along the borders, and gilt and gauffered edges. In the centre of both the front and back covers is a medallion, surrounded by a decorative motif consisting of a wreath of branches and two flower vases, which contains the initials and date 'FVO 1617'. The spine has been divided into five compartments, each decorated with a flower vase and a motif of stars in gold tooling.

PLATES: 82 engravings of plants, each one except for plate 7 bearing the Latin name of the species. This copy contains plates 1–21, 24–49, 51–61, 68–73, 75–77 (plate 76 has been numbered by hand and appears after plate 77), 83–84, plus 13 unnumbered plates. All the engravings have been coloured by hand. Written in the upper part of plate 55 is 'Amplificatio sive Dilatatio Florilegij nuper coepti auctique. Iam vero varijs atque Elegantioribus Floribus Exornati. Opera et Studio

JOHANN THEODOR DE BRY, *Florilegium novum*. Binding

Johannis Theodori de Bry. | Eruveiterung oder Vortpflantzung, des newlich angefangenen, schon Vermehrten Blumbuchs; So Jetzt mit macherley Schonen Blumen ausgebesert und gezieret worden. A° 1613'.

REFERENCES: Arber, p. 244; Pritzel 1299; Nissen 272; Hunt 197; *A Magnificent Collection*, no. 91; Blunt and Stearn, p. 102.

THIS EXTREMELY popular florilegium went through several reprintings, and was considerably expanded in later editions (see no. 16). Its author, Johann Theodor de Bry, belonged to a noted family of engravers from Frankfurt. De Bry's father, Theodor, a draughtsman, engraver and goldsmith, was born in Liège but eventually transferred to Frankfurt, where he set up a printing shop. De Bry and his brother Johann Isräel, after learning the rudiments of their trade in these modest surroundings, went on to become two of the most talented and prolific engravers of their time.

Although de Bry was already a fully mature artist when he began engraving and printing his first book of flowers, he recognized that this genre presented a difficult challenge, for to imitate the endless variety of colours in nature was a task perhaps beyond the skill of even the greatest master. But as de Bry insisted in his dedication: 'Of all the things which spring from this earth, flowers are the most beautiful for their grace and dignity, just as man surpasses every other living thing in dignity of body and soul.' Drawing on a recurring theme in the art and literature of the period, he further reflected that the flower's fragility should serve as a reminder of the ephemeral nature of all worldly things, and bids us think of the hereafter.

The publishing history of this work is quite complex, owing perhaps to the very fact that the engravers and printers were one and the same persons, and it is difficult to supply a description of an 'ideal copy'. There exists, for example, an extremely rare edition published in 1611, possibly a trial proof prepared for the great Frankfurt fair. Several editions appeared in 1612, each with a different number of plates, and its own pagination; the Oak Spring copy, for example, contains some additional plates that can be dated to 1613. The *Jardin du Roy* that Vallet had published just a few years earlier (see no. 8) provided a useful model for de Bry, who did not hesitate to copy a number of images from it with only minor modifications—leaving out the butterflies and insects, for example, and adding roots and bulbs. The illustrations of the narcissus in plate 15, although arranged somewhat differently, were clearly drawn from Vallet's work, while plate 28 shows a number of anemones adapted from it. Since the images were directly copied onto copper plates, when printed they appear as mirror images of the originals.

Many non-European plants appear in the *Florilegium novum*, such as the 'Hyacintus Peruanus', the 'Narcissus Indicus flore rubro vulgo Iacobeus' or Sprekelia, and the 'Ananas fructus Indicus Orientalis'. Other plates show examples of the floral 'monstrosities' that were cultivated in Baroque gardens, among them the 'Anemone Pavota latifolia multiplex flore miniato' in plate 32, or two Turk's-cap lilies (*Lilium martagon*)—one with a deformed stem that bloomed on 20 July 1608 in the

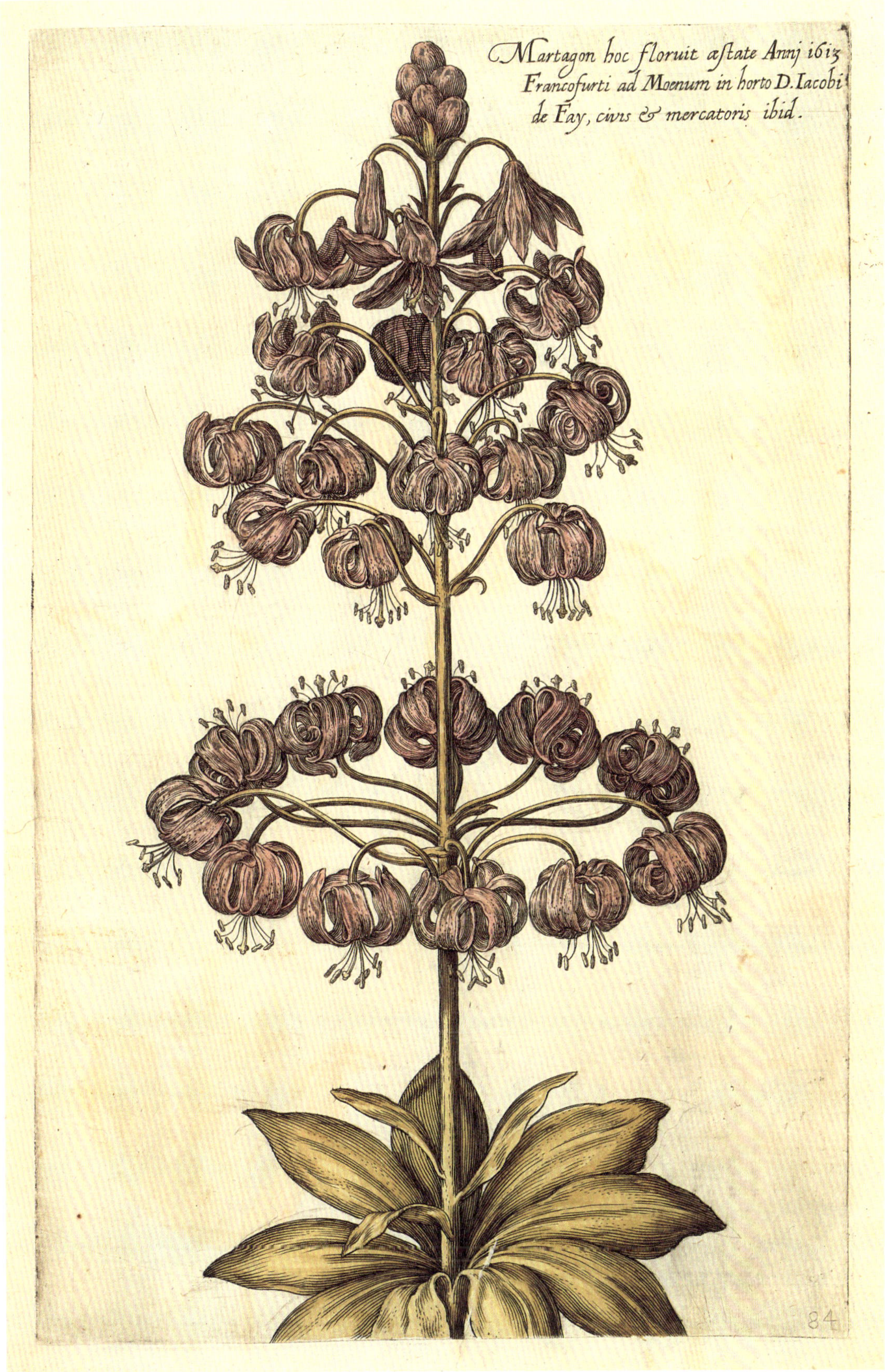

JOHANN THEODOR DE BRY, *Florilegium novum.* Martagon lily (*Lilium martagon*), plate 84

JOHANN THEODOR DE BRY, *Florilegium novum.* Jacobean-lily (*Sprekelia formosissima*) and two varieties of *Narcissus,* plate 16

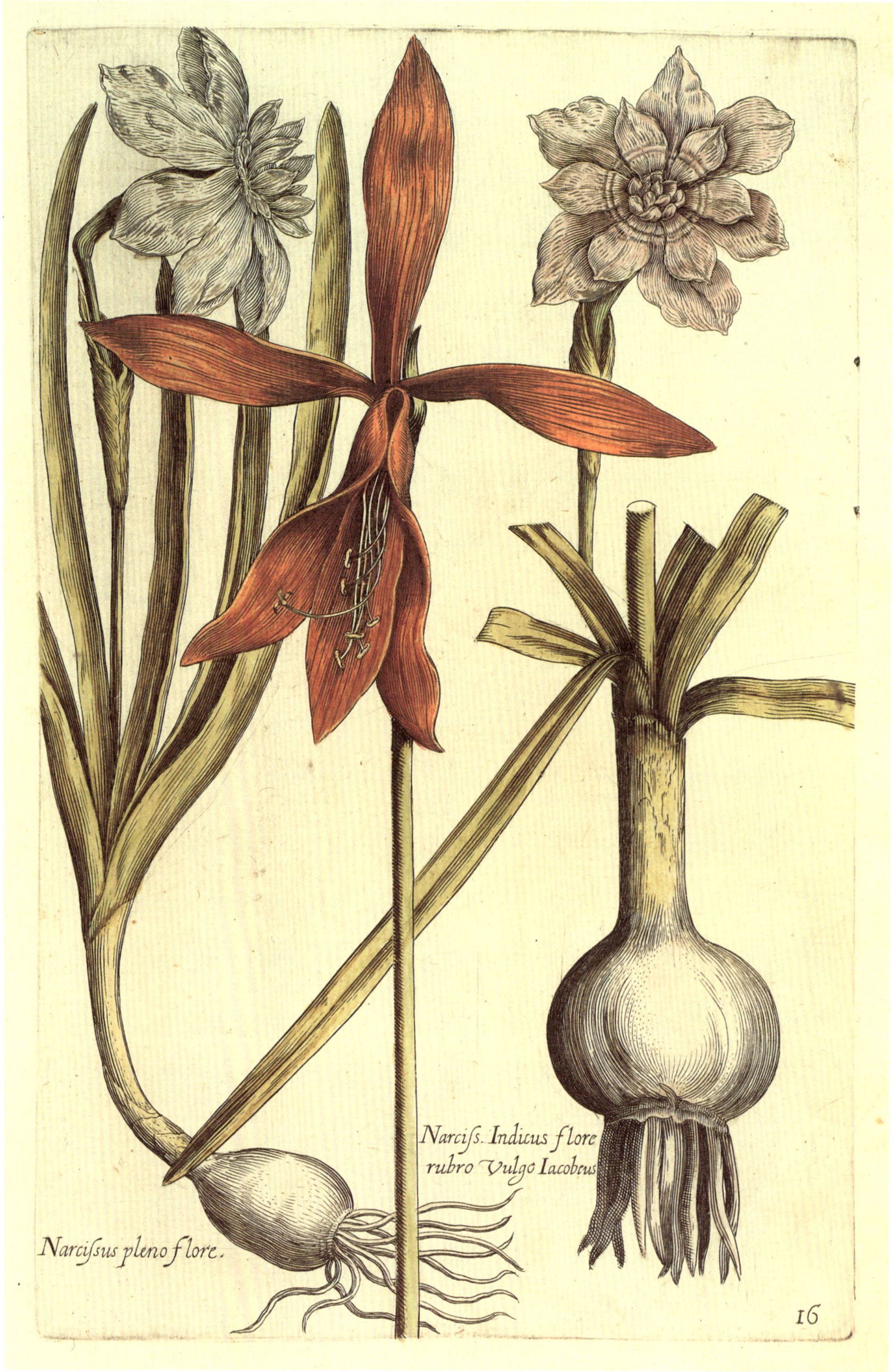

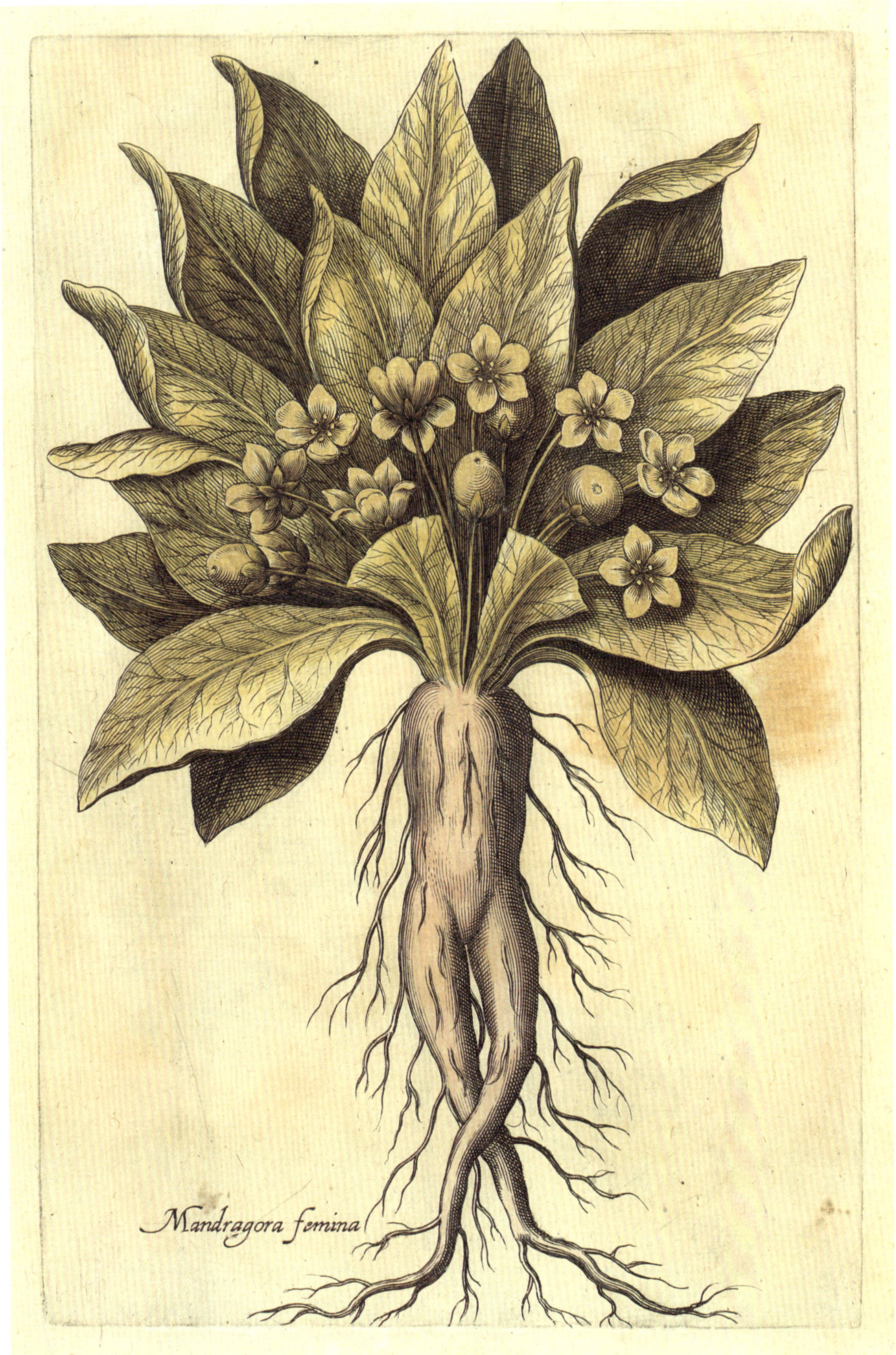

JOHANN THEODOR DE BRY, *Florilegium novum.* Mandrake (*Mandragora officinarum*)

garden of Bernahart Barth von Harmatting in the city of Beyrn (plate 68), and the other that brought forth the anomalous blooms pictured here in the summer of 1613 in the garden of Jacobo de Fay in Frankfurt (plate 84).

The unsurpassed artistry for which de Bry was renowned throughout Europe emerges clearly in the plates of this florilegium. Each has been carefully composed, and the confident lines of the engraving, with their fine shading, denote the hand of a true master. The delicate tints used to colour the plates, sometimes highlighted with touches of gold and silver, add the crowning touch to this delightful florilegium from the dawn of the seventeenth century.

11. Basilius Besler (1561–1629)

[*the title-page is dominated by a triumphal arch with the figures of Flora and Demeter and the coat of arms of Johann Cornelius von Gemmingen. Before the columns stand the Old Testament kings Solomon and Cyrus, whose names appear in cartouches at their feet. They are each flanked by two urns, one containing an aloe plant and the other a cactus. Below is a scene depicting the Garden of Eden with God, Adam, and numerous animals. At the top of the page, suspended between the double columns, is a draped cloth on which is written*] Hortus Eystettensis, Sive Diligens et accurata omnium plantarum, florum, stirpium, ex variis orbis terrae partibus, singulari studio collectarum, quae in celeberrimis viridariis arcem episcopalem ibidem cingentibus, hoc tempore conspiciuntur, delineatio et ad vivum repraesentatio opera Basilii Beslerii Philiatri et Pharmacopoei M.D C.XIII [*in the lower right*] Wolffg: Kilian August. scalpsit.

1° [Broadsheets] 55 × 45 cm. 4 parts in 2 volumes.

Volume 1:
(:)–3 (:)1)(1 χ–^{3}χ1 a (o)–o (o)1 ^{4}χ1 A–S1 ^{5}χ1 A)(–R)(1 ^{6}χ1 A2)(–R2)(1 ^{7}χ1 A3)(–N3)(1 ^{8}χ1 A4)(–o4)(1 ^{9}χ1 A5)(–P5)(1 ^{10}χ1 A6)(–H6)(1 ^{11}χ1 A7)(–17)(1 ^{12}χ1 A8)(–18)(1 2A)(1 2B2)(1 2C3)(1 2D4)(1 3A)(1 3B2)(1 3C3)(1 ^{13}χ1 2a:):(:1–20:) :(:1 ^{14}χ1 2A)(2^{1}–2Q)(2^{1} ^{15}χ1 2A2)(2^{1}–2Q2)(2^{1}

Volume 2:
^{16}χ1 2A3)(2–2K3)(2^{1} ^{17}χ1 2A4)(2–2M4)(2^{1} ^{18}χ1 2A5)(2–2M5)(2^{1} ^{19}χ1 2A6)(2–2K6)(2^{1} ^{20}χ1 2A7)(2–2M7)(2^{1} ^{21}χ1 2A8)(2–2M8)(2^{1} ^{22}χ1 2A9)(2–2M9)(2^{1} ^{23}χ1 2A10)(2–2M10)(2^{1} ^{24}χ1 2A11)(2–2K11)(2^{1} ^{25}χ1 2A12)(2–2K12)(2^{1} ^{26}χ1 2A13)(2–2L13)(2^{1} ^{27}χ1 2A14)(2–2O14)(2^{1} 2A–2G1 ^{28}χ1 ^{29}χ1 2A)(3–2N)(3^{1} ^{30}χ1 2A2)(3–2N2)(1 ^{31}χ1 2A3)(3–2I3)(3^{1} ^{32}χ1 2a–2c1 A4–G4^{1} A1

BINDING: Mottled calf, covers framed in a gilt roll, with marbled edges. On the spine of the first volume is a red calf label, lettered in gilt, 'Hortus Eystettens', and a label in green calf with 'Tom I'. The second volume bears a similar red label, and a green label with 'Tom II'.

PLATES: Title-page, followed by a frontispiece with two medallions, one bearing the author's portrait and the other his coat of arms. Followed by a total of 367 botanical plates divided among the four parts of the work (*classis verna, aestiva, autumnalis, hyberna*); each part opens with a title-page followed by a series of engraved plates numbered consecutively in manuscript. Plate number [181] and [182] of the Martagon lily joined and folded as one image.

PROVENANCE: Property stamp of 'H. Leleux' on the front free endpaper of each volume.

REFERENCES: Hunt 430 (1713 edn); Stafleu and Cowan, 1.497; Pritzel 745; Nissen 158; *A Magnificent Collection*, no. 26; Gascar and Aymonin; Keunecke; Barker; Blunt and Stearn, pp. 106–08.

BASILIUS BESLER's *Hortus Eystettensis* is generally considered to be one of the most important and most beautiful florilegia ever published. It was commissioned by the Prince-Bishop Johann Conrad von Gemmingen, a virtuoso and connoisseur of flowers whose cultural formation included

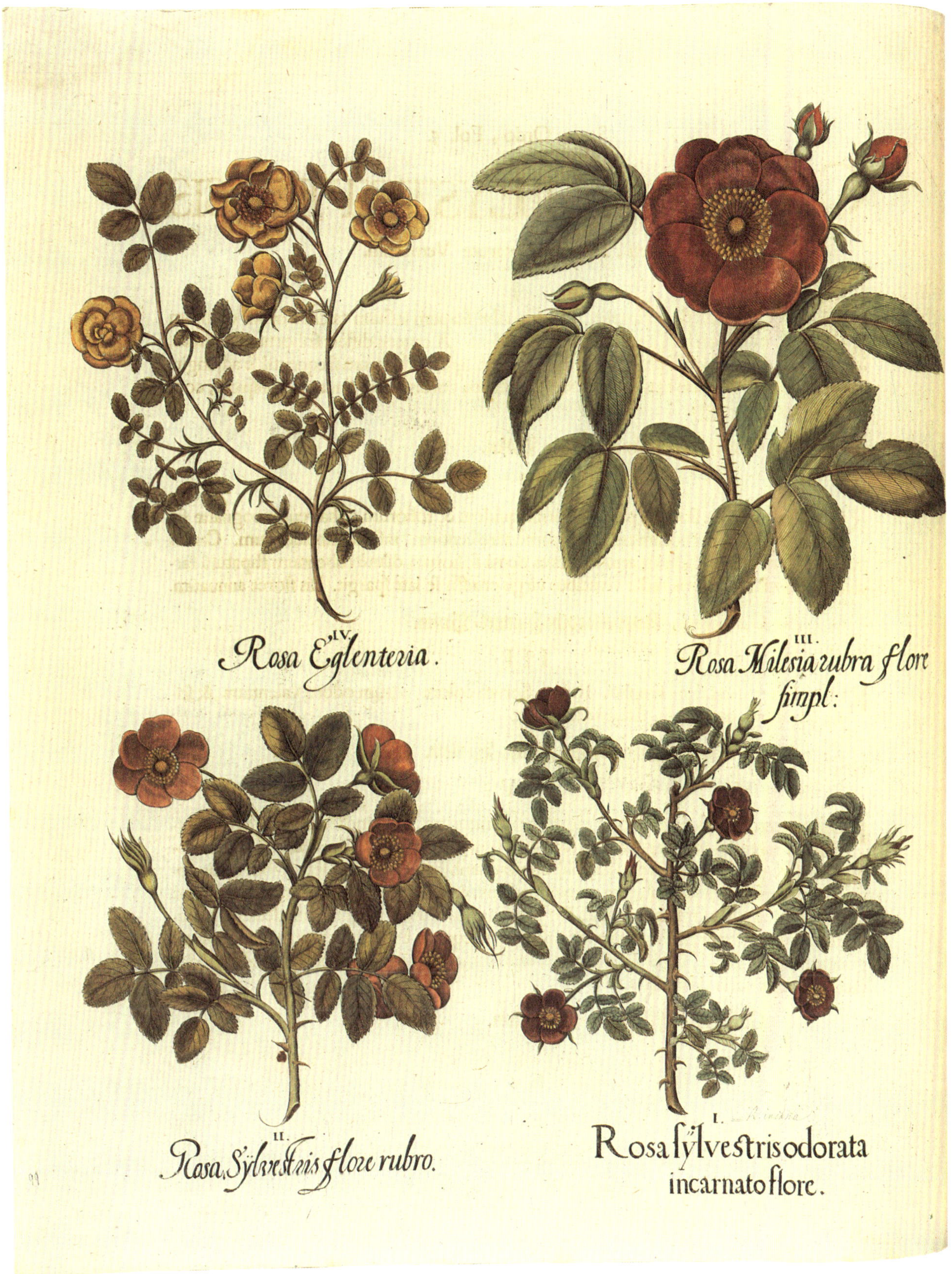

BASILIUS BESLER, *Hortus Eystettensis*. Four varieties of *Rosaceae*, 'Sextus Ordo, Fol. 6'

an extended sojourn in Italy. In 1596 Gemmingen asked the physician and botanist Joachim Camerarius the younger (1534–98) of Nuremberg to oversee the construction of extensive gardens around his castle, which stood on a hill overlooking Eichstätt, a town to the south of Nuremberg.

Camerarius died two years after beginning this project, however, and the task of completing it fell to Basilius Besler, a pharmacist from Nuremberg who also owned a fine garden and a collection of scientific curiosities. Within a few years the gardens of Eichstätt had become among the most renowned in Europe. In 1606 Gemmingen asked Besler to draw up a catalogue in the form of a florilegium, complete with illustrations, of his by now vast botanical collection. Unfortunately Gemmingen never saw the completion of this important work, dying in November 1612, just one year before *Hortus Eystettensis* was published. He was succeeded as bishop by Johann Christoph von Westerstetten, who undertook the responsibility of maintaining the castle and its gardens, albeit not at the munificent level his predecessor had.

Besler enlisted the aid of two collaborators in the compilation of his vast and ambitious catalogue —his younger brother Hieronymus (1566–1632), and a nephew of Joachim Camerarius, Ludwig Jungermann (1572–1653), who taught natural history at the University of Altdorf. Hieronymus helped Besler to write the Latin text, while Jungermann perhaps assisted in the preparation of the botanical descriptions, many of which were drawn from Camerarius's *Hortus medicus*, published in Nuremberg in 1588.

Hortus Eystettensis was published in 1613, probably in Altdorf, by the famous printing-house of Conrad Bauer. One series of copies, dedicated to Bishop Johann Conrad, was intended for the use of scientists and therefore contained the complete text, printed on the verso of the plates. However, Besler had another series of copies printed on high-quality paper, in which the verso of each plate was left blank and the letterpress was either entirely omitted, or else printed on both sides of separate leaves inserted between the pages of plates. This de-luxe edition, with its beautifully hand-coloured plates, was intended for a particularly wealthy and discriminating audience, one that could afford its prohibitive price of 500 florins (Barker, pp. 11–12, 35).

A number of the original drawings for *Hortus Eystettensis*, today conserved in the library of the University of Erlangen, were apparently executed by Besler himself; indeed, in the dedication he states that he was the author of many of them ('multae etiam istarum stirpium et planatarum mea manu sunt factae'). The task of engraving the hundreds of elegant plates, however, which took many years to complete, was entrusted to others.

On one of the opening pages of *Hortus Eystettensis* is a fine portrait of Besler set in an oval frame beneath his family's coat of arms; he is shown holding a branch of basil, in a playful reference to his first name. This engraving was executed by Wolfgang Kilian (1581–1662) of Augsburg probably after a painting, today conserved in the Göteborg Konstmuseum, by Lorenz Strauch (1554–1630) of Nuremberg (Barker, p. 13). Kilian also engraved the imposing principal title-page, while the four lesser title-pages introducing the four different parts of the work—each of which depicts a seven-

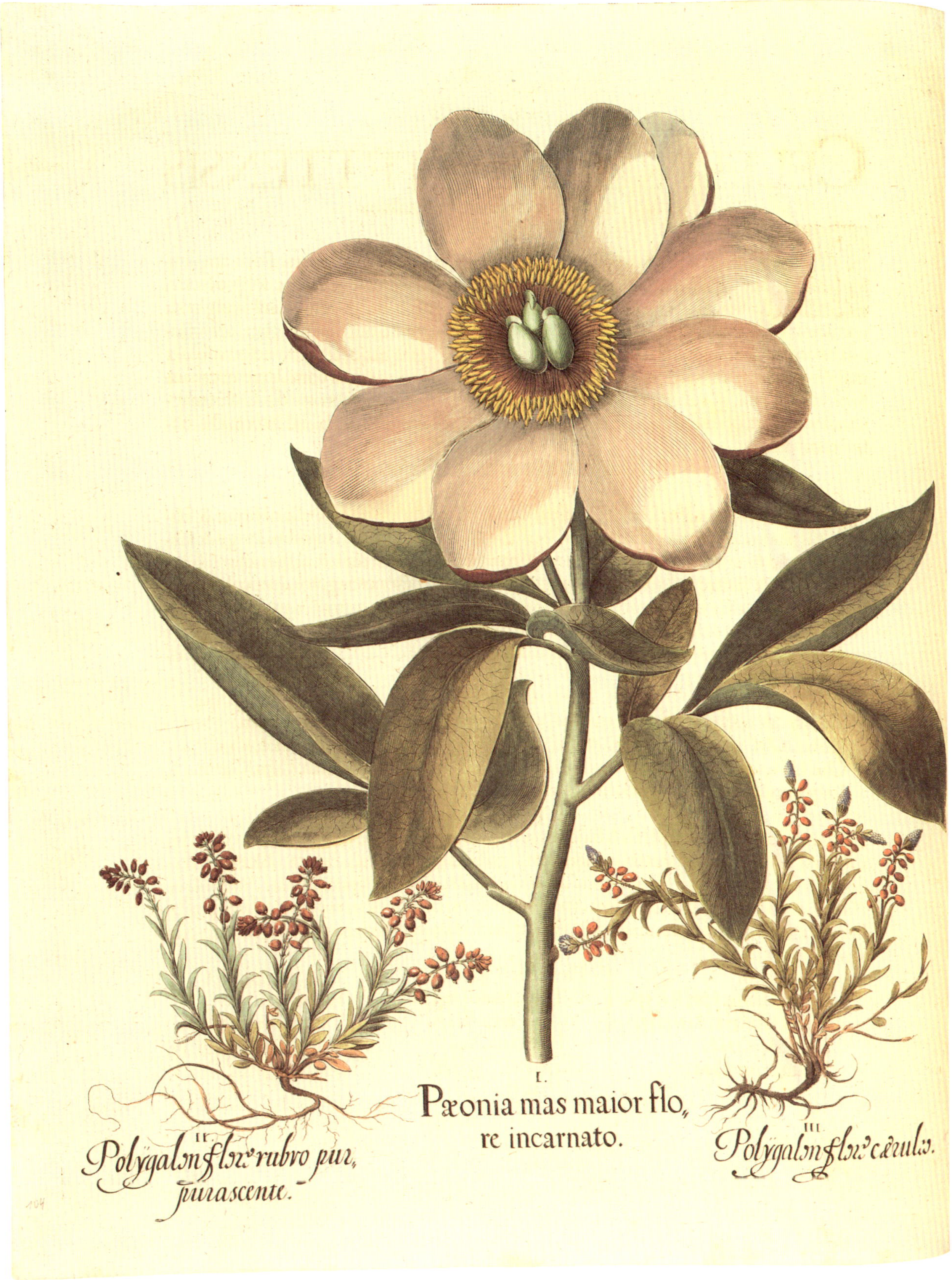

BASILIUS BESLER,
Hortus Eystettensis. Paeonia and two varieties of *Polygala*, 'Sextus Ordo, Fol. 11'

teenth-century 'cabinet of curiosities', no doubt very much like those in Besler's own collection—were engraved by Johan Leypolt (*fl.* 1607–19) of Bamberg.

In addition to the signatures of Heinrich Ulrich (1572–1625) of Nuremberg and the Dutch artist Servatius Raven (*fl. c.* 1610), the monograms of a number of other artists can be found in the plates—Hieronymus Lederer of Nuremberg; Dominikus Custos (*c.* 1560–1612), who had also worked at the court of Rudolf II; Frederich van Hulsen (*c.* 1580–1660), possibly a student of Theodor de Bry; and P. V. Isselburgh (*c.* 1580–after 1630), perhaps a student of Crispijn van de Passe (Nissen, I. 70–73).

In the Oak Spring copy of Besler's masterpiece, the engravings have been bound so that they appear on the left-hand page, while on the facing page appear the apposite botanical descriptions. In each plate one or more flowering plants are depicted. The images have been beautifully composed and arranged on the page; indeed, the artist sometimes sacrificed scientific accuracy for the sake of decorative effect. The species have, however, been carefully grouped according to the season in which they bloom, beginning with spring, and each plant is labelled with its Latin name. In the text the names of various botanists, including Rembert Dodoens, Matthias de L'Obel, Caspar Bauhin, Carolus Clusius and, naturally, Camerarius are cited. More than 1,000 plants are presented, including wild and cultivated species from Europe, Africa and the New World. According to our pharmacist-author, many of these plants had medicinal properties, which he duly describes in the text.

Although the quality of the plates is somewhat uneven—a fact that is not at all surprising considering how many different artists had a hand in their production—most of them were beautifully conceived and drawn in the Baroque style, reflecting its penchant for grandiloquent and theatrical effects, and some, such as the 'Corona Imperialis Polyantos' (I, pl. 80), the sunflower (II, pl. 24), and two poppies (II, pl. 291), have an undeniable expressive power.

Two works produced in the same period as *Hortus Eystettensis* can help to shed some light on the artistic antecedents of Besler's extraordinary florilegium. A painted florilegium that once belonged to Joachim Camerarius has recently been rediscovered; it was acquired by the German government and is now in the University Library at Erlangen (see *The Camerarius Florilegium*, sale cat. by S. Raphael and W. P. Watson, Christie's, London, 20 May 1992). This manuscript was perhaps the work of Joachim Jungermann (1561–91), the brother of the same Ludwig Jungermann who had worked with Besler on *Hortus Eystettensis*. Joachim Jungermann was a talented botanical illustrator, and he may also have collaborated with Camerarius on his *Hortus medicus*. There are many parallels between Jungermann's florilegium and *Hortus Eystettensis*, beginning with the arrangement of the species according to the seasons of the year. Even more interesting is the marked similarity in style of the paintings in the two florilegia, for in each we find the same taut, emphatic forms, beautifully epitomized in the curvilinear movement of the stems and leaves.

A second work that has recently been linked to *Hortus Eystettensis* is a florilegium known as the *Kalendarium* (today conserved in the Archives of the Royal Botanic Gardens, Kew) by another artist from Nuremberg, Sebastian Schedel (1570–1628). Barker suggests that some of its paintings may actually represent some of the first preparatory drawings for *Hortus Eystettensis*. On other pages of the *Kalendarium*, Schedel has directly pasted engravings cut from various hand-coloured copies of *Hortus Eystettensis* (Barker, pp. 32–3).

In the complicated publishing history of this work a footnote must be added, for another edition of *Hortus Eystettensis* was published at Ingolstadt sometime around the middle of the eighteenth century (even though the date on the title-page reads 1713) by Johann Georg Sthenander (or Starckmann; 1701–80). This edition was prepared for Johann Antonius I Knebel von Katznellebogen, the last Prince-Bishop of Eichstätt, whose coat of arms was added to the title-page. Some slight variations were introduced into the work, but all the illustrations were printed from Besler's original engraved plates. Oak Spring has a copy of this edition, bound in a single volume with hand-coloured plates (Barker, p. 59).

In this final edition of *Hortus Eystettensis* Besler's name no longer appears in the title and is mentioned only in the introduction. In addition, the Latin term *olim* (once) has been added to the title, for the garden at Eichstätt was no longer in existence at the beginning of the eighteenth century, having been destroyed during the course of the Thirty Years War. A final misfortune closes the publishing history of *Hortus Eystettensis*, for its engraved plates, which had been so carefully preserved, were carelessly melted down in 1817 by the royal mint in Munich.

12. CRISPIJN VAN DE PASSE *the younger* (1589–1670)

[*engraving* 14.5 × 21.5 *cm. The title is presented in the form of an inscription carved on a stone plaque, from which two allegorical figures representing the sun and the moon are drawing back a curtain; superimposed are two roundel portraits, one of Rembertus Dodonaeus and the other of Carolus Clusius. Above, a putto blows on a loose garland of flowers strung along the top of the curtain, while on the platform on which Sol and Luna stand can be seen a row of flowers (including a tulip) planted in sunken pots*] Hortus Floridus In quo rariorum & minus vulgarium florum Icones ad vivam veramq[ue] formam accuratissime delineatae. Et secundum quatuor anni tempora divisae exhibentur Incredibili labore ac diligentia Crisp: Passaei junioris Delineatae ac suum in ordinem redactae A°.1614. [*within the frame of the engraving, at bottom right*] Crisp: Passaeus execud: ulti. [*further below, but still contained within the engraved frame*] Extant Arnhemij. Apud Joannem Janssonium Bibliopolam ibid. [*the entire engraving was printed separately, trimmed, and mounted on a sheet of paper matching in size the folios of the volume*]

[*there follows a separate letterpress section of 15 leaves of Dutch text collating* π^2 $^2\pi^2$ A–F², *including a title-page in Dutch*] Den Blom-Hof, Inhovdende De Rare Oft Ongemeene Blommen Die Op Den Tegenvvoor-dighen tijdt by de Lief-hebbers in estimatie ghehouden vverden. Ghedeelt Naer De Vier Deelen Des Jaers, ende door Crispian vande Pas de Jonghe in ordre gebrocht, ende met groote moete naer het leven gheconterfeyt. Noch hier by ghevoecht de manier soomen dese bloemen naer haer eyghen coleuren ofte verven sal illumineren ofte afsetten tot dienst van alle curieuse Lief hebberen der bloemen. [*typographical ornament* 1 × 2.25 *cm.*] [*hand pointing right*] Ghedruckt tot Utrecht | voor Crispijn vande Pas, Anno 1614.

[*There are also two leaves of letterpress printed vertically accompanying the section on tulips.*]

[*letterpress title-page for the section on Summer*] Aestas Horti Floridi, Inqua Praecipui Aestatis Flores Excellenti, Cr. Passaei Stylo Ad vivum Admodum Ingeniose Exprimuntur. [*typographical ornament* 0.5 × 1 *cm.*] Ultraiecti. Ex officina Hermanni Borculoi. Et Prostant apud Joannem Jansonium Bipliopolam Arnemiensis.

[*letterpress title-page for the section on Autumn*] Autumnus. Horti

Floridi rariores Autumni Flores Summa Industria Et Labore. Crisp. Passaei Filii. In aere effictos, & in lucem recens datos continens. [*typographical ornament* 1 × 2 *cm.*] Arnhemii. Ex officina, Joannis Jansonii. 1616.

[*letterpress title-page for the section on Winter*] Hortus Floridus Hyemalis Cr. Passaei Fil. labore et Industria | Nunc Primum In Lucem Editus | Anno CI(C) DC. XIIII. | [*typographical ornament* 1.25 × 1.25 *cm.*] | Ultraiecti. / Ex Officina calcographica Cr. Passaei.

[*letterpress title-page for the second part of the volume*] Altera pars Horti Floridi In Qua Praeter Flores, Varia Etiam Comprae-henduntur Arborum Fructiferarum, Fructicum, Plantarrum quoque et Herbarum Medicinalium Genera. Per Crisp. Passaeum in lucem, edita.

18 × 27 cm. 193 leaves.

BINDING: Early 17th-century vellum, gilt fillet and central panel of floral roll with fleur-de-lis at angles, the coat of arms of John van Zuylen van de Haer (1573–1641) of Utrecht.

PLATES: All of the plates, woodcut initials and borders are hand coloured in this copy, as are the typographical ornaments on the Dutch title-pages for Summer and Autumn, the artist having followed the instructions given in Spring on how to colour the plants in the four sections of the work. ('Beschhriivinghe van de Coleuren', A1(r)–F2(v9)). Very few of the copper engravings are signed, although Crispin van de Passe did sign the engraved title-page. 'P S[monogram] f', Crispijn's brother Simon, appears on engravings nos. 15, 22 and 33 in the section on Spring, and on engravings nos. 1 and 8 for Summer; 'Willem Pass f.' appears on engraving no. 32 for Spring and nos. 3 and 9 for Summer. The monogram 'AB'—Dr A. Buchelius, the author of the first poem of the 'Ad CR. Passeum fil.'—follows the epigram in the engraving of Flora for Autumn, which is signed 'Crispin Pass. inv.' at bottom left and 'Simon Passaaeus sculp:ultr.' at bottom right. In the second part of this hortus the engraved plate of 'Cognoscite Lilia . . .' is inscribed 'Formulis Crispiani Passaei et Joannis Waldnelij'. All the plant names in Latin, and sometimes in German, Italian and Dutch, are engraved, except for no. 40 in 'Altera pars', where only the initial capitals have been completed. In addition to the engraved numbers throughout, all the pages have been numbered in a contemporary hand, the same hand that compiled a manuscript index on two leaves, bound after the final plate of the second part.

PROVENANCE: Bookplate, with a cartouche containing a ship with two armed knights, on front paste-down. On the first leaf, the right edge of which has been trimmed, is a manuscript note stating the owner to be one 'Joannes Zuijlemus Dominus de Haer', whose coat of arms appears on the cover. The date '29 Aug[ust] 1640' is also given, which indicates that de Haer purchased the book one year before his death. 'Ex libris Dominae van Teger N: Ten Dijk' appears in manuscript below the engraving attached to the title-page.

REFERENCES: Brunet, IV.414; Rohde; Savage 1923; Pritzel 6972, 6973; Nissen 1494; Hatton, p. 8; Coats, pp. 32–3; *A Magnificent Collection*, nos. 270–72; Blunt and Stearn, pp. 108–13; Gerard 1996; *An Oak Spring Hortus*, no. 62.

CRISPIJN VAN DE PASSE THE YOUNGER's *Hortus Floridus* was without question the most popular florilegium ever published. Although the printing history of this book is complicated, it has been well delineated by Spencer Savage, whose article published in 1923 remains the definitive study. The volume was produced partly by the engraver and by the printer in two cities, Utrecht and Arnhem. The plates were printed, then modified and assembled at different times between 1614 and 1617 with issues consisting of one of two distinct Latin versions of the text or of Dutch, French and English letterpress titles and prefaces. This has led to a confusing situation, in which what Savage refers to as the various 'states' (issues), not to mention individual copies, contain plates in different states with the addition of insects and backgrounds, bound in different ways, sometimes with added material. Indeed, it is almost impossible to find any two that are identical. At Oak Spring are four copies of *Hortus Floridus*: two of the Dutch issue (one coloured, the other not) and the French and the English versions. The copy described here appears to be a variant of Savage's State 2f (p. 192).

Van de Passe came from a celebrated family of engravers: his father, Crispijn van de Passe the

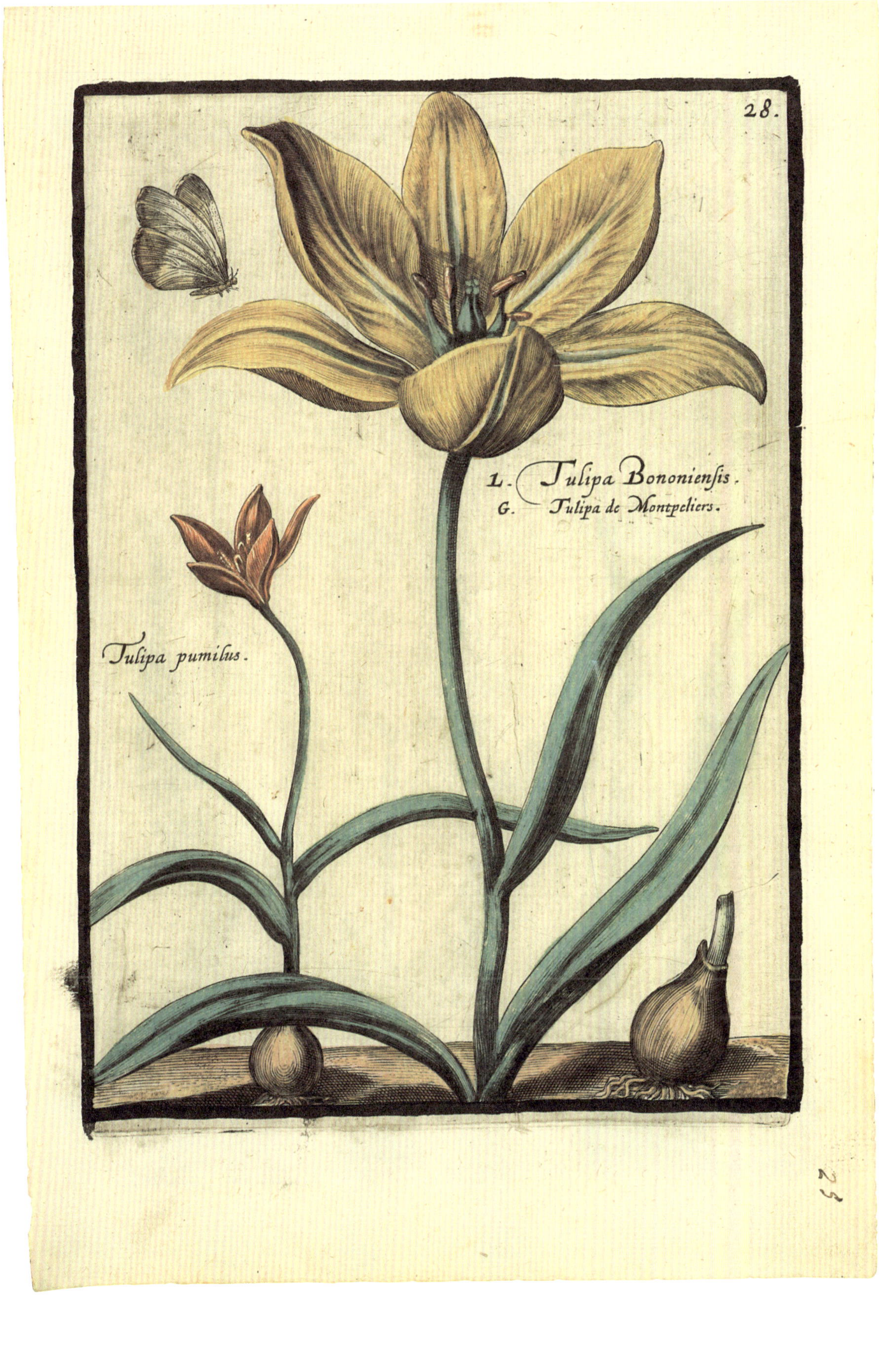

CRISPIJN VAN DE PASSE, *Hortus Floridus*. Two varieties of *Tulipa* with a butterfly, plate 28

elder (*c.* 1565–1637), was well known as an illustrator of texts and as a portrait engraver. Van de Passe himself lived from 1612 to 1617 in Utrecht, where he produced his *Hortus Floridus*. He then moved to Paris in order to take charge of his family's publishing interests there, where he gave lessons in drawing at the Maneige Académie, which was frequented by members of the court and by the aristocracy. He also continued his work as a book illustrator, as he was to do in Utrecht and Amsterdam.

Hortus Floridus is divided into two parts, containing anything from 175 to 200 leaves depending on the copy examined, mostly of engravings of flowering plants, accompanied by brief descriptions. The first, and longer, part consists of four sections corresponding to the four seasons of the year. In some of these works the signatures of two of van de Passe's brothers, Simon and Willem, appear, but they merely completed certain of the engravings based on designs prepared by van de Passe himself. Three of the four sections open with the picture of an elaborate formal garden, seen as if from a balcony. The four points of the compass are indicated in each view but, oddly enough, the four scenes do not reflect the passage of the seasons.

The second part of the work, 'Altera pars', which in almost all issues follows the first, contains engravings of somewhat lesser quality, produced by the artist in an earlier period, perhaps copied from his father's work when he was in his 'tyrocinium', or apprenticeship, in Cologne (Savage, p. 201). In fact, 'Altera pars' was issued separately, probably earlier than 1614, and is signed 'Formulis Crispiani Passaei [probably the elder van de Passe, for the title-page reads 'Crisp: Passaei junioris' and the 'Benigno florum', 'Cr. Passeus fil.'] et Joannis Waldnelij' (i.e., Woutnel, a Flemish engraver). However, some of the illustrations are not even original, having been copied from *La Clef des champs* (Hatton, p. 9; see no. 6) or from the florilegium printed by Adrian Collaert in Antwerp *c.* 1590.

The section devoted to the four seasons presents a large variety of flowers divided according to the season in which they bloom. The volume concentrates primarily on cultivated varieties, many of which had been brought to Europe around the end of the sixteenth century and the beginning of the seventeenth. Among the spring flowers can be found the narcissus, crocus, muscari (or grape hyacinth), fritillary and numerous varieties of tulip; the summer flowers include the peony, iris, lily, gladiolus and rose. The sunflower, marigold (then known as 'Flos Africanus'), morning-glory, tobacco plant, cyclamen and amaryllis (or 'Narcissus marinus exoticus', described by Carolus Clusius) are among the flowers of autumn, while winter is the season of the hellebore, snowdrop, liverwort and dog's-tooth.

Since *Hortus Floridus* contains a large amount of quite exact scientific information, it rapidly became an indispensable reference work for flower cultivators and flower-lovers. Giovanni Battista Ferrari, for example, in his magisterial treatise *Flora* published in Rome in 1633 (see no. 29), reproduced the tulip-transplanting instrument. Van de Passe intended that his work should also be used by artists, and elaborate guidance on colouring the plates is supplied, as the title-page promises:

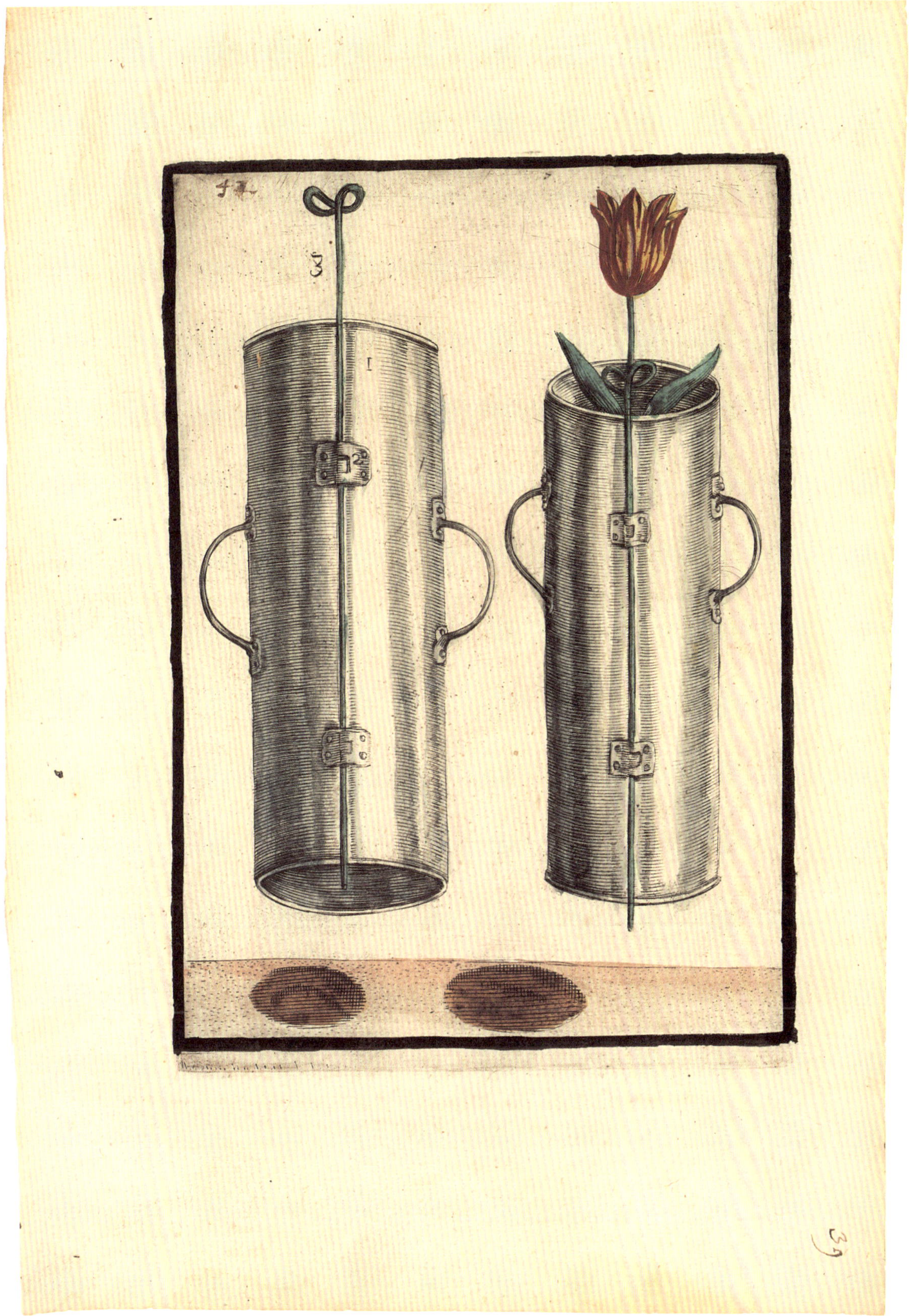

CRISPIJN VAN DE PASSE, *Hortus Floridus.* Instrument for transporting plants, plate 42

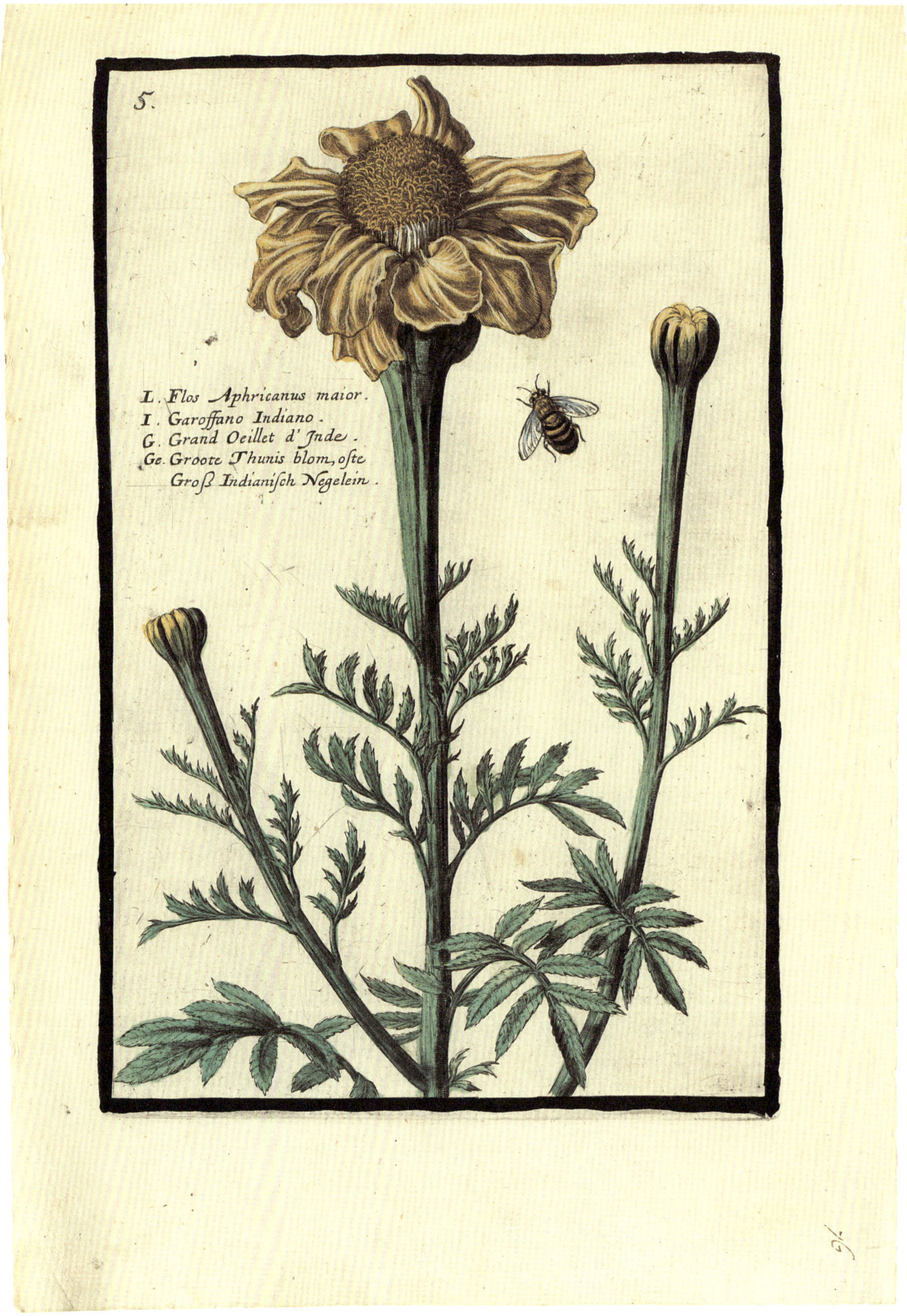

CRISPIJN VAN DE PASSE, *Hortus Floridus.* Marigold (*Tagetes*) with a yellow jacket, plate 5

'Wherein . . . is contained . . . the perfect true manner of colouringe the same with theire naturall couloures, beinge all in theire seasons the most rarest and excellentest flowers' (from the English issue of 1615). These useful directions seem to have been carefully followed by the artists who coloured the plates in this copy of the work.

Van de Passe was not merely a scientific illustrator, however, and his artistic vision found full expression in this work. Unlike the other florilegia of the period, which presented their flowers with their stems and roots floating against a blank white page, *Hortus Floridus* shows each plant growing in its natural habitat, represented by a pleasingly realistic landscape. It is worth recalling that it was in the city of Utrecht, one of the cities where *Hortus Floridus* was created, that a particular typology of still-life was developed, the floral still-life (see no. 22), for it has been noted that in some of his engravings van de Passe shows his plants together with small animals (such as the mouse in plate 23) or insects, added in later states of some of the plates, such as the common cockchafer, a juxtaposition that one also finds in contemporary works by the very first still-life artists, among them Balthasar van der Ast and Johannes Bosschaert (Bol, p. 28).

By uniting scientific illustration and the genre of the still-life in *Hortus Floridus* (Olmi 1989), van de Passe made available a precious repertory of floral images for artists such as van der Ast, Ambrosius Bosschaert and Roelandt Savery. Some of the plates of single flowers were copied for other botanical works, including Matthaeus Merian's *Florilegium renovatum* (no. 16), Hendrik Cause's *De Koninglycke Hovenier* (Amsterdam, 1676; see *An Oak Spring Hortus*, no. 65), and John Hill's *Eden* (no. 53).

13. Girolamo Pini (*fl. c.* 1610–20)

Flowers with an iris in the centre.

Flowers with a fritillary in the centre.

90 × 120.5 cm. Oil on canvas.

PROVENANCE: Ex-collection Air Marshal Sir John Tremayne, sold at auction at Croan Manor, Croanford near Bodmin, Cornwall, by Phillips, 31 October 1979 (lot 199, the pair).

THESE TWO PAINTINGS by Girolamo Pini are autograph replicas of signed and dated works by the same artist in the Musée des Arts Décoratifs, Paris; in the collection of this same institution is a third sampler, with a red peony at the centre (Gonzáles-Palacios, pp. 302–03; Gregori 1986, pp. 35–6; Gregori 1991, p. 523). The first of the two Oak Spring paintings shows forty-three different species of flowers, with an iris dominating the centre of the canvas; the second work depicts seventy-one species, with a fritillary at the centre. In both the common names of the plants are presented on a scroll in the lower part of the painting.

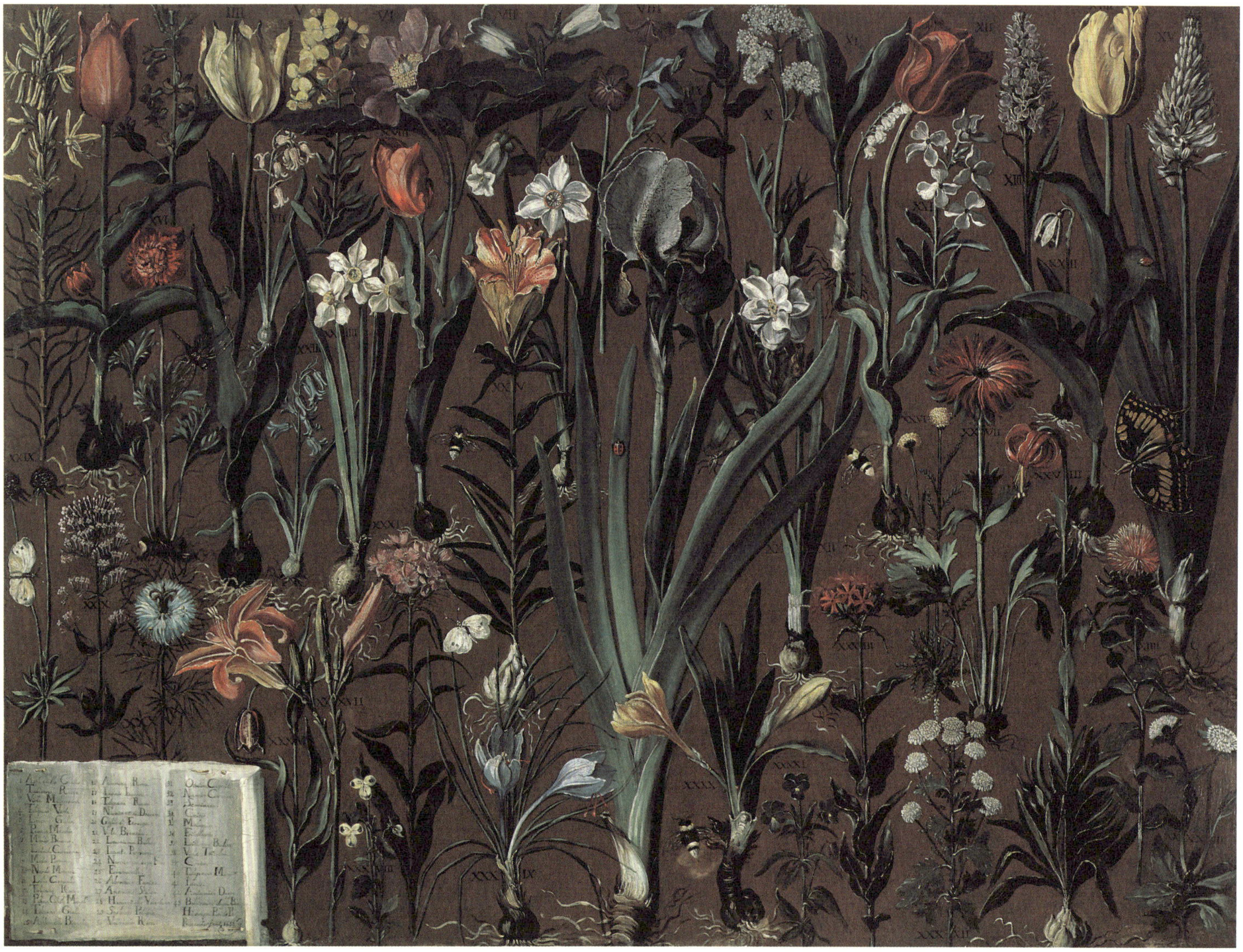

GIROLAMO PINI,
A fritillary surrounded by flowers

Nothing is known about the artist Girolamo Pini, who signed the first canvas 'Hieronimi Pini de Pis. Inventor fecit 1614' and the second 'Girolamo Pini inv.'. The third canvas—the one in Paris—bears the date '1615'. A number of studies have been written on the three kept together in Paris, but scholars have remained wholly unaware of the existence of the two Oak Spring replicas.

It is generally believed that Pini was Tuscan, originally from Pistoia close to Florence, although the Paris notation 'Pis.ro' (here 'Pis') is ambiguous and might equally well be taken to refer to the Adriatic coastal city of Pesaro (*Pisaurum*) in the region of Le Marche. However, no mention of the artist has been found in contemporary documents there, while in Florence a certain Giovanni Pini, possibly a relative, is known to have been working in the first decades of the sixteenth century as a

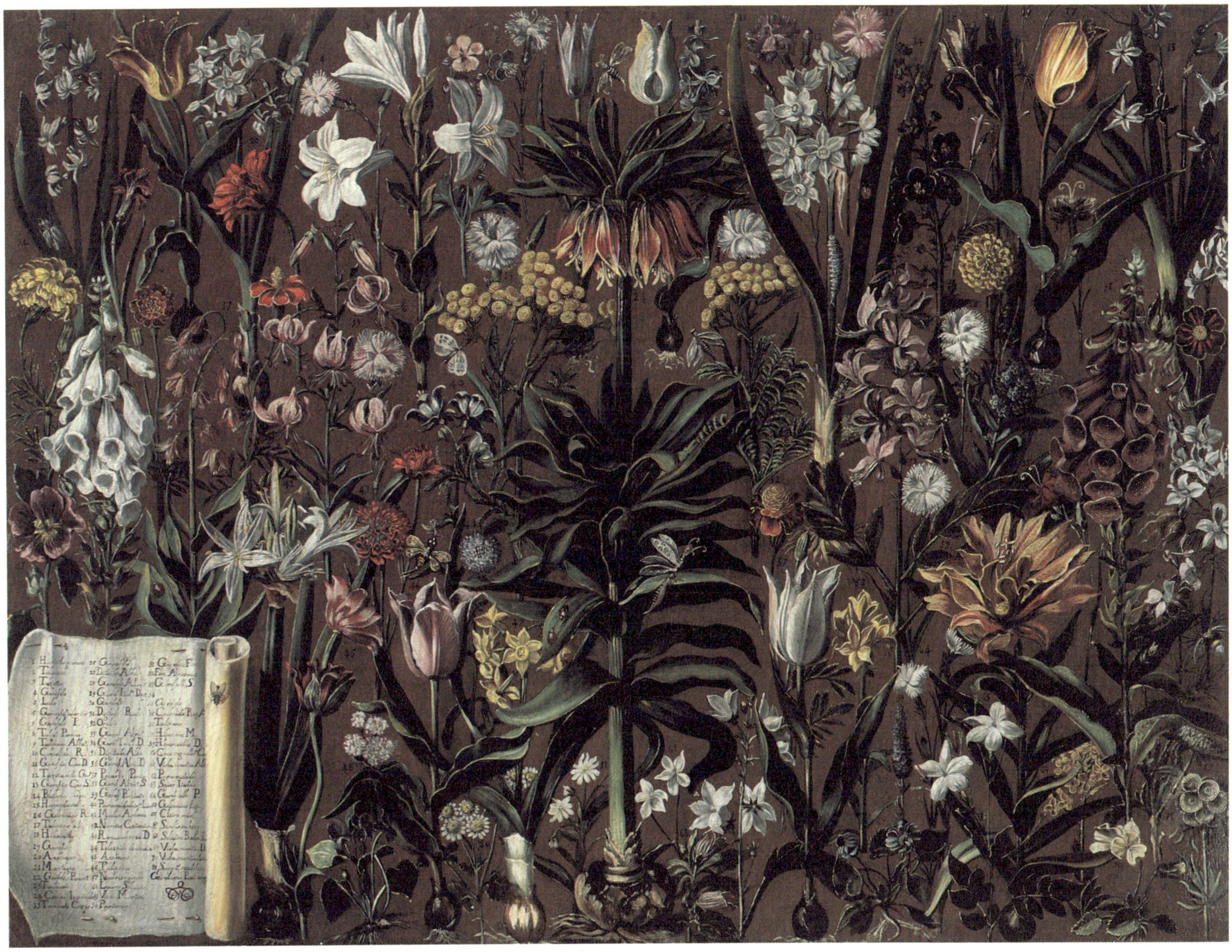

GIROLAMO PINI, An iris surrounded by flowers

still-life painter. It seems quite plausible that there was some connection between our artist and the Tuscan capital, which due to its intense intellectual activity had also become an important centre for scientific illustration. In this period many artists specializing in the genre could be found working in Florence for scientists, virtuosi and the Medici court, including the great Jacopo Ligozzi, who produced a number of magnificent botanical paintings (now conserved in the Uffizi Gallery) for the Grand-Dukes of the Medici (Tongiorgi Tomasi 1993).

Pini seems to have been influenced less by the restrained and elegant style of Ligozzi (whose compositions often focus on a single plant) than by the florilegia of the day, some of them painted by hand on paper or vellum, and some produced by the first printing-presses. His flowers, skilfully rendered with heavy strokes of colour, consist for the most part of cultivated bulbous species, which the artist often contrived to pair off in order to create charming decorative effects. The dark-brown background utilized by Pini was an unusual innovation that had been adopted by the German artist Ludger Tom Ring the younger (?1522–84) in some of his botanical paintings in bodycolour (*Daffodils and other Flowers*, cod. min. 42, fol. 166r, in the Nationalbibliothek, Vienna).

In keeping with the scientific approach of the period, the plants in each painting have been carefully numbered, each number corresponding to a name on the scroll painted in the corner of the canvas. These names, although in the vernacular and transcribed inaccurately, do not furnish any helpful clues as to the geographical provenance of the paintings. Reflecting the current typology of the florilegium, in these works a variety of insects appear. On the scroll in the painting with the fritillary there is a very lifelike fly *à trompe l'œil*, a device seized on by countless artists, beginning with Giotto, so as to demonstrate their skills at mimicking reality (see Chastel).

APPENDIX

On the scroll in the painting with the iris appears: '1. Asplendillo Gialla. 2. Tulipano Rosso. 3. Viola Maliconich. 4. Tulipano Verde. 5. Leuco Giallo. 6. Peonia Maschio. 7. Medio Bianco. 8. Lichnide Coronaria. 9. Media Purpureo. 10. Nardo Moniano. 11. Lilio Convalio. 12. Tulipano Rosso. 13. Palma Chst Megalt (?). 14. Tulipano Giallo. 15. Asphondillo Biancho. 16. Anemone Rossa. 17. Jacinto-Leuc. 18. Tulipano Rosso. 19. Narcisso di Dodoneo. 20. Giglio di Ferasne. 21. Viola Biancha. 22. Leucoio non Bulbosa. 23. Jacint. Purpureo. 24. Narciso di più fiore. 25. Emerocalle. 26. Abromato Femina. 27. Anemone Stalato. 28. Hemorecalle Valentina. 29. Scabiosa Pelagrie. 30. Valeriana Rossa. 31. Oculis. Cb. [?]. 32. Narciso Con [?]. 33. Scon [?]. 34. Cnicio. 35. Melant. [?]. 36. Fritillaria. 37. Lilium bulboso. 38. Viola Tricolori. 39. Croco. 40. Tulipano Minor. 41. Jaccs. 42. Anthemide Doppi. 43. Bellis minor Bianc [?].' The canvas is signed 'Hieronimi Pini de Pis. Inventor fecit 1614'. Among the insects one can recognize a dragonfly, three silkworm moths, two ladybugs, two butterflies, a *Pieris mannii* (Southern small-white butterfly), a bee and a *Papilio machaon* (swallowtail).

On the scroll in the painting with the fritillary appears: '1. Hiacintho pricans. 2. Tulipano. 3. Tazetta. 4. Garofolo. 5. Lilio. 6. Gariophil' color car. 7. Garofolo P. 8. Tulip.o Persico. 9. Tulipano Alba. 10. Consolida R. 11. Garof. in Car. D. 12. Tazatta della Gsore. 13. Garof. in Car. S. 14. Bletaria tęrza. 15. Hermodactilles. 16. Gelsomino R. 17. Tulipano al. 18. Hiacintho. 19. Garofo. 20. Anemone. 21. Martagon. 22. Garfolo Pavon.o. 23. Tanaceto. 24. Corona Imperiale. 25. Tanaceto Crispo. 26. Garofolo. 27. Dittamo Alba. 28. Garofalo A. Lac.o. 29. Garof. Ind.o Dop.o. 30. Garofolo. 31. Digitale Rub. 32. Orchis. 33. Garof. Afric.o. 34. Garof. Ind. D. 35. Digitale Alba. 36. Garof. Afric. D. 37. Penachio Persico. 38. Garof. Africs. S. 39. Garof. Pichiato. 40. Paramidale int. Liro. 41. Malva Arborea. 42. Narciso Costantinop. 43. Ranuncolo rossa D. 44. Telaspide de candia. 45. Anemone. 46. Tulipano. 47. Narciso giallo. 48. Lauro Silvestre. 49. Viola Martia. 50. Partemio. 51. Gromen Flori. 52. Flos Africanus. 53. Garof. alb. S. 54. [*name missing*]. 55. Garofolo. 56. Consolida Reg. A. 57. Tulipano. 58. Hipericon M. 59. Hemerocallis D. 60. Cons. reg. color Ga. 61. Viola martia Alb. 62. Paramidale. 63. Spico Italico. 64. Garof. alb. P. 65. Gelsomino hisp. 66. Cheiro min'. 67. Scabiosa hisp. 68. Solatro Bella I. 69. Viola martia D. 70. Viola martia lutea. 71. Seme di scab. hisp.' The canvas is signed 'Girolamo Pini inv.'

Among the insects depicted, a fly, a bee, five ladybugs, a butterfly, four dragonflies, a caterpillar and a silkworm moth can be recognized.

14. François Langlois (1589–1647)

[*the engraved title-page shows a formal garden bedded with young plants, and two gardeners, a male and female, carrying pots of plants and flowers. At their feet rest two large ansate vases, the first filled with narcissi and the second with fritillaries. In the foreground is an ornately decorated watering-can surrounded by flowers and insects, among which may be seen narcissus, an arum lily, a leguminous plant, a cockchafer and a butterfly. A rich festoon frames the upper part of the engraving, from both ends of which emerge cascades of columbine, narcissi, crocuses, marguerites and a* Fritillaria meleagris. *In front of the festoon floats a vase filled with carnations and jasmine. The composition is enlivened by a number of insects, among which can be recognized a dragonfly, a housefly and a species of* Diptera. *In the centre of the engraving is a panel on which is written*] Livre de Fleurs ou sont representés touttes sortes de tulippes, Narcisses, Iris, et plusieurs autres fleurs avec diversités d'Oiseaux, Mouches, et Papillons, le tout fait apres le naturel, par l'Anglois M^{e}. enlumineur. A Paris Par Jean Le Clerc rüe St Jean de Latron. Avec Privilege du Roy, 1620 [*in the lower left-hand corner of the engraving*] L. Gaultier incidit.

I^{o} [Broadsheets]. 28 × 33 cm., 26 unnumbered plates.

BINDING: Modern vellum.

PLATES: 26 engraved plates, including the illustrated title-page, with some etched details. Some of the plates measure 20 × 27, others 18 × 23 cm. The Latin names of all the birds and flowers are engraved on cartouches, while the German names sometimes appear in manuscript next to the figures. The cartouches of the final plate (fol. 25) are blank.

PROVENANCE: At the foot of the title-page is handwritten in ink 'Ma. h. Heijden [3 long s's, i.e., subscripsit]'.

REFERENCES: Nissen 1137; Hunt 205; Blunt and Stearn, p. 113; Saunders, p. 60.

THE HIGHLY ORIGINAL composition of the plates in this florilegium sets it apart from other works of the same period. Its author, François Langlois (known as 'Ciartes', for Chartres, his city of origin), worked as a printmaker and *enlumineur*, as he defines himself in the title-page to this work, as well as a bookseller and art dealer. He spent long periods of time abroad—in England, Italy and perhaps also in the Low Countries—and acquired the reputation of a connoisseur of fine *objets d'art*. In Tuscany he became a close friend of the renowned Florentine etcher Stefano della Bella, and of Jacques Stella, an etcher from Lyon who was working at the court of the Grand-Dukes.

In 1633 Langlois set up shop in Paris, where he also acted as an art agent for Charles I of England. When van Dyck was in Paris in 1641, he executed a portrait of Langlois that was subsequently engraved by Jean Pesne and Pierre Mariette. The inscription on the engraving reveals that this eclectic figure was in addition an excellent musician: 'François Langlois natif de Chartres, libraire et marchand de tailles-douces à Paris, excellent à jouer de la musette et de plusieurs autres instruments.' Printmaking was, however, Langlois's principal activity. A great admirer of Rembrandt, he executed dexterous copies of a number of the artist's works; he also transposed many of Raphael's paintings into masterly prints. In 1639 he purchased and reprinted several copper plates of religious and decorative subject-matter by Daniel Rabel, who had died two years previously (Meyer, p. 15; see no. 15).

The sophisticated title-page to this florilegium, rich in horticultural imagery, was designed by Langlois but engraved by another artist, Léonard Gaultier (1561–1641). Born in Germany, Gaultier established himself in Paris, where he engraved various portraits of members of the aristocracy and

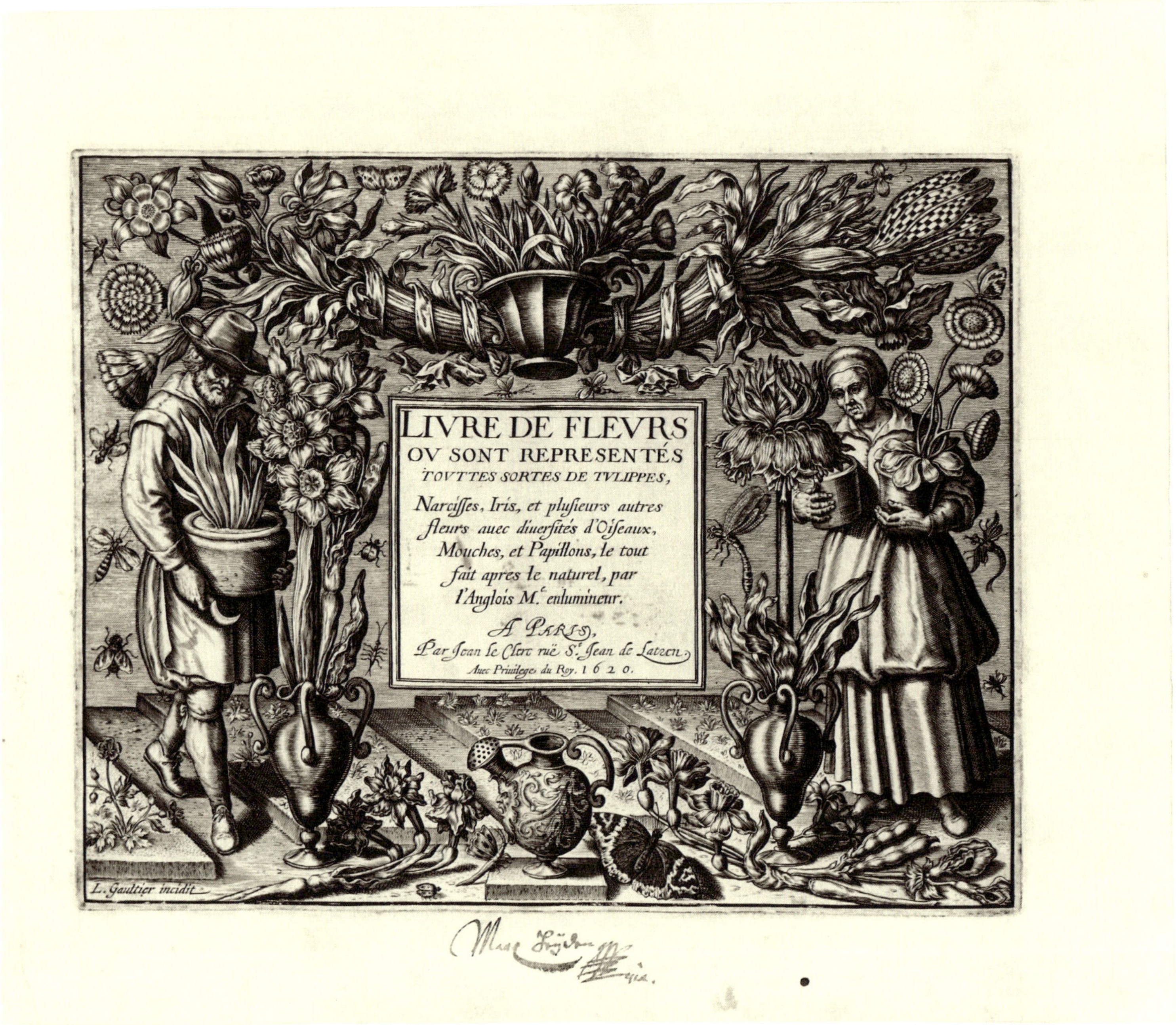

FRANÇOIS LANGLOIS, *Livre de fleurs.* Title-page

royalty, including Henri III, Henri IV, Marie de' Medici and Louis XIII. Langlois, however, undertook to work on many of the other plates himself, signing them 'Langlois fecit et Execudit. Cum Privilegio Regis'. His work reflects the influence of Crispijn van de Passe, especially in the tulip series, where, as in many of the illustrations in *Hortus Floridus* (see no. 12), the flowers are shown flourishing in their natural habitats, surrounded by other plants and herbs. In one plate the tulip on the far right is supported by an iron ring, a detail borrowed directly from the work of van de Passe. In the plates that follow, the flowers (mostly varieties of narcissus and iris) are shown together with different animals—birds (perched on branches or atop little hillocks) and insects—reminiscent of the plates in the *Quadrupedum ac volatilium, florum praeterea et fructuum* engraved by the German Johann Hogenberg in 1594. The cartouches bearing the Latin names of the various species have been adroitly incorporated into the compositions as decorative elements themselves, shown wound around a stem or dangling from a leaf. Although Langlois states on the title-page that, in keeping with the practice of the period, 'le tout [est] fait apres le naturel', it is quite clear that he was much less interested in a scientific portrayal of his subject-matter than he was in exploiting the decorative possibilities of its various elements.

This copy of Langlois's florilegium is incomplete. However, the Oak Spring collection does include two loose plates illustrating tulips, one of which bears the handwritten number '12'; they may have been taken from this very volume, which in fact contains two torn paper brackets. Oak Spring also possesses another copy of Langlois's work, comprising the title-page and sixteen mounted plates, the smaller ones glued four to a page and the larger ones two to a page. Two of the illustrations in this series are not to be found in the copy described in detail above; one depicts an owl and a fleur-de-lis, while the other shows a magpie together with a poppy and a Windsor bean-plant.

15. Daniel Rabel (?1578–1637)

[*within a double-ruled border, an engraved and hand-coloured title-page. A series of three cartouches are sustained by classical figures against a wall banked with flowers. On the first is written*] Theatrum florae [*on the second*] In quo Ex toto Orbe selecti Mirabiles Venustiores ac praecipui Flores Tanquam ab Ipsus Deae sinu proferuntur [*and on the third*] Guillelmus Théodorus Pinxit 1.6.2.4. [*on an oval step at the foot of the wall*] AN .D. | CIↃ | IↃC | XXII.

41 × 28 cm. 71 engraved and hand-coloured plates consisting of a title-page, a frontispiece and 69 numbered engravings after drawings of flowers by Rabel; plus 5 contemporary watercolours of flowers numbered 70–74.

BINDING: Straight-grained red morocco of a later date, full gilt spine.

PLATES: All hand coloured by 'Guillelmus Théodorus', who signs the title-page in gold.

PROVENANCE: At the top of the frontispiece is written, in an 18th- or 19th-century hand, 'Elizabethae Joye ex dono Jacobi Joye'.

REFERENCES: Arber, p. 245; Pritzel 10855; Nissen 1575; Vanuxem; Coats, pp. 33–35; *Printmaking in the Service of Botany*, pp. 42–3; *A Magnificent Collection*, nos. 289–90; *An Oak Spring Garland*, no. 5; *An Oak Spring Hortus*, no. 15.

THIS BOOK, printed in Paris in 1622 by Nicolas de Mathonière, was the second engraved florilegium to be published in France, the *Jardin du Roy* of Pierre Vallet having appeared in 1608 see no. 8). Published anonymously, *Theatrum florae* was an immediate success, and further editions were printed in 1627 ('Apud Petrum Firens') and 1633 ('Chez Pierre Mariette'), in both cases in

Paris. It was only in the third edition that the name of the author, Daniel Rabel, finally appeared. Confirmation of his authorship can be found in an album of botanical paintings signed by Rabel and now in the Bibliothèque Nationale (Aymonin and Schnapper); it contains drawings that formed the basis for some of the engravings in *Theatrum florae* (Stein; Schnapper 1988, p. 58; Pinault, p. 22). Rabel was a man of many talents—painter, engraver, portraitist, designer of theatre sets for ballets, creator of garden ornaments and plant-beds, engineer, and finally painter, as he signed himself in the album. However, although it is clear that Rabel was responsible for the preparatory drawings, it is not certain that he executed the engravings for *Theatrum florae*, which sometimes do not do full justice to the originals.

Rabel was born *c.* 1578 in Paris; his father, Jean, was also a painter. Rabel seems to have been the very first artist of naturalistic subjects to enjoy the patronage of Gaston, duc d'Orléans, a younger brother of Louis XIII. This prince took a lively interest in the natural sciences and in the cultivation of flowers, pastimes he could indulge to the full in the splendid gardens of his château at Blois. Paintings on naturalistic themes became fashionable in France during this period in large part due to his encouragement. Gaston commissioned Rabel to paint a number of vellums that came to form the original nucleus of the celebrated royal collection of botanical paintings today conserved in the Muséum National d'Histoire Naturelle, Paris, a collection to which Nicolas Robert also contributed a few decades later (see no. 44).

Another important patron who, from 1603 until the end of 1633, made use of Rabel's talents was the intellectual and naturalist Nicolas-Claude Fabri di Peiresc in Provence, for whom the artist completed many archaeological paintings, as well as a volume on German heraldry (Schnapper 1988, p. 178). Peiresc, speaking of him, testified: 'J'ai trouvé les desseins fort jolis et M. Rabel fort honeste homme' (Meyer, p. 8).

Active though he was in many different areas, Rabel's fame rests on his botanical and floral paintings, which inspired eulogies from many of his contemporaries, including the poets Georges de Scudéry and François de Malherbe. Scudéry wrote a sonnet in praise of some of Rabel's paintings of birds (unfortunately now lost), while Malherbe penned the following reflections on Rabel's floral painting: 'L'art y surmonte la Nature / Et si mon Jugement n'est vain, / Flore lui conduisoit la main. / Quand il faisoit cette peinture' (Hunt 212; Vezin, p. 22). To this flattering chorus may be added the words of praise in *Pensées ingégneuses* by the Jesuit Dominique Bouhours, the arbiter of good taste in seventeenth-century France.

The plates in *Theatrum florae* are indeed magnificent. Illustrating a variety of flowers, mostly bulbous species (such as the fritillary, lily and narcissus), and certain oddities greatly prized during the early part of the century, such as double and multiple blooms forced by the use of 'secret' fertilizers, the specimens have been elegantly arranged and presented on the page, each one labelled with its Latin name. While the images clearly owe much to the earlier florilegia of Vallet and de Bry (see nos. 8, 10 and 16), certain illustrations stand out due to their exceptional artistic quality and

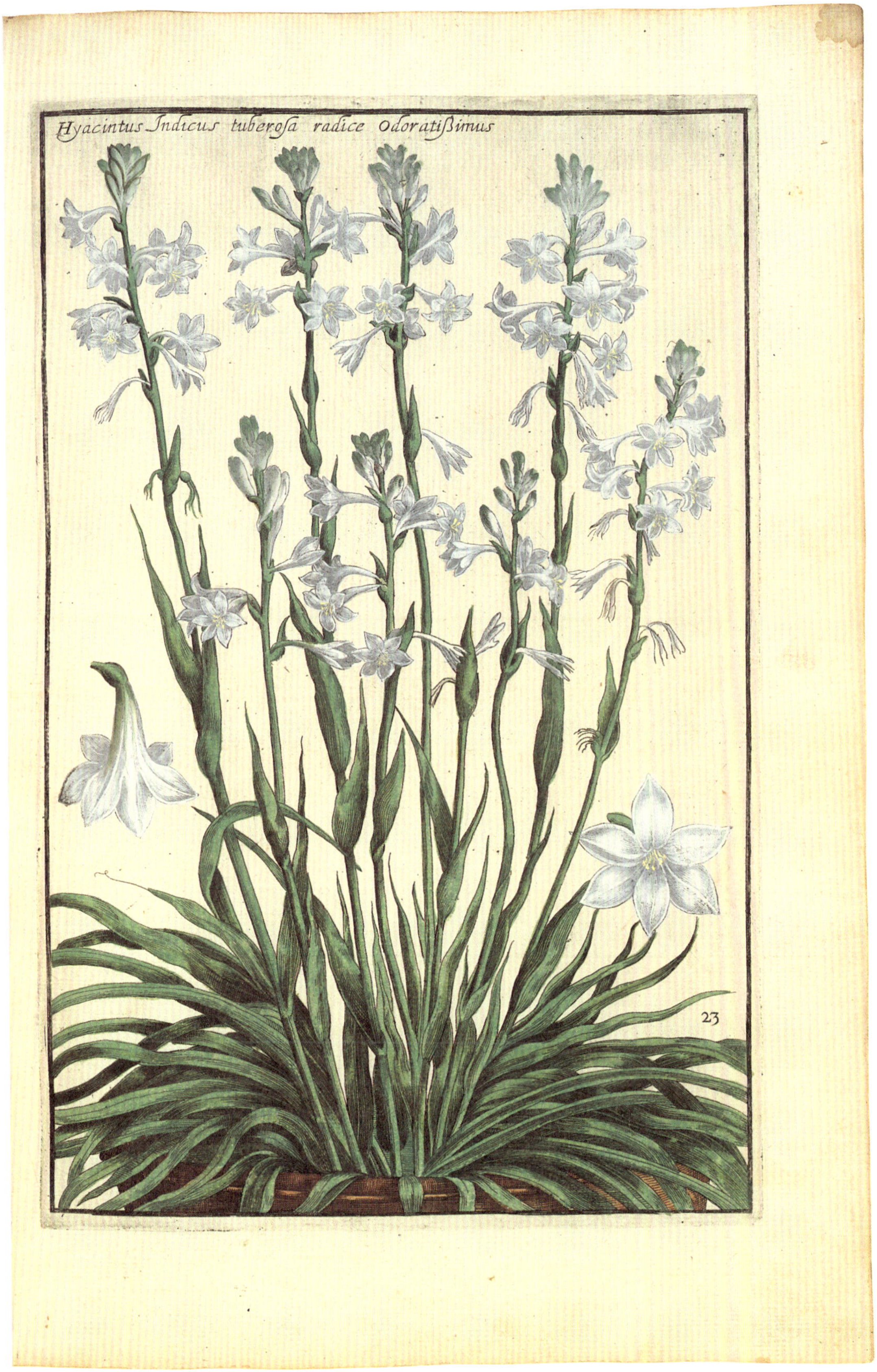

DANIEL RABEL,
Theatrum florae. Tuberoses
(*Polianthes tuberosa*),
plate 23

DANIEL RABEL, *Theatrum florae.* Three dog's-tooth-violets (*Erythronium dens-canis*) and a male moth, plate 40

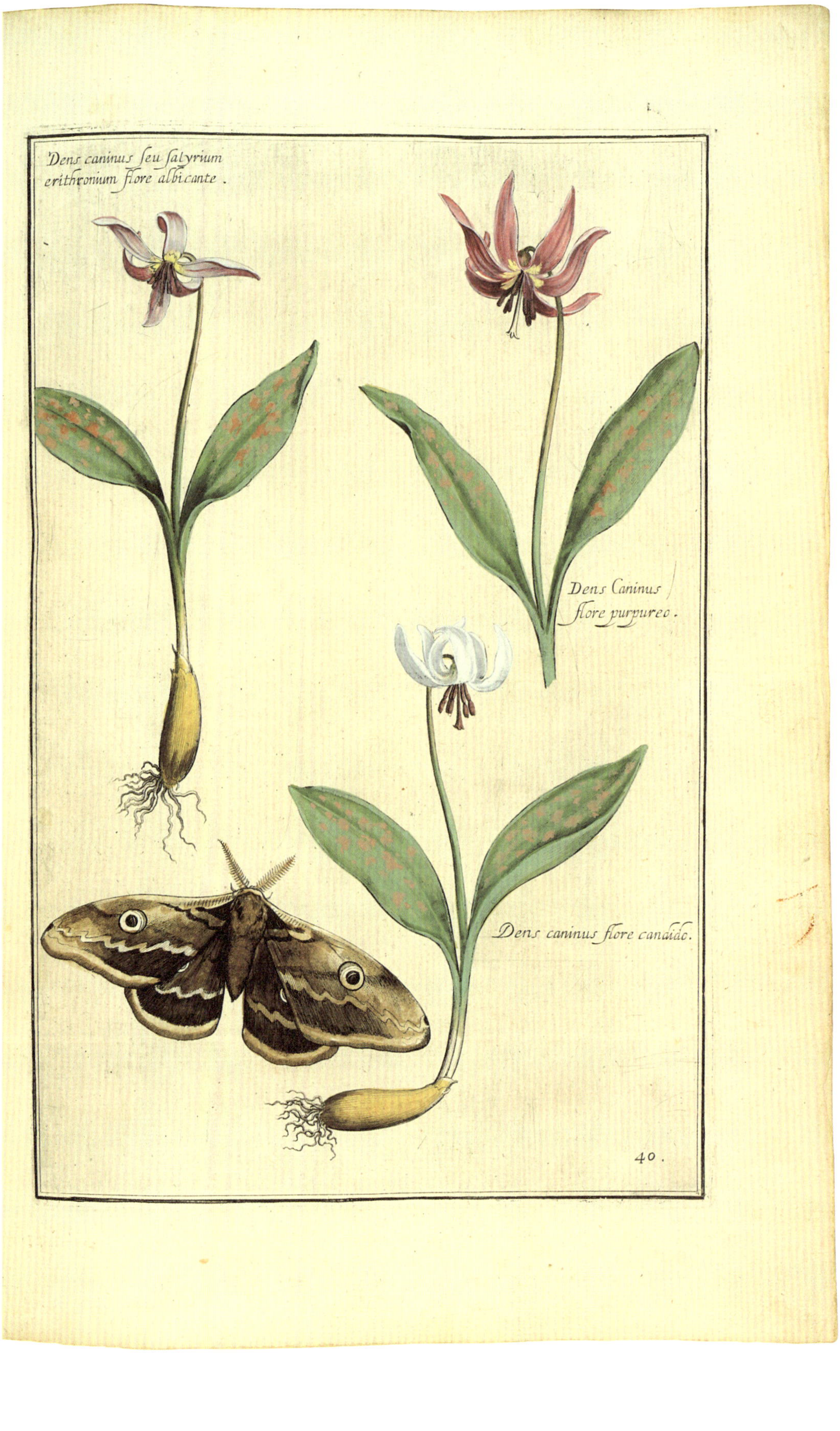

almost scientific realism: a beautiful plate showing a mass of 'Hyacintus indicus tuberosa radice odoratissimus' (*Polianthes tuberosa*) growing in an urn, for example, or another depicting various species of 'Dens caninus' or *Erythronium dens-canis* (dog's-tooth violet). This latter flower, which forms the subject of a miniature in the album in the Bibliothèque Nationale (plate XVIb), is shown here together with a life-size *Saturnia pyri* (giant silkworm moth).

Rabel also designed the title-page and frontispiece, with their sophisticated allegorical imagery. The title-page shows Apollo—the classical deity associated with botany and medicine—flanked by two female figures, perhaps nymphs, offering floral tributes. The cartouches bearing the title are also surrounded by garlands and bouquets and by female figures wearing floral wreaths. In the lower part of the engraving a male and a female figure are seated either side of an oval step, arranging flowers in large urns that are reminiscent in form of the vases in coeval Flemish flower pieces. The frontispiece, which presents the vista of an imposing formal garden seen through a green archway, forms a fitting 'entrance' to this florilegium (see *An Oak Spring Hortus*, fig. 35). In the upper part of the archway two putti hold a banner on which is announced 'DELICIÆ DOMINI / NEC. QVID SPERARE HABEBAT' (The Lord's delights. Naught else was worth hoping for). In the lower part are Flora and Vertumnus.

This copy of *Theatrum florae* is unique in the history of botanical illustration. Although it was current practice to colour by hand the plates of important botanical works, in this case the execution is of an unusually high quality, the artist having used warm and vivid tints highlighted in gold to create effects reminiscent of manuscript illumination. Nothing is known of this artist, however, who, justly proud of his handiwork, signed the title-page 'Guillelmus Théodorus'. Perhaps of Greek origin, as his surname would seem to indicate, we do not even know whether this artist worked in France or England or whether this copy appears to have migrated according the owner's notation. It is probable that 'Guillelmus' was also responsible for the five watercolours of flowers bound at the end of the volume, the style of which is closely akin to that of the hand colouring in the engravings.

16. JOHANN THEODOR DE BRY (1561–?1623) *and* MATTHAEUS MERIAN *the elder* (1593–1650)

[*added illustrated title-page: against a landscape dominated by a portico and trees, the goddess Flora and her attendant maidens are placing urns and wreaths of flowers around the herm of two-faced Janus, one of the most ancient and venerable of the Roman gods, whose role was to guard over doors and entrances—in this case the entry to a garden. Two urns of flowers stand on the steps at the base of the herm. Scattered on the ground are other flowers, while to the left appear various gardening implements. Above the arch of the portico is written*] 'Flora Sive Florum Cultura' [*and on the herm*] 'Redimitur Floribus Annus'. Florilegium Renovatum et Auctum. Prostat Francofurti apud Matthaeum Merianum. 1641 [*on the title-page*] Florilegium Renovatum et Auctum, Das ist, Vernewertes und vermehrtes Blumenbuch: von Mancherley Gewächsen, Blumen un Pflantzen, welche uns deren Schönheit, lieblicher

Geruch, Gebrauch, und manigfaltiger Unterschied angenehme machet, die nicht allein auss der von uns bekandter, sondern auch den alten unbekandter Welt, fruchtbaren Schoss, uns herfür gegeben werden: Die hierinnen auffs zierlichste und fleissigste, dem Leben nach, so viel als möglich gewesen, in Kupffer gebracht, und mit ihren Stengeln, Blettern, Blumen, Samen, Hülsen, Zwibeln, und Wurtzeln, derer Liebhabern für Augen gestellt zufinden; bey derer jedem Stück, sein eygentlicher rechter Namen, aber umb der Gewiszheit, und sicherer Erkandtnuss willen, nur in Latinischer Sprach gesetzet. [*owner's coat of arms*, 6.5 × 9 *cm.*] Frankfurt am Mayn Bey Matthaeo Merian Buchhändlern, im Jahr MDCXXXXI.

F° 30.5 × 20 cm. A–B^4 (B4 blank) *1–4* 5–14 p. Followed by 4 leaves with a manuscript index from the same period. A manuscript Linnaean index dated 1822 is loosely inserted.

BINDING: Contemporary calf with richly decorated gilt panel of rolls and ornaments with a central lozenge. Spine divided into five compartments with gilt decoration.

PLATES: 176 engraved and hand-coloured plates (some of them folded). Plates 1–32 bear engraved leaf numbers, but are not arranged sequentially. On plates 33 to 176 the leaf numbers have been written in manuscript. The plants' names appear engraved on the plates.

PROVENANCE: Coat of arms sustained by two putti, engraved and hand coloured and signed 'Martin Wevetzer, Valentins sohn, 1662' over title-page vignette. Inscribed on the front free endpaper: 'Evelyn Agnes Dingwall from her Father with his love.'

REFERENCES: Pritzel 1299 (note); Nissen 274; Hunt 237; *A Magnificent Collection*, no. 93; Blunt and Stearn, p. 102.

IN 1641, twenty-one years after the death of Johann Theodor de Bry, Matthaeus Merian the elder published an amplified version of the *opus magnum* of his father-in-law, *Florilegium novum* (see no. 10), entitling it *Florilegium renovatum*. Merian was born in Basel in 1593, and after studying with Dietrich Mayer in Zurich and living for a brief period in Paris, he settled in Frankfurt, where he married de Bry's daughter. A versatile artist, Merian not only continued the work of his father-in-law as an engraver, he also painted landscapes, portraits and historical scenes. His children, Matthaeus the younger and Maria Sibylla (see nos. 81, 82, 101), followed in his footsteps: both became artists.

Florilegium renovatum is a much richer work than the earlier *Florilegium novum*, many illustrations from other sources having been added to de Bry's seventy plates. The allegorical title-page and the first thirty-two plates—illustrating parterres, urns of flowers and gardening tools, as well as plants—were drawn from *De florum cultura* by the Jesuit Giovanni Battista Ferrari, which was published in Rome in 1633 (see no. 29). Merian reproduced Ferrari's plates, reversing them in the printing process, and sometimes adding butterflies or other features or combining figures from more than one illustration. The images in plate 32, for example, were directly copied from pages 495 and 497 of *De florum cultura*. Plate 158 instead presents, in a slightly rearranged form, the marigolds from plate 26 of Emanuel Sweerts's *Florilegium* (see no. 9). Other illustrations, such as two depicting the *Datura stramonium*, or thorn apple (pls. 113 and 114), and another (pl. 175) showing various species of peppers ('Piperi Indici'), were drawn from the work of Caspar Bauhin.

The large folded engravings in *Florilegium renovatum* are entirely new, however. They present exotic or unusual plants, some of which the artist might have seen in the gardens around Frankfurt, such as the 'Lilium liliorum sive 122 lilia ex eodem bulbo enata: floruit Francofurti ad Moenum' (The lily of lilies, that is 122 lilies grown from the same bulb. It flowered in Frankfurt am Main) in plate 116 and the 'Lilium Wran. multifolium. Floruit in horto Principissa Stukardie' (The many

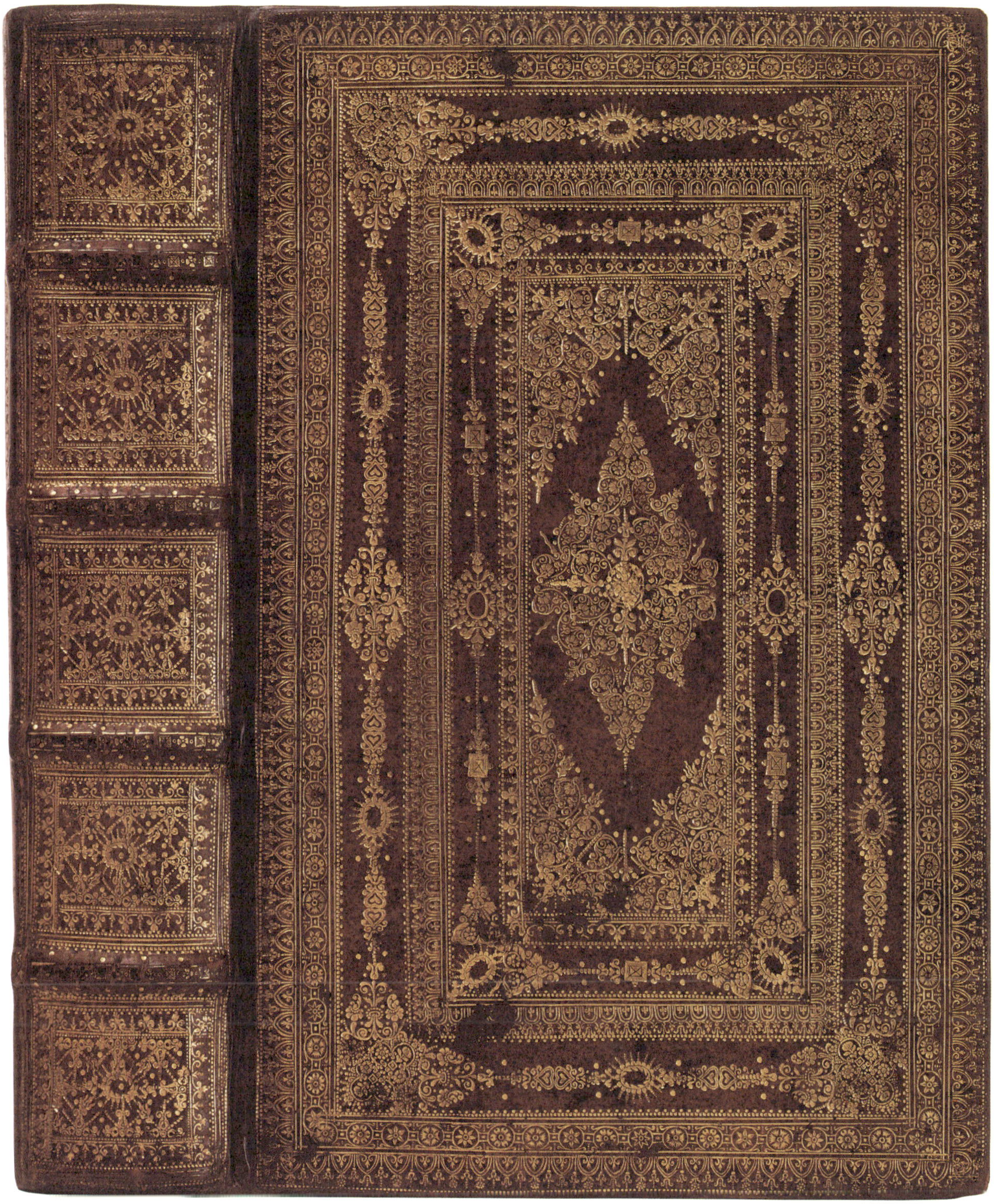

JOHANN THEODOR DE BRY and MATTHAEUS MERIAN, *Florilegium renovatum*. Binding

JOHANN THEODOR DE BRY and MATTHAEUS MERIAN, *Florilegium renovatum*. An arrangement of flowers, plate 11

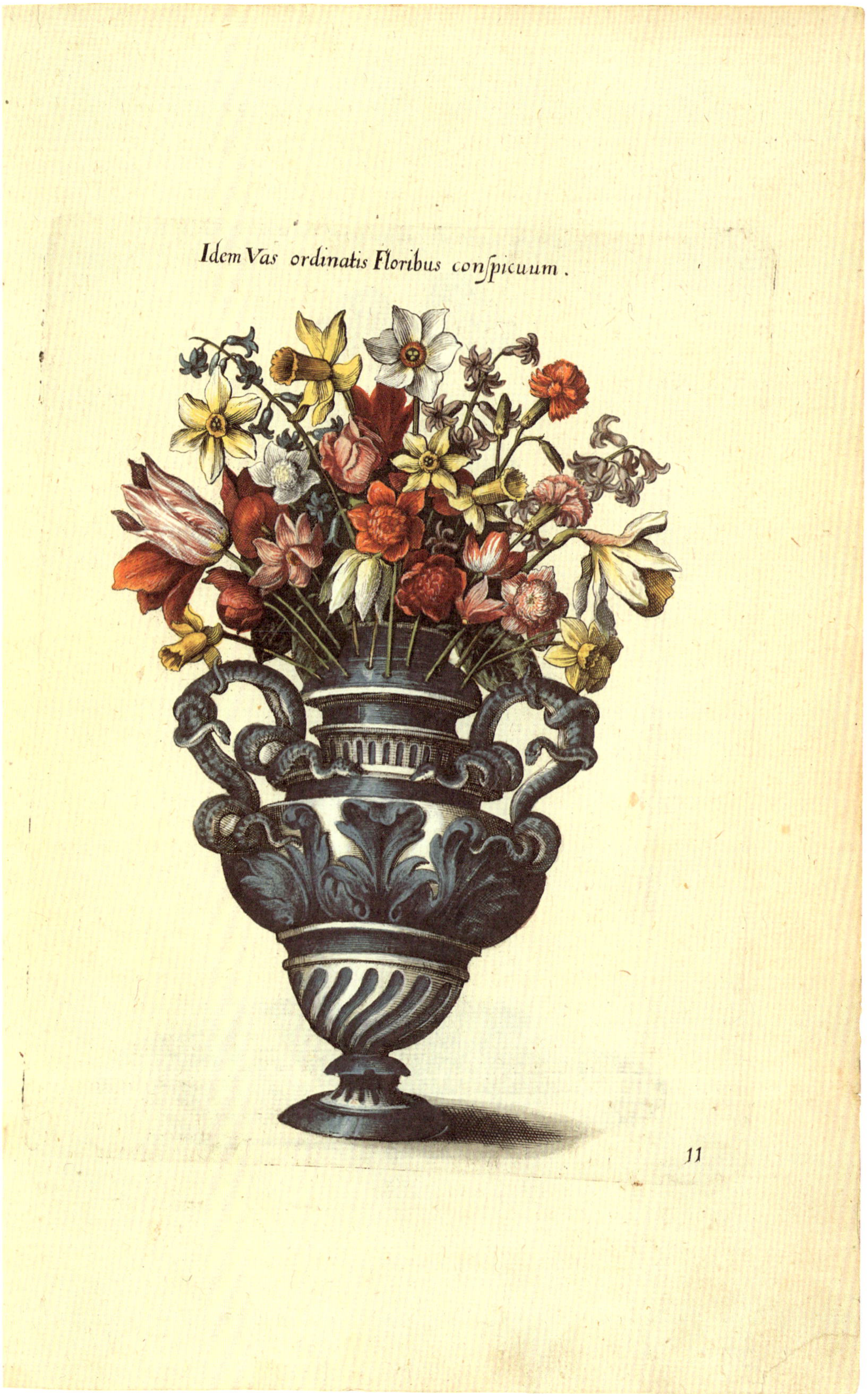

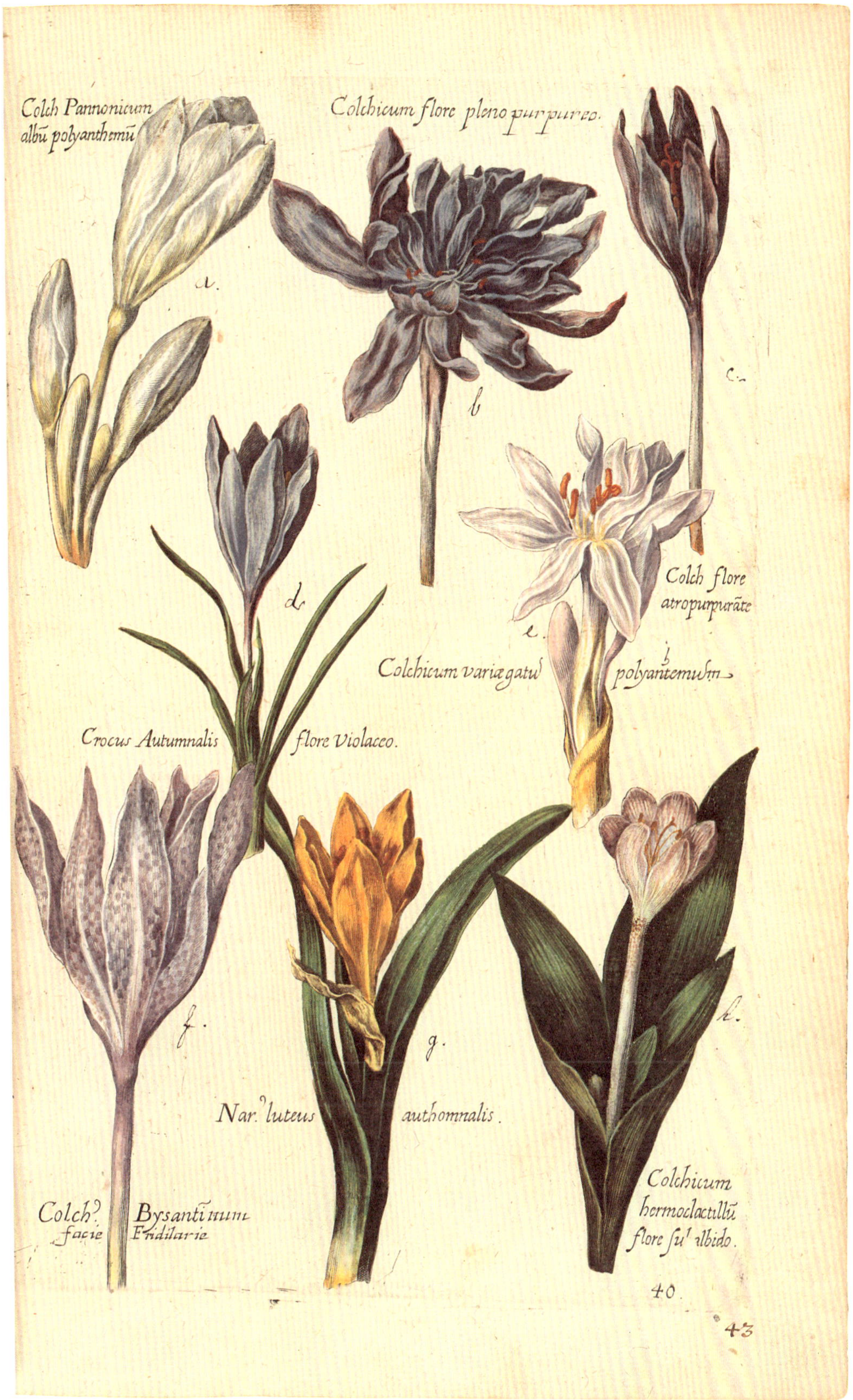

JOHANN THEODOR DE BRY and MATTHAEUS MERIAN, *Florilegium renovatum.* Two varieties of *Crocus* and six varieties of *Colchicum*, plate 40

leaved Wran lily, which flowered in the garden of the Princess of Stuttgart). Plate 124 depicts a '*Hyiucca sive Yúca Canadana*, nostris núnc Iúcca gloriosa Indorúm, ad vivum delineata' (The Hyiucca, or Canadian Yucca, now called the *Yucca gloriosa indiana*, painted from life). From the accompanying legend we learn that this plant flowered in the garden of Remigio Fesch in Basel in September 1644. An illustration of a 'monstrous' rose that bloomed in Prague (pl. 107) is dated 1647, showing that some of the plates were inserted after the title-page had been printed.

One can readily understand the economic reasons underlying the production of works such as Merian's *Florilegium renovatum*, for authors were not always able to provide original illustrations drawn from life, and they thus resorted to copies and replicas of illustrations from other works. Notwithstanding these limitations, Merian's florilegium is an admirable work, executed by an accomplished engraver. The copy at Oak Spring is further enhanced by the very fine hand colouring of its plates in vivid colours, which are quite true to life.

17. Anonymous Artist

Manuscript florilegium.

31 × 20.5 cm. Manuscript with flower paintings on 226 leaves, not numbered consecutively. The vernacular names of the plants have been added in calligraphic script. Accompanying manuscript index bound separately in vellum (20 × 12.5 cm.).

BINDING: Parchment.

PLATES: Bodycolour drawings of flowers, one on each leaf. Folio 11 is painted on the recto and verso.

THIS MANUSCRIPT, dating from the second half of the seventeenth century, depicts many of the spontaneous and cultivated species that could be found growing in Dutch gardens of the period. In addition to various narcissi, irises, ranunculi, hyacinths and carnations ('angeliern'), there are twenty-five paintings of tulips. Generally one specimen is presented on each page, shown growing out of the ground rather than with a cut stem. In a few cases different specimens of the same flower appear in a single painting, shown perhaps in different colours, such as the crocuses on folio 11 or the hepatica on folio 12.

This manuscript was probably the work of a floriculturist or an amateur rather than a professional artist. Although the flowers are painted in a somewhat naïve style and the details are sketchily treated, they nevertheless manage to furnish an immediate and quite accurate idea of the species being depicted. The very simplicity of many of these plates, such as the one depicting the 'Colcicum portugallie' (fol. 114), lends them an undeniable charm. This same flower is depicted in one plate of a florilegium by van de Passe with the name 'Colchicum lusitanicum reticulatum' (see no. 12).

The small bound index that accompanies this manuscript is of great interest, for it provides

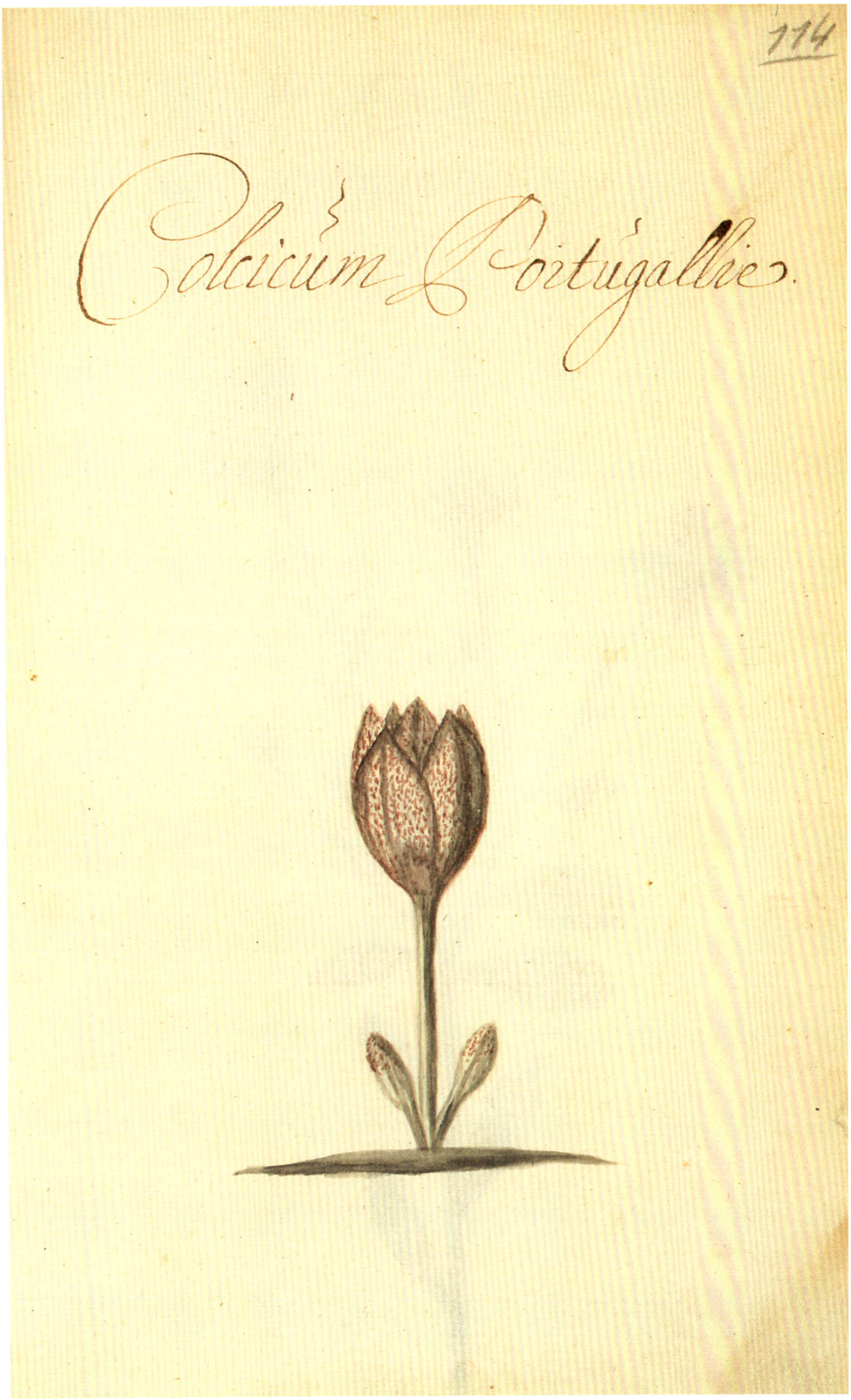

Dutch School, seventeenth century. Meadow saffron (*Colchicum autumnale*)

information on the flower trade that played such an important role in the Dutch economy during the seventeenth century. It consists of handwritten notations by an anonymous floriculturist, perhaps the very one who painted the manuscript, regarding various purchases made between the years 1678 and 1681 from horticulturists such as Thijs Jansen, Pieter Voorhelm, van de Lange and a certain Musaert. Our collector also bought plants from a group of horticulturists identified simply as the 'Italians'.

The lists of tulips are particularly interesting. For example, we find that our collector acquired 102 bulbs from Thijs in 1640, and that on two other occasions he paid fifteen *gulden* and fifteen *stivers* for thirty-four bulbs, and fifty *gulden* eight *stivers* for forty-one bulbs. A closing note shows that in this period he purchased a total of 734 tulip bulbs for 168 *gulden*.

The tulip lists are followed by others regarding purchases of hyacinths, anemones, ranunculi, irises, fritillaries, crown imperials, primroses, auriculas, carnations and other plants, such as rosemary, laurel and even some orange-trees. Usefully, these lists provide us with information regarding the names by which these varieties were known around the year 1680.

18. Pieter Holsteyn *the younger* (1612–1673)

White carnation (*Dianthus caryophyllus*). 28 × 17.8 cm. Water- and bodycolour on paper.
At the base of the plant on the left: 'PH.fe' (with the P and H connected).

Pieter Holsteyn the younger was born in Haarlem, a busy commercial city and an important centre for the cultivation of bulbous plants. His father, Pieter the elder (*c.* 1582–1662), was a draughtsman and painter on glass, while his younger brother Cornelis became a well-known painter of religious, mythological and allegorical subjects. Holsteyn learned the rudiments of drawing and painting in his father's studio, which was frequented by other young artists; in 1640 he became a member of the Haarlem guild of artists. Documents show that in 1647 Holsteyn and his brother Cornelis were in Amsterdam; the latter in fact spent the rest of his life in that city, dying there in 1658. Holsteyn married twice, first in 1662, and lived and worked primarily in Haarlem, returning to Amsterdam for extended periods in 1662 and 1671.

The *œuvre* of the younger Holsteyn and that of his father have in the past frequently been confused, and problems of attribution still remain today. It is known that our Holsteyn produced a large number of engravings, primarily portraits, including one of Isabella d'Este, Duchess of Mantua, based on a painting by Correggio (Hollstein 1949, IX.76–9). Only recently, however, has a systematic identification of his works on naturalistic themes been undertaken.

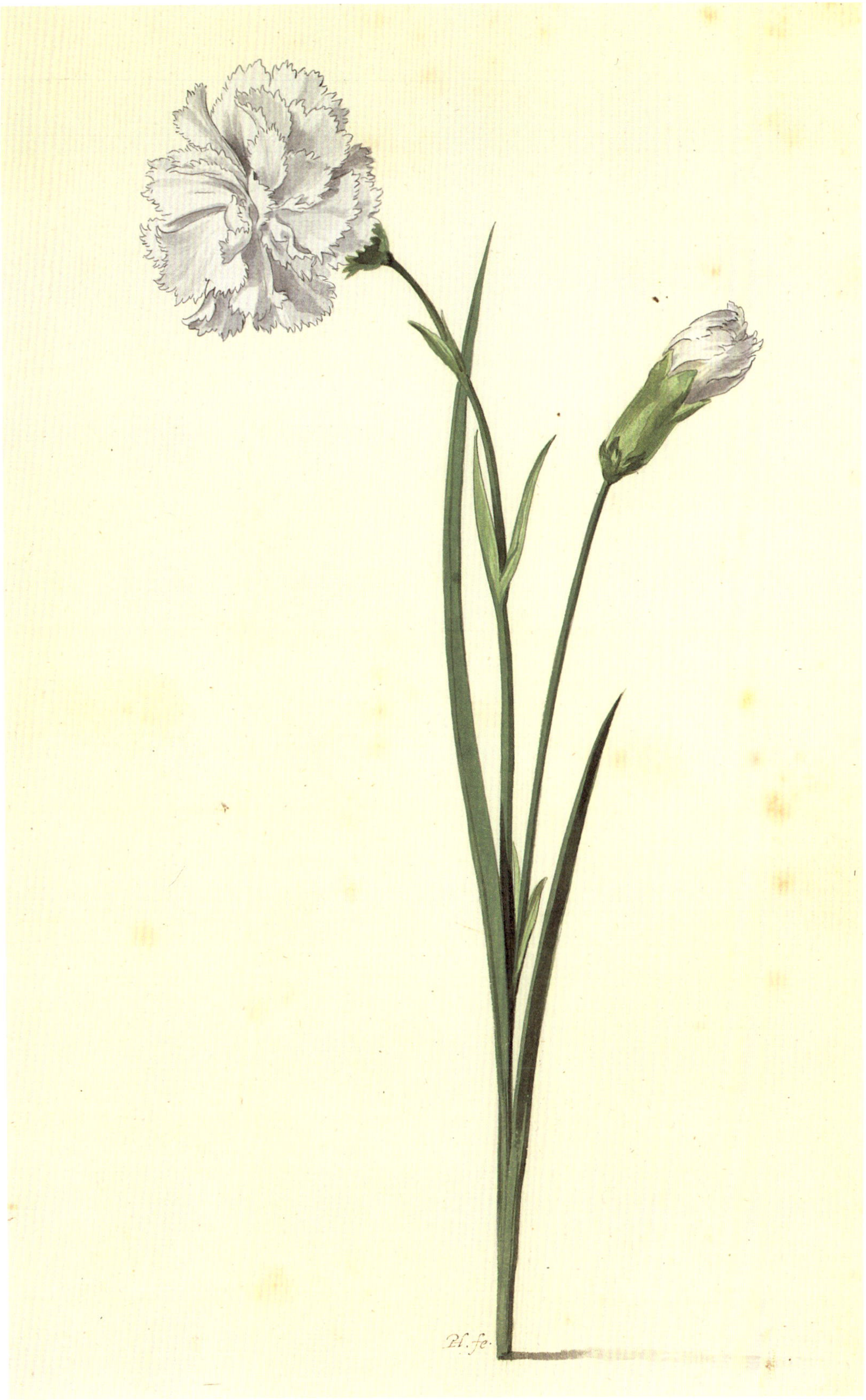

PIETER HOLSTEYN the younger, White carnation (*Dianthus caryophyllus*)

Holsteyn seems to have been particularly interested in zoological subjects, for he produced a large number of paintings of birds and insects, all of which bear his characteristic monogram—'PH fe'. Some have recently appeared on the art market. This body of works has led scholars to include him in a small group with other, slightly older, artists, such as Rochus van Veen (*fl.* 1665–*c.* 1709) and Pieter Withoos (1654–93), as the first proponents of the genre of bird and insect painting that enjoyed a great vogue in Europe during the seventeenth century. In Holland the popularity of these paintings reached its height during the last decades of the seventeenth century and the first decades of the eighteenth, with painters such as Aert Schouman (1710–92) devoting themselves almost exclusively to this subject (*Nederlandse en Vlaamse*, p. 17; *Kleuren Raffinement*, nos. 69–70).

Holsteyn also produced many works in the genre of flower painting. An album entitled *Flores a Petro Holsteyn ad vivum depicti* ('Flowers . . . drawn after nature') containing 122 bodycolour paintings—all bearing the monogram of the artist—is today conserved in the Lindley Library of the Royal Horticultural Society, London (Elliott, p. 36). An album of tulip paintings by Holsteyn once formed part of the Krelage collection, today dispersed (Krelage 1942a, p. 202, no. E), while two paintings in bodycolour by him are included in the manuscript album *Plusieurs Especes de Fleurs dessinées d'après le naturel* now in the Rijksprentenkabinet, Amsterdam.

At Oak Spring there are two flower paintings by Holsteyn, one depicting a white carnation and the other a pink and red variegated carnation. They are quite similar to two paintings (fols. 108, 109) in the London *Flores* and represent excellent examples of the artist's work in the genre of botanical illustration. The paintings are characterized by dense, opaque colours applied directly to the paper without preliminary sketches in pencil or chalk, an expedient to which many other botanical illustrators resorted, particularly in the delineation of stems and leaves. Delicate highlights have been created with touches of watercolour on the petals and buds. The leaves and closed buds spread themselves elegantly across the page, while to the right of the stems a faint grey line has been traced with a fine brush to create the merest suggestion of a background, a signature note that recurs in almost all the works in the London album.

19. Jeremias Falck (1609/10–1677)

[*in a garland of flowers*] Novae et exquisitae Florum Icones, Huius generis artium cultoribus perutiles, maxima cura delineatae, et tabulis aeneis incisae per Jeremiam Falck Hamburgi An^{o} 1662.

29.7 × 22 cm. Title-page and 11 plates.

BINDING: Sprinkled paper boards.

PLATES: Engraved title-page framed by a flower garland, numbered 1, and 11 engraved plates of flowers, numbered 2–12.

PROVENANCE: Ex-libris of the physician J. Jacobus Baierius.

REFERENCE: J. C. Block, *Jeremias Falck, sein Leben und seine Werke*, Danzig, 1890.

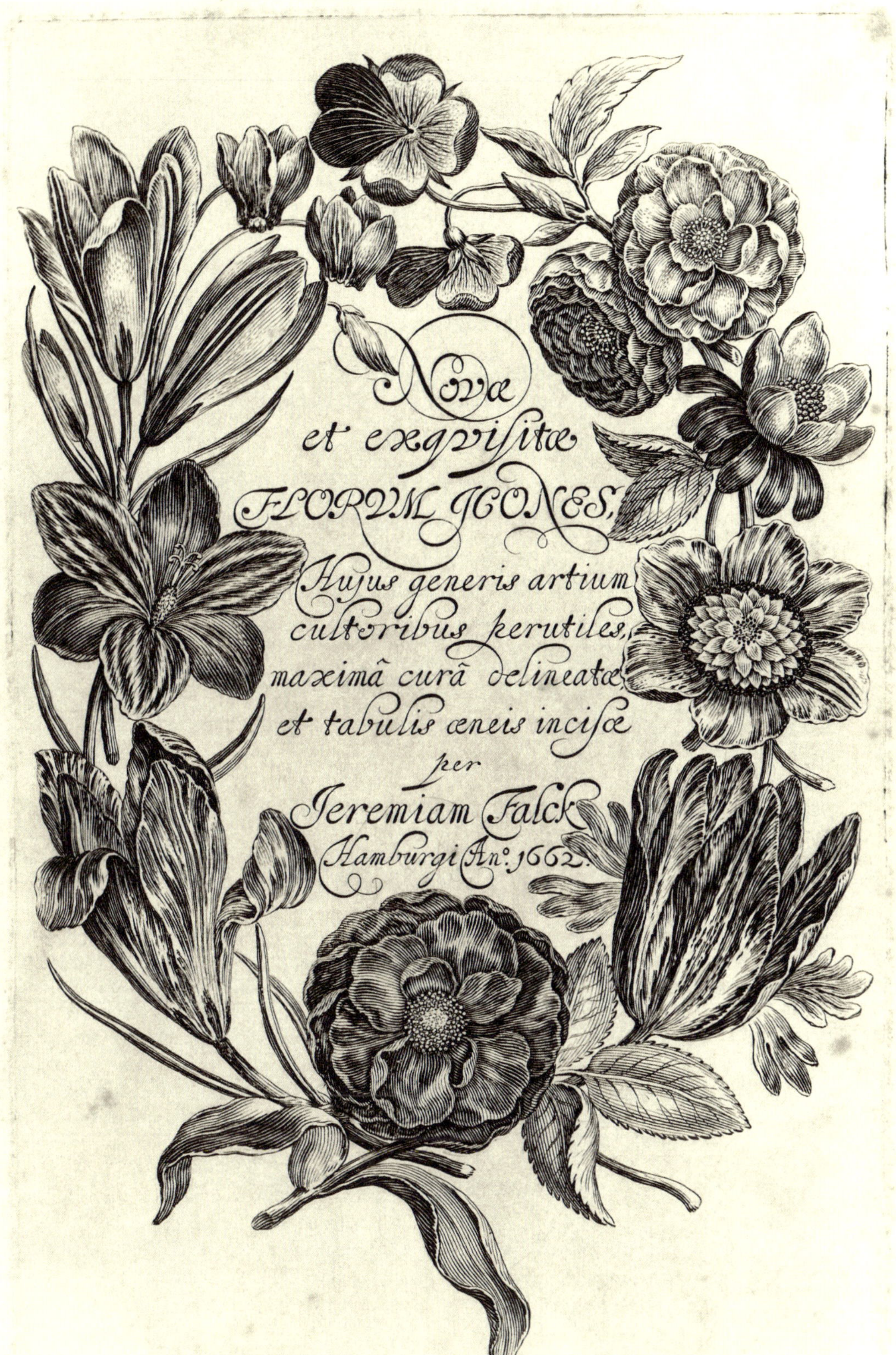

JEREMIAS FALCK, *Novae et exquisitae florum icones*. Title-page

THE DRAUGHTSMAN AND ENGRAVER Jeremias Falck was born in Danzig and had the good fortune to study with two of the greatest engravers of his time—Cornelius Bloemaert and Abraham Bosse. During his life he travelled widely, obtaining prestigious commissions in various countries. He worked in Paris, and for several years in Amsterdam, and around 1655 was appointed official engraver to Queen Christina of Sweden. During the last period of his life Falck worked in Hamburg. He was buried in his native city in February 1677.

Falck's output comprises over 300 engravings, nearly all of which are either portraits or scenes depicting mythological or religious themes. In 1662 he produced this floral series, inspired by the many engraved florilegia that were printed during the course of the seventeenth century, affirming that these plates would be of interest to 'connoisseurs of such arts'.

Falck's *Icones* was not conceived as a florilegium, however, but rather as a book of designs for the embroiderer and other artisans. The title-page is decorated with a garland of flowers, while the eleven plates show various flowering species, including the rose, lily, carnation, fritillary, iris, peony and narcissus. Each plate depicts several flowers from one or more species, presented as separate specimens rather than in formally arranged bouquets. The stem and leaves are often not shown, Falck having trimmed them off close to the blossom. The bold, simple lines of these engravings would have translated quite effectively into designs for use by other artists.

20. JULIUS FRANÇOIS DE GEEST (*c.* 1639–1699)

[*a cartouche in the form of a grotesque mask, from whose eye sockets emerge two masses of flowers composed of tulips, lilies, violets and other flowers painted in brilliant colours. Between the large open jaws is the inscription*] Jardin de Rares et curieux Fleurs faicts P Francois de Geest de Leovarde en Frise.

29.3 × 21.2 cm. Manuscript on vellum, fols. 1–102. The title-page is followed by a pen and ink drawing and 100 paintings on vellum. The title-page and paintings are protected by guard sheets, numbered 1–100, which bear a 17th-century Dutch watermark.

PLATES: 100 miniatures in bodycolour of plants, labelled with their Latin names. Preceded by a pencil drawing on vellum signed 'A Pynacker'.

BINDING: Early 18th-century Dutch binding in mottled calf. Covers decorated with a combination of gilt rolls forming a border and panels; inner panel has fleurons at corners; centre-piece of strap and arabesque work. On the spine, 7 ornate gilt compartments, and 'BLOEM BOECK DOOR P. F. GEEST' in gilt lettering.

PROVENANCE: Collection of Sir Bruce Ingram, England.

REFERENCES: Toesca; Blunt and Stearn, p. 138.

JULIUS FRANÇOIS DE GEEST was the eldest child of Hendrikje van Uylenburgh (the sister of Saskia, Rembrandt's wife) and Wybrand de Geest (1592–1659), an artist who spent four years in Rome studying the classical and contemporary art to be found there. After training as a painter under his father, de Geest moved to Antwerp (*c.* 1657) to continue with Erasmus Quellinus, a painter and man of letters, and close friend and collaborator with Rubens. The elder of his two sisters, Eva

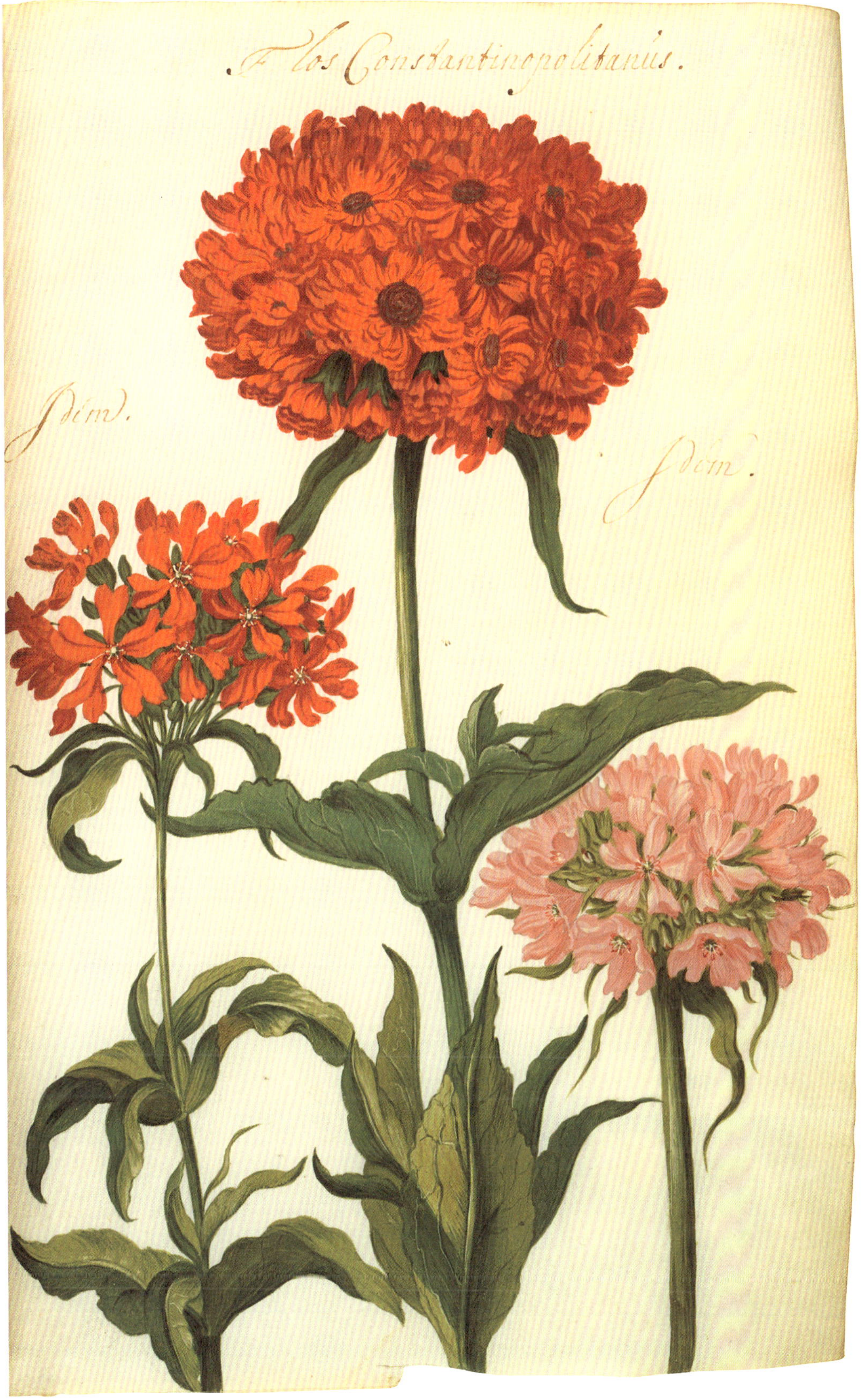

JULIUS FRANÇOIS DE GEEST, *Jardin de rares et curieux fleurs*. Three varieties of maltese cross (*Lychnis chalcedonica*), folio 31

Maria, married the landscape painter Adam Pynacker (*c*. 1620–73), who had also spent several years in Italy. De Geest eventually transferred to Leeuwarden in Friesland, where he married and remained until his death in 1699. His son, Wybrand de Geest the younger, followed the family tradition and worked as a painter.

Until recently de Geest was primarily known for a series of *c*. twenty portraits of the aristocracy, all signed and dated by him. Then, as a result of scholarly research, two important florilegia were added to his *œuvre*. The first was this manuscript, formerly owned by Sir Bruce Ingram, a plate from which was reproduced in colour in *The Illustrated London News* (25 September 1948), and the second a manuscript of flower paintings in the Biblioteca Nazionale Centrale in Rome (MS Varia 291; see Toesca). Both testify to de Geest's talent and to the excellent training he received in Antwerp, the city in which the genre of flower painting originated at the end of the sixteenth century and where it reached its fullest expression.

The Rome florilegium, bound in the same manner as the Oak Spring manuscript, bears the title 'Hortus Amoenissimus Omnigenis Floribus, Plantis Stirpibus Tam exoticis, quam indigenis, ubique locorum in hortis celeberrimis maximorum Botanophilorum summa vigilantia et selectu conquisitis . . . ab industriso et solertis ingenii Juvini Viro / D. Francisco De Geest / Wibrandi Filio Leovardiano Frisio' (see the Frans Halsmuseum *Catalogus*, no. 62). It consists of 266 leaves of paper, beginning with a title-page that shows the author offering his work to Flora. The scene is surveyed by Fame holding aloft a cartouche bearing the inscription 'Floriligium Auctore Franc.co de Geest A° 1668'. The paintings illustrate different species of flowering plants labelled with their names in Latin and Dutch. They are preceded by a pencil drawing of a garden scene dominated by a large urn of flowering plants, executed by de Geest's brother-in-law, Adam Pynacker. Many of the paintings are unfinished, and the final ones exclusively show tulips.

The Oak Spring florilegium, made probably during the same period as the one in Rome, opens with a title-page depicting a grotesque in the form of a mask decorated with flowers and inscribed with the title of the work and the author's name. Once again a signed pencil drawing by Pynacker of a garden nook (see *An Oak Spring Hortus*, fig. 16) has been inserted just before the series of plates, here comprising one hundred magnificent paintings on vellum illustrating 235 different plant species (many of which also appear in the Rome manuscript). In both works the plants are shown in flower, without their roots or bulbs, and with their Latin names added in elegant manuscript.

De Geest no doubt drew inspiration from the printed florilegia of the period, but this work stands apart due to its masterly draughtsmanship, its use of brilliant, luminous colours, and the sense of airy spaciousness that infuses each composition. As was customary at the time, many popular cultivated varieties are shown: fritillaries (fols. 4, 5, 6); narcissi (fols. 9 and 10); Asian ranunculi (fol. 11); tulips, including the famous 'Goulda' (fol. 14) and the 'Semper Augustus' (fol. 15); the 'Flos Costantinopolitanus' (*Lychnis chalcedonica*), native to the southern shores of the Black Sea (fol. 30); the 'Mirabilis Peruvianus' (fol. 99); and the sweetly scented golden 'Jasminum Americanum' (fol. 94), which Ferrari in his *Flora* (see no. 29) called 'the miracle of our world' (p. 389). Interspersed among these rare and exotic species are native wild flowers, such as the soapwort with its pale rose-coloured flowers (fol. 70), the 'Cyanus' (fol. 81) and the *Solidago virgaurea* (fol. 89).

It is not improbable that these two florilegia included flowers from the magnificent garden constructed by Count Willem Frederick of Nassau in 1648 on one of Leeuwarden's ramparts, its design perhaps inspired by the poet Jacob Cats.

This florilegium by de Geest marks an important moment in the history of botanical illustration in the Low Countries, for it is closely linked in style to the emerging compositional theme of the floral still-life. These specimens, shown without bulbs or roots, are not marshalled in the well-ordered rows of the printed florilegia, but rather on different planes, so that the leaves of one plant may cover or hang over those of another in a space that appears almost three-dimensional. As in the coeval paintings of bouquets, the flowers in de Geest's manuscript seem at first glance to have been arranged quite haphazardly, but a closer examination shows that each page was very carefully composed, with an eye to the most pleasing juxtaposition of form and colour. This manuscript at Oak Spring therefore represents a key work in the evolution of the floral still-life in northern Europe.

21. Jan Withoos (1648–1685)

Collection of flowers painted on vellum.

16.5 × 11.5 cm. 263 unnumbered plates on vellum bound in 3 volumes.

BINDING: Contemporary vellum with gilt decoration consisting of a double fillet border and a central panel made up of a double fillet with crowns at each corner. Nine-compartment spine with gilt floral ornaments. Title in black: 'J. Withoos Herbarius ad virum [*sic*] delineatum'.

PLATES: 263 bodycolours of flowers and plants on vellum.

PROVENANCE: Paulo van Uchelen (*c.* 1641–1702).

JAN WITHOOS was born at Amersfoort in the province of Utrecht. His father was Matthias Withoos, a well-known painter of cityscapes, landscapes, still-lifes and botanical and entomological subjects. During a sojourn in Italy, Matthias—dubbed 'Calzetta Bianca' (White-socks)—worked for a period in Florence as a naturalistic painter for Cardinal Leopoldo of the Medici. He taught painting to all five of his children—Jan, Pieter, Franz, Alida and Maria. Alida is known for a series of botanical paintings in bodycolour executed for Caspar Commelin, the eminent botanist and director of Amsterdam's botanical garden (Wijnands, pp. 21–2).

The eldest, Jan, followed his father's example and spent a period of time in Italy, where he became known for his paintings of insects and plants on vellum. In the Netherlands Institute for Art History in The Hague (RKD Collection), thirty paintings by him on various subjects—landscapes,

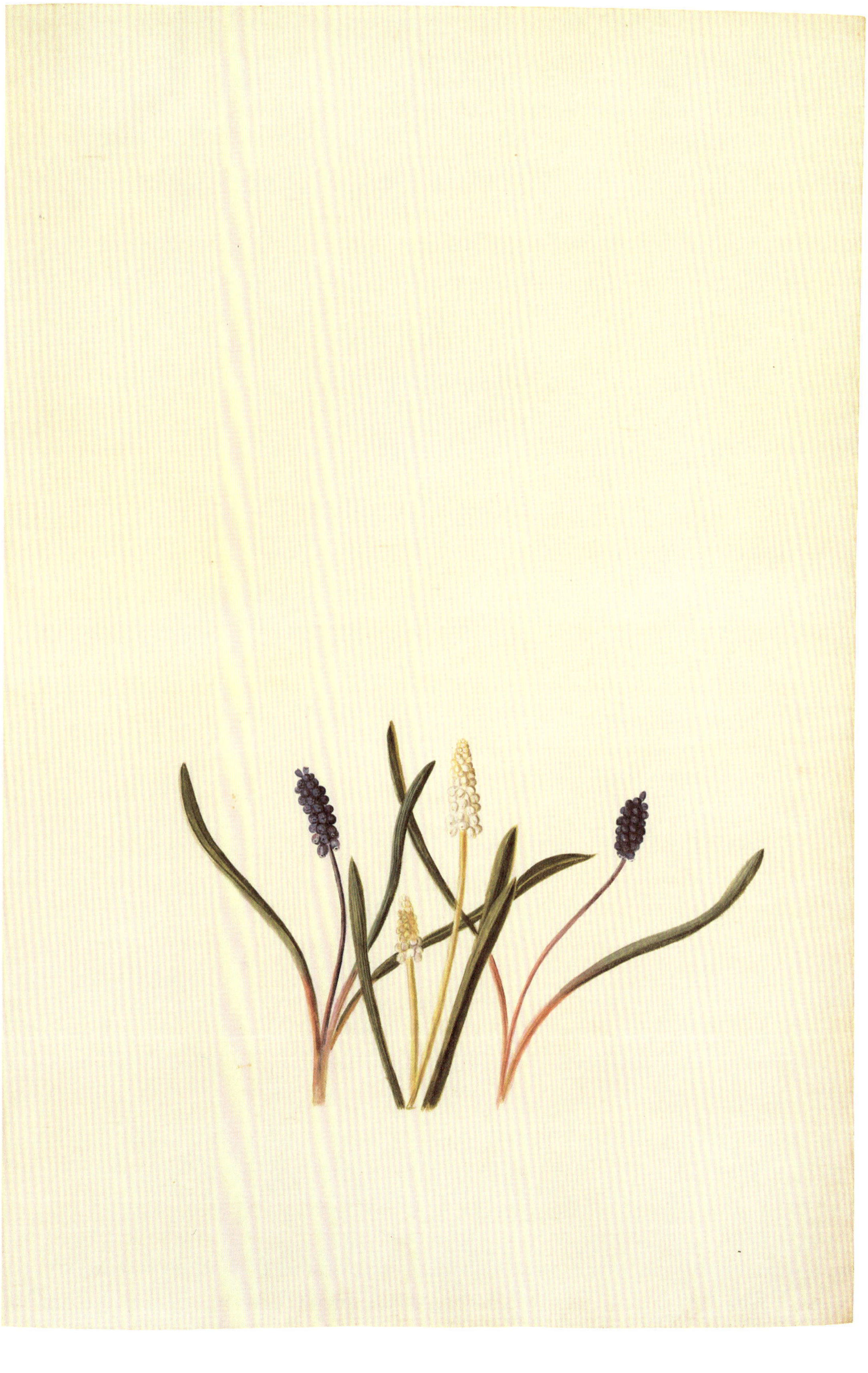

JAN WITHOOS,
A collection of flowers painted on vellum.
Grape-hyacinths (*Muscari botryoides*),
volume I, folio I

cityscapes, birds, wild turkeys, and still-lifes with flowers and insects—and twelve watercolour landscapes are documented as having passed through the hands of dealers. During an auction held by Christie's in Amsterdam on 3 November 1990, an interesting painting by Jan Withoos—bearing his monogram and the date ('J:W: A° 1678')—was sold. A depiction of 'A forest floor still-life of irises, poppies, roses, butterflies, a snake and a thistle at the foot of a tree in a landscape', it is reminiscent of the intricate forest scenes with plants, lizards and snakes painted by Otto Marseus van Schrieck (1619/20–78), an artist who was in Rome with Matthias Withoos *c.* 1650 and who also worked for the Medici court.

After Jan Withoos had returned to his homeland he was invited to the court of the Duke of Saxony-Lauenburg, where it has been assumed he spent the rest of his life. However, documentary evidence signed by Withoos shows that he was in Amsterdam in 1682, where he was involved in a tavern brawl (Bredius, p. 874). Furthermore, an inventory of his (sadly meagre) possessions was drawn up on 12 January 1688, suggesting he may have died later than has been assumed (idem, p. 873).

The collection of Withoos's flower paintings at Oak Spring is the most important body of botanical work by this artist. The collection was evidently much admired in his day because it was acquired by Paulo van Uchelen, a famous bibliophile who lived in Amsterdam during the second half of the seventeenth century. Indeed, a three-volume work containing 263 botanical paintings in bodycolour is listed in the catalogue that was prepared for the sale of van Uchelen's collection in 1703, one year after his death (*Splendidissimus librorum & iconum thesaurus uchelianus*, no. 740). Since this collector had many of the books in his library bound by the famous Amsterdam bookbinder Albert Magnus (Verwey, p. 175), it seems probable that the beautifully ornate binding of these three volumes was the work of this skilled craftsman.

Withoos's talent can be fully appreciated in this large collection of flower paintings. He combined an analytical turn of mind and scrupulous technique with an exceptional painterly style; a vibrant tactility and a vivid sense of colour imbue these paintings of anemones, hyacinths, narcissi, daffodils and carnations. Interestingly, no tulips are depicted, this once phenomenally popular flower having lost much of its appeal by the end of the seventeenth century. Withoos shows a single species in each painting, not always placed in the centre of the page, but arranged rather as his fancy took him. He always respected the natural proportions of the plant, however: when too large, he would portray it in two plates, as in the case of the acanthus (I, pls. 40, 41). Not only the most popular cultivated species of the period, but also native, spontaneous flowers are illustrated. Withoos's profound appreciation of the natural world, even in its most humble details, is particularly apparent in the third volume, where he has depicted the cyclamen and the crocus with such directness and naturalness that they appear to have just burst forth from the ground.

III · THE FLORAL STILL-LIFE IN THE FLEMISH AND DUTCH TRADITION

AS A DIRECT CONSEQUENCE of the passion for flowers that quickened at the end of the sixteenth century and the beginning of the seventeenth, an entirely new genre of painting—the floral still-life—evolved, in which the flower represented not merely a decorative element but the main subject of the work. In these paintings the artist sought to reconcile formal and aesthetic canons with a scrupulous adherence to nature. Like the florilegium, the still-life served to immortalize the ephemeral beauty of the flower, so that it could be admired even in those periods of the year when it was not in bloom.

The floral still-life developed concurrently with the genre of the botanical illustration—paintings of flowers executed in water- or bodycolour on paper or vellum by highly trained artists—and the engravings of flowers and bouquets that appeared in the printed herbals and florilegia of the period. It was from such works as these, in which the aim was to depict flowers and plants as realistically as possible (*contrefeiten* in Dutch), that the genre of the still-life drew inspiration, as well as from nature itself. Copying from such models allowed artists to work even in the depths of winter when there were no live specimens to be found. Indeed, one might see in a single still-life painting flowers that actually bloomed at entirely different seasons of the year.

That their work might be used for such purposes was anticipated by the botanical artists themselves. Pierre Vallet, for example, stated explicitly in his florilegium *Le Jardin du Roy* (no. 8) that his engravings could serve 'à l'usage de ceux qui voudraient peindre ou enluminer', while Crispijn van de Passe declared in the title-page to the edition of 1614 of his *Hortus Floridus* his intention to describe the colours of each flower shown in the plates that followed.

The floral still-life genre spread to many European countries—first Germany, then France, Italy and Spain. Its originators, however, were certainly the great Flemish and Dutch still-life artists. Indeed, according to Erwin Panofsky the profound sense of realism that imbues these works is a reflection of the quintessentially Dutch conception of the world, 'which claimed that the quality of reality belongs exclusively to the particular things directly perceived by the senses' (Panofsky, 1.8). In the Dutch Golden Age many cities became centres of intense artistic activity, where the new genre flourished. In Middelburg, Delft, Utrecht, The Hague, Amsterdam and, further south, in

Antwerp, dozens of floral painters sprang up, establishing studios to which the nobility and wealthy burghers flocked to buy pictures with which to adorn the walls of their comfortable homes.

Thus a flourishing market for paintings of flowers was sustained in this period, not only by wealthy art collectors but by floriculturists and scientists too. For example, it is known that the horticulturist Matteo Caccini of Florence (1573–1640), in a letter to the Prince-Count Charles d'Aremberg (an amateur botanist who maintained a frequent and friendly correspondence with Carolus Clusius), inquired as to how much it would cost to commission a painting depicting one hundred flowers. Aremberg responded that an excellent painter of his acquaintance asked one hundred gold *ecus* if the flowers were to be depicted as realistically as possible, but only fifty *ecus* if a less high standard of execution was acceptable (Masson 1970, p. 104).

By the mid-sixteenth century, painters in the Low Countries were depicting flower bouquets arranged in simple glass vases. According to the treatise *Het Schilderboek* (1604) by the artist and writer Karel van Mander, the very first Dutch floral still-lifes were painted by Loedwijck van den Bosch and Pauwel van Aelst (Schneider, p. 136). It was only at the beginning of the seventeenth century, however, that the formal aspects of the genre came to be codified, and that features such as butterflies and other insects—shown hovering around the bouquet or balancing lightly on the flowers—became standard elements in these compositions.

Following van den Bosch and van Aelst, whose works have unfortunately been completely dispersed, other artists from Antwerp, including Jacob de Gheyn II (1565–1629) and Ambrosius Bosschaert began to paint small floral pictures on panels or on copper. In them the flowers—usually indigenous species interspersed with a few ornamental flowers—were arranged around a central axis in carefully balanced masses. By the third decade of the seventeenth century exotic species came to predominate in these bouquets, and as a result the genre acquired a new and imposing, indeed an almost monumental quality. The simple, self-contained bouquets of the early still-life painters were gradually replaced by more sophisticated compositions made up of a wide variety of flowers. Often sprays of blossoms freed themselves from the central mass of the composition and were shown extending in various directions, lending a much more dynamic quality to the works. Earlier painters had scrupulously depicted each flower in its natural dimensions, but now artists did not hesitate to depict their magnificent blooms even larger than life. Particularly celebrated for his opulent cascades of flowers was the Dutch artist Jan van Huysum (1682–1749), who was described by his contemporaries as the 'Phoenix of flower painters'.

Next to the classical bouquet, then, with its conventional vertical composition, appeared paintings of baskets spilling over with flowers, flowers shown lying on tables or consoles, and other horizontal arrangements, as well as elaborate wreaths and garlands. Osias Beert the elder (*c.* 1580–1623) specialized in paintings of baskets of flowers, while Balthasar van der Ast (1593/4–1657) was well known both for his baskets and for his paintings of flowers shown informally heaped on tables and other surfaces. Wreaths and garlands, particularly in association with religious themes, were the trademark of Jan 'Velvet' Brueghel (1568–1625) and, especially, of the Jesuit painter Daniel Seghers (1590–1661). The garland, in fact, was often used by these artists to frame votive images, generally of the Virgin Mary with the Christ-Child, or various saints or of scenes of the Eucharist (Bergström 1956; Freedberg 1981).

In this way the vase of flowers (usually containing white lilies, irises and columbines) so familiar in the medieval and Renaissance periods from paintings of the Annunciation, had by the seventeenth century shed its Marian symbolism and become a purely secular subject. At this point, however, the flower was incorporated as a symbol in yet another process of moralizing—this time on the transience of life itself. This was entirely appropriate for, as the moralists of the period liked to point out, both the Bible and classical writings contained numerous references to the ephemeral beauty of the flower. Psalm 103 (15–16) reminds us that 'As for man, his days are as grass: as a flower of the field, so he flourisheth. For the wind passeth over it, and it is gone', while the Greek root of the word *anemone* means 'flower of the wind' (*Flowers and Nature*, p. 31). This type of symbolism was already being utilized by artists such as Joris Hoefnagel (1542–1601) of Antwerp in the 1580s. The flower also came to be associated with funereal symbology, and thus made its appearance in the still-lifes known as *vanitas* next to objects such as the lighted candle, the hour-glass, the soap bubble and the skull, symbols of the inexorable passage of time and the tragically ephemeral nature of human life (see Veca 1981).

The collection of still-life paintings at Oak Spring offers a significant overview of this fascinating genre, beginning with a small, very early painting by Ambrosius Bosschaert (dated 1606; no. 22) that depicts a nosegay of flowers in a *roemer* (a commonplace German drinking-glass), an arrangement that was to become a recurrent motif in later works by other Dutch artists. Another painting, this time by Johannes van Fornenburgh (no. 23), shows some flowers heaped casually on a stone surface, in a work that can be read as a symbolical sequence. The withered rose, the drop of water and the crumbling stone all remind us of the fleeting nature of life, while the fluttering butterfly alludes to the immortality of the human soul.

The transparent luminosity that characterizes the great masterpieces of Dutch still-life painting infuses two works by the artist Daniel Seghers (nos. 24, 25). In each, a vase containing a small posy of flowers glows against a dark background. These paintings offer us a rare example of secular works by an artist best known for the floral garlands he painted to frame devotional scenes.

Also in the collection at Oak Spring is an important work by Jan van Kessel—a series of seventeen paintings of flowers and insects on copper that were perhaps intended to be used to decorate the doors and drawer fronts of a marquetry cabinet of curiosities. Each of the panels at Oak Spring (see no. 26) is characterized by an elegant *mis-en-page* reminiscent of the work of Joris Hoefnagel, and by the artist's extraordinarily accomplished portrayal of various flowers and insects, each of which is shown life-size and in almost microscopic detail.

The painting of a *Lychnis chalcedonica* and some insects by Herman Henstenburgh (no. 27) per-

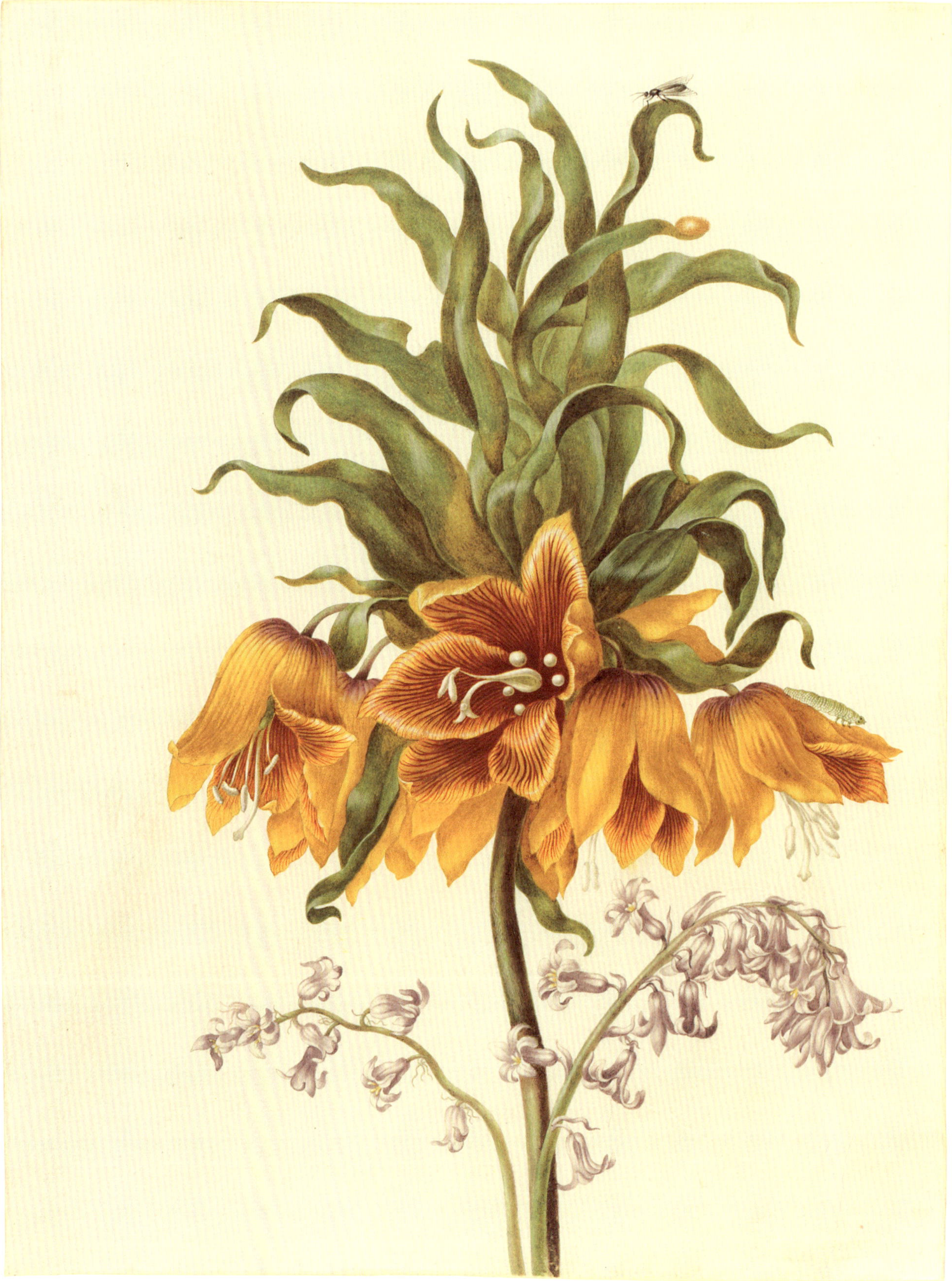

HERMAN HENSTENBURGH, Crown imperial (*Fritillaria imperialis*), two bluebells (*Scilla non-scripta*), and insects. Maria Sibylla Merian's copy after Henstenburgh of the crown imperial is in St Petersburg's Academy of Sciences mits us to appreciate the technical virtuosity of one of the most celebrated Dutch painters of the floral still-life. He was active at the end of the seventeenth century and the beginning of the eighteenth, when flower painting had assumed an almost Baroque grandeur, but this work—executed on vellum rather than on canvas—has a more intimate quality, the artist seeking merely to portray his subject-matter as faithfully as possible. Nevertheless, the elegant composition and rich colours remind us that this is the work of a master in the sophisticated style of the late seventeenth century.

22. AMBROSIUS BOSSCHAERT *the elder* (1573–1621)

Flowers in a glass vase.
20.5 × 17.5 cm. Oil on copper. Signed with the monogram 16[monogram]06, and dated 1606. Gilt frame, decorated at the four corners with a fretwork motif sprinkled with tiny flowers. 'Guilloché' style picture frame (north European, 17th century), 19th-century copy, made in the United States.

PROVENANCE: Hector Braume, 1965; Mrs Edward F. Hutton, Westbury, Long Island, New York.

EXHIBITED: London, Hallsborough Gallery, May–June 1964.

REFERENCE: Mitchell 1973, p. 54.

IN A SIMPLE, cylindrical glass vase placed on a light coloured surface against a dark background is a bouquet of flowers comprising two roses, two variegated tulips, a hyacinth, a violet, a forget-me-not, a columbine, a primrose, a lily of the valley and a marigold. A *Lasiommata maera* ('large wall-brown' butterfly) balances lightly on a white rose, while on the ground rest a sprig of forget-me-not and an insect, a common housefly. This elegant work was painted by Bosschaert the elder, one of the first artists to specialize in flower painting. His innovations were to contribute to the enormous success of the genre, which began at the end of the sixteenth century in the Low Countries and later spread across Europe.

In this period the flowers most often depicted were those in vogue among connoisseurs, unusual species that stimulated the acquisitive mania of collectors and came to form one of the underpinnings of the Dutch republic's economic prosperity (Stearn 1965, Veca 1982). The bulbous species were particularly prized, with the narcissus, lily, fritillary, iris and tulip attracting vast numbers of fanciers, who not only cultivated them in their gardens but bought pictures that captured their transient beauty in more permanent form.

Bosschaert was born in Antwerp in 1573, although soon after his family moved to Middelburg, capital of the flourishing province of Zeeland. Within the walls of this city could be found resplendent gardens, of which mention is made in the works of Jacob Cats (1577–1660), the greatest Dutch poet of the first half of the seventeenth century (Bol, pp. 15–16). The celebrated botanist Matthias de L'Obel also lived there from 1584 to 1596. His works include a scientific treatise on flowers—finally deemed a subject worthy of serious study. Bosschaert was a member of the Guild of St Luke until 1588; in 1604 he married Maria van der Ast, whose brother Balthasar was an artist who specialized in floral compositions and still-lifes. Bosschaert's three sons, Ambrosius the younger, Abraham and Johannes, all became successful painters in the same genre that their father and uncle practiced. Bosschaert was not only a talented painter, working in partnership with Jan 'Velvet' Brueghel (whose sobriquet alludes to his ability at capturing the velvety soft texture of a flower petal on canvas), for he also maintained a flourishing trade as an art dealer. Documents show, for example, that in 1612 he sent a substantial shipment of paintings by sea to England. Some scholars have suggested that this profitable activity may have distracted Bosschaert from concentrating on his art, for his total output does not exceed some fifty works, almost all of them produced during the latter part of his life.

It is not known exactly when Bosschaert began painting floral compositions. His earliest work seems to be a small still-life of a plate of fruit (now in a private collection in Sweden), signed with his characteristic monogram and dateable to *c.* 1600 (Bergström 1956, p. 57). Two known floral paintings by his hand date from 1606, one in the Cleveland Museum of Art (Stechow, p. 60) and this work at Oak Spring.

Beginning in 1607 and continuing up to his death in 1621, Bosschaert produced a series of floral paintings whose apparent simplicity belie a masterly, and indeed highly innovative, sense of design and composition. His earliest works were somewhat restricted in inspiration, each composition being made up of a limited variety of blooms arranged in a small vase or *roemer*. After 1615, when Bosschaert had abandoned Middelburg, first for Bergen-op-Zoom in northern Brabant and then for Utrecht, where he lived from 1616 to 1618, his floral repertory became much richer, and the containers holding the flowers ever more ornate. The flowers themselves were now set in shadowy niches, from which they emerged in a blaze of vivid colour. Even the zoological universe that had always formed a part of his work expanded to include butterflies, caterpillars, flies, coleopterans, grasshoppers as well as shells, often bizarre in shape and colour (Hairs, pp. 206–11).

Sometime before the end of August 1619 Bosschaert moved to Breda, where he devoted the last years of his life to his painting. It was, in fact, during this period that he produced his largest and most complex works, and introduced an important stylistic innovation: his flower arrangements were still set in niches, but these now opened onto spacious and airy landscapes stretching away to azure horizons.

The small, precisely drawn bouquet at Oak Spring is a relatively early, but already stylistically mature, painting, and thus constitutes a key work in Bosschaert's *œuvre*. To the freshness of its inspiration and the almost scientific accuracy of its execution is united the artist's great compositional skill. The components of Bosschaert's botanical microcosm are interwoven in a subtle play of light and dark: branches and leaves are cunningly concealed in the shadows, while the blooms stand out tellingly in the foreground, culminating in the most prized flower of the period, a tulip with

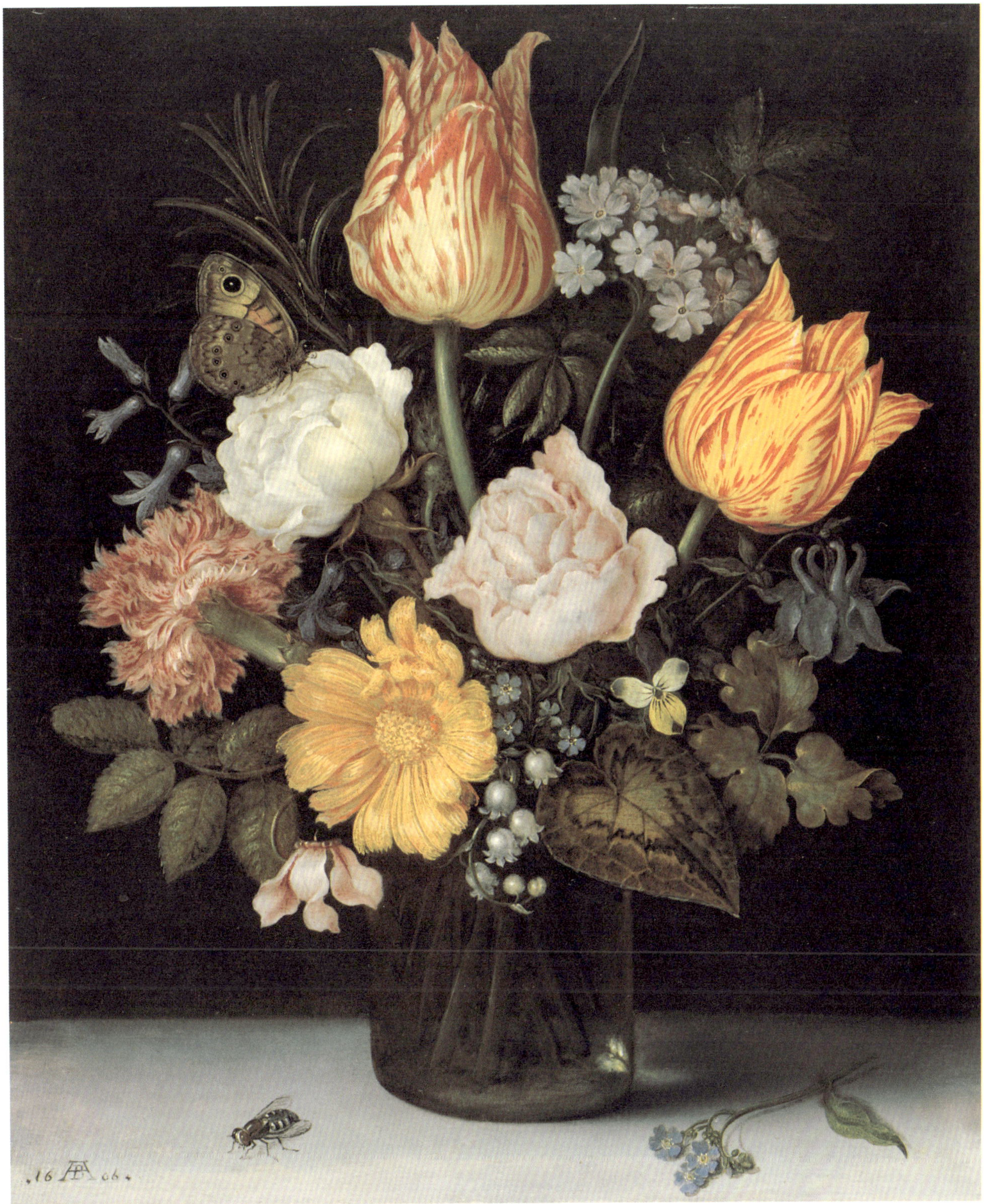

AMBROSIUS BOSSCHAERT the elder, Flowers in a glass vase

pointed, white and red variegated petals. The colours lie smooth, dense and luminous on the copper plate; the hues of the flowers are vivid, yet delicate, while on the glass can be seen a brilliant flash of reflected light. The full-blown rose, the crisply pleated leaves of the marigold and the cyclamen with its stem curving around and back on itself in a complete loop, in a characteristic insignia often repeated in later works, delight the eye, contributing to the charm of this work by one of the most sophisticated floral artists of his time.

23. Johannes Baptiste van Fornenburgh (1585–?1649)

Still-life with flowers, insects and a lizard.

30 × 36 cm. Oil on wood. Signed on the left 'IBF'.

PROVENANCE: Kunsthandel G. Cramer, The Hague.

REFERENCE: Bol, p. 89, no. 5.

On a balustrade of chipped and flaking stone—a recurrent image in the late works of Johannes van Fornenburgh—is a charming bouquet composed of a rose, a tulip, an anemone, a columbine and a sprig of what appears to be yellow jasmine. Next to the flowers a lizard basks, while a fly and a butterfly (*Vanessa atalanta*) flit in the still air overhead. As in many of his other still-lifes, the artist has painted some drops of water on the balustrade. They glisten iridescently; one seems about to run off the edge.

It is known that van Fornenburgh was living in The Hague from 1608 (the date that appears on the earliest known work by him) until 1621. His first paintings—carefully composed bouquets placed in niches and surrounded by insects and other small animals—clearly reflect the influence of the artist Jacob de Gheyn, a friend and close collaborator of the director of the botanical garden at Leiden, Carolus Clusius. In 1620 van Fornenburgh's work entered a new phase, becoming more complex and sophisticated as the artist fell under the influence of the great floral painter Ambrosius Bosschaert (no. 22); indeed, so similar in style was his work to that of Bosschaert in this period that his monogram 'I.B.F.' has sometimes been misinterpreted as being that of the son of Bosschaert (i.e., 'Johannes Bosschaert Fecit'; see Bergström 1956, p. 84; Mitchell 1973, pp. 114–15).

In general, van Fornenburgh's flower paintings consist of modest variations on a compositional formula (Gammelbo, pp. 7–8; Veca 1982, pp. 174–5). The painting at Oak Spring seems to be the work of a moment of more than ordinary inspiration, however, a spontaneous portrayal of nature captured *ad vivum*. It is possible that, like many still-lifes from the seventeenth century, van Fornenburgh's picture was also intended to be read on a symbolic level. Thus, the lizard might represent an allusion to the merciless voracity of time and the dewdrops a reminder of the brevity and fragility of

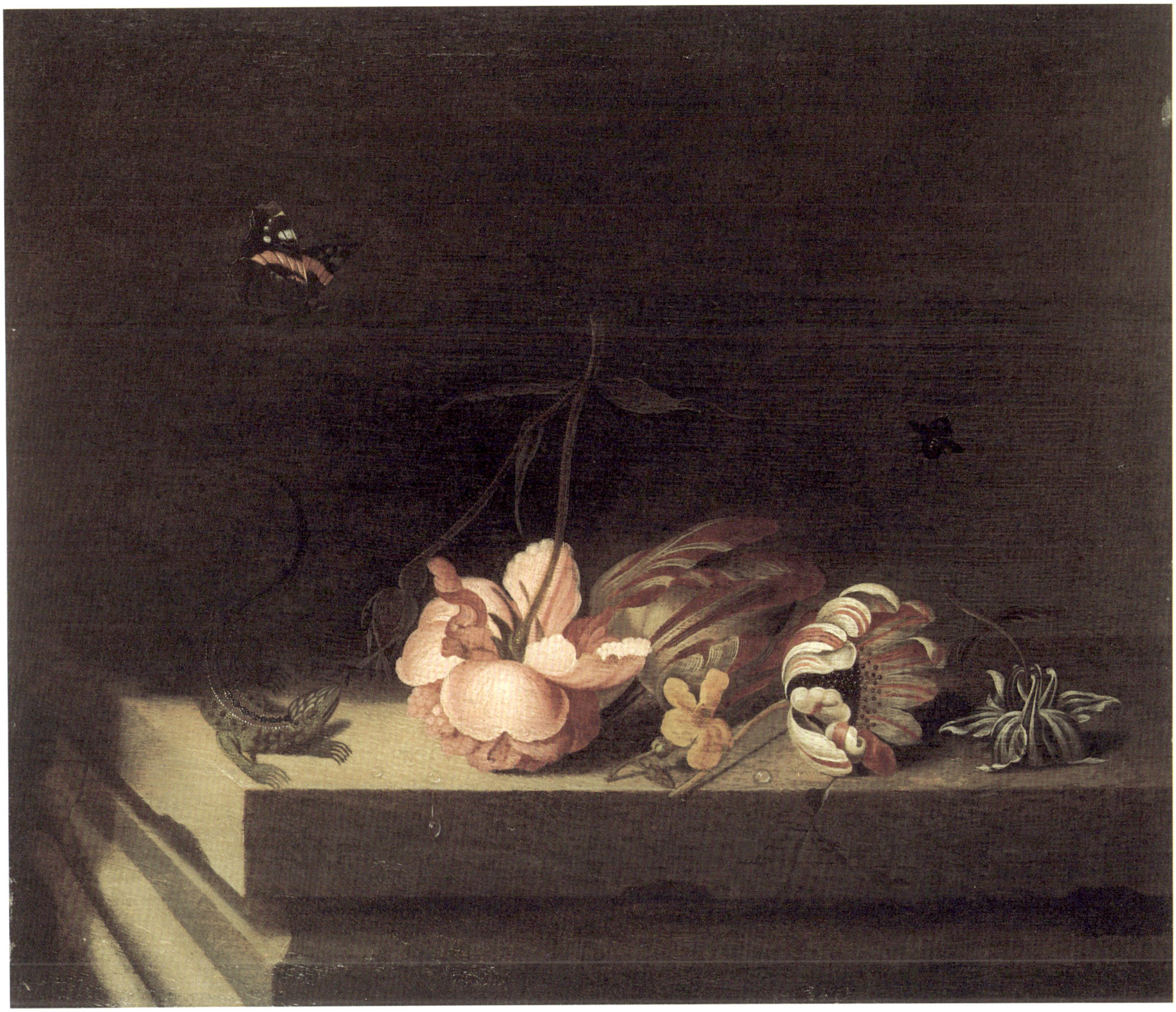

life, while the butterfly symbolizes the resurrection of the immortal soul. It is just as possible that no allegorical message was consciously intended, however, for van Fornenburgh may simply have been availing himself of images drawn at random from a repertory then in current use among painters of still-lifes and flowers.

JOHANNES BAPTISTE VAN FORNENBURGH, Still-life with flowers, insects, and a lizard

24. Daniel Seghers (1590–1661)

Flower bouquet in a glass vase with a butterfly.

44.1 × 31.5 cm. Oil on copper. Signed bottom right: 'D. Seghers Soc. tis Jesus'. Written on the left in a modern hand is 'N. 106'.

REFERENCES: Hairs, p. 134; Mitchell 1973, p. 225, no. 320; Veca 1982, pp. 170–71, no. 193.

Daniel Seghers was born in 1590 in Antwerp, where his father was a well-to-do silk merchant. He spent his childhood further north, in Holland, and received as a consequence a Protestant education. At the age of ten he already showed signs of a precocious artistic talent. Returning in 1610 to his native city, he became a student of Jan 'Velvet' Brueghel, who introduced him to the genre of flower painting and encouraged him to embrace the Catholic faith. Seghers joined the Jesuit Order as a lay brother in 1614 and took his vows in 1625. He continued to paint thereafter, but exclusively for his Order, decorating churches and other religious buildings, and painting works that were often presented as gifts to important visitors.

Seghers also spent time in Mechelen, Brussels, Rome and Ghent. His sojourn in Rome (1625–7) must have constituted a particularly stimulating experience, for in that period the interest in the cultivation of flowers and in flower painting was at its height there, and he could have met key figures such as Tobia Aldini (see no. 28) and his fellow Jesuit G. B. Ferrari (see no. 29). A sixteenth-century source informs us that 'il travailla incessament à dessineigner ce qu'il y avait de plus exquis dans les Palais et les Jardins' (Hairs, p. 120). In 1628 Seghers settled definitively in Antwerp, taking quarters in the house of the Jesuit Order, where he was visited by various illustrious personages, among them Charles II of England and Peter Paul Rubens, collaborating with the latter on the *Virgin Surrounded by Angels and Flowers* for the Church of St Ignatius in Antwerp. He was, by all contemporary accounts, a simple, modest and kindly man. He died in 1661 and was buried in the Jesuit chapel of the Church of S. Carlo Borromeo, Antwerp.

Seghers's considerable *œuvre* consists almost exclusively of paintings of bouquets and wreaths, in which the form and colour of the flowers are rendered with a skill that clearly reflects the influence of his master, Jan 'Velvet' Brueghel. Similar floral motifs could be found in the florilegia of the period, but their purpose was essentially decorative or scientific, while Nicolas Robert went so far as to associate these fragile emblems with gallant connotations in his celebrated *Guirlande de Julie*. For Seghers, on the contrary, every work was an act of devotion, the beauty of the floral emblems an echo of the glory of God. For example, around a cartouche, niche or small altar containing an image of the Virgin and Child, a sacred scene or the effigy of a saint painted in *grisaille*, generally by another artist, Seghers would construct elaborate festoons with dense symmetrical masses of blooms, from which long branches might project, sharply foreshortened, towards the observer.

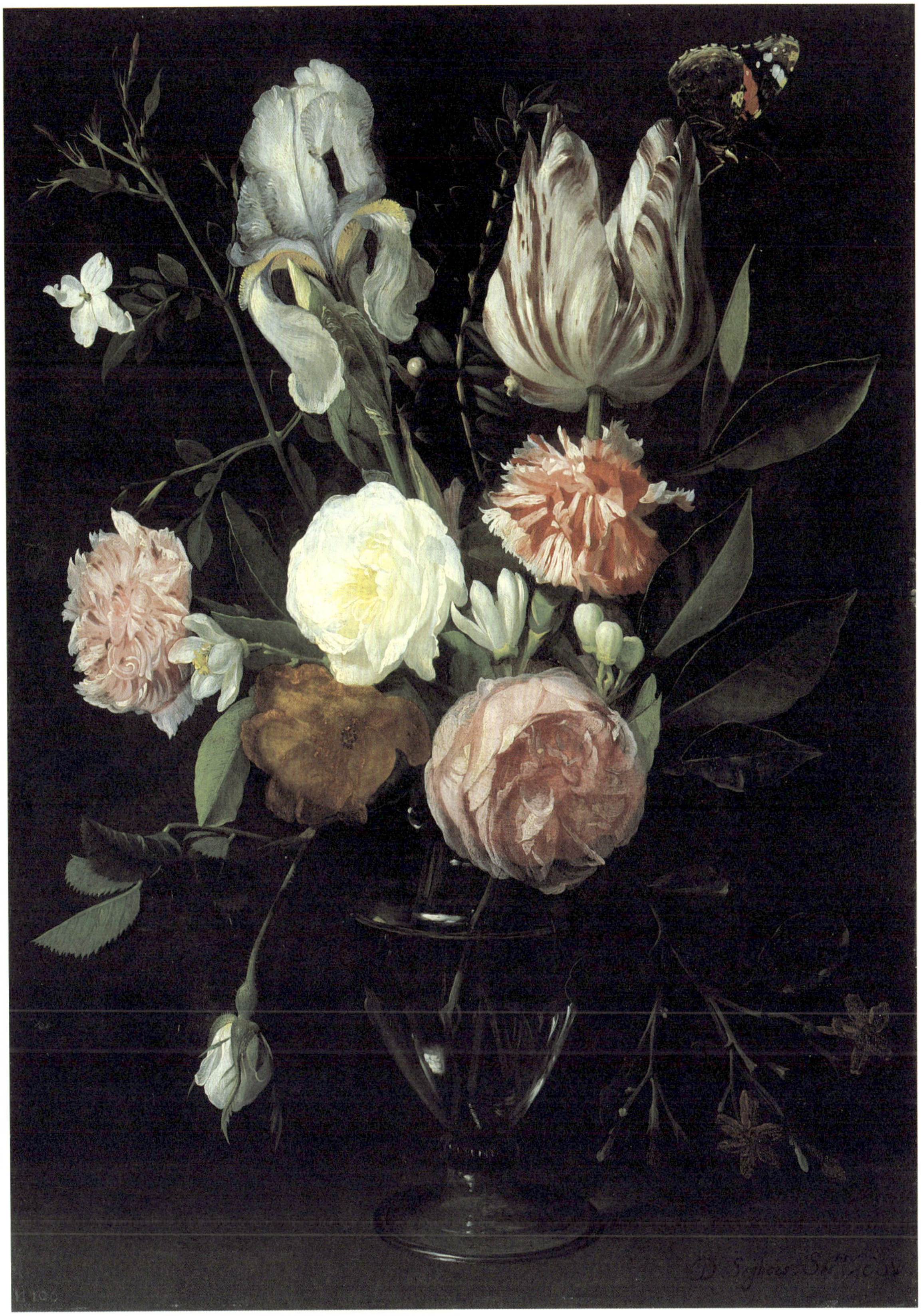

DANIEL SEGHERS,
Flower bouquet in a glass vase with a butterfly

These intricate garlands became so popular that they were also used by other artists to frame portraits and non-religious scenes.

In his paintings of floral bouquets, however, Seghers tended to eschew the sumptuous compositions favoured by Brueghel for more modest arrangements that relied for their effect on the play of light and shadow rather than on extravagant display. The painting at Oak Spring is a fine example of his sober, refined style. Against a background of dark blue, gleams an iridescent glass vase containing an iris, a tulip, two carnations, some orange blossom, four roses, a sprig of jasmine and what appears to be a *Caryophyllacea*. A variegated Red Admiral butterfly (*Vanessa atalanta*) with trembling wings is lightly poised on the tulip. The skill with which the differing tactile qualities of the tulip and the rose have been rendered are quite reminiscent of a similar painting by Seghers (signed and dated 1652) in a private collection in Belgium (Veca 1982, no. 193).

25. Daniel Seghers (1590–1661)

Flower bouquet in a glass vase.

25.8 × 18.2 cm. Oil on copper. Signed lower left: 'DS. Soc. tis Jesu'.

THIS ELEGANT PAINTING is another fine example of the work of the floral artist Daniel Seghers (see no. 24). Against a dark background the artist has set a glass vase containing a charming bouquet composed of a tulip, a sprig of orange blossom and three roses. With these few elements Seghers manages to produce a striking effect, however, with the flowers occupying nearly the entire space of the painting. A large white tulip striped with red and an overblown double rose fill the right-hand side, while on the left the light falls on another rose and the gleaming white of the orange blossom, creating a dynamic, yet balanced composition. Some of these same flowers also appear in another, more elaborate, painting by Seghers at Oak Spring. In fact, certain flowers seem to have served as leitmotifs for the artist, reappearing frequently but blooming afresh in each new context.

In this fairly late work, the vivid and dense colours of the artist's palette lend palpable life to the flowers. Unlike his master, Jan 'Velvet' Brueghel, who showed a preference for brilliant yellow and vermilion hues, Seghers's palette was dominated by softer pinks, violets and a luminous white, to which he would sometimes add a touch of ochre, as in the anemones that appear in both of the Oak Spring paintings (Hairs, p. 174).

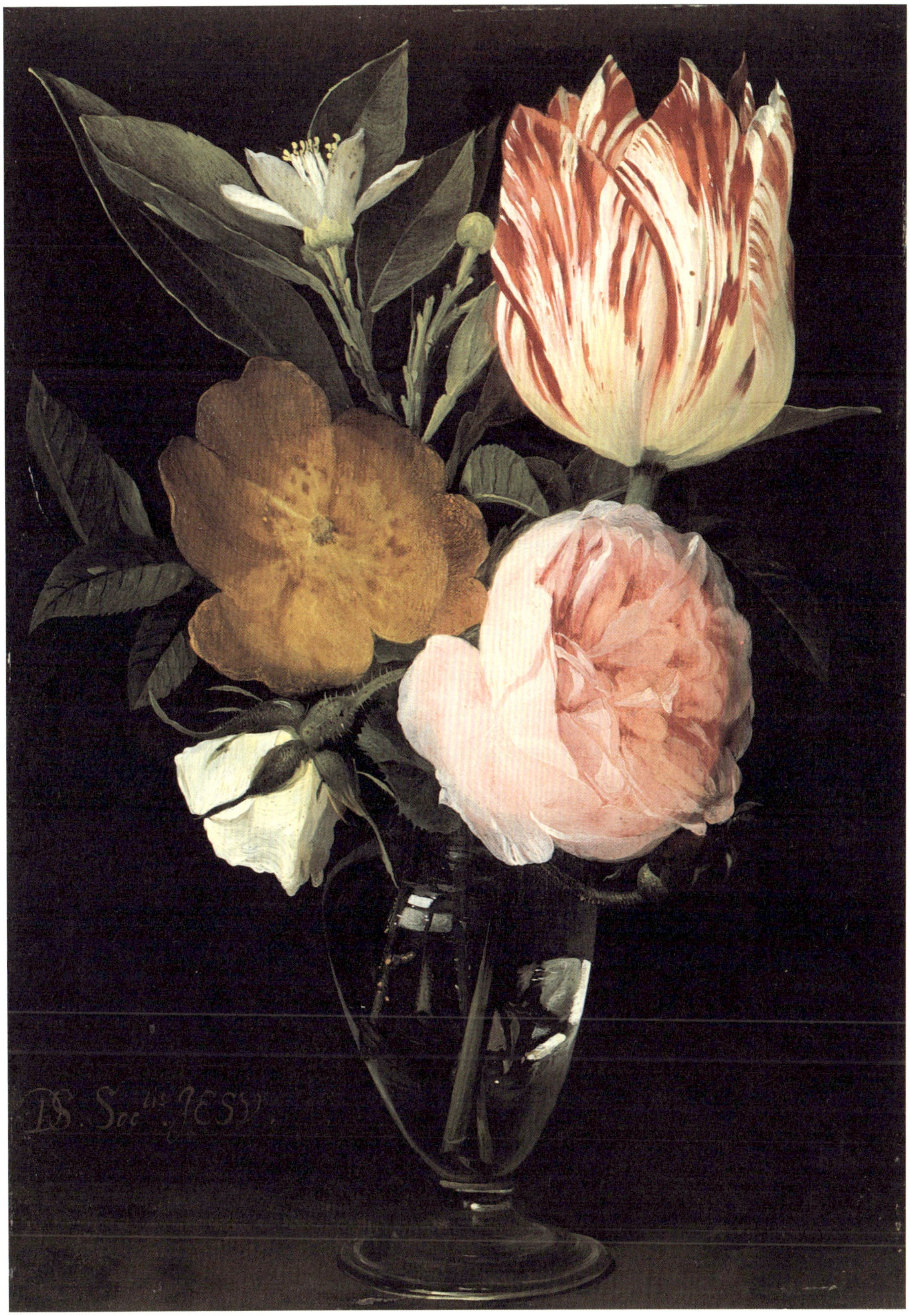

DANIEL SEGHERS,
Flower bouquet in a glass vase

26. Jan van Kessel (1626–1679)

Borage with insects.

13.2 × 18.7 cm. Oil on copper [1658]. Wooden and gilt frame.

PROVENANCE: 1923: London, Kunsthandel Gebr. Douwes; 1924: Van Valkenburg; 1935: Brussels, Fievez Firm; 1956: London, Hallsborough Gallery.

EXHIBITED: London, Hallsborough Gallery, May–June 1956.

REFERENCES: *The Connoisseur*, April 1956, p. 199; Ritterbush; *Flowers and Nature*, p. 209.

This work is one of a series of seventeen paintings at Oak Spring—all of which depict flowers, cut plants and insects *à trompe l'œil* against a light-grey background—by the Dutch artist Jan van Kessel. The artist's signature, with worms and insects forming the letters of his name, appears together with the date 1658 in one of the paintings. Van Kessel belonged to the Brueghel (or Bruegel) dynasty, for on his mother's side he was the grandson of Jan 'Velvet' Brueghel. One of his uncles was David Teniers the younger, while another, Jan Brueghel the younger, trained him as a painter. In 1635 van Kessel also began studying under Simon de Vos, a prolific artist working in Rubens's studio in Antwerp. At the age of nineteen, van Kessel's name was placed on the registry of the Academy of St Luke in Antwerp as the son of the painter Hieronymus van Kessel and as a painter of floral subjects. In 1644 he became a member, as an autonomous master, of the guild of his city and a captain of the civil guard. He had many children, two of whom—Ferdinand and Jan—also became artists.

Van Kessel's output consists mainly of small paintings in oil on copper or wood, and tempera paintings on vellum; he produced only a few works of slightly larger dimensions on canvas. All his works, however, demonstrate the same masterly confidence and technical proficiency, characterized by colours applied in thick, smooth fields against a grey background that emphasizes their brilliant hues. Only his signature varies from picture to picture, the artist utilizing in turn cursive, roman or 'insect' letters as the humour took him. His preferred subjects were animals (especially birds and insects) and flowers, generally in the form of bouquets arranged in vases or wreaths surrounding portraits executed by other artists. He also painted a small number of formal still-lifes composed of flowers, fruits, vegetables, game and drinking vessels. Other works are imbued with emblematic or symbolic meaning, reflecting the influence of 'Velvet' Brueghel; thus we can find paintings depicting Noah's ark, the Five Senses, the Four Elements and the Four Parts of the World. On this last theme there is a group of forty paintings on copper held in the Prado, Madrid.

Van Kessel's style was marked by an almost scientific approach to his subject-matter, in which he strove to create an illusionistic rendition of reality, perfect down to the last detail. In his best paintings he was clearly working *d'après nature*, and in fact we know from various contemporary biographers that he was in the habit of working outdoors in clement weather. His paintings were so much in demand—the King of Spain and the Spanish governors in the Low Countries were admir-

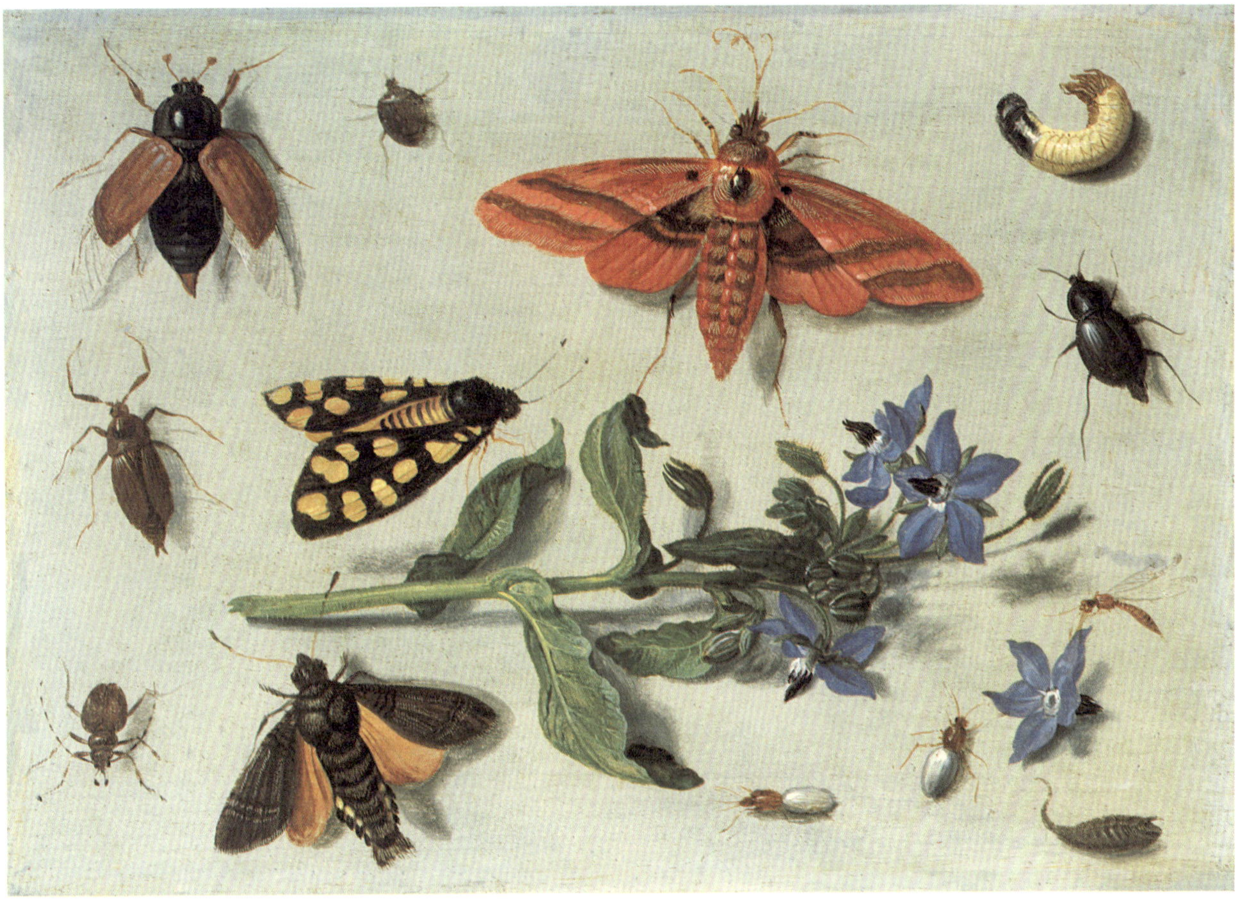

JAN VAN KESSEL,
Talewort (*Borago officinalis*)
with insects

ers of his work—that the artist had to rely on the assistance of students and collaborators, such as his son Ferdinand and the artists Erasmus Quellinus and Willem van Herp (Krempel), to produce them in sufficient numbers.

The seventeen naturalistic paintings at Oak Spring, probably produced to decorate a piece of furniture, represent an unusually fine example of van Kessel's work. An English cabinet of curiosities (*c.* 1680–90), now in the Smithsonian Institution, Washington, D.C., contains eleven panels (surely based on the Oak Spring paintings) that were wrongly assumed to be by van Kessel (Ritterbush, pp. 572–5). A similar, but somewhat less carefully executed, series of seven paintings, dated 1661, is in the Fitzwilliam Museum, Cambridge; and another three from the same year are conserved in the Uffizi, Florence. Finally, in the private collection of David M. Koester in Zurich are two works that may be linked stylistically to one of the paintings in the Oak Spring series: the first depicts a branch of redcurrant with various insects and beetles, and the second bears the date 1657 and the artist's signature formed of spiders, snakes and caterpillars (*Bruegel. Une dynastie de peintres*, p. 278).

Oak Spring's seventeen paintings constitute an undivisible *unicum* that, if studied together, allow one to appreciate fully van Kessel's refined style. The larger, horizontal panel (39 × 55 cm.) of the group must have been intended for the centre of a table-top, while the sixteen smaller works would then surround it, rather than decorate the vertical doors of a cabinet, say, or a chest of drawers. The small panel depicting a sprig of flowering borage surrounded by various insects is particularly admirable. The insects are shown in exquisite detail, and one can immediately recognize (examining the painting from left to right, and from top to bottom) a cockchafer (*Melolontha melolontha*), a five-sided stink-bug, an elephant hawkmoth (*Deilephila porcellus*), the larva of a beetle, a waterbug (*Nepa* sp.), a cream-spotted tiger-moth (*Arctia villica*), a species of darkling beetle, a long-horned beetle, a hummingbird hawkmoth (*Macroglossa stellatarum*), two dysderid spiders, a hymenopter, what appears to be an Ichneumon fly, and, finally, in the right corner, the larva of a syrphid, perhaps the flower fly (*Eristalis* sp.).

Another painting by the same artist found at Oak Spring is a small study of a sprig of rosemary and insects, dated 1653 (*Flowers and Nature*, p. 209). The great legacy of small panels wrought in van Kessel's style is testified to by an oil panel identified as 'German School, 17th century: Still-life with a Rose, Butterflies and Beetles', an oil panel of insects and a blue flower, signed 'f.v.k.'—the artist's son, Ferdinand van Kessel (1648–96)—and an unsigned study on vellum of insects and reptiles, all in the collection at Oak Spring.

27. Herman Henstenburgh (1667–1726)

Lychnis chalcedonica, aconite and insects. 37.4 × 30 cm. Water- and bodycolour on vellum.

Herman henstenburgh's career followed a somewhat unusual course, for he exercised his artistic talent both as a pastry-chef and as a painter of flowers, birds, landscapes and still-lifes in oils, body- and watercolours. Indeed, like painting, the creation of fine pastries required the dextrous mixing of ingredients and colours to create works that were a feast for the eye, however ephemeral. It was another artist-confectioner, Johannes Bronkhorst (1648–1726), who instructed him in these two subtle arts, and Henstenburgh continued to work as a pastry-chef until he had established his reputation as an artist and could dedicate himself exclusively to painting. His son, Antoine, also became a painter of insects, birds and flowers in their native city of Hoorn.

Henstenburgh's early work included a pair of portraits of the artist Nicolaas Verkolje and his wife, today preserved in a private collection in The Netherlands. In one, Henstenburgh is shown, brush in hand, presenting a painting (B.J.A. Renckens, 'Portretten van Herman Henstenburg en Zijn Vrouw Door Nicolaas Verkolje', *Oud-Holland*, LXIX/i–iv, 1954, pp. 57–8). Many other works by this prolific and eclectic artist have recently appeared at auction.

In the seventeenth century some of the greatest practitioners of the art of floral painting were

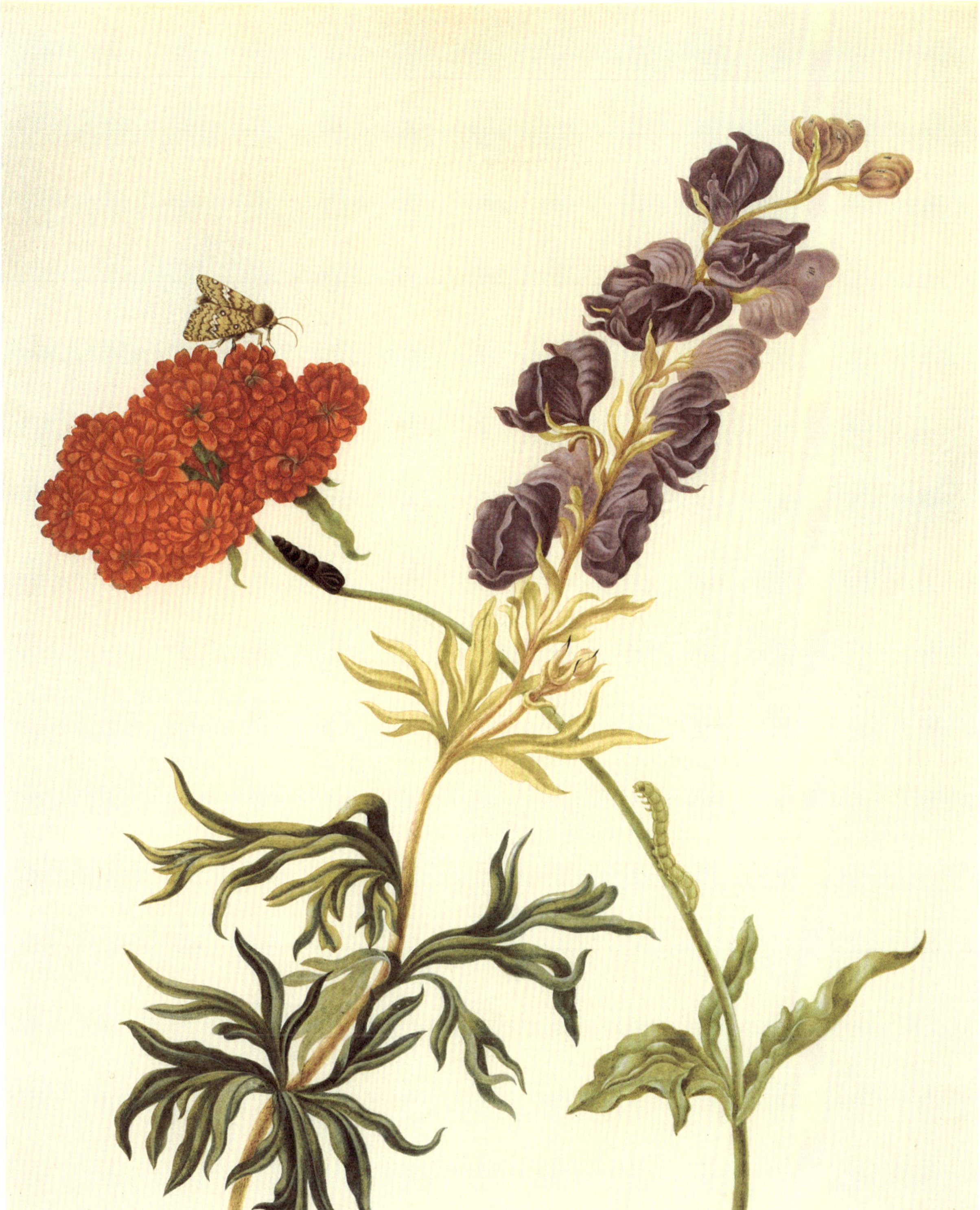

HERMAN HENSTENBURGH, Maltese cross (*Lychnis chalcedonica*), possibly a helmet flower (*Aconitum napellus*), and insects working in Holland, providing Henstenburgh with many models and sources of inspiration. However, no less a critic than Wilfrid Blunt declared one painting of three tulips by Henstenburgh, today conserved in Teylers Museum, Haarlem, to be one of the finest examples of Dutch flower painting ever produced (see Blunt and Stearn, p. 131 and pl. 36). A painting of a luxuriant *Fritillaria imperialis* in the Fitzwilliam Museum, Cambridge (PD, 644, 1973) confirms Henstenburgh's ability as a botanical painter, while three more complex works, *A Stone Bust of Flora, Garlanded with Flowers* (*Flowers of Ten Centuries*, pp. 31–2), *A Bunch of Flowers Hanging on a Dressed Stone* and *A Terracotta Vase with Flowers* (*Dutch and Flemish Artists*, exh. cat., Cambridge, Fitzwilliam Museum, 1980, nos. 17–18) demonstrate his impressive stature as a painter of floral still-lifes. Two very fine bodycolour paintings (both bearing the artist's initials), one depicting a dead Bird of Paradise (in the Uffizi's Gabinetto di Disegni e Stampe, 17896F; see Ciardi and Tongiorgi Tomasi, p. 119) and the other (Florence, Galleria Palatina) a brilliantly coloured exotic bird of unidentifiable species (*Natura Viva in Casa Medici*, p. 134), show that Henstenburgh was equally skilled as a painter of ornithological subjects.

At Oak Spring are seven floral paintings in bodycolour on vellum that permit us to study Henstenburgh's refined technique and compositional sense, and which also reveal a strong stylistic relationship with some of the watercolours by Maria Sibylla Merian preserved at St Petersburg (see no. 101). In one painting, two irises placed at divergent angles form an elegant frame for a delicate tuberose, while in the foreground the artist has placed a small sprig of muscari (grape hyacinth). In another painting a spray of aconite with violet flowers crosses the gracefully curving stem of a *Lychnis chalcedonica*, a plant brought to Europe from Asia Minor in the sixteenth century. A fragile butterfly, perhaps a *Catocala*, rests on the flower of the *Lychnis*, while a green caterpillar (an *Hesperiidae*?) climbs its stem, to which is also attached a cocoon.

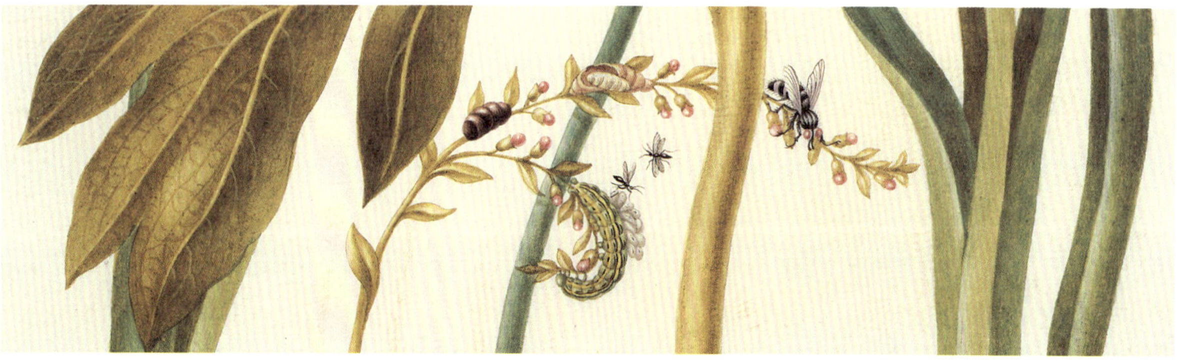

HERMAN HENSTENBURGH, Detail from one of the works on vellum showing insects amid a sprig of flowers

Henstenburgh's exceptional skill in the depiction of insects can also be seen in another drawing, this time of a peony and several stems of narcissi. In the tiny space surrounding one small sprig of flowers the artist has managed to include no less than three insects, plus a butterfly (perhaps a *Pieride*) shown in various phases of its life-cycle. He has captured with microscopic accuracy the fine hairs of the insects and the transparent scales of their wings, infusing these tiny creatures with palpable life.

IV · THE FLOWER-GARDEN AND THE RISE OF THE 'JARDINIER FLEURISTE'

THE VERITABLE PASSION for the cultivation of flowers that arose at the end of the sixteenth century and quickly swept across Europe resulted in the emergence of the flower-garden as a distinct entity, with a typology all its own. Coupled with this phenomenon was the appearance of the *jardinier fleuriste*—a connoisseur, collector and the solicitous custodian of the flower-garden.

During the second half of the sixteenth century many botanical gardens were established, first in Italy and then elsewhere in Europe, in which plants and simples were cultivated for scientific study and for medicinal purposes. Flowering plants were rarely grown, and then only unusual species considered to be collectors' items. Gradually, however, an area specifically devoted to the cultivation of flowers came to be set off from the rest of the garden. Initially designated as a *hortus coronarius* (garden of garland flowers), such sites did not differ greatly from the gardens of the medieval period, being planted with familiar annual and perennial herbaceous plants and a few bulbous plants (Segre). Interest in flowering plants grew rapidly, however, and the seventeenth-century garden was given over almost entirely to their cultivation. Beginning in Italy, these gardens became marks of prestige among the great princes, noblemen and ecclesiastics, who vied with one another to cultivate ever more rare and beautiful species in their vast *jardins d'agréments*. Their example was quickly followed by pharmacists, botanists, amateurs and *jardiniers fleuristes*, who took up on a smaller scale the hobby of cultivating rare flowers. The collections of flowers that virtuosi succeeded in establishing in their gardens served as a live pendant to the permanent collections of dried plants and botanical paintings kept in or adjacent to their cabinets of curiosities. Indeed, the English gardener John Rea, in his *Flora* (no. 31), likened the flower-garden to a 'cabinet' in whose boxes were stored 'Nature's choicest jewels'.

This veritable vogue for flowers also gave rise to a flourishing commerce, in which an exacting clientele of noblemen, virtuosi, naturalists and floriculturists vied to secure rare specimens, for which they were prepared to pay enormous sums. Indeed, the Dominican friar Agostino del Ricco of Tuscany was moved to write disapprovingly in 1595 in his manuscript treatise on agriculture (now in the Biblioteca Nazionale, Florence), 'malvagi e disleali, e poco amorevoli giardinieri no' vogliono dare di queste cipolle se non si fa gran suono di piastre d'argento' (These wicked, dishonest, and detestable gardeners refuse to sell their bulbs if one does not make a great ringing noise with many silver coins) (Maiorino *et al.*, p. 284).

Between the enthusiasts of the period a lively correspondence sprang up in which news of unusual species, advice regarding their cultivation, and boxes of seeds and bulbs were exchanged. This correspondence allows us to retrace the path of many exotic plants that were introduced to Europe and cultivated in gardens in Florence, Bologna, Rome and elsewhere. For example, between the years 1606 and 1609 the floriculturist Matteo Caccini, the owner of a splendid garden in the heart of Florence, sent to the botanist Carolus Clusius in Leiden specimens of various day lilies, ranunculi, the newly discovered *Mirabilis jalapa*, and a tulip that was subsequently named *Tulipa clusiana*. Clusius in turn sent to Caccini many varieties of hyacinths, lilies, colchicums and narcissi, as well as a painting of a rare 'green tulip' (Ginori Conti). From Bologna the naturalist Ulisse Aldrovandi (1522–1605), a prolific author and founder of the city's botanical garden, kept up a busy correspondence and exchanged seeds, plants and botanical illustrations with naturalists across Europe (Frati).

One of the first great flower-gardens in Europe was constructed in Paris, under royal patronage, by the botanist Jean Robin. Tended by his son Vespasian, this garden was chiefly given over to the cultivation of exotic species. 'Là sont dedans des pots, les fleurs d'Amérique / Avec leurs escriteux, là les plantes d'Afrique / Et tout ce que le Turc cultive en ses jardins', as a short poem declaims in the preface to the catalogue (published in 1601) of the 14,000 species that could be found growing there. Of the most popular flowers of the period, this catalogue lists fifteen varieties of roses and a large number of bulbous plants, including nineteen varieties of ranunculi, twenty-two of irises, twenty-five of hyacinths, twenty-five of anemones, twenty-six of narcissi and nine species of tulip (Schnapper 1988, p. 41). Some of these plants were subsequently depicted in the very first printed florilegium, Pierre Vallet's *Jardin du Roy* (no. 8).

Rome outshone all the other cities of Europe, however, for the size and splendour of its gardens and for the vast array of rare and unusual plants that could be found there. Sir Thomas Hanmer (1612–78), author of a mid-seventeenth-century 'Garden Book' on plant cultivation, wrote admiringly of the many beautiful varieties of anemones that he saw on a visit to Rome. Particularly imposing were the gardens belonging to the Farnese family, which occupied a considerable area of the Palatine Hill and contained many unusual species brought back by priests from the most distant corners of the globe. The curators of the Farnese gardens, Tobia Aldini and the botanist Pietro Castelli, describe some of these exotic plants in *Exactissima descriptio rariorum plantarum* (no. 28), the illustrations for which may have been prepared by Castelli as well. G. B. Ferrari also extols the gardens of Rome in his *Flora* (no. 29), the most important gardening treatise written in the seventeenth century.

Indeed, innumerable treatises and books were produced in this period on the subject of gardening and the cultivation of flowers, and they provide us with precious information on the development of the Baroque garden. Ferrari's *Flora*, for example, opens with a discussion of the correct typology for the flower-garden. The authors of all of these works agreed that the beauty of the flower-garden did not depend on an impressive architecture or ornately shaped beds, but rather on the arrangement of the flowers themselves. A well-designed garden was distinguished by a harmonious matching of colours and forms, the skilled gardener being able to achieve such effects while taking into account the compatibility of the various species.

These seventeenth-century treatises also provided the reader with very detailed instructions on how to construct a flower-garden. To begin, one must select a suitable piece of ground, and surround it with a protective wall. Then one might perhaps lay out two orthogonal paths (ideally with a fountain in the centre), dividing the garden into four spacious areas in which beds can be established. These beds must not be too complicated in form, and each should be surrounded by a narrow path for the convenience of both the gardener and those who came to admire. Small stakes could be used to mark exactly where the bulbs or plants have been set. Almost all of the authors agreed that bricks or painted wooden blocks, rather than the traditional herbaceous border, should be used to mark off the beds, while hedges of evergreen were to be avoided because they competed with plants for precious nutrients.

In their treatises, Ferrari and Filippo Arena (no. 38) both declare the ideal form of the garden to be a regular polygon, while other authors, such as Agostino Mandirola (no. 30) and John Rea (no. 31), recommend a perfect square. Pierre Morin, a *jardinier fleuriste* who supplied plants to the celebrated gardener and collector John Tradescant, whimsically constructed his own garden in Paris in the form of a flower, with a central oval surrounded by a series of smaller beds in the shape of petals (no. 33). Many renowned floriculturists and collectors visited this garden, including Cardinal Francesco Barberini and John Evelyn, the latter describing Morin as 'the most skilled and curious person of France'.

Although every writer offered his own advice on how to arrange the plants in their beds and which flowers could be placed in close proximity, all reiterated the importance of following a rigorously symmetrical plan, without which, according to Paolo Clarici in *Istoria e coltura delle piante* (no. 36), 'non vi sarà mai vaghezza' (There will never be beauty there). Clarici provides at the beginning of his work a charming bird's-eye view of the garden of his wealthy patron Gherardo Sagredo at the Villa di Marocco near Venice. To assist the reader, many authors such as Ferrari, Rea and Arena provided illustrations of designs for flowerbeds borrowed from garden architects such as Hans Vredeman de Vries (1568) and Daniel Loris (1629). They further recommend that the gardener draw a careful plan of his garden's layout on paper or ask a competent architect to do so, for such diagrams, as well as careful inventories, were essential to keep track of the plants in an orderly garden. The allure of the flower-garden could be enhanced by placing urns of flowering plants at appropriate intervals, or by constructing—as both Clarici and the anonymous author of *Connois-*

sance et culture parfaite des tulippes (no. 34) suggest—small 'floral theatres' at various points in the garden.

In all these treatises the figure of the *curieux fleuriste*, sometimes represented by the author himself, emerges clearly. The *fleuristes* contributed in many ways to the development of floriculture and botany in this period. Not only did wealthy patrons employ *jardiniers fleuristes* to tend their flower-gardens, to create new ones and to procure rare varieties, sometimes these gardeners succeeded in establishing their own nurseries, in this way further promoting the dissemination of rare plants.

The qualities of 'curiosity' and 'virtuosity' also played a central role in the birth of the experimental sciences (Olmi 1992, p. 12). The treatise *La Natura, e coltura de' fiori*, for example, which was written in the mid-eighteenth century by Filippo Arena, a Jesuit and professor of mathematics at the University of Palermo, is particularly interesting for the scientific rigour with which the author approached his subject. While his work describes the cultivation of plants, and contains a discussion of flowers from an 'aesthetic' point of view, there is also a surprisingly modern description of the cross-pollination of plants by insects, which apparently predates any other mention of this phenomenon in the botanical literature.

Therefore, the true *jardinier fleuriste* was not merely a skilled and knowledgeable gardener; he was also a man of science, a virtuoso and an *homme d'esprit*, with that passion for curiosities that has always characterized the collector, ready to make almost any sacrifice to obtain the object of his desire. As the author of *Connoissance et culture parfaite des tulippes* observed, collectors should be all the more interested in plants because a painting will always remain but a single object, while the bulb of a rare flower could be infinitely multiplied, and the most humble flower was an object of beauty far more antique than the oldest medallion. The *curieux fleuriste* must have been a conspicuous figure indeed in the intellectual landscape of the sixteenth and seventeenth centuries, for La Bruyère included him in *Caractères* as the quintessential figure of the fanatical collector, his acquisitive mania quite surpassing that of any book or medallion collector. Jean-Jacques Rousseau, too, saved some of his most cutting irony for those 'petits curieux' and 'petits fleuristes qui se pâment à l'aspect d'une renoncule, et se prostent devant les tulipes' (Schnapper 1983).

Gardening was none the less a serious and respected profession, and the treatises these *jardinier fleuristes* wrote, often on the basis of their personal experience, are full of practical and up-to-date advice. They recommend that the aspiring gardener learn the place of origin of every flower in his garden, the type of climate and soil that each required, and which ones could be planted in close proximity. They compiled detailed lists and descriptions of flowers to accompany their works, which sometimes—as in the case of Morin's *Remarques necessaires pour la culture des fleurs*—also served as sales catalogues for their collections of rare plants.

All the works published in this period reflect in one way or another the Baroque taste for the unusual, the extravagant and the bizarre. Many pages were dedicated to rare and exotic species. Hughes, for example, includes in *The Flower Garden* (no. 32) a detailed description of the sunflower

—at that time considered the most fabulous of plants—which he had seen in Florida and Jamaica. In *Twelve Months of Flowers*, Robert Furber disguised what was actually a sales catalogue by presenting it in the form of twelve lavish engravings in which 400 different flowers from his nursery (discretely numbered) have been arranged in twelve extravagant bouquets, one for each month of the year (no. 37).

Ingenious *fleuristes* such as Ferrari and the anonymous author of *Secrets pour teindre la fleur d'immortelle* (no. 35) also attempted to surpass the inventiveness of Nature herself in the creation of strange and abnormal flora. They devised 'secret' processes that they claimed would change the form, colour or scent of a given flower, and were sometimes quite reluctant to reveal how they had obtained these effects. Ferrari does describe many of his 'secret discoveries', but humorously notes that when the Roman floriculturist Polidoro Nerucci, the owner of a magnificent garden in the Trastevere, was asked how he had managed to obtain some particularly beautiful double anemones, he responded dryly, 'Secretum meum mihi' (That secret is mine and I will keep it) (p. 504). In fact, today we know that many genotypic variations may occur spontaneously, and that most of the practices suggested by these floriculturists were entirely useless, being limited to the application of special fertilizers, pruning flowers, or infusing foreign substances into the roots of the plants.

Another fascinating theme reflected in these treatises was the constantly changing taste in flowers among collectors and connoisseurs. The rose was perennially held in high esteem, but the narcissus, the carnation and other unusual flowers were each in turn celebrated by flower lovers and in literature and painting, sometimes—as in the case of the hyacinth and, most notably, the tulip—inspiring a veritable mania (see chapter Seven).

28. Tobia Aldini (*dates unknown*) *and* Pietro Castelli (*c.* 1575–*c.* 1657)

[*architectural frame composed of a classical arch (on the pediments of which are seated two putti holding flowers) surrounding a coat of arms with the fleur-de-lis of the Farnese family. At the base of the arch on either side appear the seated figures of Teofrasto (Theophrastus) and Dioscoride (Dioscorides), the former with an open book on his knees, the latter seated in a meditative attitude with a closed book in his hands. Next to Dioscorides lies a helmet, a reference to his activity as a soldier. Between the two figures are scattered distilling devices and various medical instruments*] Exactissima Descriptio Rariorum Quarundam Plantarum, qu[ae] continentur Rom[ae] in Horto Farnesiano: Tobia Aldino Cesenate Auctore Illustr.mi et Rev.mi Principis et Cardinalis Odoardi Farnesii Medico Chimico, et Eiusdem Horti Praefecto. Romae Typis Jacobi Mascardi Anno Jubilei. 1625. Superiorum permissu. L.C.F.

F° 31 × 22.5 cm. ✠[6] A–B[6] C–L[4] M[6] *i–xii* 1–100 *101–106* p. (pages 27, 40 misprinted as 37, 30 respectively.)

BINDING: Contemporary parchment. Spine title in ink: 'Aldino Descrip Plantar Horti Farnes.'

PLATES: Title plate engraved by 'L.C.' (Luca Ciamberlano), and 23 full-page engravings of plants and flowers, each with its Latin name given at the top of the page. Woodcuts showing particulars of various plants on pages 7, 16, 17, 38 and 79. Woodcut initials, head- and tail-pieces.

REFERENCES: Séguier 2; Sprengel, II.121; Pritzel 1590 (as Castelli); Nissen 13; Hunt 208; Nocchi and Pellegrini.

THIS VOLUME, dedicated to Cardinal Odoardo Farnese, is apparently a work by the pharmacist and physician Tobia Aldini of Cesena, curator of the splendid gardens of the Farnese family situated on the Palatine Hill in Rome. Beginning in the eighteenth century, however, a number of scholars (including Haller, Spregel, Pritzel) have suggested that the actual author of this work may have been Pietro Castelli, professor of medicine and botany at the University of Rome, and founder of Messina's botanical garden, for which he edited a scholarly catalogue, *Hortus messanensis* (1640). An indirect allusion to this possible authorship can be found in a poem in Latin that follows the formal dedication: here the name 'Petrus Castellus Romanus' is hidden in an acrostic (it can be spelt out by taking the first letter from each verse). Further evidence is in the printer's note, in which appears the following phrase (set out in capitals): 'In gratiam Tobiae Aldini scripsi cuncta' (With the benevolence of Tobia Aldini have I written all this), a curious detail that would seem to acknowledge the central role played by Castelli in the execution of the work. His opinions are in fact cited several times in the text, and it seems more than probable that, while Tobia Aldini was the actual author, he availed himself liberally of the advice and suggestions of the illustrious botanist, especially in the editing of the Latin text. By 1626 this fruitful collaboration had come to an end, however: after the death of Cardinal Odoardo, Aldini passed into the service of Cardinal Francesco Barberini, and in 1634 Castelli left Rome for Messina.

Regardless of its disputed authorship, the *Exactissima descriptio* remains an outstanding example of the work being produced in Rome's intellectual circles during the first decades of the seventeenth century. The fact that it was printed by Jacopo Mascardi, who also published the most significant works to emerge from the city's Accademia dei Lincei (the first scientific academy to be founded in Europe), testifies to its scientific merit. Intended for experts, it was nevertheless written in a clear and simple Latin that contrasts strikingly with the ornate and allusive prose of *Flora* (no. 29), G. B. Ferrari's *opus magnum*, published just eight years later.

In the *Exactissima descriptio* Aldini presents various rare plants from the unique collection of the Farnese family, which due to its close ties with the Jesuit Order, frequently obtained seeds and specimens of exotic plants from priests returning to Rome after long periods spent abroad. The enthusiasm with which these plants were received in Rome during the first decades of the seventeenth century is well documented: new species were avidly sought after and grown in private gardens, constituting objects of prestige to be vaunted in cultivated circles.

Exactissima descriptio is divided into sixteen chapters, each one of which is devoted to a particular plant. A complete description of the plant, as well as details concerning its medicinal and culinary properties, are provided, while elegantly engraved plates aid the reader to grasp its salient characteristics. Frequently an illustration of the entire plant follows one showing a particular detail, presented either in a full-page plate or in a more modest space on a page primarily given over to text. Although the name of the artist who made the preparatory drawings for this work is not known, the engraver can be identified as the same artist who signed the frontispiece, Luca Ciamberlano

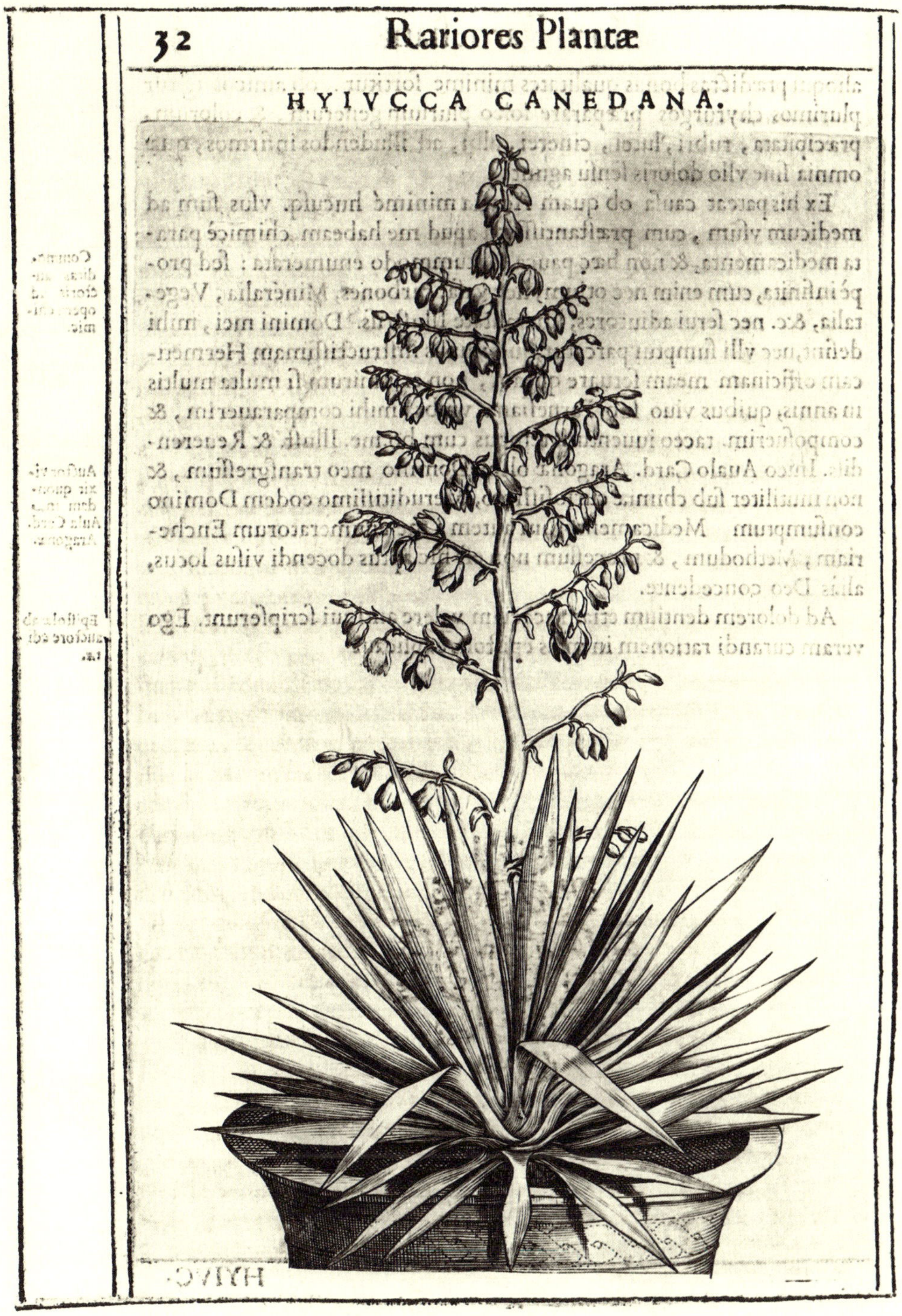

TOBIA ALDINI and PIETRO CASTELLI, *Exactissima descriptio rariorum quarundam plantarum*. Spanish-dagger (*Yucca gloriosa*), page 32

(1586–1641), then working in Rome as an engraver of religious, classical and allegorical subjects and as a designer of frontispieces. It is known that Castelli was an accomplished draughtsman with regard to plants, and it is quite possible that he was responsible for these high-quality plates, although no evidence has been found to support this.

The opening chapter of *Exactissima descriptio* is dedicated to the delicately perfumed 'Acacia farnesiana' or cassia, recently brought to Italy from the island of San Domingo and presented as a gift by the Farnese family to Ferdinand II, Grand-Duke of Tuscany. In the Farnese garden, where a seed from this tree was planted, the first branches began to appear in 1611, and three years later the tree came into flower. It continued to blossom luxuriantly twice a year, and its flowers form the subject of two beautiful plates, in one of which is also shown a detail of the seeds. Among the other plants described are the passion-flower (Maracot or Grandilla, written about at length by Nicolas Monardes, who had discovered it in the mountains of Peru), which flowered regularly in the Farnese gardens beginning in August 1619, and the 'Lauro indiano' (*Persea indica*), the 'Spinacia Fragifera' (*Bitum capitatum*) and the 'Solis flore tuberoso' (*Helianthus tuberosus*) or tobinambur, a plant with pleasantly flavoured, tuberous roots.

Among the most impressive plates are one showing the 'Lilionarcissus Rubeus Indicus', a species of amaryllis (amarillidacea) that could also be found growing in the garden of the Roman horticulturist Tranquillo Romauli, and the 'Hyiucca', brought to Europe from Canada (*Yucca gloriosa L.*). Quite different from the other species of this plant known at that time, in the Farnese garden it flowered in 1616, in 1620, and once again in 1622. A handsome engraving of this plant, shown in full flower and certainly drawn from life, is accompanied by scholarly notes, including apposite geographical-botanical observations.

29. GIOVANNI BATTISTA FERRARI (1584–1655)

[*engraved title-page within a sober frame decorated with a coat of arms and garlands, all hand coloured*] Flora overo Cultura Di Fiori del P. Gio: Battista Ferrari Sanese della Comp. di Giesù Distinta in Quattro Libri E trasportata dalla lingua Latina nell'Italiana da Lodovico Aureli Perugino [*decorative element*] In Roma Per Pier.'Ant. Facciotti. 1638. Con licenza di Superiori.

4° 24 × 17 cm. ✠–✠ ✠⁴ A–3X⁴ 3Y⁶ *i–xvi* 1–520 *521–548* p.

BINDING: Contemporary vellum. 19th-century green morocco label on the spine with 'Ferrari | Flora' stamped in gilt and '1638' in manuscript. 'Flora del P. Ferrari' written in ink on tail-edge of the volume.

PLATES: Engraved title-page and frontispiece, and 38 engraved plates depicting designs for parterres (7), different species of flowering plants (20), garden tools (5), and vases and bouquets (6). Many of the plates bear scroll captions. One of the plates is signed by Anna Maria Vaiani (active in Rome *c.* 1630). Six other plates, together with the frontispiece, were engraved by J. F. Greuter and Claude Mellan after designs by Pietro da Cortona, Guido Reni and Andrea Sacchi. They depict allegorical tales, most of them relating to the Roman goddess Flora. All the plates are fully coloured except for the 7 showing parterre designs.

The volume also contains 9 loose ink drawings of flowers copied from the book's plates, with manuscript descriptions in Latin, both probably dating from the late 17th century.

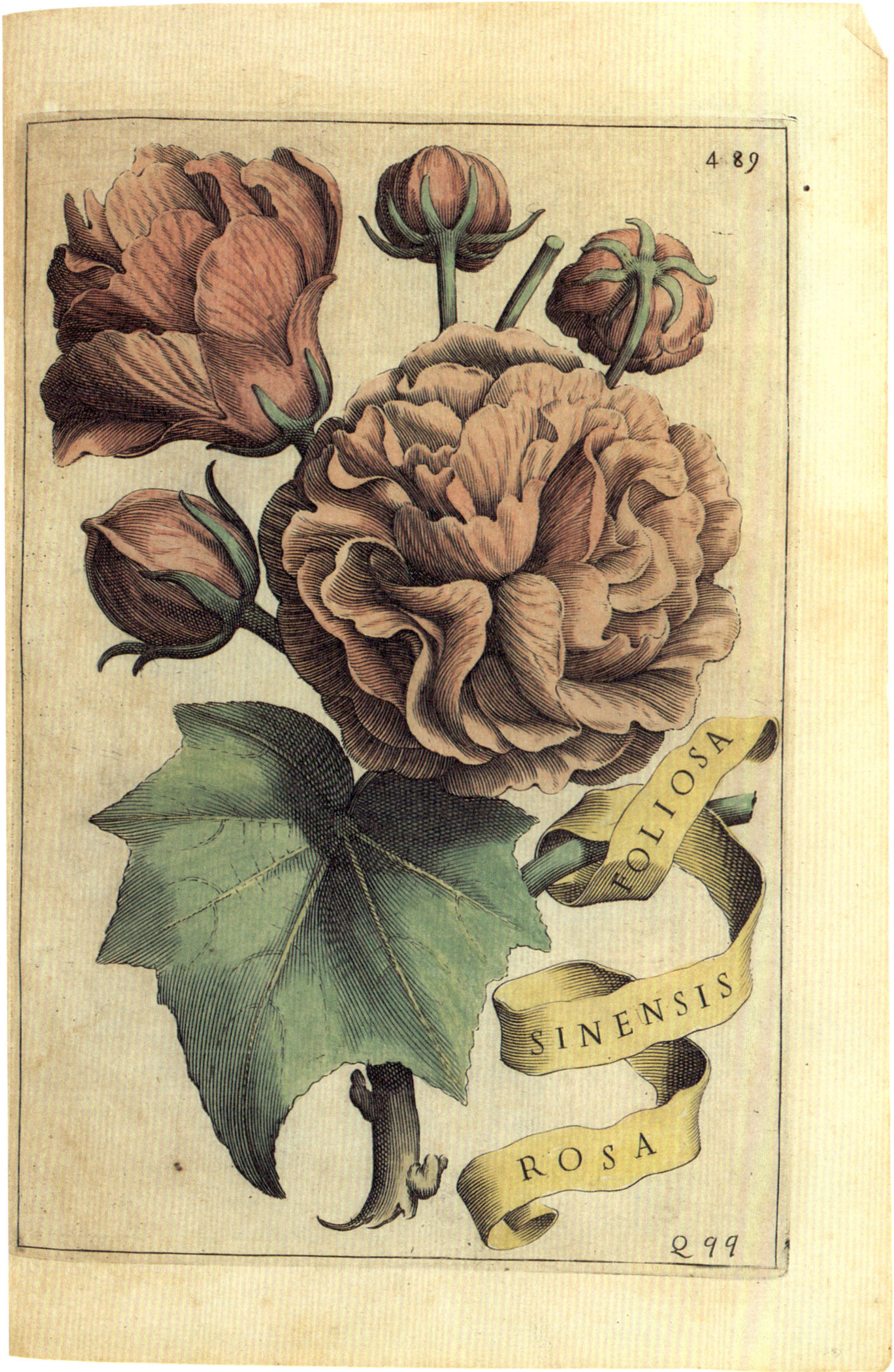

GIOVANNI BATTISTA FERRARI, *Flora overo cultura di fiori*. Chinese rose (*Hibiscus rosa-sinensis plenus*), page 489

PROVENANCE: A library shelfmark (probably from the 18th century) is written on the inside front cover.

REFERENCES: Arnold p. 246; S. Piantanida et al., *Autori italiani del '600*, Milan, 1948–51, p. 1799; Pritzel 2877; Nissen 1951; Hunt 222 (for 1st Latin edn, 1633); Masson 1972; Geissler; Belli Barsali; Tongiorgi Tomasi and Ferrari; *Age of the Marvelous*, no. 160 (for the Latin edn); Coffin 1991; *An Oak Spring Hortus*, no. 5.

GIOVANNI BATTISTA FERRARI's *Flora overo cultura di fiori* was one of the most refined and erudite works to be published in seventeenth-century Rome, then the intellectual and artistic capital of Europe. Although a serious work on the subject of botany and horticulture, it also posited itself as an object of delight (*diletto*) in accordance with the classical dictate that the acquisition of knowledge could be both effortless and diverting. Its author was a Jesuit priest, who after completing his studies in his native city of Siena moved to Rome, where he was appointed to the prestigious chair of Hebraic studies at the Collegio Romano. To his duties there, which he carried out for twenty-eight years, Ferrari added his eclectic activities as a humanist, orientalist and man of letters, publishing religious works, theatre pieces and essays on the mores of the period. His most significant output, however, reflected his special interest in the botanical sciences.

Ferrari's career coincided with five decades of unusually intense artistic activity in Rome. Furthermore, with the accession to the papal throne of Maffeo Barberini, who assumed the name Urban VIII, and the founding of the Accademia dei Lincei, Rome also became one of the most important centres of scientific activity in Europe. The study of the natural sciences flourished in this stimulating atmosphere, profiting from the generous patronage of Cardinal Francesco Barberini (the Pope's grandson), who was a connoisseur of 'cose di natura' (things of nature) and an enthusiastic collector of exotic plants, which he cultivated in the Orti Barberini on the Quirinal Hill. He had studied under the botanist Johann Faber of Bamberg at the University of Rome, and was a friend of Federico Cesi, the founder of the Accademia dei Lincei. The young Cardinal chose as one of his closest advisers the learned scholar Cassiano dal Pozzo (1588–1657), who was well known in Roman circles as a collector of both antiquities and objects of scientific interest (Solinas).

By assiduously cultivating his friendship with Cassiano dal Pozzo, Ferrari gained entry into the entourage of the Barberini and thus into the circle of naturalists who were pursuing their studies in the Orti Barberini and in the gardens of other enthusiasts, such as Francesco Caetani, Duke of Sermoneta, and Tranquillo Romauli, and those of the Lodovisi and Peretti Montalto families, all of whom possessed splendid estates in Rome.

Ferrari dedicated his very first printed work, *De florum cultura*, to Cardinal Francesco Barberini; written in Latin, it was published in Rome in 1633. Five years later an Italian edition, illustrated with the same engraved plates and entitled *Flora overo cultura di fiori*, appeared; the work at Oak Spring is a beautifully hand-coloured copy—probably unique—of this edition. In 1646 Ferrari also published *Hesperides*, a work on the cultivation of citrus species (see *An Oak Spring Pomona*, no. 67).

Flora overo cultura di fiori was intended by its author to serve as a manual on ornamental flowers

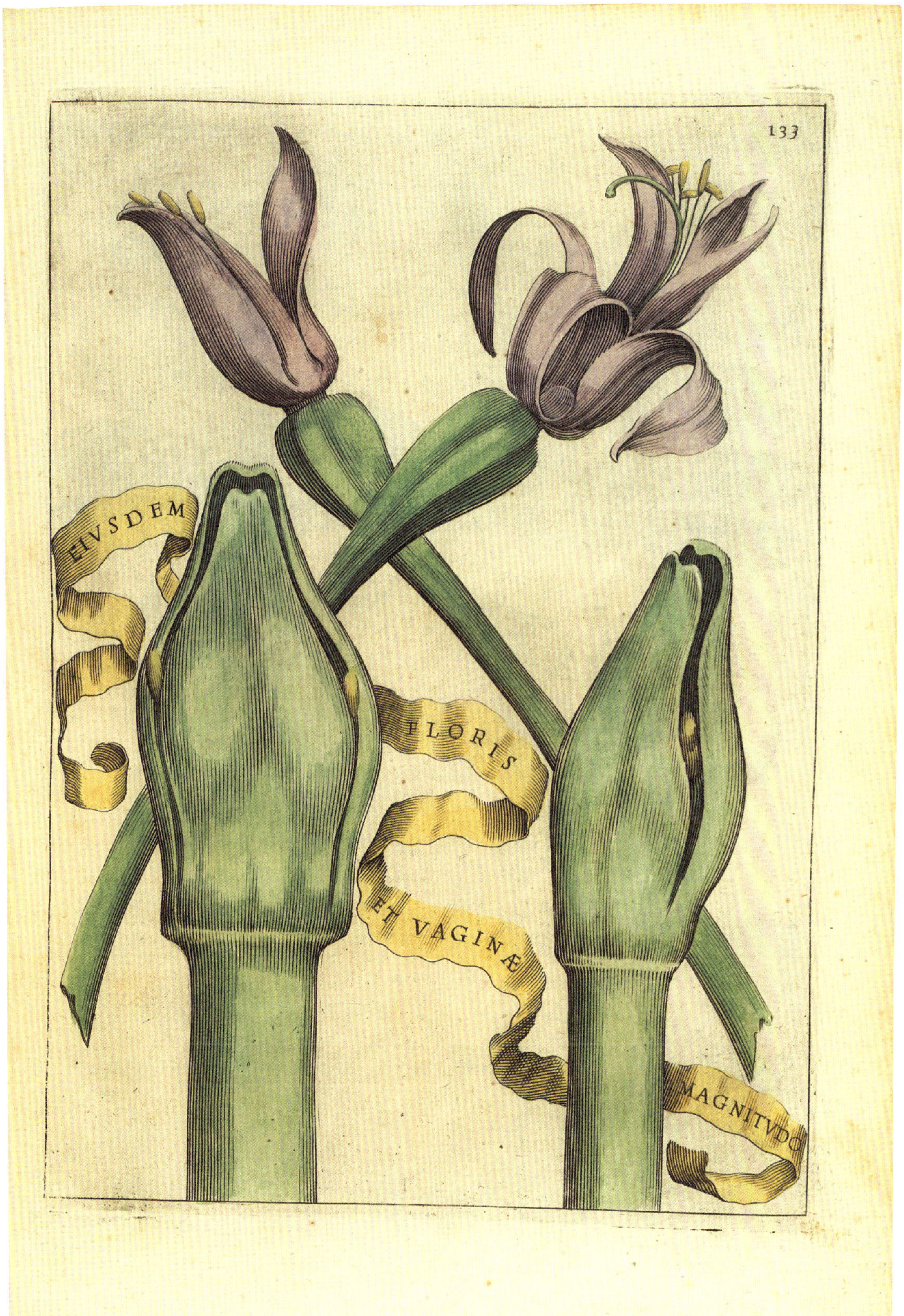

GIOVANNI BATTISTA FERRARI, *Flora overo cultura di fiori*. A type of *Amaryllis*, page 133

and their cultivation. It is divided into four parts. The first describes how to set up a garden, the second presents some of the most popular flowers of the period (the narcissus, crocus, tulip, fritillary, etc.) and their characteristics, and the third provides instructions on their cultivation. However, interspersed throughout the work and in the fourth section in particular are a variety of fascinating digressions ranging from precepts on the fitting up of a garden (see *An Oak Spring Hortus*, p. 23), to a discussion of floral 'meraviglie e artefici' (marvels and artifices), 'secrets' for tinting flowers or marking the petals with heraldic figures, and advice on how to obtain 'crespi' (puckered) blooms or flowers of unusual dimension or form.

The garden, which is described by Ferrari in a striking metaphor as a 'quasi militare alloggiamento delle fiorite schiere' (An almost military accommodation of the floral ranks), could be laid out in the form of a square, circle, rectangle or polygon (preferably an octagon), delimited by a wall or thick hedge. Special care must be taken in designing the flowerbeds, for excessively complicated forms were impractical for the gardener (Tongiorgi Tomasi 1983, p. 11). Indeed, the author declared, the beauty of the garden lay not so much in the form of the parterres as in the orderly and harmonious arrangement of the flowers within them. The beds themselves should be marked off by an herbaceous border or, even better, by a neat border of bricks. Short stakes of cane with labels could be used to mark where different flowers have been planted. Finally, the flowers when they come into bloom should not be gathered indiscriminately, for the order and beauty of the beds must be maintained and a source of seeds for the coming year ensured.

Conscious of the importance of illustrations in a text of this nature, Ferrari had many copperplate engravings prepared for his *Flora*. Some of them—most probably the ones depicting flowering plants—were drawn by Nicolas Guillaume Delafleur (d. *c.* 1670), an artist from Lorraine who was a friend of Nicolas Poussin. Delafleur was known in Rome, where he worked for Cardinal Barberini, as 'Monsù Fiore' (Monsieur Flower) in recognition of his great skill in depicting floral subjects.

Some of the plates in Ferrari's work show just the cut flower, and others the entire plant with its bulb and roots, while in each a scroll inscribed with the plant's Latin name has been wound gracefully around the stem. The life-cycle of the flower may be depicted, from the bud to full inflorescence, but the skilful arrangement of the images on the page shows that the artist's interest was not limited to the mere accurate documentation of his subject-matter. This can be appreciated in the fine drawing of a bud and flower from the 'Narcissus indicus', or in the plate showing an amaryllis, and in the splendid illustration of the 'Foliosa Sinensis Rosa' (Chinese rose) or hibiscus, a malvaceous plant from the Far East that was first cultivated in Rome by Ferrari himself. Another extraordinary plate shows a seed from this same plant as seen through 'quella sorta d'occhiale a cannello, che piccolissimi corpicciuoli fa parere assai grandi' (That sort of lens in the shape of a tube, which makes the tiniest objects appear quite large; p. 478); this is a clear reference by the author to the microscope, an instrument which the members of the Accademia dei Lincei were just beginning to use in their studies around the year 1626.

Other plates in *Flora* illustrate ground-plans for parterres, gardening tools, containers for transporting plants, vases containing elaborate bouquets and urns and frames for the construction of handsome floral centrepieces. One of these was drawn and engraved by Anna Maria Vaiani, a Roman artist of some renown whose name can be found in Galileo Galilei's correspondence.

Although in his introduction Ferrari describes his work as 'entirely new', the botanical plates for *Flora* in fact follow quite closely the typology of the florilegia being published at that time in central and northern Europe. Entirely novel, however, were the six magnificent plates engraved by Johan Friedrich Greuter and Claude Mellan from drawings by some of the most renowned artists of the period, including Pietro da Cortona (who also designed the imposing frontispiece), Guido Reni, Andrea Sacchi and Giovanni Lanfranco. These illustrate various allegorical or mythical tales scattered throughout the text by Ferrari in order to divert the reader during his perusal of this sometimes excessively specialized work (Freedberg 1989, p. 45). One charming vignette depicts the story of Limace and Bruno, two clumsy and dishonest gardeners in the service of Flora, one of whom was transformed by the goddess into a slug (*limaccia*) and the other a caterpillar (*bruco*). This tale appears at the end of a paragraph dealing with animal pests noxious to flowering plants.

Thus, although Ferrari's *Flora* was an important text on gardening and horticulture, it also reflects the culture of allegorical humanism that flourished in seventeenth-century Rome. In this context it is interesting to recall that in his youth Ferrari wrote a brief Latin oration in verse, *Aetas florae*, composed entirely of floral images linked to the conventional themes of beauty and love but also to more melancholy associations, the flower in its ephemeral beauty being transformed by the precocious author into a *memento mori* and a symbol of the transience of human life.

The success of Ferrari's *Flora* was as much due to its splendid illustrations as it was to the text. It is no coincidence that the 1641 edition of the famous florilegium by Johann Theodor de Bry, published in Frankfurt am Main by Matthaeus Merian, contains at the beginning thirty-two plates which are either exactly copied from or directly inspired by illustrations from *Flora* (see no. 16).

30. Agostino Mandirola (*dates unknown*)

Il Giardino De' Fiori Diviso in tre Libri, Nelli quali brevemente s'insegna Nel primo il modo di conoscere, e cultivare li Fiori di Bulbi più rari. Nel secondo la cognitione, e cultura delli Fiori di Radiche più riguardevoli. Nel terzo il modo di moltiplicare, cultivare, e conservare gl'Agrumi. Di F. Agostino Mandirola Da Castel Fidardo. Dell'Ordine Min Con di S. Francesco Dottore di Sacra Theologia. Dedicato All' Illustriss. Sig. Conte Girolamo Rossetti. [*typographical ornament*] In Ferrara, Per Giuseppe Gironi Stampator Episc. Con licenza de' Superiori 1650.

12° 17 × 10 cm. ✠[6] A–F[12] *i–xii* 1–142 *143–144* p.

BINDING: Contemporary paper boards; title in manuscript (in ink) on the spine; manuscript shelf mark on the spine and front cover.

PLATES: [*engraved half-title*] On a broad pedestal flanked by two rose bushes and, in the distance, by two trees, is the coat of arms of the Rossetti, Counts of Ferrara, surmounted by a crown and an imperial fritillary, and surrounded by six flowers (a carnation, an anemone, a tagates, a lily, and two tulips). On the pedestal is written: 'Il' Giardino De Fiori All' Ill:mo Sigr: Conte Girolamo Rossetti.' In the lower left-hand corner of the page: 'Jac. P.' Woodcut head-pieces, initials.

PROVENANCE: Bookplate: 'Fuerstlich Auerspergsche Fideicommisbibliothek Zu Layback' and in ink: 'V. M. 17'. On the half-title in ink: 'Wolfg. Engelb: S.R.I. Com: ab Aursperg Sup: Cap: Carnia'. Catal. Inser: Anno 1655'.

REFERENCE: Cleveland, no. 211.

THE AUTHOR of this early flora and manual on flower gardening was the Franciscan brother Agostino Mandirola, a native of Castelfidardo in the region of Le Marche. It was first published in Macerata in 1646 with the title *Manuale di giardinieri*, and reissued in 1658. The original edition was dedicated to the Marquis Francesco Ricci of Macerata. This little 'pocket-book' was evidently an extremely popular and highly valued work, for it continued to be published and read until almost the end of the eighteenth century. Oak Spring's copy of *Il Giardino de' fiori* would appear to be the rare 1650 edition, which was published in Ferrara and dedicated to Count Girolamo Rossetti, a wealthy dilettante and the author of a treatise on hydraulics. In a florid dedication the printer, Giuseppe Gironi, presents him with this work containing 'i fiori . . . fatti dalla stampa immortali' (Flowers . . . immortalized by the printer's art). This edition includes a fine engraved half title-page signed 'Jac. P.', most probably Jacopo Piccini (or Pecini or Pizzini) of Venice (1617–69), an engraver of sacred, mythological and allegorical scenes as well as portraits, mostly of other artists.

Ferrara was one of the principal cities under the dominion of the enlightened Este family, and an important centre of intellectual activity, particularly for the botanical sciences. The names of celebrated physicians and naturalists such as Leonicenio (1428–1524), Manardo (1462–1536) and Alfonso Pancio (d. before 1579) are associated with the city. Pancio was private physician to Duke Alfonso II d'Este and the director of Ferrara's botanical garden; he was also personally acquainted with many of the most important naturalists of his day, including Ulisse Aldrovandi of Bologna and the Dutch botanist Carolus Clusius (Masson 1972, p. 66).

During the seventeenth century, interest in the cultivation of flowers became widespread and many works were written on the subject. Mandirola's treatise appeared not long after Ferrari published his famous *Flora*. Ferrari wrote his work in Rome when the papal city was at the very height of its glory under the rule of the cultivated Barberini family, a splendid centre of wealth and power, attracting artists and scholars from all over Europe.

In contrast to such erudite works as Ferrari's *Flora*, Mandirola's *Giardino de' fiori* is an eminently practical manual of flower gardening, full of advice based on the direct experience of the author, although in his introductory note Mandirola assures us that he has also consulted the works of Pliny, Dodoens (perhaps his *Florum, et coronariarum odoratarumque nonnullarum herbarum historia* of 1568; see no. 39), a certain 'Chisio' (no doubt a misspelling by the typesetter of the name 'Clusius') and, of course, Ferrari. Nevertheless, Mandirola reckons it particularly fitting that a member of a relig-

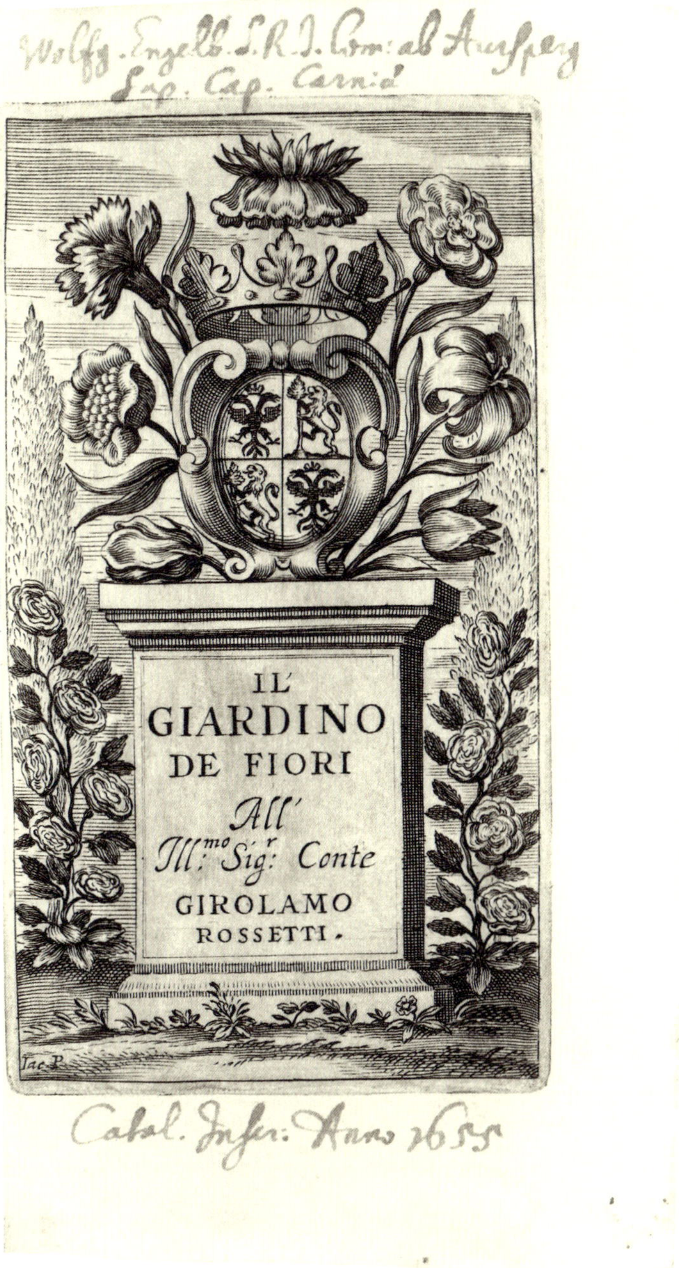

AGOSTINO MANDIROLA, *Il Giardino de' fiori*. Title-page

ious order such as himself present a treatise on the subject of gardening, since the garden was itself a symbol of Paradise.

Il Giardino de' fiori is divided into three 'books', or parts, which describe the laying out and planting of a garden and the cultivation of various flowering plants and fruit trees. The first book opens with detailed instructions on how to design and construct a garden. The gardener must first acquire a thorough knowledge of the site and the quality of the soil, and of the flowering plants that he wishes to establish there, whether they be bulbous or root species. He should therefore be 'young, conscientious, diligent, [and] assiduous' (p. 1), with some notion of the principles of architecture in order to be able to design the layout of the garden, and with a considerable knowledge of different types of flowers so as to be able to arrange them suitably. He must know how to cultivate the various species, and be equipped with all the necessary tools—a spade, shovel, hoe, rake and measuring rule, at least two screens for sifting soil, a watering-can, a wheelbarrow, several brooms, etc. He should also have a box in which to keep brushes, scoops and bundles of numbered sticks (for marking where the rare plants have been set).

Ideally, the garden site should be slightly sloping, so that rainwater does not accumulate there, with a sunny corner where plants may be placed during the winter, and a shady corner to protect them during the hottest months. Mandirola recommends a rectangular ground-plan, this being the easiest to design and lay out. The 'aiette' or flowerbeds should be designed to allow for the placement of the largest possible number of plants. Furthermore, the author advises the gardener to draw the layout of the garden and its beds on a sheet of paper, so that the flowers that are planted there during the course of the year can be noted down. The pathways may be lined with box or myrtle unless the garden is very small, in which case they should be marked off with white tiles embedded in the soil. Hedges of wild thyme are to be avoided as they dry the soil and may cause damage to bulbs. The garden should have two types of soil—rich earth where plants with roots will flourish, and sandy soil for bulbous plants. The soil in the beds should be completely replaced once every three years.

The gardener must pay particular attention to the arrangement of the plants in his flowerbeds; different species should not be placed too close together, particularly the ranunculi, which will not grow next to other plants. Jonquils and the narcissi can be planted along the borders, while tulips and anemones will grow best if placed in the centre of the bed. Furthermore, bulbs should always be planted an equal distance apart, for which purpose it is useful to mark the ground with a measuring rule.

Flowers may also be planted in glazed terracotta pots. The pots must be as wide as they are tall, with a hole in the base that the gardener should cover with small stones so that it does not become blocked with soil. It can then be filled with rich soil for plants with roots, or sandy soil for bulbs; the latter should be planted at a distance of at least four fingers below the rim of the pot. Once the plants have firmly taken root and come into flower, the vases may be placed around the garden following 'quell'ordine d'architettura, che il migliore sarà giudicato per l'abbellimento del Giardino' (That architectural order which will be judged the best for the embellishment of the Garden; p. 9). Further advice is provided on sowing, watering and ridding the garden of pests.

Mandirola then supplies detailed instructions on how to cultivate a number of the most highly prized flowers of the period (which are listed in alphabetical order, with their page numbers, in the index preceding the text). The remainder of the first book deals with bulbous flowers, including different varieties of narcissi, jonquils, hyacinths, martagoni lilies and tulips, which 'like innumerable chameleons, transform themselves in a thousand colours' (p. 49). The second book describes the cultivation of root plants, such as the rose, the double-blossom peach tree and many types of jasmine, including (p. 85) a variety from America 'called Quamoclit' (*Ipomoea quamoclit*). The third book is dedicated to the citruses, which were frequently used to construct espaliers along garden walls and which, together with bulbous flowers and the varieties with double blooms, constituted the species most sought after by the seventeenth-century gardener.

In comparison to Ferrari's sophisticated and richly illustrated *Flora*, Mandirola's treatise is a far more modest work, its sole embellishment being an engraved half-title. Its text, however, couched in a simple and direct language wholly unlike the complex, metaphorical idiom used by Ferrari, provided concrete, up-to-date advice that must have served as an invaluable guide to generations of gardeners.

31. John Rea (d. 1681)

Flora: seu, De Florum Cultura. Or, A Complete Florilege Furnished with all requisites belonging to A Florist. In III. books. by John Rea, Gent. London, printed by J. G. for Thomas Clarke, 1665. [*but with the added engraved title-page showing Flora seated on a wall with a pier on either side, each supporting an urn of flowers. Before them are two allegorical figures, Ceres and Pomona, who stand on a broad, low platform. Between the piers hangs a draped cloth, on which is written*] Flora Ceres & Pomona by John Rea Gent: London printed for Richard Marriott & are to bee sold at his shopp in fleetstreet under ye kings head Taverne. 1665. [*on the base of the platform is engraved*] FLOS speculum vitae modo vernat et interit aura. | FLORIS imago fugax rapidi nos admonet aevi. | FLORI par juvenis tener est crescentibus annis. | FLOREM si ostendet, feret ipso tempore fractum | O FLOS sic vernans juvenili aetate, pudorem | A FLORE accipias, et honeste vivere discas [*The flower, mirror of life, blooms and dies in an instant. The fleeting image of the flower reminds us of the passing years. Tender youth is like the flower when it grows, showing its bloom and immediately thereafter its fruit. Oh virginity, that blossoms at an early age, adopt the modesty of the flower, and learn to live honestly*] [*beneath the platform*] D: Loggan sculp.

F° 29 × 19.5 cm. π^4 B^4 b^4 C–Z^4 2A^4 (2A3 + χ1) 2B–2I^4 2K^2 *i–xxiv* 1–239 *240–244* p. This copy lacks the preliminary leaf and title-page. Bound at the end is a broadsheet from the seed catalogue of the nurseryman Edward Fuller (*fl.* 1660–1720).

BINDING: Bound in quarter red morocco and beige cloth by W. N. Keynes in 1905 (signed on the lower turn-in). Decorated endpapers.

PLATES: 8 engraved plates of flowerbeds, numbered 1–16. Woodcut initials.

PROVENANCE: Armorial bookplate of Edward Howard, Duke of Norfolk (1686–1777). On a label on the first front free endpaper: 'Belonging to the Library bequeathed by the Will of Edward Duke of Norfolk to remain in his Family. Henry Howard & Thos Eyre Esq^rs Executors.'

REFERENCES: Pritzel (2nd edn) 7445; Hunt 301; Henrey, I.195–8 and no. 325.

THE FLORICULTURIST John Rea was the author of one of the most important gardening books to be published in England during the second half of the seventeenth century. His *Flora: seu, de florum cultura* first appeared in 1665. It was republished in 1667 with some additional plates depicting flowers and evidently continued to serve as a manual for gardeners for many years, for it was issued again in 1702. Rea titled his work after that of the most famous florilegium of the century, G. B.

Ferrari's *De florum cultura*, published in Rome in 1633 (see no. 29), although the scope and nature of his work was quite different. Ferrari's *Flora* is a product of the high Baroque period, a sophisticated florilegium interwoven with classical references and arcane digressions couched in an elaborately metaphorical language, whereas Rea's *Flora* is a genuine gardening manual, designed not 'to amuse the Readers with Romantick Inventions' but 'to acquaint the unskilful with such Rules and apt Forms, as may be fit for the planting and disposing of the best Flowers; and in so plain and easie a method, that every person of any capacity may be enabled thereby to be his own Gardener' (p. 2).

Rea's *Flora* opens with two letters of dedication, the first addressed to Charles Gerard, fourth Baron of Gerard's Bromley, Staffordshire, whose garden was designed by Rea. The second was addressed to Sir Thomas Hanmer, the author of a projected 'Catalogue of choice plants', which, however, was not published in his lifetime. Rea appears to have had a long and close connection with Hanmer, who furnished him with many plants for his collection and encouraged him during the writing of *Flora*. Rea also gracefully included two poems dedicated to the wives of his two illustrious patrons.

Rea was a 'florist', a term that in the seventeenth century was used to refer not to one who sold flowers as a profession but to those who took up the study and cultivation of plants as a hobby or to satisfy their thirst for knowledge. He was well known in his day to floriculturists all over England for his expertise and for his collection of rare plants; as Hanmer affirmed in a letter to John Evelyn, among florists 'the most skillfull and best stored in these parts is Mr. Rea of Kinlet in Shropshire, whose florilege is in print' (Goodchild, p. 101).

In the epistle addressed 'To the Reader' Rea explains his motivations for writing *Flora*. John Parkinson's *Paradisi in sole*, the most authoritative English florilegium and gardening book extant (see no. 40), had been published in 1629, but as Rea observed, many of the plants listed in it were already 'by Time grown stale, and for Unworthiness [to be] turned out of every good Garden', while it was lacking many of those 'noble things of newer discovery' without which no florilegium could be considered complete. Thus, he decided to write a thoroughly updated and modern work in 'the form of a *Florilege*, furnished with all requisites belonging to a *Florist*, [rather] than continued in the old method of an *Herbal*, and instead of old names, uncertain places, and little or no virtues, to insert some other things much more considerable'.

Rea's *Flora* originally was to have been illustrated with numerous plates of flowers, but at some point he abandoned this idea, apparently with little regret, for he wrote 'As for the cutting the Figures of every Plant . . . as Mr. *Parkinson* hath done, I hold [it] to be altogether needless; such Artless things, being good for nothing, unless to raise the Price of the Book, serving neither for Ornament or Information'. Rea did have a very fine title-page prepared, with three allegorical figures shown against an architectural backdrop, and a poem in Latin, every line of which begins either with *flos* or an inflected form of it. This page was engraved by David Loggan, a German artist of some renown, who had worked for a time in the Low Countries before moving to England.

Each of the three books that make up *Flora* is dedicated to one of the three goddesses who appear together on the title-page. The first book, offered to Flora herself, describes how to construct a garden, recommends which trees, shrubs and flowers to plant, and explains various gardening techniques. The second, dedicated to the harvest goddess Ceres, describes the cultivation of various annuals and biennials, while the third, offered to Pomona, the goddess of tree-fruits, invites the reader 'to a Banquet of the best Garden-Fruits our cold Northern Country will afford'.

Perhaps the most interesting and unique aspect of Rea's *Flora* is the introductory section to the first book, entitled 'Of the making and planting of Fruit and Flower-gardens, with a Nursery for Fruits and Flowers to furnish both'. In it the author provides step-by-step directions for the layout and construction of the two 'Gardens of delight'—the fruit and the flower-garden—which should grace the 'Habitation of every Gentleman'. These written instructions, together with the book's eight engraved plates of designs for flowerbeds, make it possible for us to recreate a seventeenth-century English garden complete in every detail.

After some thoughtful reflections on the true origin of a beautiful garden—money alone cannot create one, for only under the loving care of a skilled gardener will flowers and trees flourish 'though planted in a barren Desart'—Rea begins instructing the aspiring gardener. Both the flower-garden and the fruit-garden should be set out on the south side of the house, to provide plants with the benefit of both shelter and sun. Thirty yards square was ample for the flower-garden of a nobleman's seat, and twenty yards square for a 'private Gentleman', for, as Rea observes, too large a garden is commonly 'ill furnished, and worse kept'. Rea included 'draughts' or layouts of various sizes to that 'every one may take that which best agrees with his ground', seeking to maintain the garden's ideal proportions by using a single 'division' (ranging from two feet to one yard, depending on the size of the draught) to determine the width of every border, walk and flowerbed.

Both gardens should be surrounded by a strong brick wall at least nine feet high, with a lower wall or paling where the two gardens adjoin. Every border should be lined with a 'rail' assembled by an 'understanding Joyner' to the exact length required, and carefully set so that it was straight and level. Rea provides remarkably precise directions for the construction of these rails, which he considered to be the best system for defining the garden's beds and borders. They should be of fine seasoned wood, one and one quarter inches thick and five inches high for the fruit-garden or four inches high for the flower-garden, with strong hardwood stakes nailed at intervals to fix the rails firmly in the ground and keep them from warping.

Rea describes in some detail how to establish a fruit-garden—what trees, shrubs and flowers to plant, how to prepare the soil for planting, and how to set the turf, walks and alleys—before passing in the next section to the subject of the flower-garden, which, to use his charming imagery, should be fashioned 'in the form of a Cabinet, with several boxes fit to receive, and securely to keep, Natures choicest jewels' (p. 6).

After providing further directions for the construction of rails, as well as a 'Lattice-frame' for the

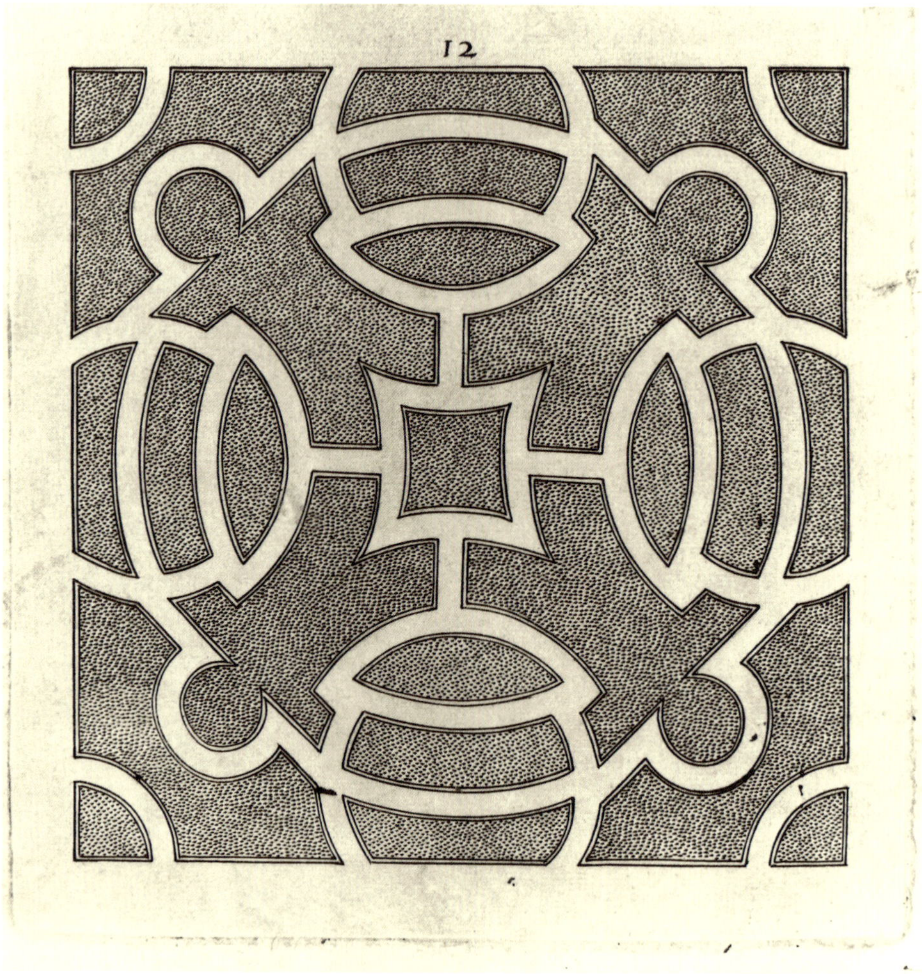

JOHN REA,
Flora: seu, de florum cultura.
Design for a flowerbed,
number 12

training of roses, Rea explains how the flower-garden should be laid out. At its centre was to lie the Fret, twenty-three divisions wide (in a garden thirty yards square, a division should measure two feet six inches), containing the beds and circumscribed by walls and paths. Along the surrounding garden wall, which was to be planted with '*Auricula, red Primroses, Hepaticaes, double Rose-champion, double Nonsuch, double Dames-violet*, the best *Wall-flowers, double Stock-gilliflowers*, and many other things', there should run a 'Great Alley' eight feet nine inches wide. The inner side of this Great Alley was to be lined with a 'latticed Border' against which 'several sorts of fine *Roses*' could be trained to form a thick hedge, while 'The border next to this *Rose-hedge* is chiefly intended to place Pots upon, on all sides, with the best *Gilliflowers, Auricula's, Myrtles, Oleanders*, and all other potted Plants, onely the edge next the Rail is to be set with all the several sorts of *Crocus*' (p. 9). Next to this inner border should run an 'inner Alley', both border and alley being one division wide.

The space within the Fret could be laid out in a geometrical pattern of flowerbeds following one of the designs presented by Rea in the eight engravings that illustrate this section. In deciding which

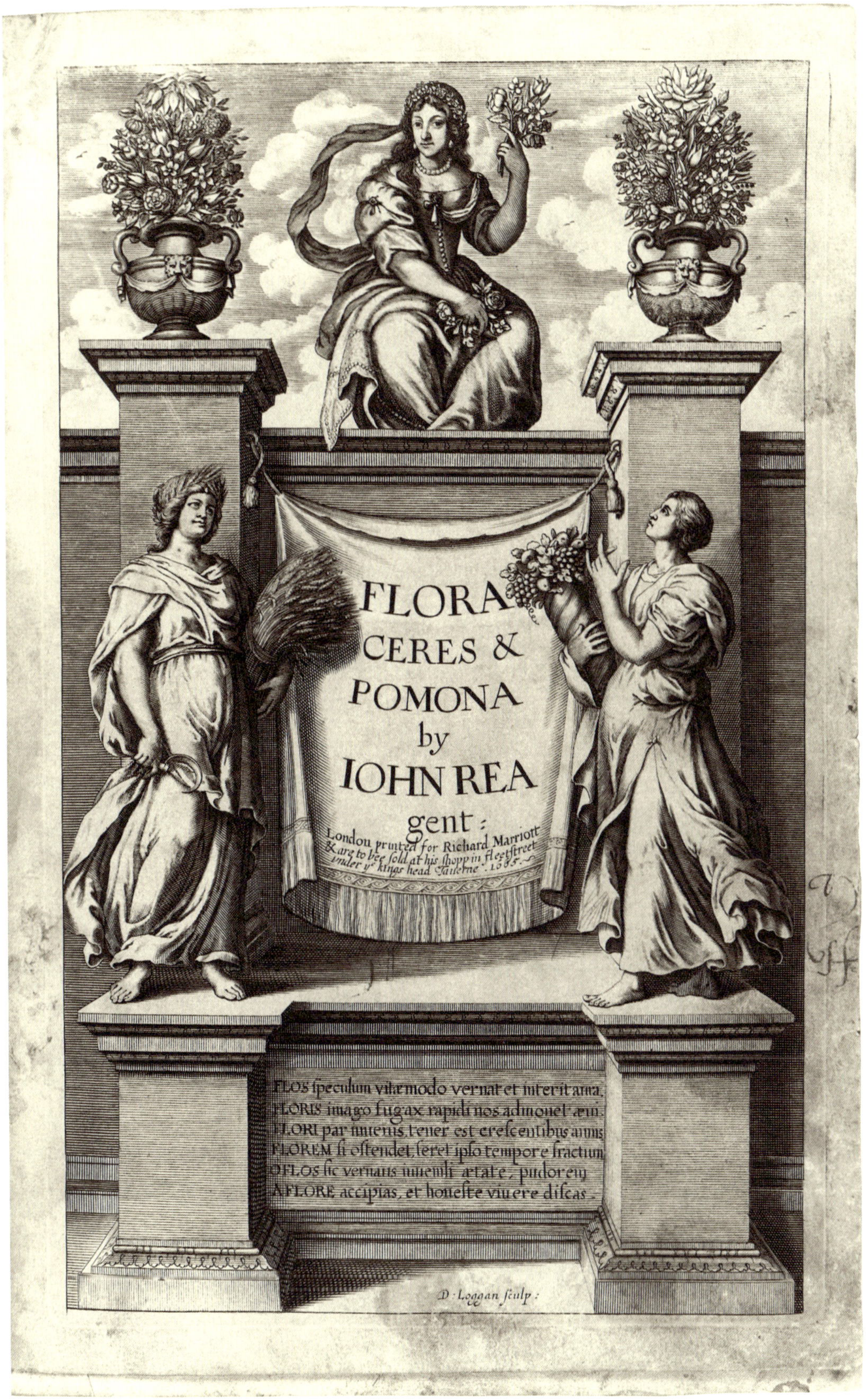

JOHN REA,
Flora: seu, de florum cultura.
Title-page

flowers to plant, the knowledgeable gardener should take care to place in close proximity only those varieties that will not harm one another. At the corners of each bed '*Crown-Imperials, Lilies, Martagons*, and such tall Flowers' could be set, while in the centre he recommends 'great Tufts of the best *Pionies*, and round about them several sorts of *Cyclamen*; the rest with *Daffodils, Hyacinths*, and such like: the streight Beds are fit for the best *Tulips* . . . : *Ranunculus* and *Anemonies* also require particular Beds; the rest may be set all over with the more ordinary sorts of *Tulips, Fritillaries, bulbed Iris*, and all other kinds of good Roots' (p. 9). Every year the beds could be rearranged according to the gardener's taste as his stock of plants increased.

Rea then recommends building into one of the garden walls 'a handsom Octangular Somerhouse, roofed every way, and finely painted with Landskips, and other conceits, furnished with Seats about, and a Table in the middle, which serveth not onely for delight and entertainment, to sit in and behold the beauties of the Flowers, but for many other necessary purposes' (p. 9). He goes on to describe how to set up a nursery, with stoves to 'gently attemper the Air in time of hard Frosts', and in the concluding section provides practical advice on such matters as how to water one's garden, how to plant from slips and cuttings, and how to construct a hot bed. More detailed instructions regarding the various plants, flowers and trees mentioned in this section are provided in the two subsequent books.

32. WILLIAM HUGHES (*fl.* 1655–1683)

The Flower Garden. Shewing Briefly how most Flowers are to be ordered, the time of Flowering, taking of them up, and of planting them again. And how they are increased by Layers, Off Sets, Slips, Cuttings, Seeds, &c. With other necessary observations relating to a Flower Garden. Whereunto is now added. The Gardiners or Planters Dialling, (viz.) how to draw a Horizontal Diall, as a Knot in a Garden, on a Grass-plot, or elsewhere, the like before not extant. By William Hughes. London, Printed for William Crook at Green Dragon without Temple Barr, 1672.

12° 14 × 7.5 cm. A⁶ B–E¹² F⁶ (F3 missigned F5) *i–xii* 1–100 [i.e., 102: p. 95–96 repeated in the pagination sequence; with the first p. 96 misprinted 94] *101–106* p.

BINDING: Mottled calf.

PLATES: Woodcuts: 4 folded plates, plus 5 illustrations in the text, and head-pieces and initials.

PROVENANCE: Ex-libris of 'The Right Hon[ble] Thomas Lord Viscount Weymouth Baron Thynne of Warminster 1704' pasted to the verso of the first leaf.

REFERENCES: Henrey, I.206; Desmond 1994, p. 363.

WILLIAM HUGHES was a gardener to Countess Conway at Ragley Hall, Warwickshire. It is known that he spent many years in Jamaica, and that he visited Florida *c.* 1652. He was the author not only of this volume on the cultivation of flowers, which was very successful and went through several reprintings, but also of the *The Compleat Vineyard* (1665) and *The American Phy-*

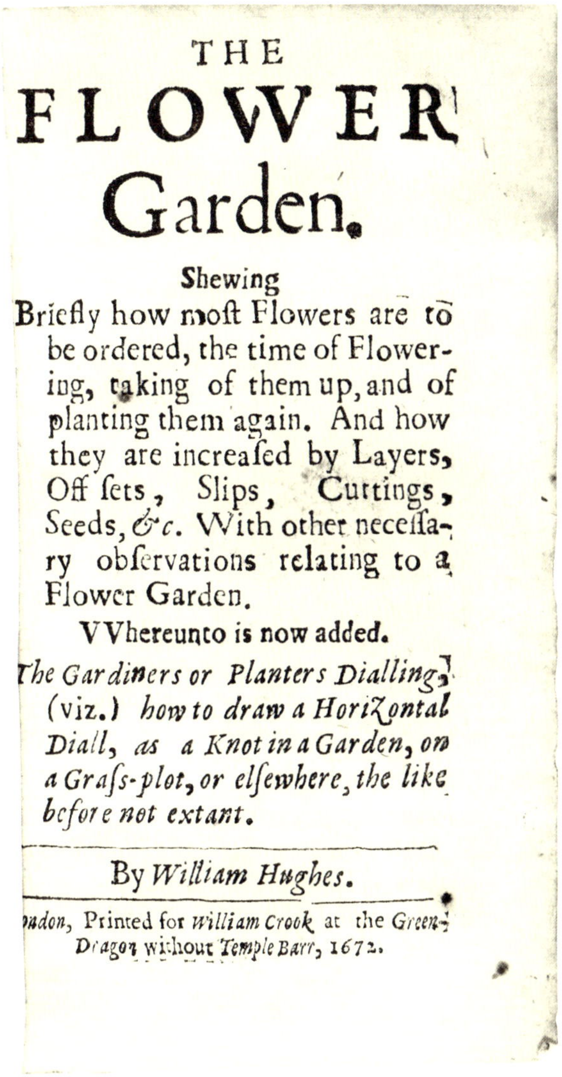

THE
FLOWER
Garden.
Shewing
Briefly how moſt Flowers are to be ordered, the time of Flowering, taking of them up, and of planting them again. And how they are increaſed by Layers, Off ſets, Slips, Cuttings, Seeds, *&c.* With other neceſſary obſervations relating to a Flower Garden.
VVhereunto is now added.
The Gardiners or Planters Dialling, (viz.) how to draw a Horizontal Diall, as a Knot in a Garden, on a Graſs-plot, or elſewhere, the like before not extant.

By *William Hughes.*

[Lo]ndon, Printed for *William Crook* at the *Green-Dragon* without *Temple Barr,* 1672.

WILLIAM HUGHES, *The Flower Garden.* Title-page

sitian; or, a Treatise of Roots, Plants, Trees, Shrubs, Fruit, Herbs, &c. Growing in the English Plantations in America (1672).

In *The Flower Garden* Hughes offers his readers an epitome—the fruit of his vast personal experience—on the cultivation of a flower-garden. One must begin, he affirms, with a correctly constructed hot-bed for the germination of certain seeds, among which he cites the French marigold, amaranth, bindweed and the 'Merveille of Peru' or *Mirabilis jalapa,* all extremely popular flowers in this period. He then briefly discusses those plants that may be grown from seed without the benefit of a hot-bed, and those that can be planted from 'layers, slips and cuttings'. Hughes also specifies their flowering times and what must be done in terms of maintenance. Concerning the asphodel, for example, 'the most part of them Flower in *May* and *June,* they are increased by taking them up once in two or three years, and parting the Root when the Stalk is dry, and then quickly planting them orderly again' (p. 23). In contrast, '*Sun-flowers* grow very tall, and do for the most part Flower in *September*, and are yearly raised from seed, sown or set in *March* or *April* under a warm Wall' (p. 42).

This useful, if somewhat cursory, work concludes with detailed instructions, accompanied by diagrams, on how to construct a sundial, which the author affirms could serve in a flower-garden 'as any other Knot for ornament'.

33. PIERRE MORIN *the younger* (*c.* 1600–1690)

Remarques necessaires pour la Culture des Fleurs. La maniere avec laquelle il les faut cultiver, & les Ouvrages qu'il faut faire selon chaque Mois de l'Année. Avec une Methode facile pour faire toutes sortes de Palissades, Bosquets, & autres Ornemens qui servent à l'embellissement des Jardins de plaisir; & un Catalogue des Plantes les plus rares; Le tout diligemment observé par P. Morin, Fleuriste. Nouvelle Edition. Augmentée d'un Traité des Oeillets, & de la maniere qu'il les faut cultiver. A Paris, Chez Charles De Sercy, au Palais, au Sixiéme Pilier de la Grand' Salle, vis à vis la Montée de la Cour des Aydes, à la Bonne-Foy couronnée. M. DC. LXXVIII. Avec Privilege Du Roy.

12° 16 × 9.5 cm. $\tilde{a}^6$ ($\tilde{a}6 + \chi 1$) A–O (in alternating 8s and 4s) P^4 *i–xiv* 1–171 *172–176* p.

BINDING: Sprinkled calf binding. Gilt spine with lettering 'Rema Des Fleu'.

PLATES: Engraved title-page showing a scene with Flora offering a wreath and a bouquet of flowers to a crowned female figure—probably representing France—who holds a banner displaying the title of the book. At the top of the page is another female figure sprinkling water on the scene below; she is accompanied by a putto blowing gentle breezes. Head-pieces and vignette.

PROVENANCE: Unidentified coat of arms with a crown and fleur-de-lis on the front pastedown and on the verso of the title-page. On the title-page in manuscript 'Mr Jeanbaptiste Durand avocat'.

REFERENCES: Pritzel 6454 (for 1st edn, 1658); Hunt 300 (for 1665 edn); Schnapper 1985, p. 358; Leith-Ross.

PIERRE MORIN THE YOUNGER (also known as Pierre Morin III) belonged to a family of floriculturists in Paris who became quite well known during the course of the seventeenth century. 'Curieux fleuristes', they were not simply gardeners, but connoisseurs who collected as well as sold rare and unusual flowers. In 1619 Pierre Morin married Françoise de La Brosse, a cousin of Guy de La Brosse, personal physician to Louis XIII and later the first Intendant of the Jardin Royal des Plantes Médicinales, as the Jardin des Plantes in Paris was originally called (see *An Oak Spring Hortus*, no. 18).

During his trips to Paris in 1644 and 1651, the diarist John Evelyn visited the Morin gardens and collections. In 1649 and 1669 the English royalist Richard Symonds also toured their gardens. Symonds's account of his visit in 1649 to Pierre Morin's garden in the Faubourg Saint-Germain is augmented by a sketch-plan he made on the spot, revealing that it was fancifully designed in the form of an oval flower, with box-edged beds representing the petals (see Leith-Ross, fig. 1).

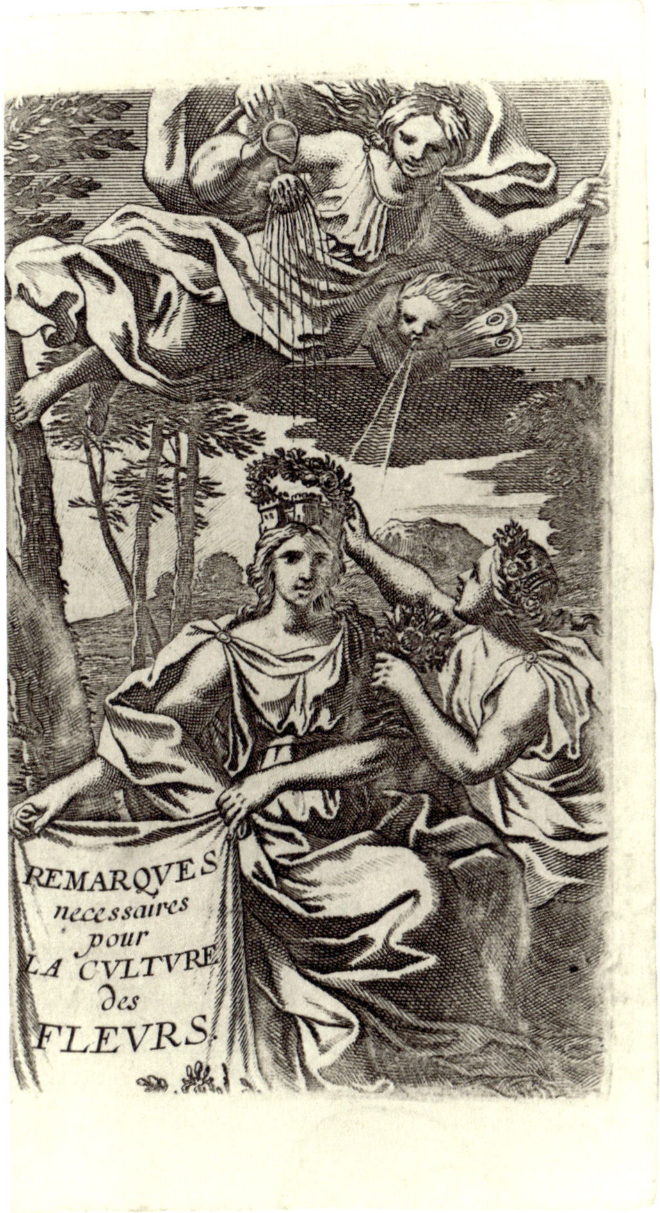

PIERRE MORIN,
Remarques necessaires pour la culture des fleurs.
Frontispiece

The first catalogue of the plants available from the Morin gardens was published in 1621 by Pierre's older brother, René. In 1651 Pierre produced a catalogue of the flowering plants to be found in his own garden in the Faubourg Saint-Germain, plants that were also available for export. In his catalogue the most popular flowers of the period were amply represented; among those listed we find one hundred varieties of tulip, twenty-four species of ranunculi, seventy-one different irises and twenty-seven species of anemones, many with double blooms. In his house, which was located within the grounds of his garden, Pierre Morin had assembled a cabinet of natural curiosities—sea-shells, corals, minerals and insects (he had planned to write a natural history of insects)—as well as a collection of fine porcelain and prints by Dürer, Lucas van Leyden and Jacques Callot.

In his *Remarques* our eclectic author not only presents a new catalogue of his collection, augmented by various species recently introduced from North America, but also offers detailed observations and specific advice for those interested in the cultivation of flowers. Published in 1658 by Charles de Sercy, the work was very well received and went through several reprintings, including this edition of 1678, which appeared together with a brief treatise on the carnation.

The first part of *Remarques* describes the tasks that every gardener must carry out in the different months of the year, from January through to December. It is followed by detailed lists of the flowering plants that are sensitive to the cold, those plants requiring rich damp soil, those preferring thin dry soil, etc. Notes on the season in which each plant comes into flower are included. There is also a list of the flowers most prized for their fragrance, headed by 'l'oreille d'ours' or *Primula auricula*, a native flower that was soon to become extremely popular. This somewhat dry presentation of gardening rules is followed by a brief and agreeable treatise on the carnation, which opens with a curious apologue describing a battle between the carnation and the rose 'pour sçavoir si l'un sera le Roy, & l'autre la Reyne dans la petite Monarchie des Fleurs' (p. 66). Morin describes the virtues of the plant, the different colours in which the blooms might be found and the curious names the plant has been given at various times, and also provides advice regarding its cultivation.

At the end of the work is a catalogue of the flowering plants that could be obtained from the author's garden, including many varieties of tulip, then celebrated as the 'King of Flowers'. There is also a note informing readers that many 'plantes des Indes', such as the 'Couronnes Imperiales à grandes Fleurs', 'Tulipes de la Chine' and 'Muscaris nouveaux de differentes couleurs des Anciens', could be procured through him. This practical, yet elegant, little volume testifies to Morin's competence as a gardener and his love for his profession. The opening 'Avis aux Curieux' reflects the new attitude adopted by the intellectuals of the Enlightenment: 'On dit que c'est un bonheur tout particulier, quand la satisfaction de notre esprit est jointe à l'utilité que notre Profession nous apporte: Ayant reçu cette faveur du Ciel, je croirois être ingrat, de la posseder, si je ne fasois part à ceux qui aiment l'Agriculture du peu de connoissance que j'en ay tirée par ma propre experience'.

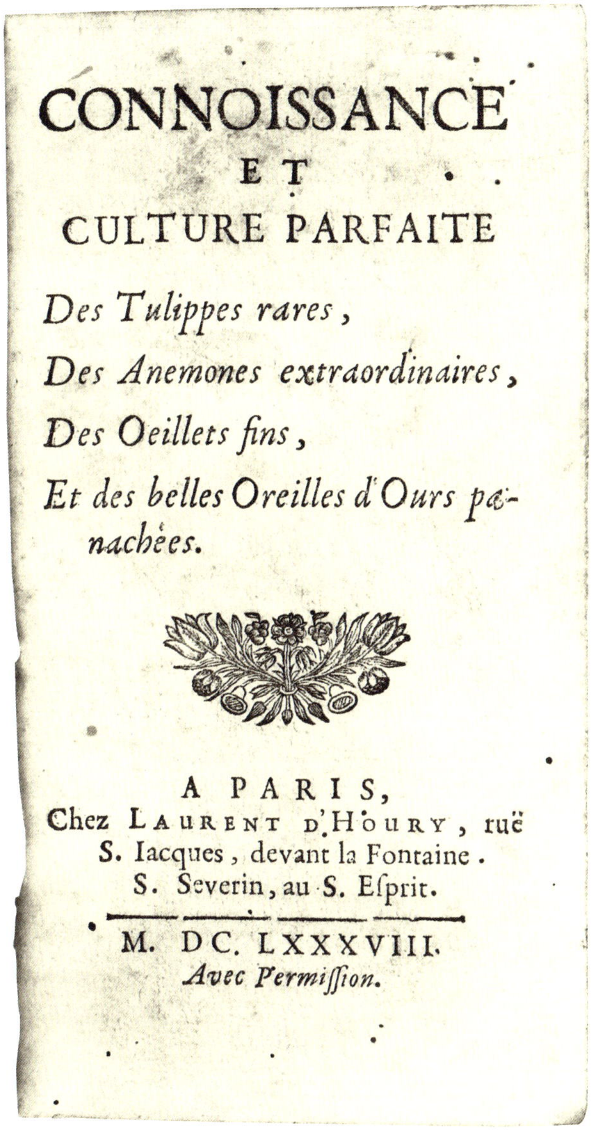

CONNOISSANCE
ET
CULTURE PARFAITE
Des Tulippes rares,
Des Anemones extraordinaires,
Des Oeillets fins,
Et des belles Oreilles d'Ours panachées.

A PARIS,
Chez LAURENT D'HOURY, ruë
S. Iacques, devant la Fontaine.
S. Severin, au S. Esprit.

M. DC. LXXXVIII.
Avec Permission.

Anonymous author,
Connoissance et culture parfaite des tulippes rares.
Title-page

34. Anonymous Author

Connoissance et Culture Parfaite Des Tulippes rares, Des Anemones extraordinaires, Des Oeillets fins, Et des belles Oreilles d'Ours panachées. [*typographic ornament*] A Paris, Chez Laurent d'Houry, rüe S. Jacques, devant la Fontaine. S. Severin, au S. Esprit. M. DC. LXXXVIII. Avec permission.

12° 15 × 8.5 cm. ã^4 ^{2}ã^8 ẽl A^4(–A4) B–M (in alternating 8s and 4s) *i–xxvi* *1* 2–142 p.

BINDING: Mottled sheep; red letter spine label with gilt lettering 'Des Tulippes'; date at the foot of the spine.

PLATES: Woodcut head- and tail-pieces.

REFERENCE: Goldsmith C1382.

THIS SOPHISTICATED little volume discusses four of the most highly prized flowers of the period, and gives precise instructions as to their cultivation. Although its author chose to remain anonymous while dedicating his work to one of the most celebrated figures in his *métier*, André Le Nôtre—'Contrôleur general des Bâtimens . . . Dessignateur des Jardins de Sa Majesté' and creator of the gardens of Versailles—he nevertheless proudly affirms: 'Je puis dire sans vanité ya peu de personnes en France qui connoissent mieux que moi les belles fleurs.' Antoine Schnapper (1988, p. 44) has posited the author of this work to be the 'curieux fleuriste' Valnay, but since this surname is cited by the author many times in the text together with those of many other celebrated florists, this attribution is somewhat questionable.

In the *Avvertissement* the author observes that, while God created many flowers, all of which can testify to His glory, he himself wished to concentrate on the four he considers truly exceptional, the 'tulippe rare', the 'anémone extraordinaire', the 'œillet fin' and the 'oreille d'ours panaché', flowers unknown to most gardeners, and which could be fully appreciated only by 'gens d'esprit'. In fact, our author himself was clearly an *homme d'esprit*, a virtuoso in the fullest sense of the term, for in this book he not only demonstrates his vast experience and knowledge of flowers, but discourses with ease on poetry and painting, observing in passing, for example, that in the city of Paris a painter as great as Nicolas Poussin could barely earn enough to live on. In his own opinion the infinite variety of nature was much more rewarding to the discriminating connoisseur than any work of art, be it a painting, a medallion or the most delicate piece of porcelain statuary, for 'un tableau est toujour un, un ognoin se multiplie'. An artist could never compete with nature's palette of colours, and a medallion, however old, remained 'modern' next to the eternal beauty of the flower.

The author opens his work with the tulip, which he too declares to be the king of flowers. Indeed, in this period many French floriculturists, such as Morin (see no. 33), Robeau, Vireau, Brunet, Lombard and Laure, dedicated themselves to the difficult art of multiplying bulbs; according to our author, Lombard in particular succeeded in developing many varieties that were exceptional for their colour or size and were thus eagerly sought after by wealthy collectors. The perfect tulip, the author continues, should be neither too tall nor too short, of fine shape, and with thick petals, so that the flower would last at least eleven or twelve days. Its colours should be unusual and variegated, without a predominance of red. After furnishing detailed instructions on their cultivation, the author describes how these striking flowers could be used to create 'Theatres de Fleurs de Tulippes', an idea conceived by the same Monsieur de Valnay. Such theatres comprised five or six terraced steps covered with a green cloth on which tulips of different form and colour placed in phials of water could be arranged to make designs or even pictures.

The author then passes to the anemone, which he states was brought to France from the Indies by Monsieur Bachelier, who was also responsible for introducing the Indian chestnut to France. Apparently, Bachelier managed to cultivate not only the simpler varieties, but double anemones too, in a wide range of brilliant hues. These flowers were much less difficult to grow than the tulip and were greatly prized for their size and for the velvety texture of their petals. Thus, many floriculturists of the period, including Destrange, Caboud, Valnay and Descoteaux, devoted themselves to its cultivation. After describing the culture of anemones, the author passes to the third of his preferred flowers, the carnation, of whose origin he admits himself to be ignorant. Although both Morin and Charpentier, the Lieutenant-General de Compiègne, had written works on the subject, our author felt that much remained to be said, particularly with regard to its cultivation, which was not nearly as difficult as these two authors claimed it to be. Among the floriculturists who specialized in the growing of carnations he cites Descoteaux, Valnay, Breard and Caboud. The perfect carnation must be large in size and variegated in colour, with a large number of petals without excessive dentation, and many leaves. The author describes with admiration an unusually large variety that had been successfully cultivated and which was named 'Le Nouveau Monde'.

The final section is devoted to the auricula, which, the author observes, could be found growing spontaneously in France, although the wild flower was small in size and pale in colour when compared to the cultivated varieties '[qui] ont ces qualitez desirables dans les fleurs qui sont plaisir à voir' (p. 122). The multicoloured varieties were still so new in the author's day that when, in 1685, the Chevalier de Saint Mory, 'cet illustre & grand Curieux', sent some specimens to the King, the latter could not hide his surprise that there existed in France flowers of such exceptional beauty. The corolla of the flower must be large and open if one is to fully appreciate the beauty of the auricula, the author expounds, and the stem must be sufficiently strong to bear the weight of the bloom. As with the tulip, anemone and carnation, this flower should possess those qualities of satiny texture, brilliant colour and bizarre features that were then so much sought after.

35. Anonymous Author

Secrets Pour Teindre La Fleur D'Immortelle En Diverses Couleurs, Avec La Maniere De La cultiver. Pour Faire Des Pastes de differentes odeurs fort agreables. Et Pour Contrefaire du Marbre au naturel, propre pour toute sorte d'ouvrages figurez. Par F. L. D. T. R. [*typographical ornament*] A Paris, Chez Charles De Sercy, au Palais, au sixiéme Pilier de la Grand' Salle, vis-à-vis la montée de la Cour des Aydes, à la Bonne Foy Couronnée. M. DC LXXXX. Avec Permission.

12° 15 × 8 cm. A–F (in alternating 8s and 4s) G^2 *1–9* 10–72 75–77 *78*p.

BINDING: 17th-century calf. On the spine in gilt 'La Fleur d'Imm'.

PLATES: Woodcut initials, head- and tail-pieces.

PROVENANCE: On the verso of the title-page is the coat of arms of 'His Excellency Alexander Lord Polwarth, Eldest Son of Patrick Earl of Marchmont, &c. His Majesties Ambassador Extraordinary & Plenipotentiary to Congress at Cambray, Lord Clerk Register of Scotland, & Lord Lieutenant of Berwickshire. Anno 1722'.

THIS RARE VOLUME of unknown authorship belongs to a minor, somewhat recondite genre of literature that emerged at the end of the 1600s and continued to flourish for well over a century. Its object was to make public the 'secret' practices of artists and artisans, which, in keeping with

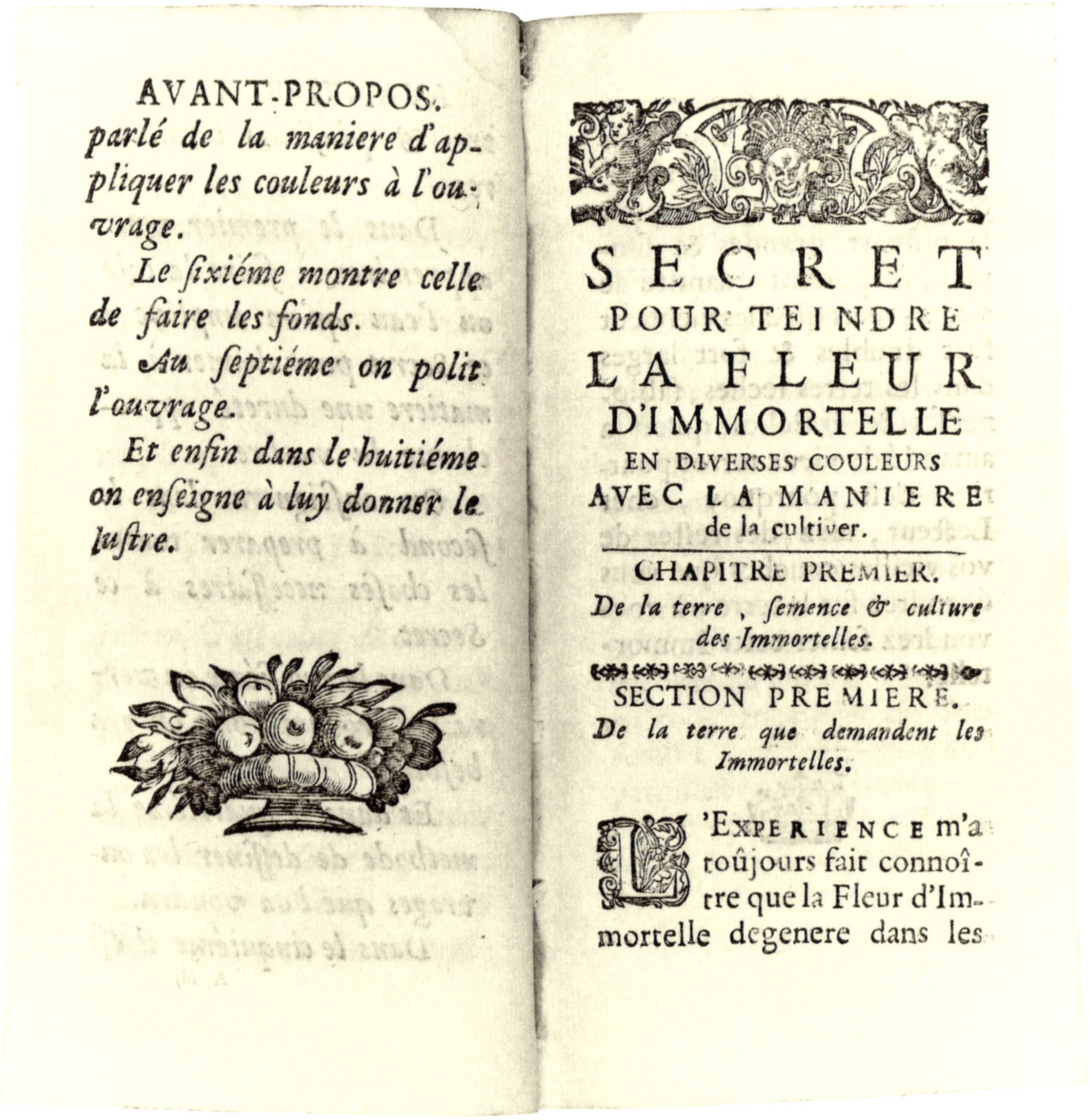
AVANT-PROPOS.
parlé de la maniere d'appliquer les couleurs à l'ouvrage.
Le ſixiéme montre celle de faire les fonds.
Au ſeptiéme on polit l'ouvrage.
Et enfin dans le huitiéme on enſeigne à luy donner le luſtre.

SECRET
POUR TEINDRE
LA FLEUR
D'IMMORTELLE
EN DIVERSES COULEURS
AVEC LA MANIERE
de la cultiver.

CHAPITRE PREMIER.
De la terre, ſemence & culture des Immortelles.

SECTION PREMIERE.
De la terre que demandent les Immortelles.

L'EXPERIENCE m'a toûjours fait connoître que la Fleur d'Immortelle degenere dans les

Anonymous author, *Secrets pour teindre la fleur d'immortelle en diverses couleurs.* Tail-piece and head-piece on pages 8 and 9

a tradition going back to the Middle Ages, were jealously guarded in the workshop or studio and carefully handed down from father to son. The scholars of the Enlightenment were vehemently opposed to this custom and declared that these secrets should be brought out into the open. That such close practices were widespread is demonstrated by the fact that Diderot himself, in the 'Discours Préliminaire' to the *Encyclopédie* made a direct appeal to all artists, artisans and other persons whose work involved particular techniques, to divulge their secrets for the benefit of all.

Although, when considered in this light, *Secrets pour teindre la fleur d'immortelle* seems to herald a new epoch, its contents are firmly rooted in the Baroque period. The seventeenth-century love of artifice and its obsession with the imitation and transformation of materials influenced every sphere of the arts and sciences. Botany and its related activities were no exception: garden architecture, floriculture and floral painting and decoration lent themselves with particular felicity to the expression of the Baroque aesthetic. In his famous treatise *Flora overo cultura di fiori* (see no. 29) G. B. Ferrari provided various 'secret' recipes that he claimed would modify the colour, size or form of certain flowers. In *Secrets* the author, identified only by his initials F.L.D.T.R., reveals how to tint the petals of the immortal or everlasting flower (*Xeranthemum annum*), so named because it could be dried without losing its colour (which varied from white to purple) or form. The author also describes the seeds, how the plant should be cultivated and the flowers gathered, and, finally, the best way of arranging the dried flowers in pleasing bouquets.

According to the author, the everlasting flower could be tinted by infusion with various substances, such as vitriol spirits, oil of tartar, essence of sulphur or aqua vitae, all of which could be procured from the pharmacist in rue Nueve de la Cadrerie. Common household items, such as vinegar, salt, quicklime or ink, could also be used. The procedure itself was quite simple: one merely had to immerse the stems of the flowers for a certain length of time in a solution prepared according to the author's instructions, and then leave the flowers to dry. Sulphur would colour the flowers deep black; cream of tartar would tint them a pale yellow; and vinegar would dye them a beautiful cherry red. We also learn that Flanders glue mixed with water and applied to the tips of the stems will protect the flowers from insects. The brightly coloured dried flowers could then be used to make handsome floral arrangements in vases made of clay, plaster or paper.

36. Paolo Bartolomeo Clarici (1664–1725)

[*title-page in red and black*] Istoria e coltura Delle Piante Che sono pe'l Fiore più ragguardevoli, e più distinte per ornare un Giardino in tutto il tempo dell'Anno, Con un copioso Trattato degli Agrumi. Di D. Paolo Bartolomeo Clarici. Opera postuma Consacrata à Sua Eccellenza il Sig. Gerardo Sagredo Procurator di S. Marco. [*vignette of putti with lemon-tree in a vase, watering-can and tools*, 5 × 7.5 *cm.*] In Venezia, MDCCXXVI. Presso Andrea Poletti. Con Licenza De' Superiori, e Privilegio.

4° 23.5 × 17 cm. a–c^4 d^6 A–5B^4 5C^6 *i–xxxvi* 1–761 *762–764* p.

BINDING: Old full vellum; dark-red morocco spine label with lettered gilt.

PLATES: Engraved portrait of the author by F. Zucchi on a4, and one folding plate by F. Zucchi after Gio. Filippini of 'Prospetto e topografia del Palazzo e Giardini Sagredo nella Villa di Marocco' between signatures [d4] and A1; woodcut decorations and initials throughout.

REFERENCES: Arnold, p. 158; Pritzel 1728; Masson 1961, pp. 238–41; Puppi, p. 110; *DBI*, XXVI.136–7; *An Oak Spring Hortus*, no. 10.

CLARICI's *Istoria e coltura delle piante*, primarily devoted to the cultivation of flowers, had a great influence on garden architecture during the course of the seventeenth and eighteenth centuries in Italy. The work was published posthumously by the author's brother, Domenico Maria Clarici, who also appended some 'Notes on the life of the author'. Thus we learn that as a young man

Clarici studied theology and the liberal arts at the Collegio Nazareno in Rome. He moved to Padua, seat of one of the oldest universities in Italy, at the age of twenty-three, and there attracted the attention of Cardinal Giorgio Correr. This learned prelate became his patron, encouraging him in his studies and introducing him into various intellectual circles in Padua and Venice.

Clarici's interest in geography and topology led him to the study of botany, which became an absorbing passion. He constructed his own garden and dedicated himself to the cultivation and study of plants and flowers. He was ordained at the age of fifty-three, and at this point began to write *Istoria e coltura delle piante*. He was particularly encouraged in this project by Gherardo Sagredo, a procurator of S. Marco, whose family owned a villa with a splendid garden at Marocco, near Venice. The first plate in *Istoria*, in fact, presents a view of this garden. Clarici spent the last years of his life in Udine, where he had been invited by Federico Cornaro, a deputy of the city. There, in recognition of his great learning, Clarici was immediately accepted as a member of the Accademia degli Sventati. In Udine he wrote many works on historical and topographical subjects, but died unexpectedly in Venice in 1725, before they could be published. Fortunately, the author's brother undertook the task of publishing *Istoria*, thus providing us with a fitting monument to the accomplishments of this learned ecclesiastic.

The work is divided into four sections: the first provides the reader with a brief history of garden architecture and instructions on how to construct a proper garden; the second is dedicated to cultivation techniques; the third to botanical history and the cultivation of certain prized flowering plants; and the fourth to a dissertation on citrus plants.

In the first part Clarici, drawing on his considerable personal experience, provides detailed instructions on how to set out a flower-garden. In addition to beds laid out in pleasing designs, he recommends there should be 'four small platforms or pyramids oriented towards . . . the East, West, North and South' respectively, built of earth or timber. Plants and flowers placed on these raised structures could then automatically find their most favourable orientation for growth. He protests against the design and construction of a certain style of staircase that was apparently quite the fashion in his day, but which served, he thought, only to encumber the landscape.

Flowerbeds, Clarici stated, must be laid out so that the gardener can move about and work with ease. It was also desirable to place those flowers with longer stems behind those with shorter stems, so that the latter are not constantly thrown into the shade. The borders should be marked off with low boards, or hedges of box or some other unscented shrub. Plants such as thyme, sage or southernwood are to be avoided, as they are too highly perfumed, and some grow for only a brief season before dying and then spring up again in unexpected places, thus destroying the order of the garden, 'whose greatest beauty lies in the perfection of its symmetry, without which, even if the most lovely flowers are to be found there, . . . it will never be pleasing to the eye' (p. 5).

The second and third parts of the work then deal with the 'History and cultivation of those plants that are the most beautiful to look at because of their flowers, and that provide the most

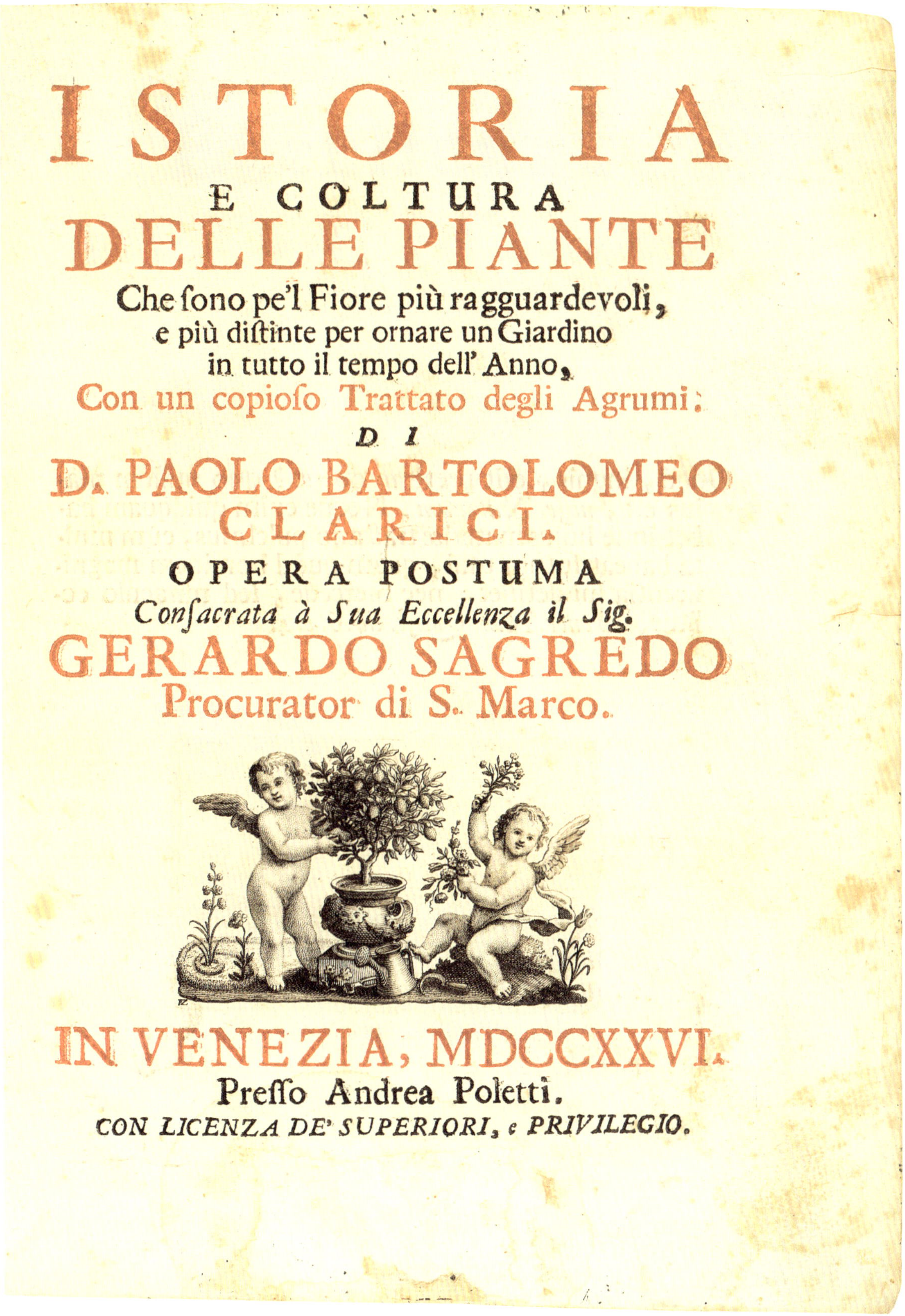
ISTORIA
E COLTURA
DELLE PIANTE
Che ſono pe'l Fiore più ragguardevoli,
e più diſtinte per ornare un Giardino
in tutto il tempo dell' Anno,
Con un copioſo Trattato degli Agrumi:
D I
D. PAOLO BARTOLOMEO
CLARICI.
OPERA POSTUMA
Conſacrata à Sua Eccellenza il Sig.
GERARDO SAGREDO
Procurator di S. Marco.

IN VENEZIA, MDCCXXVI.
Preſſo Andrea Poletti.
CON LICENZA DE' SUPERIORI, e PRIVILEGIO.

PAOLO BARTOLOMEO CLARICI, *Istoria e coltura delle piante*. Title-page elegant ornamentation for a Garden'. After the soil has been suitably prepared, one must be ready to plant 'the bulbs, roots, and flowers that grow from seeds', not haphazardly, but in a rational and scientific manner. If more than one species is to be planted in a single flowerbed, careful note must be made of what has been planted each time. The serious gardener should know the provenance of every plant in his garden, and the climate and soil conditions characteristic of its native habitat. Bulbs and roots should be counted and planted, allowing for a sufficient amount of space between each. Great care must be taken in the choice of which flowers are to be placed in close proximity. For example, the ranunculus and the anemone should not be planted in the same bed for they require the same type of compost, and the ranunculus, being the stronger plant, will quickly absorb most of the nutrients, to the detriment of the anemone. Furthermore, different plants with large roots or bulbs, such as the double jonquil, tulip and ranunculus, should be planted in separate beds.

Directions are also given on how to gather and store seeds and the 'secret' of how to force 'very large blooms' from them, following the procedure described by Ferrari (see no. 29)—that of choosing the largest seeds, cleaning them carefully and planting them in wax tubes in the earth.

Clarici then embarks on a detailed discussion of the appropriate use of urns for planting. In his opinion they constituted the most beautiful 'ornament to rejoice the eye'. Since they can be continually replenished with plants in full flower, one can create the impression of a garden in perpetual bloom. The most suitable urns for the garden are of glazed terracotta: the most fitting colours are white, green, red or light blue, but never black. They should be well-proportioned—that is, as high as they are wide, with drainage holes in the sides, not at the bottom. Such urns must first be immersed in water for one day to draw off the odours of the fired clay, which are noxious to seeds. Once his plants have taken root in these containers, the wise gardener will take care to weed them regularly, for weeds are not only unsightly but may steal nutrients from the plants themselves.

The third part of *Istoria* opens with some general remarks in praise of flowers and their beauty—their delicate fragrance, and the infinite variety of colours and forms that a bountiful Nature has seen fit to create. Clarici then embarks on a rigorously scientific discussion of the morphology of flowers and the latest theories concerning the sexual reproduction of plants. He cites the opinions of Marcello Malpighi, Joseph Pitton de Tournefort and Giulio Pontedera (professor of botany at the University of Padua and director of its famous botanical garden), thus demonstrating his perfect mastery of the scientific aspects of his subject.

Istoria e coltura delle piante was a work of serious scholarship, one based on a careful reading of the most recent scientific works on botanical subjects as well as the author's own research. Thus Clarici left behind him a useful legacy that was to influence generations of garden architects and floriculturists both in Italy and elsewhere.

37. Robert Furber (*c.* 1674–1756)

Twelve Months of Flowers.

1° [Broadsheets] 59 × 47 cm. Engraved list of subscribers and 14 engraved and hand-coloured plates.

BINDING: Contemporary half calf and marbled paper boards. On the spine is a green label: 'Casteels' Flowers and &'. Bound with the author's *Twelve plates with figures of fruit* of 1732 (12 engraved and hand-coloured plates depicting fruits).

PLATES: Bound as the frontispiece is a leaf bearing 4 engraved plates of flower vases. This is followed by an engraved leaf consisting of a list of subscribers in alphabetical order, surrounded by a frame of flowers (each labelled with its common name). At the end of the list: 'To Their Royal Highness's Frederick Prince of Wales, the Princess Royal, To the Most Noble, R^{t} Honble, and the other Generous Subscribers to These Twelve Months of Flowers. This plate is most humbly Dedicated, by their Most Obedient, and Oblig.^{d} humble Servants. Robert Furber, Peter Casteels, Henry Fletcher'. Each of the 12 plates for the months of the year carries the inscription 'From the Collection of Robert^{t}. Furber Gardiner at Kensingtn. 1730. Desig'd by P^{tr}. Casteels Engrav'd by H. Fletcher'.

A plate depicting the foliage and flowers of 3 trees follows, on which is written: 'These 3 Trees Flower'd at the Earle of Pembrokes. The Yellow Tulip Tree at 19 years old in the year 1720 it bore above 509 Tulips; The White Tulip with ye Fibers green the Leaf ever green with an Aramatic Smel flower'd in 9. years but never grows near so big as other Tulip tree. The Silke Cotton Tree flowered in 11 years, very beutifull out of the Green Pods thewhite, not as the Linnen Cotton, but bright Shining as Silke, the Cotton when all expanded out of one of these little Pods takes as much room as a large Walnut, this southern Silke Cotton tree is the first that flowere in Europe. Furber Gardner over against high Parke Gate near Kingsinton'.

Followed by 12 engraved and hand-coloured plates of flower bouquets.

PROVENANCE: On front cover is the armorial bookplate of Henry Grey, Duke of Kent (1671–1740), and, on front pastedown, that of his descendant Thomas Philip, second Earl de Grey (1781–1859).

REFERENCES: Dunthorne 113; Nissen 674; Henrey, III.733; *An Oak Spring Pomona*, no. 26.

THIS IS THE FIRST EDITION (1730) of a celebrated flower catalogue published by Robert Furber, a nurseryman from Kensington, then on the outskirts of London. Entitled *Twelve Months of Flowers*, its twelve plates present almost 400 different flowering species, grouped according to the month in which they flower. Reflecting in style the grand manner of the Baroque period, the flowers are arranged in elaborate bouquets and placed in elegant urns. Next to each flower appears a number, which corresponds to a name in the key printed at the bottom of the plate on either side of a cartouche inscribed with the name of one of the months of the year.

Twelve Months was conceived as a flower catalogue, but its commercial function was adroitly veiled and the artistic quality of its illustrations distinguishes it from the more modest pamphlets generally produced by floriculturists, including Furber himself, in this period. For the work Furber sought the collaboration of Pieter Casteels (1684–1749), an artist from Antwerp who had achieved great renown in England for his paintings of birds and flowers, which decorated the walls of many aristocratic homes. Casteels designed the series of twelve floral plates, which were then engraved by Henry Fletcher, an artist well known for his reproductions of famous paintings and portraits. The plates were designed to be afterwards coloured in by hand, as they are in the Oak Spring copy. Preceding the plates is a page containing a list of wealthy subscribers, the title of the work and the

ROBERT FURBER,
Twelve Months of Flowers.
Flowers blooming in the month of July

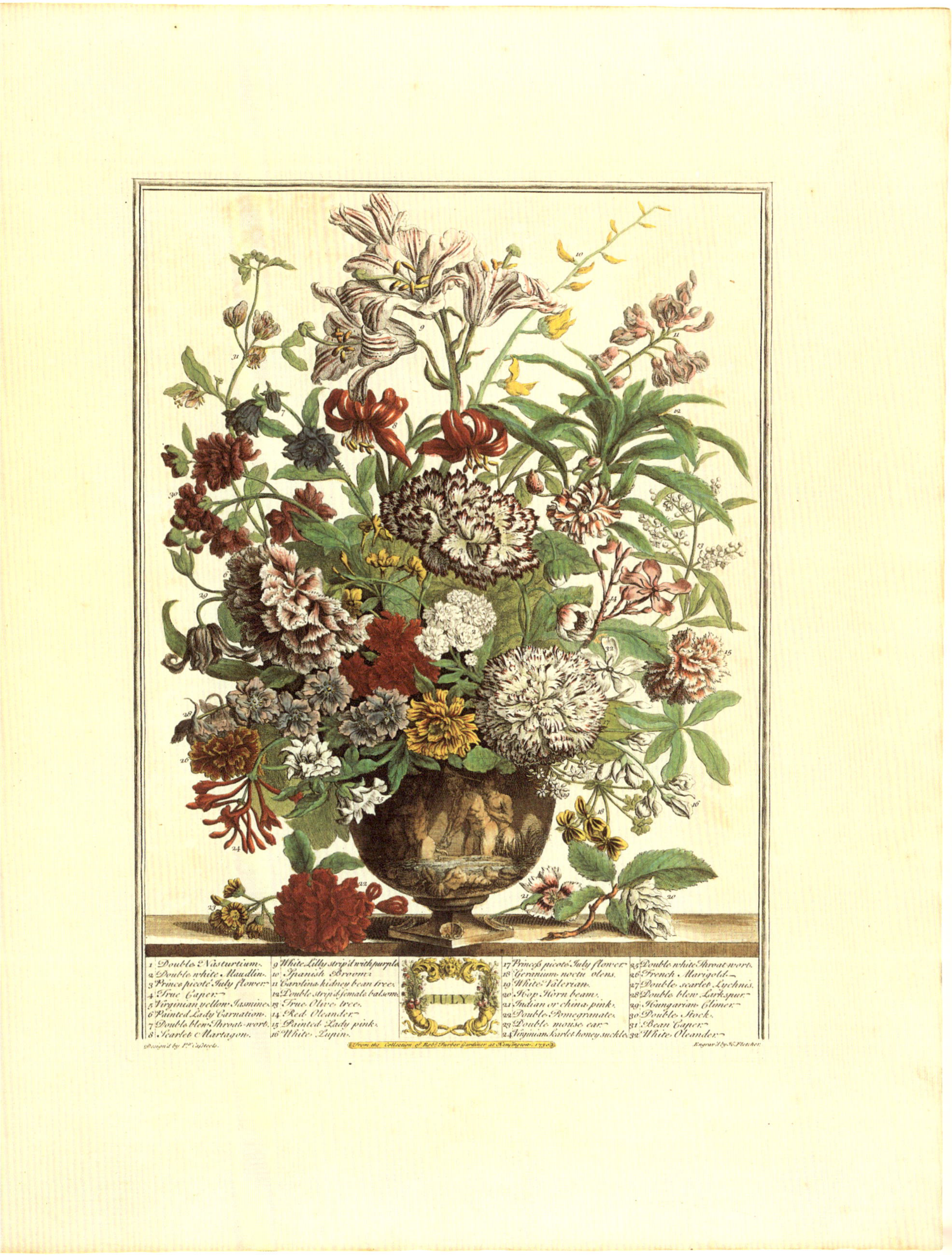

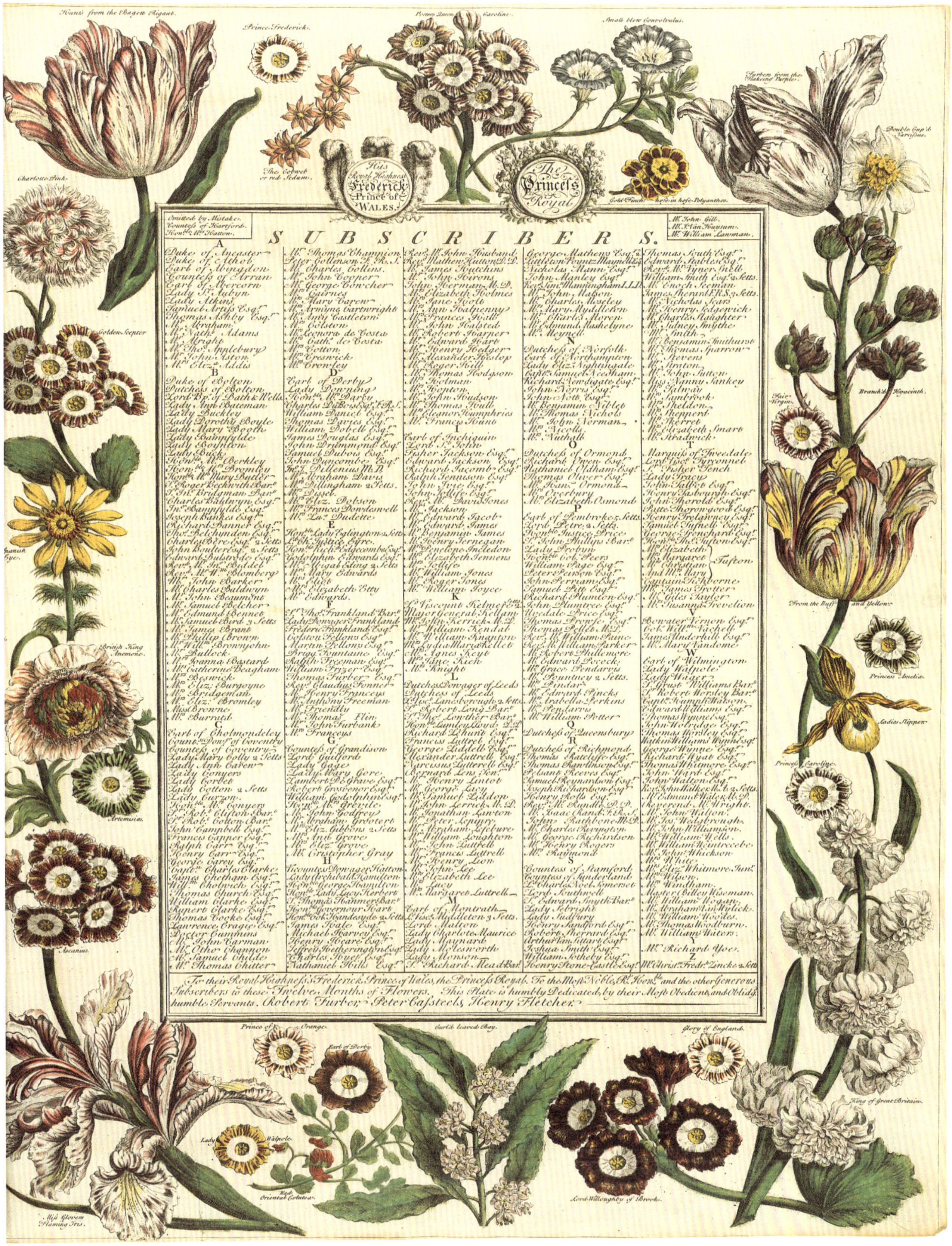

ROBERT FURBER,
Twelve Months of Flowers.
Subscribers' page

names of its authors. The page is framed by a wreath of auriculas, a very popular flower at the beginning of the eighteenth century. Many new species were cultivated and, like the tulip a century earlier, named after a famous contemporary or classical figure, such as the 'Earl of Derby', 'Artemisia' or 'Ascanius'. In 1726 Richard Bradley, then professor of botany at Cambridge, noted in his *New Improvements of Planting and Gardening* that 'numerous varieties of Auriculas are all distinguished by names of titles of great men'. (The 1719–20 and 1739 editions of *New Improvements* are also at Oak Spring.)

Despite their great profusion, every flower in these compositions is readily identifiable. They range from native species to exotic specimens from the Americas: Furber records that he received various specimens from Mark Catesby, 'a very curious gentleman, from Virginia' (Dunthorne, p. 14). Thus we can see the blossoms from an 'American flowering maple' (*Acer saccharinum L.*) and a 'Virginian flowering maple' (*Acer rubrum L.*) in the plate for the month of March (nos. 12 and 15), a 'Virginian Scarlet honeysuckle' (*Lonicera americana rubella*) in the plate for July (no. 24), and a 'Virginian Birthwort' (*Aristolochia durior*) in the plate for September (no. 10).

Twelve Months was such a success that in 1732 Furber published another series, this time depicting the various fruits that come into season in each month of the year, with plates designed once again by Pieter Casteels. In fact, a copy of the fruit series can be found bound together with the Oak Spring copy of *Twelve Months*. In the same year Furber published a smaller, more economical edition of the floral series, entitled *The Flower-Garden Display'd*, this time furnishing it with a brief text on the cultivation of flowers. In its introduction Furber explains that 'As the First Impression of the Monthly Flower-Pieces have been so well receiv'd by the Publick, we thought a particular Description of the Flowers, and the nature of their Culture, which could not be express'd in the Plates themselves, might hope for the same Success.' Oak Spring has this edition of 1732 as well as the 1734 issue. In 1733 Furber published another small catalogue of fruit and flowers entitled *A Short Introduction to Gardening*, which boasted an even longer list of subscribers than the one in *Twelve Months* (see *An Oak Spring Pomona*, no. 27).

We do not know whether the *Twelve Months of Flowers* actually increased the sales of seeds and plants from the Kensington nursery, but the catalogue itself went through innumerable reprintings during the course of the eighteenth century. The images were reversed during the engraving of new plates for some of the subsequent editions, and were heavily coloured after printing to hide the numbers; thus the work's original purpose was forgotten, and it came to be admired for its aesthetic qualities alone. Oak Spring also possesses a hand-coloured copy of the edition of 1749 engraved by Thomas Bowles, J. Clark and Parr and printed by John Bowles with the title *Flora, or, A curious collection of Ye most beautiful Flowers as they appear in their greatest Perfection each Month of the Year.*

38. Filippo Arena (1708–1789)

Tomo I. La Natura, e Coltura de' Fiori fisicamente esposta in due trattati Con nuove ragioni, osservazioni, e sperienze. A vantaggio de' Fioristi, de' Fisici, de' Botanici, ed Agricoltori. Per il P. Filippo Arena Piazzese della Compagnia di Gesu, Professor di Mattematica nell'Imperial Collegio di Palermo. Tomo Primo. [*etched vignette*] In Palermo Appresso Angelo Felicella MDCCLXVII. Impr. Del Castillo V.G. Impr. Natoli R.C.P.

Tomo II. Della Natura, e Coltura de' Fiori fisicamente esposta Trattati due del Sac. Ignazio Arena da Piazza Dottore in Sacra Teologia, e Canonico dell'Insigne Collegiata nella sua Patria. Tomo Secondo. [*etched vignette*] In Palermo MDCCLXVIII Appresso Angelo Felicella. Impr. Del Castillo V.G. Impr. Natoli R.C.P.

4° 25 × 19 cm. 3 volumes bound in 2, 65 folded leaves of plates. π^4 A–3I^4, π^4 A–3F^4 ^{2}A–Y^4 (volume 1: 2V2 missigned 2T2). Imperfect copy: lacking the title-page to volume 3; the plates of the atlas of illustrations of volume 3 are divided and bound at the end of volumes 1 and 2.

BINDING: Decorated paper bindings. On a leather label attached to the spine of volume 1: 'P. Arena Coltura de Fiori. To I.'; and on the label to volume 2: 'P. Arena Coltura de Fiori. To. II.'

PLATES: 65 engraved plates; etched title vignettes; woodcut initials, and head- and tail-pieces.

REFERENCES: 'Some Early Italian Gardening Books', *JRHS*, 1923, pp. 183–4; Nissen 48; Pritzel (2nd edn) 223; Plesch, no. 16; *A Magnificent Collection*, no. 9; *DBI*, IV.79–81.

FILIPPO ARENA was born in Sicily at Piazza Armerina in 1708. He entered the Jesuit Order at a very young age and eventually became professor of mathematics and philosophy at the University of Palermo. His real passion, however, was for the botanical sciences, and he conducted meticulous studies on hundreds of plant species, including many that were native to Sicily. He was very widely read, and familiar with the work being done by his colleagues elsewhere in Europe, including the great Swedish botanist Linnaeus. This enabled him to compile a systematic work on his studies that was as up to date as anything being published in his day. But because *La Natura, e coltura de' fiori* was printed in Sicily, an island far from the principal intellectual centres of Europe, his work became known only to a handful of contemporaries, and his significant contributions to the history of botany were never fully recognized.

Arena states in the introduction to the first volume of his work that it was intended not only for 'florists' but for scientists too, in particular 'botanists, physicists and agriculturists', since the knowledge contained therein was the fruit of direct 'observation and experience'. Volume One is devoted to a general discussion of botany, and to the 'anatomy of the flower and the function of each of its parts'. In fact, it contains a remarkably advanced dissertation on the sexual generation of plants, including the function of pollen and the importance of its transmission by insects, a discovery that has traditionally been attributed to the German botanist Joseph Gottlieb Koelreuter (1733–1806), director of the botanical garden in Karlsruhe.

The second volume deals with horticulture, in particular the cultivation of flowers. In the first sixty pages Arena describes in considerable detail how to set up a flower-garden. One must first seek a flat, sunny piece of ground, well sheltered from harsh winds, and blessed with good soil and an abundant source of water. He declares the elaborate French parterre to be quite impractical and a

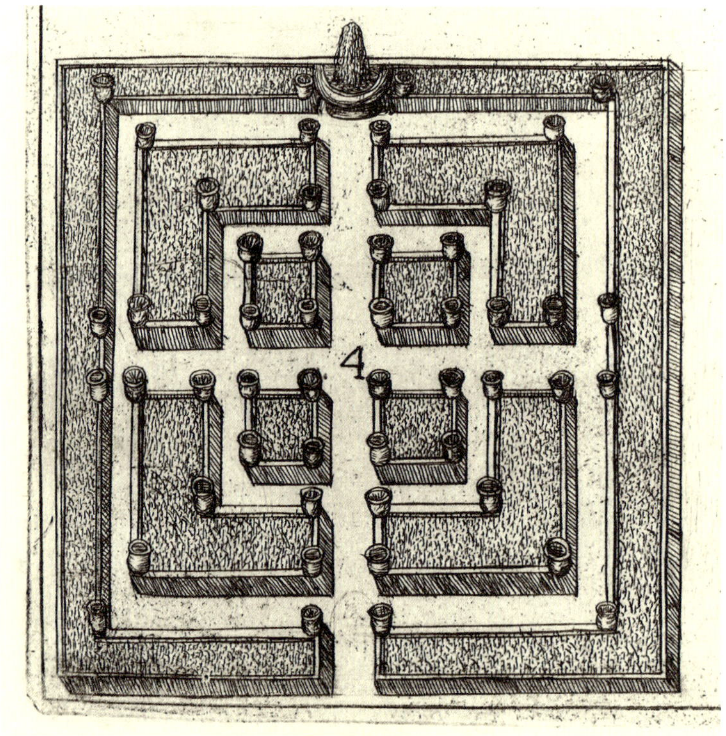

FILIPPO ARENA,
La natura, e coltura de' fiori.
Detail of flowerbeds
from volume I,
plate 3

hindrance to the proper cultivation of plants. Ideally, a flowerbed should be laid out in the form of a regular polygon and surrounded by a narrow path, permitting the flower lover 'full enjoyment of all the flowers' and at the same time allowing the gardener access to every plant. He recommends setting up a fountain in the centre of the garden and benches or seats along the main paths. Urns of flowers might be placed on columns, pedestals and flights of steps, but an excess of such decorative effects was to be avoided. The garden should be surrounded on all sides by a wall lined with citrus espaliers. In the second plate the author provides five examples of designs for gardens that the architect or surveyor might draw on for ideas. The flowerbeds themselves should be lined with bricks firmly driven into the earth, or painted blocks of hardwood (oak or chestnut), or evergreen shrubs such as thyme, myrtle or box. He even suggests materials that might be used to cover the paths, for example, small shells mixed with lime.

Part Two contains a discussion of flowers and their cultivation. The first chapter in this section tells how to force semi-double and double blooms from the seed, and offers a genetic explanation of this phenomenon. He then describes the proper use of the seedbed, and explains when and how young plants should be transplanted (taking into account various factors, including the disposition of the stars), and how to care for the plants once they have established themselves, protecting them from inclement weather, plant diseases and noxious insects. A catalogue of various 'noble' species of flowers closes the second volume.

Although Arena spent most of his life in Sicily, his perspective and training were anything but

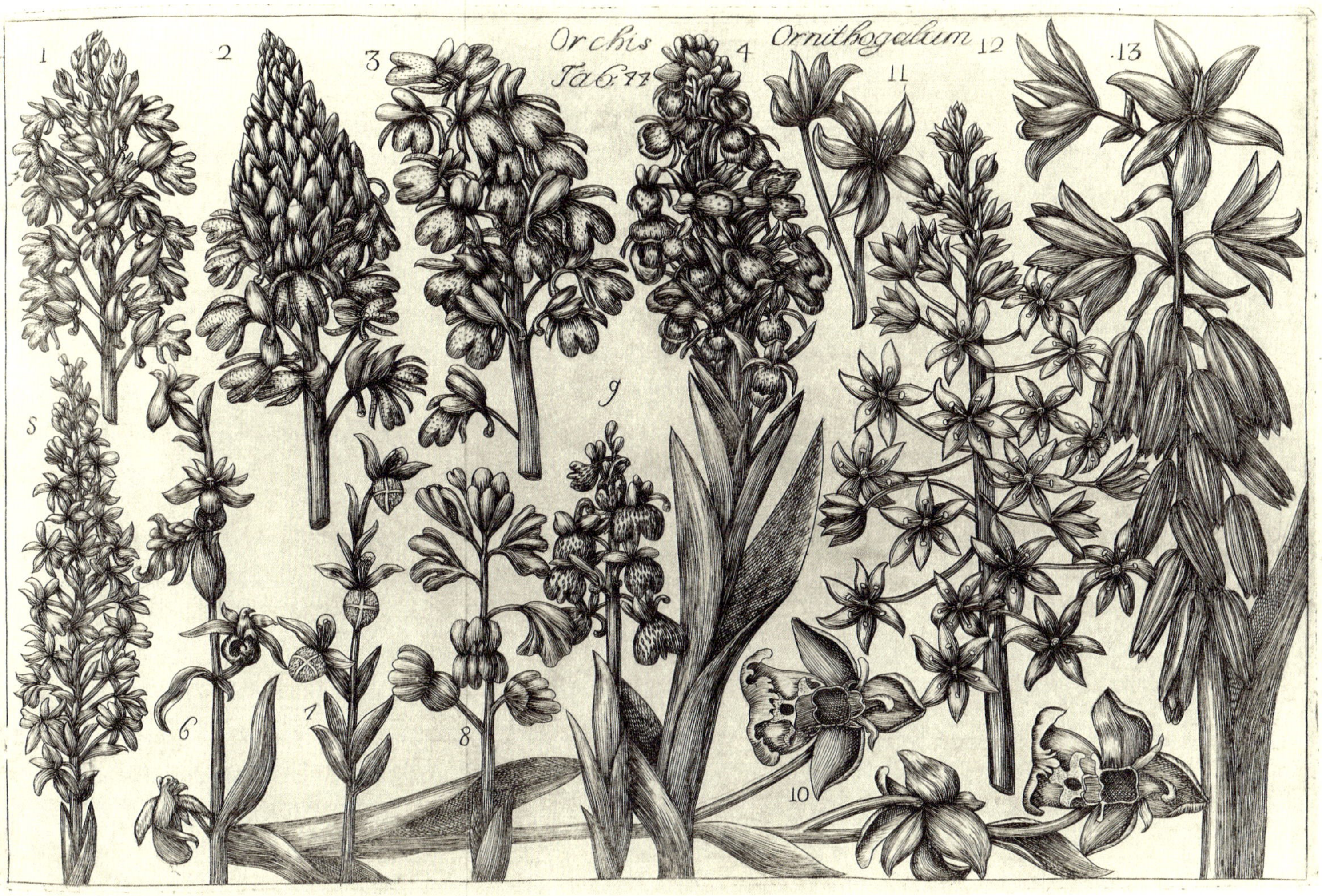

Volume II, plate 44: 1. Soldier orchid (*Orchis militaris*); 2 and 3. *Orchis purpurea*; 4. Common spotted orchid (*Dactylorhiza fuchsii*); 5. Fragrant orchid (*Gymnadenia conopsea*); 6. Early spider orchid (*Ophrys sphegodes*); 7. *Ophrys crucigera*; 8 and 9. Green-winged orchid (*Orchis morio*); 10. Late spider orchid (*Ophrys holoserica*); 11–13. Varieties of *Ornithogalums*

provincial, and he was thoroughly acquainted with the European literature on his subject. He avows that he has diligently studied the most important botanical texts published in France and Italy over the past 200 years, and, for example, cites with great admiration the work of G. B. Ferrari (see no. 29), the edition of 1684 of *Manuale di giardinieri* published in Venice by Agostino Mandirola (see no. 30), and *Istoria e coltura delle piante* by P. B. Clarici (see no. 36). He also mentions French authors, including Louis Liger, who in 1704 published the *Jardinier fleuriste*, Jean Paul de Rome d'Ardène, author of the *Traité des renoncules* of 1746 (see no. 76), the poet and theologian René Rapin, and Abbé Noël Antoine Pluche, author of the controversial work *Le Spectacle de la nature* (1732).

Arena dedicated himself with equal energy to the illustration of his work. In the introduction he states that he felt the need for a third volume, this one to consist entirely of illustrations (in the Oak Spring copy these illustrations have been divided and bound at the ends of volumes One and Two) because, as he says, 'I have always judged incomplete those books of botanists, and of florists, which have accurately described the plants and flowers without placing before one's eyes their drawn image'. To this end he decided to 'engrave with his own hands' sixty-five plates for *La Natura, e coltura de' fiori*. He enlisted the aid of a confrère of the Jesuit Order, Mario Cammareri, to help him in this project; their signatures appear together in the first engraving, an allegory showing Flora, crowned with a garland of flowers, offering floral tributes to three figures—Botany (shown holding an alembic), Physics (who is looking through a magnifying glass) and Pomona (who holds a spade and a basket of fruit).

In the preparation of his plates Arena drew heavily upon *Phytanthoza iconographia*, a work containing more than 1,000 botanical illustrations published in four folio volumes between 1737 and 1745 by the German botanist Johann Wilhelm Weinmann. In some of his plates Arena copied floral images from different pages by Weinmann, arranging them in original compositions that reflect his own considerable aesthetic taste. The anemones that appear in plates 6 and 7, for example, were copied from plates 119–26 of volume I of *Phytanthoza*, while the lilies in plates 36 and 37 were drawn from plates 653–62 in volume III. Plate 42, which shows an amaryllis, and plate 53 illustrating a hibiscus, were instead copied from Ferrari's *Flora*. Arena noted that the illustrations, which he intended to arrange to be coloured by hand in some copies, would be of use not only to the botanist, but to painters, sculptors, embroiderers and weavers as well.

Even if *La Natura, e coltura de' fiori* did not win the recognition that it certainly deserved, it represents a paradigmatic work from the era of the Enlightenment. With such works as these, which were based on direct experience rather than on a reinterpretation of the classics, science entered the modern era. As Arena significantly points out in the closing paragraph of his introduction: 'Reason, and Experience are two sisters who must never separate, for one serves as the escort of the other', and indeed the science of floriculture could not have advanced as it did without the guidance of both.

V · FLOWERS FOR MONARCHS, VIRTUOSI AND MEN OF SCIENCE

THE FIRST scientific text dedicated exclusively to flowers, *Florum, et coronariarum odoratarumque . . . historia* by Rembert Dodoens (no. 39) appeared in the second half of the sixteenth century, but it was only at the beginning of the seventeenth century that the image of the flower acquired a specific typology, evolving out of the purposes for which it was created and modulated by the personality of the artist who was called on to depict it. Rembert Dodoens's work, for example, was published in the Low Countries at a moment when scientific interest in botany was united with a commercial interest arising out of the new and highly successful floriculture industry established there. Thus, the images in *Florum, et coronariarum* sought to adhere to a high standard of scientific objectivity, and even though the artist who produced the preparatory drawings, Petrus van der Borcht, was famous for his own work in other genres, he sought to adapt his style to the tradition of the contemporary illustrated herbal, and to depict each plant as realistically as possible. He and other artists were seriously hampered at first, however, by the graphic conventions of this tradition and by the severe limitations of the woodcut, which produced a harsh, unsupple line with no *chiaroscuro* effects.

It was primarily the great monarchs of Europe who, during the course of the seventeenth and eighteenth centuries, encouraged the study of botany by establishing academies and botanical gardens, and who generously supported both the scientists and the botanical artists on whom the former relied to assist them. Thus, the principal capitals of Europe—Paris, London and Vienna, as well as the papal city of Rome—became important centres for botanical studies.

One of the most singular works to be produced in this period was *Paradisi in sole* by the English herbalist, pharmacist and gardener John Parkinson (no. 40). This book, according to its author, was not only a 'speaking garden' and a 'feminine work of flowers' (being dedicated to Queen Henrietta Maria, the consort of Charles I); it was above all a 'visual garden' and a floral anthology, presenting hundreds of indigenous and exotic plants.

It was in France under the reign of Louis XIV that politically influential figures began to sponsor projects in all the areas of the natural sciences—not only botany but zoology and mineralogy. Thus, Gaston, duc d'Orléans (1608–60), initiated a magnificent collection of botanical paintings, known as *vélins* because they were painted on the very finest vellum. These works, which combined scientific accuracy with a high degree of artistic and technical refinement, were inherited in 1660 by Gaston's nephew, Louis XIV. Unlike his uncle, however, Louis XIV did not consider these *vélins* to be a private possession created solely for his personal enjoyment. He realized that they constituted a unique resource, as well as a monument to the glory of the French monarchy, and placed them on display at the Bibliothèque du Roi, where they were universally admired. (They can be seen today in the Muséum National d'Histoire Naturelle, Paris.)

By Louis's express order, and that of his successors, the collection continued to be added to for many years by artists working in the service of the crown under the close supervision of qualified scientists. Hundreds of botanical illustrations were produced by artists that included Nicolas Robert (no. 44), Jean de Joubert, Claude Aubriet (no. 47), Madeleine Basseporte (who was particularly admired by Louis XV—he enjoyed visiting her studio to watch her paint) and Gerard van Spaendonck (no. 58), all following the same format and using the same technique of bodycolour on vellum. Louis XIV reckoned this project to be of sufficient importance to justify the creation of a special position, 'Peintre du cabinet du Roi pour la miniature', a title that was not suppressed during the Revolution, but merely modified to that of 'professeur d'iconographie naturelle au Jardin des Plantes' (Vezin, p. 41).

Colbert, Louis XIV's chief minister, also asked Nicolas Robert to prepare the illustrations for a monumental *Histoire générale des plantes*, to be compiled under the auspices of the Académie Royale des Sciences. This work was never published in definitive form, but Robert did manage to complete several hundred plates before his death in 1685 (see no. 43). Colbert was not slow to realize that these naturalistic illustrations, which were being produced for strictly scientific purposes, could also be used as models in the applied arts. Thus, a purely decorative form of flower painting arose, which was further encouraged by Charles Le Brun (1619–90), 'premier Peintre du Roi', director of the royal tapestry works of Gobelins, powerful patron of the arts, and undisputed arbiter of taste, pomp and artistic fame in the second half of the seventeenth century in France. The works of Jean Baptiste Monnoyer represent perhaps the most complete expression of this sumptuous new style of 'peinture des fleurs'. Monnoyer not only produced magnificent paintings in oils, he also worked on large engraved plates, prints from which were widely disseminated (no. 46); they served as models for other artists, especially the tapestry weavers of Aubusson, who were celebrated throughout Europe for their *verdures*.

This grandiloquent and theatrical approach to botanical illustration, characterized by rich formal and chromatic contrasts, was continued by Gerard van Spaendonck, an artist who did not conform to the more 'scientific' tradition of his predecessors even though he worked exclusively at the Jardin des Plantes for forty-eight years, and in close collaboration with such eminent botanists as Jussieu, Buffon, Lamarck and Geoffroy de Saint-Hilaire. Spaendonck produced paintings of great visual impact, which were displayed and admired in numerous Parisian salons. In contrast to these somewhat ostentatious works, a sketchbook by Spaendonck at Oak Spring (no. 58) offers us a rare and intimate glimpse of the artist's working technique.

At the same time that official institutions were sponsoring large-scale botanical works, many more modest texts illustrated with interesting plates were being produced by scientists without the benefit of patronage from the wealthy or powerful. Some were illustrated by the authors themselves, but most were the fruit of a close collaboration between the botanist and the artist. It had, in fact, become common practice for naturalists to invite artists to accompany them on their expeditions, so that they could document their observations and discoveries *in loco*. Claude Aubriet, for example, travelled through the Levant from 1699 to 1702 in the company of Tournefort, while Nikolaus Joseph von Jacquin took Franz Anton von Scheidel along with him on many of his Alpine expeditions.

The very appearance of the illustrated book underwent significant modifications in this period, for advances in the printing process during the eighteenth century allowed the insertion of ever more numerous and elegant decorative elements—vignettes, head- and tail-pieces, title-pages and frontispieces. They even made their appearance in scientific works, which were now being produced in ever larger numbers.

The typology of the florilegium also changed considerably during the course of the seventeenth and eighteenth centuries. The Dutch botanist and professor Abraham Munting, for example, introduced a radical innovation in his *Phytographia curiosa* (no. 45) by placing his plants against landscape backgrounds filled with learned references to the classics or other sources. Later, the eclectic scientist John Hill (no. 53), who was also a competent botanical artist, was to publish a lavishly illustrated (although scientifically somewhat dubious) catalogue of flowering plants, while L'Héritier de Brutelle (no. 57) created an entirely new typology for the illustrated monograph.

A refined Rococo elegance characterizes many other works from the eighteenth century now in the collection at Oak Spring, such as the hand-painted florilegium that Guillaume van Blarenberghe and his son Louis-Nicolas—two artists not known hitherto for their work in this genre—prepared for a pharmacist and 'amateur' of Lille (no. 49). A similar charm invests the bodycolour paintings of the German Christoph Ludwig Agricola (no. 48), and a floral painting by the English artist Thomas Robins the elder (no. 52). The works of the younger Robins (no. 52), on the other hand, are marked by a more rigorously scientific approach, while an album of paintings by Franz Anton von Scheidel (no. 54) reflects the influence of both the traditional florilegium and the contemporary, more decorative approach to flower painting.

The works of Georg Dionysius Ehret, which were created for both scientists and connoisseurs (nos. 50, 51), are distinguished by the unique vision of one of the greatest artists in the history of botanical illustration. Having worked in his youth for some of the most eminent botanists of the period, including Linnaeus, Ehret utilized his profound knowledge of plant structure to create flower paintings in which the subject-matter dominated the painting with truly impressive force.

Inspired by his example, or benefiting directly from his teaching, younger artists such as Simon Taylor (no. 55) and various amateurs (including the many aristocratic English ladies whom Ehret counted among his students; see no. 56) were to carry on the tradition he had established with such magisterial authority.

At the end of the eighteenth century and the beginning of the nineteenth, as the Romantic period came into its own, the genre of flower painting adapted itself to the shift in taste and became more popular than ever. One need only recall the figure of Pierre Joseph Redouté, perhaps the most celebrated flower painter of all time. Working for a succession of royal patrons—Marie-Antoinette as well as the Empress Joséphine and Marie-Louise of Austria (the first and second wives of Napoleon)—Redouté was instrumental in shaping the taste of the period; indeed, for many years his exquisite style and technique completely dominated the genre.

At Oak Spring are several works by another artist from this period who is almost unknown today but who certainly merits consideration—the Italian Baldassare Cattrani. This talented artist worked for many years as a botanical illustrator in the prestigious botanical garden at Padua, and many of his paintings of plants found their way into the collection of Eugène Beauharnais, probably via his mother, the Empress Joséphine, who had invited Cattrani to join the group working in the gardens of her country estate at Malmaison. The paintings at Oak Spring shed valuable light on the *œuvre* of this now almost forgotten figure (no. 59).

In contrast to his contemporary Redouté, the great English botanical illustrator James Sowerby (no. 60) worked almost exclusively with scientists. His efforts contributed to the establishment of one of the most important and, to this day, indispensable instruments for the dissemination of botanical information, the scientific journal.

39. Rembert Dodoens (1517–1585)

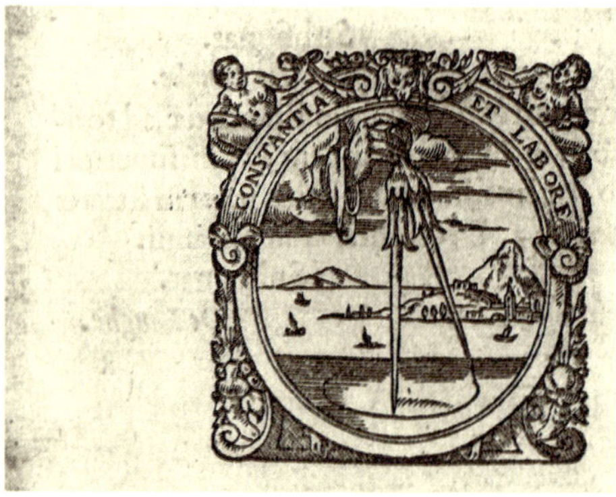

REMBERT DODOENS, *Florum, et coronariarum.* Christophe Plantin's printer's device on the title-page

Florum, et Coronariarum Odoratarumque Nonnullarum Herbarum Historia, Remberto Dodonaeo Mechliniensi Medico auctore. Altera Editio. [*printer's device*] Antverpiae, Ex officina Christophori Plantini M.D. LXIX.

8o A–V^8 *1* 2–309 308 309 [i.e., 311] *312–320* p. Bound with *Frumentorum, leguminum, palustrium et aquatilium herbarum* . . . , Antverpiae: Ex Officina Christophori Plantini, 1566. A manuscript index of the plants on the front and back pastedowns continuing on the free endpapers. Manuscript notations in ink throughout. On the title-page in manuscript: '27. Libris 4r Pemptadis 2 um Editi Lat. Fol. 1 [. . .] respondentur'.

BINDING: Contemporary vellum. On the spine in ink: 'Remberti Dodonaei Historia Florum et Coronariarum'.

PLATES: Woodcuts by Petrus van der Borcht (1545–1608).

PROVENANCE: Three library property stamps, all nonidentifiable.

REFERENCES: Pritzel (2nd edn) 2347; Nissen 514; Stafleu and Cowan, no. 1486; J. Van Meerbeeck, *Recherches historiques sur la vie et les ouvrages de Rembert Dodoens (Dodonaeus)*, Utrecht, 1980; *Botany in the Low Countries*, no. 104.

THIS IS the second edition of a work on flowering plants by the Dutch physician and botanist Rembert Dodoens, which was first published in 1568 by the printer Christophe Plantin (*c.* 1520–89) of Antwerp. Plantin himself was an amateur botanist who possessed a fine garden at Berchem (*Botany in the Low Countries*, no. 14). From his presses emerged some of the most important botanical texts of the sixteenth and seventeenth centuries, written by authors of the stature of Carolus Clusius and Matthias de L'Obel, as well as Dodoens; their work contributed significantly to the development of the modern system of botanical taxonomy.

Dodoens was born in Leeuwarden, although his family soon afterwards moved to Malines. where his father, a physician, decided to practice. Dodoens also took up the study of medicine, graduating from the University of Louvain in 1535. He began working in Malines in the very period when the plague and a number of other infectious diseases were ravaging Europe. His professional activity was not limited to the practice of medicine, however, for he had a lively curiosity that embraced all areas of science, and wrote numerous books on a wide range of subjects. His earliest work, *Pauli Aeginetae de febribus . . .*, published in Cologne in 1546, is a treatise on fevers. It was followed by *Cosmographica*, a compendium on the subject of cosmology and astronomy, published in Antwerp in 1548.

In 1553 Dodoens's Dutch translation of the seminal work *De historia stirpium* appeared, which had been published by the German botanist Leonhart Fuchs just one year earlier. Instead of presenting the plants in alphabetical order, as the author had done, Dodoens attempted to group them systematically by botanical class. This project was followed by a treatise on seeds and leguminous plants, two botanical atlases, and a translation of Clusius's *Histoire des plantes.* In 1554 the first important botanical work written by Dodoens himself, *Cruydeboeck*, was published by Jan van der Loe of Antwerp. An updated edition appeared in 1563, and an English edition entitled *A nievve Herball* appeared in 1578.

In 1565 Dodoens began working on a monumental herbal that was initially published by Christophe Plantin in parts consisting of small volumes. The work *Florum, et coronariarum*, which was published in 1568, actually constituted the second volume of this ambitious project; it appeared immediately after a volume on legumes, grains and aquatic plants. Work on the project was interrupted when Dodoens moved to Vienna to assume the post of personal physician to Emperor Maximilian II. He subsequently transferred to Prague when Rudolf II succeeded Maximilian. Finding the atmosphere at Prague's royal court less than congenial, however, in 1577 Dodoens moved to

Cologne, where he spent some years and wrote several medical works, including a treatise on melancholy and one on the use of wine in the treatment of illnesses. In 1582 he was appointed professor of botany at the University of Leiden, and one year later his major work, *Stirpium historiae pemptades sex*, illustrated with 1,341 woodcuts, was published by the Officina Plantiniana.

Indeed, Plantin played a key role in the history of botanical illustration, as well as in the history of botany, for at his celebrated printing-house in Antwerp he set up a permanent artists' workshop, where the plates to illustrate his scientific texts could be prepared. One of the artists working for him was Petrus van der Borcht, a gifted young draughtsman who specialized in maps and cityscapes. A native of Malines, van der Borcht had already executed the title-page and a portrait of the author for Dodoens's *Cruydeboeck*, and it would have been natural for Dodoens to recommend him to Plantin. Van der Borcht produced many botanical illustrations for Plantin, which were then engraved by artists that included Gerard Janssen van Kemen (*c.* 1540–90), Arnaud Nicoläi (*c.* 1525–*c.* 1590), Antoine van Leest (*c.* 1545–*c.* 1592) and Cornelis Müller. In order to counter the high costs of preparing these illustrations, Plantin frequently utilized the same plates in different works. Furthermore, if images of certain plants were lacking, he did not hesitate to have his artists copy suitable ones from works already published, one of his sources being Fuchs's *De historia stirpium*. Nevertheless, in Plantin's workshop many illustrations were drawn *ex novo* from live specimens by van der Borcht and others. Numerous original woodblocks and drawings by van der Borcht are now in the Musée Plantin–Moretus, Antwerp.

Dodoens's *Florum, et coronariarum* was devoted to those flowers, ranging from herbaceous annuals and perennials to rare bulbous plants, that were traditionally used in the sixteenth and seventeenth centuries to form garlands and wreaths, *coronae*; in fact, a garden planted with such flowers was sometimes referred to in this period as a 'hortus coronarius'. Among the flowers he describes are the rose, violet, lily, marguerite, marigold, amaranth, iris, narcissus, anemone and carnation and sweet-smelling herbs such as thyme, marjoram, lavender and lavender cotton.

In the second edition of his work, which was published in 1569, Dodoens included descriptions of the tulip (p. 107), the 'Flos africanus' or tagetes (p. 62), and the 'Chrysanthemum perunianum' or sunflower (pp. 305–6), which, as the author explained, is crowned with a flower of 'exceptional size' and is native to 'Peru and other American regions'. Dodoens then informs the reader that this rare flower had already been cultivated in the gardens of the royal palace in Madrid and in the botanical garden at Padua, while in the Low Countries the virtuoso and amateur botanist Jean de Brancion had managed to coax a moderate-size plant to grow from a seed in his garden in Malines, although the cold of the approaching winter prevented it from coming into flower.

The edition of 1568 of *Florum, et coronariarum* was illustrated with 108 woodcuts that cost Plantin more than eighty-two guilders to have prepared, and could be purchased for the price of six *stuivers* (*Botany in the Low Countries*, no. 41). The edition of the following year contained 109 engravings, of which five were entirely new (four were used to replace four illustrations in the first edition). The

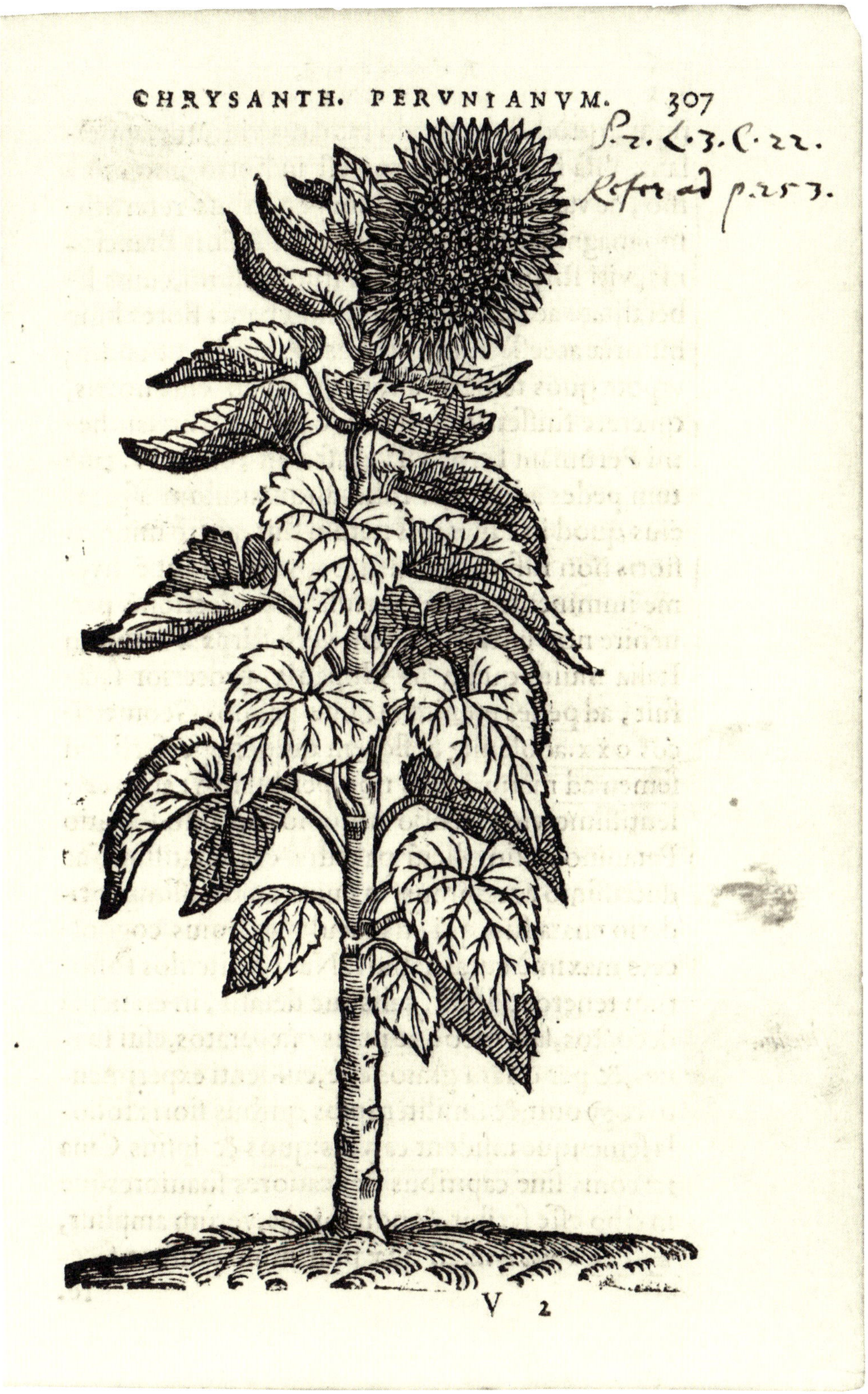

REMBERT DODOENS,
Florum, et coronariarum.
Sunflower
(*Helianthus annuus*),
page 307

original drawings were all the work of Petrus van der Borcht, who had also illustrated the first volume on cereal grains, legumes and marsh plants. We know from contemporary documents that the artist received a payment in advance in June 1567 for his work on the latter volume.

40. JOHN PARKINSON (1567–1650)

[*woodcut title-page showing the Garden of Eden in an oval frame from which rays emanate. The frame is flanked above by two winged putti representing the winds and below by two urns containing flowers. A stream runs through a wooded landscape in which various native and exotic plants can be seen looming over the tiny figures of Adam and Eve. At the bottom of the page on a heart-shaped tablet is written*] Paradisi in Sole Paradisus Terrestris. or A Garden of all sorts of pleasant flowers which our English ayre will permitt to be noursed Up: With A Kitchen garden of all manner of herbes, rootes, & fruites, for meate or sause used with us, and An Orchard of all sorte of fruitbearing Trees and shrubbes fit for our Land together With the right orderinge planting & preserving of them and their uses & vertues Collected by John Parkinson Apothecary of London 1629 [*at the very foot of the page, in two columns, is written*] Qui veut parangonner l'artifice a Nature Et nos Parcs a L'Eden indiscret il mesure Le pas de l'Elephant par le pas du ciron, Et de l'Aigle Le vol par cel du mucheron, A Switzer.

[*in the colophon on 3G2v*] London, Printed by Humfrey Lownes and Robert Young at the signe of the Starre on Bread-Street hill. 1629.

2° 33 × 21 cm. ⋆⋆^{6} A–3F^6 3G^2 *1–12* 1–612 *1–16* p.

BINDING: Calf, with spine title 'Parkinson's Herbal 1629' in gilt lettering.

PLATES: Woodcut title-page, woodcut portrait of the author (16 × 10 cm.), one woodcut of flowerbeds, 108 full-page woodcut botanical plates, 3 small woodcuts in the text. Decorated head-pieces and initials.

PROVENANCE: On the front free endpaper in manuscript 'John Hay pre 1–10–0'. Ex-libris of Glyndebourne.

REFERENCES: Rohde, 142–62; Arber, pp. 135–6; Pritzel (2nd edn) 6933; Nissen 1489; Hunt 215; Anderson 1977, pp. 227–34; Blunt and Stearn, p. 87.

THIS IS THE first edition of one of the oldest and most highly regarded books on gardening and flowers to be published in England. Very little is known about the author, John Parkinson, except that he was a herbalist and gardener born in Nottinghamshire, who in 1616 was actively practising as a pharmacist in London. He received the title of 'Apothecary' from James I, and after the publication of his magisterial herbal, Charles I had him named 'Botanicus Regius Primarius' (First botanist to the King).

Parkinson possessed a private garden at Long Acre in London 'well stored with rarieties', as he boasted in his *Theatrum botanicum* (p. 609), where he cultivated a large number of flowers and medicinal plants. Many of these are described in *Paradisi in sole paradisus terrestris*. He was a friend of the editor Thomas Johnson, who revised the 1633 edition of John Gerard's famous *Herball*, first published in London in 1597, and he knew John Tradescant the elder, a widely travelled naturalist who possessed a large and eclectic museum across the Thames at Lambeth. He also frequented the home of the celebrated physician Sir Theodore Mayerne; the preface to *Paradisi* takes the form of a eulogy dedicated to Mayerne.

In 1640 Parkinson published an herbal in English entitled *Theatrum botanicum*, in which he

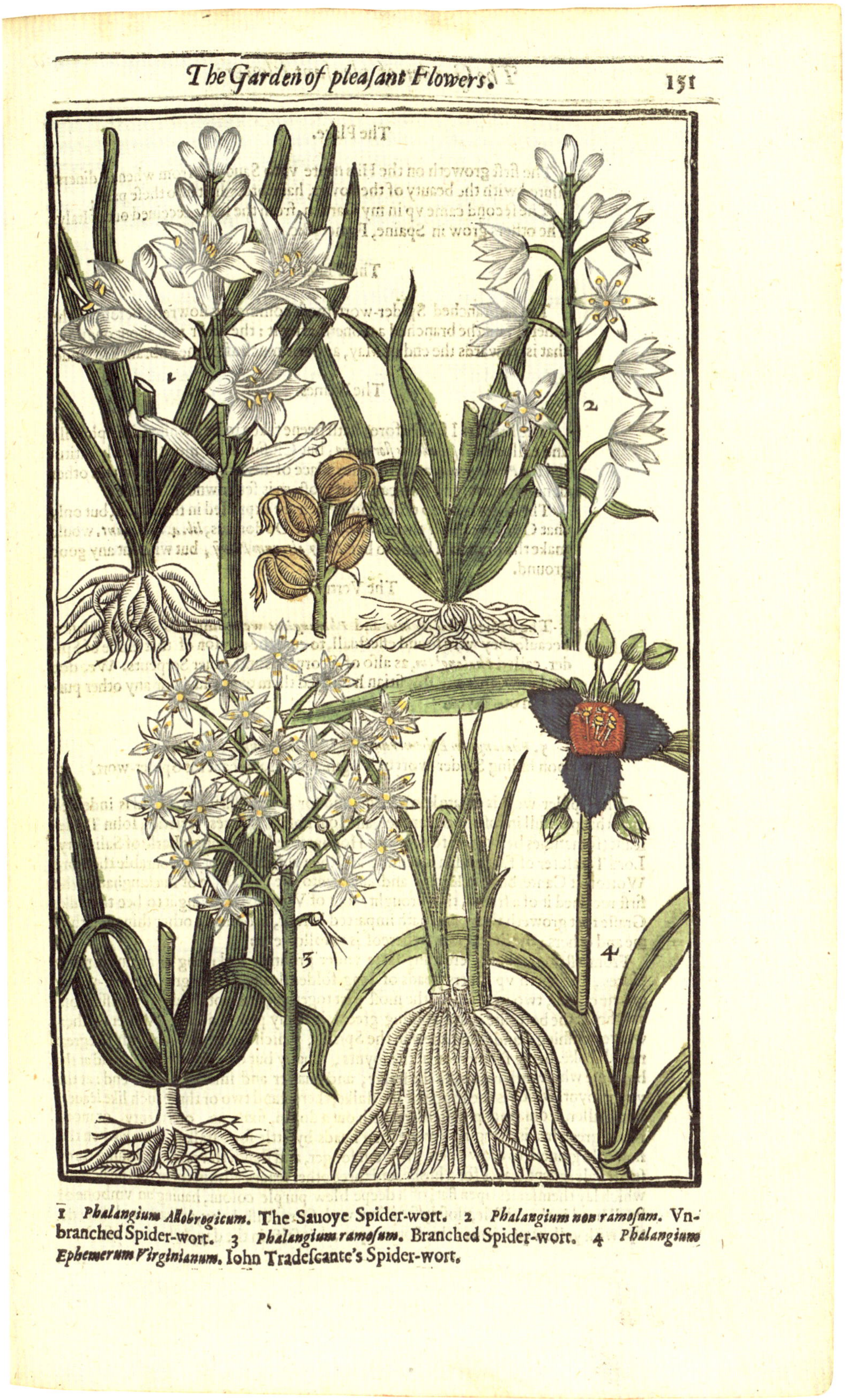

JOHN PARKINSON, *Paradisi in sole*, page 151. An *Asphodelus*, two *Ornithogalum*, and a common spiderwort (*Tradescantia virginiana*)

described more than 4,000 plants, most of them with medicinal properties. In the compilation of this herbal he availed himself liberally of the notes and unpublished material left by the French botanist Matthias de L'Obel, who died in 1616. L'Obel had spent the last years of his life in England, living on the edge of north London at Highgate and overseeing the gardens of Edward, eleventh Baron Zouche, at Hackney.

Paradisi in sole, which the author describes as a 'speaking Garden', was dedicated to Queen Henrietta Maria. In keeping with the prevailing taste for anagrams and other forms of wordplay, the first three words of the Latin title represent a punning form of his name: PAR IN SO. In his introduction Parkinson defines the plant world as one of the most wonderful expressions of divine creation, and he tells us that in *Paradisi* he will describe 'all sorts of pleasant flowers which our English ayre will permitt to be noursed up'. Although conceived as a text on gardening, in which the layout and structure of gardens of various types (ranging from the kitchen-garden to the orchard) are presented, the ample space dedicated to flowering plants also makes it one of the most important books on flowers to be produced in the early Stuart period. The text is written in a simple but effective style, and is full of useful digressions and historical notes. Its main interest lies in the 108 large plates illustrating almost 780 plant species. While less refined than the copper engravings generally used to illustrate the florilegia of the period, these woodcuts have an expressive power of their own. They were executed by A. Switzer, a German artist based in England during the first part of the seventeenth century.

The title-page consists of an elaborate woodcut showing the Garden of Eden planted with a variety of exotic species only recently introduced to Europe. They are depicted in curiously stylized, almost fantastical form, equalling or exceeding in height the diminutive figures of Adam (shown gathering fruit from a tree) and Eve (who is grasping a strawberry plant). Among the unusual species one can recognize a cactus, a pineapple, a banana tree and various flowers, much magnified, including a tulip in the centre foreground. A small 'Vegetable Lamb' may be discerned in the middle ground; this very peculiar species also appears in the title-page to the *Theatrum botanicum*. Known also by the names 'Boramet' (or 'Borametz') and 'Agnus sciticus', this 'plant–animal' was first described in medieval times in Sir John Mandeville's *Travels*. It was subsequently reproduced, with very few variations, in a number of herbals, including one by Castore Durante published in Rome in 1585. This anomalous species was also included in the plant kingdom by the Italian physician and naturalist Cesare Scaligerio in his encyclopaedic work *De subtilitate*, published in 1557, where the 'lamb' was presented complete with head, ears, eyes, legs and downy fleece. The plant was believed to grow wild in Tartary, its thick stem springing directly from the earth and into the umbilicus of the 'lamb-fruit'. The source of this legend has been traced to a species of fern, the *Frutex tartareus*, whose hairy roots were cunningly sculpted into animal forms for merchants who then sold them in the markets of Europe as rare and highly prized curiosities.

The second plate shows a portrait of the author. In keeping with the typology of fifteenth- and sixteenth-century portraits of botanists, he is shown holding a flower, which can be identified as a *Lychnis chalcedonica*.

A vast array of wild and cultivated flowering plants (including almost fifty varieties of anemones and more than 150 different tulips) are described in detail by the author and illustrated, perhaps by Switzer, in the woodcut plates. Several plants are presented on each page, each one very close to the next, suggesting an effect almost of *horror vacui*. Most are shown with their bulb or roots, while at the foot of the page the Latin and English names of each species is given. Some of the plates are original, while others were inspired by, or directly copied from, contemporary botanical texts by eminent botanists such as Clusius and L'Obel. Worth noting are the illustrations on page 151: among the various lilies depicted one can recognize a white asphodel and two ornithogalums. A *Tradescantia virginiana* also appears on this page; Parkinson named it 'John Tradescante's Spiderwort' in honour of his celebrated friend and colleague, who provided him with many new species for his collection.

The plates in this copy of *Paradisi in sole* have been hand tinted in bodycolour, not particularly carefully but in vivid and brilliant hues, which the artist attempted to match as closely as possible to the natural colours of the plants, and in fact his somewhat slapdash style complements very well the simple lines of the woodcuts.

41. Nicolas Robert (1614–1684)

[*engraved title on a shield with floral decorations and the protome of a lion, 15.2 × 10 cm.*] Fiori diversi Novame[n]te posti in luce Intagliati da Nicolo Rubert Francese ad Instanza di Gio. Battista de Rossi in Roma in P. Navona. An. MDCXXXX Deddicati al Mag.co Giovanni Orlandi Amico cariss.mo [*at the foot of the page*] In Roma Appresso Gio. Battista de Rossi, con licenza delli Superiori.

20.5 × 14 cm. 25 leaves of plates.

BINDING: Sprinkled calf. On a paper label attached to the spine: 'Robert. Fiori Diversi. 1640'.

PLATES: 25 etched plates, including the title-page, numbered in pen and ink, with the names of the flowers printed in Italian. The traces left by the border of the plate vary from 14 × 9 cm. to 15.5 × 10.5 cm.

PROVENANCE: Previous owner's signature removed from title-page.

REFERENCES: Bultingaire 1926; Nissen 1643; Blunt and Stearn, p. 118.

NICOLAS ROBERT was one of the most illustrious artists of the seventeenth century to specialize in floral illustration. His fame is based on three bodies of work of particular importance. The most famous is the elegant *Guirlande de Julie*, justly described as 'une des plus illustres galantéries qui aient jamais été faites'. Commissioned by the marquis de Montausier for the fascinating Julie d'Angennes, daughter of Mme de Rambouillet, the *Guirlande* is an album of sixty-one madrigals by the most celebrated poets of the period (Conrart, Chapelain, Scudéry, Racan and others), com-

NICOLAS ROBERT,
Fiori diversi. Two varieties of *Rosaceae* and insects

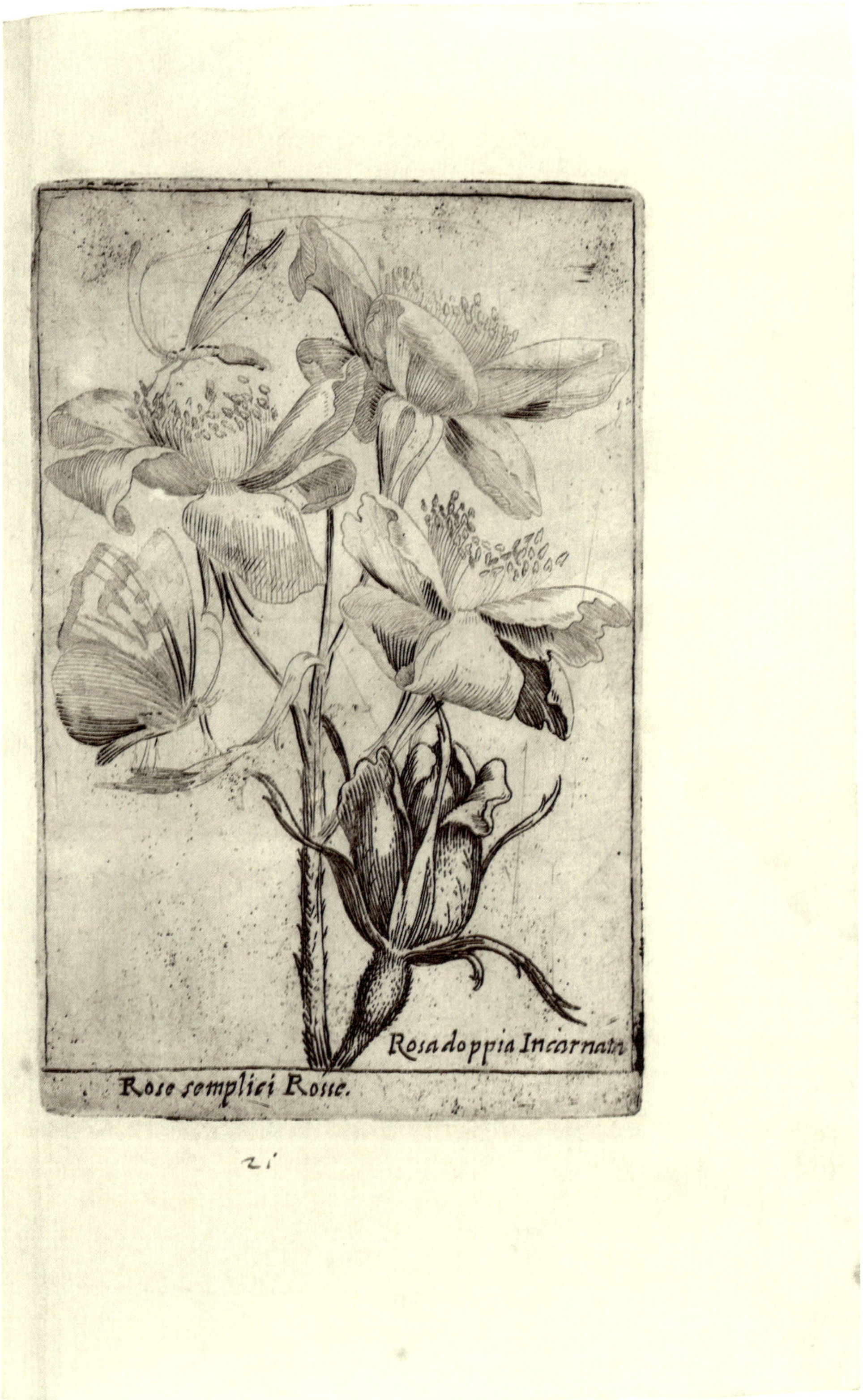

piled in 1641, and decorated by Robert with splendid floral miniatures. Robert was also responsible for a series of more than 700 miniatures of flowers painted on vellum, a project begun for Gaston, duc d'Orléans, brother of Louis XIII, and continued under the patronage of Louis XIV. This magnificent series of illustrations formed the nucleus of what was to become a vast collection of botanical illustrations now in the Muséum Nationale d'Histoire Naturelle, Paris. In addition, Robert was an inspired engraver, producing a substantial body of works ranging from decorative pieces to scientific illustrations (see nos. 42–4).

Fiori diversi is the earliest known work by Robert, completed during a sojourn in Rome in 1638–40. Since no earlier works by Robert are known, it seems probable that he took up the genre of flower painting during this period, influenced by the cultural ambience and the passion for gardens and horticulture he found in the city. During the first half of the sixteenth century the eminent figures working in Rome included the botanist Tobia Aldini (see no. 28), G. B. Ferrari (see no. 29) and Nicolas Guillaume Delafleur of Lorraine. In fact, in 1638 Delafleur published an engraved florilegium that the young artist might very well have seen and drawn inspiration from. Robert soon came to excel in the art of botanical illustration, becoming in 1666 'Peintre ordinaire du Roy pour la miniature'. *Fiori diversi* itself was dedicated to the Roman engraver, printer and connoisseur Giovanni Orlandi, and was published by Giovanni Battista de' Rossi, from whose presses in the Piazza Navona issued forth a constant stream of prints designed by the most celebrated Italian engravers of the period as well as by the many foreign artists based in Rome.

The small etchings to be found in *Fiori diversi* already show, by their composition and pagination, unmistakable signs of a refined artistic sensibility. Not satisfied with the simple enumerative approach that characterized most of the florilegia to date, Robert introduced an entirely new approach to the genre of flower painting, emphasizing the decorative aspect without, however, sacrificing the realism that was still the primary object of these works. His elegant style can already be discerned in such plates as no. 8, depicting a wallflower ('Violle Fiore Colombino'), or no. 21 showing two species of roses ('Rose semplici Rosse' and the 'Rosa doppia Incarnata'), in which the silken texture of the petals is rendered with delicate, sure lines by the etcher's needle, while butterflies and insects flit across the page or rest on leaves and petals.

This series of floral etchings was so well received that Robert decided to apply himself to the production of prints too. Around 1660 he completed a new and more complex series of illustrations, this time combining the techniques of etching and engraving. Entitled *Variae ac multiformes florum species* or *Diverses fleurs*, it was first printed in Paris by the celebrated François Poilly (see no. 42), and then went through several reprintings, including one in Rome in 1665 by Giovanni Giacomo de' Rossi, who had by this time transferred the family's printing works to the Temple of Peace.

42. Nicolas Robert (1614–1684)

[*engraving with the title on a shield surrounded by a garland of flowers tied above and below with ribbons. In the lower left: n. 1*; 20.5 × 15 *cm.*] Variae ac Multiformes Florum Species appressae ad Vivum et aeneis tabulis incisae. Authore N. Robert. Diverses Fleurs dessinees et Gravees d'Apres le Naturel par N. Robert. Avec Privil. du Roy A Paris F. Poilly excudit rüe St. Jacques a limage St. Benoist. [*preceding the title-page is a manuscript title-page followed by two manuscript pages*] Les Leçons Royales.

27 × 20.3 cm.

BINDING: Modern red morocco. On the spine in gilt lettering: 'Robert Diverses Fleurs'.

PLATES: 31 plates including the illustrated title-page produced by a combined process of engraving and etching (20.5 × 15 cm.). Each plate bears the following (also etched): a number from 2 to 30 (the same number being repeated in manuscript under the plate), the Latin name of the plant, and the monogram of the artist (NR) followed by 'cum pri. re.' Alternating with the plates are 35 manuscript leaves (first and last are blank) of Catherine Perrot's *Les Leçons royales*.

REFERENCES: Nissen 1645; Hunt 282; Coats, pp. 37–8; *A Magnificent Collection*, no. 306; Blunt and Stearn, p. 122.

THIS SERIES of plates on floral subjects was designed and etched by Nicolas Robert, and was published *c.* 1660 by the renowned printer François Poilly, whose shop was located in rue Saint-Jacques 'à l'image de Saint-Benoist' in Paris. In comparison to *Fiori diversi*, produced by the same artist twenty years earlier in Rome (see no. 41), *Variae ac multiformes florum species*, or *Diverses fleurs*, clearly shows Robert's increased mastery of the problems of technique and composition. The earlier plates are characterized by a certain harshness of line, the flowers standing out starkly against the page. In this later work Robert made much more creative use of the various elements at his disposal, incorporating the stems and leaves of the plants in sophisticated compositions that anticipate the quintessentially Baroque motif of the formal flower bouquet.

Butterflies and other insects are also much more numerous than in the artist's previous work. For example, in plate 7 Robert depicts two crocuses (one with a double bloom) and a cornflower together with two insects, on the right a tussock-moth (*Dasychira pudibunde*) caterpillar and on the left what is probably a carpenter bee (*Xylocopa violacea*), while plate 27 shows two very realistic grasshoppers at the base of a bouquet composed of a cornflower, a buttercup and an *Iris tuberosa*.

Having mastered the art of miniature painting during his work on the famous *Guirlande de Julie* and the *vélins* for the court of Louis XIV, Robert strove for a similar degree of detail in his illustrations, combining the use of etching and engraving to create a subtle spectrum of tones reminiscent of the delicate play of actual colours. It is worth noting that the title-page is written in a garland, as in the famous manuscript dedicated to the noble Julie d'Angennes.

The Oak Spring copy of *Diverses fleurs* is particularly interesting because alternating with the plates are the eighteenth-century manuscript pages of a singular work by a pupil of Robert, Catherine Perrot. Also known as Mme Horry (her husband was C. Horry, notary to the Archbishop of Paris), Catherine studied under Robert and became an able painter of flowers and birds in her own

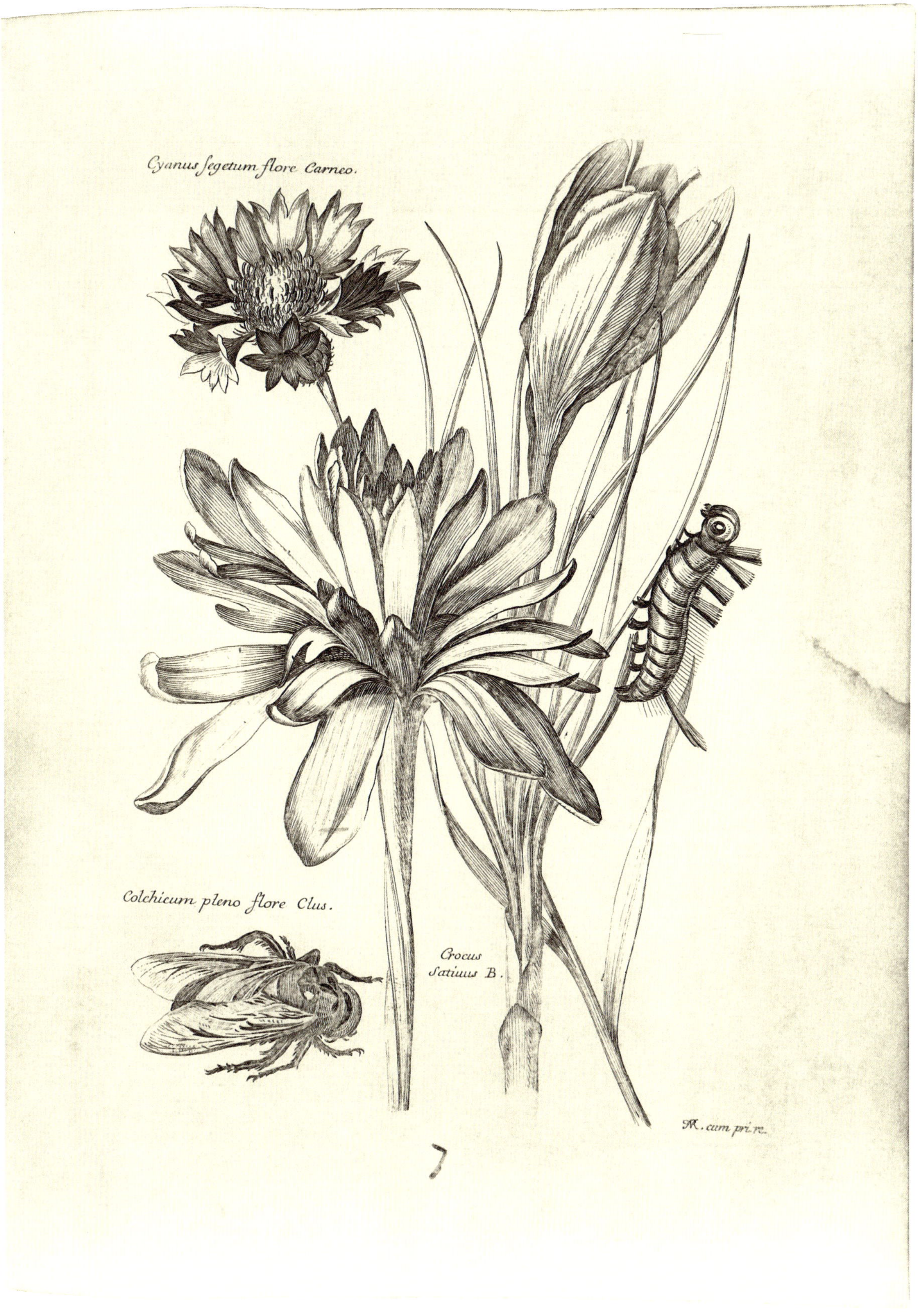

NICOLAS ROBERT,
Variae ac multiformes florum species. Cornflower (*Centaurea cyanus*), *Colchicum* sp., saffron crocus (*Crocus sativus*), and insects

NICOLAS ROBERT,
Variae ac multiformes florum species. Manuscript text on folio 7

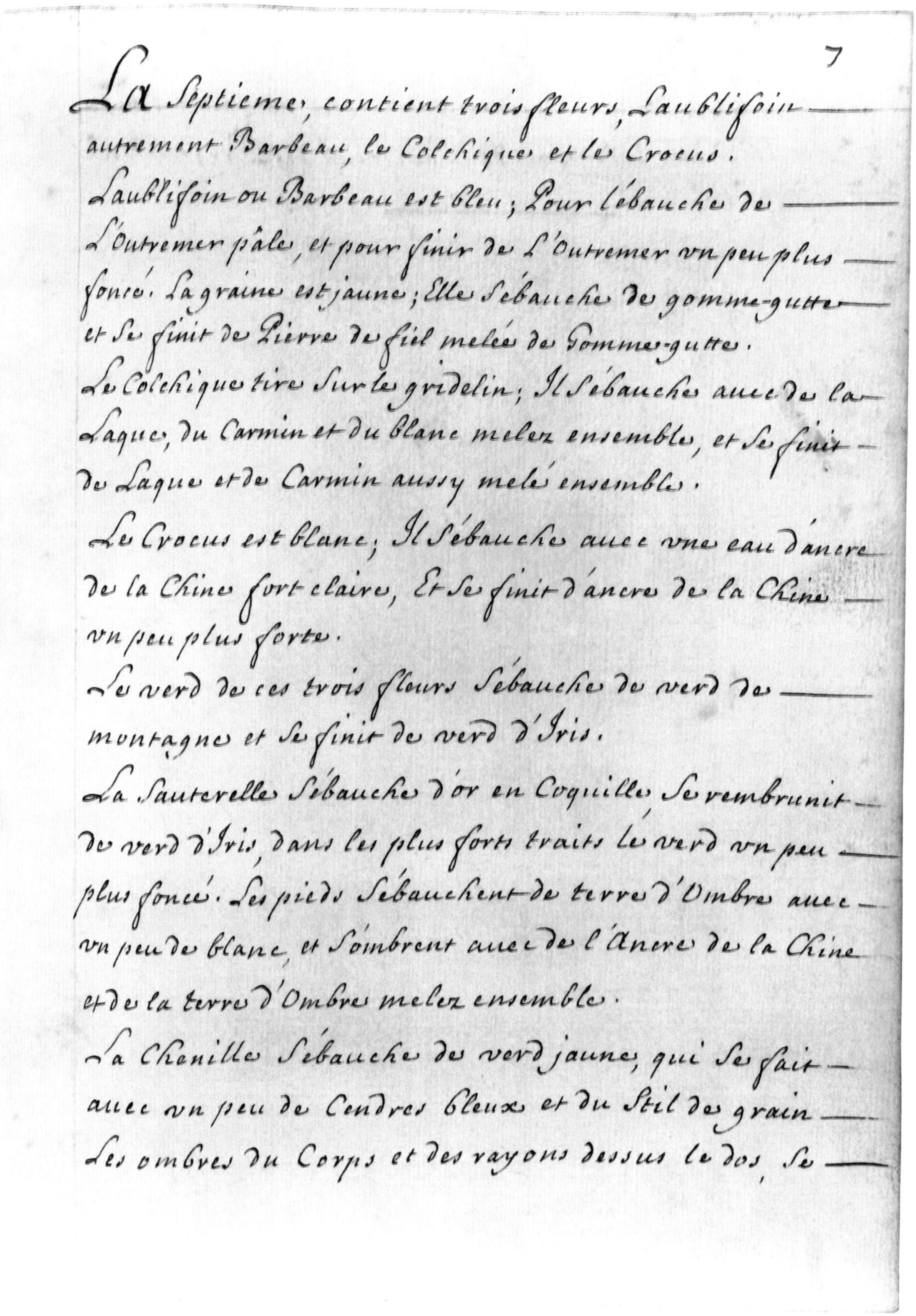

7

La Septieme, contient trois fleurs, Laublifoin
autrement Barbeau, le Colchique et le Crocus.
Laublifoin ou Barbeau est bleu; Pour lébauche de
L'Outremer pâle, et pour finir de L'Outremer vn peu plus
foncé. La graine est jaune; Elle Sébauche de gomme-gutte
et Se finit de Pierre de fiel melée de Gomme-gutte.
Le Colchique tire Sur le gridelin; Il Sébauche auec de la
Laque, du Carmin et du blanc melez ensemble, et Se finit
de Laque et de Carmin aussy melé ensemble.
Le Crocus est blanc; Il Sébauche auec vne eau dancre
de la Chine fort claire, Et Se finit dancre de la Chine
vn peu plus forte.
Le verd de ces trois fleurs Sébauche de verd de
montagne et Se finit de verd d'Iris.
La Sauterelle Sébauche d'or en Coquille, Se rembrunit
de verd d'Iris, dans les plus forts traits le verd vn peu
plus foncé. Les pieds Sébauchent de terre d'Ombre auec
vn peu de blanc, et Sombrent auec de l'Ancre de la Chine
et de la terre d'Ombre melez ensemble.
La Chenille Sébauche de verd jaune, qui Se fait
auec vn peu de Cendres bleux et du Stil de grain
Les ombres du Corps et des rayons dessus le dos, Se

right. It seems most probable that this copy was intended to be used for teaching purposes. The transcription of each chapter of the *Leçons* precedes the plate to which it refers, beginning with the frontispiece ('La premiere feuille . . . contient une Couronne de fleurs, à la tete de laquelle est une Imperiale . . . Couleur Orangé') and ending with a plate depicting a cluster of 'Viola lutea' : '. . . plusieurs Violettes ensembles', as Catherine Perrot states, 'Elles Sébauchent d'une eau de Laque et d'Outremer melez ensemble plus d'Outremer que de Laque, et ses finissent du méme melange'.

At Oak Spring is a copy of another edition of *Les Leçons royales*, dedicated 'à Madame la Dauphine' and published in Paris by Jean Nego in 1686. Indeed, as Catherine Perrot testifies, this collection of Robert's flower engravings became immensely popular and went through several reprintings, not only in Paris but also in Rome (1665), providing useful models for painters, decorators and embroiderers. Some of Robert's plates were borrowed from by the Bolognese engraver Francesco Curti for an undated florilegium *Variae ac multiformes florum species* published in Bologna by the printer Giuseppi Longhi, while Maria Sibylla Merian utilized several of Robert's plates in her *Histoire des insectes de l'Europe*, published in Amsterdam in 1730, as did the unknown author of *Nederlandsch Bloemwerk* of 1794, also published in Amsterdam (two copies of which are at Oak Spring).

43. Nicolas Robert (1614–1684), Abraham Bosse (1602–1676) *and* Louis de Chatillon (1639–1734)

[Plantes du Roi: Paris, 1685–1701, 2 vols.]
[*first volume: title-page in manuscript*] Roberti Icones Plantarum. Tom. I.
[*second volume. manuscript index of plants* (fols. 1r–3r); *title-page in manuscript*] Roberti Icones Plantarum. Tom. II.

54 × 37.5 cm. Vol. I, plates 1–154; vol. II, plates 155–327.

BINDING: 18th-century Russia; gilt borders with coronets; red and blue labels on spine bearing the handwritten titles 'Roberti—Icones Plantarum Tom I' and 'Roberti—Icones Plantarum Tom II.'

PLATES: 327 engraved plates of different plant species, almost all of them with the Latin name engraved and the Linnaean name in manuscript. Many of the plates contain at the foot of the page the name of the engraver, whether Robert, Bosse or Chatillon. Eight plates, as listed in manuscript on a loose page slipped into the second volume, are present in two states (plates 16 and 262; plates 71 and 227; plates 72 and 169; plates 88 and 310; plates 94 and 250; plates 111 and 298; plates 125 and 259; and plates 220 and 325).

PROVENANCE: Both volumes have the bookplate of Chillingham Castle, Northumberland. These copies originally belonged to John Stuart, third Earl of Bute (1713–92), a politician who was also a patron of the arts and an enthusiastic botanist and horticulturist; he is generally considered to be the founder of the Royal Botanic Gardens, Kew. Bute's extensive and specialized library, which contained more than 3,000 works, was auctioned off by Leigh and Sotheby's in 1794. This work, listed in the catalogue as lot 969, was acquired for £33 by Charles Bennett, fourth Earl of Tankerville (1743–1822), proprietor of Chillingham Castle (*Catalogue of the botanical and natural history part of the library of the late John, Earl of Bute*, May 1794, p. 42).

REFERENCES: Laissus 1969; Coats, pp. 59–60; Blunt and Stearn, pp. 121–2.

THIS IS THE first edition of a series of engravings produced by Nicolas Robert for a 'history of plants' that the newly created Académie Royale des Sciences in Paris intended to publish. The original idea for this encyclopaedic undertaking was conceived by Perrault and the proposal was enthusiastically received by Colbert, minister to Louis XIV, although it appears to have actually begun to take shape only when the botanist Denis Dodart (1634–1707) joined the Académie in 1673. His work, *Mémoires pour servir à l'Histoire des Plantes*, which was intended to form the introductory volume to this series, appeared in 1675 and contained thirty-nine plates engraved by Robert. In it Dodart lamented the fact that none of the engravings could be in colour, but at least, he assured the reader, all the illustrations in the new series would be drawn directly from life, the artist making every effort to present the plants in their actual dimensions.

Putting to one side his regular activity of flower painting on vellum in order to concentrate on this project, Robert managed to produce an enormous number of engravings. At the time of his death in 1684, however, the work was far from complete, although it is known that 319 finished plates were deposited with the Imprimerie Royale in 1692. It fell to Abraham Bosse, a talented engraver of various subjects, including scientific texts, who also wrote a treatise on engraving, to continue his work. The project was finally completed by the engraver and painter on enamel, Louis de Chatillon. Both he and Bosse based almost all of their engravings on the detailed drawings, now preserved in the Muséum Nationale d'Histoire Naturelle in Paris, which Robert had prepared in red chalk.

A limited number of plates, without any text, was published for the first time in 1701. Here the plants were presented in alphabetical order, an arrangement that necessitated the mixing of works by the three different artists. These engravings were incorporated, with certain adjustments to the nomenclature as demanded by the Tournefort system of classification then in vogue, into a three-volume work that was finally published in 1788. Neither this work nor the preliminary set of engravings, however, were ever made available for sale.

The *Recueil des plantes* was judged by the master flower painter Gerard van Spaendonck to be the most beautiful botanical work ever published. Most of the engravings show the plants in flower, complete with their roots, and many are accompanied by enlarged drawings of certain details. This work was clearly the result of a close collaboration between the scientist and the artist, both seeking to portray the plant as accurately and recognizably as possible. The plates engraved by Robert are instantly distinguishable by the lightness of his touch and by the sensitive modulation of the various tones of grey and black in the leaves and flowers. His virtuosity can be seen in the splendid plates of the 'Althoea sive Ibiscus' (pl. 24), the 'Clematis Americana' or bignonia (pl. 109), and the passion-flower (pls. 108 and 110), as well as one showing a plant described as 'Anonyma, Anapodaphyllon quibusdam'—'Plante sans nom, appellée par quelques uns Anapodaphyllon' (pl. 45). This 'anonymous' plant is actually the May-apple (*Podophyllum peltatum*), a tree native to Virginia, which Robert depicted with its delicate white spring blossoms.

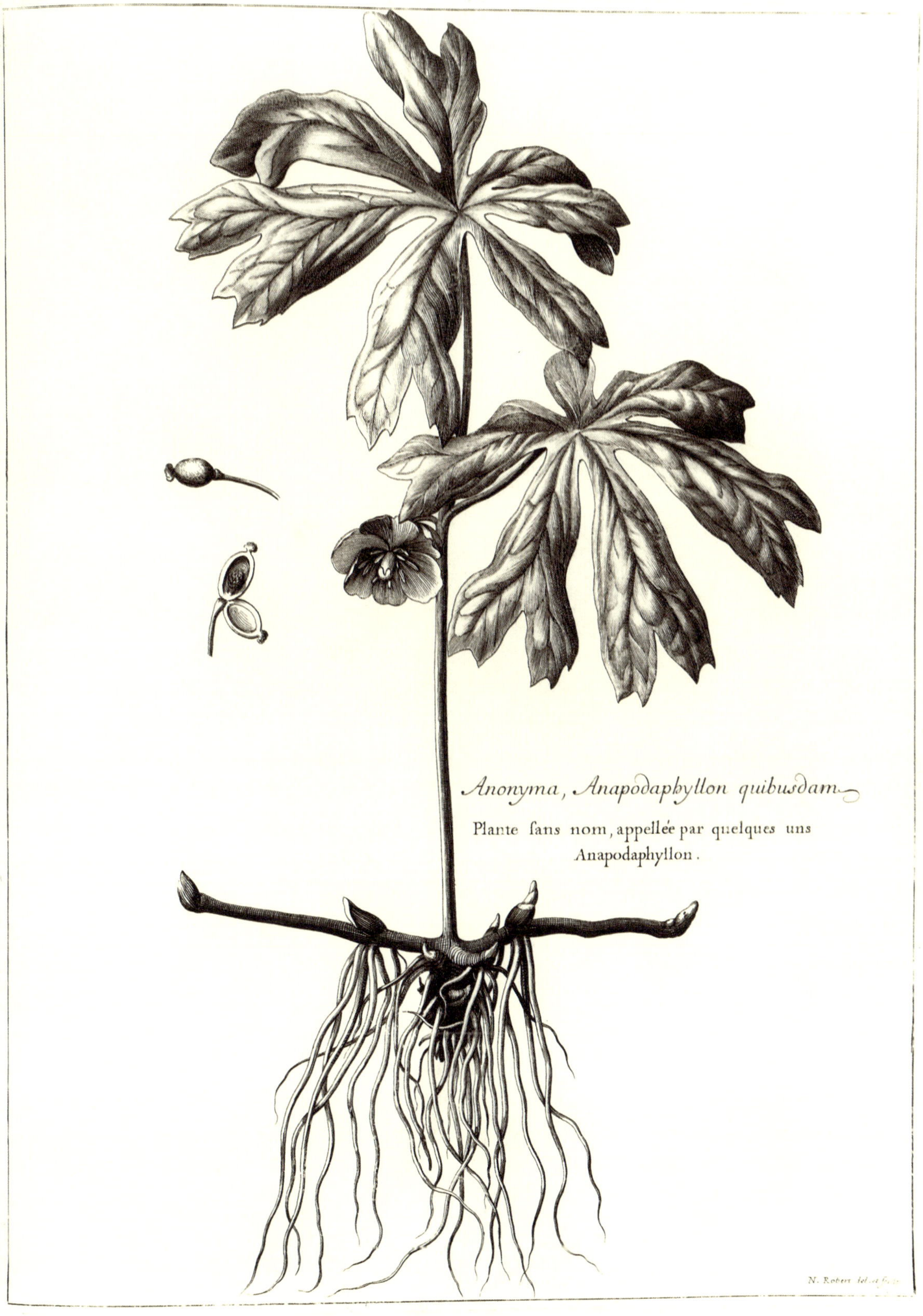

NICOLAS ROBERT, *Plantes du Roi*. May-apple (*Podophyllum peltatum*), volume I, plate 45

Bosse and Chatillon were fully competent to carry on the work of Robert, as may be seen from the illustrations of the 'Gentianella Alpina' (pl. 148) and the 'Mandragora mas' (pl. 230) engraved by Bosse, or the 'Ananas Acostae' (pl. 32) and the 'Caltha palustris' (pl. 125) by Chatillon. Chatillon also produced the wonderful engravings of two 'Plantes inconnues venues de Siam', which close the second volume of the series (pls. 326 and 327).

The Oak Spring edition, 'Plantes du Roi', contains the complete series of engravings in their first state, as well as eight plates in other states. They were almost certainly produced before the re-engraving of 1719. Some, such as those not provided with captions (these, in fact, first appeared in the 1701 edition), probably represent preliminary proofs executed by Robert himself (see, for example, plates 124, 180 and 255).

44. Nicolas Robert (1614–1684)

Scilla maritima (*Urginea maritima*) with bulb.

45.2 × 33.2 cm. On vellum, within a gilt border (0.5 cm.).

There are traces of writing on the left side of the painting that have been covered with white paint.

This painting depicts the flower of the *Urginea maritima* (a species of lily native to the warm littoral zone of the Mediterranean), together with an enlarged detail of its bulb, the medicinal portion of the plant. The painting is very similar in style and dimension to the *vélins* painted by Robert for Gaston d'Orléans and Louis XIV.

In 1645 Robert, who was already celebrated for his masterly illumination of the *Guirlande de Julie*, was invited to Blois by Gaston d'Orléans. This royal château had a large botanical garden in which many rare plants and exotic flowers were cultivated. Robert was asked to continue a project begun by Daniel Rabel (see no. 15)—the painting on vellum of the rare plants and flowers in Gaston's collection. It was perhaps the Scottish botanist Robert Morison, director of the gardens at Blois for ten years, who encouraged Robert to travel there and devote himself to the art of botanical illustration (Blunt and Stearn, p. 121).

After the death of Gaston d'Orléans in 1660, this collection of botanical paintings was inherited by Gaston's young nephew, Louis XIV, who had them transferred to the Jardin des Plantes in Paris. Thanks to the patronage of Colbert, the powerful chief minister of the King, who had Robert named the King's 'peintre ordinaire . . . pour la miniature', the artist was able to continue his work on this project up until his death. He produced several hundred miniatures of plants, all of them shown in flower and at least 700 bearing his signature. Robert also completed various paintings of birds from the royal aviary (Stein; Laissus 1980).

This matchless collection of *vélins*, to which numerous other artists successively contributed

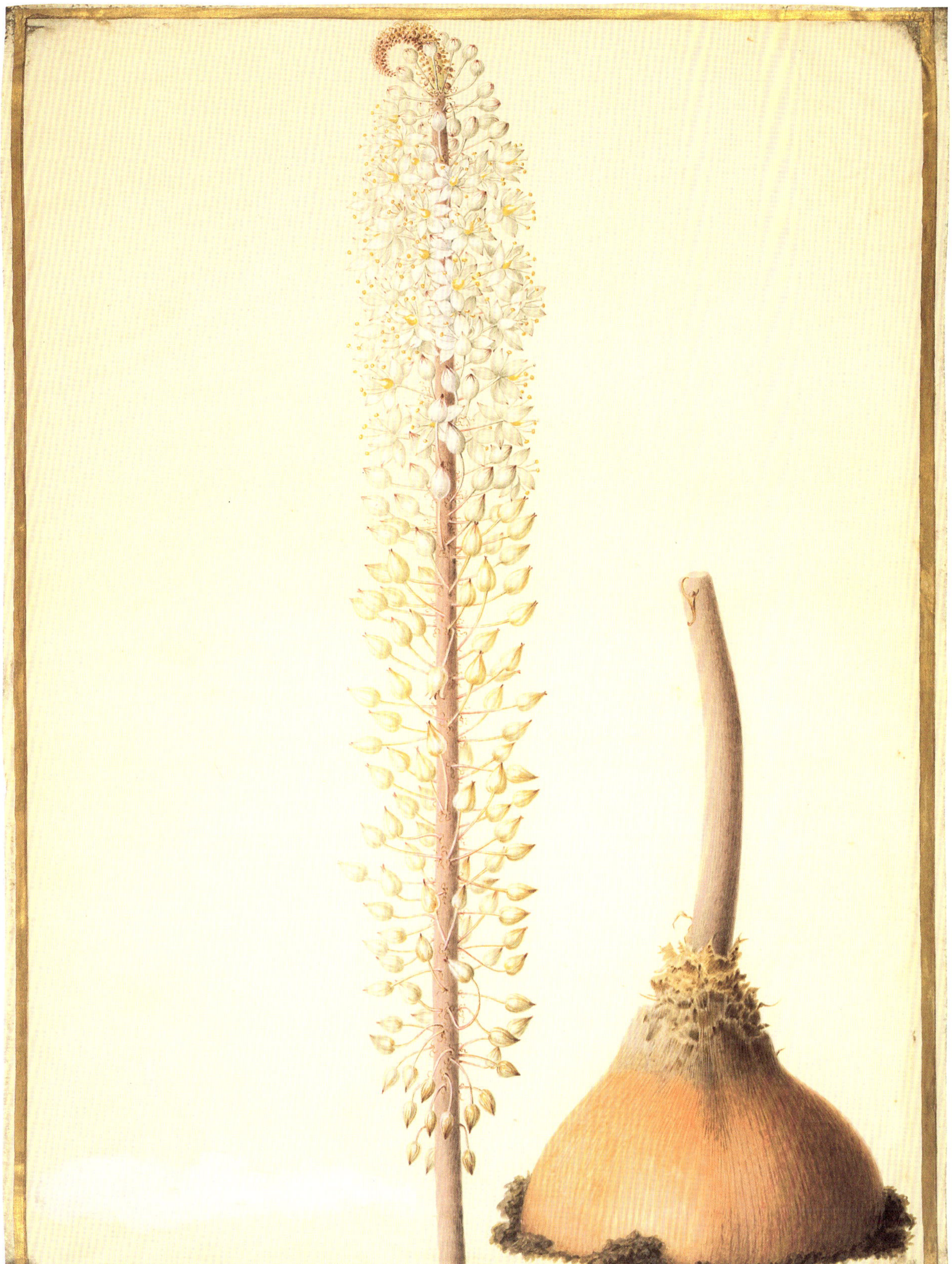

NICOLAS ROBERT,
Squill (*Urginea maritima*)
with bulb

during the course of the eighteenth century, and which came to comprise forty-eight bound volumes, was transported in 1793, by the decision of the Convention Nationale, to the Muséum Nationale d'Histoire Naturelle, where they can be found today. In the history of botanical illustration, the paintings by Robert constitute a singular accomplishment. The artist, in his rendition of each flower, was able to unite the precision of the scientist with the freshness of inspiration of the artist. His paintings could thus serve both as a reference tool for the scientist and as a source of aesthetic pleasure for the flower lover.

45. Abraham Munting (1626–1683)

[*printed in red and black*] Abrahami Muntingii, Medicinae Ac Botanices Professoris: Phytographia Curiosa, Exhibens Arborum, Fruticum, Herbarum & Florum Icones, Ducentis & Quadraginta Quinque Tabulis ad vivum delineatis ac artificiosissime aeri incisis. Varias earum denominationes, Latinas, Gallicas, Italicas, Germanicas, Belgicas, aliasque, ex probatissimis Anthoribus [*sic*], priscis ac neotericis, desumptas Collegit & Adjecit Franciscus Kiggelaer, Botanophilus. [*printer's device* 5 × 6 *cm.*] Amstelaedami, Apud Rod. & Gerth. Wetstenios HFF MDCCXIII.

4° 41.5 × 25 cm. π^4 ★–★★★^{2} A–M^2 ★★★★–★★★★★★^{2} (Oak Spring's copy has bibliography and index to plates and descriptions bound in at end) *1–20* 1–47 *48 21–32* p.

BINDING: Vellum.

PLATES: 245 engraved and etched hand-coloured plates of plant species. Engraved and hand-coloured initial plate with a coat of arms and the dedication: 'Nobilissimo, Amplissimo, Spectatissimo Consultissimoque Viro D. Benjamino Fagelio, Domino A' Ter Weer, Curiae Hollandiae, Zelandiae, Frisiaeque Senatori Prudentissimo, Praepotentium Ordinum Hollandiae Nomine ad Causas Ecclesiasticas, Synodumque Hollandiae Septentrionalis Delegato Dignissimo, Etc. Etc. Summo Bonarum Artium, Elegantissimorumque Librorum Cultori; Possessori Peritissimo Numismatum, Mineralium, Conchyliorum, Insectorum, Florum, Herbarum, Plantarum Exoticarum, Ac Rariorum Quorumvis Viro Amantissimo, Curiosam Hanc Abrahami Muntingy Phytographiam Summâ Cum Veneratione Offerunt Nobilitatis Tuae. Humillimi Clientes P. vander AA. Et F. Halma.' Illustrated half-title depicting Nature being offered gifts, with the inscription: 'Abrahami Muntingii Phytographia curiosa', designed by J. Goeree and engraved by J. Baptist; head- and tail-pieces.

PROVENANCE: Unidentified armorial bookplate.

REFERENCES: Nissen 1429; Hunt 404.

THE TITLE *Phytographia curiosa* designates a series of Latin editions, the first one published in 1702, which reproduce (with some variations) a Dutch florilegium that appeared in 1696 in Leiden and Utrecht, *Naauwkeurige Beschryving der Aardgewassen*. This in turn was an updated version of an earlier work, *Waare Oeffening der Planten*, which was published in Amsterdam in 1672 and in 1682. The author, Abraham Munting, was an eminent professor of botany and chemistry at the University of Groningen. Encouraged in his medical and botanical studies by his father, Heinrich, as a young man Abraham also travelled extensively, visiting England, Germany and Italy, and spending a considerable amount of time in France, where he became greatly renowned for his learning. Munting eventually returned to Groningen, where he taught at the university for more than

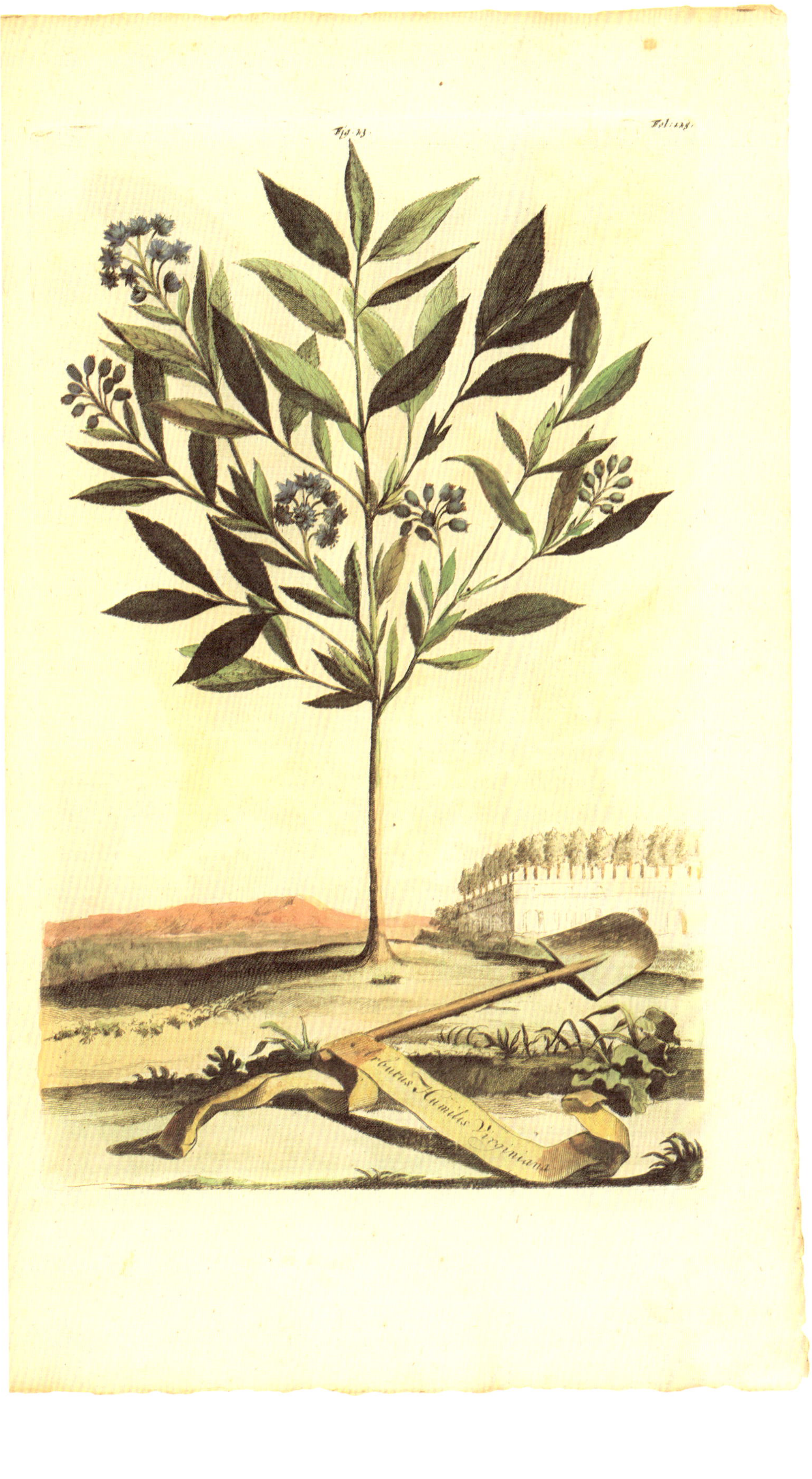

ABRAHAM MUNTING,
Phytographia curiosa.
Strawberry tree
(*Arbutus unedo*),
plate 23

twenty-four years and founded one of the most extensive botanical gardens of the period, known as the 'Paradise of Groningen', which he directed from 1658 up until his death in 1683.

Munting wrote a number of works on medical–botanical topics, but his posthumously published *opus magnum*, the *Naauwkeurige*, enjoyed particular success, at least in part due to the novelty of the plates, which, in a radical departure from the iconography of the traditional florilegium, presented its plant species against a charming series of landscape backgrounds. The first Latin edition, edited by the botanist Franz Kiggelaer (d. 1722), opens with the funeral oration on Munting delivered in February 1683 by one of his colleagues, Johannes Mensing, a professor of elocution. Kiggelaer included all the plates used in the Dutch original, but replaced the descriptive text with a simple enumeration of the plants illustrated.

The illustrations are remarkable for their elegance and originality. The sophisticated title-page to the first edition, which also appears as the frontispiece to Kiggelaer's *Phytographia curiosa*, was designed by an artist of considerable merit, Jan Goeree (1670–1731), who had been a student of Gérard de Lairesse. The plate was etched by a certain 'Jan Baptist', identified as the French engraver Jean Baptiste Adam. It depicts a procession of classical figures bearing floral tributes to be laid at the feet of Nature—'Rerum magistra' (Mistress of all things), as is written on the marble pedestal against which she leans. Medallions containing classicist vignettes, including one of Narcissus gazing at his own reflection, have been inserted in the decorated border.

Each plate shows a different plant in flower, including many exotic species from America and other distant lands. The plant dominates the foreground, filling the entire page, often with a detail of the fruit or flowers presented on a smaller scale. In some cases the plants are presented *à trompe l'œil*, while in others they have been arranged in decorated urns. Sometimes gardening tools are depicted as well. The name of each plant appears written on an elegantly fluttering ribbon or cartouche, or on a crumbling marble plaque.

The originality of the work lies, however, in the small landscapes that have been inserted into the background of the plates. Here the artist gave free rein to his imagination, delineating scenes that in reality bore little relation to the actual habitat of the plant. Pastorals with animals and figures alternate capriciously with vistas of walled cities and landscapes containing classical statues and ruins. It is not known who actually conceived this idea of combining botanical illustration with landscape scenes; it may have been Munting himself, or perhaps the artists who were commissioned to do the illustrations. Only one plate (201) is signed; it bears the name of Joseph Mulder (1659–1710), an engraver who had studied with Hendrick Bogaert and who was working in Amsterdam as an illustrator of historical and geographic works. Mulder also depicted various unusual plants and insects from Surinam for Maria Sibylla Merian (see no. 101).

The hand colouring in this copy of *Phytographia curiosa* generally owes more to the imagination of the artist than to scientific fact, as may be seen in the illustration of the 'Arbutus Humilis Virginiana' (*Arbutus unedo* or strawberry-tree), which bears bizarrely tinted flowers and light-blue berries.

Such odd details, however, add a certain charm to the plates. The two final illustrations (244 and 245) are supplementary to the original edition. One shows a date-palm and Christ's entry into Jerusalem, the other a 'Caryophyllus Arbor' (*Eugenia* or clove).

A copy of the first edition from 1696 and an edition published by Peter de Coup in 1727 in Amsterdam are also at Oak Spring.

46. Jean Baptiste Monnoyer (*c.* 1636–1699)

Collection of engraved and etched plates, most of them executed by Jean Baptiste Monnoyer.

51.5 × 36 cm.

(I) Etched title-page, numbered 1: 'Livre De Toutes Sortes De Fleurs D'apres Nature [*on the pedestal*] Baptiste Monnoyer Pinx. Vauquier f [*and lower down*] A Paris ruë St Jacques chez N. Poilly a la belle Image Avec Privilege du Roy'. Eleven etchings numbered 2–12.

(II) 4 unnumbered etchings, two on each leaf.

(III) 4 unnumbered etchings, two on each leaf.

(IV) 4 unnumbered etchings, two on each leaf.

(V) 2 unnumbered engravings.

(VI) 12 numbered etchings, two on each leaf: title-page numbered 1: 'Diverses Fleurs mises en boucquets dessignées et Gravées par J. Bailly Peintre du Roy. Qui se vendent à Paris aux Galleries du Louvre chez led. Bailly Avec Privil, de sa Maj.té'.

(VII) 4 double-page etchings. On the first is written: 'Livres de Plusieurs Paniers de Fleurs dessiné et Gravé Par J. Baptiste. A Paris chez N. De Poilly ruë S. Jacques à la belle Image. avec privil. du Roy.'

(VIII) Etched double-page title-page: 'Livres de Plusieurs Vaze de Fleurs faicts d'Apres le Naturel Par J. Baptiste. avec Privil. du Roy et ce vend A Paris chez N. De Poilly ruë St Ja. a la belle Image.' [*and above*] 'Nobil.mo Viro D.D. Carolo le Brun Equiti, Proto Pictori Regis Christian.mi Regijs Picturis et Operibus Praeposito, necnon Regiae Pictori Academiae Cancellario Meritatissimo. Humilissimus Servus J. Baptista', followed by 8 double-page etchings.

(IX) 3 etchings on double pages; on the second is written: 'Livres de Plusieurs Corbeilles de Fleurs dessiné et Gravé par J. Baptiste. A Paris chez N. De Poilly ruë S. Jacques a la belle Image avec privil. du Roy'.

(X) 2 etchings on double pages. On both are written: 'Guirlande de Fleurs dessiné et Gravé par J. Baptiste. Se vendent A Paris chez N. De Poilly rue S. Jacques a la belle image avec Pr. du Roy'.

(XI) 2 etchings on two pages, on both of which are written: 'J. Baptiste Monnoyer Pinxit. Vauquier sculpsit cum Privil. Regis'.

BINDING: Parchment. On the spine in manuscript: 'Z' and 'N 12'.

PLATES: (I) Title-page consisting of a pedestal framed by a floral garland and a ribbon tied in an ornate bow, etched by François Vauquier from a design by Monnoyer, followed by 11 plates with ruled frames depicting bouquets.

(II) 4 etchings of baskets of flowers, with ruled frames, designed and executed by Monnoyer.

(III) 4 etchings with ruled frames, each containing 2 plates depicting flower urns standing on horizontal surfaces, etched and printed by Monnoyer.

(IV) 8 etched plates depicting bouquets (2 on each page), signed by Monnoyer.

(V) 4 engraved plates depicting flowers (2 on each page, unsigned.

(VI) Title-page consisting of a bouquet with the title written on one of the leaves, followed by 11 etchings of bouquets, 2 on each page (the first on the same page as the title-page), designed and executed by Jacques Bailly.

(VII) 4 etched plates on double pages with ruled frames, depicting baskets of flowers.

(VIII) 9 double-page etchings comprising a title-page with a dedication to Charles Le Brun, whose coat of arms appears on the base of the urn, followed by 8 plates of flower urns on double pages with ruled frames, designed and etched by Monnoyer.

(IX) 3 plates depicting baskets of flowers, on double pages with ruled frames, painted and etched by Monnoyer.

(X) 2 garlands of flowers on double pages with ruled frames, painted and etched by Monnoyer.

(XI) 2 plates with ruled frames showing urns of flowers, painted by Monnoyer and etched by Vauquier.

PROVENANCE: Ex-libris Mildred Bliss, Dumbarton Oaks, Washington, D.C.

REFERENCES: Pavière, p. 12; Mitchell 1973, p. 178.

JEAN BAPTISTE MONNOYER was born in Lille, and completed his artistic apprenticeship in Antwerp, an important artistic centre with a well-established tradition of floral painting. He began his career in Paris, where he was immediately taken on by Charles Le Brun, 'Premier peintre du Roy', to help with the painted decoration at Versailles, the royal château at Marly and other buildings, such as the Trianon and the Ménagerie. He also collaborated in the creation of designs for the great tapestry *ateliers* of Gobelins and Beauvais. In 1662 Monnoyer became a member of the Académie Royale de Peinture, and in 1673 four of his paintings were exhibited at the Académie's Salon. At the age of forty-four, on the invitation of the British ambassador at Paris, Ralph, first Duke of Montagu, Monnoyer transferred to England, although he continued to make frequent trips back to the Continent. Monnoyer decorated the Duke's London residence (later to become the first home of the British Museum), and undertook commissions for a number of other prestigious clients in England, including the Queen.

Monnoyer's works, which range from small paintings to fresco decorations for palaces, all bear the stamp of his inimitable style, combining grandeur of composition with an inexhaustible inventiveness. His beautiful flowers, almost all of them cultivated species, fill baskets or fancy urns, or are woven into dense garlands. Despite their decorative function, these compositions are highly naturalistic and never fall into mere repetitive patterns. The immense popularity of his designs encouraged Monnoyer to produce them in multiple copies by means of etchings. He himself prepared the drawings, which were then etched and printed either under his direction or that of François Vauquier (see no. 62) at the shop of Nicolas de Poilly, 'à la Belle Image' in the rue Saint-Jacques. They were then sold in Poilly's shop and in that of Jacques Bailly (1629–79) in the Galerie du Louvre. Monnoyer signed all of his etchings, often simply with the name 'Jean Baptiste' by which he was universally known.

The etchings in this volume generally occupy a double page and feature tulips, chrysanthemums, peonies, carnations, lilies and other flowers. Arranged in urns, bowls and baskets, they form asymmetrical compositions that stand out with striking decorative effect against the white background of the page. The vast range of greys and blacks achieved by the adept use of the etching needle combined with skilful printing echo the strongly contrasting hues of Monnoyer's paintings.

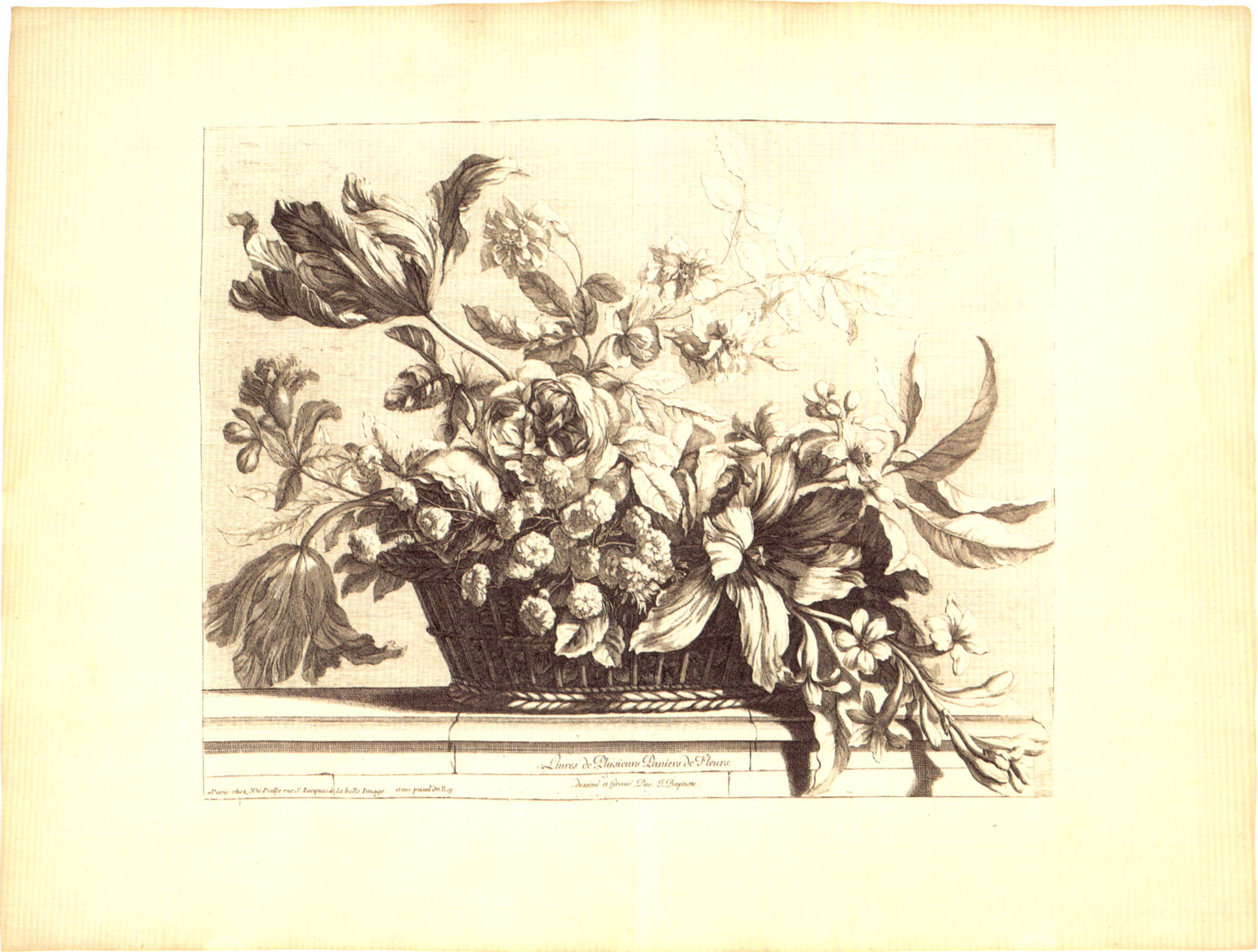

Of particular interest is the series of etchings (VIII) dedicated to the great Le Brun, who had helped the young artist at the outset of his career.

This volume also contains twelve etchings by Jacques Bailly (mentioned above), who produced floral compositions in a variety of media (oils, bodycolour, engraving), and who became a member of the Académie in 1664. The frontispiece and the last two plates in the volume were etched by Vauquier after designs by Monnoyer.

JEAN BAPTISTE MONNOYER, Collection of engraved and etched plates. A basket of flowers in 'Livres de Plusieurs Paniers de Fleurs'

47. Claude Aubriet (1665–1742)

[*title in red, blue and gold manuscript lettering within a gilt frame*] Papillons, Plantes Fleurs et Animaux, peints par Aubriet.

[*on the front endpaper in manuscript:*] 'Rien de plus beau que cetet ouvrage de Claude Aubriette Peintre et Dessinateur du Cabin du Roi. Je le recommande a mes enfants comme infiniment précieux. Il contient 53 feuillets, dont 28 représentent des plantes avec leurs fruits et les 25 autres 97 papillons mâles et femelles, plus un assez grande nombre en coques, en chenilles et in chrysalides. P.'

42.5 × 29 cm.

BINDING: Contemporary red morocco, with gilt triple fillets. All edges gilt, and a gilt floriated spine: 'Plantes et Papillon Peints par Aubriet'.

PLATES: 53 watercolours on vellum and one (fol. 3) on paper. Each plate has a gold ruled frame (37.5 × 27 cm.) and is protected by a guard leaf. Presented in two suites: (1) wild flowers, plants and fruits (fols. 1–28); (2) moths and butterflies (fols. 29–54). Paintings 29–54 are signed 'Aubriet pinx'. At the foot of each plate is the Latin name of the plant, often followed by a reference to a published work in which a fuller botanical description can be found.

PROVENANCE: From the collection of Louis Jean Gaignat (1697–1768; the manuscript was sold in 1764, listed as no. 1603 in the *Supplement a la bibliographie instructive ou Catalogue des livres du cabinet de feu M. Louis Jean Gaignat*), and the duc de La Vallière (sold in 1783 as no. 1677 in *Catalogue des livres de la bibliotheque de feu M. Le Duc de la Vallière*).

REFERENCES: Nissen, pp. 98–100; *An Oak Spring Garland*, no. 15; Blunt and Stearn, pp. 124–6.

CLAUDE AUBRIET, 'Peintre du Cabinet du Roy pour la miniature', followed Jean Joubert, the successor to Nicolas Robert, in the production of *vélins* depicting the royal botanical collections in Paris (see no. 44). Aubriet early acquired a considerable reputation as a painter of plants and insects, and the botanist Joseph Pitton de Tournefort commissioned him to produce engravings for his *Elémens de botanique* (1694), one of the most important botanical texts to be published at the end of the seventeenth century; a Latin edition entitled *Institutiones rei herbariae* appeared in 1700. For this project Aubriet accompanied Tournefort to the Levant in 1700 on one of the many scientific expeditions commissioned by Louis XIV for the purposes of naturalistic, historical and archaeological study. During the course of this three-year expedition Aubriet produced many sketches in monochrome that, due to their unembellished style, were not always appreciated by Tournefort. Many of them later served, however, as a useful basis for the illustrations to two works on the flora of the Near East published in the nineteenth century, the first by René Desfontaines and the second by Hyppolithe François Jaubert and Eduard Spach. In 1706 Aubriet was named 'Peintre du Cabinet du Roy', a post he was to occupy until 1735, when he was succeeded by one of his students, Madeleine Basseporte. Aubriet also collaborated with another distinguished French botanist, Sébastien Vaillant, in the production of 300 illustrations for the second edition of *Botanicon parisiense*, a work describing the flora native to the region of Paris, which was published in Leiden in 1723.

Most of Aubriet's works are today to be found in Paris (Muséum Nationale d'Histoire Naturelle), Göttingen and London (Lindley Library).

The volume of bodycolour paintings on botanical and entomological subjects at Oak Spring appeared, together with two other manuscript volumes by the same artist, at auction when the collection of the Conseilleur-Sécretaire du Roi Honoraire, Louis Jean Gaignat, was sold in Paris in

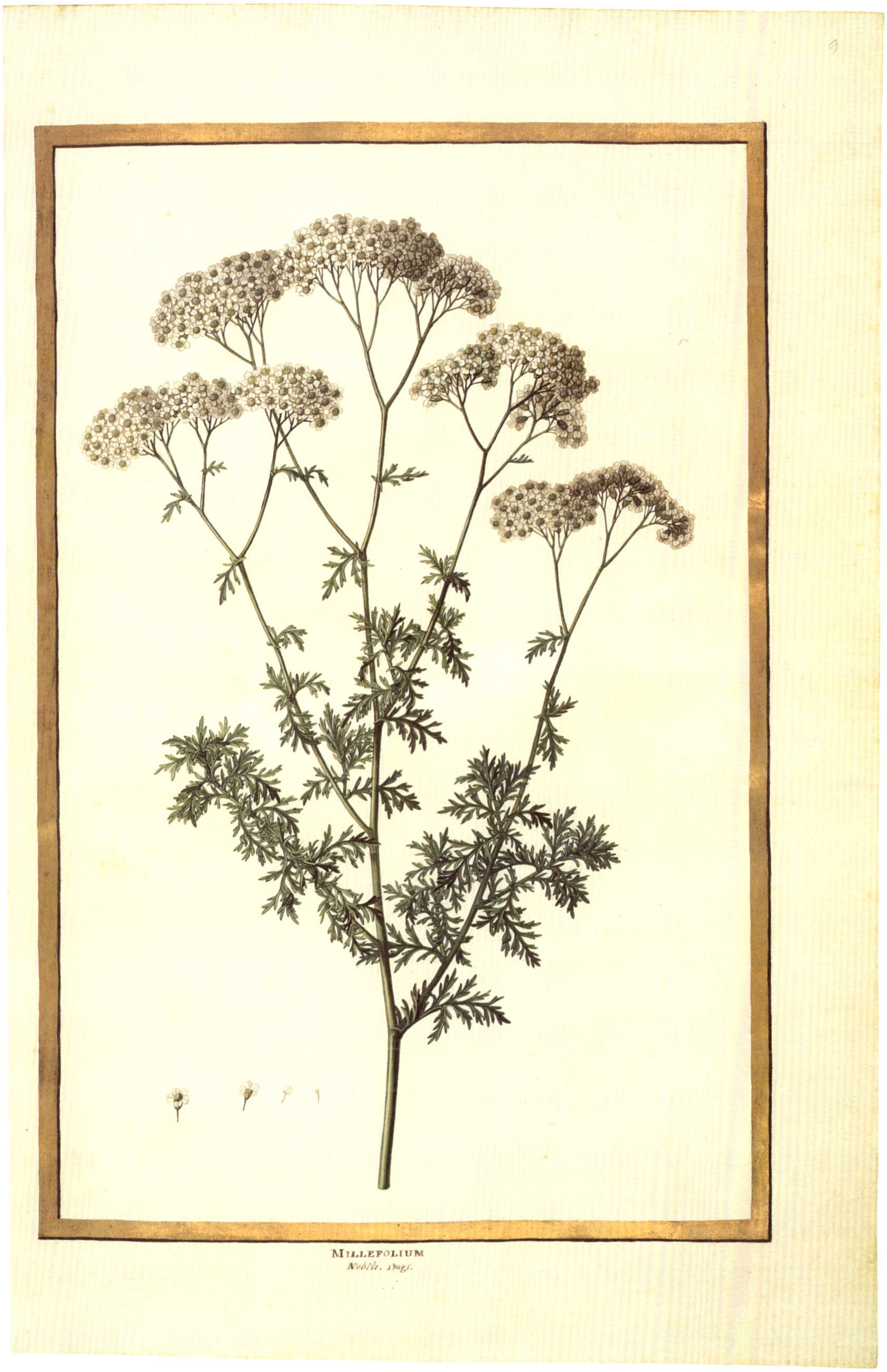

CLAUDE AUBRIET,
Papillons, plantes, fleurs et animaux. Common yarrow (*Achillea millefolium*), folio 9

1764 (G.-F. de Bure (*le jeune*), *Bibliographie instructive*, Paris, 1764, II, no. 1603, p. 348), and later as lot 1677 in the sale of the collection of the duc de La Vallière in 1783 (G. de Bure (*fils aîné*), *Catalogue des livres*, Paris, 1783, I, no. 1677, p. 490). The paintings in the Oak Spring manuscript are clearly the work of a master in the iconographic and stylistic tradition of Nicolas Robert. Aubriet's images are scientifically impeccable—the smallest details of the flowers and fruit are perfectly distinguishable—and his technical execution so perfect as to border sometimes on an excessive formalism. Nevertheless, these luminous paintings, with their forms defined by dense, vivid touches of colour and atmospheric effects created by fine shadowing, undeniably mark him as one of the most important botanical artists of the eighteenth century.

Among the plates, many of which depict exotic species in flower, are some that are particularly striking: the *Tamarindus* (fol. 1), a pomegranate with brilliant red flowers (fol. 5), a *Nicotiana* (fol. 6), and a yellow-flowered African *Ficoides* (*Mesembryanthemum*) (fol. 7). The *Achillea millefolium* (fol. 9) demonstrates Aubriet's skill at painting flowers on a smaller scale with the same brilliant precision. Equally impressive are his entomological plates of moths and butterflies, many shown in two views (dorsal and ventral), together with their caterpillar, cocoon and chrysalis.

48. Christoph Ludwig Agricola (1667–1719)

Rosa gallica versicolor. 29.5 × 21.5 cm. Bodycolour on vellum. 'Agricola' is written in ink in the centre of a pen, ink and bodycolour cartouche on the original mount.

CHRISTOPH LUDWIG AGRICOLA, a painter and engraver from Regensburg, did not receive any formal training as an artist, preferring to develop his hand and eye by drawing directly from nature. A visit to Italy during the course of a Grand Tour marked a decisive turning-point in his life and work, however. In Rome he was able to study the works of Nicolas Poussin, and he eventually decided to settle in Naples, where he remained for many years. There he made a comfortable living by painting views of the city and the surrounding Campania, works that were much sought after by wealthy foreign travellers, particularly the English, who liked to take them home as souvenirs of the places they had visited.

Agricola never lost his interest in natural history, however, and continued to paint still-lifes with birds and flowers; he also completed a series of sixteen paintings of butterflies, unfortunately now lost. The painting of a pink rose in the collection at Oak Spring forms a pair with one that depicts a gardenia and butterfly. Both works have been painted against a striking black background. The considerable skill of the artist can be seen in his rendition of the silky petals curling around their bright golden stamens, and the green leaves with their serrated edges and delicate veining, all of which have been depicted with superb realism.

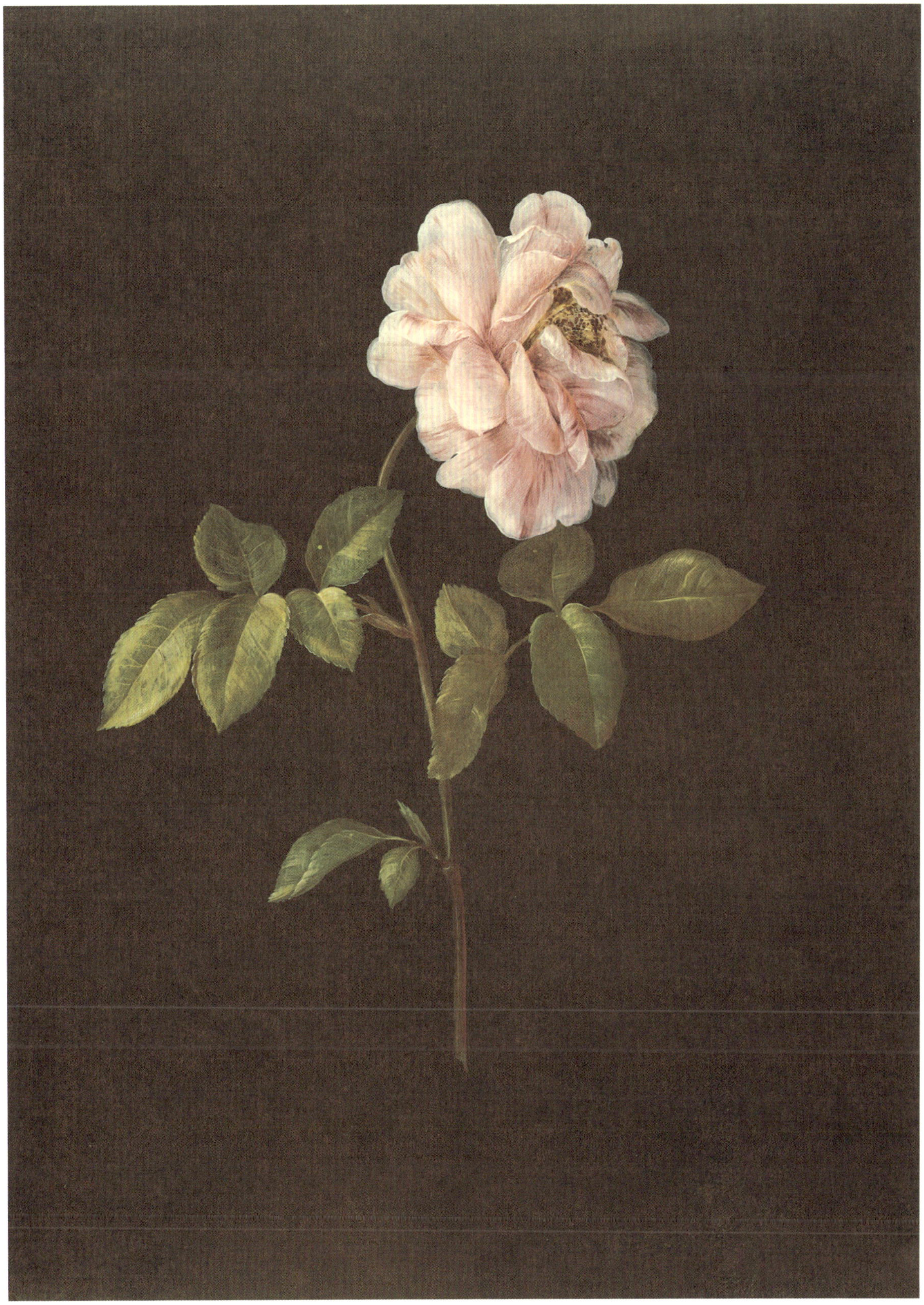

CHRISTOPH LUDWIG AGRICOLA. French rose (*Rosa gallica versicolor*)

49. Jacques Guillaume van Blarenberghe (1679–1742) *with* Louis-Nicolas van Blarenberghe (1716–1794)

[*manuscript lettering within a watercolour border of garlands and vignettes of garden scenes*] Horti herbarii Domini Lamberti Michaelis Winckelman (Dum viveret) Pharmacopolae Lillensis, et Botanophili. Plantae quaedam, ad vivum pictae. a D: D: Vanblaremberghe. Faucquette scripsit Insulis 1755.

37 × 23 cm. After the frontispiece the leaves are numbered in manuscript 1–275, followed by 7 unnumbered leaves of manuscript index.

BINDING: Contemporary calf; full gilt spine with the title 'PLANTAE AD VIVUM PICTAE'.

PLATES: The manuscript title-page credits the van Blarenberghes (the 'm' in the name on the title-page is an error) with the authorship of this collection of watercolour drawings; the dedication mentions the son, Louis-Nicolas, by name, adding 'cum patre suo' (referring to his father, Jacques Guillaume). The calligrapher, Faucquette, executed the lettering for the title-page, the dedication pages, and the 13 pages of index. The 250 water- and bodycolour paintings of plants that follow (each specimen labelled with its Latin name) are numbered by hand 3–275. Pages 107, 110, 111, 114, 115, 118, 119, 122, 123, 127, 130, 131, 134, 135, 138, 139, 142, 143, 146, 147 and 203 are blank.

PROVENANCE: Ex-libris 'I F J Hatton Procuratoris' (1712–58); sold Brussels, 20 November 1957.

REFERENCES: Marcel Decroix, 'Les Peintres van Blarenberghe: à propos d'un manuscrit botanique lillois illustré par eux', *Bulletin de la Société de Botanique du Nord de la France*, XI/2, 1958, pp. 61–5; Lauterbach, pp. 82–6; *An Oak Spring Hortus*, no. 30.

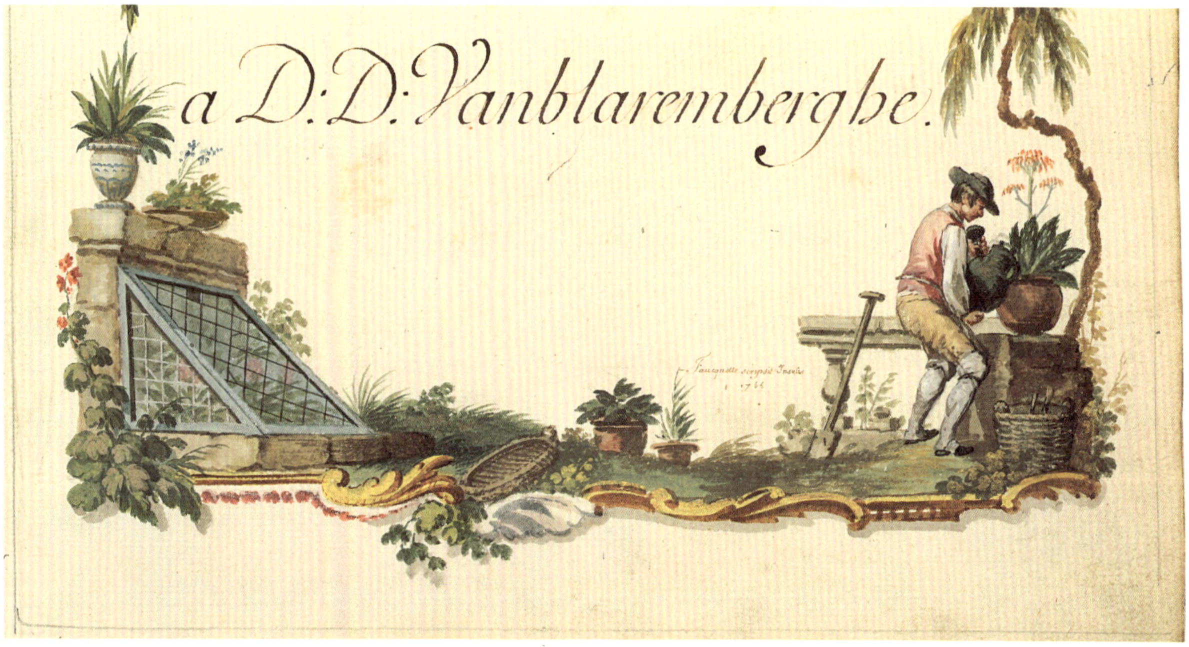

JACQUES GUILLAUME VAN BLARENBERGHE and LOUIS-NICOLAS VAN BLARENBERGHE, *Horti herbarii*. Detail from the frontispiece: a gardener and pots of plants

THIS MANUSCRIPT HERBAL opens with a Latin dedication addressed to all lovers of botany ('Botanophilis'), written by the procurator Ignace François Joseph Hatton. Hatton had received from the children of the pharmacist and botanist Lambert Michel Winckelman (or Wickelman) of Lille (who was also Hatton's father-in-law), the gift of a large number of botanical paintings

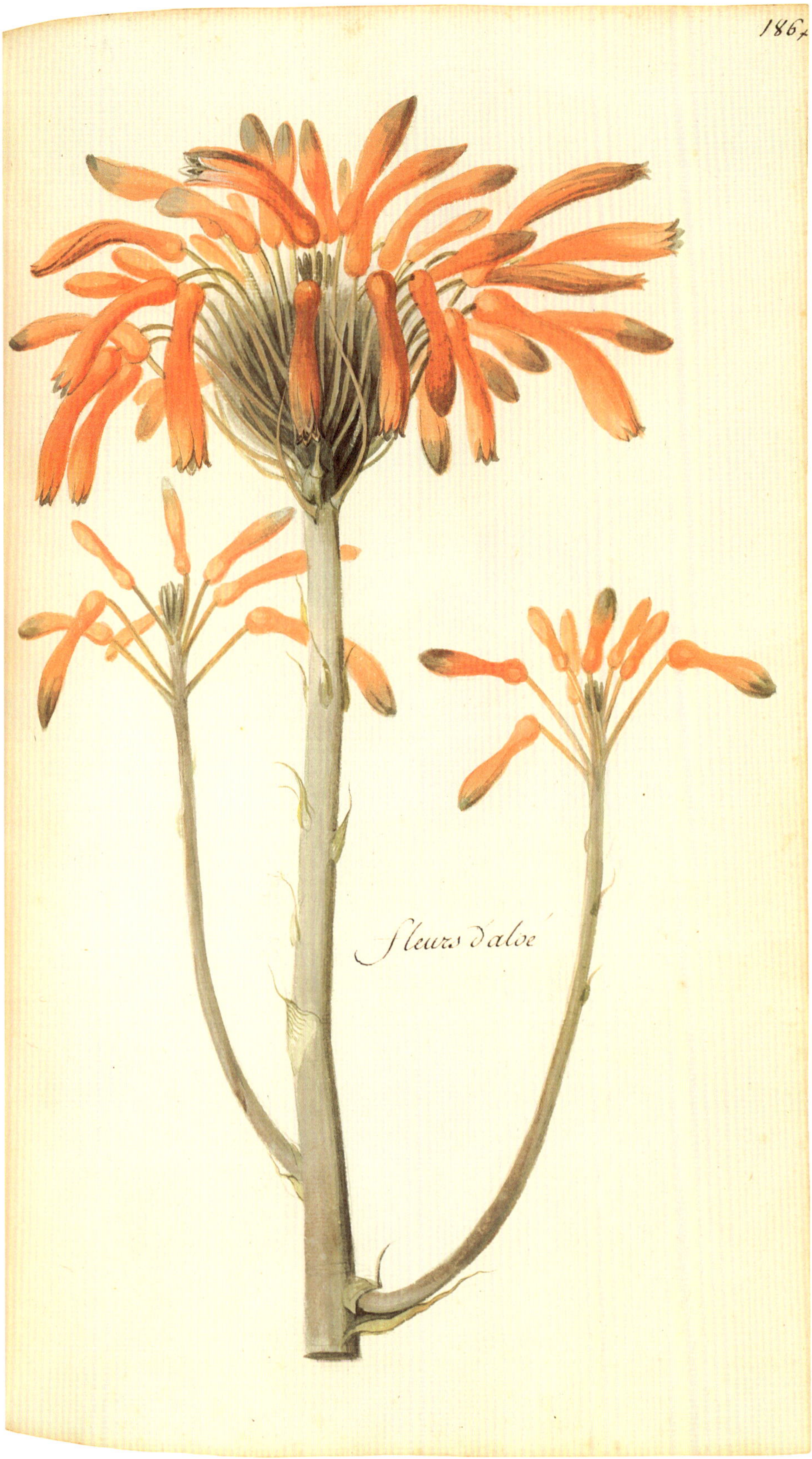

JACQUES GUILLAUME VAN BLARENBERGHE and LOUIS-NICOLAS VAN BLARENBERGHE, *Horti herbarii*. Candelabra aloe (*Aloe arborescens*), folio 186

JACQUES GUILLAUME VAN BLARENBERGHE and LOUIS-NICOLAS VAN BLARENBERGHE, *Horti herbarii*. Snowdrop (*Galanthus nivalis*), folio 26

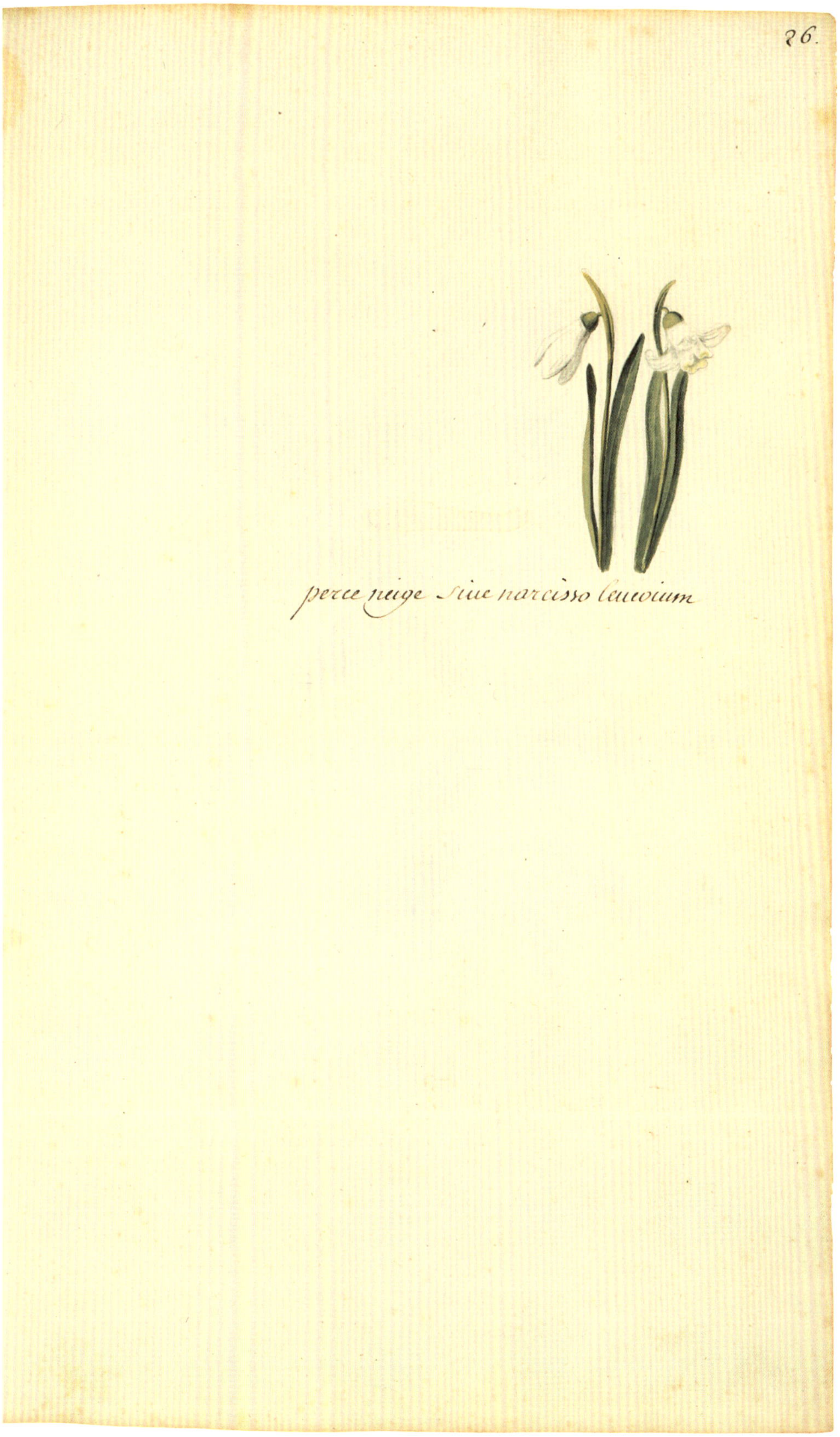

commissioned by Winckelman from the artists Jacques Guillaume and Louis-Nicolas van Blarenberghe. In his dedication Hatton explains that Winckelman's original intention had been to compile a detailed herbal of native and exotic plants, but his premature death in 1755 interrupted this project, and the text, as well as many of the illustrations of rare plants mentioned in the dedication, were unfortunately dispersed.

In order to preserve at least those paintings that had fallen into his hands, Hatton had this manuscript prepared and bound. In addition to the paintings it contains a preliminary leaf decorated with a garden landscape, a frontispiece with a pleasant series of gardening scenes, a dedication, a closing tailpiece showing putti engaged in gardening tasks and, to conclude, a useful index. Most of the botanical illustrations in the manuscript were executed by Jacques Guillaume, who was engaged in this work up to the year of his death in 1742. His son Louis-Nicolas then completed the project, and the frontispiece (dated and signed by the calligrapher 'Faucquette scripsit Insulis 1755') and tailpiece were probably designed by him.

Jacques Guillaume was the son of Henri van Blarenberghe, a Dutch painter of miniatures. Jacques Guillaume eventually settled in Lille, where he became quite renowned as a painter of battle-scenes. His sons, Henri-Desiré and Louis-Nicolas, followed in his footsteps, the latter becoming a well-known miniaturist and *gouachiste* who worked in Paris for the duc de Choiseul as a painter of naval battle-scenes until 1750.

Given the well-established tradition of flower painting in the Low Countries, it is not surprising that Winckelman turned to two artists of Dutch origin to prepare the illustrations for his herbal. And, in fact, the van Blarenberghes gave proof of considerable skill and artistic sensibility in the execution of these botanical paintings, most of them illustrating wild species in flower. The primary scope of the work being scientific, Winckelman had demanded that certain details be focused on: thus the miniaturists provided enlarged views of the seeds, buds and, most importantly, the leaves of each plant. They managed nevertheless to arrange the various views quite elegantly and harmoniously on the page. There is, for example, a very fine representation of the aloe, separated into two parts; on folio 185 the leaves and part of the trunk are shown, while on the following folio (186), its extravagantly beautiful red flower is depicted.

50. Georg Dionysius Ehret (1708–1770)

[*title printed at head of the first plate*] Plantae et Papiliones rariores, depictae et aeri incisae a Georgio Dionysio Ehret, Palat. Heidelb. 1748

55 × 38.5 cm. 15 hand-coloured etchings. Plate XV is followed by a leaf of descriptive letterpress.

BINDING: Original marbled boards with grey paper spine.

PLATES: 15 numbered, hand-coloured etchings. Plate XIII is numbered I. At the foot of every plate the name of the plant has been engraved, and on some of the pages there are extensive descriptions. Plates II and XV carry a short text in English.

Ehret's name generally appears, with a date varying from 1748 to 1759. Plates IV and VIII are descriptive tables. At the foot of the explanatory text to plate XV is a small etched vignette of a *Rondeletia* (6.7 × 11.2 cm.). Bound in the back are two additional hand-coloured etchings by Ehret, a 'Walkeria' (29.8 × 14.3 cm.) signed in the lower right, and a 'Theobroma foliis ferratis Linn.' (18.8 × 13.5 cm.) signed and dated in the lower right 'G. D. Ehret 1762'. 'Tab I' appears in the top right-hand corner of the latter.

REFERENCES: Dunthorne 109; Nissen 583; Henrey, II.62–4, and no. 677; Calmann, pp. 93–4; Blunt and Stearn, p. 166.

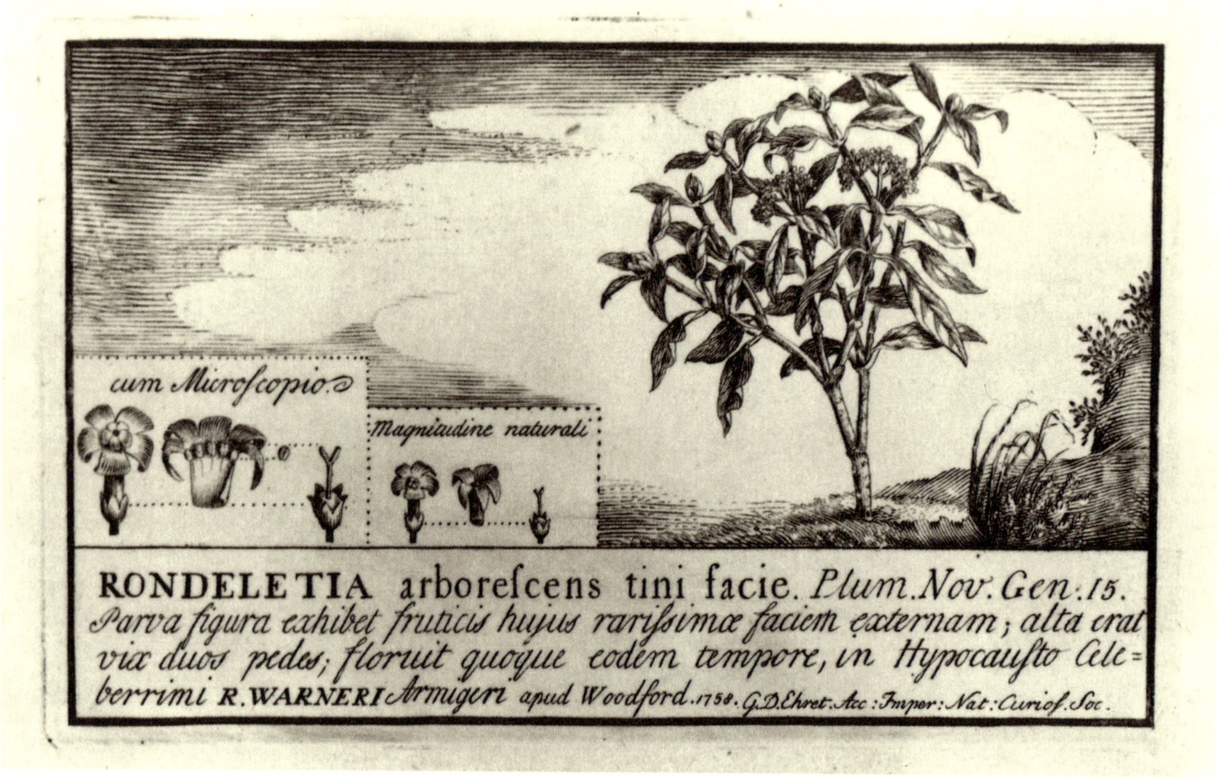

GEORG DIONYSIUS EHRET, *Plantae et papiliones rariores*. A tail-piece of a *Rondeletia*, a tropical evergreen, on tab XV, 'The History and Analysis of the Parts of the Jessamine . . . July 1758'

GEORG DIONYSIUS EHRET was a remarkably gifted botanical painter, yet one whose name is little known today beyond a small circle of specialists. His *œuvre* includes a large number of botanical paintings on vellum (signed and dated), most having survived and some of which can be seen in London (Victoria and Albert Museum; Royal Botanic Gardens, Kew; Natural History Museum), Cambridge (Fitzwilliam Museum) and at the University of Erlangen. *Plantae et papiliones rariores* was the first and most important engraved work by Ehret, and perhaps the only one that can truly compare with his magnificent paintings.

Ehret's work, as Blunt noted in 1950, greatly influenced the art of botanical painting in a period when the science of botany was itself undergoing radical transformations (see nos. 55 and 56). Ehret was personally acquainted with many of the greatest naturalists of his day, including Bernard de Jussieu and Sir Hans Sloane. He also had the good fortune early in his career to collaborate with the great Swedish botanist Linnaeus. We know something about the career of this exceptional artist from some autobiographical notes set down by him in the latter part of his life (see the 'Memoir', trans. E. S. Barton, *Linnean Society Proceedings*, 1894–5, pp. 41–58).

Ehret was born at Heidelberg into a modest family of market gardeners; he received some drawing lessons from his father, but on his death the young Ehret was forced to take work as an apprentice gardener. While he was in the employ of the Elector of Heidelberg he attracted the attention of the Margrave of Baden, who took him into his service as a gardener. He soon left this position, however, and set off to seek his fortune. After a less than satisfying experience in Regensburg, where the apothecary Johann Wilhelm Weinmann commissioned many drawings, later published in *Phytanthoza iconographia*, and where Ehret also worked for the banker Loeschenkohl, who set him to work copying and colouring the plates in printed botanical works, such as *Hortus judicus malabaricus*, published in Amsterdam between 1678 and 1703, Ehret finally had the good fortune to meet Christoph Jacob Trew (1695–1769), a wealthy physician and enthusiastic bibliophile who became not only Ehret's most important patron, but a lifelong friend as well. It was Trew who first encouraged the artist to study plants from a scientific as well as an artistic perspective (Calmann, p. 27).

In 1733 Ehret began to travel more widely, visiting various countries and sending Trew many hundreds of botanical drawings and paintings, which the latter subsequently included in his *Plantae selectae* (1750–73). Ehret stopped in Paris, where he made the acquaintance of the great botanist Jussieu, and also visited Geneva, Lyon, Montpellier and London, where he met Sir Hans Sloane (the president of the Royal Society) and Philip Miller, curator of the Physic Garden in Chelsea (the only botanical garden in London at that time) and author of the monumental *Gardeners Dictionary*. Holland, naturally, and in particular Leiden with its famous botanical garden, represented an important stop on Ehret's itinerary. While there in 1736 he visited Hartekamp near Haarlem, the estate of the banker and amateur horticulturist George Clifford, where Linnaeus was staying as botanist in residence. Clifford bought many paintings from Ehret and encouraged him to assist in the illustration of *Hortus Cliffortianus*, a work that Linnaeus was preparing on the rare plants at Hartekamp, and which was published in 1737.

Linnaeus's scientific work had a profound effect on Ehret's art: under his tutelage Ehret learned to analyse the structure of plants before attempting to depict them. The Swedish botanist explained to Ehret his revolutionary method of identifying plant species on the basis of their reproductive organs. Ehret immediately grasped the importance of this new classification system and produced a beautifully engraved *Table* to illustrate it (a rare copy of the original drawing for it is in London's Natural History Museum). The *Table* was subsequently included, unbeknown to Ehret, by Linnaeus in the 1737 edition of his *Genera plantarum*, published in Leiden.

By the end of 1736 Ehret had decided to settle in London, where he eventually married the sister-in-law of Philip Miller, and where he remained until his death in 1770. He became quite well known and acquired a number of important patrons. His work was admired by Richard Mead, physician to the royal family and the possessor of many works of art as well as a fine collection of natural specimens. Mead commissioned paintings from Ehret and introduced him into various cultivated and aristocratic circles in the capital. The highly influential Duchess of Portland also became one of his patrons and did much to further his career.

Ehret continued to produce botanical paintings and engravings, collaborating in the illustration of the *Transactions of the Royal Society*, and illustrating some travel books, including Richard Pococke's *A Description of the East* (1743–5), Griffith Hughes's *The Natural History of Barbados* (1750) and Patrick Browne's *The Civil and Natural History of Jamaica* (1756). Like many other prominent artists, he became a successful and popular teacher of flower painting. Once Ehret's fame had spread he was even invited to Denmark to teach, although he refused as he had no wish to leave England. In fact, the artist never strayed far from London during the second half of his life, except for a brief period spent at Oxford's botanical garden. He was elected a member of the Royal Society in 1757, where he delivered three papers. He continued to paint actively for another decade, only slowing down when he began to suffer serious problems with his eyesight.

Plantae et papiliones rariores was published as a volume of plates without any text; even the title was simply printed at the top of the first illustration. Plates IV and VIII are tables with minute drawings illustrating various plant characteristics according to the Linnaean system (with a few departures), accompanied by brief explanatory phrases in Latin. Each of the thirteen other plates illustrates one or more different flowering species. These engravings are marked by a perfect unity of artistic vision, for a single hand made the drawings, engraved and printed the plates and coloured the final images, a fairly unusual procedure in this period. The illustrations, while drawn with scrupulous attention to scientific fact, are pleasantly varied in composition, and sometimes enlivened with a Rococo touch. Butterflies flit across the pages, and in some of the plates, such as those of the *Martynia annua* (pl. I) and the 'Deccan hemp' (*Hibiscus cannabinus*, pl. VI), below the magnificently drawn flower is a tiny drawing of the entire plant, shown growing in the earth or in a pot.

Ehret managed to imbue even his table of plant characters (pl. VIII) with unusual charm, depicting *à trompe l'œil* a drawing of an *Abutilon* or glade mallow (together with its male and female states), attached casually to the table itself, while in the lower-right corner there is the sketch of a peanut. Some of the plates carry acknowledgements by the artist: the Carolina allspice (*Calycanthus floridus*) in plate XIII is dedicated to the Earl of Macclesfield, and the ginger plant (*Amomum zingiber*) in plate XIV to his beloved patroness the Duchess of Portland.

Some of the original drawings for *Plantae et papiliones rariores* are today in the Natural History Museum. These include a preparatory watercolour sketch of the *Hibiscus cannabinus* that appears in plate VI. In Ehret the vision of the scientist and that of the artist were mingled in perfect harmony. Possessed of an unusually thorough grounding in botany, his images were models of scientific accuracy; at the same time the delicate outlines and exquisite colouring of his illustrations betray a fine artistic sensibility.

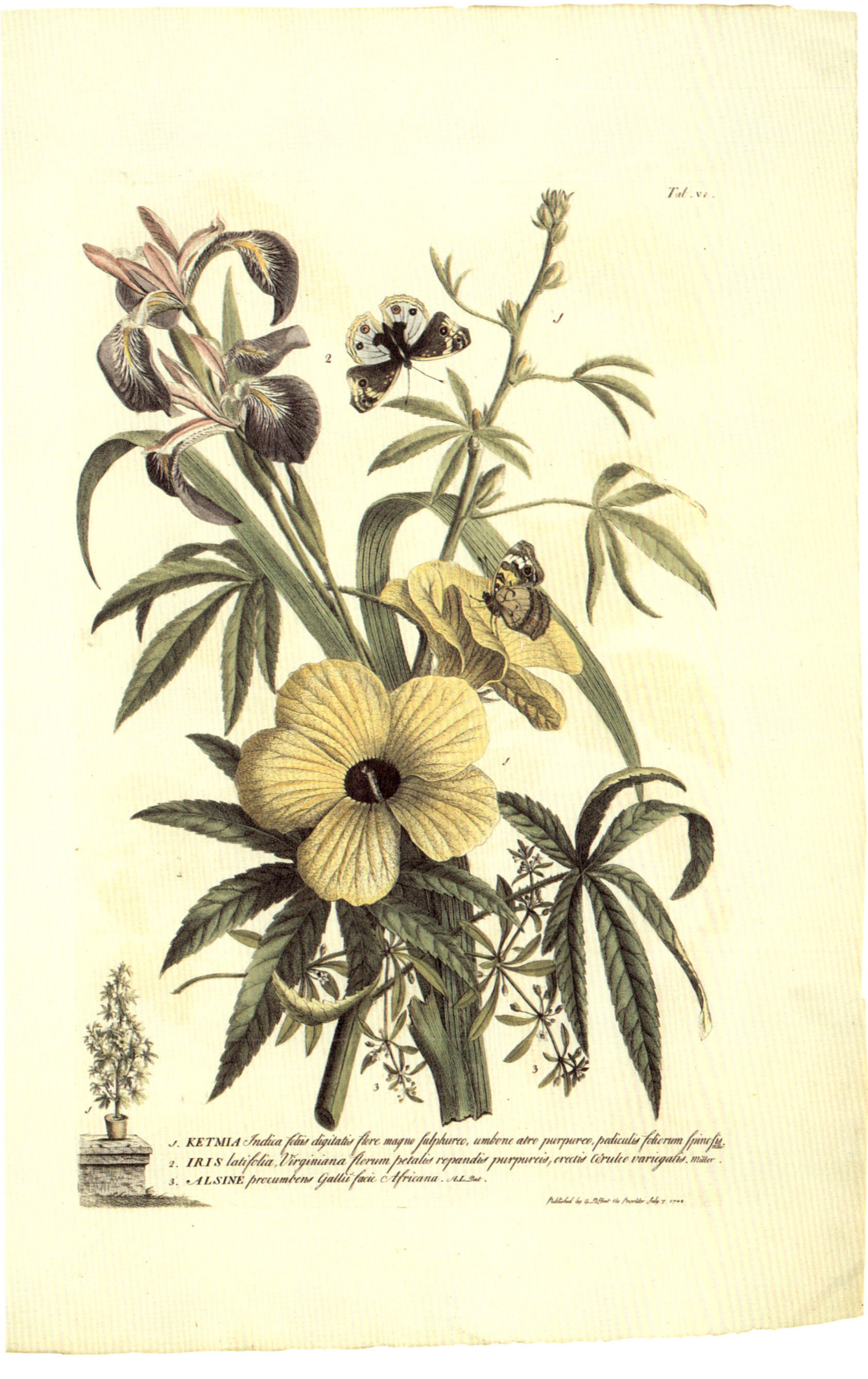

GEORG DIONYSIUS EHRET, *Plantae et papiliones rariores*. Plate VI: 1. Indian hemp (*Hibiscus cannabinus*); 2. Southern blue flag (*Iris virginica*); 3. Carpet-weed (*Mollugo verticillata*) with butterflies

The more closely one studies his works, the more anomalous it appears that Ehret is still almost unknown to art historians and is rarely mentioned in their dictionaries (a point made in Mitchell 1973, p. 103). The vast number of drawings and paintings he left allow us to follow the evolution of his style and to ascertain that very few artists in the history of botanical illustration could match the consistently high level of his work, either scientifically or artistically. With his unerring sense of form and colour, elegant style and unusual mastery of technique, Ehret managed to create a unique and quite personal vision, so accurate and realistic that each painting became an almost idealized version of the actual subject.

The passion that Ehret brought to his work is beautifully expressed in a portrait in oils by George James (Linnean Society, London). The artist, who was fifty-nine years of age when he sat for this painting, is shown informally dressed, with his white shirt open at the collar, and his eyes fixed attentively, one might almost say tenderly, on a fine specimen of day-jasmine. Pencil in hand and a blank sheet on the table before him, he is about to begin yet another one of his miraculous plant 'portraits'.

The collection at Oak Spring also includes many bodycolour paintings by Ehret; in addition to the imposing *Magnolia grandiflora* (see no. 51), there is an important collection of thirty-seven paintings on vellum, *Flowers, Moths, Butterflies and Shells*, produced by the artist between the years 1756 and 1769, and ten botanical paintings on vellum, including one of a *Convallaria* (lily of the valley) signed and dated 1762, that was subsequently engraved by A. L. Wirsing for Trew's *Hortus nitidissimis* (pl. 130).

51. Georg Dionysius Ehret (1708–1770)

Magnolia grandiflora. 75.6 × 55 cm. Bodycolour on vellum. Signed in the ribbon at the bottom right: 'G. D. Ehret Pinxit'. Framed.

The Heidelberg-born Ehret decided at the age of twenty-eight to settle in England, where he lived for the rest of his life. As we learn from his 'Memoir' he arrived there in August 1763, and once settled in London every day used to walk the three miles that lay between his home in Fulham and the Physic Garden in Chelsea, in order to observe *en route* the coming into flower of a magnificent magnolia-tree then growing in the garden of Sir Charles Wager at Parsons Green. In this way Ehret could study at leisure every phase of the flowering process, from the first appearance of the bud through its opening into the beautiful, wide-petalled flower, to the formation of the seeds. He made a number of drawings and engravings of these various stages 'in order to publish a perfect botanical study of it' (see Calmann, p. 62). Some of these engravings he sent to his

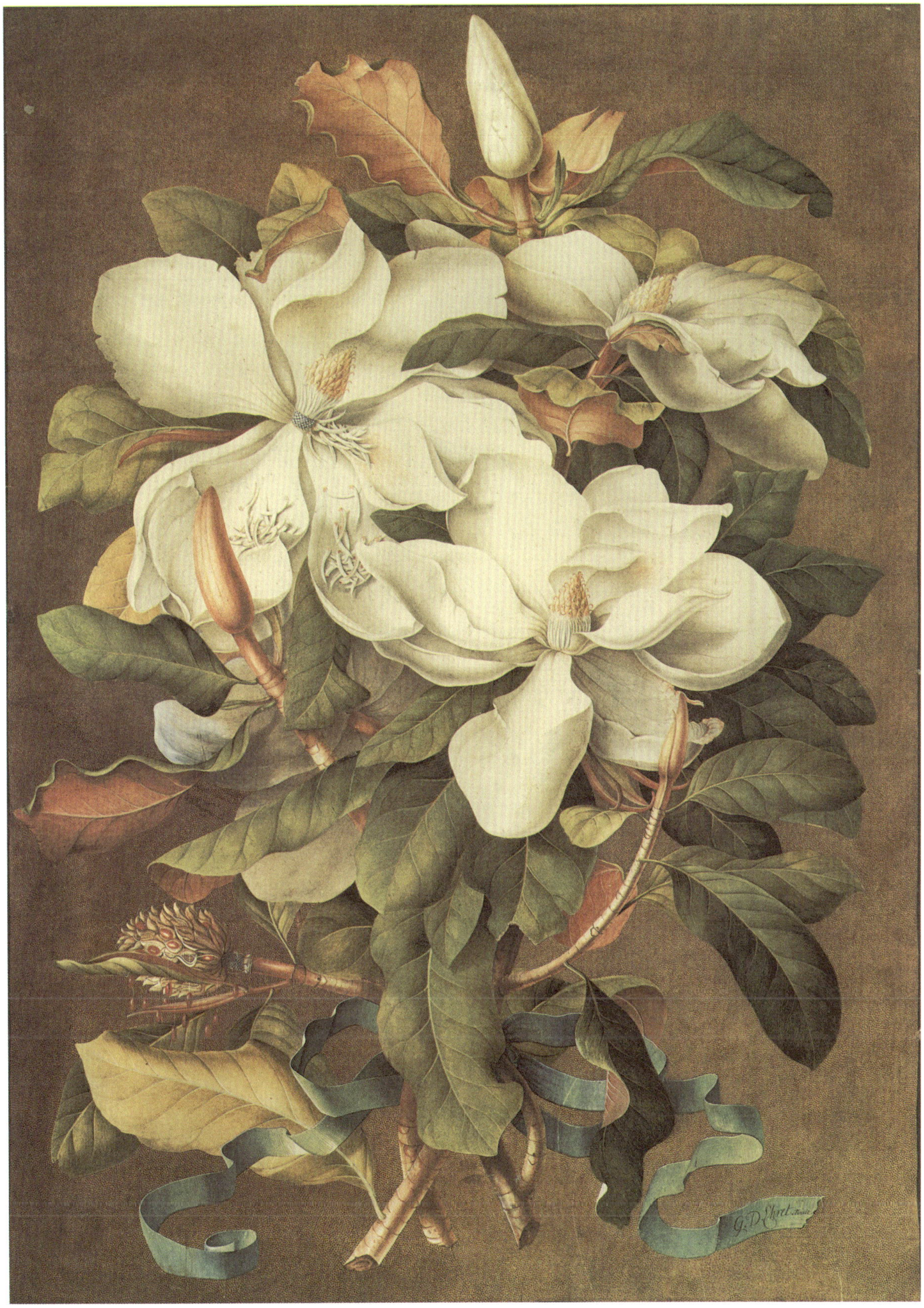

GEORG DIONYSIUS EHRET, Southern magnolia (*Magnolia grandiflora*)

patron and friend, the naturalist Dr Christoph Jacob Trew, who subsequently described the flower in *Commercium litterarium*, the journal of the Nuremberg Society of Physicians, prefacing his article with a eulogy to the artist. The pharmacist Johann Ambrosius Beurer, another close friend of Ehret's and a cousin of Dr Trew, sent one of the artist's engravings to the botanist Bernard de Jussieu in Paris, who professed himself to be delighted with it.

Ehret's illustrations of the magnolia flower were an immediate success. Trew utilized them for two plates in the fourth volume of his *Plantae selectae*, published in 1754—one a splendid representation of the flower (pl. XXXIII) and another (pl. XXXV) showing details, including the follicles and seeds. Both plates were accurately engraved by the German artist Johann Jacob Haid from the original drawings. Two illustrations of the flower also appeared in the second volume of the *Natural History of Carolina* by Mark Catesby, a celebrated work on the flora and fauna of the New World published in 1743. Every single plate in this lavishly illustrated work is by Catesby except for these two of the magnolia, whose unusual beauty evidently greatly impressed the author. (A smaller bodycolour painting of just one magnolia flower, executed somewhat later by Ehret—it is dated 1743—can be seen in the Victoria and Albert Museum.)

The large bodycolour painting at Oak Spring is undated (perhaps *c.* 1737) and represents an interesting, because atypical, example of his work. In its creation Ehret seems to have allowed artistic considerations to take precedence over scientific ones. While the other works in this series were careful botanical studies of a single blossom, here we find a complex and dynamic composition—four flower-laden branches arranged in a bouquet and bound with a light-blue ribbon on which the artist's signature appears, 'G.D. Ehret pinxit'. This is no mere botanical painting, but an impressively sophisticated work in the Rococo style.

Against a dark-brown background, uniform in colour but textured to create a pleasant dappled effect, seven flowers—each in a different stage, from the tightly closed bud, to the flower in its moment of greatest splendour, to the bare capsule with its seed-bearing follicles—are dramatically silhouetted. Ehret made brilliant use of his palette to heighten the visual impact of the work: the upper sides of the leaves have been painted in a range of intense greens, while their lower surfaces are a warm brown. In the midst of the pure white fleshy petals, the brilliant yellow of the ovaries stands out, while the vivid red seeds have been painted just as they are falling to the ground. In this way Ehret underlines the temporality, and the fleeting beauty, of the living flower. This painting is thus a mine of scientifically accurate information, but couched in the language of the artist, one of the greatest painters in the history of botanical illustration.

52. THE ROBINS FAMILY OF BATH

THOMAS ROBINS *the elder* (?1715–1770)
Wild flowers, a butterfly and a fly.
29.8 × 22 cm. Water- and bodycolour on paper.

THOMAS ROBINS *the younger* (?1743–1806)
Agapanthus umbellatus, Coronilla varia and butterflies.
35.5 × 24 cm. Water- and bodycolour on paper, glued to a sheet of grey cardboard, on which is written: 'N° 10. Vol: 2d'. The flowers are labelled 'a' and 'b', and the butterflies from 1 to 4; a manuscript key with the corresponding scientific and common names of the species is provided on another sheet. Inscribed in pen and brown ink in the lower right: 'Painted from Nature by Thos Robins the younger Bath 1787'.

PROVENANCE: In 1967 'Pictures of Exotic Plants and Insects' by Robins the younger were placed on sale by the Bath & West & Southern Counties Agricultural Society.

THOMAS ROBINS THE ELDER OF BATH, 'the limner of the city' in the mid-eighteenth century, is best known for his paintings of landscapes, country houses and gardens, to which he often added a border of highly naturalistic motifs, such as the honeysuckle vines framing his *View of a Gloucestershire Country House* now at Oak Spring (see Harris). Robins also painted many botanical pictures, marked by a highly decorative style and an unusual technique that combined body- and watercolour. At the Royal Horticultural Society's Lindley Library, for example, there are three interesting paintings by him of mushrooms, insects and butterflies (Elliott, p. 50). Robins's *œuvre* also includes a number of impressive floral images and bouquets. At the Fitzwilliam Museum in Cambridge is a sketchbook entitled 'Drawings of Flowers; Ehret and Robins' (Broughton Collection, PD 115, 1973; see *Flowers of Three Centuries*, p. 66) that contains about one hundred paintings by Robins, some of them signed and one of them dated (1762), as well as a painting by Ehret (fol. 38) and works by other artists. The bouquet of mallows, geraniums, orange blossom and other flowers bound with a blue ribbon on folio 39 of the Fitzwilliam sketchbook is stylistically quite similar to the *Wildflowers, a Butterfly and a Fly* at Oak Spring. Furthermore, the bouquet in folio 41 includes a rare flower—the *Campanula canariensis* or Canary bell-flower, just recently arrived in Europe—that can also be seen in the Oak Spring painting.

There are four other known floral paintings (all in the Fitzwilliam) by Robins the elder (two of which are signed), all of which reflect the artist's highly decorative sense of composition. Three depict floral bouquets and the fourth an arrangement of *Lonicera*, roses and convulvulus shown against a Gothic ruin (*Flowers of Three Centuries*, nos. 88, 89, 90).

Wildflowers, a Butterfly and a Fly constitutes an unusual example of Robins's work because here he has placed his bouquet against a background of dark-brown bodycolour. The mass of flowers is held by a pink ribbon knotted in a simple bow, while the stem of a white cyclamen dips gracefully over the bow and into the lower part of the composition. The bouquet contains a number of wild and cultivated flowers, including some hellebore, snowdrops, a cyclamen, a Canary bell-flower, an *Anemone hepatica*, an *Arabis*, an *Anthemis* and a *Herantis hiemalis*, all depicted in remarkable detail. The leaves of these flowers have been arranged in a decorative frame around the bouquet, reminis-

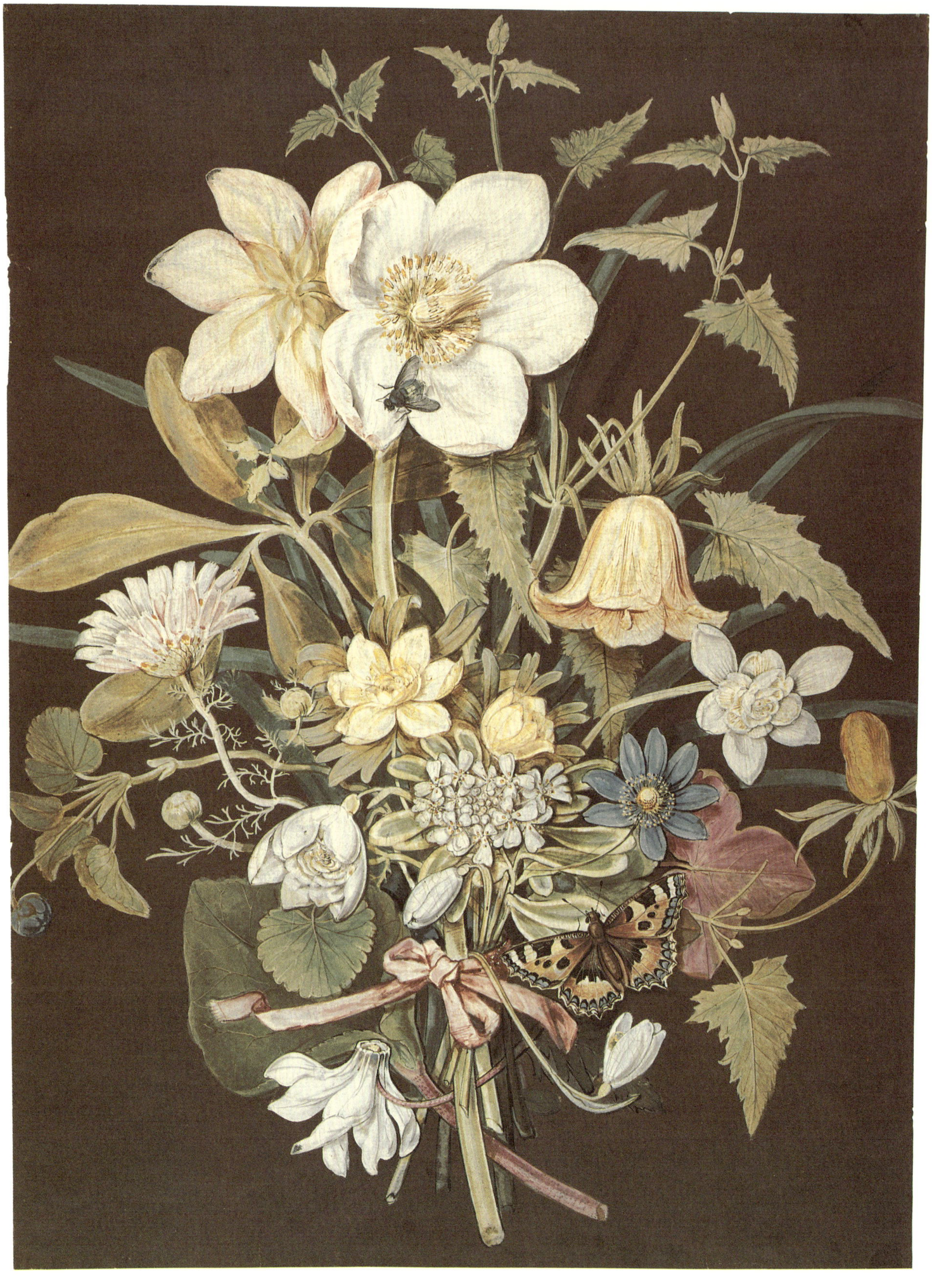

THOMAS ROBINS the elder, Wildflowers, a butterfly and a fly

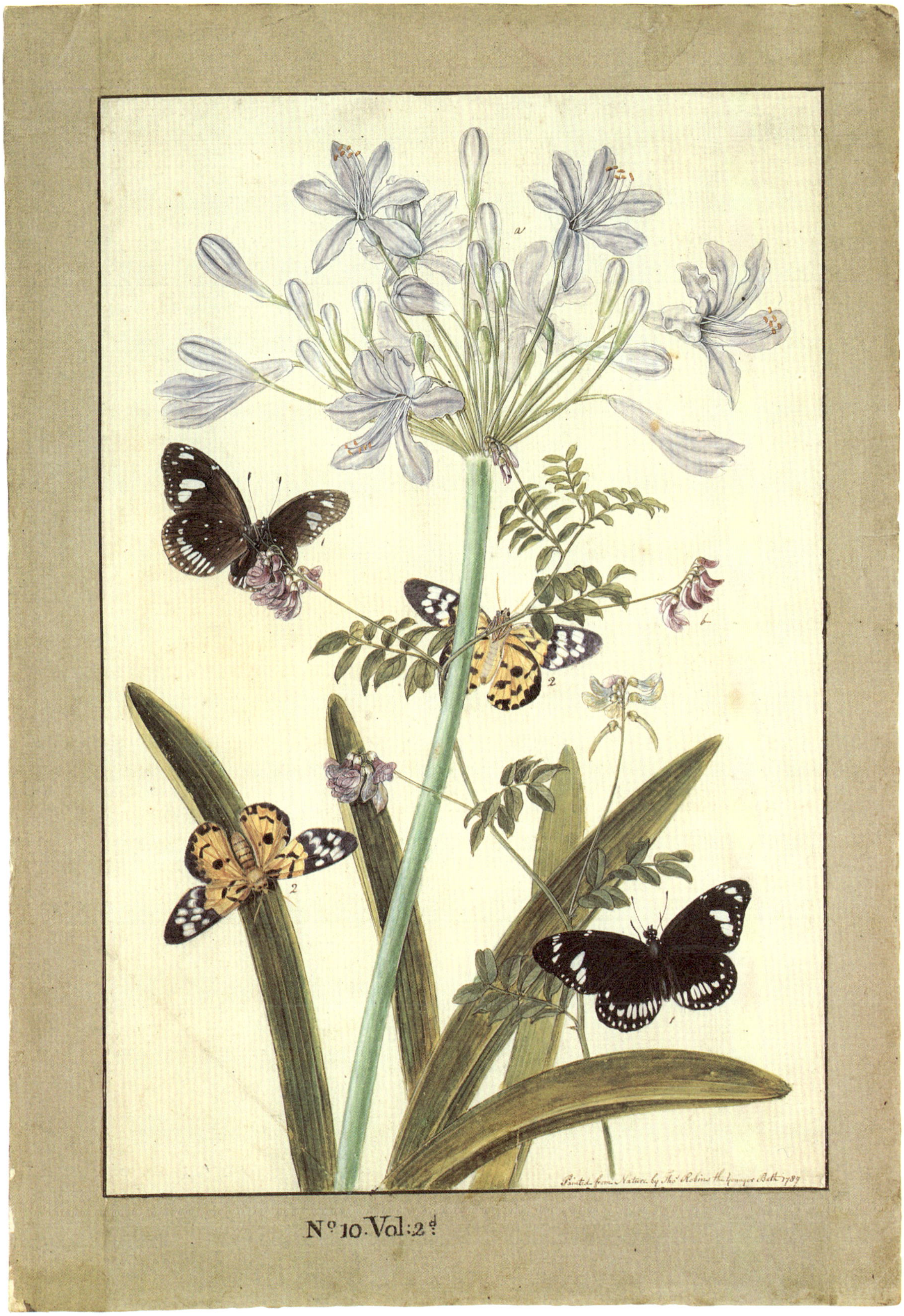

THOMAS ROBINS the younger, Blue-African-lily (*Agapanthus africanus*), crown vetch (*Coronilla varia*), and butterflies cent of the plant borders that surround Robins's landscape paintings. A variegate Small Tortoiseshell butterfly (*Aglais urticae*) and a fly poised *à trompe l'œil* against the white petal of a hellebore further animate this sophisticated Rococo composition.

At Oak Spring there are also seventeen paintings in bodycolour of flowers and insects and one painting of butterflies (all of the same dimensions) by Thomas Robins the younger, one of which is the *Agapanthus africanus* discussed below. They originally formed part of the second of the two bound volumes of paintings that the artist placed on sale at his house in Bath in 1787; shortly after his death in 1806 these volumes were acquired by the Bath & West & Southern Counties Agricultural Society. Under each of the seventeen paintings is a handwritten inscription with 'Vol: 2d' and the page number of the painting as it appeared in the original volume. Each work has been signed by the artist in an inscription that includes a date between 1785 and 1788 and the notation 'Painted from Nature'. This series allows us to better analyse the style and technique of this talented artist, who left behind him a relatively small number of works. A painting in the Fitzwilliam, *Flowering Branch of Plumeria against a Landscape* (PD 913, 1973; see *Flowers of Three Centuries*, p. 66), signed 'Thomas Robins' and dated 1769, has been attributed to him, as has a series of modest watercolours depicting orchids in the Lindley Library (*Treasures of the RHS*, pl. 15).

In contrast to the highly decorative style of his father, Robins the younger utilized a more rigorously scientific approach in his work. His paintings of flowering plants, ranging from the most common cultivated varieties to the latest exotic species, interspersed with butterflies and other insects, are impressive in their realism. Nevertheless, Robins had also clearly inherited his father's decorative sense for his works are skilfully composed, and he managed to interweave opaque strokes of bodycolour with the lighter, fresher hues of watercolours to create tonalities in keeping with the eighteenth-century taste for rich chromatic effects.

One of the most interesting works in the Oak Spring series is the painting of an *Agapanthus africanus*, an exotic plant that was introduced and widely cultivated in Europe towards the end of the eighteenth century. Around its stem the artist has wound a sprig of *Coronilla varia*, while in the air hover two different species of Indian butterflies; each is depicted twice, once in the dorsal and once in the ventral view. The black and white butterfly is clearly identifiable as a *Euploea sylvester*, while the yellow one appears to be a somewhat fanciful version of a *Castniidae*.

Two other paintings from the same manuscript volume can be found in the Pierpont Morgan Library, New York (*Flowers in Books and Drawings*, no. 117).

53. JOHN HILL (?1714–1775)

Eden: or, A Compleat Body of Gardening, Both in Knowledge and Practice; Directing the Gardener in his Work, for every distinct Week in the Year, In the Kitchen-Ground, Flower-Garden, and Nursery; And Shewing the Proprietor those Shrubs, Plants and Flowers, which are most usually cultivated in Gardens; or most worthy to be so; With the History, Description, and Figure of every Kind; and their true Culture and Management; The raising of Flowers according to English, French, and Dutch Florists. The Management of Fruit Trees in the present improved Method: with a Catalogue of Fruits as they come in Season. The Culture of the Kitchen Ground; the making of Hot Beds, &c. And The raising of the Products, in the natural, and artificial Manner. Illustrated With Figures of about Four Hundred of the finest Shrubs, Flowers, and Plants. By J. Hill, M.D. Member of the Imperial Academy [*armorial vignette* 4.5 × 5 *cm.*] The Second Edition, corrected; And Enlarged With the Addition of Twenty Folio Plates of new Plants, now first raised in the Royal Gardens at Kew. London: Printed for the Author in St. James's Street; And sold by all Booksellers. M.DCC.LXXIII.

2° 43.2 × 27.5 cm. π^2 (π2 + 1) a1 B–8S^2 χ^2 8T1 *i–ii* i–iv i–ii 1–716 *713*–714 p.

BINDING: Modern three-quarter calf and marbled boards. On the spine is a red leather label with 'Hill's Eden' in gilt lettering.

PLATES: 82 engraved plates. Preceding the title-page is a mezzotint portrait of the author. Inserted after the first gathering is an etched plate showing the Genius of Botany. At its foot: 'The Genius of Botany explaining to the Gardner the Characters of Plants, while Flora & Pomona offer him their choicest Products, as Rewards of his Labour'. 80 engraved and hand-coloured plates of plants, each with its English name.

REFERENCES: Dunthorne 129; Nissen 880; Sitwell and Blunt, p. 59; Hunt 559; Plesch, pp. 261–2; Henrey, II.90–98 and no. 805.

JOHN HILL was a somewhat controversial figure in the history of botany in England. Although the author of a vast number of works on an astonishing range of subjects, some doubts have been raised as to the real merit of his scientific achievements. He was born either at Spalding or at Peterborough in 1714 or 1716. He received his initial training in a pharmacy and then passed into the service of Charles Lennox, second Duke of Richmond, and Robert James Petre, eighth Baron Petre of Writtle, as a gardener and also as curator of their collections of dried plants. Hill soon abandoned this line of work, feeling it to be insufficiently renumerative, and took up acting, a step that did little to improve his economic situation. He then opened a pharmacy in Covent Garden, but it was his translation from the Greek of a brief treatise on mineralogy by Theophrastus that brought him scientific recognition and a certain degree of financial security.

Hill had discovered his vocation; from that moment he became an indefatigable author, turning his hand to works on almost every subject, from the sciences to theology and naval history; he even wrote a few plays and novels. His most impressive achievement was a twenty-six-volume work, *The Vegetable System*, for which he personally designed and etched 1,600 plates. In this way he became quite renowned despite a somewhat difficult and caustic personality, which procured him not a few enemies. In 1750 Hill received a medical degree from the University of St Andrews; he later obtained the title 'Knight of the Swedish Order of Vasa', and, by a concession on the part of George III, as a result he was allowed to assume the rank and title of an English knight. It also appears that in 1760 he participated in the groundbreaking work for what was to become the great botanical garden of Kew, and also worked as a gardener at Kensington Palace.

Hill is famous for having introduced the Linnaean system of botanical nomenclature to England, even if he himself did not make any original contributions in this area. Indeed, his *History of Plants* (the second of three volumes on a general natural history 'of the world'), published just two years before Linnaeus's revolutionary *Species plantarum* (1753), contained nothing that went beyond the generally accepted canons of the day. Hill incorporated some aspects of the new system in his *British Herbal*, published in 1756–7, but he did not refrain from expressing a number of criticisms of Linnaeus in it, which he reiterated in *The Sleep of Plants* (1757), a treatise taking the form of an open letter addressed to the Swedish botanist. In 1756–7 Hill also published the first edition of *Eden*, illustrated with sixty plates. Many of these were designed and engraved by Hill himself, although other artists participated in the work, including the flower painter C. A. Edwards, the portrait painter B. Cole, the Dutch painter Jacobus Caspar Philips (author of various works on perspective), Edward Dalton, and an artist who signed himself 'Darley'. The allegorical frontispiece to *Eden* was designed by Samuel Wale, one of the founder-members of the Royal Academy, who died in 1786. It was engraved by Charles Grignion (1717–1810), the scientific illustrator who executed the plates for Albinus's work on anatomy. The frontispiece shows the Genius of Botany explaining the nature of his science to a gardener, while Flora and Pomona offer gifts. This engraving is stylistically quite similar to the frontispiece designed by Wale for Hill's *British Herbal*, where the Genius of Health is shown receiving tributes, which will be duly offered to the reader, from the continents of Europe, Asia, Africa and America.

Eden is arranged as a calendar, its aim being to guide the gardener week by week, beginning with the last week of August, in the care of the flower-garden, the kitchen-garden and the nursery. Illustrations of different plants, generally in flower, are provided, together with a description and a brief history of each. The sixty plates in the first edition all show six or seven different plants, realistically portrayed and artistically arranged on the page. The twenty plates added to the second edition each illustrate a single species.

It has already been noted by previous scholars that Hill drew heavily from published works for many of the illustrations in *Eden*. Some were copied from the famous *Hortus Floridus* by Crispijn van de Passe or from Parkinson's *Paradisi in sole* (Henrey, II.98). The author also borrowed from other sources: the two exotic species portrayed in plate 11—the 'Short thick leav'd Aloe' and the 'Long fruited Trichosanthes'—were inspired by two plates in *Catalogus plantarum horti Pisani*, published in Florence by Michelangelo Tilli in 1723. Tilli's preparatory drawings were almost certainly in London by the beginning of the eighteenth century; Hill may well have used them (Tongiorgi Tomasi 1991, p. 192). The very fine tulips in plate 34 were engraved by Hill from a drawing by one of the most famous flower painters of the period, Jan van Huysum.

The charmingly hand-coloured copy of *Eden* at Oak Spring is a second edition, published in 1773 with the twenty additional plates. Its title-page bears a fine portrait of the author in mezzotint, executed by the Irish engraver Richard Houston (*c.* 1721–75), who was famous for his prints after

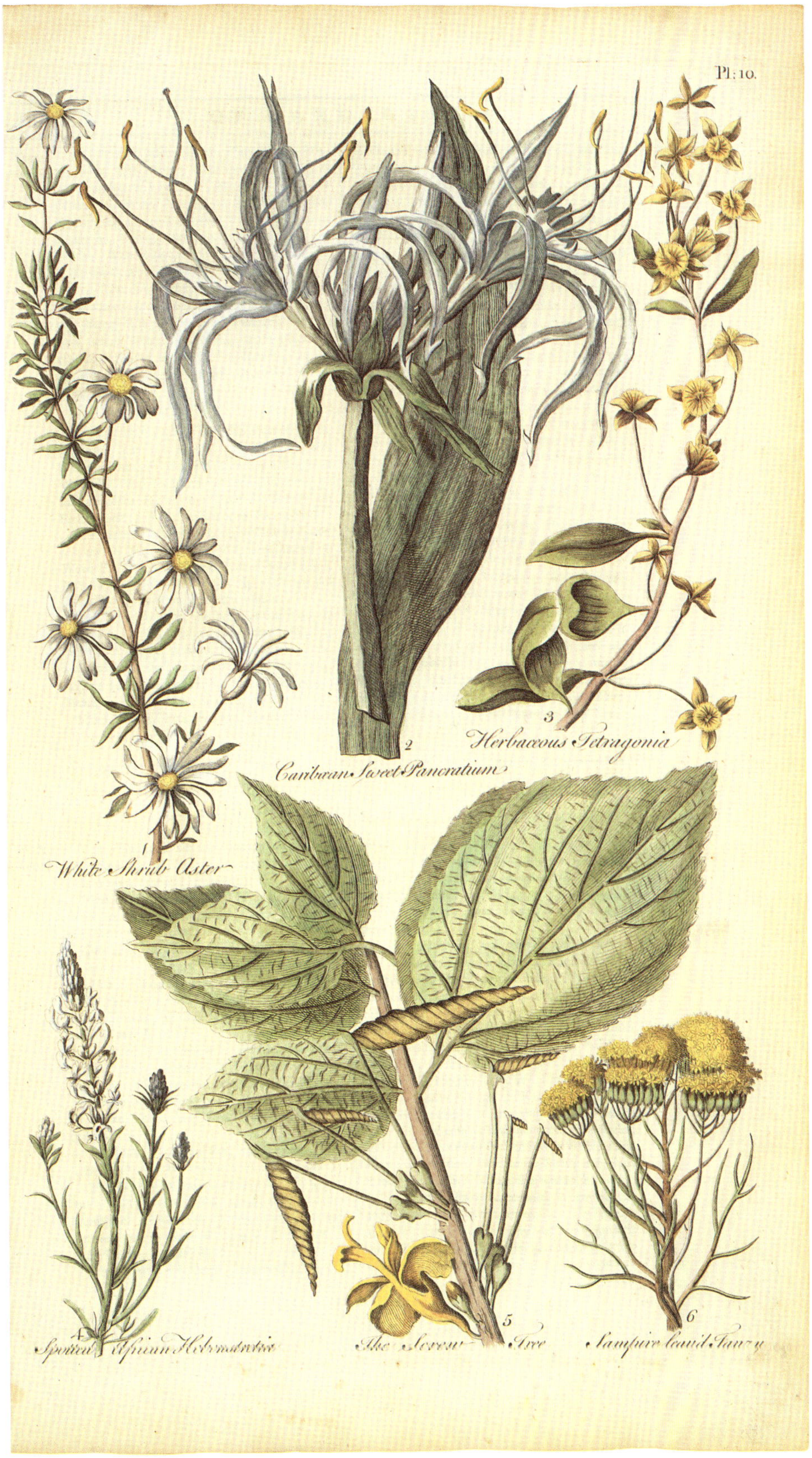

JOHN HILL, *Eden: or, A Compleat Body of Gardening*. Plate 10: 1. White Shrub Aster (*Olearia* sp.); 2. Caribaean Sweet Pancratium (*Hymenocallis caribaea*); 3. Herbaceous Tetragonia (*Tetragonia herbacea*); 4. Spotted African Hebenstretia (*Hebenstretia comosa*); 5. The Screw Tree (*Helicteres isora*); 6. Sampire leav'd Tanzy (*Tanacetum crithinifolium*)

JOHN HILL, *Eden: or, A Compleat Body of Gardening*. Nine varieties of *Fritillaria*, plate 35

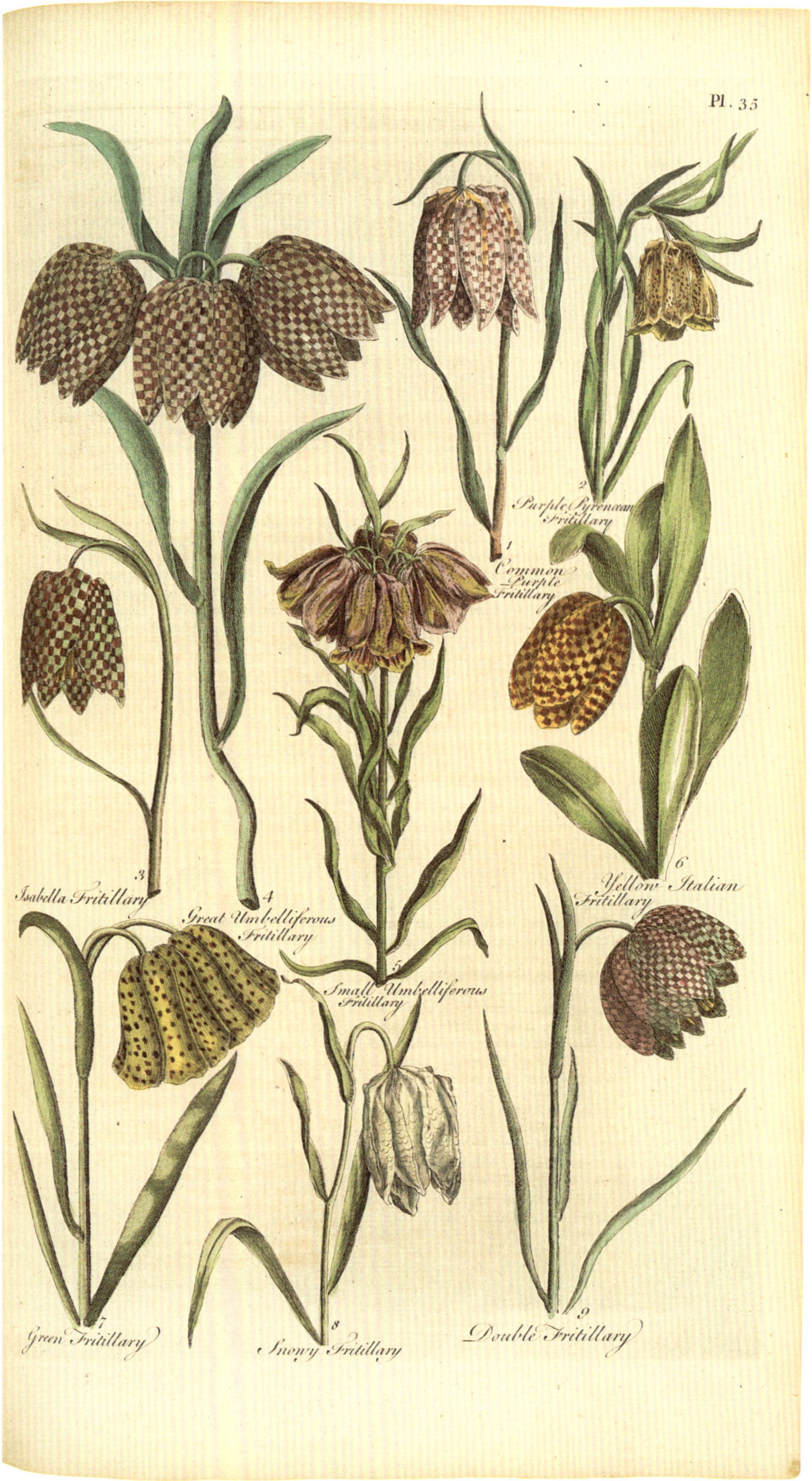

portraits by Reynolds. The portrait of Hill, based on a drawing made in 1757 by Francis Cotes (1725–70), appears in an oval frame set in a stone pedestal; against the soft background shading of the mezzotint emerges the intelligent and determined head, in three-quarters profile, of this complex figure, who undeniably played a conspicuous role in the intellectual history of eighteenth-century England.

54. Franz Anton von Scheidel (1731–1801)

Album of flowers, plants, vegetables and fruits.

51 × 36 cm. 162 leaves (the last a double-page), irregularly numbered, with paintings in bodycolour of flowers, plants, vegetables and fruits.

BINDING: Mottled half calf and decorated paper boards. Gilt spine with red label: 'Scheidtl. Blumen Pflanzen und Früchte'.

PLATES: 160 bodycolours of flowers, plants, vegetables and fruits. In many of the plates the images are numbered, with a manuscript key (apparently in different hands) provided, generally on the facing verso page. However, the botanical name(s) given in Latin and/or German in the key do not always correspond to the plant depicted. Inscriptions by a single hand appear at the bottom of most of the tulip plates. On folio 50: 'Tulippen. Von Pater Norbert Hazl ordinis Eremittarum auf dem Kaltenberg Floriret Anno [1]765 (Tulips. Cultivated by Father Norbert Hazl of the Augustinian order in Kaltenberg in 1765)'. On folios 53, 55, 57, 58 60, 61, 63, 66, 67, 68, 70, 72, 73, 74, 75: 'Tulippen, Welche Ao: 1765 in Lissenberg floriret haben (Tulips, which in the year 1765 in Lissenberg bloomed)'. On folios 54, 59, 69: 'Tulippen. Von Pater Florian ordinis Eremittarum zu Maurberg Floriret Anno 1765' (Tulips. Cultivated by Father Florian of the Augustinian order in Maurberg in the year 1765). On the verso side of folio 7: 'Die Welche Keine Nummer hamb, seind Isabel Rossen (The ones that have no number are Isabel Rossen)'.

FRANZ ANTON VON SCHEIDEL (or Scheidl or Scheidtl) was an outstanding naturalistic painter who worked in Vienna during the second half of the eighteenth century; he collaborated with the botanist Nikolaus Joseph von Jacquin (1727–1817) on many of the latter's most important projects. Jacquin was born in Leiden, but having moved to Vienna to pursue his medical studies he spent the rest of his life there. While still a student, he so impressed Franz I with his abilities that the Emperor appointed him curator of the gardens of Schönbrunn, and under his care they became among the most famous in all Europe. Jacquin filled Schönbrunn with exotic species brought back by him from expeditions to the West Indies and Central America. He eventually became professor of chemistry and botany at the University of Vienna.

Jacquin was a great appraiser of the art of botanical illustration and a more than competent artist in his own right. However, his scholarly activities left him little time for painting and he generally relied on collaborations with professional artists. One of the most talented of these was Scheidel, who, according to Jacquin, was 'in pingendo celeberrimus' (very successful in painting); the pair went on many expeditions together. Scheidel illustrated two of Jacquin's most important works—the three-volume *Hortus botanicus vindobonensis*, produced between 1770 and 1776, and the extraordinary five-volume *Florae Austriacae* (1773–78)—both of which were published in Vienna under the patronage of the Austrian royal family. For the latter, Scheidel also drew the charming landscape

FRANZ ANTON VON SCHEIDEL, Album of flowers, plants, vegetables and fruits. Six varieties of *Papaveraceae* and six varieties of holly-hock (*Alcea*), folio 6

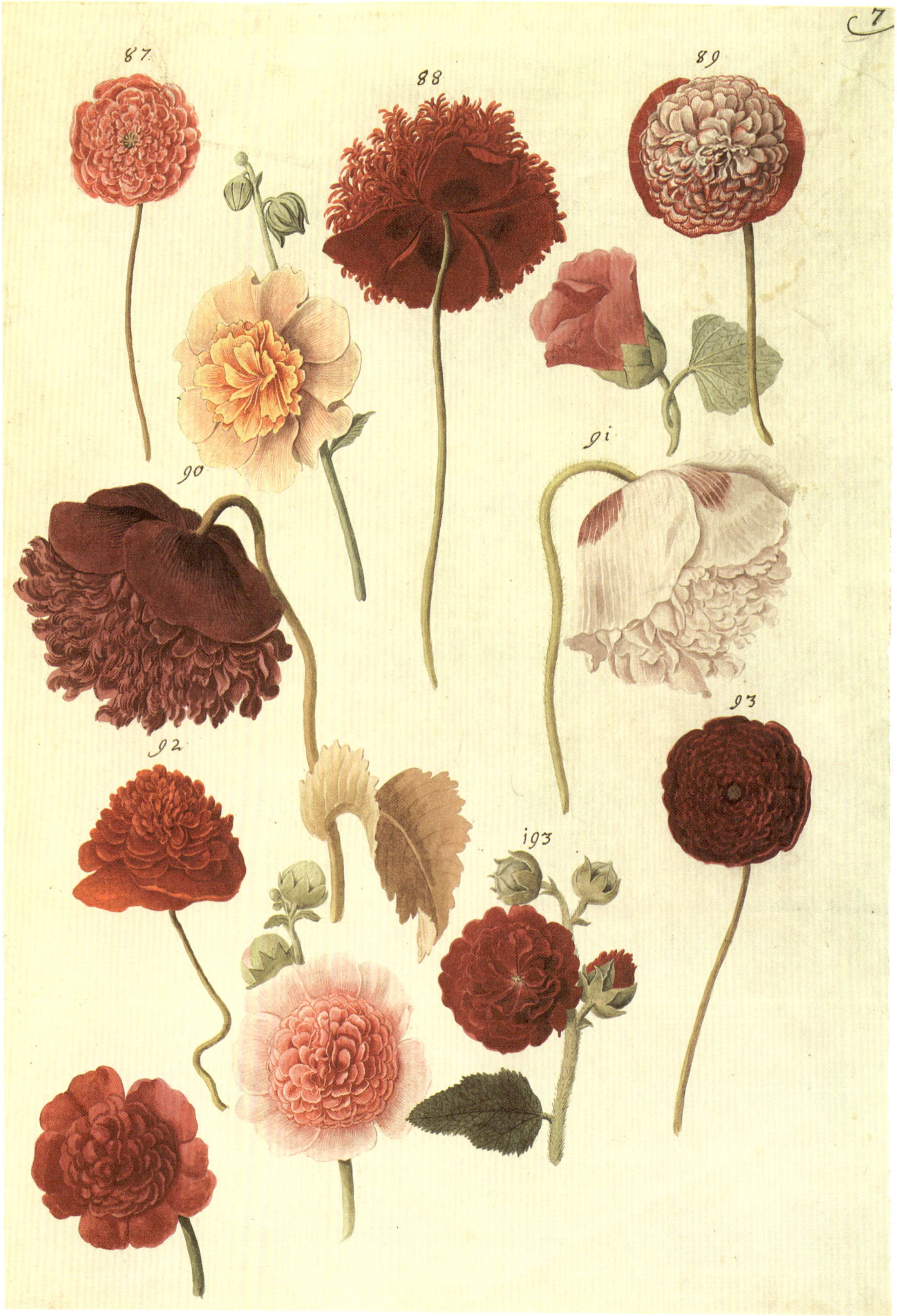

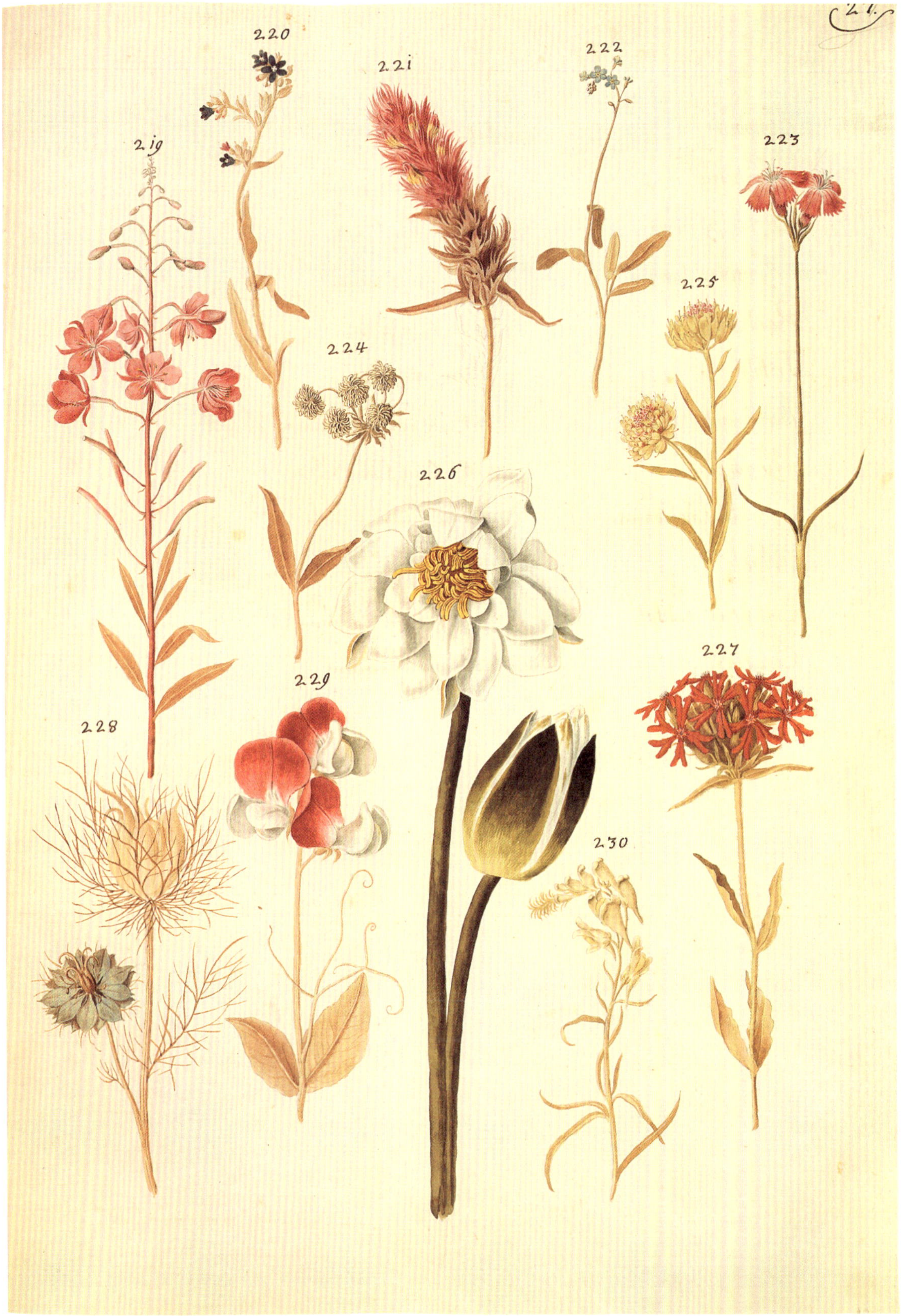

FRANZ ANTON VON SCHEIDEL, Album of flowers, plants, vegetables and fruits. Folio 98: 219. Fireweed (*Epilobium angustifolium*); 220. Bugloss (*Anchusa arvensis*); 221. Cow-wheat (*Melampyrum* sp.); 222. Forget-me-not (*Myosotis* sp.); 223. Pink (*Dianthus* sp.); 224. Clover (*Trifolium* sp.); 225. Candytuft (*Iberis* sp.); 226. European white water-lily (*Nymphaea alba*); 227. Maltese cross (*Lychnis chalcedonica*); 228. Love-in-a-mist (*Nigella damascena*); 229. Everlasting pea (*Lathyrus latifolius*); 230. Toadflax (*Linaria* sp.)

vignettes for the title-pages to volumes III and IV, which were subsequently engraved by Mansfeld. Together with the artist Johann Scharf (1765–94), Scheidel completed the illustrations for Jacquin's work on the gardens of Schönbrunn, *Plantarum rariorum horti caesarei Schönbrunnensis descriptiones et icones*, which was published between 1797 and 1804.

Although he specialized in botanical painting, Scheidel was equally adept at other types of scientific illustration—drawing animals, birds and minerals with the same deft virtuosity. The album of bodycolour paintings at Oak Spring gives ample proof of his versatility. Its first plates reflect the typology of the seventeenth-century florilegium, with each page presenting different varieties of a single cultivated species. Here we can find the iris, amaryllis, fritillary, double narcissus and double and semi-double anemone, all beautifully depicted in a variety of colours. There are also pages showing a vast array of tulips and carnations, including many striated varieties. An even larger number of the plates in this album follow the typology of the eighteenth-century scientific treatise. In some of them Scheidel presents small branches from different varieties of the same species of plant (or closely related species), carefully arranged and numbered, often with the names of the plants written by hand on the facing (verso) page. Other paintings depict a variety of cacti, some of them in pots. Reflecting the eighteenth-century vogue for Chinoiserie, Scheidel has included a charming plate showing four 'Chinese' baskets filled with flowers and fruit (fol. 80). There is another series of paintings of different vegetables, whose rounded forms are depicted with loving attention to detail. The album closes with the painting of an enormous pumpkin, which takes up an entire double sheet.

55. SIMON TAYLOR (1742–1796)

Botanical paintings for the Earl of Bute.

61 × 47.5 cm. 46 botanical plates in water- and bodycolour.

BINDING: Red morocco with floriated borders. Arms of John Stuart, third Earl of Bute, in gilt on covers. On the spine in gilt: 'Plants by Taylor'. Encased in a box case of vellum and decorated paper.

PLATES: 46 leaves in water- and bodycolour on vellum (43.2 × 33.5 cm.) mounted on sheets of paper with borders ruled in watercolour.

PROVENANCE: John Stuart, third Earl of Bute (1713–92). This volume of 46 watercolours was included in a lot of 15 volumes, containing 690 botanical paintings by Simon Taylor, sold by Leith and Sotheby at an auction of the Bute collection in 1794 (*Catalogue of the Botanical and Natural History Part of the Library of the Late John Earl of Bute Sold by Auction by Leigh and Sotheby*, London, 8 May 1794, pp. 60–63).

THIS SUPERB VOLUME contains forty-six paintings in water- and bodycolour of various flowering herbs and shrubs belonging to the families of the docks, sages, heaths, smilaxes and saxifrages, primroses, gentians, antholyza and digitalis. Each plant is labelled at the bottom of the page with its Latin name. They depict just a small part of the collection of plants in the botanical garden that John Stuart, third Earl of Bute, constructed in 1763 at Luton Hoo, Bedfordshire. Bute

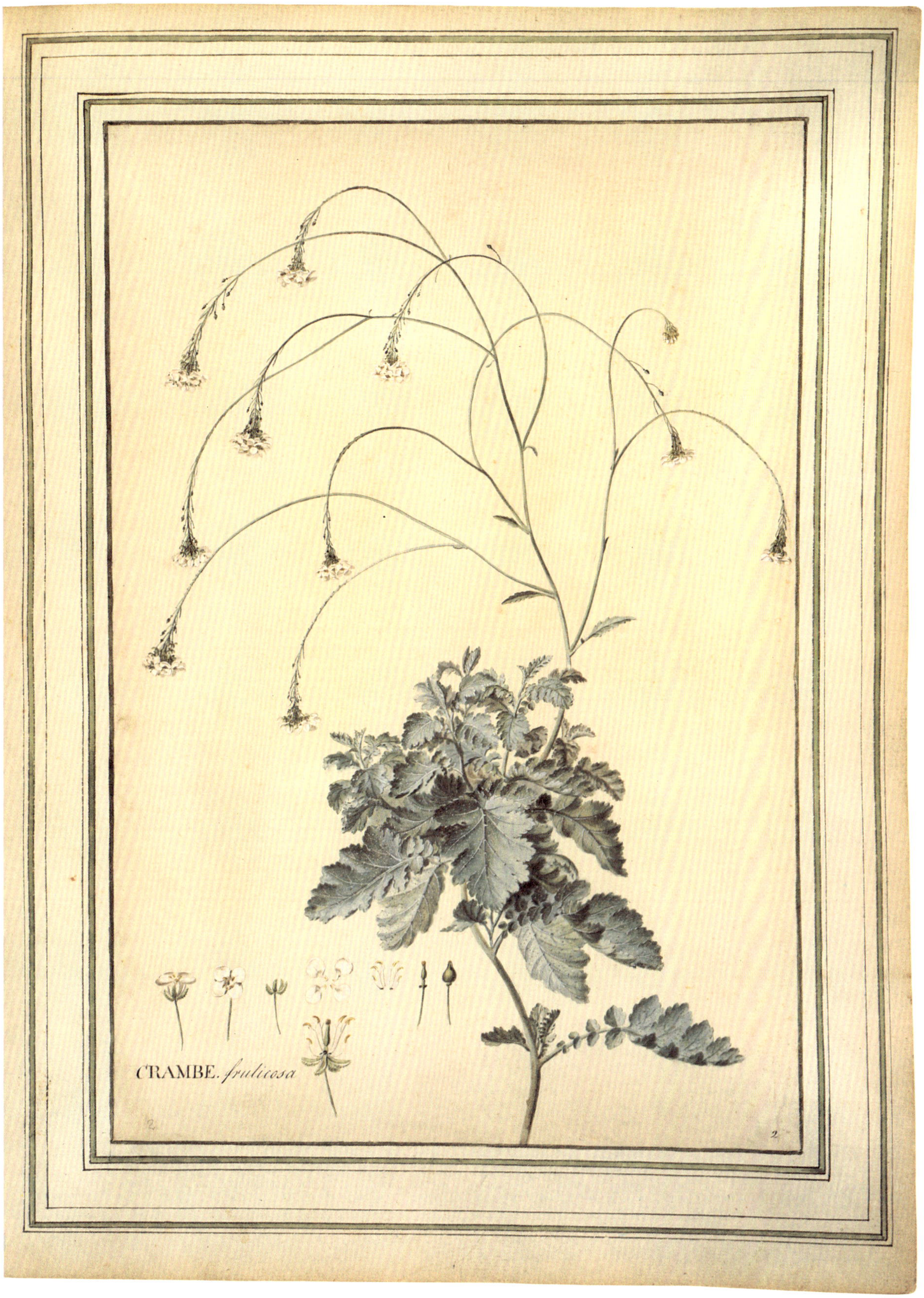

SIMON TAYLOR,
Botanical paintings for the Earl of Bute. Sea kale (*Crambe maritima*)

was a prominent politican (he served as prime minister in 1762–3) and friend and adviser to George III, but he was also a botanist, landscape gardener and generous patron of the botanical sciences. He underwrote William Curtis's *Flora Londinensis*, a magnificent illustrated work describing the plants that could be found growing within a radius of ten miles of London. He is best known, however, as the founder of the Royal Botanic Gardens at Kew. Bute realized the importance of having a visual record made of his botanical collection, and commissioned works from some of the most important illustrators of his time, including Johann Müller (1715–*c.* 1790) (Le Rougetel, p. 108) and Simon Taylor. Both of these artists had worked with Ehret (see nos. 50, 51) when he was engaged in depicting the botanical collection of Dr John Fothergill (1712–80) at Upton House near Stratford, on the eastern edge of London (Calmann, p. 64).

Müller was born in Nuremberg, but left his native city and settled in London, where he changed his name to John Miller. He executed the plates for *An Illustration of the Sexual System of Linnaeus*, which was published in London in 1770–77 and which the eminent Swedish author expressed himself to be greatly pleased with. Miller also produced the illustrations for the splendid *Botanical Tables* that were privately printed by Bute in 1785 in just twelve to sixteen copies for the astounding sum of £12,000. The set of the *Botanical Tables* at Oak Spring was once owned by the Duchess of Portland and later by the Marchioness of Bath.

Simon Taylor studied painting in London at William Shipley's school, and in 1760 was already working in Bute's artistic entourage. A series of 103 watercolours of wild flowers, mounted on paper with an engraved border simulating a carved frame and attributable to Taylor and Miller, is also at Oak Spring. Taylor was a fine artist, not perhaps comparable to the brilliant Ehret but one nevertheless possessed of a perfectly respectable talent, as the naturalist John Ellis affirmed in a letter in 1770 to Linnaeus: 'I suppose you know that Ehret is dead. We have nobody to supply his place in point of elegance. We have a young man, one Taylor who draws all the rare plants of Kew garden for Lord Bute; he does it tolerably well' (Calmann, p. 109). Ellis himself had one of Taylor's paintings engraved by John Miller to illustrate his work *An Historical Account of Coffee*, published in London in 1774 (a copy of this title is in the collection at Oak Spring).

In this fine volume of paintings we can recognize the influence of Ehret: the skill with which the plants have been arranged on the page, and the practice of showing the parts of the plant in detail at the bottom of the page, reflect Ehret's distinctive approach. Certain idiosyncracies in Taylor's technique allow us to distinguish his work from that of his master, however. Ehret's colours were applied in almost transparent layers, while Taylor laid his paints on much more densely, his brush often leaving thick blobs of bodycolour, including generous amounts of lead white. In this way he created a richly tactile effect, less refined than that of Ehret, but equally effective.

A work drawn by Taylor and engraved by Maleuvre appears in L'Héritier de Brutelle's *Geraniologia* (see no. 57). Other paintings by Taylor are conserved in London at the Natural History Museum and the Kew Herbarium.

56. Anonymous Artist

Album of garden-flower studies.

25.5 × 20 cm. 150 bodycolour paintings on vellum loosely inserted between blank leaves.

BINDING: Contemporary calf with clasps and a gilt spine.

PLATES: 150 paintings in bodycolour on vellum. The paintings vary in size from 18 × 13.5 cm. to 24 × 18 cm. On most of the plates the common name of the plant has been written by hand at the bottom of the page, and the Latin name on the verso side.

PROVENANCE: Together with this album is a letter dated 1 December 1923 from E. W. Stevens, a bookseller in Bridgwater, Somerset, stating that this album came from a sale 'some years ago' of Vice-Admiral R. W. Stopford's effects at Iwerne Courtney, Dorset.

TO JUDGE from their style, this noteworthy collection of paintings in bodycolour of flowering plants was produced in England sometime during the second half of the eighteenth century. The manuscript notations that appear at the foot of each plate cite works published as late as 1750, showing that the artist, or rather the patron who commissioned the album, was *au courant* of the very latest developments in the science of botany. The earliest works mentioned are *De plantis* by Andrea Cesalpino (1583), John Gerard's *The Herball* (1597), *Rariorum plantarum historia* by Carolus Clusius (1601), *Hortus Eystettensis* by Basilius Besler (1613) and *Paradisi in sole* by John Parkinson (1629). Many works from the latter part of the seventeenth century and the beginning of the eighteenth are also referred to, including *Elémens de botanique* (1694) and *Institutiones rei herbariae* (1700) by Joseph Pitton de Tournefort, *Synopsis methodica* by John Ray (1696), *Index alter plantarum* by Herman Boerhaave (1720), *Historia plantarum rariorum* by John Martyn (1728–37), *Historia botanica practica* by Giovanni Battista Morandi (1744) and, finally, *Hortus Cliffortianus* (1737–8) and *Hortus Upsaliensis* (1748) by Linnaeus.

The provenance of this album provides some clues as to the artistic milieu in which it was created. It was found among the possessions of Rear-Admiral Robert William Stopford when he died in 1911. Stopford spent the latter part of his life at Shroton House, Blandford, in Dorset, not far from the magnificent estate of Merly once owned by Ralph Willett (1720–95), a wealthy sugar plantation owner. Willett purchased Merly in 1751 and spent much of his time there, particularly in the summer. He was an enthusiastic student of botany, and commissioned many artists to depict his collection of plants. Simon Taylor (see no. 55), who had worked for Dr Fothergill and Lord Bute, and Peter Brown (*fl.* 1758–99), floral painter to the Prince of Wales, both went down to Merly to paint. (At Oak Spring are four bodycolour paintings by Brown.) Willett also knew the great botanical illustrator Georg Dionysius Ehret (see nos. 50, 51), who frequently stayed with him at Merly. Willett enjoyed going on long nature walks with Ehret, although the latter was wont to say of his host: 'Mr. Willett loves plants, but is no botanicus' (Calmann, pp. 65–6). In addition, in a neighbouring village, Hanford, lived two artists—father and son, both named Henry Seymer—who were well known for their still-life paintings of birds, shells and insects.

This fine album contains 150 paintings in bodycolour on vellum of different wild and cultivated flowers, as well as a few unusual fruits and vegetables, such as the pomegranate (pl. 30) and the tomato (pl. 77), shown together with their flowers. The confident style and skilful *mise-en-page* of these paintings show them to be the work of a talented artist who had evidently made a careful study of the works of the great botanical illustrators of the period, in particular Ehret. Like the German artist, he used thin bodycolour applied in successive layers to suggest subtle modulations in colour, and sought to depict the smallest particulars of the stems, shoots, flowers and leaves to create an effect of heightened realism. This artistic heritage is clearly reflected in his series of paintings of the iris (fols. 22–6), with long, blade-like leaves, fine veining and arabesque flowers. However, the two most distinctive qualities of Ehret's work, his minute and passionate interest in his subject-matter and the almost tactile quality of his paintings, are missing from the works in this album.

Other stylistic similiarities confirm the existence of a link between the two artists. Ehret's paintings are scientifically impeccable, reflecting the meticulous training he had received in his youth under the tutelage of various eminent botanists, and the paintings in the English album show the same striving for an almost documentary realism. Following the example of Ehret the various parts of the plant, including the flower and fruit, are presented as clearly as possible, and the paintings have been carefully labelled, with the English name of the plant written on the recto and the Latin name on the verso of almost every painting.

Although the quality of the paintings in this album is uneven—the last five, for example, are rather modest works—many are very fine indeed, such as the spiderwort (*Tradescantia virginiana*) (fol. 122), the 'Double flowered poppy' (*Papaver flore-pleno*) (fol. 57), and the 'corn marigold' (*Chrysanthemum segetum*) (fol. 70), which is shown together with a butterfly from India, a *Hypolimnas misippus*, whose colours—black, white and deep blue—contrast strikingly with the bright yellow of the flowers.

Interestingly, at Oak Spring there is also a set of 161 bodycolour paintings of flowering plants on loose, unnumbered sheets. They are accompanied by two labels on which have been stamped in gold lettering, using a typeface datable to the mid-nineteenth century: 'Drawings by Dame Ann Hamilton Née Heathcote 1762 to 1766. Vol I; Vol II'. Only five of the sheets are dated: 1752, 1753, 1753, 1765 and 1766. The paintings in this collection are so similar in style to those in the English album that they almost seem to be by the same hand.

Both groups of paintings have been executed on modest sized sheets of vellum, with the English name of the plant written at the bottom of the page, and the Latin name on the back. On numerous occasions, for a plant depicted in both collections, the same bibliographic references (from Clusius to Linnaeus) have been cited in an identical fashion. Many other similarities can be traced between the work of Ann Hamilton and the paintings in the English album featured here, and thus to the influence of Ehret himself. For example, like Ehret, Hamilton sometimes shows her plants growing out of the ground with stones or clumps of moss around the base of the stem, as in her plate of the

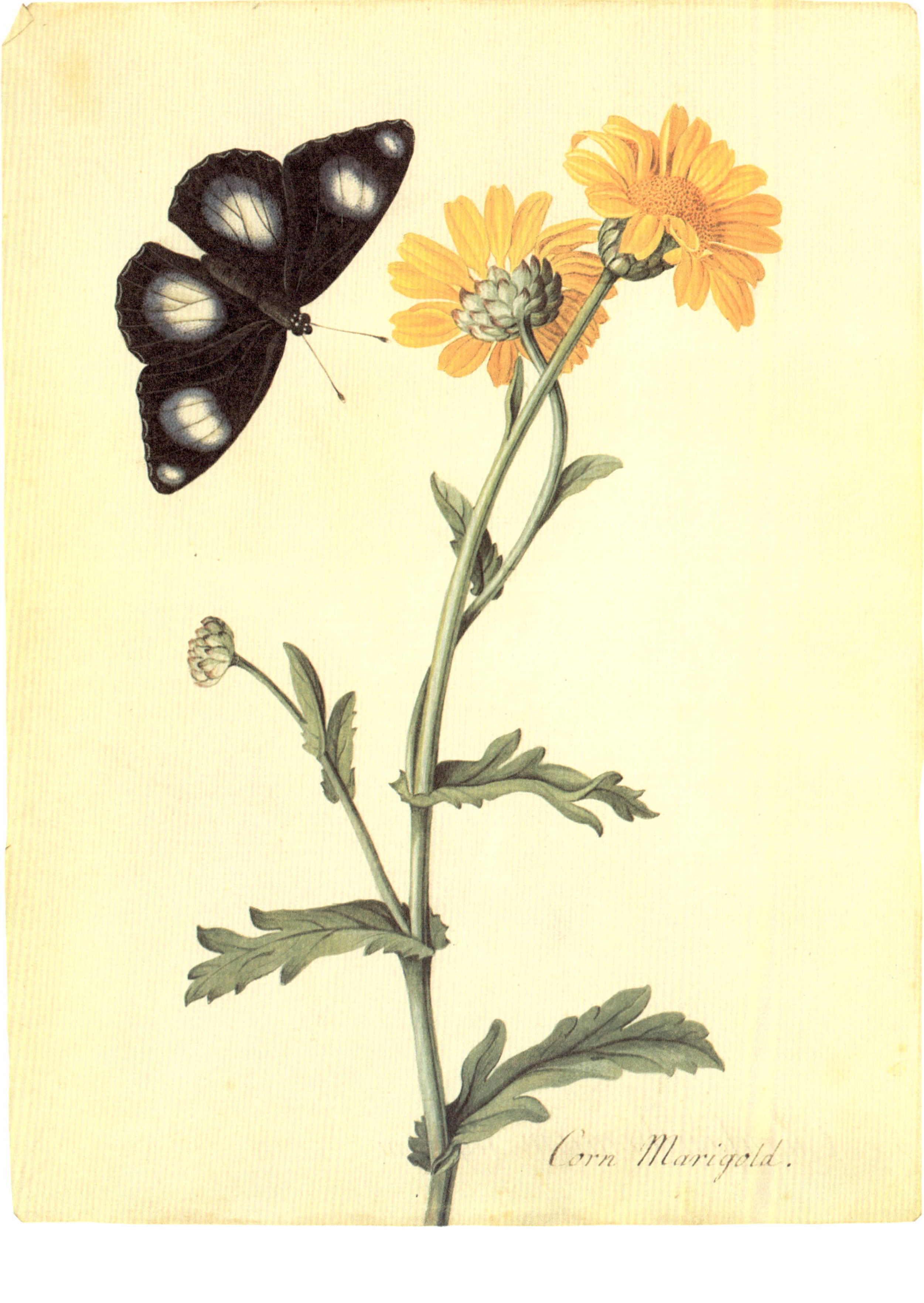

Anonymous artist,
Album of garden-flower studies. Corn marigold (*Chrysanthemum segetum*) and butterfly

Anonymous artist,
Album of garden-flower
studies. Trumpet gentian
(*Gentiana clusii*)

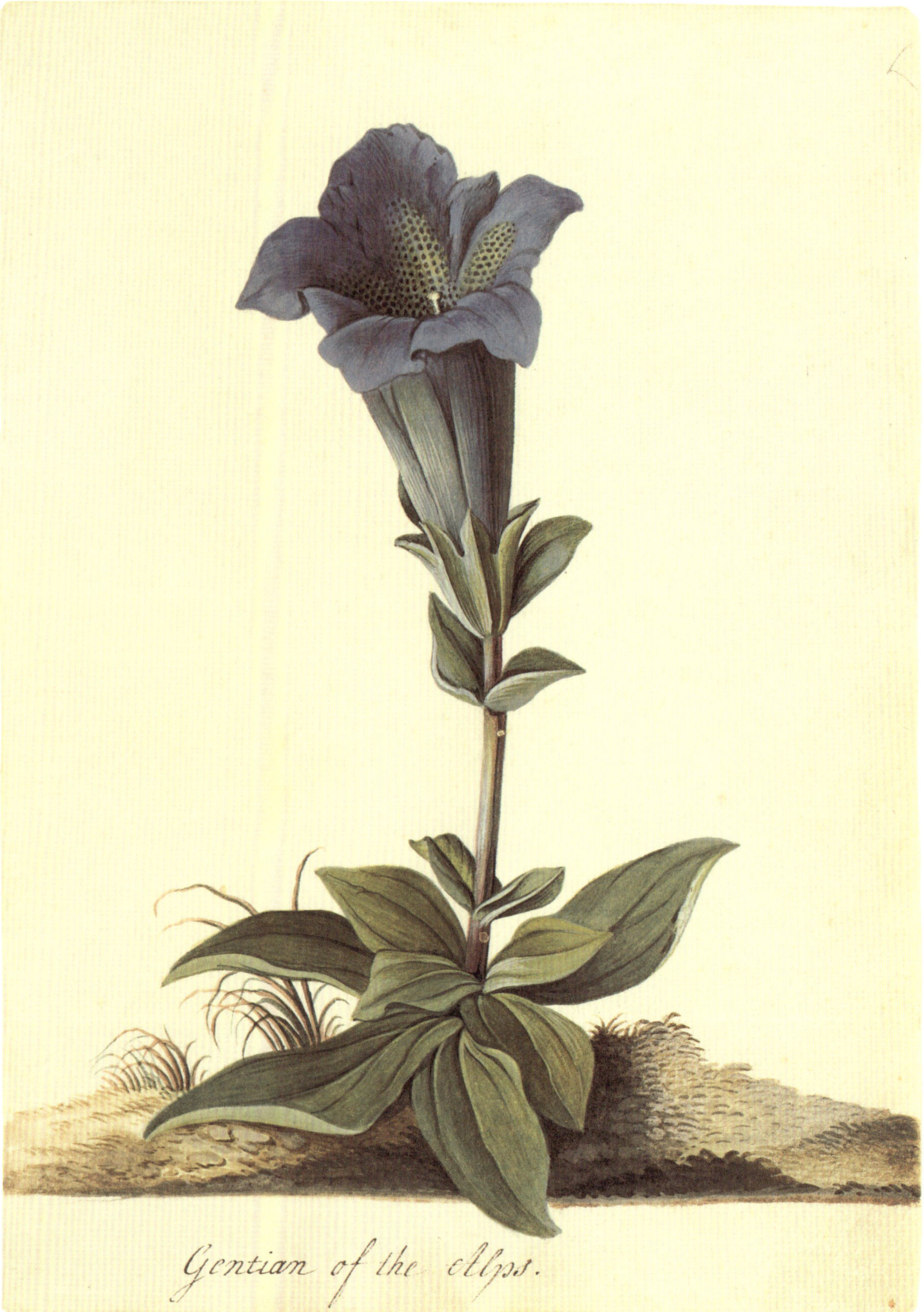

'Gentian of the Alps'. Other plants, such as the 'Yellow Alpine Alysson', are shown together with gaily coloured butterflies, or with their fruit, as in the painting of 'The Scarlet been' (*sic*), whose handsome, spearhead-shaped leaves and delicate flowers have been juxtaposed with the curving pod and brightly coloured bean to form an elegant composition. Here clearly was an artist gifted with an observant eye and unusual talent; her paintings are at once wonderfully accurate and objective records, and beautifully composed pictures in the tradition of Ehret.

Nothing at all is known of Ann Hamilton 'née Heathcote'. However, among the many pupils who had the good fortune to study under Ehret were four of the five daughters of Sir John Heathcote, Bt (1689–1759), of Normanton Park in Rutland (Calmann, p. 82). In fact, beginning in 1755 when he left Oxford, Ehret undertook a new activity, that of teaching floral painting to young ladies of good family. This sort of work was quite renumerative and he was evidently very successful at it, for he became a popular teacher and acquired many students. In his 'Memoir' he informs us that he was quite proud of having taught his well-born students not only how to paint flowers but how to conserve their work as well, carefully labelling each painting with the botanical name of the species depicted. Among his students he counted such illustrious figures as Catherine, Duchess of Norfolk; Mary, Duchess of Leeds; the two daughters of the Duchess of Bridgwater; the two daughters of Margaret, Duchess of Portland; and Bridget, the wife of James Douglas, fourteenth Earl of Morton, and President of the Royal Society from 1764 until his death in 1768. This last pupil was in fact one of Sir John Heathcote's daughters (Calmann, p. 81), and probably a sister to our Ann Hamilton. Another very talented pupil of Ehret's was Lady Frances Howard, who executed a series of nearly one hundred bodycolour paintings of wild flowers, all of them signed and dating from *c.* 1760, which can also be found in the collection at Oak Spring (see *An Oak Spring Garland*, no. 20).

57. Charles Louis L'Héritier de Brutelle (1746–1800)

Car. Lud. L'Héritier, Dom. De Brutelle in Aula Juvam. Paris. Reg. Consil. Geraniologia, seu Erodii, Pelargonii, Geranii, Monsoniae et Grieli Historia Iconibus Illustrata. Parisiis, Typis Petri-Francisci Didot. Prostat Parisiis, apud Ludovicum Nicolaum Prévost, Theophilum Barrois, via quam vocant Quai des Augustins. Londini, apud Petrum Elmsly, Viennae, apud Rudolphum Graeffer, Argentorati, apud Amandum Koenig, Bibliopolas. 1787–1788.

54 × 37 cm. 1 leaf, 44 leaves of plates. Bound with L'Héritier's *Cornus. Specimen botanicum sistens descriptiones et icones specierum corni minus cognitarum.* Parisiis, Typis Petri-Francisci Didot, 1788.

BINDING: Calf binding with a border of gilt fillets and rolls. On the spine in gilt 'L'Heritier Geraniolo & Cornus' and a shelfmark in manuscript.

PLATES: 44 engraved plates, principally after drawings by Pierre Joseph Redouté and James Sowerby.

REFERENCES: Pritzel (2nd edn) 5267; Nissen 1188; Sitwell and Blunt, p. 65; Plesch 451; Stafleu and Cowan, no. 686; *A Magnificent Collection*, no. 213.

CHARLES LOUIS L'HÉRITIER DE BRUTELLE was a wealthy jurist and celebrated amateur botanist who lived through the tumultuous period of the French Revolution and its aftermath. He courageously commanded a battalion of the National Guard, and when the Revolution came to an end was imprisoned. On release, L'Héritier found himself entirely destitute. He managed to obtain a modest post at the Ministry of Justice, which brought him a measure of security, but in April 1800 his life came to a tragic end when he was murdered, quite by chance, on a Paris street.

A more than competent scientist, L'Héritier formed part of the botany section—together with Jussieu, Adamson, Desfontaines and Ventenat—of the Classe de Mathematique et Physique at the Institut National, which was created in 1795 to take the place of the Académie des Sciences.

L'Héritier placed great weight on the role of the artist in the study of the natural sciences. It is not surprising, therefore, that when Pierre Joseph Redouté arrived in Paris, L'Héritier was the first to recognize the young artist's extraordinary gifts and as a result guided him in the direction of what was to become Redouté's life's work. L'Héritier invited Redouté to his home, allowed him free use of his vast library, and was the first to instruct him in exactly what was required of a botanical illustration that would afterwards be studied by other scientists (see no. 61). Fifty-four of the plates in L'Héritier's first published work, *Stirpes novae* (1784–91), were engraved after paintings by Redouté (a copy is at Oak Spring). In 1786 the artist joined L'Héritier in London to work with him on *Sertum Anglicum*, a book on the rare plants in Kew Gardens that was published in 1788 and which the author intended to present to his friends and colleagues on his return to France. It was L'Héritier's recommendation that procured for Redouté his first prestigious position, that of official draughtsman to the Cabinet of Marie-Antoinette, thus launching him on what was to be a long and immensely successful career.

L'Héritier was also apparently the first to discover and promote the career of another celebrated botanical artist, and the founder of a family of naturalists and scientific illustrators, the Englishman James Sowerby (see no. 60). Sowerby prepared the first illustrations for L'Héritier's monograph on the geranium, *Geraniologia*, in 1786–7 during the latter's stay in England. Many other French and English artists collaborated on this project, although it was never finished due to the outbreak of the Revolution in France. *Geraniologia* was finally published in an incomplete form in 1792 (even if the date that appears on the title-page is 1787–8).

The work consists of forty-four plates depicting various types of geraniums—crane's-bills, stork's-bills and heron's-bills (*Geraniums, Pelargoniums, Erodiums*). Six of the plates are based on paintings by Sowerby, while thirty-one were prepared after Redouté and three after the artist B. Pernotin. Louis Fréret, Simon Taylor (see no. 55) and Redouté's brother Henri Joseph each contributed one illustration to the work, while one plate is perhaps based on a vellum by Claude Aubriet (see no. 47). The plates were engraved by J. B. Guiard, F. Hubert, P. Maleuvre, L. Bouteleu, Baron, Milsan and Voyard. The flowers, with their distinctive leaves and tiny blossoms, have been

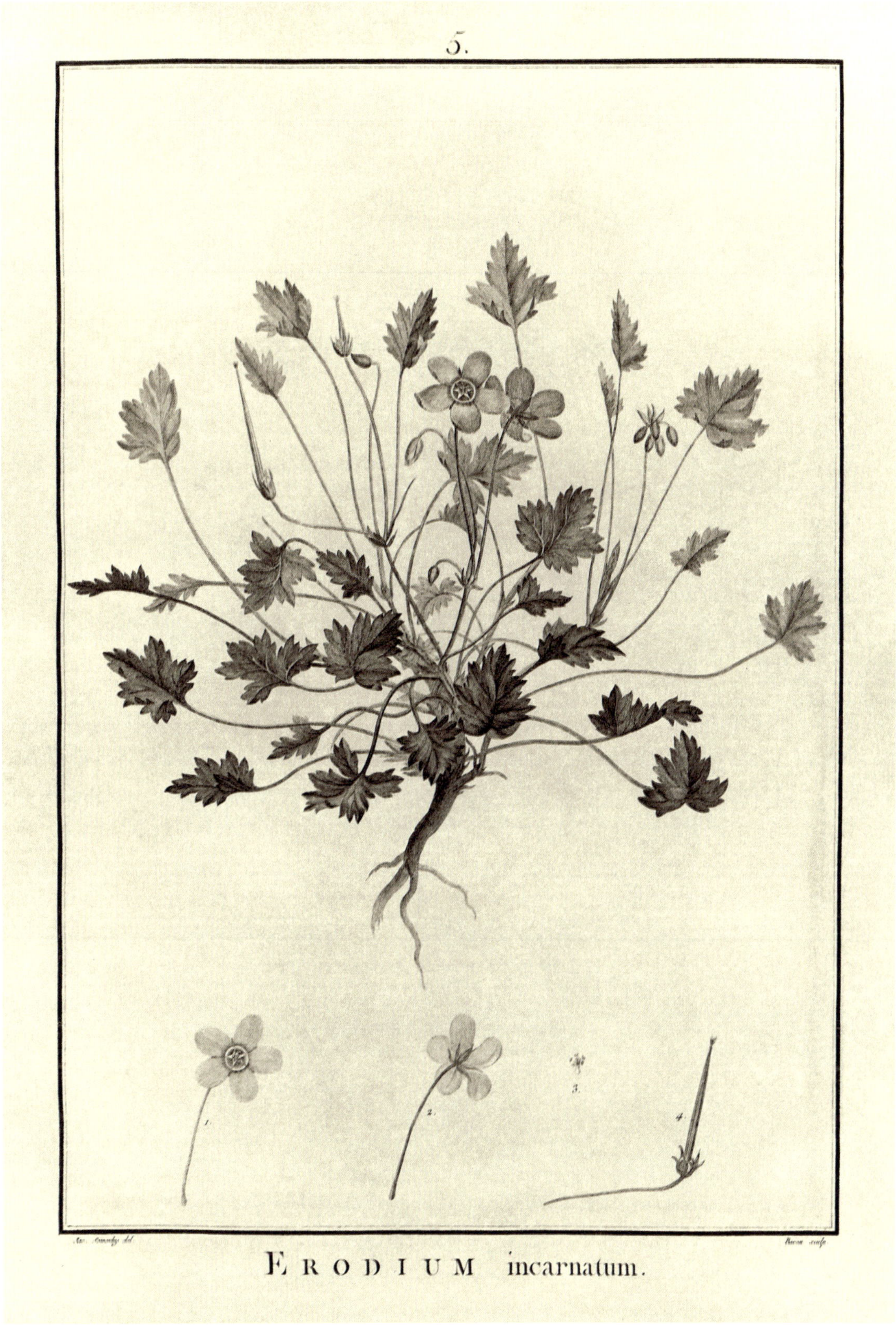

CHARLES LOUIS
L'HÉRITIER DE BRUTELLE,
Geraniologia. Stork's-bill
(*Erodium* sp.), plate 5

portrayed with great fidelity by L'Héritier's artists, and the drawings skilfully transferred by his engravers. In each plate enlarged details of the plant are also presented.

L'Héritier chose to leave his illustrations uncoloured, perhaps influenced by the example of the splendidly clear plates of the *Recueil des plantes* by the seventeenth-century botanical illustrator Nicolas Robert (see no. 43). L'Héritier probably felt that if his engravings were coloured, much of the sharpness and precision of the images would be lost. The plates in *Geraniologia* by Sowerby and Redouté are particularly fine, and provide us with an example of the early work of these two botanical illustrators, when they were still under the influence of the great French tradition established by Robert.

58. Gerard van Spaendonck (1746–1822)

Album of flower studies.

17.5 × 11 cm. 79 unnumbered leaves.

BINDING: Bound in wallet-style limp cloth. On the cover in manuscript 'II G. van Spaendonck'. Housed in a white leather and pink cloth case.

PLATES: 79 plates in black chalk, drawn on the recto; plates 77 and 78 also have drawings on the verso.

REFERENCES: Hardouin-Fugier and Grafe 1989, p. 365 (fol. 68, depicting a rose branch, is reproduced).

This collection of sketches by the Dutch artist Gerard van Spaendonck—jottings in black chalk of ideas for floral motifs—constitutes a genuine *album d'atelier* and offers a fascinating glimpse into the creative processes of one of the most important botanical painters of the eighteenth century. Van Spaendonck was born in Tilburg, but like many a young and ambitious artist, soon set off for Paris to seek his fortune. The Dutch school of floral painting was then in vogue and Spaendonck, who had been a pupil of Jacob Herryns and had studied the works of Jan van Huysum, quickly established a name for himself. He spent the rest of his life in the French capital, an intensely active artist who received many marks of recognition during the course of his long career.

When he first arrived in Paris, van Spaendonck found work painting floral motifs on snuff boxes. By 1774, however, he was designing decorative motifs for the royal porcelain factory at Sèvres. In the same year he was appointed a 'peintre du Roy'. Further titles and positions were showered upon him, particularly after his work was first shown at the annual Salon in 1777. There he demonstrated his versatility in different media, presenting watercolours and miniatures as well as paintings on floral themes. His work was appreciated by numerous intellectuals and connoisseurs, among them Denis Diderot. In 1780 van Spaendonck succeeded Madeleine Basseporte as Professeur de Peinture de Fleurs at the Jardin des Plantes; Pierre Joseph Redouté was one of the many students who studied under him. He became a member of the Académie de Peinture in 1791 and continued to exhibit works at the annual Salon until his death.

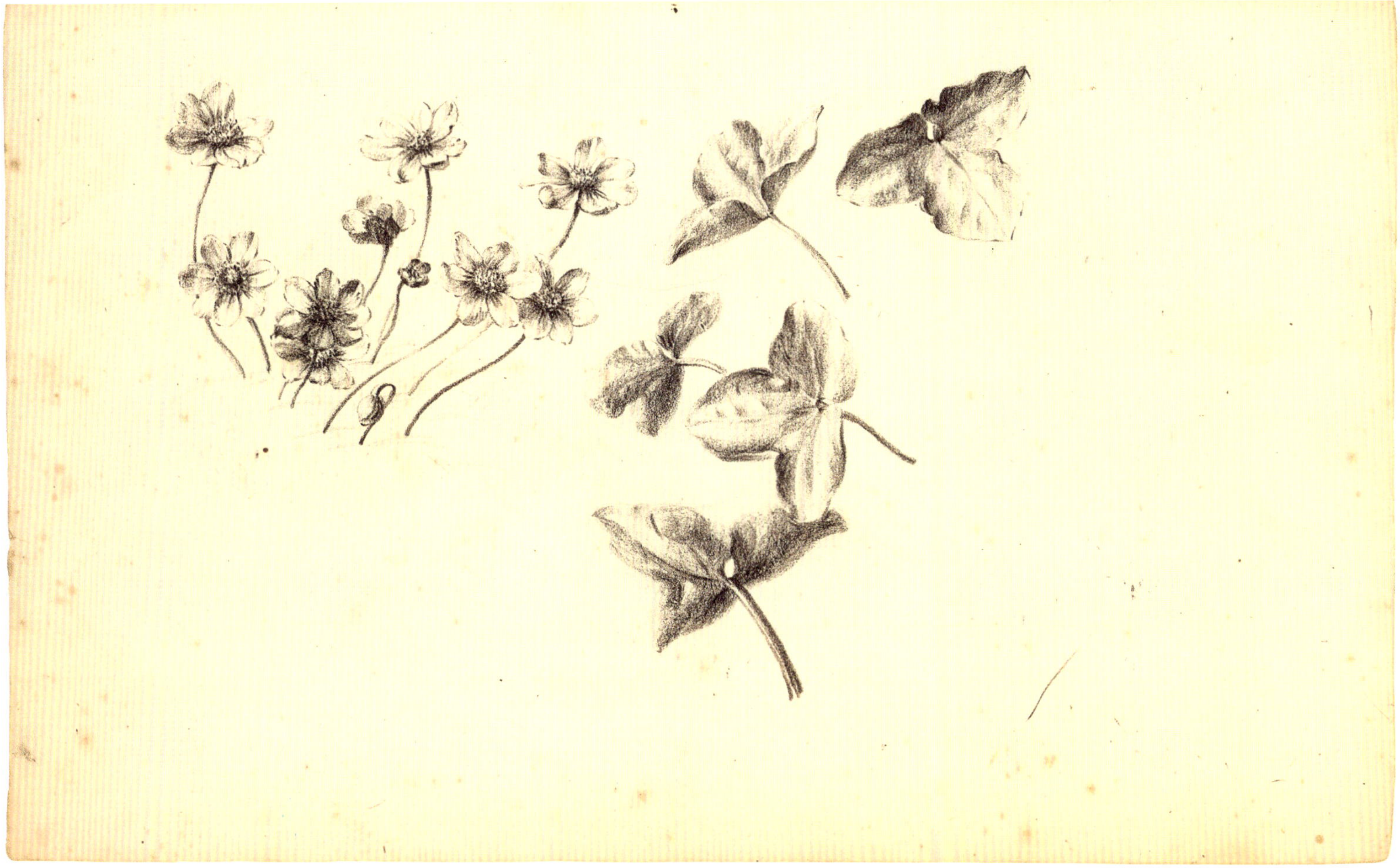

GERARD VAN SPAENDONCK, Album of flower studies. Flowers and leaves of a *Hepatica*

Among Spaendonck's contributions are the fifty paintings, which include a particularly fine bodycolour painting of a Virginian hibiscus, he made to the collection of *Vélins du Roy* begun by Daniel Rabel (see no. 15) and Nicolas Robert (see no. 44). Despite the fact that he bore the official title of Peintre Ordinaire du Roi pour la Miniature, he did not suffer persecution during the French Revolution, and continued to teach undisturbed at the Jardin des Plantes. There he received the appointment of professor of 'iconographie naturelle' in 1793. Many important naturalists worked with Spaendonck during his long tenure at the Jardin, including Bernard de Jussieu, Lamarck and Geoffroy de Saint Hilaire (Vezin, pp. 41–2).

When the Académie des Beaux-Arts was created, Spaendonck became one of its first members. Quatremère de Quincy, who supplied the artist's funeral oration, declared that he was worthy to join the select company of great French painters on the Mount Olympus of the arts.

A part of Spaendonck's *œuvre* comprises the large floral paintings he produced for a ready market of wealthy Parisian art collectors. In them he managed to unite the realism of the Dutch school with the suave elegance of the French school, as exemplified by the most celebrated French floral painter of the seventeenth century, J. B. Monnoyer (see no. 46). In this way he created a pleasing, showy style that became something of a formula, the artist repeating *ad infinitum* a compositional motif in which ornate vases, placed on pedestals or tables with rich fabrics artfully arranged around their bases, are filled to overflowing with luxuriant blooms, mostly double poppies, roses and peonies. Some of Spaendonck's best work is to be found among his engravings and miniatures, where his refined technique had fullest expression and where he perhaps felt that he could allow himself a more direct and spontaneous approach to his subject-matter. Among these works are twenty-one superb plates produced using the new technique of stipple engraving and destined for a manual entitled *Fleurs dessinées d'après nature* published *c.* 1800, a copy of which is at Oak Spring. Also at Oak Spring are some fine paintings, including 'Six tulip studies', a charming bouquet of 'Roses, a tulip, auricula, forget-me-not, double hyacinth and syringa', and another of 'Roses, delphiniums and carnations'.

When one considers the polished, sophisticated works for which Spaendonck is best known, this modest album of sketches in black chalk becomes all the more interesting, for here we can see the artist at work, pausing to study with delight the complex convolutions of an iris petal, the compact form of a rosebud, or the leaves and petals of a humble ranunculus. They reflect a side of him far removed from the celebrated painter of Salon pieces, for here his art is entirely intimate and true to life. There is even a self-portrait, sketched in a few quick, deft strokes, of the artist sitting before his easel (fol. 57). Some individual flower drawings by Spaendonck that can be compared stylistically to this album are preserved in various public and private collections (Boven and Segal, nos. 102, 109, 110, 114 and 115).

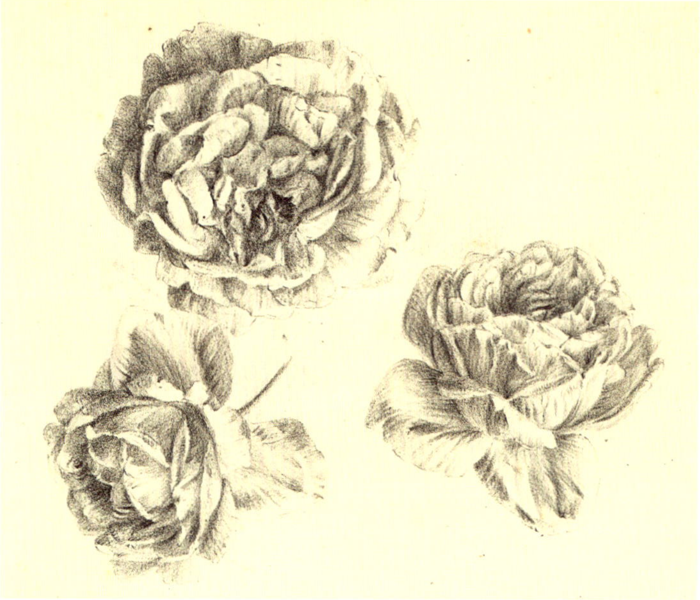

GERARD VAN SPAENDONCK, Album of flower studies. Three sketches of *Rosaceae*

59. BALDASSARE CATTRANI (*fl.* 1776–1810)

Stirpium Imagines LXXV quas ex Horto Patavino delineavit et vivis coloribus ad naturam fidelissime expressit Balthassar Catrani Romanus curante et dirigente Joanne Marsilio Botanices Professore eiusdemque Horti Prefecto. CIↃIↃCCLXXVI. 1776.

54 × 38 cm. 75 watercolours of flowers and fruits, preceded by a leaf with the title in manuscript and another leaf of a manuscript index of the plants.

BINDING: Modern three-quarter green morocco and green cloth. On the spine in gilt: 'Catrani. Stirpium Imagines LXXV. Padua 1776'.

PLATES: 75 bodycolours of flowers and fruits on paper, within borders double ruled in ink. Plates numbered in roman numerals.

PROVENANCE: Probably from the botanic garden at Padua.

THE COLLECTION at Oak Spring includes a large number of works by the Italian artist Baldassare Cattrani; several can also be found in the Garden Library at Dumbarton Oaks, Washington, D.C., while others have appeared at auction (Christie's, Florence, 1977, and New York, 1996). They permit us to reconstruct the personality and *œuvre* of an artist almost entirely unknown today—his name is not mentioned in any catalogue, nor have his works ever been published, even though he surely merits consideration in any history of botanical illustration. At Oak Spring, in addition to this manuscript, there are seven paintings on vellum of flowering plants, all of similar dimensions (53.5 × 38 cm.), and sixty-three loose paintings on paper of various indigenous and exotic plants. The title-page to the latter collection reads *Exoticarum atque indigenarum plantarum sthegmata ad naturam forma ac coloribus expressa collegit Baldasar Cattrni* (sic).

In May 1935 a sale was held in Zurich of the collection of books and manuscripts left by Eugène Beauharnais (1781–1824), the son of Joséphine Beauharnais and stepson of Napoleon I, who served as viceroy of Italy and who later received the title of duc de Leuchtenberg. Twenty-four volumes of botanical paintings in bodycolour by Cattrani were included in this collection. The catalogue description of lot no. 71 reads as follows: 'Cattrani, Baldassare, Collection de Botanique. Recueil de 1640 aquerelles originales, peintes sur vélin et représentant des fleurs et des plantes. Grandeur des feuilles: 53.5 : 38 cm. 24 vols. Peint en 1806. In folio. Dos orné en or et a froid. Magnifique oeuvre de botanique. Chaque feuille est encadrée d'un double filet noir et porte en bas le nom de la plante, dans quelques volumes les ont omis. Le premier volume contient plusieurs aquerelles peintes sur papier et en général celles de ce volume ont un peu suffert par l'humidité. Les aquerelles sont exécutées avec beaucoup goût et de soin et l'artiste a su fixer avec son pinceau l'éclat des couleurs et les nuances que que nous admirons dans les fleurs. Commes les liliacées de Redouté, cette oeuvre aussi a été peinte a Malmaison sur les ordres de l'imperatrice Josephine' (*Bibliothèque Eugène de Beauharnais (vice Roi d'Italie et Duc de Leuchtenberg) et des Ducs de Leuchtenberg provenant du Château de Seeon en Bavière. Vente aux Enchères 23 et 24 Mai 1935 à Zurich Zenfihaus zur Meise par Braus-Riggen bach Basel et Ulrico Hoepli Milano*, p. 21). From the dimensions and other details provided in the catalogue it can be deduced that the seven paintings on vellum at Oak Spring, as well as 300 works in bodycolour on the same material loosely inserted in five in-folios and a series of 172 paintings in bodycolour on vellum now at Dumbarton Oaks, originally came from this important group of manuscripts.

The catalogue also states that the paintings were executed for the Empress Joséphine at her country estate, Malmaison, in 1806, and thus during the very period in which the garden was being expanded under the Empress's personal supervision. Malmaison, a magnificent estate just north of Paris, had been acquired by Joséphine in 1798. It is well known that Cattrani invited various artists to come and paint her magnificent collection of flowers (Jouanin). Redouté was there in 1806, preparing his voluminous work *Les Liliacées* (1802–16), the original watercolours for which were inherited by Eugène Beauharnais and appeared as lot 82 at the same auction in Zurich. Actually, no sources testify that Cattrani did actually work directly under Joséphine's patronage, since in the same year (1806) the artist was working once again in the botanic garden at Padua, this time for the director, Giuseppe Antonio Bonato (1753–1836), the successor to Giovanni Marsili. For Bonato, Cattrani painted seventy bodycolours on vellum that were sold at auction (Christie's) in Florence in 1977 and are now dispersed. In the title-page to this manuscript, framed by an elegant flower garland, we read: 'Stirpium Imagines . . . quas ex horto Patavino delineavit et vivis coloribus ad naturam fidelissime expressit Balthasar Catrani Romanus curante et dirigente Joanne Marsilio Botanices Professore eiusdemque Horti Prefecto'—'Die Vigesimo Junii Anno Millesimo Octingentesimo Sexto'. Stylistic relationships have finally allowed the attribution to the artist of more than 225 botanical plates, hitherto recorded as anonymous, painted in water- and bodycolour on paper (51.5 × 36.5 cm.), now kept in the library of the botanic garden at Padua. The plates are inserted in a miscellaneous manuscript with the title *Piante dell' orto botanico di Padova*, probably bound at the time of the same 'prefetto', Bonato.

The seventy-five bodycolour paintings of native and exotic flowering plants contained in the Oak Spring manuscript *Stirpium imagines* were painted in Padua in 1776 at the botanic garden, then under the direction of the alert director Marsili (1727–95), who was also an astute collector of paintings. Although they probably represent an early undertaking by Cattrani, these beautifully composed and carefully executed paintings show that at this point he was already an artist of great accomplishment. In each plate the specimen dominates the centre of the page, while individual parts of the plant or microscopic details appear enlarged in the margin. Next to common European species such as the *Digitalis purpurea* (pl. XXV) are depicted such exotics as the *Crinum africanum* (*Agapanthus umbellatus*) (pl. XXI) and the *Dodecatheon meadia* (pl. XXVI), a plant recently introduced to Italy from America.

Particularly interesting in terms of botanical history is plate XXXII, which depicts the so-called 'Mugherino del Granduca di Toscana', actually a cultivar of the *Jasminum sambac Soland*. This exotic plant from Goa, India, was seen for the first time in Italy when the King of Portugal presented a

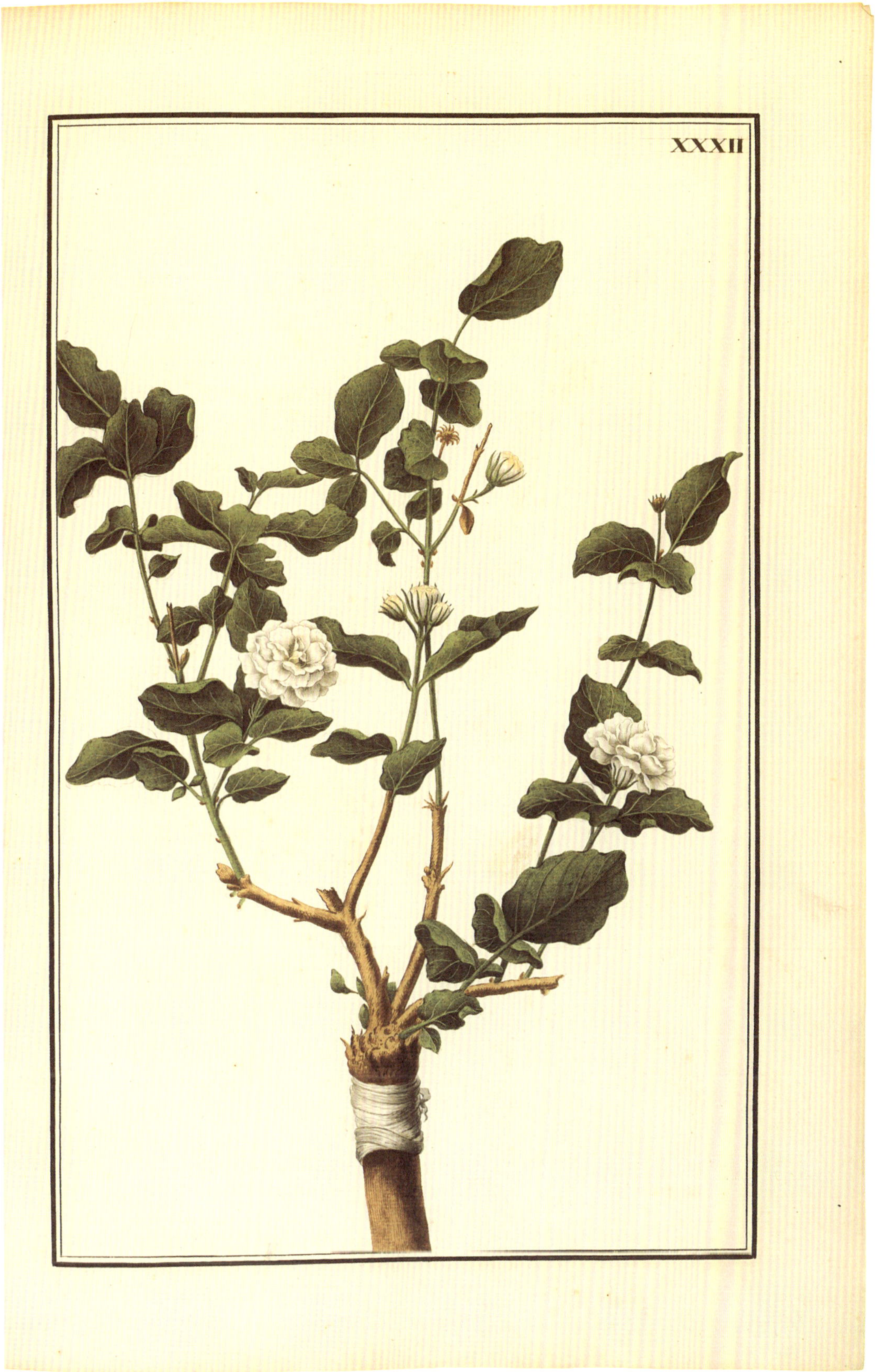

BALDASSARE CATTRANI, *Stirpium imagines.* Arabian jasmine (*Jasminum sambac*), presented to the Grand-Duke of Tuscany, folio XXXII specimen to Cosimo III, Grand-Duke of Tuscany. So rare was this plant that Cosimo III asked his most talented naturalistic painter, Bartolomeo Bimbi (1648–1729), to paint it immediately; the picture, which shows the plant in full flower, can be seen today at the Palazzo Pitti in Florence (*Floralia*, pp. 95–7).

The Grand-Duke jealously guarded his plant, setting severe proscriptions against any grafting or layering from it. These restrictions were officially raised only in 1791 by the Grand-Duke Pietro Leopoldo, although in reality the *Jasminum sambac*, which is mentioned in a poem by Lorenzo Magalotti, was being cultivated in botanical gardens all over Italy long before this date. A description and an engraving of the plant appear in the *Catalogus plantar um horti Pisani*, published in Florence in 1723 by Michelangelo Tilli, the director of the botanical garden at Pisa (Tongiorgi Tomasi, 1991, p. 197). We also learn from contemporary sources that a specimen of *Jasminum sambac* cultivated in the botanical garden at Padua came into flower in 1776. Cattrani's watercolour shows the plant still bearing its grafting bands.

Even though we have almost no biographical data on Cattrani, aside from a brief note on the manuscript title page of *Stirpium imagines* and the manuscript sold in 1977 that states that the artist was Roman by birth, given the volume and quality of his output Cattrani is certainly an important figure in the history of late eighteenth- and early nineteenth-century botanical illustration. His singular talent led to permission to study and paint in Padua's botanical garden, one of the oldest and most prestigious in Europe, and support from the Empress Joséphine, patroness of one of the most productive circles of artists and plantsmen in the history of botany. Thus, the high scientific, as well as artistic, quality of Cattrani's work comes as no surprise. Every painting is marked by his clear and assured style, with each plant elegantly arranged on the page and depicted with a minute attention to detail.

60. James Sowerby (1757–1822)

Chinese magnolia. 33.3 × 20.5 cm. Pencil and watercolour. Inscribed on the verso: 'Magnolia from China Feby. 15 Bulstrode. By J. Sowerby'.

PROVENANCE: Collection of Iolo A. Williams.

This restrained yet sophisticated sketch was executed by one of the most important and prolific English artists in the history of botanical illustration, many of whose descendents became naturalists and scientific illustrators. Sowerby first studied at the Royal Academy, and then with the marine painter Richard Wright. He began his career as a painter of portraits and landscapes, but eventually decided to specialize in botanical illustration, perhaps with the encouragement of the naturalist L'Héritier de Brutelle, who was the first to utilize his paintings to illustrate his books

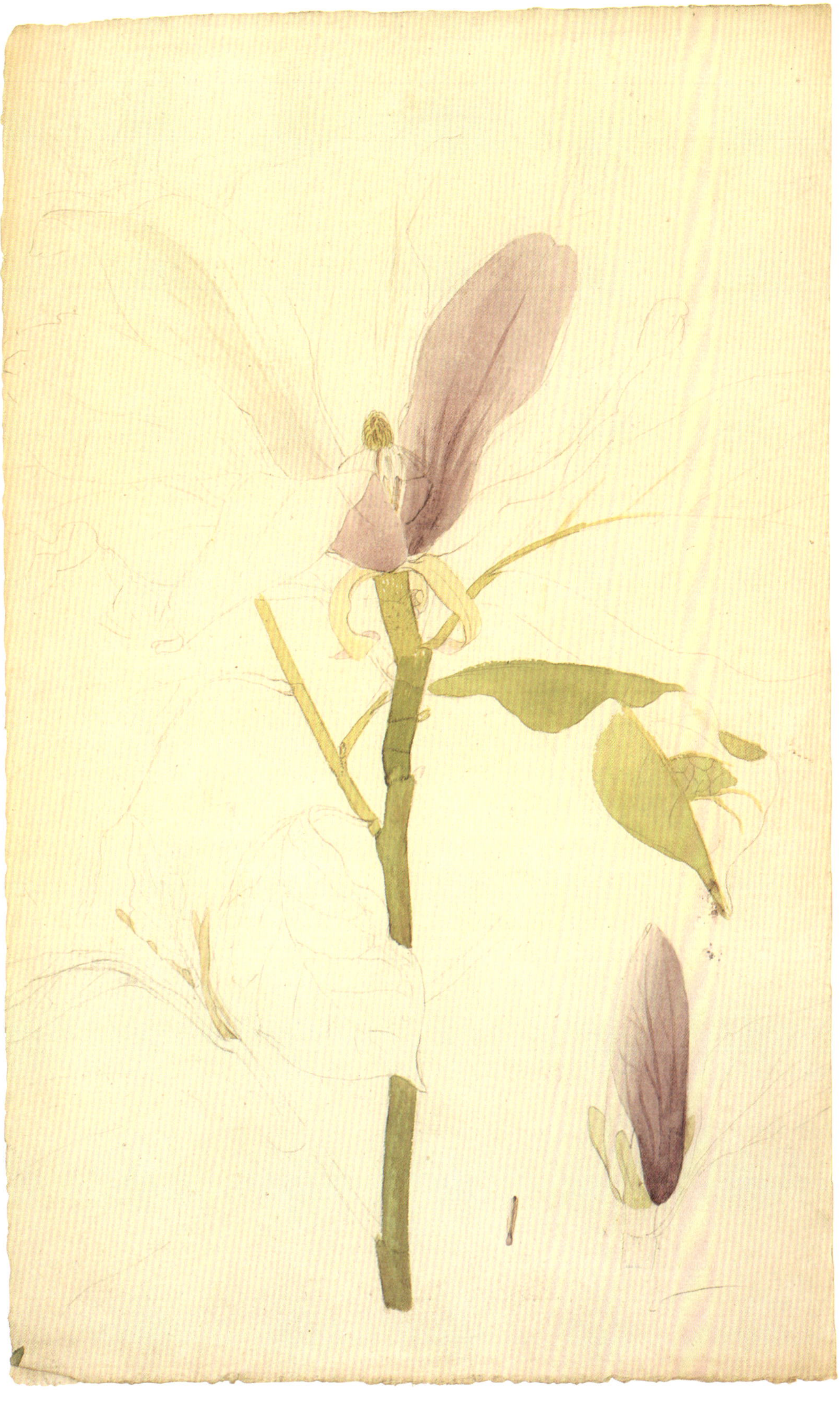

JAMES SOWERBY,
Chinese magnolia

(see no. 57). In 1787 Sowerby was engaged by William Curtis (1746–99), director of the Chelsea Physic Garden in London, to produce drawings and copper-plate engravings (later hand-coloured) for *The Botanical Magazine*, the first journal on the subject of botany to be published in England. A painting by him of an elegant *Iris persica*—'highly esteemed by all lovers of flowers' in the words of the editor—formed the opening plate to the very first issue of the journal, which sold for one shilling (Desmond 1987, pp. 26, 30). Many other works by Sowerby (some of them unsigned) appeared in subsequent issues.

Sowerby was a cultivated artist with an extensive knowledge of botany, but he was also quite versatile; for example, in the last years of his life he undertook a study of zoology and geology in order to be able to depict animals and minerals. He also founded his own journal, *English Botany* (1790–1814), publishing a total of thirty-six issues with nearly 2,500 plates. Collaborating with him on this project was the naturalist Sir James Edward Smith (1759–1828), founder of the Linnean Society of London, who wrote most of the text that accompanies the artist's plates. Sowerby's expertise in the formal aspects of botanical illustration also enabled him to produce *A Botanical Drawing-Book; or, An Easy Introduction to Drawing Flowers according to Nature*, which was published in 1788 and reissued in 1791 (a copy of the edition of 1807 is at Oak Spring), as well as *Flora luxurians, or, The Florist's Delight*, published between 1789 and 1791. He also collaborated on Curtis's unfinished project *Flora Londinensis* (1777–98), *Medical Botany* (1790–94) by William Woodville, and *Exotic Botany* (1804–05) by Sir James Edward Smith. He produced illustrations of many popular flowers, such as the carnation, tulip, auricula and hyacinth, for *The Florist's Delight*, and prepared both the text and the 440 plates for *Coloured Figures of English Fungi or Mushrooms*, published between 1797 and 1815. Finally, he engraved the plates—from drawings by Ferdinand Bauer—for John Sibthorp's *Flora Graeca*, which was published in parts from 1806.

At Oak Spring are nine watercolours by Sowerby, among which the *Lily and Bulb* (*An Oak Spring Garland*, no. 29) and *Spray of Three Lilies* are particularly noteworthy for their elegant composition, harmonious colours, and scientific accuracy. The *Chinese Magnolia*, although a quick sketch in comparison to the eight other paintings, should not be dismissed *tout court* as an unfinished work, for it gives us an opportunity to study the remarkable technique of this artist–naturalist. As Wilfrid Blunt observed; 'in his rapid sketches, his pencil line is swift and vivid' (Blunt and Stearn, p. 220), and here we can retrace the confident pencil outline with which Sowerby began each of his works before applying delicate strokes of watercolour to recreate the tints of the petals and leaves.

61. Pierre Joseph Redouté (1759–1840)

[*cover title*] Le Bouquet Royal Oeuvre Posthume de P.-J. Redouté Dédié à Sa Majesté la Reine des Français. Paris, chez les Marchands de Nouveautés. MDCCCXLIII [*on the verso of the half-title*] Paris: Typographie Lacrampe et Comp., Rue Damette, 2.

2° 45 × 31 cm. 2 leaves, 5 leaves of plates.

BINDING: Three-quarter green morocco on marbled paper with the original wrappers bound-in. On the spine, gilt rose ornaments and 'Redouté Le Bouquet Royal'.

PLATES: A portrait of the author, lithographed by François-Fortuné-Antoine Férogio from a drawing by Marie Eléonore Godefroy, forms the frontispiece. It is followed by four colour plates of roses, stipple engraved by N. Rémond from drawings by Redouté and retouched by hand.

REFERENCES: Dunthorne 236; Pritzel 7457 (cites *Le Bouquet royale*, 1844); Nissen 1590; Sitwell and Blunt, p. 76; *A Catalogue of Redoutéana*, no. 63; *A Magnificent Collection*, no. 293.

PIERRE JOSEPH REDOUTÉ is widely thought to be the greatest floral artist of all time, and at Oak Spring is a complete collection of his printed works as well as a number of original drawings and paintings in watercolour. The published works include the eight-volume *Les Liliacées* illustrated with 486 plates, the three-volume *Les Roses* with 168 plates, *Choix des plus belles fleurs . . .* with 144 plates, *Le Cours de fleurs du Jardin des Plantes* with forty-eight plates and the *Jardin de la Malmaison* by Etienne Pierre Ventenat with 120 plates. Every work was published in folio form, with the engravings made after drawings by Redouté. The Oak Spring collection also includes seven of the artist's original watercolours for Ventenat's *Jardin*, two for the partially completed *Figures de la flore des Pyrenées* by Jean François La Péyrouse, a bodycolour painting of tulips and roses executed in 1811, another entitled *Carnations, Hyacinths and Campanula* and dated 1812, a third painting, *Pink Roses*, from 1833, and an early watercolour (1793) of a *Dodecatheon meadia* (Shooting star). The elegant simplicity of this last work contrasts strikingly with the somewhat grandiose style of the artist's mature work (see *An Oak Spring Garland*, no. 23).

Redouté counted three successive queens among his patronesses, and also taught painting for many years to the wives and daughters of the French aristocracy. His auspicious career, combined with his prodigious output, helped to create his perhaps somewhat exaggerated reputation, for in his own day Redouté was considered to be the 'Raphael of flower painting'. His life and career spanned one of the most tumultuous periods in the history of France. Born in the Ardennes region (now part of Belgium) into a family of painters going back many generations, Redouté, together with his two brothers, began his artistic training very early, and by the age of thirteen was already working as an itinerant decorator. He passed a significant formative period in Holland, where he was able to study the many sixteenth- and seventeenth-century Dutch masters of flower painting, in particular the works of Jan van Huysum and Rachel Ruysch. In 1782, with his elder brother Antoine Ferdinand (1756–1809), Redouté made his way to Paris. He began painting at the Jardin du Roi and there met the famous botanical artist Gerard van Spaendonck (see no. 58), who took him on as a student and taught him the art of painting in watercolours on vellum.

In Paris Redouté was also introduced to the botanist and connoisseur Charles Louis L'Héritier de Brutelle, who commissioned many works from him (see no. 57). The first was a series of more than fifty paintings for a book published by L'Héritier in 1784–5, *Stirpes novae*. When this project was finished, the artist set off to join L'Héritier in London, where he collaborated with him on *Sertum Anglicum*, a work describing the rare plants at Kew. While in London Redouté frequented artistic circles and met various well-established figures, including the engraver and botanical painter James Sowerby (no. 60), who prepared some of the plates for *Sertum Anglicum*, and the great Florentine engraver Francesco Bartolozzi. Bartolozzi had settled in London many years before, and there perfected the art of stipple engraving, which immediately became the preferred technique for the reproduction of paintings in books (Watson, pp. 12–13). On returning to France, L'Héritier introduced Redouté to his friends at court, and in 1787–8 Redouté was made official draughtsman to the *cabinet* of Marie-Antoinette. The Queen adored flowers, but the temper of the times did not allow her to indulge this passion to the full. Indeed, history has recorded a poignant episode in which the Queen, during her imprisonment, sent Redouté a night-blooming cactus with the request that he portray it for her (Blunt and Stearn, pp. 196–7).

In 1793 Redouté and his brother Henri Joseph (1766–1852) were appointed to the staff of the Muséum d'Histoire Naturelle—as the Jardin and Cabinet du Roi were renamed following the Revolution. Redouté was engaged to paint botanical subjects, and his brother zoological ones. Henri Joseph was later invited to accompany Napoleon on his Egyptian campaign in order to make a visual record of all findings of scientific interest. Redouté too found his star once more on the rise under the new regime. Empress Joséphine, Napoleon's first wife, became one of his most enthusiastic admirers and a generous patron. He was included in the small group of scholars and artists invited by Joséphine to pursue their work at Malmaison, her estate to the north of Paris. The gardens there were stocked with rare native and foreign plants tended by the botanist Etienne Pierre Ventenat, and in 1804 a superb collection of roses was also established.

Redouté prepared many of the illustrations for Ventenat's *Jardin de la Malmaison* (1803–4); in addition, he contributed to the illustrations for *Les Liliacées* (1802–16), the first in a series of gargantuan and celebrated works by the same author. In *Les Liliacées* Redouté perfected the refined style that was to become his trademark and the chief reason for his phenomenal success. After Joséphine's death in 1814, Redouté began working for Napoleon's second wife, Marie-Louise of Austria, to whom he also gave some lessons in the art of flower painting. When Spaendonck died in 1822 Redouté was considered to be his natural successor, and he took on many of Spaendonck's students, including Pancrace Bessa, Antoine Pascal and Ernestine Panckoucke (see Hardouin-Fugier).

Les Roses was published between 1817 and 1824, a work that Redouté would have dedicated to Joséphine had she not died before it was completed. Although not one of his finest productions, and conceived on so ambitious a scale that it brought its author to the brink of financial ruin, *Les Roses* was an enormous success. The art of flower painting, once considered a minor genre, was by now

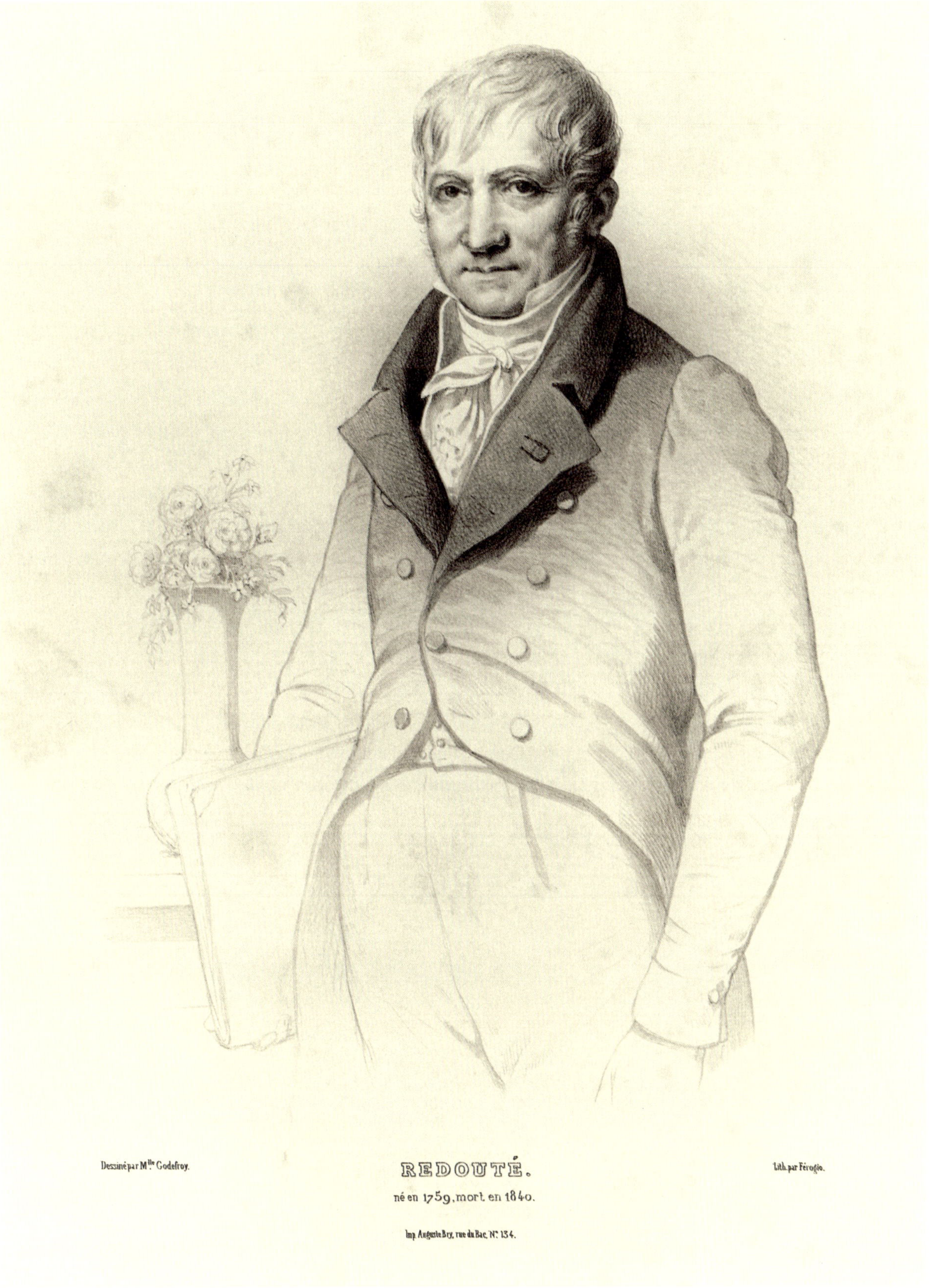

PIERRE JOSEPH REDOUTÉ, *Le Bouquet royal*. Frontispiece portrait of the author greatly appreciated by many connoisseurs and collectors. Paintings in water- and bodycolour on vellum, with their fine outlines and delicate hues, and stipple engravings with their rich tonal effects, were particularly sought after. The prestigious annual Salons, to which Redouté began sending pictures in 1796, contributed even further to the popularity of this genre.

Redouté's work is characterized by immense charm and a perfect mastery of technique, which allowed his many admirers of the time to gloss over the scientific inaccuracies and a certain superficiality in conception. Redouté was an indefatigable artist; he even produced sixty-five plates for an edition of Jean-Jacques Rousseau's *Essais élémentaires sur la botanique*, which was published in 1805 as *La Botanique de J. J. Rousseau*. Unavoidably, the endless stream of paintings produced by Redouté for dozens of published works formed at once the source of his fame and set limits to his artistic achievement. In Redouté's 'elegant period', as the last part of his life has been aptly described (*A Catalogue of Redoutéana*, p. 27), his watercolours were no longer painstakingly painted *d'après nature*, but were the fruit of a consummate technique.

It is for this reason that a lesser-known work by the artist was chosen for inclusion here. *Le Bouquet royal* was a posthumous publication, dedicated to the Queen by Redouté's widow and his daughter. It contains engravings of four varieties that were not included in the artist's monumental monograph on the rose, each variety bearing an appropriately feminine name—'Amélie', 'Hélène', 'Clementine' and 'Adelaide'. The delicate modulations in hue of the original watercolours are captured with great skill in Rémond's colour-printed stipple engravings. The title-page consists of a sober portrait of the artist engraved after a drawing by the celebrated portraitist Marie Eléonore Godefroy. Redouté is shown holding an album, while a vase of roses stands on a table by his side. The dignified mien and meditative expression presented to us in this idealized image bear little relation, however, to the awkward and inelegant figure described by Redouté's contemporaries (Blunt and Stearn, p. 194).

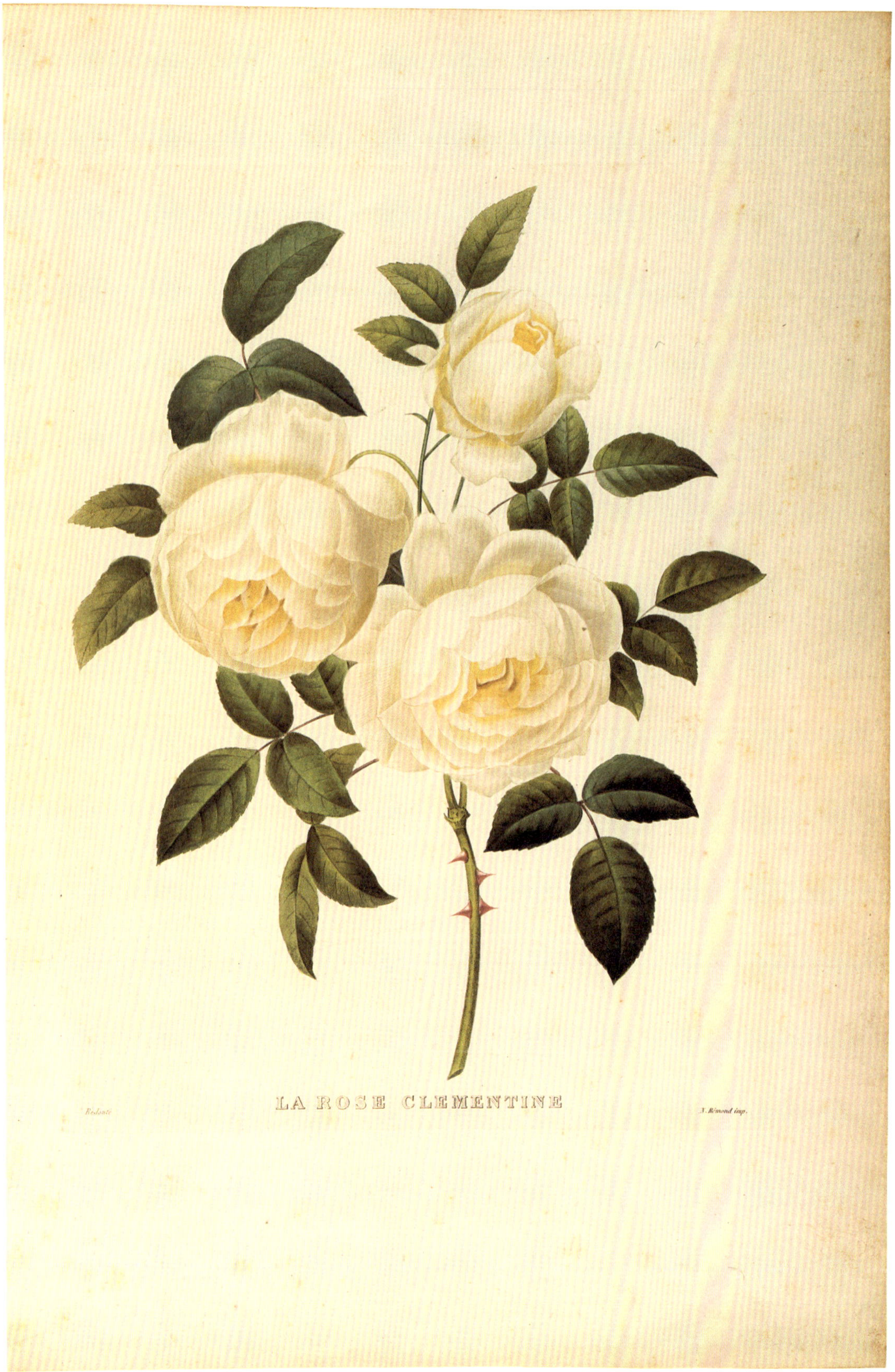

PIERRE JOSEPH REDOUTÉ, *Le Bouquet royal.* White roses ('La Rose Clementine'), plate 4

VI · FLOWERS IN THE DECORATIVE ARTS AND MANUFACTURING

FLOWERS have not simply constituted a recurrent theme in naturalistic illustration and the figurative arts, for they have remained a cherished motif in the applied and decorative arts in both Eastern and Western art since antiquity. Floral designs have been used to decorate almost every object created by human hands, from the simplest household implement to the most precious *objet d'art*. To this end, unlike the botanical artist, whose primary concern was scientific accuracy, the designer could give free rein to his imagination, creating forms and images that were the epitome of refinement and elegance.

The flowers depicted by the botanical illustrator were usually presented in a strictly frontal view, often with their leaves, roots or bulbs, and fruit and seeds. The designer, on the other hand, could arrange his flowers and bouquets as he pleased, even inventing forms or views not to be found in nature. These motifs, often highly stylized, were admirably adapted to serial and geometric repetition and thus to the decoration of cloth, tapestries, carpets and paper. Nevertheless, botanical illustration and the decorative arts have always been bound by the closest of ties. The Italian artist Jacopo Ligozzi (1547–1626) provides a case in point; he belonged to a celebrated family of embroiderers, and not only was he an extraordinary naturalistic painter, he also created beautiful designs for embroidery, elaborate blown-glass cups and vases and *pietre dure* marquetry (Tongiorgi Tomasi 1993, pp. 32–3). Many other great artists applied their talents to the decorative arts. Jean Joubert, for example, who succeeded Nicolas Robert as 'Peintre du Roi pour la miniature', also produced designs for the decoration of furniture (*Flowers into Art*, p. 17), while Gerard van Spaendonck (no. 58) began his career as a modest painter of floral designs on snuff boxes. The noted nineteenth-century botanical illustrator and lithographer Walter Hood Fitch (nos. 104 and 107) first worked as a designer of 'oriental' floral patterns for a calico manufacturer.

In the seventeenth century the printed florilegium served as a source of designs for artisans working in all the decorative arts, from jewellery and ceramics to marquetry and, above all, textiles (*Book of Flowers*, pp. 65–132). Floral motifs, already popular in the medieval period, became ubiquitous in embroidered fabrics and in the fine cloths and tapestries woven for the wealthy, who demanded ever more splendidly elaborate attire and a new luxuriance and beauty for their domestic

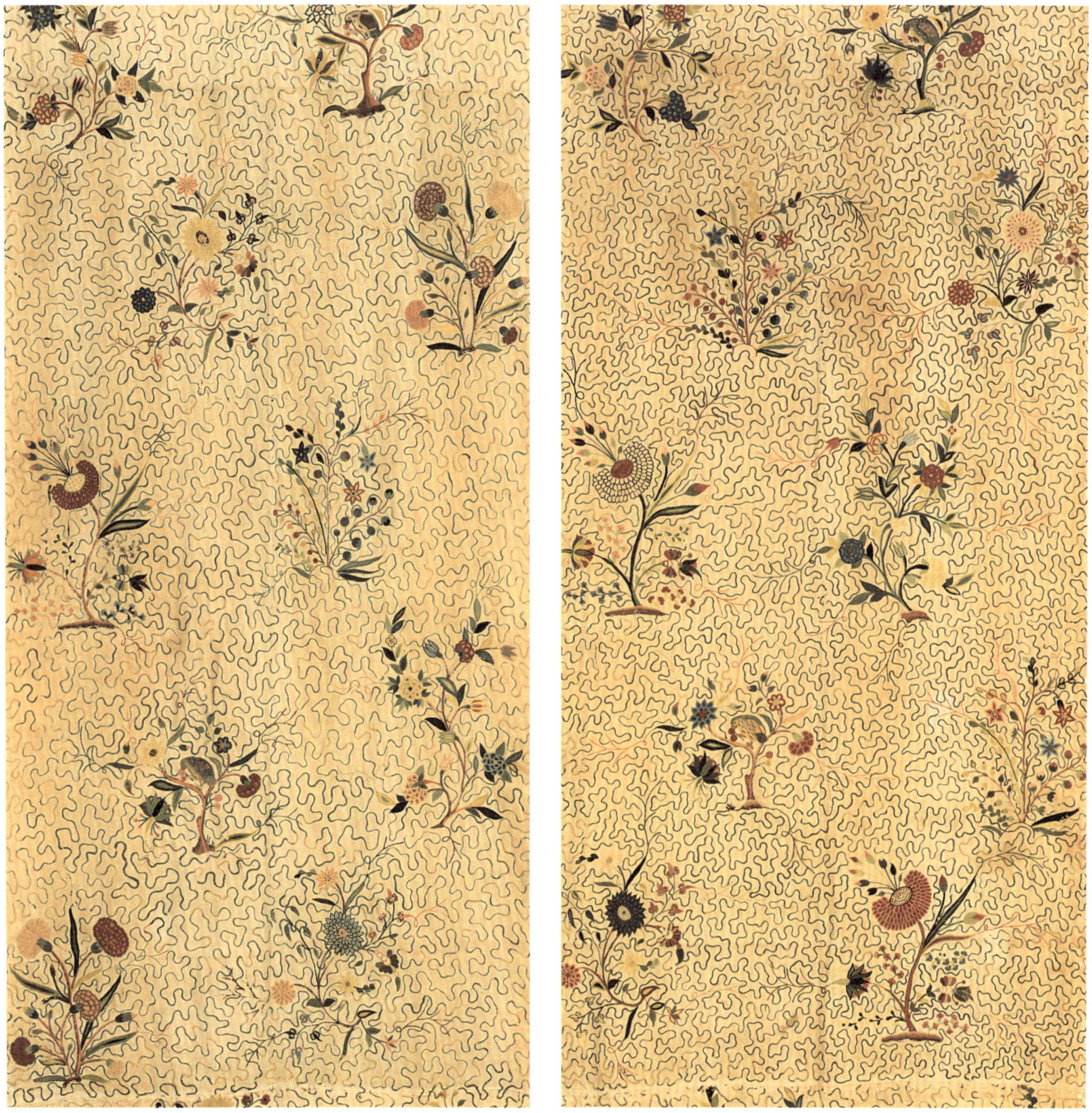

Details from two eighteenth-century embroidered Indo-Portuguese fabrics

furnishings. In the seventeenth century Venice and Genoa became famous for their *velluti giardino*, velvet cloth decorated with brilliantly coloured garden motifs. Floral designs thus became an important element in the production of most handiwork, and the richness of the design a reflection of the possessor's wealth and position.

As we have already seen, Jacques Le Moyne de Morgues (no. 6) and Pierre Vallet (no. 8) were quite aware that their paintings could also serve as a source of inspiration for artisans. Vallet, who had been granted the title of 'Brodeur ordinaire du Roy', dedicated one of his works to Marie de' Medici, the consort of Henri IV, who had introduced the use of floral motifs in embroidery to the

French court. The self-portrait that opens Vallet's florilegium is framed not only by the traditional tools of the painter, the architect and the engraver (palette and paintbrush; T-square and compass; gouge and the burin) but by the scissors, needle and bobbin of the weaver and embroiderer, thus indicating the parity of these arts in Vallet's eyes. Daniel Rabel, author of *Theatrum florae* (no. 15), also produced *Dessins des parterres* (1630), a series of patterns that could be used, as the author himself pointed out, interchangeably, either for embroidery or when designing a flower-garden.

The florilegia of Emanuel Sweerts (no. 9), Johann Theodor de Bry (no. 10), Basilius Besler (no. 11) and Crispijn van de Passe (no. 12) were created, as we have seen, for botanists or connoisseurs, although artisans sometimes looked to them for inspiration. Only later did artists such as François Langlois (no. 14), Jeremias Falck (no. 19) and the young Nicolas Robert (nos. 42, 43) begin to produce works with the kind of accentuated decorative qualities that favoured their use as patterns or models. At Oak Spring is a bound volume, with 'Fleurs' stamped on the spine, containing many designs by different artists (no. 62), including a series of engravings of flowers and insects by François Vauquier that were inspired by etchings carried out by the most famous floral decorator of the period, Jean Baptiste Monnoyer (no. 46). Many celebrated printing-houses in Paris, such as François Poilly's 'à la Belle Image', Nicolas Langlois's 'à la Victoire' and, later, Jean Mariette's 'aux Colonnes d'Hercule', all situated in the rue Saint-Jacques, furnished a selection of *feuilles ornées* that could be acquired by the numerous artisans working in the capital.

During the course of the eighteenth century the decorative arts reached new heights of prestige and popularity. Like the other arts, they faithfully mirrored the changes in taste of their patrons. Thus, the magnificent splendour—*le grand goût*—that marked the reign of Louis XIV had given way before mid-century to the charming grace of the Rococo, which in its turn was replaced by the more sober style that we know today as Neoclassicism. In the Rococo period artisans delighted in the creation of small objects of exquisite refinement in rare or precious materials. The architect, painter, decorator, cabinet-maker, bronze sculptor, *tapissier* and ceramicist all sought to excel one another in the exquisiteness of their materials, in the perfection of their workmanship, and in the invention of elegant and exuberant decorative motifs, most of them inspired by naturalistic themes. Designs featuring plants, flowers, birds and animals, as well as charming landscape scenes (many in the popular Chinoiserie style), appeared on every decorated surface and in every material, from fine porcelain, furniture and wall-panelling to carpets, tapestries and gowns. In 1754 the Society for the Encouragement of Arts, Manufactures and Commerce—the very first organization of its kind—was established by William Shipley in London. The aim of the Society (which in 1908 became the Royal Society of Arts) was to encourage craftsmanship of all kinds and the development of new and improved techniques. And in the same period, in his *Encyclopédie*, Diderot was encouraging artists to abandon 'pure' art and dedicate themselves to its practical applications in manufacturing.

An exquisite album at Oak Spring by Charles Germain de Saint-Aubin (no. 63) reflects all of the gay charm of the Rococo. The more than 250 works in this album range from beautiful, quite realistic, flower paintings to charmingly witty 'inventions', and we are not surprised to discover that Mme de Pompadour, who passionately adored floral motifs, participated in the creation of one of the watercolours in the collection. Germain, who was 'Dessinateur du Roi pour la broderie et la dentelle', affirmed that his paintings were quite worthy of comparison to the celebrated *vélins* of the royal botanical collection. At Oak Spring there are also several pieces of embroidered cloth of Indo-Portuguese origin dating from the second half of the eighteenth century. Intended for decorative use, these lengths of cotton cloth are colourfully embroidered in silk thread with various traditional Eastern motifs, among them the tree of life with flowers (including carnations and tulips), lanceolate leaves, and birds pecking at fruit and berries.

The work of the English artist John Edwards (no. 64) and the French artist Jean Louis Prévost (no. 65), both active *c.* 1800, reflect the change in taste in the decorative arts imposed by the new Romantic aesthetic. They produced designs in which the delicate Rococo motif of single flowers was replaced by bouquets and other compositions. Jean Baptiste Huet the elder was not only a painter of pastoral landscapes, genre scenes and flowers, such as the fine *Iris germanica* in the collection at Oak Spring (no. 66); he also created designs for the famous tapestry works of Gobelins and Beauvais. The artist Antoine Berjon (no. 67) worked closely with the silk manufacturers in his native city of Lyon, providing them with designs, while, as a teacher of floral decoration at the local Ecole des Beaux-Arts, training many artists who subsequently went off to work for textile manufacturers in various parts of France. The Lyon school of decoration became quite renowned after the publication of two manuals, the *Méthode Grobon frères*, published in the late 1840s in Paris and London, and Chabal Dussergey's *Etudes et composition des fleurs*, which appeared twenty years later.

Flowers were popular not just in textile manufacturing, but in the production of ceramics and porcelain too. During the eighteenth century the most celebrated porcelain works in Europe, including Vincennes and Sèvres in France, Chelsea in England, and Meissen in Germany, all produced services in fine porcelain decorated with a variety of floral motifs. Even delicate cups and vases in the form of flowers were made. Many of the patterns were inspired by the work of botanical illustrators such as Ehret (no. 50), van Spaendonck (no. 58) and Redouté (no. 61). At the same time, as can be seen from a Meissen pattern-book at Oak Spring produced by Johan Samuel Arnhold and Johann Stephan Capieux (no. 68), lesser-known artists also produced many beautiful designs, adapting themselves with great ingenuity to the relatively small surface areas at their disposal by the use of motifs such as small bouquets with short stems.

The fashioning of artificial flowers out of fine silk, hammered metal, paper or wax was yet another example of the use of floral motifs in the decorative arts. Already quite popular in the eighteenth century, such fripperies became all the rage in the succeeding one, and *plumassiers* and *bouquetiers* collaborated with dressmakers to create ever more elaborately decorated gowns for their exacting clients in the great capitals of Europe (*Flowers into Art*, p. 118).

In England, Emma Peachey revived the seventeenth-century art of wax sculpting. By her efforts

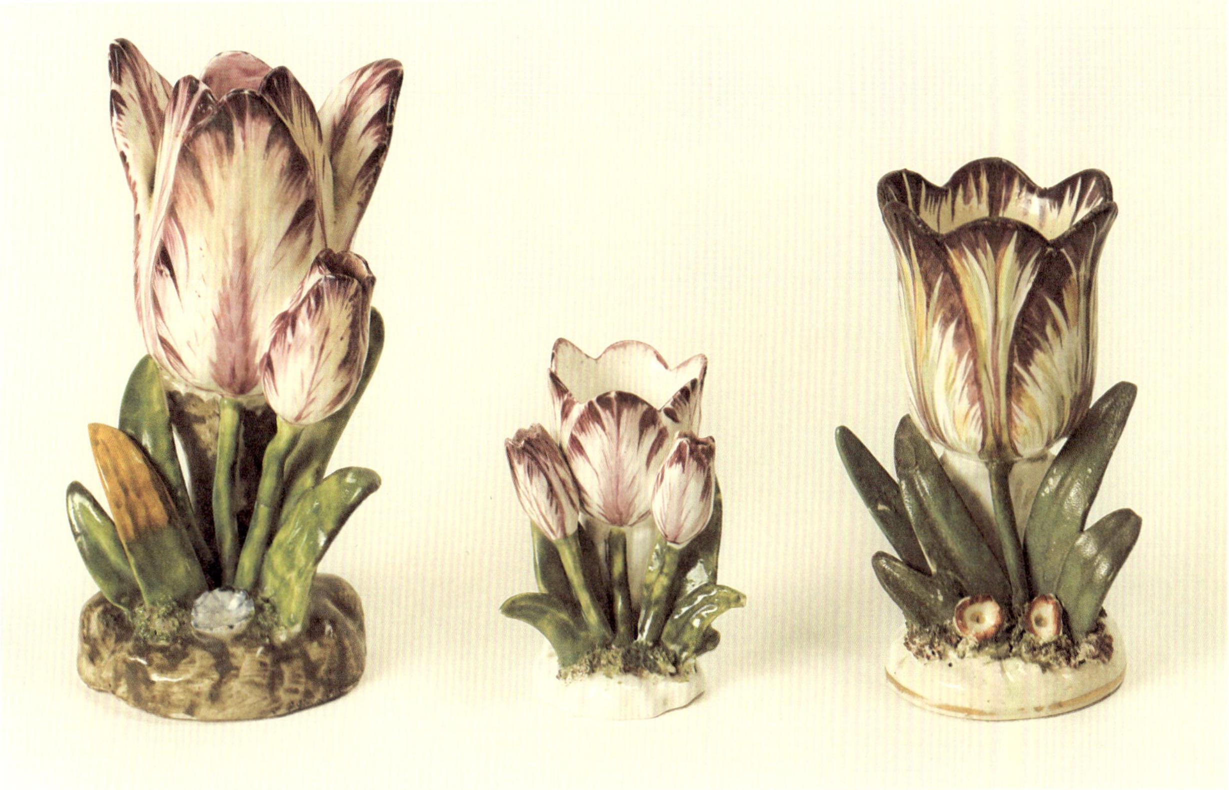

Staffordshire tulips, 1830s

—she taught at her home in London, wrote a manual on the subject (no. 70), and created highly theatrical fruit and flower compositions, some of which she had hoped to display at the Crystal Palace during the Great Exhibition of 1851—she contributed to the immense popularity of this art, which, as she pointed out, required as careful a study of botany as did the most exacting botanical illustration. T. C. March (no. 71) was another quintessentially Victorian artist, who designed elaborate tiered *epergnes* (table-centres) out of glass, ceramics or metal, in which flowers, fruits and other elements (trailing vines, osprey feathers, etc.) could be arranged to create handsome centrepieces. These objects, which were manufactured by the London firm of Dobson & Pearce, became a great commercial success after they were shown at the International Exhibition held in the capital in 1862.

The universality of the floral motif is reflected in the objects produced by simple artisans of other cultures too. At Oak Spring are various examples of handiwork from the American colonial period, among them a painted wooden box dating from 1776 and a highly stylized painting of a vase of flowers by John Ingersoll (no. 69). There is also an unusual collection of sixty small cigarette cards, with their own protective folder, issued by the tobacco manufacturers J. Wix & Sons of London in 1923. On each card is mounted a small piece of cloth bearing a machine-embroidered flower; the name of the flower is supplied, as is a short text and in some cases a few lines of verse. These cards were issued free in packets of *Kensitas* cigarettes, and keen collectors could obtain the folder for them

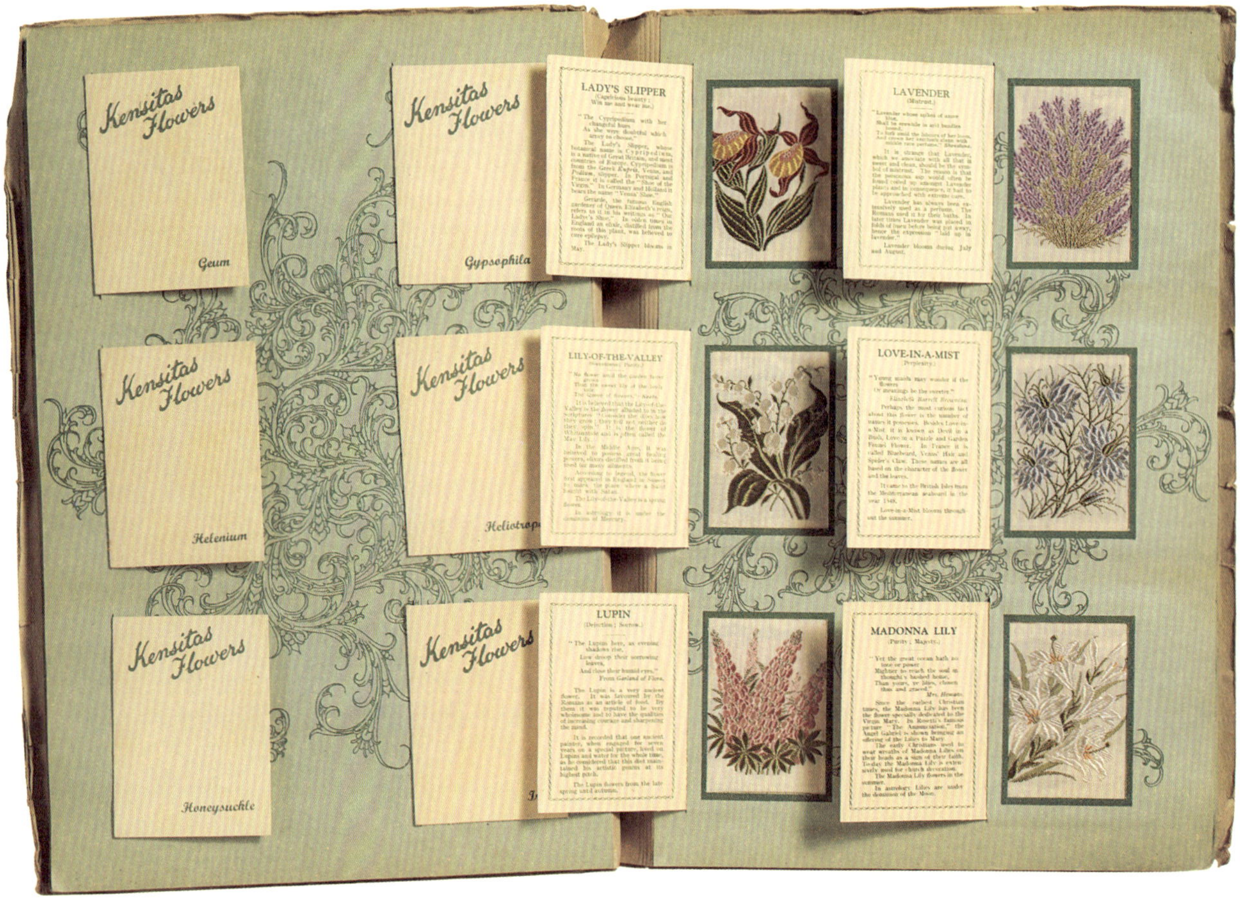

Collection of *Kensitas* cigarette cards, 1923

by post from Wix & Sons, who, on the back of each card, remind the curious that 'Tobacco is a plant and we want you to realize that only the finest Virginia Tobacco is used in Kensitas'.

This brief *excursus* on flowers in the decorative arts closes with a very unusual work, *Pot de fleur* (18.5 × 22 cm.), made in 1960 by the French artist Jean Schlumberger (1907–87). In his works, which consist mostly of designs for jewellery and other precious objects, Schlumberger, who lies buried on the beautiful island of San Michele in the lagoon of Venice, created a poetic world woven from dreams, fantasy, history and nature. This painting in bodycolours (20.2 × 12.7 cm.) depicts the modest terracotta vase, a gift to the artist from Rachel Lambert Mellon, that he transformed into a rare object by ornamenting it with various precious stones—square-cut emeralds, garnets and a sapphire. The bejewelled *Pot de fleur* itself is now on loan from Mrs Mellon to the Virginia Museum of Fine Arts, Richmond (*Schlumberger*, pl. 43). Schlumberger also designed the lead flower urn that today tops the roof of the greenhouse pavilion in the garden at Oak Spring.

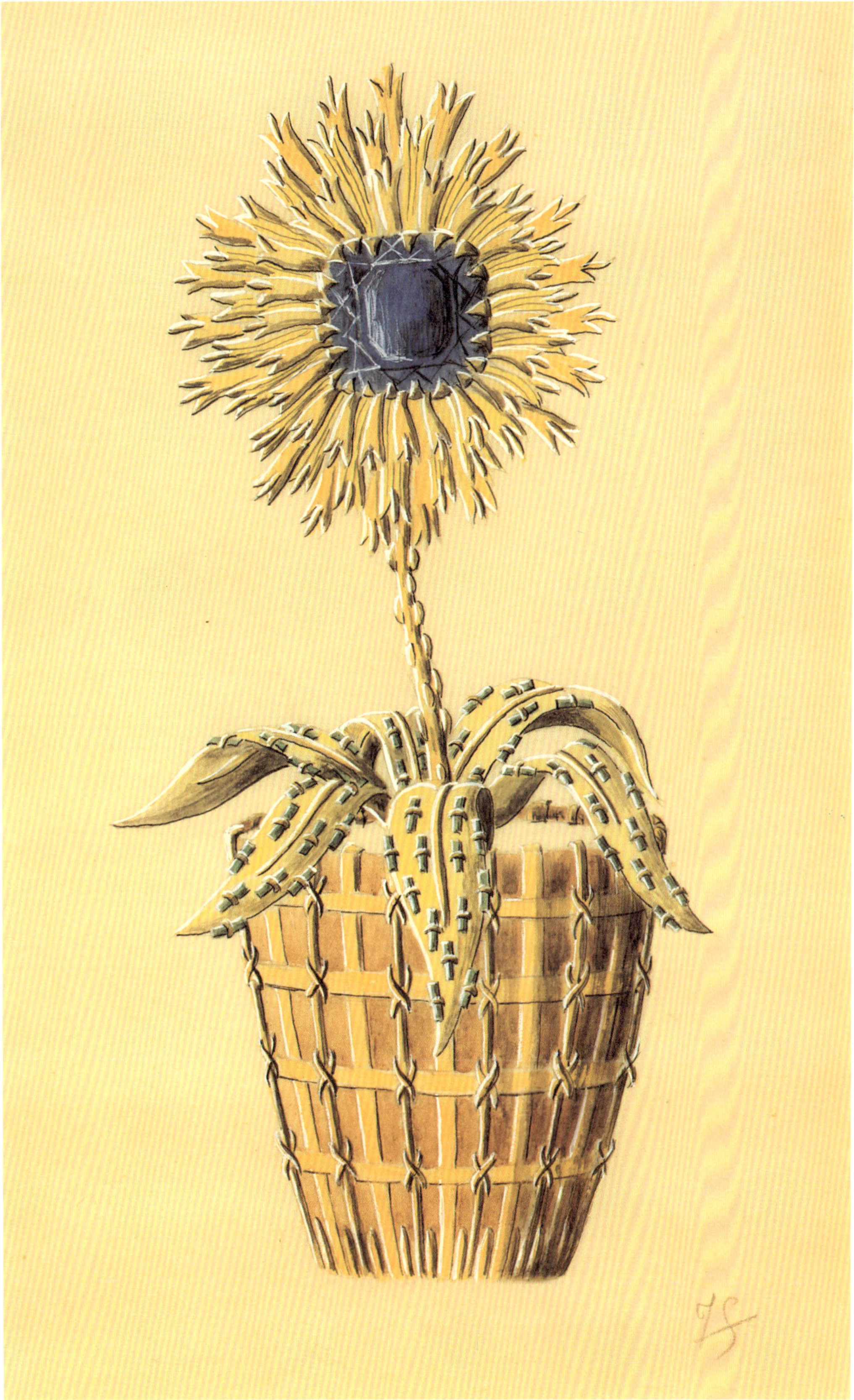

JEAN SCHLUMBERGER,
Pot de fleur,
bodycolour

62. François Vauquier (1621–1686) *and other artists*

Collection of engraved and etched prints of flowers and insects.

29 × 20 cm. 97 leaves composed of XV series as follows: (I) title-page and 11 plates numbered 1–11; (II) title-page and 6 plates numbered 2–7; (III) 6 plates numbered 1–2, 2 *bis*, 3, 3 *bis*, 4; (IV) 6 plates numbered 1–6; (V) 2 plates numbered 1–2; (VI) 22 unnumbered plates; (VII) 6 plates numbered 1–6; (VIII) 11 plates numbered 2–12; (IX) 5 plates numbered 2–5, 5 *bis*; (X) title-page numbered 1, followed by 5 plates numbered 2–6; (XI) 3 unnumbered plates; (XII) 8 unnumbered plates; (XIII) 1 unnumbered plate; (XIV) 2 plates numbered 2 and 4; (XV) 1 title-page.

BINDING: Mottled calf with gilt spine bearing the title 'FLEURS'.

PLATES: (I) Engraved title-page showing a balustrade with 'Livres de Fleurs' written on the central pier, from which hangs an elegant floral wreath. At the foot of the balustrade: 'Par Vauquier. A Paris chez De Poilly rue St Jacques a limage St Benoist. C.P.R.' This series consists of 11 engraved plates of bouquets and urns of flowers (pls. 1, 3, 4, 6, 7, 8, 10, 11 with ruled frames).

(II) Engraved title-page with a flower urn on a balustrade decorated with flower festoons, on which is written: 'Vases de Fleurs Propre pour Peintres, Brodeurs & Dessinateurs'. At the bottom: 'Vauquier sculp. De Poilly ex. cum Privil. Regis. rue St Jacques, à l'Image St Benoît.', followed by 6 engraved plates of urns of flowers, all with ruled frames. In plate 5 the name of each flower is given.

(III) 6 etched plates with ruled frames, each depicting an urn of flowers with at the bottom 'A Paris chez J. Mariette rue St Jacques aux Colonnes d'Hercule'.

(IV) 6 etched plates of flower urns, with ruled frames and 'A Paris chez Mariette rue S. Jacques aux Colonnes d'Hercule'.

(V) 2 engraved plates of flower bouquets, signed by Vauquier, with 'De Poilly ex. cum Privil. Regis rue St Jacques, à l'image St Benoît'.

(VI) 22 engraved plates of bouquets with ruled frames, most of them signed by Vauquier, with 'A Paris chez N. Langlois rüe St Jacques a la Victoire Avec Privilege du Roy'.

(VII) 6 etched plates of flower bouquets, unsigned, with 'Chez J. Mariette rue St Jacques aux Colonnes d'Hercule'.

(VIII) 11 engraved plates of bouquets, with ruled frames, most of them signed by Vauquier, with 'De Poilly ex. C.P.R.' All the plates also give the names of the flowers.

(IX) 4 engraved plates with ruled frames, unsigned.

(X) Engraved title-page with insects. In a cartouche 'Muscarum Scarabeorum Verminumq[ue] Varie Figure & Formae omnes primo ad vivum coloribus depictae & ex Collectione Arundelian. Hermann Weyen execudit A Paris rue St Jacque à lemage S. Benoist pres la Poste. Henri le Roy fecit Æ 72 1651'. Followed by 5 etched plates of insects.

(XI) 3 engraved plates of insects, cut out and laid down.

(XII) 8 engraved plates of flower urns, bouquets, and a garland, signed by Franz Ertinger and C. Galle, cut out and laid down. One plate bears the date '1676'.

(XIII) 1 etched plate with a ruled frame, cut out and laid down. It is signed 'I. Baptiste [Monnoyer]' with the inscription 'Poilly rue St Jacques a la belle image'.

(XIV) 2 etched plates of flower urns with ruled frames, unsigned, with the inscription 'A Paris chez J Mariette rue St Jacques aux Colonnes d'Hercule'.

(XV) Engraved title-page from series (X) with butterflies and insects, and the title 'Naade Naatúúr Getekende Beessiens. Nieúwlyks úytggeven door Cornelis Danckerts tot Amsterdam met Privelegie' presented between the two butterflies at the top of the page. Laid down on the inside back cover is an engraving of a cartouche with an allegorical figure representing Autumn and a ribbon on which is written 'Autumnus'. Below appears: 'M. Heylbrouck jnventor et Sculpsit'.

REFERENCE: Berlin, Staatliche Museen, Kunstbibliothek, *Katalog der Ornamentstichsammlung der Staatlichen Kunstbibliothek*, Berlin, 1939, cat. 4432.

FRANÇOIS VAUQUIER was an engraver, and perhaps also a goldsmith, from Blois; his brother Robert was a painter on enamel. Vauquier lived at a time when the genre of flower painting in France was dominated by Jean Baptiste Monnoyer, an artist whose work was so popular that it was also made available to a wider public in the form of engravings (see no. 46).

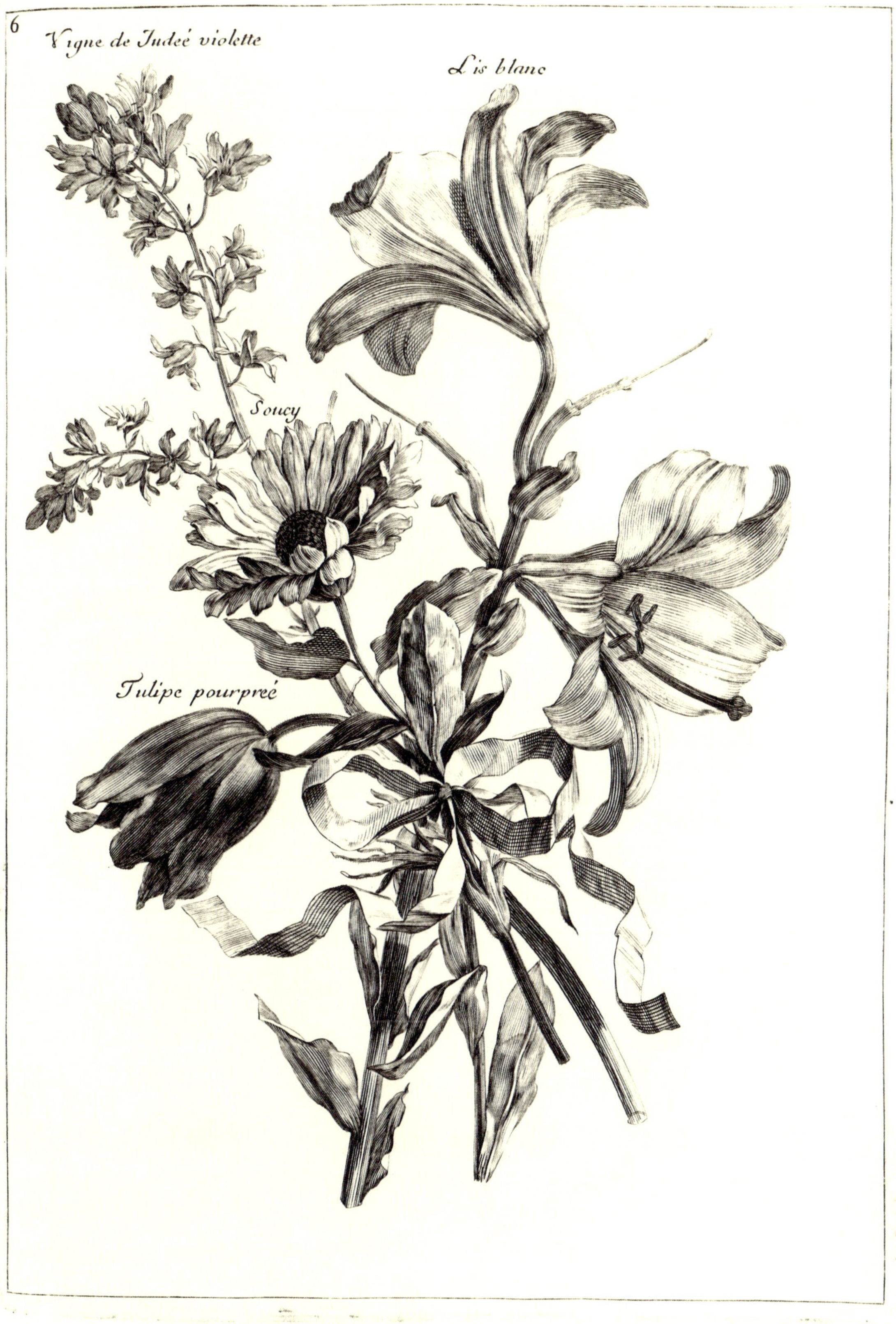

FRANÇOIS VAUQUIER *et al.*, Collection of engraved and etched plates. Bouquet of flowers with a ribbon, plate 6

FRANÇOIS VAUQUIER *et al.*, Collection of engraved and etched plates. Bouquet of *Dianthus* and *Tagetes* with a ribbon

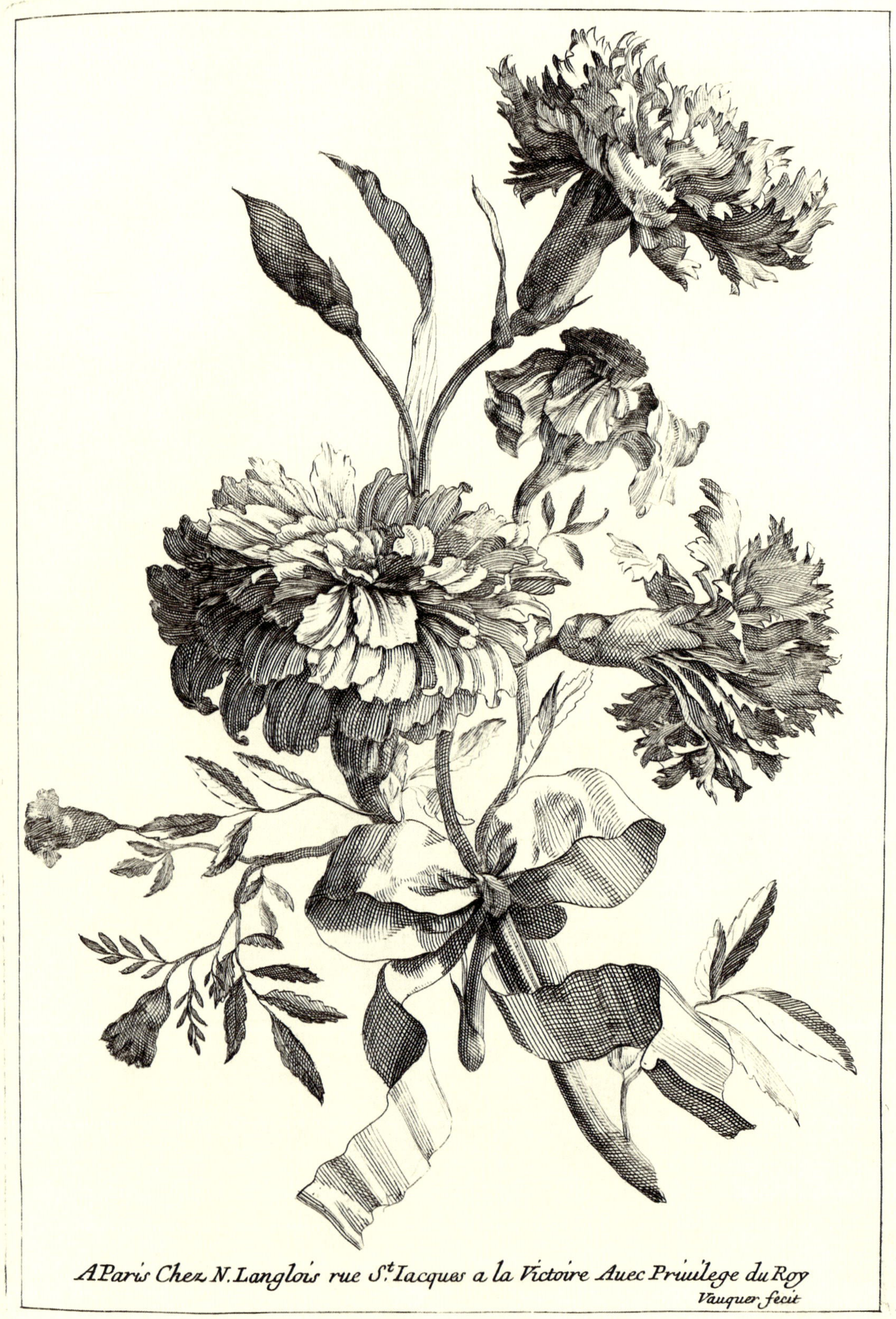

In order to provide connoisseurs, artists and artisans (such as embroiderers, upholsterers and painters on ceramics, who were constantly searching for novel floral motifs) with affordable images of flowers for their work, Vauquier produced many series of engravings, taking advantage of the skills he had acquired engraving plates under the supervision of the great Monnoyer. His own works were issued and sold by some of the most important printsellers in the rue Saint-Jacques in Paris, among them Nicolas de Poilly (1627–96), the brother of François, at the sign 'à la Belle Image', and Nicolas Langlois (died 1640), whose brother was also called François, at the sign 'à la Victoire' (see no. 14).

The engravings in this volume reveal Vauquier to be a technically accomplished artist with considerable compositional skill. The flowers in his elegant bouquets, each one bound with a ribbon, have been exquisitely engraved, Vauquier managing to convey the complex curving lines of each leaf and petal. There are other images in this volume that reflect the influence of Monnoyer, engraved (perhaps at a later date) by Jean Mariette (1660–1742) at his shop 'aux Colonnes d'Hercule' in the rue Saint-Jacques. These are followed by a series of engraved prints depicting flower bouquets in urns against landscape backgrounds by the German artist Franz Ertinger (d. 1710), who worked in Paris for many years. Prints of this kind would have provided yet another useful source for artists and artisans seeking fresh motifs.

Since insects were a popular component in the floral compositions of the period, a series of plates on entomological subjects (series x) was included in this volume. They were engraved by Henri Le Roy (1579–after 1651) and seen through the press by Hermann Weyen, a printer originally from Holland who died in Paris in 1672. Also bound in this volume is a frontispiece with insects (series xv) by the Dutch artist Cornelis Danckerts (see no. 72), who based his design on certain prints by a more famous artist, Wenceslaus Hollar.

This collection of prints, with its miscellaneous floral and insect motifs, represents an interesting example of the volumes of floral images that were produced as a corollary to the immensely popular genre of flower painting in seventeenth-century France.

63. Charles Germain de Saint-Aubin (1721–1786)

[*framed by a floral garland, in gold lettering on vellum*] Receuil [*sic*] De Plantes Copiées D'Aprés [*sic*] Nature Par De Saint Aubin Dessinateur Du Roy Louis XV. 1736–1785.
36.5 × 24.5 cm. Two manuscript leaves tipped-in with biographical information, in part by the hand of the same artist. 258 numbered leaves on paper. On the verso of plate 258 further biographical information in manuscript. A manuscript index of the plates on two leaves (fols. 259–60, recto and verso).

BINDING: Mottled brown calf. Gilt floriated spine with gilt lettering: 'Sepecet De La Natur'.

PLATES: 258 plates of plants, flowers and decorative floral motifs in bodycolour, watercolour, pen and ink, wash, or red chalk. The recto side of each plate is numbered by hand in the upper right corner (fols. 1–258); the name and a description of the plant appears in manuscript, as well as a date between 1740 and 1786. 2 watercolours are pasted in, on fols. 131 and 209; etchings have been pasted on fols. 231v and 232–8.

PROVENANCE: On the death of Charles Germain in 1786, this manuscript passed into the hands of his eldest daughter, Mme Marie Françoise Dounebecq, in whose possession it remained until her death in 1822. In her will she bequeathed it to the engraver Pierre Antoine Tardieu, the husband of Mme Dounebecq's niece, Eugénie Isabelle de Bernard. The manuscript later reappeared in the possession of the Parisian collector Hippolyte Destailleur, in whose auction in 1893 it was sold for 3,000 gold francs.

REFERENCES: Goncourt, 1.410–11; Thieme-Becker, XXIX, 318; Faré, pp. 389–90.

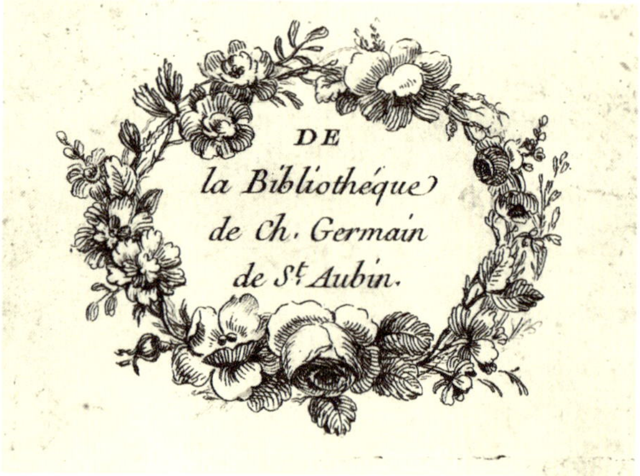

CHARLES GERMAIN DE SAINT-AUBIN, *Recueil de plantes*. An etched bookplate on folio 232

FOR MORE THAN two centuries, from the end of the seventeenth century to the beginning of the nineteenth, the Saint-Aubin family constituted a veritable dynasty of embroiderers, engravers and artists serving the French monarchy. Charles Germain de Saint-Aubin was the eldest son of the Brodeur du Roi, Gabriel Germain, and himself became 'Dessinateur du Roi pour la broderie et la dentelle' to Louis XV, as the title-page of this manuscript proclaims. Other works by Charles Germain at Oak Spring, such as *L'Art du brodeur* and *Premier Recueil de chiffres*, both published in Paris in 1770, confirm that he was one of the most celebrated floral decorators of his time, as do the prints in both *Bouquets champêtres dédiés à Madame la Marquise de Pompadour* and *Bouquets champêtres, dédiés à Madame le Maréchale de Biron*, both published (without date) in Paris by F. Chereau.

Recueil de plantes is absolutely unique, however, for it represents a collection of more than 250 watercolours, bodycolours, sketches and drawings by the artist, gradually accumulated over a period of almost fifty years. Below a frieze of flowers on folio 131 the artist has written: 'Un des quarante mille Dessins de Broderies de Charles Germain de St. Aubin'. The first works in this collection (fols. 257, 258) were produced in 1736, when Germain was only fifteen years of age, while the last paintings date from 1785, just one year before his death. Gifted with a fertile imagination and a bent for naturalistic subjects, in these works we can trace his growing artistic versatility and technical mastery. The documentary value of this manuscript is further enhanced by the autobiographical notes added in the artist's own hand, and other notes on the family supplied later by the relatives who inherited this precious volume.

The Goncourt brothers fully appreciated the importance of *Recueil de plantes*, and described it in great detail in their work on the history of eighteenth-century French art. They noted that a portrait of the artist, drawn in pencil by his brother Augustin and dated 1767, was pasted to the title-page. Unfortunately, both the title-page and the portrait were removed from the volume after its sale by the collector Destailleur for the sum of 3,000 francs in 1893. The two detached leaves are today in the Pierpont Morgan Library (see *Flowers in Books and Drawings*, pp. 96–7, and *French Master Drawings*, p. 136). The original title-page was decorated with a fine Rococo cartouche containing trophies and a variety of musical instruments, artist's tools and gardening implements as well as the inscription 'à la Nature'. It was replaced by the present title-page, decorated with a rather more conventional floral garland.

Recueil de plantes appealed greatly to the refined taste of the Goncourt brothers, and they unerringly focused on some of its most beautiful works, such as the bodycolour painting on folio 46, in which 'à la tige d'un jasmin [actually a wallflower] s'attache, a moitié enroulé, le trompe l'oeil d'une petite estampe en couleur, d'un papier de musique' (p. 410), and folio 68 depicting a convolvulus, which, as they note, bears an annotation by the author: 'Madame la marquise de Pompadour à travaillé à cet bouquet en 1757'.

Germain in fact, enjoyed the patronage of the powerful Mme de Pompadour. She appears to have collaborated with him on the painting of a bouquet of convolvulus, as we can read below the image on folio 68 of this *Recueil*. She even had a special box of watercolours from China sent to him as well as various pieces of porcelain. From this shared love of Chinoiserie must have originated the idea for the engraved *Livre de fleurs chinoises et de caprices*, which the Goncourts cite among the many works by the artist. Germain's talent was greatly appreciated by the aristocracy: he designed a dinner service for Mme du Barry, while a note at the foot of the plate depicting a *Myosotis* (fol. 110) explains that the King's daughter, Mme Clotilde, had sent him a spray of these flowers to paint for a 'portefeuille destiné pour le Roi'.

The works in *Recueil* reflect Germain's wide-ranging interests and his constant experimentation with new styles and techniques. The floral motifs are delicate and refined, as befitted their intended use as embroidery designs; yet an unusual vein of playfulness runs through the collection as well. On the verso of folio 2, for example, the artist mischievously notes that the flowers depicted on folios 140, 149, 156, 215 and 257 are pure inventions, and not real botanical specimens. Nevertheless, the trained eye of the artist is evident in each plate, and the various parts of every flower have been portrayed with minute exactness.

There is a captivating variety to the work in *Recueil*; although it consists for the most part of exquisite studies of single, and occasionally two or three, flowering plants against a blank page, there are also many skilfully composed bouquets, as well as some very striking paintings of flowers against darker backgrounds. Some of the works were executed in wash, and others in pen and wash, or in red chalk. A few include butterflies and other insects, birds or rare sea-shells. There are also six enchanting paintings in which the artist has depicted next to his flower a *trompe l'œil* 'page' bearing

CHARLES GERMAIN DE SAINT-AUBIN, *Recueil de plantes*. Double carnation (*Dianthus caryophyllus*), folio 117

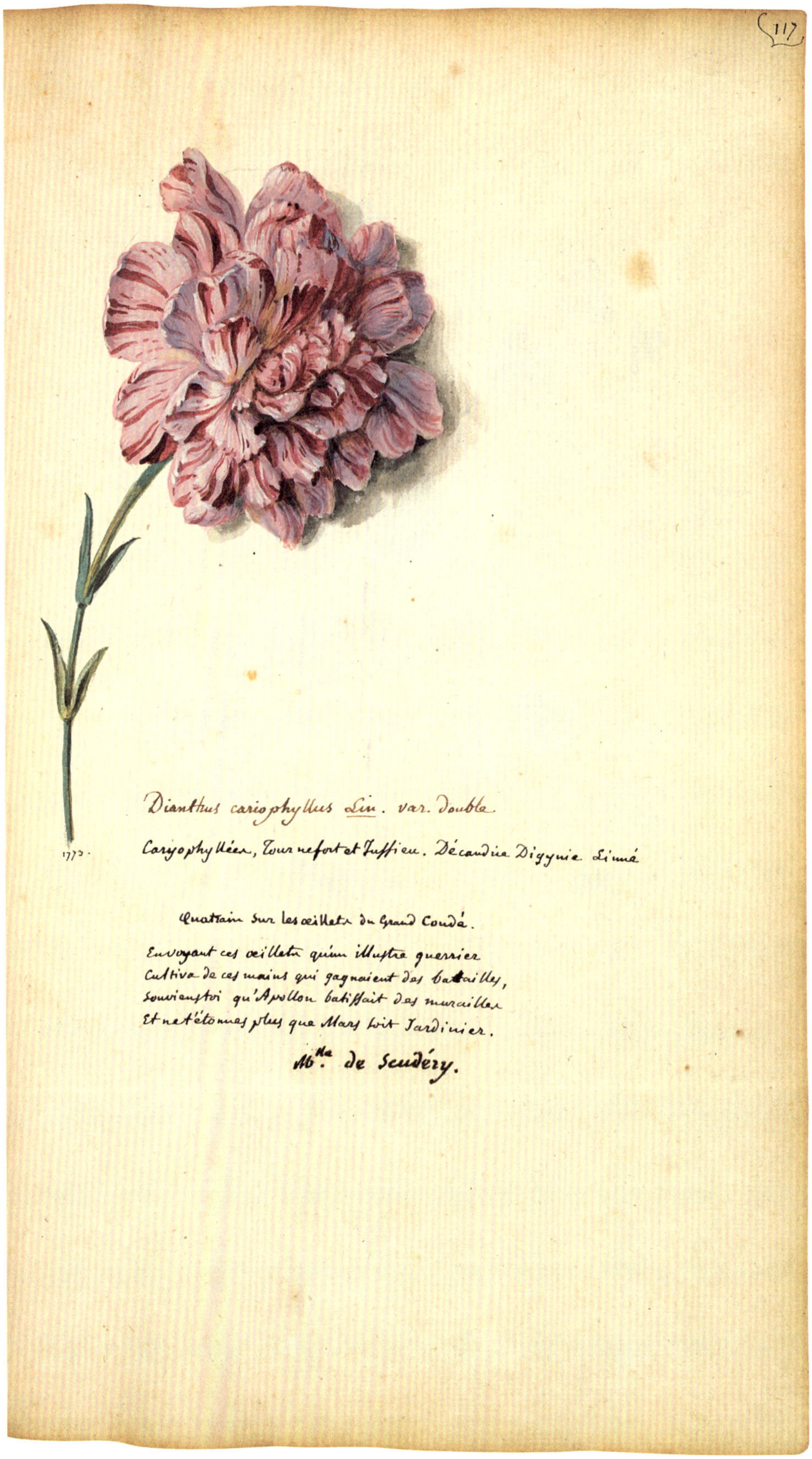

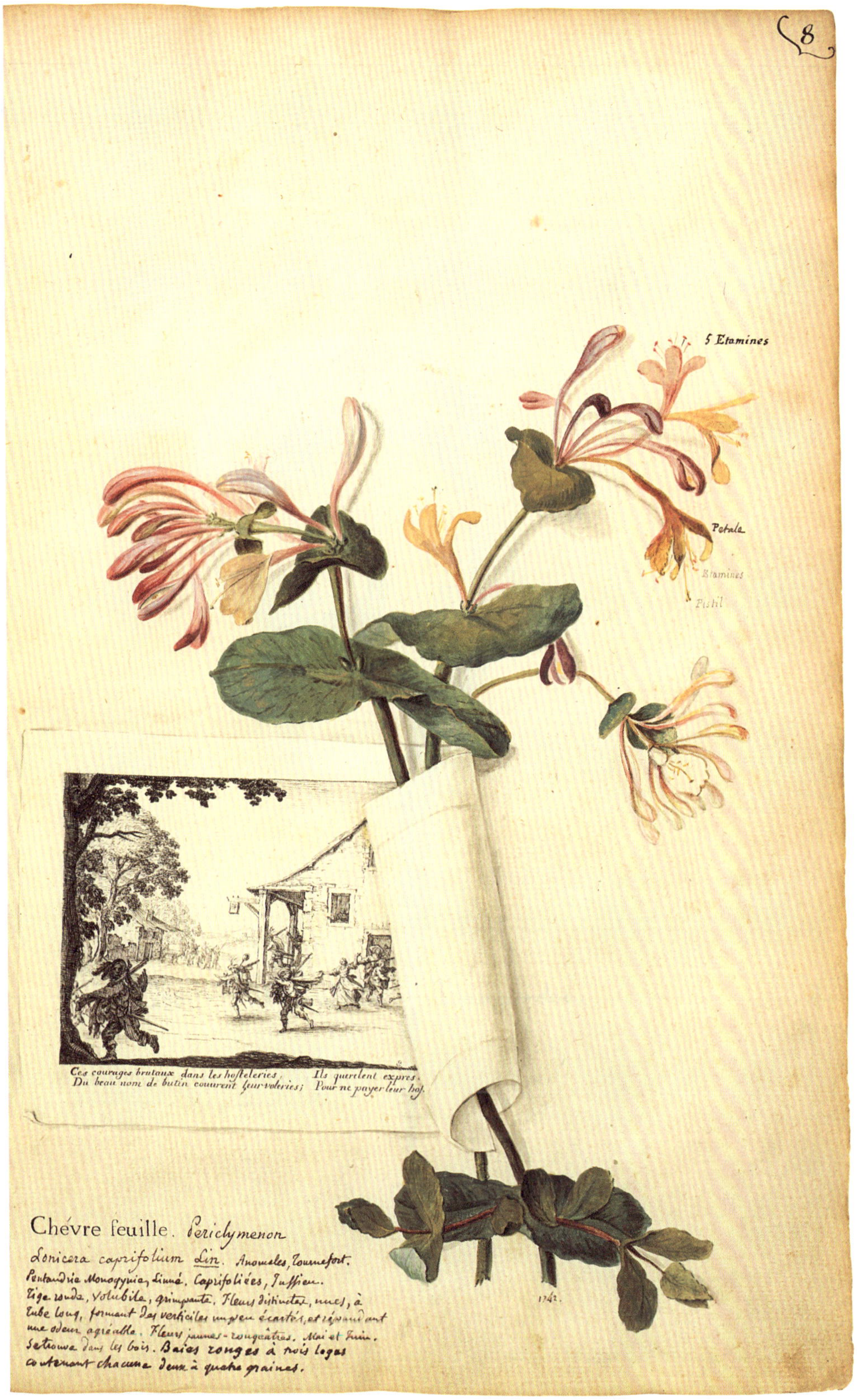

CHARLES GERMAIN DE SAINT-AUBIN, *Recueil de plantes*. Honeysuckle (*Lonicera caprifolium*), with a *trompe l'œil* drawing of a print, folio 8

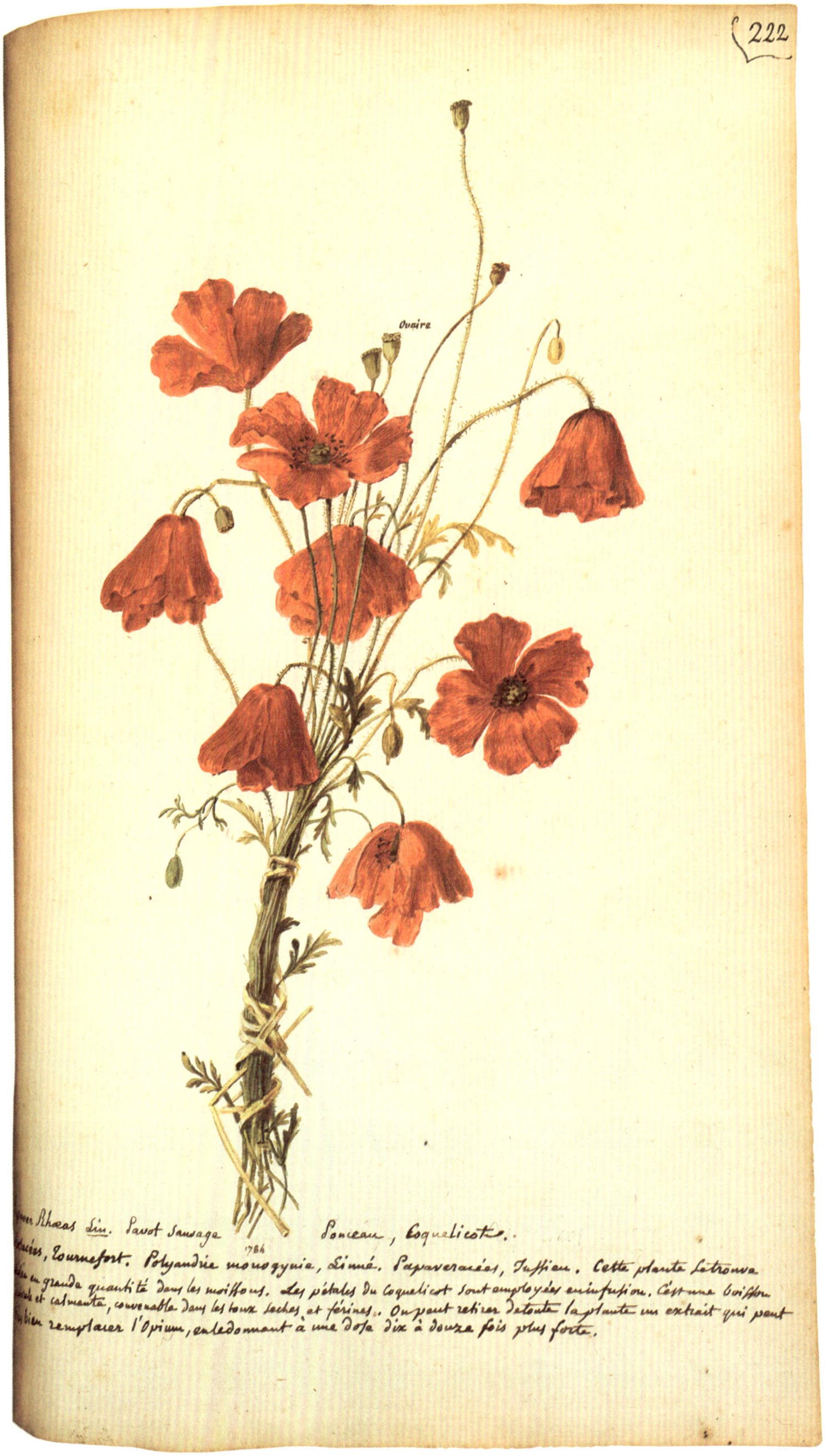

CHARLES GERMAIN DE SAINT-AUBIN, *Recueil de plantes*. Common field poppies (*Papaver rhoeas*), folio 222

another sketch or engraving. On folio 2 he has added a landscape in watercolours of the château of Choisy; on folio 8 there is a scene showing soldiers pillaging an inn in a *trompe l'œil* 'engraving' reminiscent of *Les Misères de la guerre* by Jacques Callot. Folio 46 has a sheet of music partially rolled around the stem of a wallflower (the composition that so pleased the Goncourts), while on folio 55 he has introduced a 'pen and ink sketch' of a Rococo vase. Folio 56 has a 'watercolour' depicting part of the garden of the hôtel de Lesdiguieres in Paris (see *An Oak Spring Hortus*, fig. 33), and, finally, folio 209 includes a depiction of a monkey, wolf and fox, perhaps illustrating a fable by La Fontaine.

On folio 79*v* there is a cipher alphabet, while examples of Germain's engraved work have been glued to other pages—two elegant ex-libris (fols. 231*v* and 232), and some small etchings from his work *Fleurettes* (fols. 233–8). In one plate depicting two insects (*Ephippigere ephippiger*), the artist has fancifully traced the path of their flight, recalling another curious work by him, *Essai des papilloneries humaines*, which would later inspire the work of Grandville (see no. 99).

The final pages of this 'floral diary' contain some penned reflections by its author, who states without false modesty that many of his works could be considered quite the equal of any contained in the Cabinet des Estampes du Roi by the greatest French masters of the art of naturalistic painting. He also confides that the time spent working on the paintings in this volume could be counted as among the happiest moments of his life: 'Il y a 60 volumes in folio de fleurs et plantes Peintes sur vélin en belle mignature par Joubar, Aubriet, Robert, et madame Basseporte au Cabinet des Estampes du Roy. Sans vanité quelques unes de celles cy sont aussi belles. J'ai pendant quarante ans travaillé à ce volume comme on peut juger par les dattes de chaque bouquet, seulement à mes moments de loisir et de desoeuvrement et quand jay eù la nature sous la main; quoy que ces études ne vallent pas grand chose je n'ai guerre passé de moments plus tranquilles et plus délicieux que ceux que j'ai donné à cet petit ouvrage.'

64. JOHN EDWARDS (1742–*after* 1812)

[*depicted in an oval frame against a pale grey background is an urn of flowers standing on a rectangular pedestal. Three winged putti are hanging a floral garland, while on the ground lies a palette with brushes. On the right some trees can be seen in the distance. Carved into the pedestal, in an oval framed by a laurel wreath, are the words*] A Collection of Flowers, drawn after Nature, & disposed in an Ornamental & Picturesque Manner, by J. Edwards, F.S.A. [*at the foot of the frame*] John Edwards invt delt et sculpt Published as the Act directs Jany [. . .]

54.8 × 37.5 cm. 80 leaves of plates not sequentially numbered (some are unnumbered). Guard leaves.

BINDING: Brown calf with gilt borders of acanthus leaves. On the spine in gilt lettering 'Collection of Flowers Edwards'.

PLATES: Stipple engraved title-page, followed by 79 etched plates of flowering plants and ornamental elements, sometimes presented in oval frames, all hand coloured in bodycolour. Each plate is signed and bears a date somewhere between 1783 and 1795. All of the plants are labelled with their English names. On plate I: 'John Edwards fecit. Published as the Act Directs Septr 1st 1783'.

REFERENCES: Dunthorne, 105; Nissen 579; Henrey, III.673.

JOHN EDWARDS was an engraver and printer and one of the most famous English artists to specialize in the genre of flower painting. From the very beginning of his career he devoted himself to floral and ornamental subjects. His work was exhibited at the Royal Academy, and at the Society of Artists, of which he was a member, as the initials FSA following his name on the title-page indicate. Edwards lived in London until *c.* 1778, when he moved to Surrey. In addition to his flower paintings and botanical illustrations, he produced designs for textiles, especially the cotton calico that was manufactured in great quantities in England and exported all over the world. Some of his designs were included in an exhibition held at the Victoria and Albert Museum in London in 1960 (Henrey, II.19).

In 1769 Edwards published his most famous work, the *The British Herbal.* It was reissued in 1770 as *Edwards's Herbal* and again five years later as *A Select Collection of One Hundred Plates.* In its graceful illustrations the artist allowed his imagination free rein, feeling no need to consult a botanist or to tailor his drawings to the exacting requirements of science. This highly personal approach can also be seen in *A Collection of Flowers, drawn after Nature*, in whose plates, despite the claims of the title, a whimsical and purely decorative vision prevails.

In the traditional herbal, specimens were carefully presented in a rigidly frontal position to allow one the clearest possible view of every detail. In the plates to *A Collection of Flowers*, which were all drawn, etched and coloured by Edwards himself, the flowers are gathered in tempestuous bouquets tied with brightly coloured ribbons, or else placed on the page in unusual, yet graceful, arrangements. A charming example is provided by the illustration of 'Eastern poppies', whose serpentine stems seem hardly able to bear the weight of the flowers' heavy white corollas bordered with red, or that of the *Dodecatheon meadia* (Shooting-star), an American plant brought to Europe in 1744, whose numerous tiny violet flowers with their pointed styles hang gracefully from long pedicels that spring in a cluster from the slender stem of the plant. In many cases the plant is depicted in an oval frame—an extremely popular motif at the end of the eighteenth century—against a light-grey, stipple-engraved background.

The decorative nature of *A Collection of Flowers*, evident in the flower studies, emerges even more clearly in the numerous monochrome plates, in which classicizing elements—tripods, urns, vases and architectonic structures—appear in fanciful compositions together with flower garlands, birds, gardening tools and elements from the arts, all hand coloured in bright hues. Thus John Edwards depicted the natural world from the perspective of the artist, revelling in its infinite variety of forms and colours, in a vision of Neoclassical elegance with a lingering trace of the Rococo.

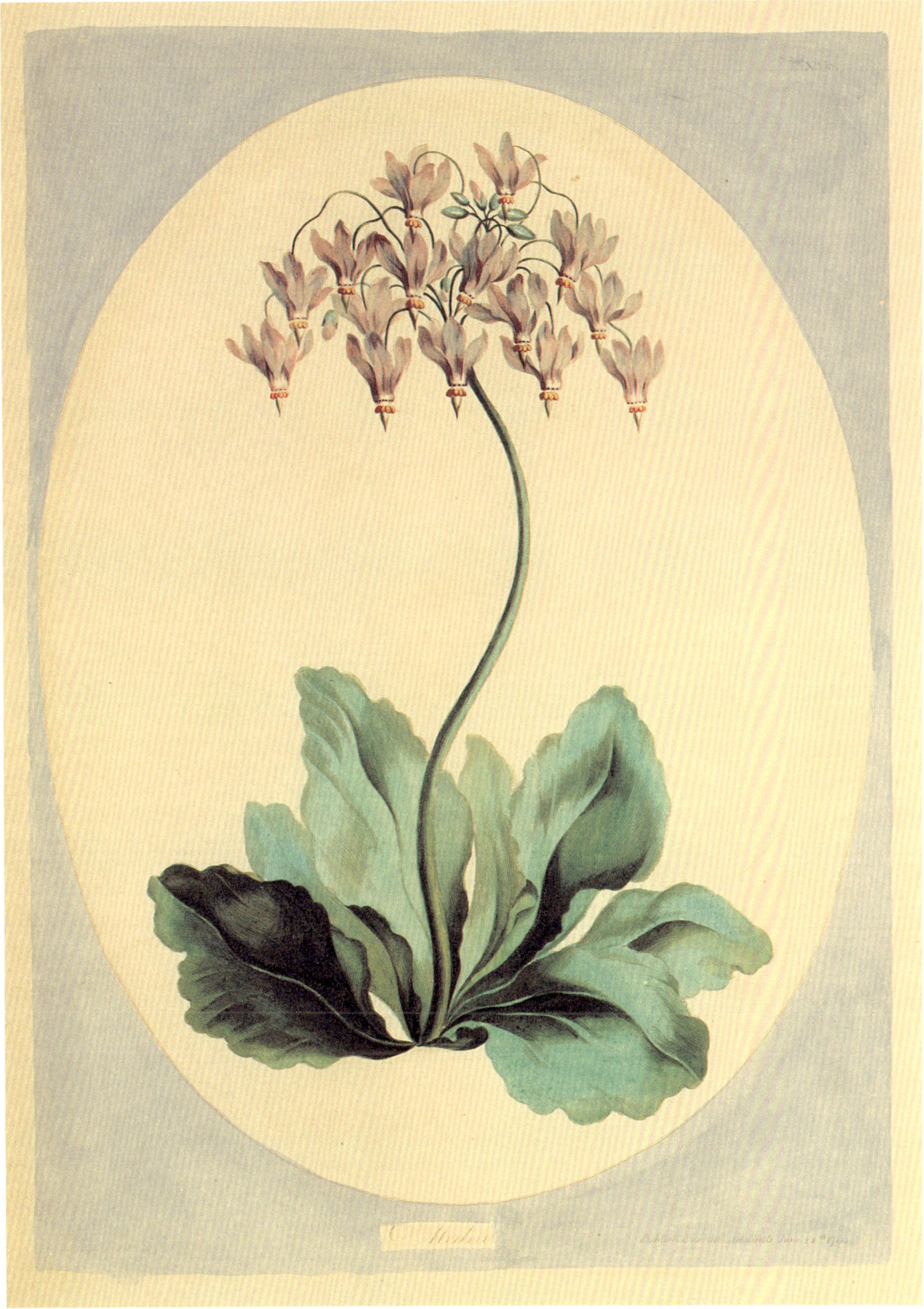

JOHN EDWARDS,
A Collection of Flowers.
Shooting-star or American cowslip (*Dodecatheon meadia*)

65. JEAN LOUIS PRÉVOST *the younger* (*c.* 1745–*c.* 1810)

Collection des Fleurs et des Fruits, Peints d'après Nature, par Jean-Louis Prevost, et Tirés de son Porte-Feuille, Avec un Discours d'Introduction sur l'usage de cette Collection dans les Arts et les Manufactures, suivis d'un Précis historique sur l'Art de la Broderie, et d'une Vue générale de toutes les Manières de peindre depuis l'Antiquité jusquà nous; Par M. Gault-de-Saint-Germain. [*engraved vignette*] Paris, Chez Defer, Marchand d'Estampes, quai Voltaire, N° 19. De L'Imprimerie de Chassaignon.

55.5 × 37 cm. *i–iv* i ii–iv 1–19 *20–22* p. 48 leaves of plates.

BINDING: Half-bound in red buckram and leather, gilt spine with the title 'Collection des Fleurs and Fruits'. Bound with Gerard van Spaendonck [*Fleurs dessinées d'après nature*, 1801], 24 stipple engraved plates, without text. Also with three additional plates by Pancrace Bessa (1772–1835) and four by Anne Vallayer Coster (1744–1818).

PLATES: 48 stipple-engraved colour plates numbered 1–48. At the bottom of each plate: 'J. L. Prevost invenit' (or 'inv.'), 'L. C. Ruotte sculp.', and 'A Paris, chez Valquin, Md d'Estampes, grande cour du Palais du Tribunat, N- 20.' Guard tissues.

REFERENCES: Dunthorne 229; Pritzel 7332; Nissen 1568 (these three references give imprint date Paris, Vilquin, 1805); Pischel; Coats, p. 66; Faré, pp. 285–98; Stafleu and Cowan, no. 8319; *A Magnificent Collection*, no. 287.

IN FRANCE during the Napoleonic period many talented floral painters flourished in addition to the celebrated Redouté. Among these the Prévost brothers, Jean Jacques the elder and Jean Louis the younger, merit particular attention. Jean Louis Prévost was born at Nointel, close to Paris, and studied with the floral painter Nicolas Bachelier. He became a member of the Académie de Saint-Luc, an institution that sought to rival the Académie Royale de Peinture, and together with his brother Jean Jacques showed many pictures there.

While still a young artist, and in collaboration with his brother, Jean Louis Prévost undertook the project of illustrating in twelve manuscript volumes all the species of flowers that could be found growing in the garden of the Château de Celles. The proprietor, Monsieur Roussel, paid them 20,000 *thalers* for this work, which in 1782 was acquired by the Muséum d'Histoire Naturelle in Paris. Prévost was a gifted botanical illustrator, although he considered himself primarily a painter and decorator. Flowers and plants were in fact a favourite source of decorative motifs in the nineteenth century, and many of his bodycolours were intended to serve as designs for the production of porcelain, textiles and embroideries. The fact that they were destined for such uses meant that the artist's primary concerns were aesthetic rather than scientific. Thus in his exuberant and sophisticated bouquets Prévost did not hesitate, for example, to bring together flowers that in nature bloomed in entirely different seasons.

In addition to numerous flower paintings, skilfully composed and often imbued with symbolic meaning (Faré, p. 288), Prévost produced the original watercolours for *Collection des fleurs et des fruits*, which was published *c.* 1805. All the paintings for this work were brilliantly engraved in stipple by Louis Charles Ruotte (1754–*c.* 1806) with the exception of plate 5, which was undertaken by Alexandre Chaponnier (1745–1806). Ruotte was a highly versatile artist who had worked in

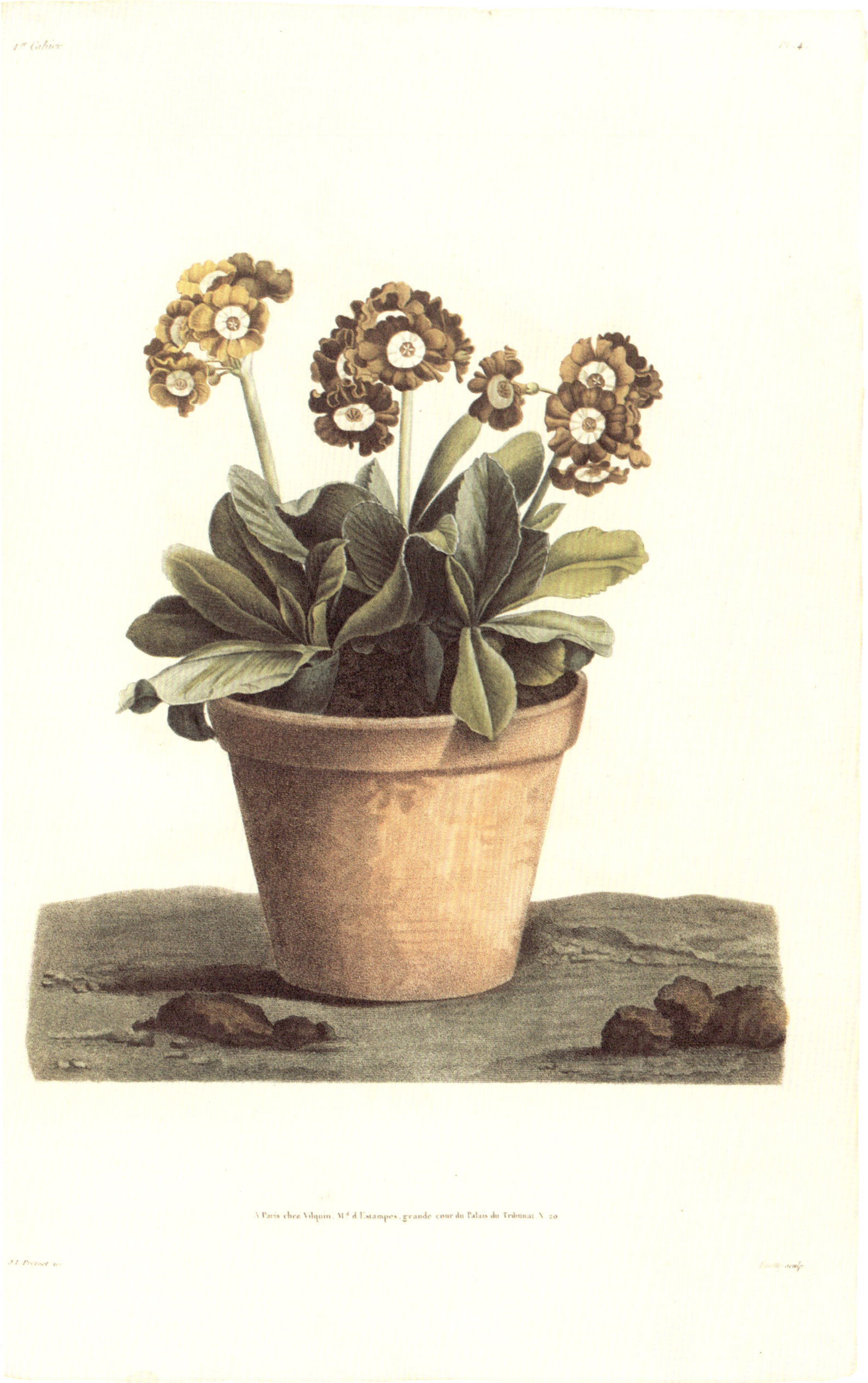

JEAN LOUIS PRÉVOST the younger, *Collection des fleurs et des fruits.* Three auriculas (*Primula auricula*) in a clay pot, plate 4

JEAN LOUIS PRÉVOST the younger, Bodycolour drawing of two anemones (*Anemone coronaria*), two *Primulas*, and a lily-of-the-valley (*Convallaria majalis*)

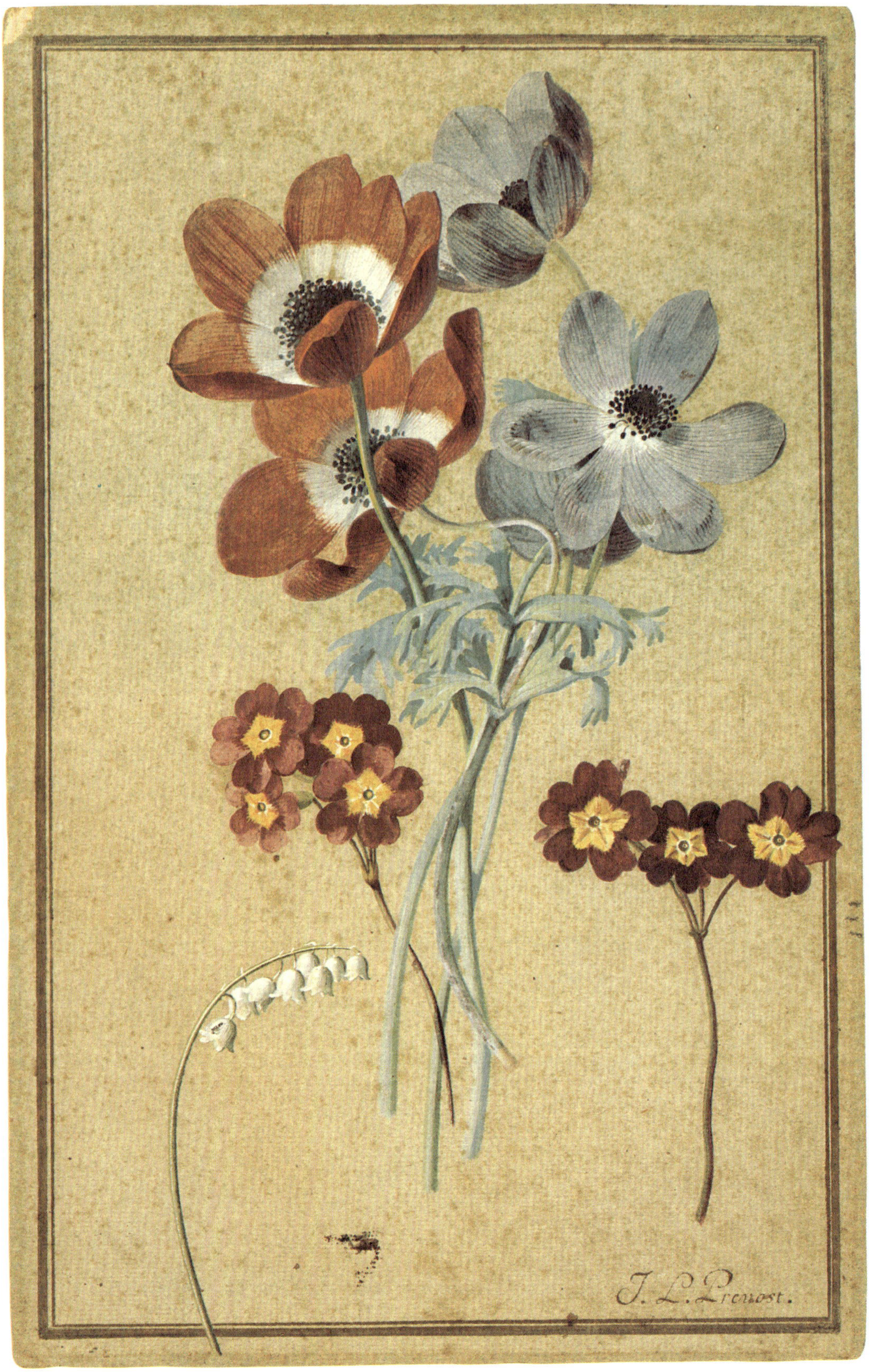

London with Bartolozzi; he was also a master of the new techniques of stipple engraving and colour printing.

The forty-eight plates in *Collection des fleurs et des fruits* comprise a pleasantly varied series of floral bouquets, and arrangements of soft fruits on plates. Each work is skilfully composed and imbued with a transparent luminosity, culminating in the iridescent drops of water that seem about to roll off the leaves. The illustration of an amber-coloured auricula (pl. 4), a flowering alpine plant that was extremely popular in this period, is masterly in its realism, while the stippled plate creates a fine interplay between the luminous highlights and velvety shadowing.

The collection is preceded by an introduction written by the portrait painter and miniaturist Anna Rajecka, subsequently Mme Gault de Saint-Germain (*c.* 1760–1832). She observes that, in the art of floral painting, artists such as Prévost 'à été forcé à suivre la marche de la nature pour en sesir les variétés; il a cherché a en pénétrer les secrets; il est devenu naturaliste, et il s'est appliqué à découvrir les espèces'. 'L'Oeuvre de Prévost', Mme Gault de Saint-Germain continues, 'offrira . . . des modèles précieux pour les manifactures de porcelaines, de teintures et d'étoffes d'ameublemens et de parures'. Aware of the practical ends for which this work was destined, she also wrote three short treatises to accompany the plates, a *Précis historique sur l'art de la broderie*, a *Vue génerale sur toutes les manières de peindre depuis l'antiquité à nous* and, finally, a *Catalogue des couleurs propres à peindre à l'eau et à l'huile.*

Collection des fleurs et des fruits was originally published in twelve parts, each comprising four plates, and could be bought from the printseller Vilquin at 20 Grand Cour du Tribunat. The collection at Oak Spring also has a signed painting in bodycolour by Prévost, *Anemones, Primrose, and Lily of the Valley.*

66. JEAN BAPTISTE HUET *the elder* (1745–1811)

Iris germanica. 33 × 25 cm. Oil on paper, laid on stiff paper.

JEAN BAPTISTE HUET THE ELDER probably received his first lessons in painting from his father, Nicolas Huet the elder, an artist who specialized in still-lifes and paintings of animals. He then studied under C. Renou, and in 1764 was recognized as the most promising student of J. B. Leprince. Before setting himself up as an independent artist he also worked for a period with François Boucher and with the animal painter Charles Dagomer. Success came quickly, for Huet's paintings, which ranged from airy landscapes to *scènes galantes*, were very popular with wealthy art collectors in Paris. He was accepted into the Académie de Peinture in 1769 at the age of twenty-four, and showed his works regularly at the annual Salons until 1802. Huet was an accomplished artist in a variety of media, from drawing to etching and from painting in watercolours to painting in oils.

JEAN BAPTISTE HUET the elder, Garden or bearded iris (*Iris germanica*)

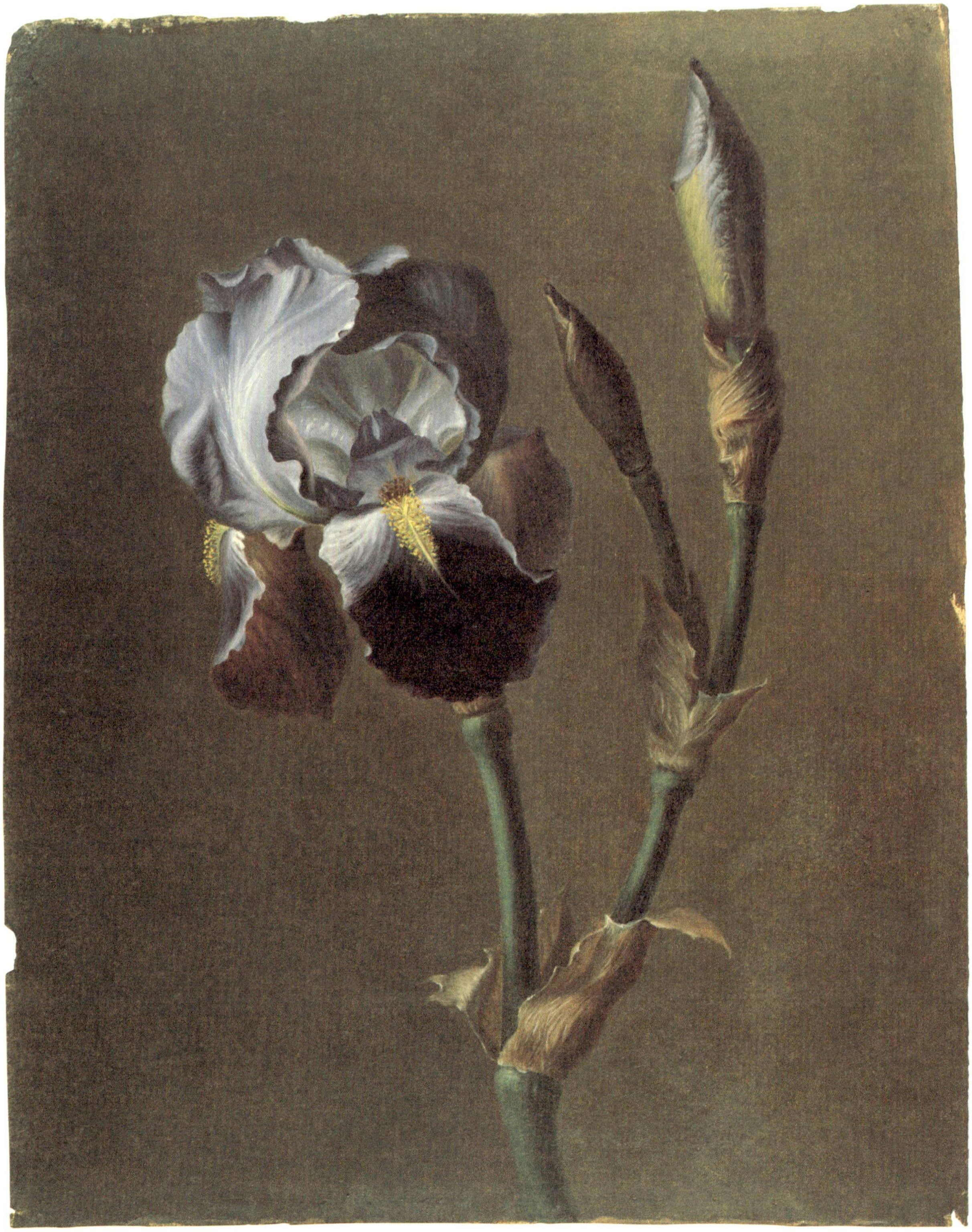

In 1770 he also began to produce designs for the great tapestry manufacturers, Gobelins and Beauvais, for his elegant still-lifes and landscapes were eminently suitable for translation into tapestry designs.

In this painting of the *Iris germanica*, Huet demonstrates his mastery of the difficult technique of painting in oils on paper. A magnificent flower and two buds are shown highlighted against a brown background. With dense, deft brushstrokes the artist has managed to recreate the velvety texture of the blue-violet petals, the papery bracts, and even such details as the minute, hair-like barbs at the ends of the petals. This plant, which since antiquity had been known for its medicinal properties, was adopted in the medieval period as a symbol of the Virgin Mary; a fine example may be seen in Dürer's *Madonna with an Iris* (National Gallery, London). Huet instead concentrates on the purely decorative aspects of the flower, exhibiting his technical bravura in a formal exercise of colour and form, without, however, neglecting to depict with careful precision the smallest details of each flower. Several branches of lilac painted also in oil on paper, dated 1786, represent another important example of the artist's work in floral painting. (Mitchell 1992, p. 39).

67. Antoine Berjon (1754–1843)

Bouquet of zinnias and tagetes. 47.5 × 31 cm. Pastel on blue paper. Laid on a paper frame with gold borders, matted, and housed in a white leather and decorated paper case with three other pastel drawings.

The work of Antoine Berjon is of particular interest because it was closely tied to the textile manufacturing industry that was one of the glories of nineteenth-century France. Born in Lyon and originally destined for the medical profession, Berjon soon abandoned this path to study art with the sculptor Michel Perrache. Lyon, a prosperous city renowned since the Middle Ages for its silk industry, was also historically one of the most lively intellectual centres in France, and boasted an art school that during the nineteenth century shone with particular lustre (Guillot).

Berjon first worked in a silk factory as a *dessinadier*, or designer, of textile patterns. He visited Paris a number of times in connection with this work, and there formed close ties with members of various artistic circles. He was thus inspired with larger ambitions and moved to the capital, hoping to make his fortune. For several years he scratched a bare living turning out flower pictures and still-lifes in oils, pastels, watercolours and drawings 'aux trois crayons', occasionally showing his work at the annual Salons. He did not, however, manage to make his mark in Paris, and, tired of living in penury, returned to Lyon, where he immediately found a position at a factory producing embroidered textiles.

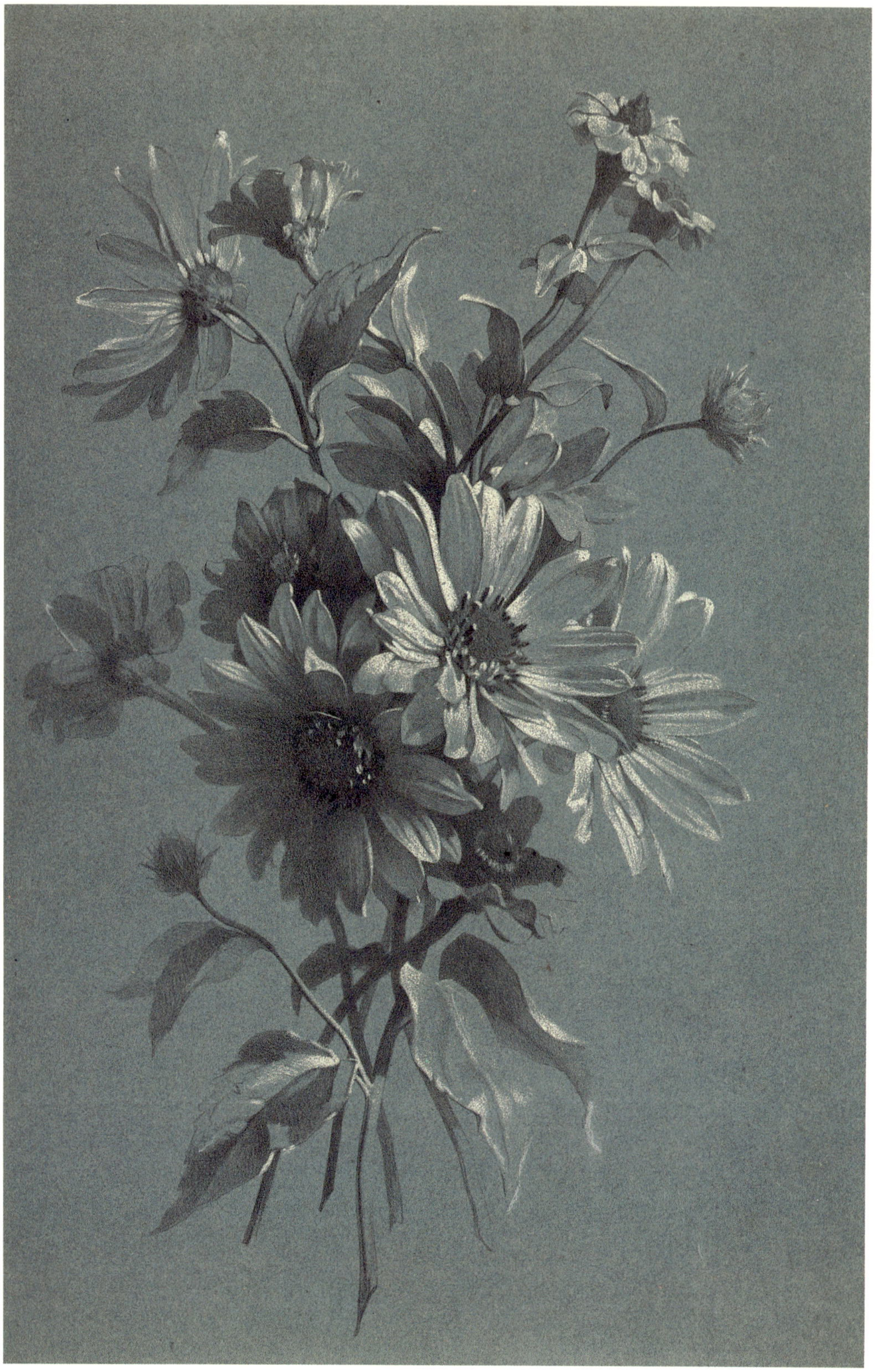

ANTOINE BERJON,
Bouquet of *Chrysanthemum* and *Tagetes*

In 1810 Berjon was appointed professeur at the local Ecole des Beaux-Arts, and he enthusiastically dedicated himself to teaching for the next thirteen years. In this period the Ecole specialized in training young artists for the textiles industry, in particular in the creation of floral motifs and the technique of *mise en carte*, the design and preparation of squared pattern cards for the industry's mechanized looms (see Hardouin-Fugier and Grafe 1989, pp. 25–6). Berjon's teaching contributed to the formation of a series of artists whose abilities were much appreciated, both in Lyon and abroad. Berjon himself was an eclectic artist who also painted portraits, pictures of animals and costume designs; he even experimented with a new engraving technique. Contemporary sources describe him as an unpleasant character, however, and his death was barely noted in the local newspapers.

In addition to his stylized floral patterns for textiles, Antoine Berjon is known for his floral still-lifes, which drew upon the tradition of the great seventeenth-century French floral painters, but which he infused with a freshness of inspiration and a figurative vocabulary all his own (Faré, p. 357). The four pastel drawings at Oak Spring are stylistically quite similar to an exquisite bouquet in black chalk conserved in the Musée des Beaux-Arts, Lyon (Hardouin-Fugier and Grafe 1989, p. 25), and constitute an important example of this talented artist's work in pastels. The bouquet of chrysanthemums and tagetes, in particular, stands out with a strikingly decorative effect against the blue of the paper. The delicate petals seem to vibrate with colour under the bright light falling from the right, shimmering with reflection highlighted in small, sure touches of white paint.

68. Johan Samuel Arnhold (1766–1828) *and* Johann Stephan Capieux (1748–1813)

Meissen pattern-book.

24 × 18.5 cm.

BINDING: Modern calf binding.

PLATES: 43 etched, numbered, and hand-coloured plates, comprising 37 of flowers, 5 of birds and 1 of fishes. Every figure in each plate is also separately numbered. Each plate is signed 'Arnhold del.' and 'Capieux sculps.' (or 'sc.'), followed by the date '1798' or '1799'. Interleaved with the etchings are leaves containing numbered manuscript notations for the figures.

THIS PATTERN-BOOK was compiled for the Meissen porcelain factory by one of their *Zeichenmeistern*, Johan Samuel Arnhold, who executed the drawings for the 226 floral and animal motifs and was perhaps also responsible for the hand colouring of this copy. His designs were transferred to plates by Johann Stephan Capieux, a talented etcher and engraver who illustrated many works of anatomy, botany and minerology. The names of both artists appear on all the plates, which were produced between 1798 and 1799. Arnhold received his initial artistic training at the school established by the porcelain factory, and then worked with the portrait painter Christian Lindner, who was also a *Zeichenmeister* for Meissen. Arnhold eventually became quite well known as a landscape painter and for his paintings on porcelain.

The Meissen porcelain works, situated just twelve miles from Dresden, was founded in 1710. It quickly became famous for dinner sets, snuff boxes and perfume bottles of fine white porcelain made from the local kaolin and exquisitely decorated by hand. The floral patterns created at Meissen were particularly admired. These initially featured highly stylized oriental motifs (*Indianische Blumen*), but were gradually replaced by more naturalistic patterns incorporating their charming *Deutsche Blumen*. One of the first great *Zeichenmeistern* for Meissen was the painter Johann Gottfried Klinger, who worked for them from 1726 to 1746 and provided the models for many of their first *Deutsche Blumen* (W. B. Honey, *German Porcelain*, London, 1951, p. 13).

The 'German flower' motifs designed at Meissen were soon copied by porcelain manufacturers all over Europe, including Vincennes in France and Chelsea in England. The Chelsea porcelain works was already celebrated for its typically English floral patterns, many of them probably inspired by flowers to be seen in Chelsea's famous botanical garden and others drawn from the plates designed by Ehret for two works, the *Plantae et papiliones* (see no. 50) and Philip Miller's *Figures of Plants* (P. Synge-Hutchinson, 'G. D. Ehret's Botanical Designs on Chelsea Porcelain', *The Connoisseur*, CXLII, October 1958, pp. 88–93). Indeed, this period coincided with one of the most glorious moments in the history of botanical illustration, and illustrations for texts by some of the greatest artists of the day must have provided an inexhaustible source of models for designers. Many of the Meissen designs, for example, can be traced directly to *Phytanthoza iconographia* by the pharmacist Johann Wilhelm Weinmann, which was published in four volumes in Regensburg in 1737–45 (Baer, p. 83).

After the Seven Years War (1756–1763), the Meissen porcelain factory entered a period of slow decline, despite the efforts of Count Camillo Marcolini, who took over the direction of the works in 1774 and struggled valiantly for four decades to restore Meissen to its once pre-eminent position. Although it continued to produce exquisite porcelain with beautifully painted floral motifs, its fame was eclipsed by the more modern porcelain works at Berlin, Sèvres and Vienna. Sèvres, for example, produced with great éclat a 'Service de déssert de Liliacées' for the Empress Joséphine, inspired by the celebrated work of Redouté.

Towards the end of the eighteenth century the Rococo style enjoyed a brief revival, and this pattern-book by Arnhold and Capieux was drawn up for Meissen's in-house artists to be used as a guide to the new porcelain designs. On every page five or six flowering species, such as the rose, peony, lily, tulip, iris and lily of the valley are presented, sometimes arranged in small bouquets, and always shown with very short stems. They have been elegantly depicted, but since they were intended for use as decorative elements, Arnhold often sacrificed scientific accuracy for purely aes-

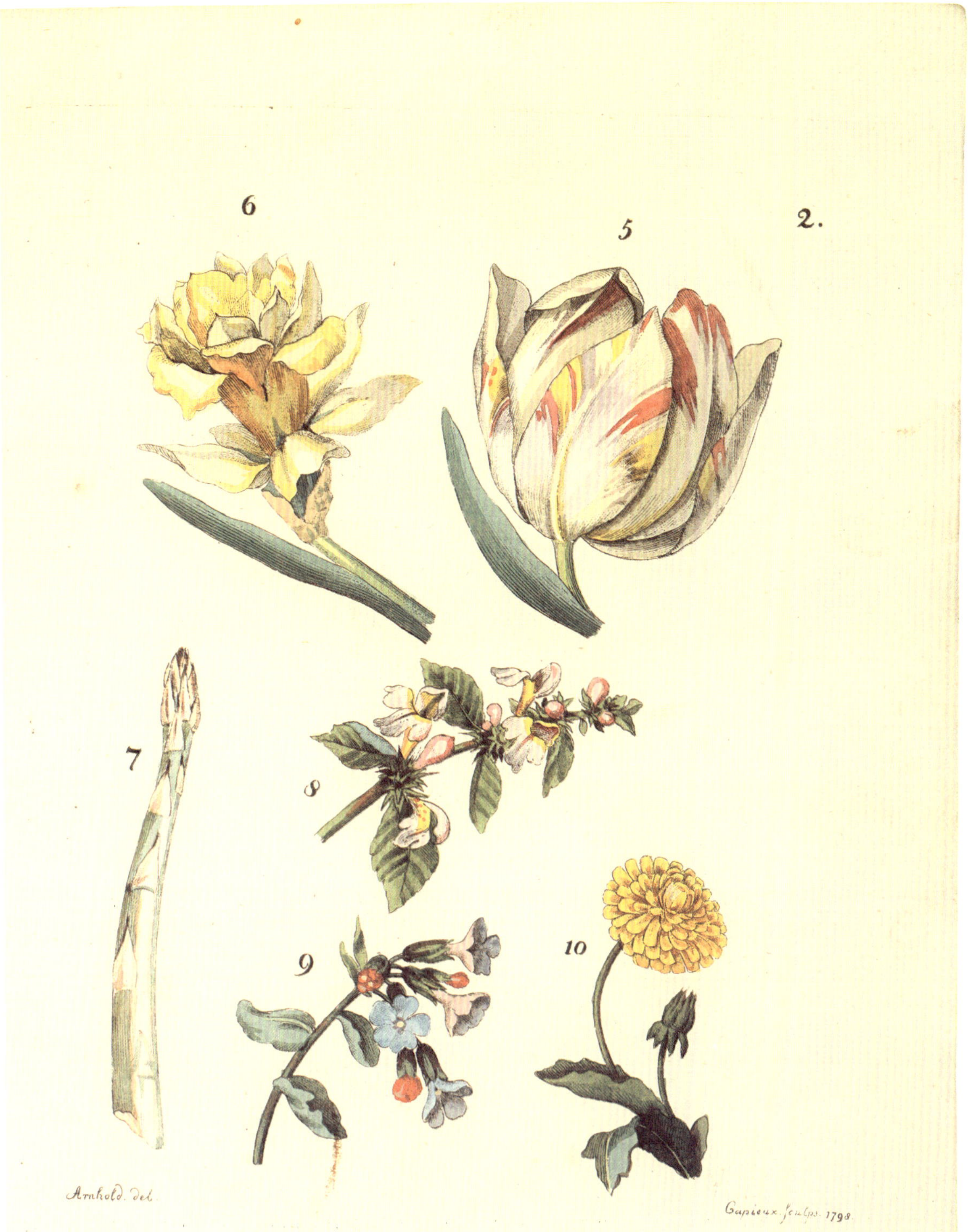

Meissen pattern-book.
Plate 2: 5. *Tulipa* sp.;
6. double *Narcissus*;
7. *Asparagus officinalis*;
8. honeysuckle (*Lonicera*);
9. *Primula* sp.; 10. dandelion (*Taraxacum officinale*)

JEAN GEORG WALCHER,
Tickseed (*Coreopsis tinctoria*)

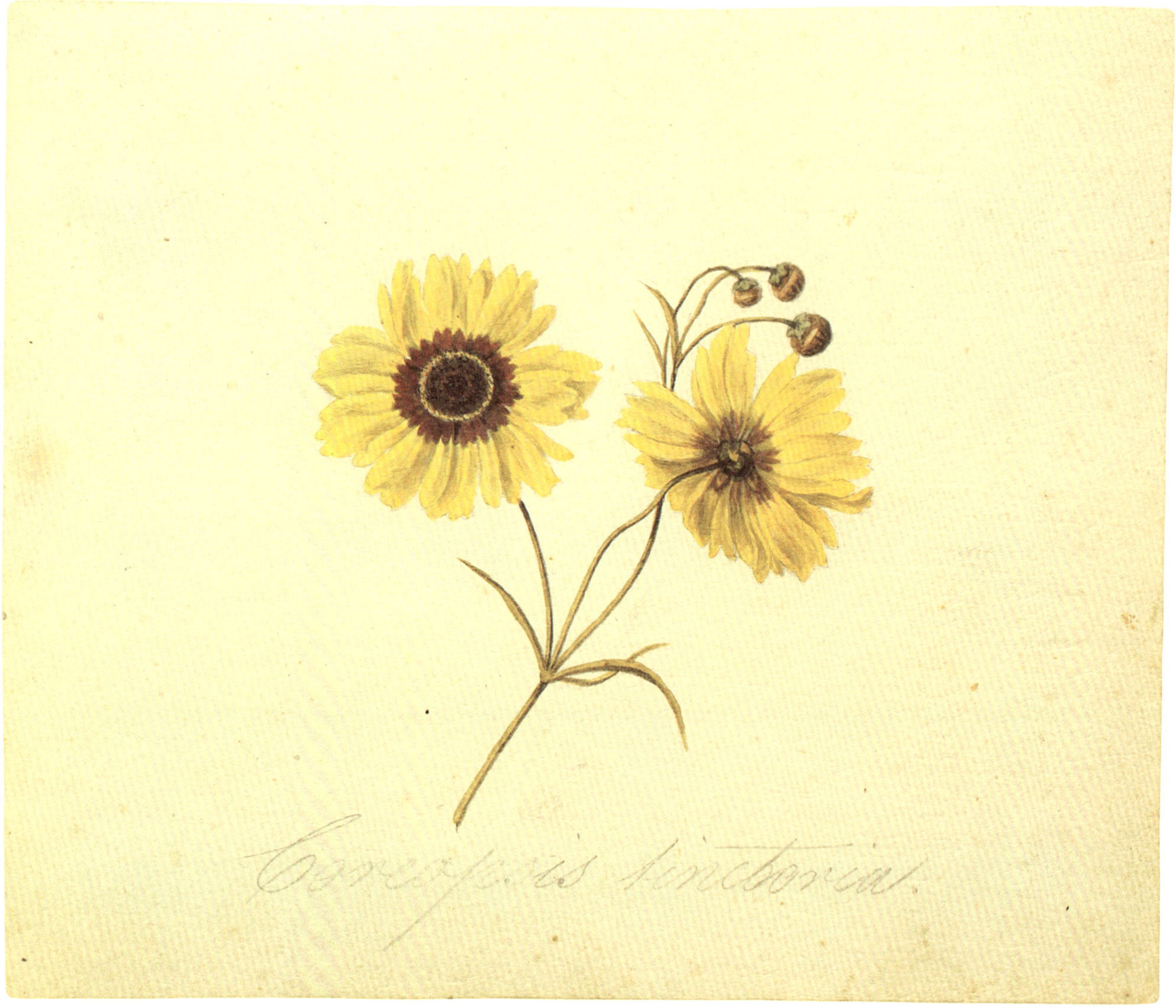

thetic considerations of form and colour. Every image is numbered, with each number corresponding to an entry on the manuscript page following the plate. Next to the common name of the flower (in German), in each entry appears a two-number code (one written in upper case and the other in lower case ciphers) referring to the specific pattern or series into which this motif was to be incorporated.

At Oak Spring there is also a noteworthy collection of flower paintings executed by the German artist Jean Georg Walcher (1785–1862) for the Meissen factory. It consists of forty-nine works in bodycolour preceded by a sober self-portrait in an oval frame. Walcher came from a noted family of artists specializing in sculpture and in painting on porcelain. Born in Niederweiler, he worked for the porcelain factory of Sèvres and then that of Ludwigsburg, where he died in 1862. Like Arnhold,

Walcher concentrated on common flowers in this collection of models, in his case presenting only one per page. They have been meticulously painted, although here again scientific accuracy has sometimes been sacrificed to decorative considerations. The Latin name appears in manuscript at the bottom of many of the paintings.

69. John Ingersoll (1788–1815)

Flowers in a vase.

9 × 23 cm. Watercolour on paper, in a gilt wood frame. With the artist's signature in the top-right corner, and his initials in the lower right.

This watercolour was painted by the American artist John Ingersoll, who signed the work and inscribed his initials on the vase. He was married to the somewhat better known artist Ann Warren, who had studied at the Academy of Fine Arts of Pennsylvania and in Paris. Little is known of Ingersoll except that he worked in Haverhill, Massachusetts.

Ingersoll's composition echos floral motifs widely used in embroidery and the decorative arts, with its bouquet of flowers—flattened and simplified into almost abstract forms—fanning out gracefully from a tiny, footed urn. The five disproportionately large blooms include two tulips that droop gracefully on either side of the vase, a cornflower, two pansies on a single stem and, at the centre, an enormous blossom, perhaps a carnation. A tiny bird hovers to the right of these giant flowers, while a winged insect has alighted on one of the pansies.

The artistic roots of this work can be traced back to the colonial period, with its naïve reinterpretation of the pictorial tradition of European art, echoes of which periodically reached America through artists who had studied in Europe, such as Ann Warren. The flower constituted a perennial theme in American painting (Anderson 1980, p. 8), not only among professional artists such as Rembrandt Peale (1778–1860), author of the celebrated portrait *Rubens Peale with a Geranium* (1801; National Gallery of Art, Washington, D.C.), but also among the itinerant artists who, in the colonial period, travelled from town to town executing decorative paintings of the simplest kind. So distinctive and characteristic was this theme, originally inspired by the seventeenth-century Dutch still-life, that the catalogue for the Whitney Museum's exhibition on American folk art devoted an entire section to it called 'Flowering'.

At Oak Spring is also a charming wooden box, painted black, with white insets on three sides decorated with floral and bird motifs. The base of one of these decorated insets is signed 'Eva. Anno 1776'. The artist utilized motifs from a repertory familiar to the anonymous creators of the embroideries, tapestries and quilts of the period.

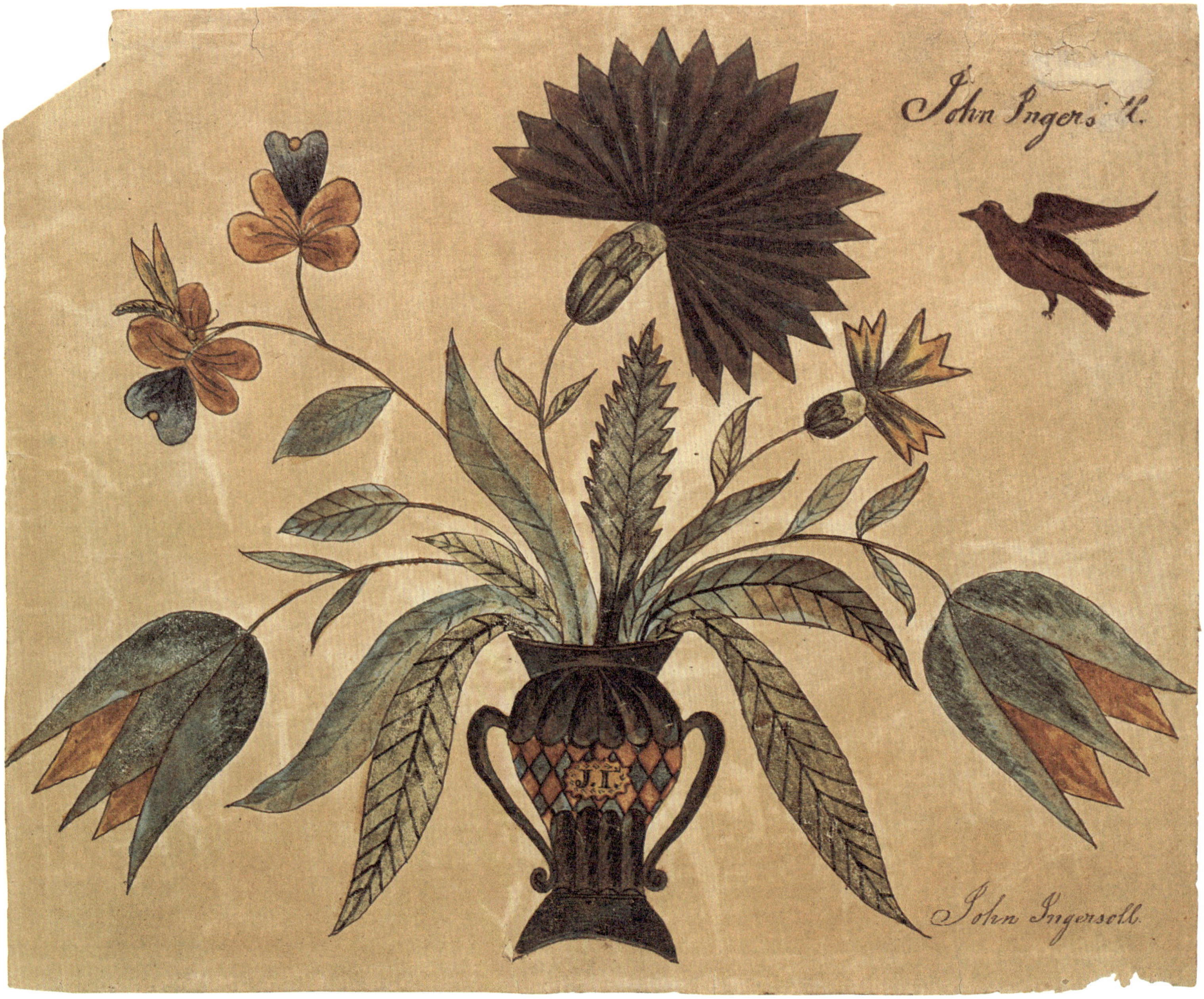

JOHN INGERSOLL,
Flowers in a vase

The still-life, with its flowers, insects and birds representing a microcosm of God's wondrous universe, remained a favourite subject with artists from colonial times until the mid-nineteenth century. The popularity of this theme also gave rise in the 1820s to a vogue for floral dictionaries, literary annuals and gift-books illustrated with botanical paintings.

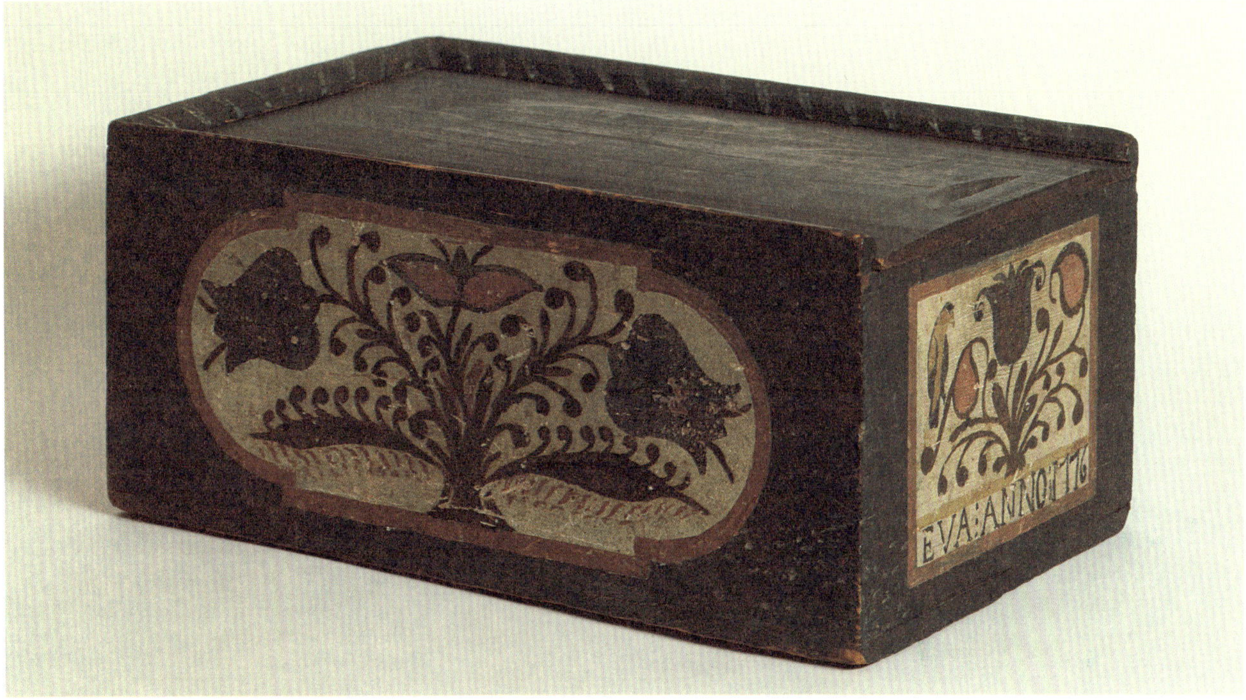

An American-made wooden box decorated with flowers, 1776

70. Emma Peachey (*fl. mid-19th century*)

The Royal Guide to Wax Flower Modelling. By Mrs. Peachey, Artiste to Her Majesty. [*poem*] 'For *not alone* to please the sense of *smell*, | Or charm the sight, are flowers to mankind given,— | A thousand sanctities do them invest, | And bright associations hallow them! | Which to the cultivated intellect | May give delight, and all the heart improve.' London: Published and sold by Mrs. Peachey, Artiste to Her Majesty, and sold by all Booksellers. MDCCCLI.

22 × 14 cm. *i–ix* x–xvi *1* 2–72 p.

BINDING: Embossed publisher's cloth binding. Title in gilt on front cover and spine: 'Peachey's Royal Guide to Wax Flower Modelling'.

PLATES: 4 lithographed plates signed 'J. Gardner & Co Zinc. [2 have 'Lith.'] 85, Hatton Garden'.

THIS *vade mecum* on the sculpting of flowers in wax was written by Emma Peachey, a self-styled 'artiste' who sought to revive a technique that had been introduced to England in the latter part of the seventeenth century by Mary Beatrice, the second wife of James II. Peachey gave lessons in wax flower modelling at her home, 35 Rathbone Place, London, where many examples of her work were also on display.

In 1851, the year of the Great Exhibition, Peachey found herself at the centre of a brief flurry of controversy: she had indignantly turned down the modest space offered to her in the Galleries

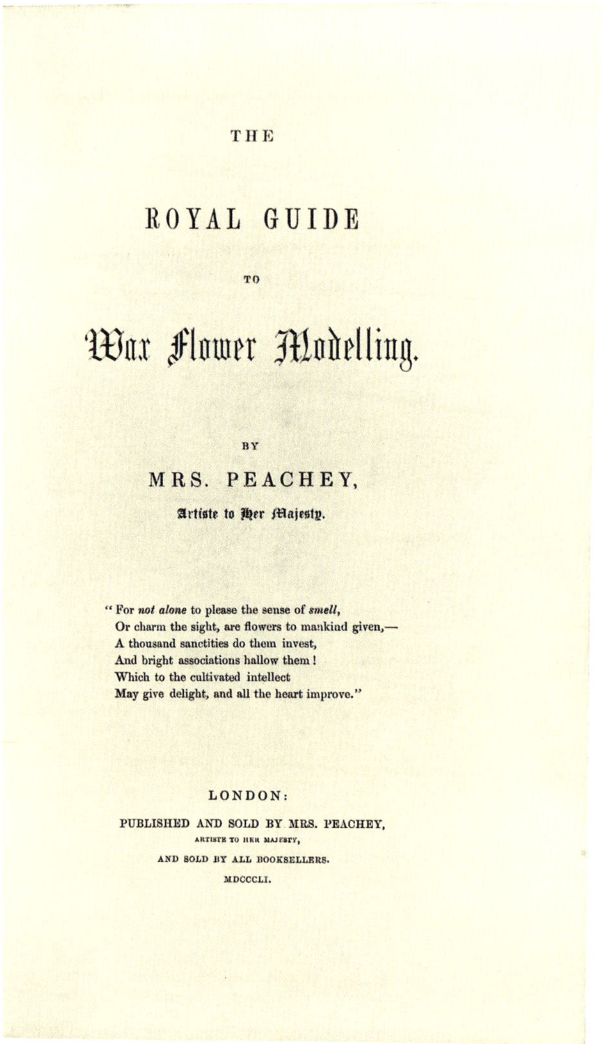

THE

ROYAL GUIDE

TO

Wax Flower Modelling.

BY

MRS. PEACHEY,

Artiste to Her Majesty.

"For *not alone* to please the sense of *smell*,
Or charm the sight, are flowers to mankind given,—
A thousand sanctities do them invest,
And bright associations hallow them!
Which to the cultivated intellect
May give delight, and all the heart improve."

LONDON:

PUBLISHED AND SOLD BY MRS. PEACHEY,
ARTISTE TO HER MAJESTY,
AND SOLD BY ALL BOOKSELLERS.
MDCCCLI.

EMMA PEACHEY,
The Royal Guide to Wax Flower Modelling. Title-page

instead of the position in the Crystal Palace she felt she deserved for the display of her sculptures. The most prominent newspapers of the period, including *The Times*, *The Morning Post*, *The Illustrated London News* and *The Sunday Times* all dedicated ample space to the affair and to the artist's work. *The Morning Post*, for example, described in admiring detail the pieces that Mrs Peachey had created for the Exhibition: 'They consist principally of an enormous bouquet of flowers and a colossal vase of fruit, both of which have been executed upon a scale never previously attempted in this country.' In one basket Mrs Peachey had arranged flowers 'from the simple honeysuckle . . . to the rarest and most valuable exotics of the East', in such a way that they appeared to be suspended over a pool of water in which several magnificent 'Victoria regia' (*Victoria amazonica*) were floating.

The Royal Guide was written in the same year as the Great Exhibition, and therefore constituted not only a manual for what rapidly became a phenomenally popular art form, but a proud affirma-

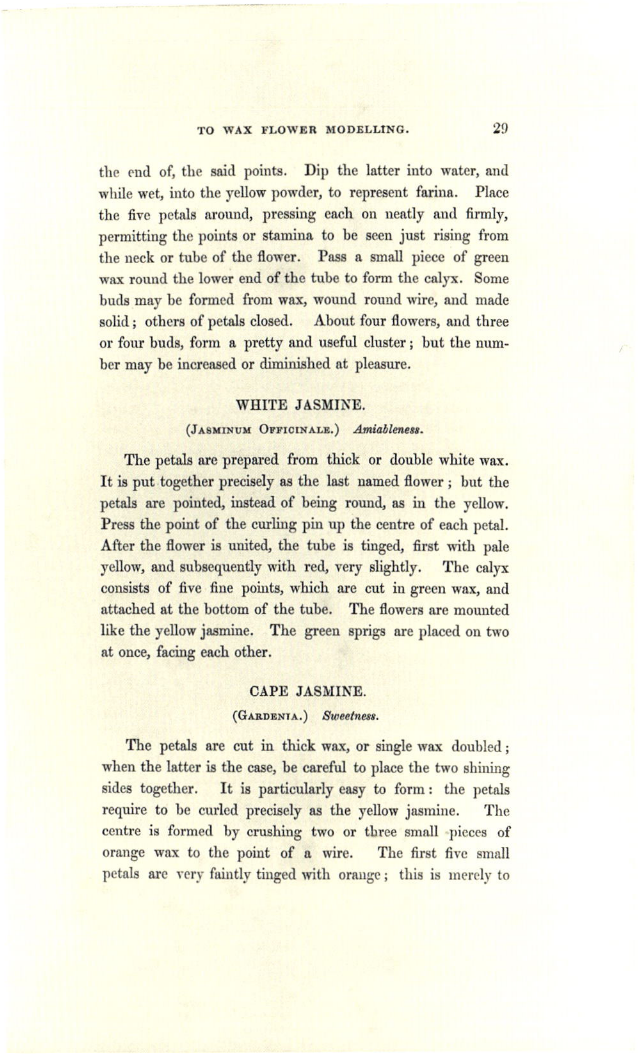

TO WAX FLOWER MODELLING. 29

the end of, the said points. Dip the latter into water, and while wet, into the yellow powder, to represent farina. Place the five petals around, pressing each on neatly and firmly, permitting the points or stamina to be seen just rising from the neck or tube of the flower. Pass a small piece of green wax round the lower end of the tube to form the calyx. Some buds may be formed from wax, wound round wire, and made solid; others of petals closed. About four flowers, and three or four buds, form a pretty and useful cluster; but the number may be increased or diminished at pleasure.

WHITE JASMINE.

(JASMINUM OFFICINALE.) *Amiableness.*

The petals are prepared from thick or double white wax. It is put together precisely as the last named flower; but the petals are pointed, instead of being round, as in the yellow. Press the point of the curling pin up the centre of each petal. After the flower is united, the tube is tinged, first with pale yellow, and subsequently with red, very slightly. The calyx consists of five fine points, which are cut in green wax, and attached at the bottom of the tube. The flowers are mounted like the yellow jasmine. The green sprigs are placed on two at once, facing each other.

CAPE JASMINE.

(GARDENIA.) *Sweetness.*

The petals are cut in thick wax, or single wax doubled; when the latter is the case, be careful to place the two shining sides together. It is particularly easy to form: the petals require to be curled precisely as the yellow jasmine. The centre is formed by crushing two or three small pieces of orange wax to the point of a wire. The first five small petals are very faintly tinged with orange; this is merely to

EMMA PEACHEY,
The Royal Guide to Wax Flower Modelling. Directions on how to make a model of a 'White Jasmine', page 29

tion by the author of the artistic value of her work. All the articles that had appeared in the press regarding the controversy were punctiliously reproduced in the appendix. In response to the question posed by *The Manchester Examiner* among others, regarding the possible toxicity of the materials she used, Mrs Peachey explained that a German chemist had helped her to prepare her colours and that all of them had been shown to be quite harmless.

The art of wax flower modelling naturally required some knowledge of botany and, in particular, a thorough understanding of plant structure. Mrs Peachey decided that the aspiring artist should visit the botanic gardens of Kew and Regent's Park, where he or she would find an endless source of models to study and copy. After providing a list of the materials that were required in order to begin, she offers general instructions on the mixing of colours and on the use of curling pins and scissors to model various forms. The second part of *The Royal Guide* then explains how to sculpt specific flowers, including such favourites as the violet, anemone, tulip, narcissus, jonquil and white jasmine, and exotic blooms such as the *Victoria amazonica*, which had been taken to England from Guyana in 1837. The author even provides her female readers with instructions on how to sculpt flowers that could be used to adorn the hair. This work was clearly a product of the Romantic period, for the text on each flower includes a short poem and a description of the flower's attributes —the violet symbolizing 'Modesty', the anemone 'Sickness', the daisy 'Innocence', etc.

This fascinating example of Victoriana is illustrated with four colour lithographs of flower bouquets, one for each season of the year, undertaken by Gardner's firm in London's Hatton Garden.

71. T. C. March (*fl. second half of the 19th century*)

Flower and fruit decoration: with some remarks on the treatment of town gardens, terraces, &c.; and with many illustrations of colour and contrast applicable to both subjects. By T. C. March. London: Harrison, 59, Pall Mall, Bookseller to the Queen, 1862.

22.5 × 15 cm. *i–v* vi–vii *viii 1* 2–108 *1–2 109* 110 *111* 112 *1–3* 4–16 p., 1 plate.

BINDING: Decorated paper boards. On the spine: 'Flower & Fruit Decoration March'.

PLATES: Steel-engraved frontispiece. Steel engravings in the text, some of which are hand coloured.

THIS CHARMING WORK, illustrated with steel engravings based on photographs, provides practical instructions on how to create compositions out of fruit and flowers, an immensely popular form of decoration in upper- and middle-class Victorian households. The author, T. C. March, was an artist employed by the London firm of Dobson & Pearce, which specialized in the production of table centrepieces, glass ornaments, ceramics, lustreware and gaslight chandeliers.

The idea for the book grew out of the First Competition for Table Decoration, held in London in 1861. The designs showed by March attracted particular attention and were subsequently produced by Dobson & Pearce: 'by adopting various improvements, especially as to fitting the pieces together, they succeeded in making an Epergne, either all glass, or of metal mixed with glass, which, besides being moderate in price, could be easily moved and packed' (p. 3). The firm presented these ornamental pieces at the International Exhibition held in London at Kensington in 1862. Filled with fruits and flowers, they made impressive centrepieces and striking sitting-room decorations, particularly adapted for country houses. They were an instant success and, as the firm of Dobson & Pearce complacently observed, for a long time their factory could barely produce enough to keep up with the popular demand.

These receptacles, as March pointed out, did not have to be filled with rare or costly flowers; an

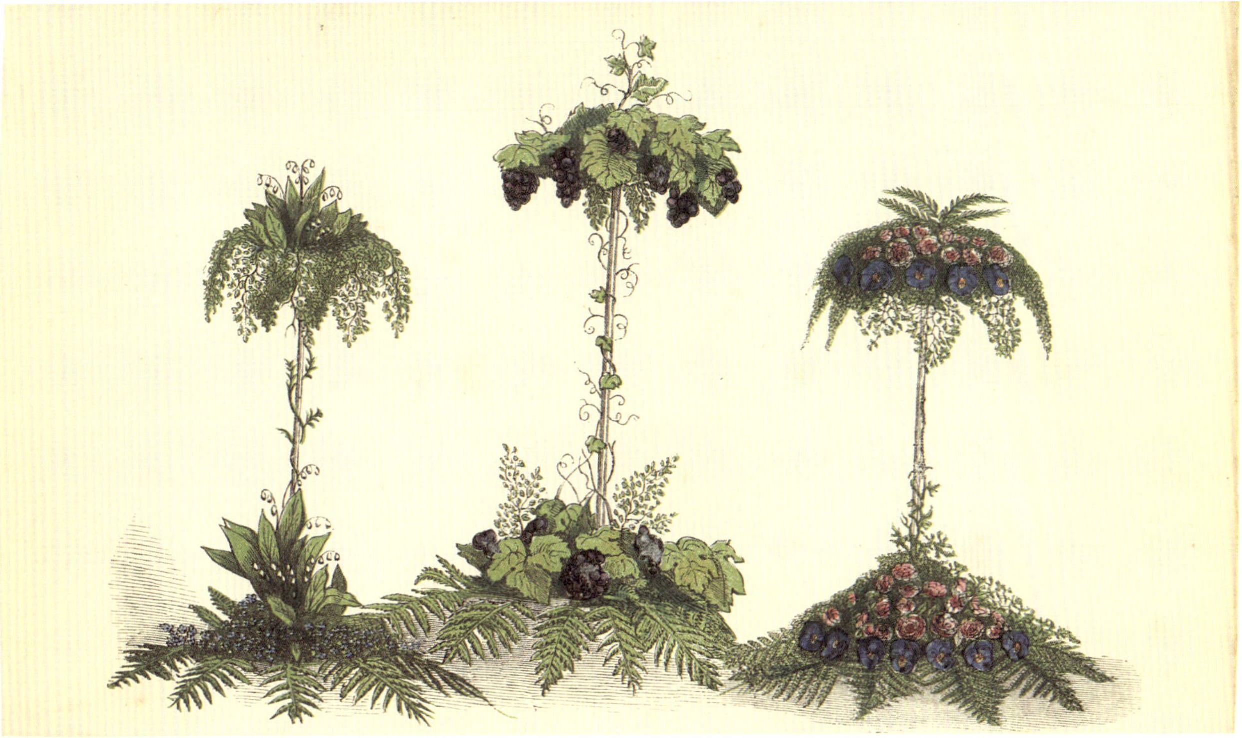

T. C. MARCH,
Flower and Fruit Decoration. Frontispiece

equally pleasing effect could be achieved with the simple flowers to be found in every garden in and around London, such as the lilac, laburnum, yellow azalea, pink May and wisteria, or summer flowers such as the geranium, rose, verbena and dahlia. Each piece could be used alone, or assembled in more complicated constructions, as March shows in diagrams accompanied by advice on how to stabilize the more towering edifices. He devotes an entire chapter to the problem of matching and contrasting colours; in general, he says, all one need do is follow the same rules that every gardener applies in arranging his flowerbeds, adapted to the smaller scale of the table decoration, of course. March also offers advice on how to keep the flowers and leaves in one's composition fresh, and, in chapter VIII, a list of 'fifty different groups of flowers and fruits recommended from experience'.

Although this volume focuses on the use of flowers for centrepieces, March also has something to say about gardening. Since few Londoners could aspire to more than a tiny plot of land or a terrace with a few flower-pots, he recommends various inexpensive plants that put on a very handsome show when in flower. Furthermore, as he points out, at the market in Covent Garden it was possible to find a wide variety of cut flowers, green branches and other items for one's flower arrangements or garden, all at moderate prices.

VII • TULIPOMANIA AND OTHER FLOWER MANIAS

WHICH FLOWER most deserved the title of King or Queen of Flora's realm? This apparently futile question was continually posed by florists and horticulturists and has received varying answers down through the centuries. Yet the Milanese poet Buonvesin de la Riva, who lived in the second half of the thirteenth century, wrote a Latin *contrasto* (dialogue poem) entitled *Disputatio rosae cum viola* (A debate between the rose and the violet). In it, the lily has to choose between the two flowers; the lily chose the modest violet, considering the rose a mundane symbol of vices. Nevertheless, the rose dominated not only in the classical period, but also from the Middle Ages to the sixteenth century. For historical reasons, as well as the result of changing whims in taste, different flowers were at different times admired with an enthusiasm sometimes amounting to a mania.

The rose was the undisputed queen of flowers in classical antiquity and in the Middle Ages. However, from the mid-sixteenth century, the botanical landscape of Europe began to undergo a radical transformation as large numbers of hitherto unknown flowering plants arrived from the Near and Far East and from the New World. These novelties quickly replaced the more familiar species in the gardens of wealthy connoisseurs; they were avidly sought after and became status symbols reflecting the power and wealth of their possessors. Thus, during the seventeenth century, botanists and flower lovers were entirely taken up with various new cultivated species, particularly bulbous plants. Indigenous and medicinal plants were relegated to the botanical gardens of the universities, and to the private gardens of botanists and herbalists.

One of the most striking examples of the excesses that the passion for flowers could lead to is provided by the case of 'tulipomania', a curious and fascinating socio-cultural and economic phenomenon that initially appeared in the Low Countries, and then spread to England and France. So pervasive was this phenomenon that it became the target of innumerable moral tracts as well as satirical texts, paintings and engravings.

First brought to Vienna from Constantinople in 1554 by Augier Ghislain de Busbecq, the ambassador of Ferdinand I of the Habsburgs to the court of Suleiman the Magnificent, the Sultan of Turkey, the tulip soon reached Holland and from thence spread across Europe. Botanists were quite intrigued with this flower when they first saw it, and Gessner, Dodoens, Clusius and L'Obel all studied and described the plant in great detail. Floriculturists dedicated themselves with enthusiasm to the creation of new varieties, many of which—unbeknown to them—were actually the result of a virus that only in our own century has been isolated and identified. Most of the new varieties were named after prominent persons—the floriculturist who first grew the flower, an important figure from antiquity, or some conspicuous personality such as a general or a widely admired beauty of the period. This convention was subsequently adopted in the naming of new varieties of other popular flowers, such as the hyacinth, the auricula, the carnation and the rose.

In this way the gardens of wealthy collectors came to be filled with increasingly unusual and bizarre exemplars of the tulip, and its bulbs were sold at ever more dizzy prices per *asen*—a unit of measure equivalent to about one-twentieth of a gram (Murray; Krelage 1942a, pp. 15–141; Blunt 1950; *Flowers and Nature*, pp. 44–5; Segal 1993). The observations of the English herbalist and apothecary John Parkinson in his work *Paradisi in sole paradisus terrestris* (no. 40) on the tulip (of which he presents 122 different varieties) reflect the great admiration for this flower, whose 'sundry diversities of colours . . . found out in these later dayes by many [of] the searchers of natures varieties . . . have not formerly been observed: our age being more delighted in the search, curiosity, and rarities of these pleasant delights, than any age I think before' (p. 45).

The tulip also appeared frequently in the florilegia of the period; for example, Crispijn van de Passe dedicated eighteen plates to the flower in his *Hortus Floridus* (no. 12). Some of these blossoms, apparently too heavy for their long, slender stems, are shown sustained by a special rod, while in another plate he illustrates a container designed to transport the plant without damaging it.

Collectors and floriculturists also commissioned artists to produce manuscripts depicting their collections, in order that they might show them to fellow collectors or to potential buyers at those times of the year when the flower was not in bloom. Thus, painters specializing in *tulpenboeken*, such as Anthony Claesz. (*c.* 1607–49), Judith Leyster (1609–60), various members of the van der Vinne family, and above all Jacob Marrel (nos. 73–5), began to appear. They produced beautiful album catalogues, often complete with price-lists, and illustrated with bodycolour paintings of tulips in an amazing variety of forms and colours—tulips with oval, pointed, fringed or 'parrot's crest' petals, or flowers with double blooms, depicted in an infinite range of colours, including many variegated effects. From these florilegia and catalogues the tulip passed directly into the floral repertory of the still-life painters of the period.

Until *c.* 1620 in Holland, only botanists and a restricted circle of flower lovers from the aristocracy and the wealthy burgher classes could permit themselves the luxury of collecting tulips, which were sold by a very small group of floriculturists. At this time, however, the market began to open up and a large number of new flower growers entered, thus significantly modifying the conditions of distribution. Eager hordes of more modest buyers from the middle and lower classes contributed to swell the market, leading to spiralling prices and, eventually, to uncontrolled speculation (see

Schama, pp. 350–66). In the first months of 1637 a 'Semper Augustus' reached the incredible price of 10,000 florins, while an 'Admiral Van Enckhuyzen' cost 5,400 florins and a 'Viceroy' 4,600. To give some idea of the relative value of the florin in the seventeenth century, a weaver typically earned about seven florins a week and a skilled artisan in Amsterdam earned six to eight florins weekly. The staggering sums spent on tulips were often paid in land, houses, herds of animals or other goods. Speculation reached its height in the period 1634–7, before the market abruptly collapsed and brought ruin on countless buyers and sellers.

Many paintings and emblem books of the period allude to this disaster, and the tulip became a symbol of the unpredictability and the fleeting nature of fortune, and of the infinite folly of man. In a painting by Hendrik Gerritsz. Pot (*c.* 1585–1657), *Floraes Mallewaagen* (Haarlem, Frans Halsmuseum), which was copied in an engraving by Crispijn van de Passe the younger, Flora is shown in her sail-powered cart holding in one hand a cornucopia filled with tulips and in the other a bouquet of some of the most prized varieties of the period. Behind her a motley crowd of tulip lovers seeks to clamber onto the cart, oblivious to the fact that it is about to roll into the sea.

This collector's mania, which continued until the end of the century, also caused a reaction—a rejection of the tulip that came to be expressed with ever more ferocious irony. There is an amusing contemporary account of the physician Everhardus Vorstius (1566–1624), a professor at the University of Leiden who could not bear the sight of a tulip, and if he passed one would beat it mercilessly with his walking-stick. The *fleuriste*, who, as we have seen, represented one of the most striking figures in La Bruyère's *Les Caractères*, is vividly described by the moralist as he tends his garden:

> He seems to have been planted there, and to have taken root among his tulips, in front of *La Solitaire*: he opens his eyes wide, rubs his hands, bends down, examines it more closely, he has never seen it look more beautiful, his heart overflows with delight; he leaves it to gaze at *L'Orientale*, thence he goes on to *The Widow*, next to *Cloth of Gold*, and from this to *Agathe*, returning at last to *Solitaire*, where he stands transfixed, grows weary, sits down, forgets his dinner: . . . This rational man, who has a soul, a creed and a religion, goes home exhausted and starving, but delighted with his day: he has looked at tulips. (trans. J. Stewart, Harmondsworth, 1970, pp. 250–51)

During the course of the seventeenth century and the beginning of the eighteenth, other flowers enjoyed periods of great popularity, for example the carnation, which Louis II, Prince de Condé (1621–86), cultivated during his imprisonment in the Bastille, or the ranunculus, which was prized for the wide variety of delicate colours in which it could be found. Anemones in particular were collected in Florence by Cardinal Giovan Carlo de' Medici (1611–66), who used to assign them strange names: *Bella Parigina, Diavolessa, Santa Barbara.* As these flowers became commonplace in every collector's garden, however, they lost their fascination and were replaced by other, rarer plants, although they would continue to form the subject of treatises by floriculturists such as Jean Paul de Rome d'Ardène (no. 76).

By the eighteenth century the tulip itself was no longer in fashion, the sophisticated tastes of collectors having led them to search for ever more rare and unusual species. In *La Nouvelle Héloïse* (1761), Rousseau, who was also an amateur botanist, describes with scathing wit an imaginary conversation between an English floriculturist and certain connoisseurs anxious to vaunt their knowledge of the latest fashions: 'Je leur confessai humblement comment ayant voulu m'évertuer à mon tour, et hasarder de m'extasier à la vue d'une tulipe dont la couleur me parut vive et la forme élégante, je fus moqué, hué, sifflé de tous les savants, et comment le professeur du jardin, passant du mépris de la fleur à celui de panégiriste, ne deigna plus me regarder de toute la séance' (Schnapper 1983, p. 180).

One of the most popular flowers in the eighteenth century was the hyacinth (no. 77), which was cultivated with particular enthusiasm in the period 1720–36 when a veritable 'hyacinth-trade' reigned (Krelage 1942a, pp. 142–96). Certain horticultural varieties of the *Hyacinthus orientalis* were introduced to Europe *c.* 1560. As with the tulip, the botanist Clusius was one of the first to study this flower and to send bulbs to fellow naturalists. Next to the original blue variety, soon a white hyacinth was bred, while a double blossom was successfully induced in 1612. At first this flower was not greatly appreciated, due to the oddly irregular shape of its petals; in fact, gardeners used to root up hyacinths when they found them in their gardens. The popularity of this flower began to grow when the floriculturist Pieter Voorhelm of Haarlem (d. 1728) found a double hyacinth in his garden that he did not destroy, but instead named 'Marie' and started to cultivate and show them to fellow botanists.

Between the years 1700 and 1720 a large number of books and treatises on the hyacinth were published, and nearly one hundred double varieties were cultivated. In addition to the blue and white varieties, floriculturists managed to produce a pale pink flower, then a deep pink one, and finally a red flower. In 1709 the 'Koningin van Sheba' (Queen of Sheba), which boasted a magnificent white flower cluster with a pink heart and thirty-eight claws, reached a price of 140 guilders per bulb, while the blue and white varieties cost between thirty and fifty guilders. Over the next two decades, growing demand led to inflated prices and rampant speculation, which reached its peak between 1733 and 1737. For a single mature bulb of the 'Passé non plus ultra' with its younger bulbs, one investor paid the staggering sum of 1,600 guilders. Growing alarm at the prospect of this speculation gave rise in 1733–4 to a series of moralizing pamphlets that echoed the cautionary tracts printed at the height of the tulipomania. For example, on 14 August 1733 in the *Hollandschen Spectator*, a fictitious letter appeared in which a certain 'Japik Schietspoel' complained that his son-in-law had completely lost his wits due to his obsession with hyacinths. Beginning in 1737, however, the price of these bulbs began to fall and the phenomenon passed before it had reached tulipomania's catastrophic proportions.

Another flower that became quite popular in the eighteenth century was the yellow Alpine primrose or 'auricula' (from its Latin name *Auricola ursi*, or bear's ear). In the mid-sixteenth century, horticulturists working in the service of the Habsburg emperor began to cultivate this flower in the Imperial gardens in Vienna, and soon many beautiful varieties had been created, a fact of sufficient interest to be noted by Parkinson in chapter Thirty-three of his *Paradisi in sole.* Not only red and purple, but also striped varieties were obtained, which John Rea describes admiringly in his *Flora* (no. 31). Interest in this species reached its height during the first half of the eighteenth century, when numerous treatises were written on the subject. One of the first was the anonymous *Traité de la culture parfaite de l'oreille-d'ours ou auricole, par un curieux de province*, published in 1732 (Moreton, pp. 25–6). But perhaps the most significant tribute to this flower appeared in Robert Furber's magnificent *Twelve Months of Flowers* (no. 37); twenty-six different varieties of auricula (each one named after a prominent aristocrat or a figure from antiquity) are presented in the plates, while an elegant frame composed of these flowers surrounds the page listing the work's subscribers. Finally, the artist and decorator Jean Louis Prévost included a plate depicting a vase of auriculas in his album *Collection des fleurs*, published *c.* 1805 (no. 65).

The rose, whose place in culture goes back to ancient times, represents one of the eternal favourites of mankind, its unique beauty and perfume having been associated with myth and legend ever since antiquity, in both the East and the West. More frequently than any other flower it has been vested with symbolic meaning, and portrayed in paintings and sculptures and in the decorative arts. At the end of the eighteenth century the rose once again became the favourite flower of both the élite and the masses, and would remain so for more than one hundred years.

In the Middle Ages, the rose in its simple, indigenous forms (the *Rosa gallica*, the *Rosa alba* and the wild variety of the *Rosa canina*) was cultivated in both the gardens of monasteries and on the estates of the wealthy. The *Rosa damascena* probably reached Europe during the period of the Crusades. The Flemish botanist Matthias de L'Obel (1538–1616) was the first author to furnish an accurate, scientific description of those roses that were being cultivated at the end of the sixteenth century. His *Plantarum seu stirpium historia* of 1576 contains descriptions and plates of the *Rosa centifolia*, the *Rosa gallica*, the *Rosa canina*, the 'Rosa cinnamomaea', the *Rosa eglanteria*, the *Rosa spinosissima*, the *Rosa foetida*, and three forms of the 'Rosa moschata' (Krüssmann, p. 51). Over the next hundred years many new species were to be added to this list, including the 'Rosa virginiana', the first rose from America to reach Europe, which is described by Parkinson in his *Theatrum botanicum* (1640). Many important species from Asia also arrived during the course of the seventeenth and eighteenth centuries (Shepherd, p. 10).

Oddly enough, it was not until 1799 that the first book exclusively dedicated to this prized flower—*A Collection of Roses from Nature* by the English flower painter Mary Lawrance (no. 78)—was published. Splendidly illustrated with large colour plates, this instantaneous success was quickly followed by many other works, among them *Die Rosen* (Leipzig, 1802–20) by Karl Gottlob Roessig, illustrated with sixty plates by Luise von Wagenheim; *Roses* (London, 1805–28) by Henry Charles Andrews with 129 plates; *Histoire des roses* (Paris, *c.* 1818) by Charles Malo with twelve plates designed by Bessa (no. 79); *Rosarium monographia* (London, 1820) by John Lindley with eighteen plates; and *Album des roses* (Paris, *c.* 1830) by Pancrace Bessa with twenty-three plates.

Without a doubt, however, the most exceptional work to be produced on this cherished flower was *Les Roses*, published in three volumes in Paris in 1817–24 with 167 plates by Pierre Joseph Redouté (no. 61) and a text by Claude Antoine Thory. This work was inspired by the extraordinary collection of roses at Malmaison, the country estate of the Empress Joséphine (d. 1814), which contained many new varieties developed by André Dupont, Hardy, Descemet, Cels and Vilmorin, the most celebrated rose breeders of the period. The rose-garden at Malmaison, which any floriculturist in France could visit, together with the artistic genius of Redouté, contributed to make the rose the preferred flower of the Romantic period.

During the second half of the nineteenth century the orchid began to seize the imagination of botanists and flower lovers. In his *Genera plantarum*, published in 1737, Linnaeus lists eight genera and sixty-nine species of orchids. In *Species plantarum* (1753) he describes thirty species of epiphytic orchids under the designation *Epidendrum*. Today, perhaps 800 genera and 35,000 species, excluding hybrids, have been identified (Fisher, p. 293). The orchid is indigenous to practically every region of the world, but the most admired species have always been those from the tropical and sub-tropical zones. Indeed, it has long been a customary practice for botanists to send orchid hunters to search the length and breadth of Central and South America, Asia, Madagascar and Africa for flowers with unusual colours or forms (although the latter in fact merely represent variations on the plant's very simple inherent structure).

In the 1830s orchid growing became a fashionable pastime among the aristocracy, and orchid auctions and flower shows became common events, thronged not only by specialists but also by those who came merely to admire the unusual beauty of these strange flowers. Many nurseries specializing in the cultivation of orchids were also established, such as that of James Veitch & Sons in England.

The first book dedicated to the orchid, B. S. Williams's *Orchid Grower's Manual*, published in 1850, was so successful that it was reprinted many times (*History of Orchids*, p. 1). Among the many other important works on the orchid published in the nineteenth century was one written by Charles Darwin, *On the Various Contrivances by which British and Foreign Orchids are Fertilised by Insects, and on the Good Effects of Intercrossing* (London, 1862), and Charles L. de Blume's *Collection des orchidées les plus remarquables de l'Archipel Indien et du Japon*, published in Amsterdam in 1858. Darwin's book incorporated a wealth of corroborative detail in support of the theory of natural selection he had advanced in *The Origin of Species* published three years earlier. Many rare specimens of orchid were sent to him from Kew by Sir Joseph Dalton Hooker (no. 104), though Darwin found himself intrigued just as much by those he encountered growing wild in the countryside near his home at Downe in Kent. Works by other botanists describe the adventurous, sometimes perilous, expeditions that they organized or participated in, such as Albert Millican's *Travels and Adventures of*

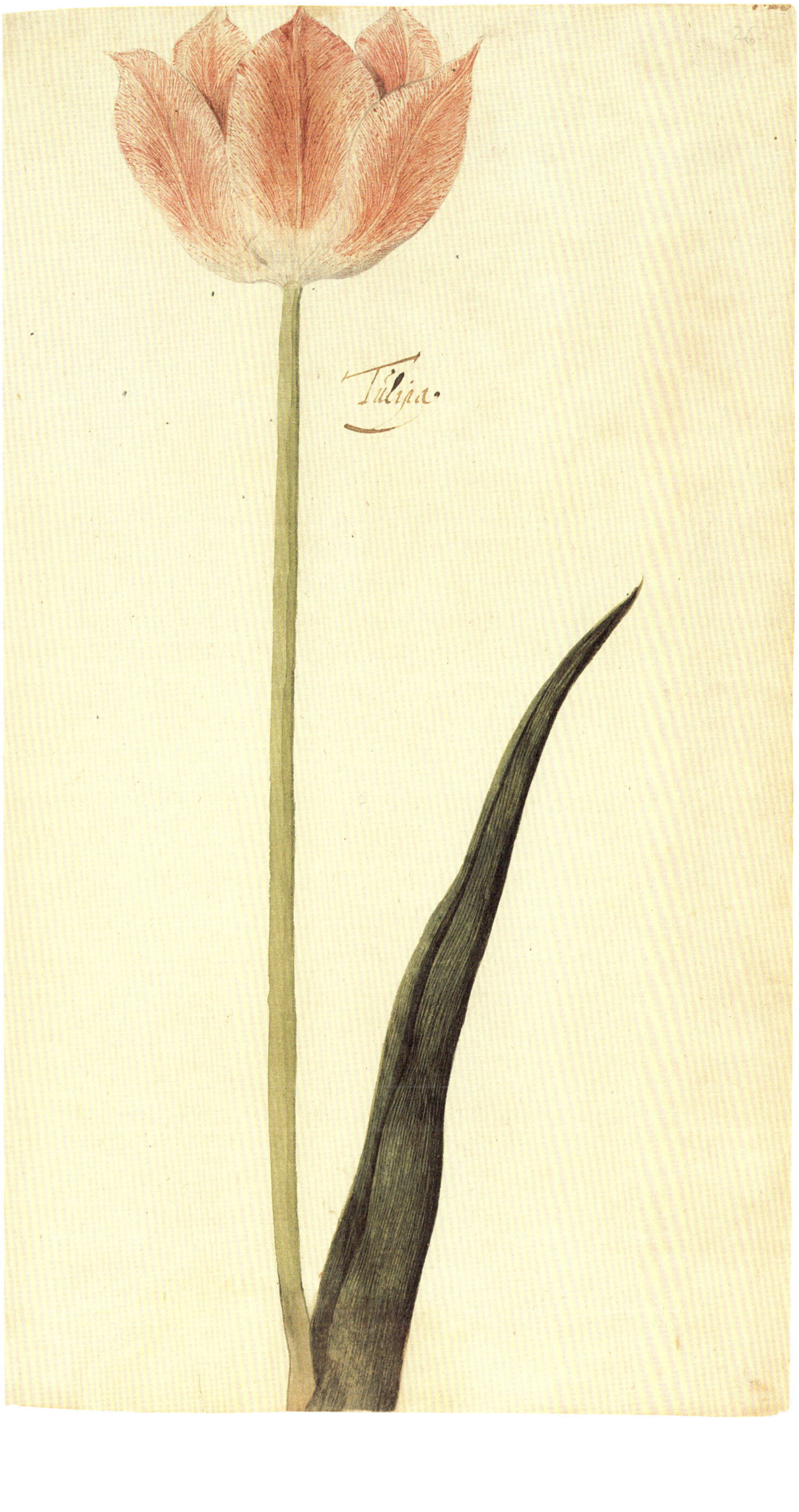

German School, watercolour studies of flowers, *c.* 1630. *Tulipa* sp., folio 17

an Orchid Hunter: An Account of Canoe and Camp Life in Colombia, while Collecting Orchids in the Northern Andes (London, 1891) or Ashmore Russan and Frederick Boyle's *The Orchid Seekers: A Story of Adventure in Borneo* (London, 1893).

One of the most eminent orchidologists of the period was John Lindley (1799–1865), professor of botany at the University College in London and for many years the moving spirit of the Royal Horticultural Society, one of the institutions most active in promoting the acclimatization and cultivation of these plants. Lindley was the author of many important works on the orchid, including *Sertum orchidaceum* (1837–41), which was illustrated with forty-nine plates painted by Ms S.A. Drake and lithographed by Maxime Gauci. William Curtis's celebrated *Botanical Magazine* also contained many illustrated articles on the orchid (Desmond 1987, pp. 150–56). Finally, James Bateman, a protégé of John Lindley and an important private collector, published a remarkable elephant folio in 1837–43 dedicated to the orchids of Guatemala (no. 80).

The orchid, with its elaborate and fantastic forms, has also provided an important font of inspiration for artists (particularly during the Art Nouveau period) as well as for photographers and scientific illustrators. The painting *Orchids and Hummingbird* by Martin Johnson Heade (1819–1904) in the Museum of Fine Arts, Boston, dated *c.* 1865 (*A World of Flowers*, p. 155; *Martin Johnson Heade*, nos. 47, 48, 49), and the many works painted more recently from life in the Amazon forest by Margaret Mee (no. 109), provide examples of the irresistible hold that this flower continues to exercise on the imaginations of artists.

72. Pamphlets and Broadsides on Tulipomania

Collection of 43 Dutch pamphlets and broadsides on tulipomania, followed by 33 bodycolour paintings of tulips. On the front free endpaper, in manuscript 'Dit bij een gebragte, waarschijngljk in de tulpen woede zelve, wierd door mij onder vele onwaardige boeken gevonden, bij een kraamer op de brug van de Bloemmarkt te Amsterdam. B. v. Voshol' (This collection, probably formed during the period of tulipomania, was found by me among a heap of old books in a secondhand dealer's shop on the Bloemmarkt bridge in Amsterdam. B. v. Voshol). A not entirely accurate manuscript list of the pamphlets follows.

21.3 × 16.5 cm.

Bound with 33 paintings in bodycolour on vellum, and preceded by a paper leaf on which is written 'Dit was een staal kaart tot verkop in tij den de tulpen niet bloeide B. v Voshol' (This was a sale catalogue for those periods when tulips were not in bloom B. v Voshol). 'Afbeeldinge der Voornaamste en Schoonste Tulpen van de Jaaren 1636 en 1637. Na't Leeven getekend door J. Marrel' (Pictures of the most Important and most Beautiful Tulips from the Years 1636 and 1637. Drawn from Life by J. Marrel). With a 'Register der Naamen van de Tulpen' (List of Tulips) and, on the back, a manuscript list of other tulips.

BINDING: Half calf and speckled paper over boards. Gilt spine with the remnants of a shelfmark.

PLATES: 2 etched plates on no. 1 (1637; 42 × 53.5 cm) by Cornelis Danckerts, and no. 43 (1763; 16.5 × 23 cm). Vignettes and decorated initials. 33 paintings in bodycolour on vellum by Jacob Marrel (1614–81), numbered 7–39, each with a guard leaf; some are signed and dated 1636 or 1637. The plates have been trimmed.

PROVENANCE: Gerard Beeldsnijder van Voshol en Vrije Nes.

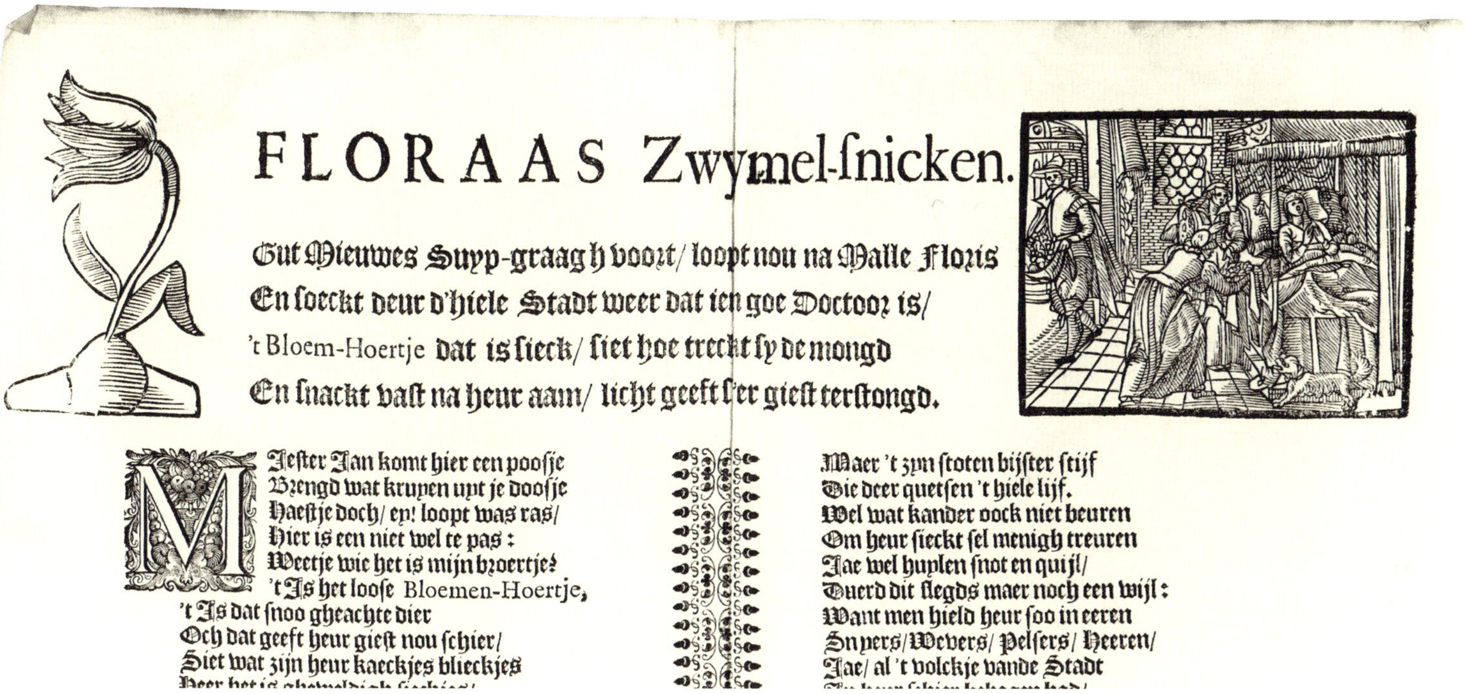
FLORAAS Zwymel-ſnicken.

Gut Mieuwes Suyp-graagh voort/ loopt nou na Malle Floris
En ſoeckt deur d'hiele Stadt weer dat ien goe Doctoor is/
't Bloem-Hoertje dat is ſieck/ ſiet hoe treckt ſy de mongd
En ſnackt vaſt na heur aam/ licht geeft ſ'er gieſt terſtongd.

Meſter Jan komt hier een pooſje
Brengd wat krupen uyt je dooſje
Haeſtje doch/ ey! loopt was ras/
Hier is een niet wel te pas:
Weetje wie het is mijn broertje?
't Is het looſe Bloemen-Hoertje,
't Is dat ſnoo gheachte dier
Och dat geeft heur gieſt nou ſchier/
Siet wat zijn heur kaeckjes blieckjes

Maer 't zyn ſtoten bijſter ſtijf
Die deer quetſen 't hiele lijf.
Wel wat kander oock niet beuren
Om heur ſieckt ſel menigh treuren
Jae wel huylen ſnot en quijl/
Duerd dit ſlegds maer noch een wijl:
Want men hield heur ſoo in eeren
Snyers/Wevers/ Pelſers/ Heeren/
Jae/ al 't volckje vande Stadt

A broadside on tulipomania (no. 19): *Floraas Zwymel-Snicken*, with a vignette of Flora on her deathbed

THIS IS PERHAPS the most complete collection extant of broadsides and pamphlets pertaining to the tulipomania that seized Holland in the mid-seventeenth century and subsequently spread to other countries (see Appendix). Another collection of thirty-two broadsides and pamphlets in the possession of the Dutch scholar E. H. Krelage (see Krelage 1942a) was broken up and sold at auction in Amsterdam in 1948. All but seven of the forty-three pamphlets at Oak Spring are dated 1636 or 1637, and were therefore printed, most of them in Haarlem, at the very height of tulipomania. Subsequently bound together, this curious volume, as a manuscript note states, was found in the shop of a second-hand dealer in Amsterdam by Gerard Beeldsnijder van Voshol (1791–1853), a collector of plants and botanical books. (Another series of tulip paintings in bodycolour, dated 1634, by Jacob Marrel at Oak Spring also once belonged to van Voshol; see no. 73.)

This collection of tracts represents an invaluable source for scholars studying tulipomania, a fascinating historical lesson on the consequences of uncontrolled financial speculation, in this case sparked off by the acquisitive mania of the collector. The severely moralizing tone that characterizes most of the art relating to tulipomania (*Flowers and Nature*; Tongiorgi Tomasi 1995), likewise infuses the pamphlets and broadsides in this collection. They obsessively reiterate the same dire warnings regarding the dangers of speculating in this ephemeral market, which can only bring financial ruin on the foolish buyer, and point out that many flowers other than the tulip are quite as beautiful and worthy of cultivation (see 3, 4, 9, 14).

A recurrent figure in these tracts is the goddess Flora, who because of her preference for the tulip above all the other flowers in her kingdom has been transformed from a benevolent figure into a perfidious one; in some of the pamphlets she is represented, as she sometimes is in classical mythology, as a prostitute (19, 20, 23, 29). The true goddess Flora is shown slumbering, or stricken with illness, or dying, quite unaware of the disaster that threatens her realm. Only her most faithful servants—wise floriculturists who did not give themselves up to speculation in tulips—can perhaps save her (7, 16, 17, 21, 24, 27, 37, 41). Some of the broadsides are embellished with woodcut vignettes designed to capture the reader's attention and impress him with the dangers tulipomania posed. These range from the simplest of outlined images of flowers (13, 16, 19, 21, 23, 30) to scenes depicting a tulip market (22), or Flora on her deathbed (19), or even her funeral (27). In one amusing vignette (26) a wife can be seen mercilessly beating her husband, whose reckless speculation threatens to bring ruin on the family.

Two very fine etchings introduce and close this collection; the first, *Floraes Gecks-Kap*, was published in *T'Zamenstraeken*, a volume consisting of a series of 'conversations' between two weavers, Warmondt and Gaergoedt, on the subject of tulipomania. Engraved by Cornelis Danckerts from a drawing by Peter Nolpe, it bears the legend, here translated, 'Flora's fool's cap, or a picture of the odd year 1637, when one fool deceived the other, then people were rich without goods, and wise without mind'. A tavern in the form of an enormous fool's cap dominates the scene; within, a busy trade in bulbs is going on, while in the background, Flora, mounted on a mule, is being driven from the city. Nothing remains for the impoverished populace but to dispose of baskets and wheelbarrows full of now worthless bulbs. On the left a sinister, winged figure holds up a long rod on which innumerable promissory notes have been strung—bits of paper with which many a hapless citizen sold himself into penury. An admonitory legend exhorts the reader not to follow their example.

The second etching, *Uytlegging van de Plaat* by an anonymous artist, was engraved for a broadside (43) that appeared in 1763, another disastrous year for Holland, when numerous persons were ruined by speculating in shares, in a direct echo of the tulipomania experience. And in fact, in a small medallion set into the arch that frames the picture, Flora can be seen, seated on a pig.

The thirty-three bodycolour paintings on vellum by the floral painter Jacob Marrel form a fitting coda to this collection. The dates that appear beside the signature or monogram of the artist in most of the works indicate that they were produced in the very years of the tulipomania in Holland. One of the first tulips illustrated is the celebrated 'Viceroy' with its magnificent deep red petals (fol. 7); this is followed by the popular 'Goulda', an elegant white tulip with red striations (fol. 8). Other magnificent varieties depicted include the 'Engels Admirael' (fol. 12), the 'Semper Augustus' (fol. 18), the 'Branson Kaar' (fol. 31) and the 'Anvers' (fol. 38). Marrel liked to embellish his floral paintings with minutely observed and beautifully painted insects; thus we find caterpillars,

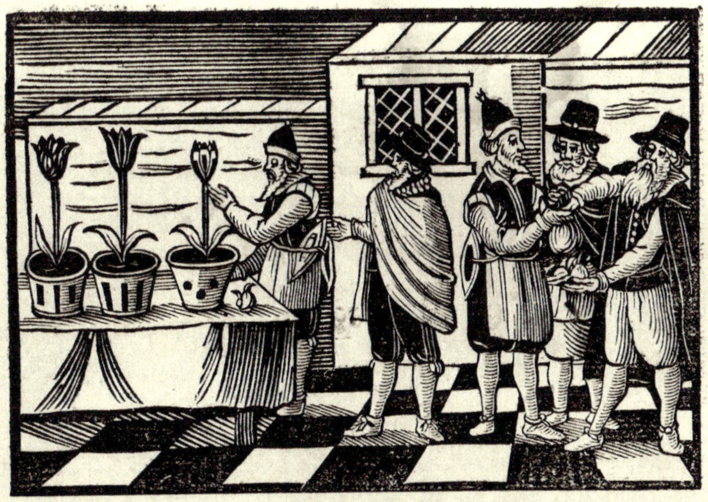

VVt Antvverpen den 8 Februarius.

Erweerdighe Vrient ende Heer,

Ick en can niet nae laten/om u. l. door deze mijn schrijvens te veradviteren / hoe dat de Bloemmisten alhier hebben op Vrydagh een solemnele Misse doen singhen in St. Joris kerck / op dat de Tulpen wel mochten voort comen / ende hebben de Goddinne Flora doen opstellen / met allerhande vercierde Roosen ende ghevlochten Cranssen met den naem ghevende (ô uytmuntende ende der hooghster Goddinne Flora) t'welc met Goude letteren boven haer hooft gheschreven stont.) Sijn daer nae in goeden ghetale in een voorname Herberghe met malcanderen gegaen/ daer sy goede ciere gemaeckt hebben/ende onder den anderen gemaeckt Officieren ende Rechters over de questien die mochten voorvallen van weghen de Coopmanschap over de Tulpen ende Bloemen onder heurlupden ghedaen/etc.

VVt Haerlem den 11 Februarius.

MOnsieur: Ick hope dat wy haest wederom V.E. sullen connen gerieven met Manufacturen/also de gene die ick binnen 14 dagen herwaerts hebbe sien Ridderē te Peerde en in Kalessen/ nu weder op t'houte Peert (zijnde t'Weef-tou) geraeckt zijn/ so dat wel groote veranderinghe in de Tulpisten/ (anders ghenoemde Bloemmisten) ghecomen is/sien veel sleght by haer Neus neer/datse van groote Marghands sonder Gheldt / in een Maent herwaerts weder soo haest op t'Weef-tou ghecomen zijn/ so dat de Proffecien van de haters der Bloem-sotten haest vervult is/die propheteerden dat de gheen die daer uyt ghereden waren/wederom in gaen souden/ende die haer hantwerck door dartelheyt ofghelept hadden/ door armoet het selve weder aennemen soude moeten/daer de oude vermaerde Bloemisten haer speculatie in hebben (soo ick hoore.)

Volght een clught van een Boer / die tot een Pekelharingh drie Tulpbollen op adt/waert 7000 gulden/ghesciet tot Haerlem.

VErleden Vrijdagh smorgens/comt een Boer met sijn Schou voor een Brouwery/om zijn Schou te laden met Draf/de Brouwer yets nodigh te doen hebbende belast hem over een cleen uurtjen weder te comen/de Boer ondertussen gaet ende coopt een Pekelharingh ende gaet doen weder nae de Brouwery toe / daer hy int Cantoor drie Tulpbollen siet legghen / als de Semper Augustus, Bruydt van Haerlem ende den Admirael, die op den derden Januarius ghecoght waren voor seven duysent gulden / die den Boer in zijn hant nam (meenden dat het Uyen waren) die hy tot de Pekelharingh op adt/ de Brouwer opt madt comende/ ende sagh de schade door den Boer begaen/begon te roepen ende te tieren als een uytsinnigh Mensch/ de Boer dit siende liep ten huyse uyt ende keech niet eens om/liet zijn Schouw voor de Brouwery legghen/daer de Brouwer zijn vermaeck inde plaets vande Tulpen innemen magh/wantse deur ghenoegh te gheef is.

A broadside on tulipomania (no. 22): *VVt Antwerpen, den 8 Februarius*, with a vignette of a tulip market butterflies, spiders, flies, ants and dragonflies decorating these pages. On folio 17 a snail can be seen creeping up a tulip leaf.

Although Marrel generally painted his flowers from life, some of the works in this series reproduce, with only minor variations, paintings from the manuscript of tulip paintings in bodycolour of 1634 once owned by van Voshol (see no. 73). The 'Viceroy' on folio 7 in fact corresponds exactly to the one depicted on folio 19 of the earlier manuscript, while the 'Manassir' with a grasshopper hanging from its stem (fol. 18) repeats the bodycolour painting on folio 57.

APPENDIX

The collection of Dutch pamphlets and broadsides on Tulipomania at Oak Spring Garden Library:

1. FLORAES GECKS-KAP . . . 1637. Broadside, etching by C. Danckerts. 44.4 × 54.6 cm.
2. MANDAMENT OP ENDE JEGHENS DE HEYDEN ENDE TURCKSCHE TULP-BOLLEN. No date. Broadside. 36 × 26.4 cm.
3. DROOM-DICHT VAN HET PROCES...ENDE DE TULIPA. Haarlem, Vincent Kasteleyn, 1612. Broadside. 36.2 × 24 cm.
4. NOODLIJKE VERDEDINGE VAN DE TULIPA... Haarlem, Vincent Kasteleyn, 1611. Broadside. 35.6 × 25.8 cm.
5. CLARE ONTDECKINGH DER DWAESHEYDT... The Hague, Zacharias Cornelissz., 1636. Pamphlet, 16 pages. 20 × 15.3 cm.
6. TULPA-LOF. No date. Broadside. 29.5 × 16.2 cm.
7. EEN NIEUW LIEDEKEN VAN DE FLORISTEN... No date. Broadside. 39.2 × 27 cm.
8. EEN NIEUW LIEDEKEN VAN DE FLORISTEN... No date. Broadside. 36.1 × 28 cm.
9. PROPHETYE... [1636]. Broadside. 19.8 × 18.3 cm.
10. EEN NIEUW LIEDEKEN / TEGHEN DE VERACHTERS DER FLORISTEN... Haarlem, Thomas Fonteyn, 1637. Broadside. 38.5 x 28.5 cm.
11. AAN DE GODDINNE FLORA. Amsterdam, Joost Hartgers, 1646. Broadside. 37.6 × 25 cm.
12. TWEESPRAECK VAN DE FLORISTEN. No date. Broadside. 38.8 × 29.3 cm.
13. EEN GHEDIGHT AEN ALLE LIEFHEBBERS / DER EDELE GODINNE FLORA. No date. Broadside (with tulip woodcut at top) attached to a sheet of paper with poem in manuscript. 28.7 × 18.6 cm.
14. EEN KLAEGH-LIEDT, WEGHEN DE LELIE-NARCISSE / OFTE TULIPA... No date. Broadside. 38.6 × 29 cm.
15. LOF-DICHT VAN CALLIOPE. OVER DE GODDINNE FLORA... Haarlem, Hans Passchiers van Wesbusch, 1637. Broadside. 38.8 × 30 cm.
16. EEN NIEU LIEDEKEN / VANDE SIJNE FLORISTEN... Haarlem, Thomas Fonteyn [?1632]. Broadside with tulip woodcut at top. 39.5 × 29.6 cm.
17. EEN NIEUW ROUW-MANTELS LIEDT / VOOR DE FLORISTEN... Haarlem, Thomas Fonteyn, no date. Broadside. 38.8 × 29.2 cm.
18. DEN SCHAT DER BLOMMISTEN. 1637. Broadside. 26 × 18.4 cm.
19. FLORAAS ZWYMEL-SNICKEN... No date. Broadside with two woodcuts. 39.2 × 29 cm.
20. SAMEN-SPRAECK / TUSSCHEN WAERMONDT ENDE GAERGOEDT... Haarlem, Adriaen Roman, 1637. Pamphlet, 24 pages. 20.1 × 15.5 cm.
21. BUYRE-PRAETJE TOT VERTROOSTINGE VAN IANTIE FLORAAS. 1637. Broadside with tulip woodcut at top. 38 × 28.4 cm.
22. VVT ANTWERPEN DEN 8 FEBRUARIUS. No date. Broadside with woodcut of tulip market. 30 × 16.3 cm.
23. KLACHTE, VAN IOOST VAN CORTRIJCKE OVER DE BEDRIEGERYE VAN FLORA. FLORAAS ZWIJMEL-SNICKEN. EEN CLAEGH-LIEDT VAN FLORA. Haarlem, Thomas Fonteyn, 1637. Broadside with tulip woodcut at top. 37.2 × 29.2 cm.
24. LAATSTE SNICKEN VAN EEN GEEST-GEVENDE HAERLEMMITER. Haarlem, no date. Broadside. 37.3 × 28.5 cm.
25. AEN MONSIUER S.V.S.W. RIJCK-SINNIGE POET, OP SYNE LAETSTE SNICK-GEEST-GEVENDE. Amsterdam, Nicolaes van Ravensteyn, 1637. Broadside. 37.2 × 28.3 cm.

26. EEN NIEU LIEDEKEN VAN SOETERS NAE-SMAECK... 1637. Broadside with woodcut of wife beating husband. 31.3 × 17.4 cm.

27. DOOD-ROLLE ENDE GROES-MAAL VAN FLOORTIE-FLORAAS... No date. Broadside with two woodcuts. 39.1 × 29 cm.

28. COPYE VAN EEN NOTORIAEL ACCOORT, GEMAECKT TUSSCHEN DE GECOMMITTEERDE DER FLORISTEN... Amsterdam, no date. Broadside. 38.5 × 28.8 cm.

29. DEN ONDERGANCK OFTE VAL VANDE GROOTE THUYN-HOER / DE BOESS-GODDIN FLORA. No date. Broadside. 38.2 × 28.4 cm.

30. APOLOGIA: OFTE VERANTWOORDINGHE / VAN FLORA... No date. Broadside with tulip woodcut at top. 39.5 × 29 cm.

31. TESTAMENTS OPENINGHE EN UYTDEELINGHE / GHEDAEN BY DE VRIENDEN ENDE ERSGHENAMEN WIJLEN FLOORTJE FLORAAS. No date. Broadside. 39.7 × 28.7 cm.

32. OPWECKINGH VAN DE SLAPENDE FLORA... Delft, Jan Pietersz. Waalpot, 1637. Broadside with woodcut emblem. 39.2 × 28.3 cm.

33. DE VERSTOORDE / EN NOYT GESTORVEN FLORA. No date. Broadside with woodcut at top. 39 × 28.7 cm.

34. DE RECHTE BLOEM-PRIJS... Kluppel and Enchuysen, Albert Wesselsz., 1637. Broadside. 39.7 × 28.9 cm.

35. TWEEDE SAMEN-SPRAECK / TUSSCHEN WAERMONDT ENDE GAERGOEDT... Haarlem, Adriaen Roman, 1637. Pamphlet, 24 pages. 20 × 14.5 cm.

36. AENLEYDINGH' TOT OPMERCK VAN 'T MISBRUYCK / EN RECHTE GHEBRUYCK DER BLOEMEN... Kluppel and Enchuysen, 1637. Broadside. 39.3 × 29.7 cm.

37. D'ONTLOOKE FLORA. No date. Broadside. 19.7 × 17.7 cm.

38. COPYE. DE STATEN VAN HOLLANDT... Amsterdam, 1637. Broadside with woodcut emblem. 39.4 × 28.3 cm.

39. TOONEEL VAN FLORA... Amsterdam, Ioost Broersz., 1637. Pamphlet, 28 pages. With engraved vignette on title-page of Flora in a garden, signed 'PBV fecit 1631'. 20 × 15 cm.

40. REGISTER VAN DE PRIJSEN DER BLOEMEN... Haarlem, Adriaen Roman, 1637. Pamphlet, 24 pages. 20.2 × 15 cm.

41. TROOST-BRIEF, AEN ALLE BEDROEFDE BLOEMMISTEN / DIE TREUREN OVER 'T STERVEN OFT 'T OVERLIJDEN VAN FLORA. Haarlem, Hans Passchiers van Wesbusch, 1637. Pamphlet, 24 pages. 19.8 × 15.3 cm.

42. COPYE VANDEN APPOINCTEMENTE OFTE APOSTILLE... 1638. Broadside with woodcut emblem at top. 37.3 × 29.5 cm.

43. UYTLEGGING VAN DE PLAAT. Rotterdam, Maronier en Vis; Leiden, Honcoop; Amsterdam, H. van Werven; The Hague, van Thol en van Os; Gouda, de Vry; Haarlem, J. Bosch; Zutphen, van Bulderen, 1763. Broadside with large vignette of people with a carriage in a city square and the god Mercury. 50.5 × 31.2 cm.

73. Jacob Marrel (1614–1681)

'Tulpenboek'.

17.5 × 13.5 cm. Manuscript composed of 91 miniatures of tulips, 90 on vellum and one on paper (fol. 14), interleaved with guard sheets. Each plate has been numbered by hand (2–92). A manuscript index with the name of each flower and its corresponding leaf number has been laid into the last two leaves.

BINDING: Calf binding decorated with gilt fillets forming a border and panel with intersections; gilt arabesque and floriated centrepiece. Gilt spine with a label on which a number (now illegible) is written.

PLATES: 91 bodycolour miniatures. Each depicts a tulip, sometimes together with an insect or another flower. The leaves were cropped before being bound together. Folio 2 bears a handwritten inscription 'Jacobii Marell'us fecit 1634', which is repeated on folios 12 and 35. The initials of the artist appear on folios 3–11, 13, 15–16, 18, 19, 21, 24, 27–31, 33–34, 37–38. On the first vellum leaf is a cartouche, coloured in yellow and pink.

PROVENANCE: The manuscript originally belonged to George Clifford (1685–1760), who owned an estate, De Hartekamp, at Heemstede close to Haarlem, where Linnaeus, the

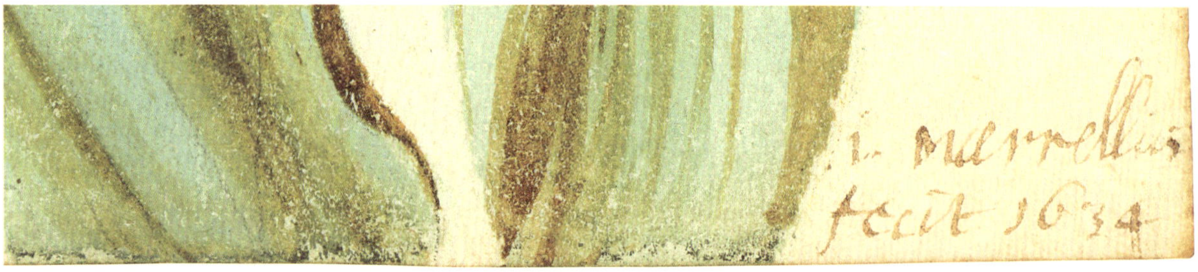

JACOB MARREL, Tulpenboek, 1634. Artist's signature and date, detail from folio 35

great botanist and physician, worked from 1736 to 1738. Clifford's name appears on the second leaf: 'Dit Boek is van George Clifford P: Z. 1756. Geteekend door J Marrel 1634 zie pag. 2'. It is also repeated on the first front free endpaper, in a note written by Gerard Beeldsnijder van Voshol en Vrije Nes (1791–1853), a naturalist and member of many European scientific societies, who subsequently came into possession of this manuscript. The note reads 'Daar de tulpen handel eenige weinige jaren duurde en de handel altijd doorging zoo wel des zomers als des winters, zoo was nodig een monsterboek of staalkaart. Een der geenen die in die tijd gebruikt wierd, is nevens gaande. Dit boek is zeker in handen van Linnaeus geweest daar deze grote botanicus in die tijd bij de Heer George Clifford in betrekking was. Beeldsnijder van Voshol' (Since the commerce in tulips lasted only a few years and was active only in the summer and winter, it was necessary to have a sample book or catalogue such as those which were used and sent out in that period. This particular book was certainly known to Linnaeus, as the great botanist was working in the service of Mr George Clifford at the time. Beeldsnijder van Voshol).

REFERENCE: *An Oak Spring Garland*, no. 9.

As emerges clearly from a biography of the artist by Joachim von Sandrart in *Teutsche Academie* (1675–9), Jacob Marrel (or Moral, Murel, Marellus) was one of the most widely esteemed flower painters of the seventeenth century. He was also a draughtsman, engraver and dealer in paintings and flowers. Born into a family of French Protestants, he and his parents fled their native country and settled in Frankfurt in 1624. There Marrel began his artistic apprenticeship under Georg Flegel. He moved to Utrecht to study with Jan Davidsz. de Heem, marrying and remaining there for many years. When his wife died in 1649 Marrel returned to Frankfurt, where in 1651 he married Johanna Sibylla, widow of the engraver Matthaeus Merian the elder. He taught his step-daughter Maria Sibylla, who later herself became a painter of botanical subjects (see nos. 81, 82 and 101). In the company of Maria and another one of his students, Abraham Mignon, Marrel revisited Utrecht in 1664. The family lived for a brief period in Nuremberg before returning to Frankfurt in 1679, where Marrel died in 1681.

Marrel was an extremely precocious artist, and although German by adoption and education, he united what he had assimilated of the German tradition of flower painting from Flegel with the tradition of the Dutch school as represented by de Heem and Seghers, creating a highly individual style of his own. A painting now in a private collection (see Mitchell 1973, p. 186, no. 258) of a bouquet of flowers in a niche, signed and dated 1635, shows that by the age of twenty-one he had already reached a high degree of proficiency; the four tulips, rendered with a minute attention to detail, are particularly striking. This unusual aptitude can only have been the fruit of years of assiduous practice. The manuscript collection at Oak Spring represents one of these efforts; executed in 1634, when Marrel was just twenty years old, it is his earliest known 'book of flowers'. Even though only

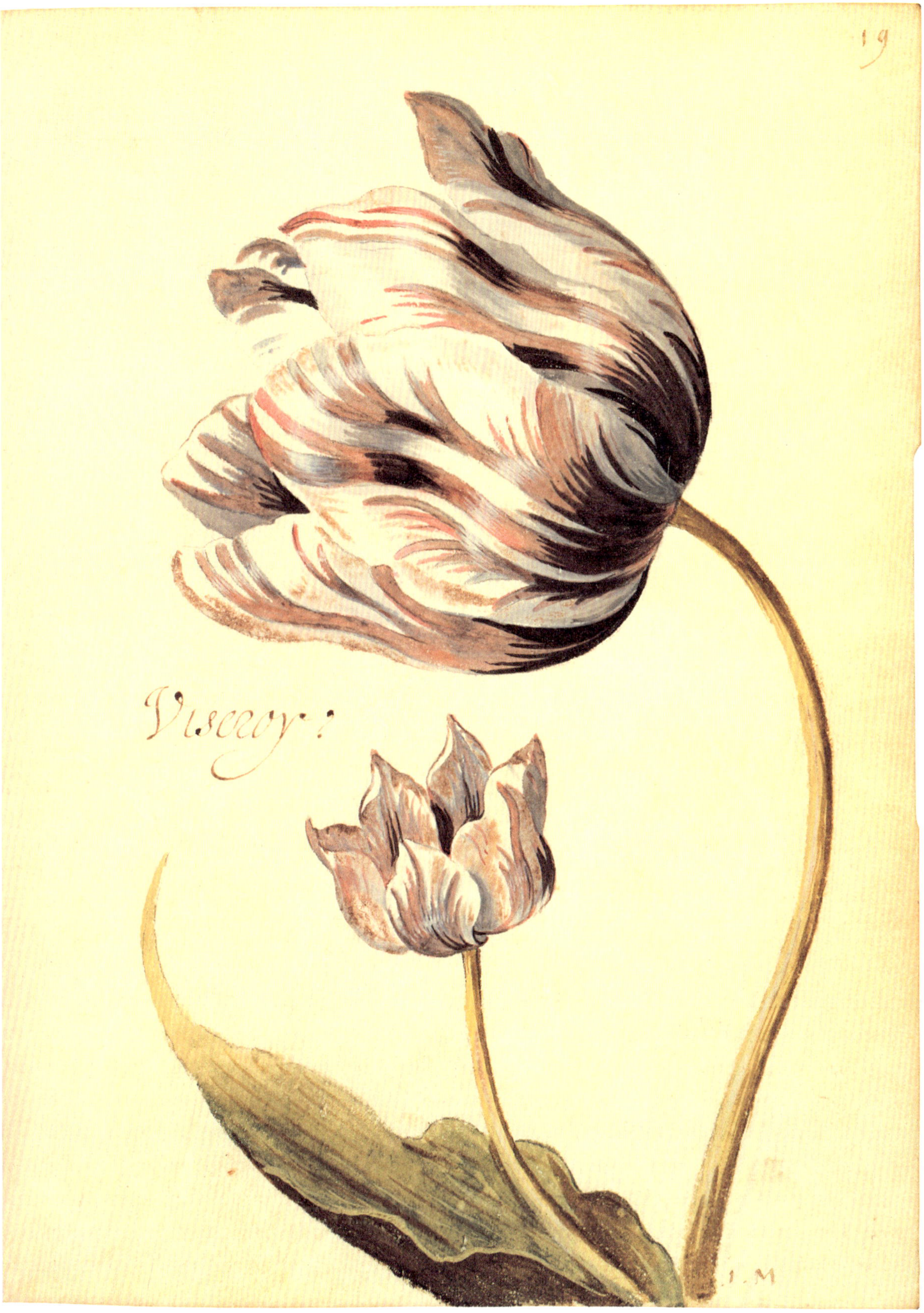

JACOB MARREL,
Tulpenboek, 1634.
Tulipa sp., 'Viceroy',
folio 19

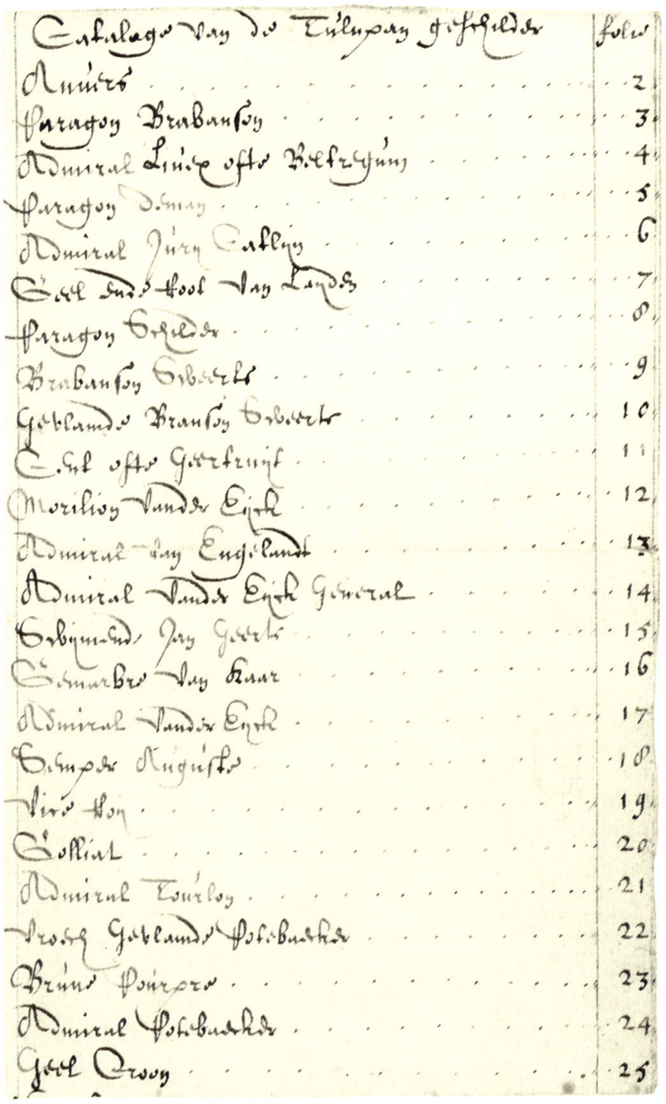

Catalogo van de Tulipan geschilder	folio
[illegible]	2
Paragon Brabanson	3
Admiral [illegible] ofte [illegible]	4
Paragon [illegible]	5
Admiral [illegible]	6
Geel ende Rool van Leyden	7
Paragon Schilder	8
Brabanson [illegible]	9
Gevlamde Branson [illegible]	10
Geel ofte [illegible]	11
Morillion vander Eyck	12
Admiral van Engelandt	13
Admiral vander Eyck General	14
[illegible]	15
[illegible] van Haar	16
Admiral vander Eyck	17
Semper Augusto	18
Vice Roy	19
Golliat	20
Admiral [illegible]	21
[illegible] Gevlamde Potebacker	22
Bruine Pourpre	23
Admiral Potebacker	24
Geel Croon	25

JACOB MARREL,
Tulpenboek, 1634.
Manuscript index

twenty-seven of the sheets bear his signature or initials, all ninety-one miniatures can be ascribed to Marrel. Each one exhibits the fine detail and exquisite colouring that are the hallmarks of his style. The insects that decorate many of the pages have been depicted with the same skill; there is a handsome stag beetle on folio 60, while a spider with an insect trapped in its fine-spun web appears on folio 88.

Many *Tulpenboeken*, or catalogues illustrated with paintings of tulips, were commissioned by floriculturists in this period from a number of artists in addition to Judith Leyster and Marrel. It seems probable that this manuscript represents just such a catalogue, for—unlike the traditional florilegium, whose scope was primarily aesthetic—a complete index is s upplied and each flower has been carefully labelled. The vast array of tulips depicted include the red and white variegated

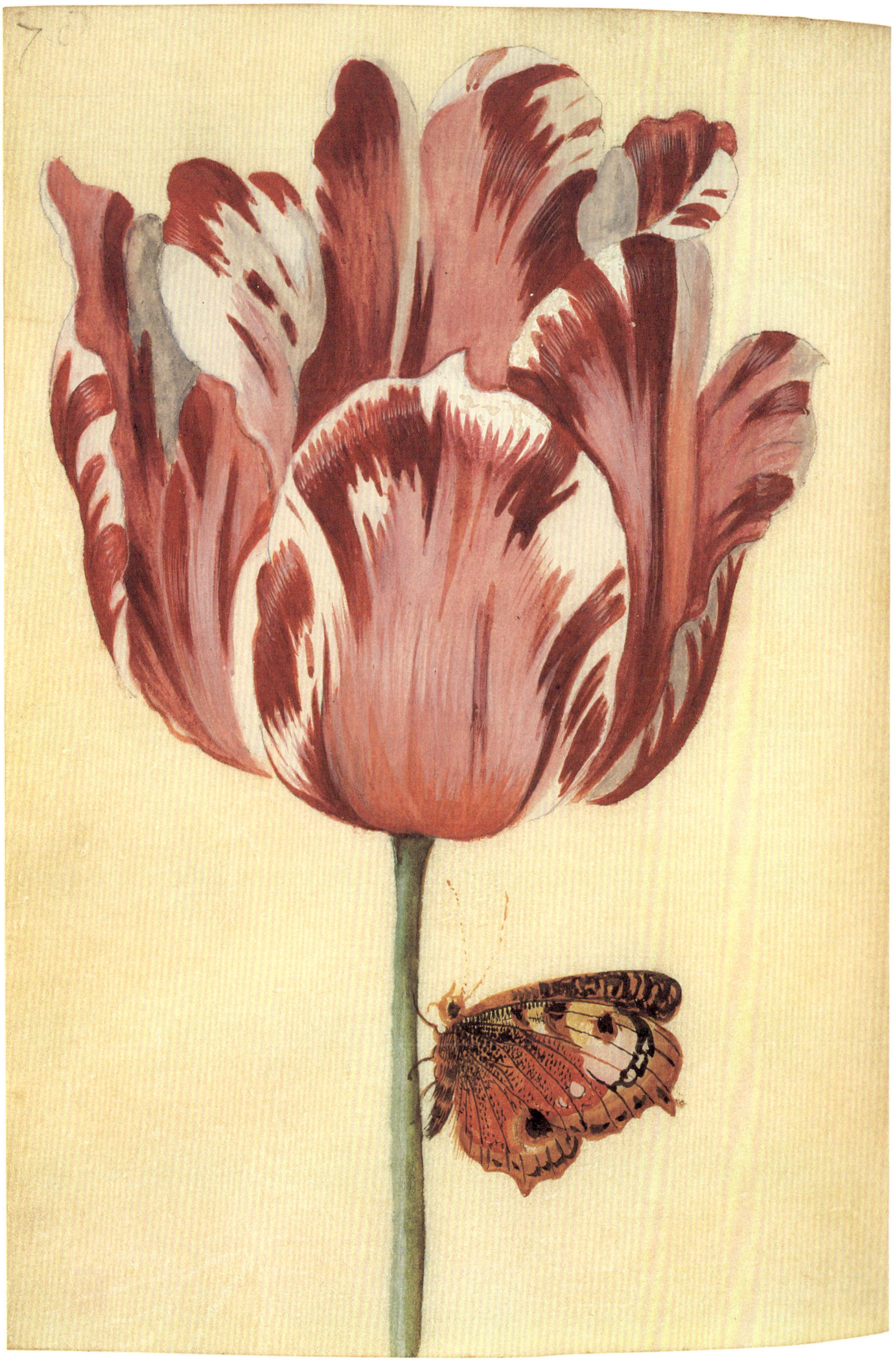

JACOB MARREL,
Tulpenboek, 1634.
Tulipa sp. and butterfly,
folio 78v

'Semper Augustus' (fol. 18), the handsome white and violet 'Viceroy' (fol. 19), and a magnificent 'Admiral Pottebacker' with yellow petals streaked with red (fol. 24).

Oak Spring has the largest known collection of tulip paintings by Marrel, although three other albums of miniatures have been documented (Bergström 1984). Among them is a work in bodycolour on paper attributed to Marrel, representing two *Iris germanica*. A *tulpenboek* compiled between 1637 and 1645 is in the print room of the Rijksmuseum, Amsterdam; another, commissioned by a Portuguese floriculturist living close to Haarlem, eventually passed from the Krelage collection to a private collection in Paris; while a third album is in a private collection in Germany. Some loose pages of tulip paintings in bodycolour can also be found at the Museum of Fine Arts, Boston, and at the Metropolitan Museum of Art, New York (Frans Halsmuseum *Catalogus*, pp. 17–18; *Flowers of Three Centuries*, p. 42).

74. Jacob Marrel (1614–1681)

'Tulpenboek'.

[*manuscript lettering framed by a garland formed of two living vines, around which are strewn gardening tools, an overturned vase containing a tulip, and some large round seeds*] Jacobus Marrellus Franckendalensis delineavit ad vivam [*sic*], hodie floru[m] sculptor Traiectensis, Anno CIƆ IƆ CXLII [1642].

35.5 × 23 cm. 96 unbound leaves with a manuscript title-page laid-in blank leaves. Following the title-page is a folded leaf, unbound, containing a manuscript list of the plants illustrated. Neither the loose plates nor the leaves of the volume bear leaf numbers.

BINDING: Contemporary sprinkled calf with green satin ties.

PLATES: 95 bodycolour paintings of tulips (31.5 × 20 cm.), with the name of each flower in manuscript.

REFERENCE: Bergström 1984.

ALTHOUGH PAINTED ON PAPER rather than his preferred material, vellum, this catalogue of tulips is undoubtedly also the work of Jacob Marrel. The title-page consists of a carefully composed still-life surrounding an inscription with the artist's name and a biographical note. The note informs us the Marrel was originally from Frankfurt and had worked as a floral engraver in Utrecht, an important centre of flower painting where, in fact, it is known that he lived from 1634 to 1649. This was the period when the tulip was the most popular flower of all the rare and exotic blooms that were being enthusiastically cultivated, the celebrated herbalist John Parkinson states in his *Paradisi in sole paradisus terrestris*.

Every new variety of tulip created by Dutch horticulturists was named after some famous personage—an important general, a well-known beauty, or a figure from antiquity. Avidly sought after by collectors, the prices paid for rare or novel varieties spiralled to dizzying heights, until the market suddenly collapsed in 1637 due to overproduction. From the papers of the painter Daniel Seghers (see nos. 24, 25) we can obtain an idea of the prices that were being asked: in 1635–6 one could acquire four 'Semper Augustus' bulbs for 43,000 florins, an 'Admiral Van Enckhuyzen' for

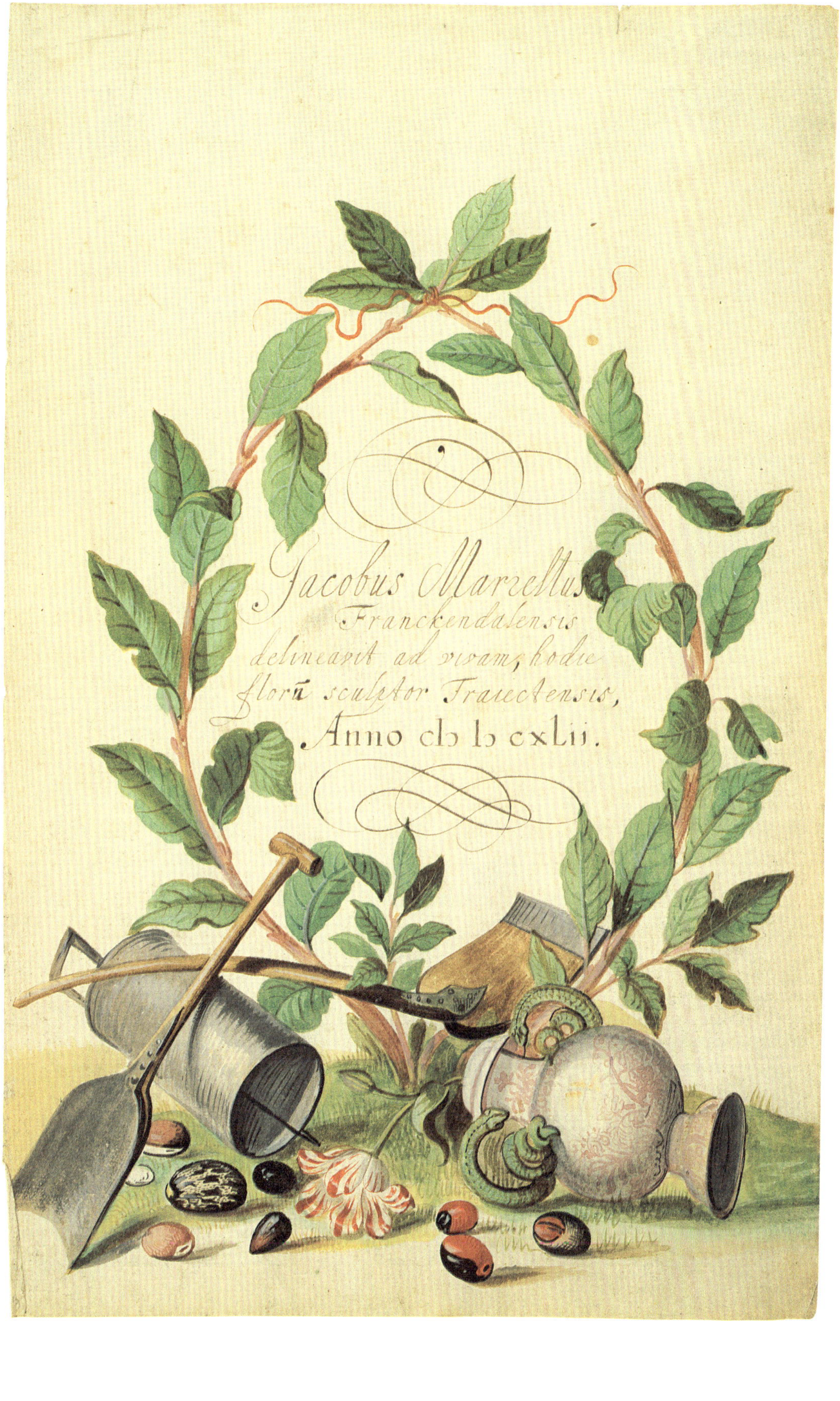

JACOB MARREL,
Tulpenboek, 1642.
Title-page

JACOB MARREL,
Tulpenboek, 1642.
Tulipa sp., 'Doncker root en Stroo-geel',
folio 9

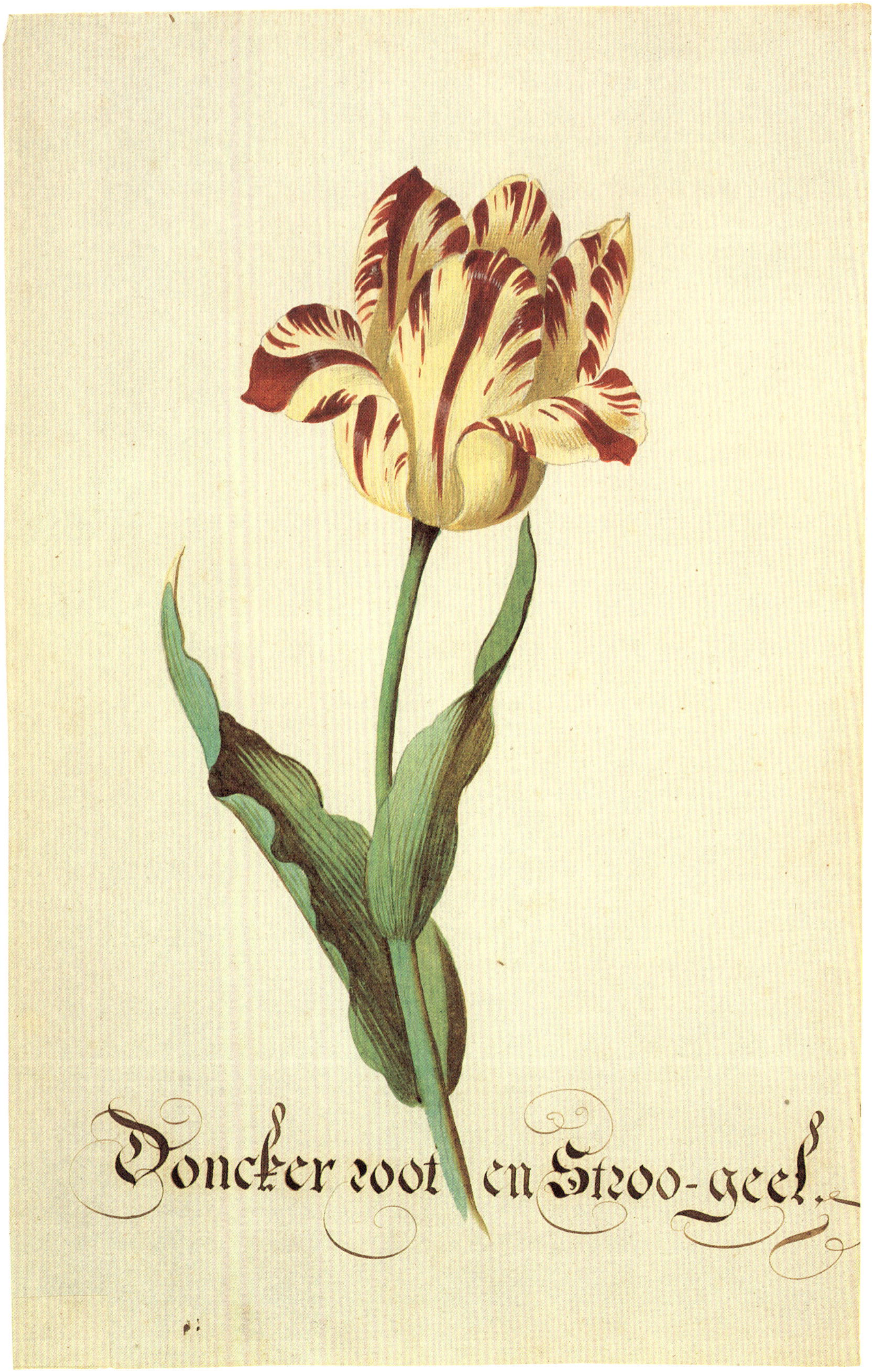

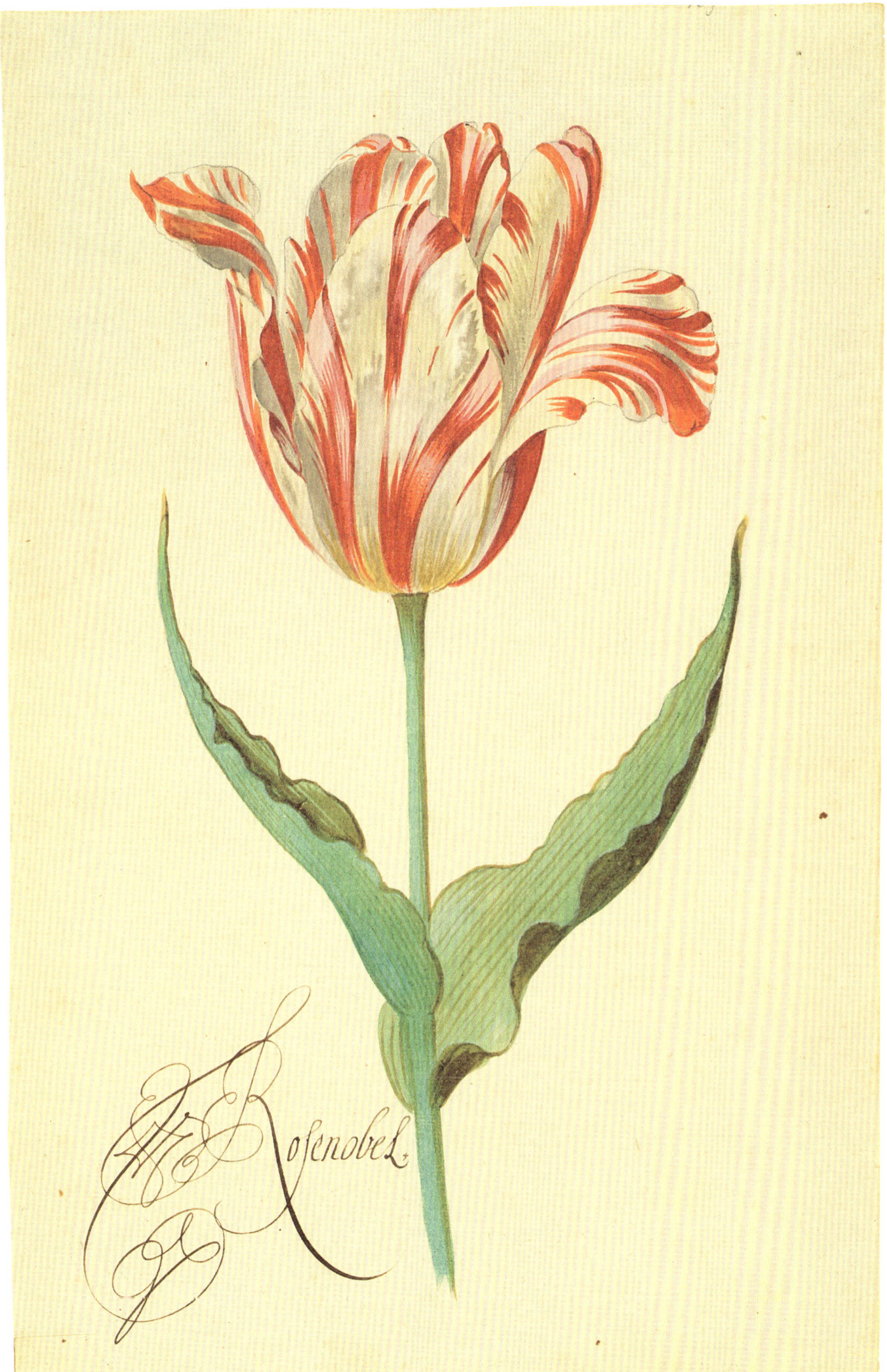

JACOB MARREL,
Tulpenboek, 1642.
Tulipa sp., 'Rosenobel',
folio 80

5,400 florins, and a 'Viceroy' for 4,200. This painted manuscript was clearly intended as a catalogue for tulip fanciers; in it Marrel depicts a vast assortment of tulips in many different forms and colours. At the foot of each he has added the name of the flower in elegant calligraphic lettering with elaborate flourishes. Of particular interest is the register that has been inserted into the volume, listing the flowers illustrated together with their prices based on weight. It opens with the following remarks: 'Register of ninety-nine sorts of the most famous tulips painted from life in the year 1642 and thus preserved from decay through the artistic brush of Jacob Marelles to the memory of the extravagant trade, carried on in them in 1635, 1636 and 1637, adding the weight of their bulbs and the prices at which they were being sold at that time, as far as I have been able to recover them in the "Op-en ondergang van Flora" [Rise and fall of Flora] printed in Amsterdam in 1643'.

Bulbs in this period were in fact sold by the *azen*, one *azan* corresponding to 0.048 grams or 10,240 *azen* to one pound. The practice of selling tulip bulbs by weight and dimension is satirized in a seventeenth-century Dutch painting now in the Musée des Beaux-Arts in Rennes. In another painting, *Persiflage on Tulipomania* by Jan Brueghel the younger (Frans Halsmuseum, Haarlem), a group of monkeys is shown huddled over various activities relating to the cultivation and sale of tulips.

The manuscript opens with an 'Admiral d'Hollande', depicted with its stem curving elegantly to the left and its long, lanceolate leaves painted a brilliant, intense green. It is followed by the celebrated 'Viceroy', much prized for the unusual dark purple streaks on its petals, and by the more modest, but equally beautiful, 'Brabanson'. The 'Bruine purper' is marked by black striations, testifying to the attempts by floriculturists to obtain the elusive 'black tulip'. A rose, the 'General Standaert', is shown with its leaves nibbled away at the edges by snails. The prices in the Register show that the value of tulip bulbs had fallen precipitously by 1642, so that it was possible to acquire a 'Viceroy' for a mere 2,700 florins. However, prices are not given for some of the most valuable species, such as the 'Admiral d'Hollande', 'General d'Hollande', 'Grand Tameralan', 'Merveille d'Hollande', 'Marcus Aurelius' and 'Semper Augustus'. Although this manuscript was originally intended to be no more than a catalogue for the selling of tulips, Marrel's talent made of it a work of art, a beautiful example of the genre of flower painting by an outstanding artist.

75. Jacob Marrel (1614–1681) [*attributed*]

'Tulpenboek'.

38 × 27 cm., manuscript, 4 front free endpapers, 65 bodycolour paintings on vellum, each with a guard sheet bearing an unidentified watermark, followed by 9 endpapers. On a slip of paper a manuscript list (on the recto and verso) precedes the first plate. On a slip tipped-in before the first front free endpaper, in manuscript: 'Drawings of Tulips. Sixty-six Drawings on vellum, with the name of each flower, beautifully executed by a Dutch artist, bound in Dutch vellum. About

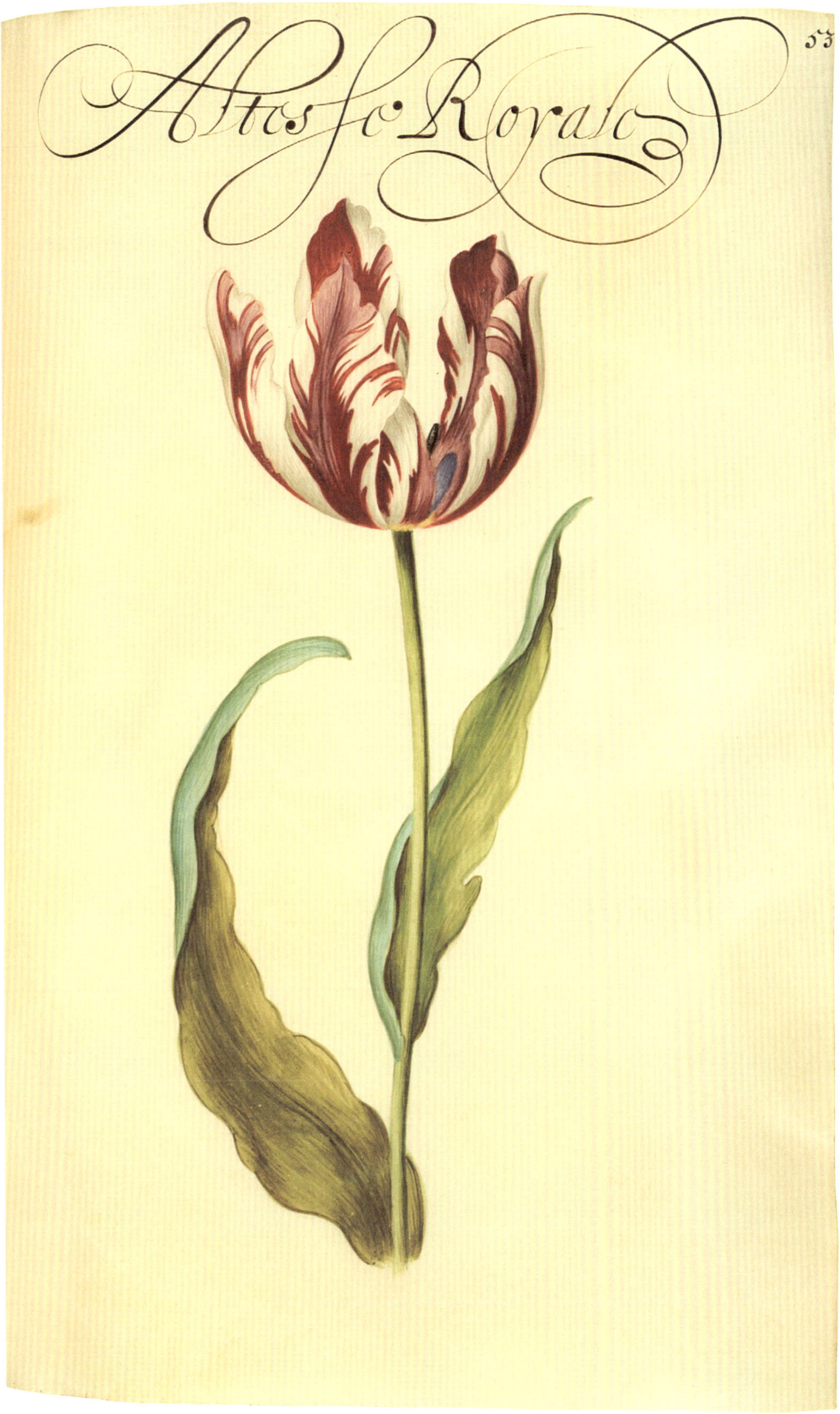

JACOB MARREL, attributed, Tulpenboek. *Tulipa* sp., 'Altesse Royale', folio 53

JACOB MARREL,
attributed, Tulpenboek.
Price-list

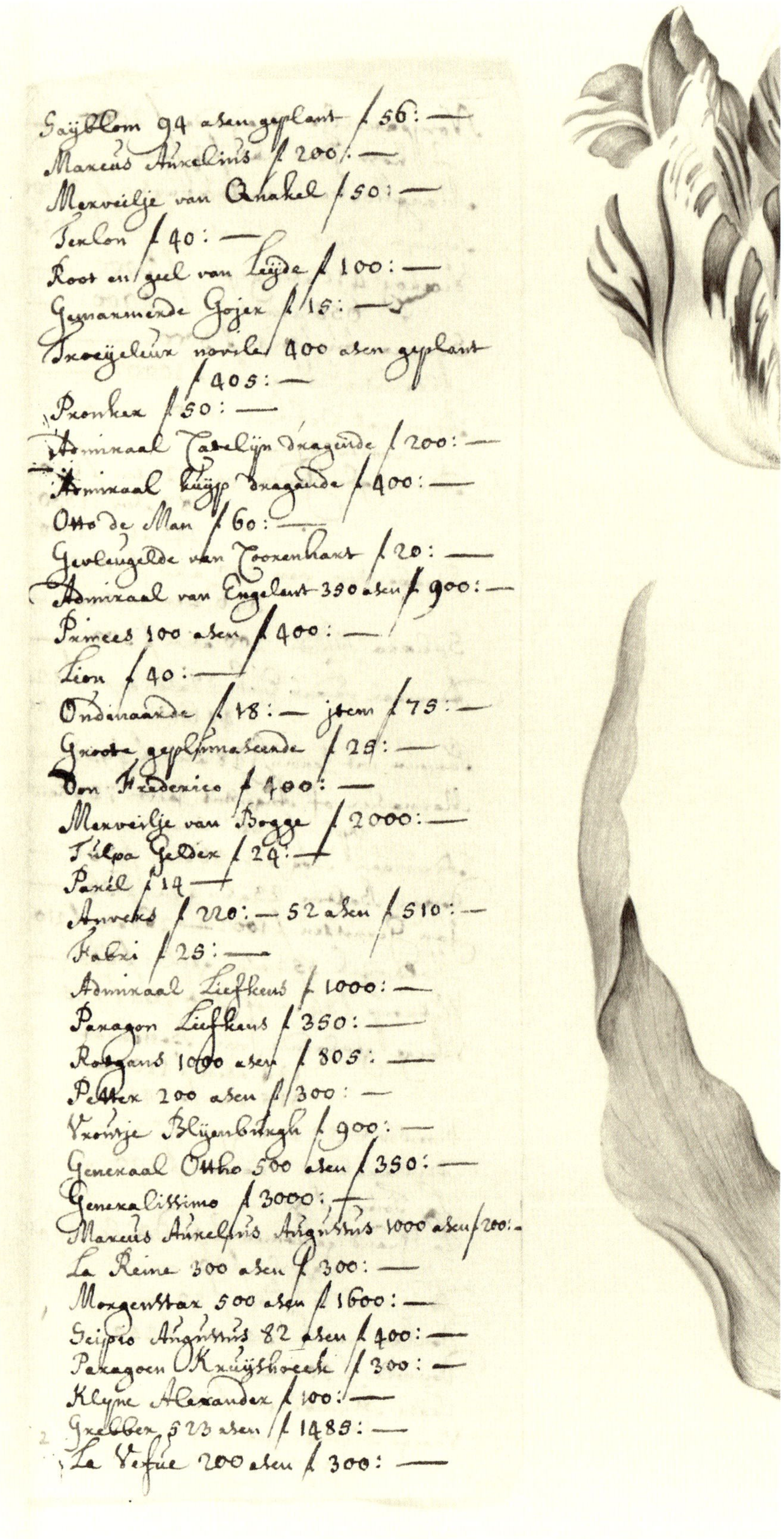

Gaijblom 94 asen geplant f 56:—
Marcus Aurelius f 200:—
Merveilje van Quakel f 50:—
Senlon f 40:—
Root en geel van Leyde f 100:—
Gemarmerde Gojer f 15:—
Vroeijeleux novile 400 asen geplant
f 405:—
Pronker f 50:—
Admiraal Catelijn dragende f 200:—
Admiraal Ruijs dragande f 400:—
Otto de Man f 60:—
Gevlamgelde van Tooxenhart f 20:—
Admiraal van Engelant 350 asen f 900:—
Princes 100 asen f 400:—
Lion f 40:—
Oudenaarde f 18:— item f 75:—
Groote geplumaseerde f 25:—
Don Frederico f 400:—
Merveilje van Bogge f 2000:—
Tulpa Gelder f 29:—
Parel f 14—
Anvers f 220:— 52 asen f 510:—
Fabri f 25:—
Admiraal Liefkens f 1000:—
Paragon Liefkens f 350:—
Rotgans 1000 asen f 805:—
Petter 200 asen f 300:—
Vrouwtje Bluijenburgh f 900:—
Generaal Ottho 500 asen f 350:—
Generalissimo f 3000:—
Marcus Aurelius Augustus 1000 asen f 200:—
La Reine 300 asen f 300:—
Morgenstar 500 asen f 1600:—
Scipio Augustus 82 asen f 400:—
Paragoen Kruijshoeck f 300:—
Klyne Alexander f 100:—
Grebber 523 asen f 1485:—
La Vefue 200 asen f 300:—

1680'. Between the first two front hand papers another tipped-in slip on which is written in manuscript 'N.B. Numbers 3, 11, 13, 42, 62, 64, 65, 71 appear to have been stolen from this Tulip Book. H.'

BINDING: Contemporary vellum with panels of double gilt fillets, and a crown at each corner of the central panel. On the spine in manuscript: 'Tulpe-Boeck'. The binding and tooling are very similar to those of the three volumes of vellum drawings by Jan Withoos (see no. 21), attributed to the Amsterdam bookbinder Albert Magnus.

PLATES: 65 plates of tulips in bodycolour. Plates 3, 11, 13, 42, 62, 64, 65 and 71 are missing. In every plate the name of the flower has been added in manuscript.

PROVENANCE: Ex-libris 'Philip Lord Harwicke Baron of Harwicke in ye County of Gloucester', and that of Paul Mellon. In addition, there is a small label, 'Heirloom 105'. On the first front free endpaper is the bookplate of Crewe Hall (Cheshire).

THIS REPRESENTS a fine example of the numerous *Tulpenboeken* that circulated widely in the Low Countries in the seventeenth century, produced for the delectation of flower lovers and as catalogues for the sales of plants and bulbs (see no. 73). Indeed, the paintings are preceded by a manuscript list of the prices asked for the flowers (see no. 74). In these paintings the tulips, with their elegantly elongated stems and leaves, stand out vividly against the white vellum folios. They testify to the exceptional skill of the artist, who surpassed himself in the rendition of the finest details, from the delicate tracery of veins in the petals and leaves to the irregular spots and holes left by ravaging insects. In a few paintings (fols. 16, 21, 36, 52) other flowers are also depicted, but the regal tulip invariably dominates the design. Among the most impressive paintings are the 'Admiral van der Eyck' (fol. 12), 'General Standaert' (fol. 22), 'La Belle Gabrielle' (fol. 28) and the 'Altesse Royale' (fol. 53).

As in the autograph manuscript by Marrel at Oak Spring (no. 73), each folio in this work bears the name of the flower written in fine calligraphy, this time in the upper margin. The ornate capital letters and elegant script add the appropriate finishing touch to these splendid paintings. The prices cited in the manuscript list at the front of the work are significantly lower than those to be found in the manuscript of 1642 (no. 74), suggesting that it was compiled at a later date. In fact, a 'Viceroy' bulb, the famous white tulip striated with red that in 1642 cost 2,700 florins per *azen*, in this list is quoted at 3,000 florins for 410 *azen*.

The unusual quality of these paintings, which denotes the hand of a true master in the art of flower painting, and their marked stylistic similarities to the works in the autograph manuscript, suggest that this work may be attributed to Jacob Marrel, who is known to have specialized in the painting of tulips, and who was active right up until 1681, the year of his death.

76. Jean Paul de Rome d'Ardène (1689–1769)

Traité des Renoncules, Dans Lequel Outre Ce Qui Concerne ces Fleurs, on trouvera des Observations Physiques, et Plusieurs Remarques utiles, soit pour l'Agriculture, soit pour le Jardinage. [*printer's decoration*] A Paris, Rue Saint Jacques, à la Vérité, Chez Ph. N. Lottin, Imprimeur-Libraire, et Augustin-Martin Lottin, Fils, Libraire. M. DCC. XLVI. Avec Approbation & Privilege du Roi.

8º 21 × 12.5 cm. π^4 (π^4 + 1) A–R^8 S^4 χ2 *i–x* 1–279 *280–284* p., 7 leaves of plates (4 folded).

BINDING: 18th-century mottled calf binding; floriated gilt spine with red title label: 'Trait Des Renon'.

PLATES: Woodcut title vignette, initials, head- and tail-pieces. Six etched plates. Etched half-title page, designed by Le Sueur and engraved by B. Audran, depicting Flora with an angel overhead bearing a trumpet and a banner which reads 'Les Renoncules'. Beneath the engraving: 'Flore au milieu de ce Jardin | Qui se ranime à sa présence, | À la fleur qu'elle tient en main | Donne une juste préférence.'

REFERENCES: Pritzel (1st edn) 263; Hunt 526.

AROUND the middle of the eighteenth century the French priest and botanist Jean Paul de Rome d'Ardène resigned his post as *Supérieur* of the Collège de Marseille and retired to the Château d'Ardène in the diocese of Sisteron, where he established a garden and devoted himself to botany. He wrote a series of treatises, each one dedicated to the cultivation of a particular flower prized by the botanists, floriculturists and connoisseurs of the period. The treatise on the ranunculus, published at Paris in 1746, was the first, to be followed by others on the hyacinth (1759), the tulip (1760) and the carnation (1762), all published at Avignon. In 1767 Ardène also published, again at Avignon, *Abrégé des instructions sur le jardinage, qui font partie de l'année champêtre.*

Traité des renoncules opens with a fine etched title-page designed by Nicolas Le Sueur (1691–1764) and engraved by Benoît Audran the younger (1698–1772), both of whom were members of celebrated artistic families. The winged figure at the top of the page represents Fame, who has perenially guided the fortunes of this popular flower. In the background one can glimpse a restless sea with two ships, from one of which proudly waves the flag of the Crusades, a reminder that the first ranunculi were brought to Europe from Persia by the Crusaders. Further beyond is the island of Candia (Crete), where the flower could be found growing wild. In the foreground Flora strolls through a garden with a basket of flowers under one arm, her gaze fixed on the ranunculus she holds up in her right hand.

As originally planned, *Traité des renoncules* was to provide a fully illustrated catalogue of the best-known species of ranunculi, correctly designated by their scientific names. The author sought without success a contemporary 'Aubriet' to collaborate with him in this task—that is, an artist with a talent to equal that of the great Claude Aubriet, who had accompanied Tournefort on his voyage to the Levant and who subsequently became a painter at the French court. Ardène was also dissatisfied with the limits of the techniques of engraving and etching, which could not reproduce the colours that formed part of the very essence of a flower; thus, such images were destined to remain, as he puts it, 'mute'. The author therefore resigned himself to producing a more limited work,

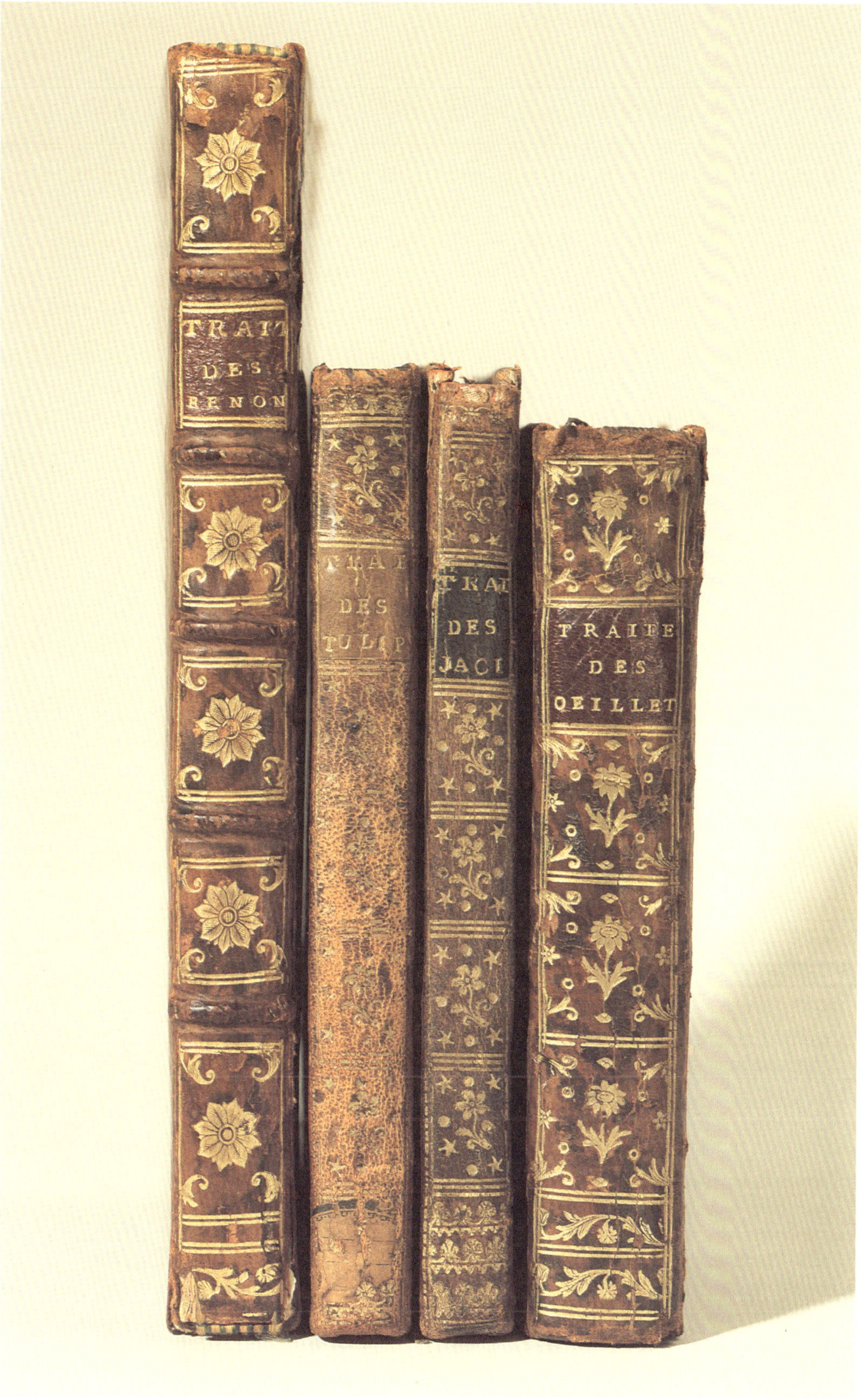

JEAN PAUL DE ROME D'ARDÈNE. Spines of the four titles: *Traité des renoncules*, *Traité des tulipes*, *Traité sur la connoissance et la culture des jacintes*, and *Traité des oeillets*

which he divided into three parts. The first supplies a brief history of the plant, in which Ardène demonstrates his vast erudition. In the second part he describes how the plant should be cultivated, drawing on his personal experience and studies. The third part contains six plates depicting various ranunculi together with enlarged details accompanied by brief explanations.

Ardène's work was addressed not only to florists and gardeners, but to scientists, flower lovers and cultivated persons generally, and is therefore full of information by turns erudite and practical. Moreover, every page of *Traité des renoncules* is infused with the author's profound love for his subject: as he muses at the close, 'au milieu d'elles habitent l'innocence & la candeur; les noirs chagrins, les remords cuisans osent rarement approcher des parterres, ou en sont bientot bannis. Cherchez donc dans ces aziles fortunés l'inestimable paix, la joie tranquille que vous désirez, & que vous ne trouvez que difficilement ailleurs' (p. 251).

Traité des renoncules was a great success: by 1763 it was in its third edition, while a German translation was published in 1754 by G. L. Hulth in Nuremberg. Thus encouraged, Ardène turned his attention to other flowers, producing *Traité sur la connoissance et la culture des jacintes* and then *Traité des tulipes* and *Traité des oeillets*, each of which contains detailed historical observations, concrete advice on the cultivation of the plant and a list of earlier works on the same subject. Each of these treatises was reprinted in numerous editions. At Oak Spring there is a copy of the edition of 1762 of *Traité des oeillets* and copies of the editions of 1765 of both *Traité sur la connoissance et la culture des jacintes* and *Traité des tulipes*.

77. Pieter van Loo (1731–1784) *and* Cornelis van Noorde (1731–1795)

[*Choix de Jacintes*] manuscript made up of 13 bodycolour paintings on paper of double hyacinths, dated 1765 and 1769.

2° 48 × 29 cm. a–c² d1 e–k² 31 unnumbered leaves.

BINDING: Red morocco with gilt floral frame and dentelles; large gilt-tooled hyacinth on each cover. Spine gilt-tooled with bearings from the arms of Soubise.

PLATES: 13 bodycolour paintings of double hyacinths, with the names of the varieties lettered at the top and signatures and dates below (the remaining leaves are blank):

1: *François Premier*, signed at the lower left 'Pr Van Loo; ad vivum del; 1769'.

2: *Marechal De Soubise*, signed at the lower left 'Pr Van Loo ad vivum del; 1769'.

3: *Fontaine Bleau*, signed at the lower right 'C. V. Noorde ad viv: del: 1765'. Bodycolour painting on paper, mounted on a page of the manuscript.

4: *Gloria Florum Suprema*, signed at the lower left 'Pr van Loo; ad vivum del; 1769'.

5: *Ophir*, signed at the lower right 'Pr: Van Loo. Fecit. 1769'.

6: *Comeet*, signed at the lower left 'Pr Van Loo ad vivum del; 1769'.

7: *Prince de Piedmont*, unsigned.

8: *Flavo Superbe*, signed at the lower left 'Pr Van Loo Fecit 1769'.

9: *Pythagore*, signed at the lower left 'Pr Van Loo ad vivum del 1769'.

10: *Heroine*, signed at the lower left 'Pr Van Loo ad vivum del; 1769'.

11: *Rose Virginale*, signed at the lower left 'Pr Van Loo; ad vivum del; 1769'.

12: *Parhelie Solaire*, signed at the lower left 'Pr Van Loo; ad vivum del 1769'.

13: *Goud-Myn*, signed at the lower left 'C. V. Noorde, fecit'.

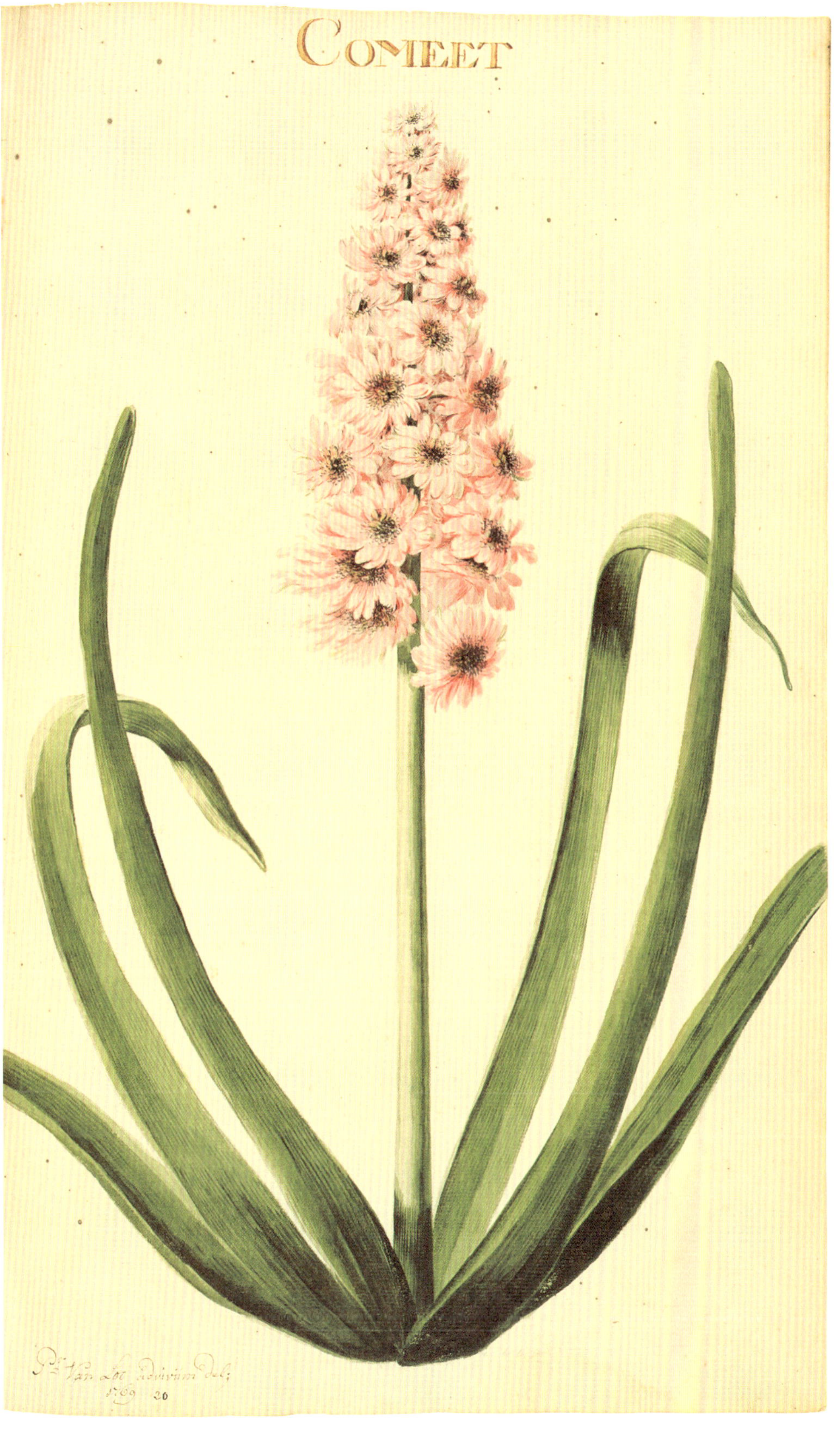

PIETER VAN LOO and
CORNELIS VAN NOORDE,
Choix de jacintes.
Double pink *Hyacinthus*,
'The Comeet',
folio 6

And on lower right '1765'. Bodycolour painting on paper, mounted on a page of the manuscript.

PROVENANCE: This manuscript was prepared for Charles de Rohan, Prince of Soubise (1715–87), a favourite of Mme Pompadour. He led the French to defeat in the Seven Years War, but was nevertheless later named Marshal of France. The insignia that decorates the spine of the volume testifies to his title, as does the imposing blue hyacinth bearing his name pictured on plate 2. This volume once formed part of the collection of Rachel McMaster Miller Hunt, whose ex-libris is attached to the front pastedown.

REFERENCE: Hunt 613.

PIETER VAN LOO (or LOON), the principal author of this remarkable manuscript, was born in Haarlem in 1731 and passed the whole of his life there, where he was officially registered in the city guild as a 'painter of flowers', although he was also active as a landscape painter and a designer of tapestries. Some bodycolour paintings by him, of flowers arranged in vases, are in the Albertina, Vienna, and the Fitzwilliam, Cambridge (*Flowers of Three Centuries*, p. 41); they show the influence of the celebrated artist Jan van Huysum. A painting by van Loo, *Hyacintus orientalis flore pleno*, was etched and included in the famous work *Icones plantarum rariorum* (Haarlem, 1793) by the Haarlem nurseryman George Voorhelm Schneevoogt (1775–1850). Another signed painting of a hyacinth, *L'Admiration*, quite similar in style to the works in the Oak Spring manuscript, was sold in New York at Sotheby's in 1993.

Eleven of the paintings in *Choix de jacintes* have been signed by van Loo, but two others, dating from 1765, are the work of Cornelis van Noorde, a draughtsman, engraver and drawing-master who also lived in Haarlem. The manuscript itself was probably collated *c.* 1769, the date that appears on van Loo's works, and van Noorde's two paintings were inserted at that time, being pasted onto two blank pages in the volume. Described in Hunt as the 'apotheosis of the Double Hyacinth' (p. 340), this volume presents the great reputation of the flower, especially in the second and third decades of the eighteenth century (Krelage 1942a, pp. 142–96).

The hyacinth, with its increasing number of available specimens with greater colour nuances and regularity of form, was praised for its unique characteristics. The Haarlem florist George Voorhelm (1711/12–87) in his first treatise, *Traité des jacintes* (Haarlem, 1752), mentions eight points to underline the excellence of the hyacinth and its sublimity over other flowers: 1. After winter the hyacinth shows itself a harbinger of spring. 2. Its smell is to be praised. 3. The single-flower hyacinth shows a perfect bouquet at its branch. 4. The hyacinth is faithful (unlike the tulip), only seldom does the colour change from blue towards white. 5. The hyacinth doesn't degenerate in fifty years (unlike the carnation, which is worthless after four to five years, and the anemone after twenty). 6. It can be grown without difficulty at either end of Europe. 7. Its flowering can be advanced at an earlier time (up to three months), either in earth or in water. 8. The hyacinth is less commonly known, so it is more rare than other flowers.

According to the same author, the hyacinth had to have a bulb of moderate size, the branch should be stout and regular in height, the leaves upright, but also horizontal. The little flowers, completely filled out like little roses, are to be attached to the branch, water-levelled, the biggest at

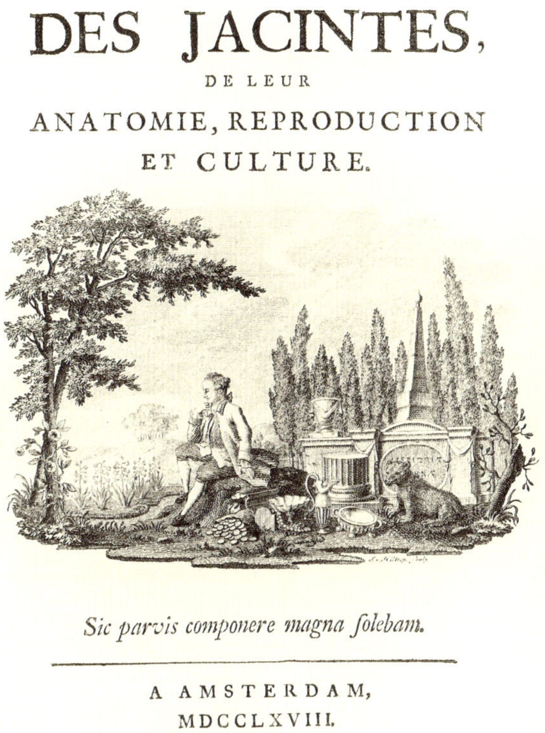
DES JACINTES,
DE LEUR
ANATOMIE, REPRODUCTION
ET CULTURE.

Sic parvis componere magna solebam.

A AMSTERDAM,
MDCCLXVIII.

MAXIMILIEN-HENRI DE SAINT-SIMON, *Des Jacintes, de leur anatomie, reproduction et culture*, Amsterdam, 1768. Title-page

the bottom, the little branches getting short towards the top. The bunch of flowers should form a pyramid, whereas the top flower should stand upright. All these qualities are present in the flowers painted in this manuscript by van Loo and van Noorde, as well as in those painted by other artists, such as Jan Laurensz. van der Vinne II (*Flowers in Books and Drawings*, no. 94, pl. 15).

The hyacinth was much praised in every European country, but especially so in France. After the brief speculation period, Mme de Pompadour, whose favourite, Charles de Rohan, was once the owner of this manuscript, admired the flower very much. Between 1745 and 1765 she motivated Louis XV to buy flowers, plants and trees for their gardens, especially for those at Choisy. Each year the King ordered bulbs from the Haarlem florists worth between 6,000 and 8,000 Dutch *ponds*. This stimulated others in France to make purchases, and the Haarlem florists enjoyed golden years.

We owe to another Frenchman, the polygraphic marquis Maximilien-Henri de Saint-Simon (1720–99), the most important treatise of the eighteenth century on hyacinths: *Des Jacintes, de leur anatomie, reproduction et culture*, published in Amsterdam in 1768; a copy of the edition of 1794 is at

Oak Spring. Saint-Simon spent part of his life in Holland, at Haarlem, where he greatly admired the views to be had of the fields of hyacinths: 'Le spectacle en est si brillant dans les environs d'Haarlem qu'un Amateur mème est trop ébloui du premier coup-d'œil, pour pouvoir jouir des charmes singuliers de chaque espece ou varieté. On y voit des arpens entiers couverts de jacintes doubles et simples, sans nouvelle intervalle, que celui des sentiers nécessaires pour leur culture' (*Des Jacintes*, 1794, p. 5).

Going beyond the standard treatises on hyacinths published in the eighteenth century, which generally limit themselves to a simple enumeration of the different species known, Saint-Simon approached his subject from a genuinely scientific perspective, discussing the complex problems of anatomy, generation and classification. The plates that illustrate his work had to be scientifically accurate in every detail, and therefore they lack much of the aesthetic charm of van Loo's works. The title-page, however, bears a humorous vignette etched by J. Kiltrop showing a gentleman gazing raptly at a vast field of hyacinths, oblivious to the heap of precious objects at his feet. The motto accompanying the vignette states: 'Sic parvis componere magna solebam' (Thus was I used to join great things to lesser ones).

78. Mary Lawrance (d. 1830)

[*printed in calligraphic style*] A Collection of Roses from Nature. London. Published by Miss Lawrance, Teacher of Botanical Drawing, &c. N° 86 Queen Ann Street, East, Portland Place, 1799. Vincent fer. & feu.

40 × 32 cm. Frontispiece, calligraphic title-page, calligraphic dedication page, and two leaves of letterpress.

BINDING: Morocco with gilt decoration. On the floriated spine 'Roses by Miss Lawrance' and, at the base, 'London 1799'.

PLATES: 90 etched and hand-coloured plates, some of them with stipple engraving, numbered 4 to 90. Plate 14 is lacking. Etched, stipple-engraved and hand-coloured frontispiece with a garland of roses surrounding the word 'Frontispiece' printed in calligraphic script. Every plate is signed 'M. Lawrance del. et sculp.' The dates of the plates vary from 1796 to 1799.

REFERENCES: Dunthorne 215; Pritzel 5100; Nissen 1151; Sitwell and Blunt, p. 26; Henrey, II.580–81 and III.948; *Printmaking in the Service of Botany*, no. 33.

THIS IS THE EARLIEST known published work entirely dedicated to the rose, a flower that enjoyed an extraordinary vogue towards the end of the eighteenth century. The author, also known by the name of Mrs Kearse, taught flower painting in London. Her lessons, priced 'at ½ guinea a lesson & a guinea entrance', had many takers. She also produced a number of flower books, among them the charming *Sketches of Flowers from Nature*, published in 1801. However, her fame as an artist is linked to one work in particular, the *Collection of Roses from Nature*. This large illustrated volume, begun in 1796 and printed in thirty parts, was finally published in its entirety in 1799 together with a dedication to Queen Charlotte. It met with such a warm reception that the artist felt

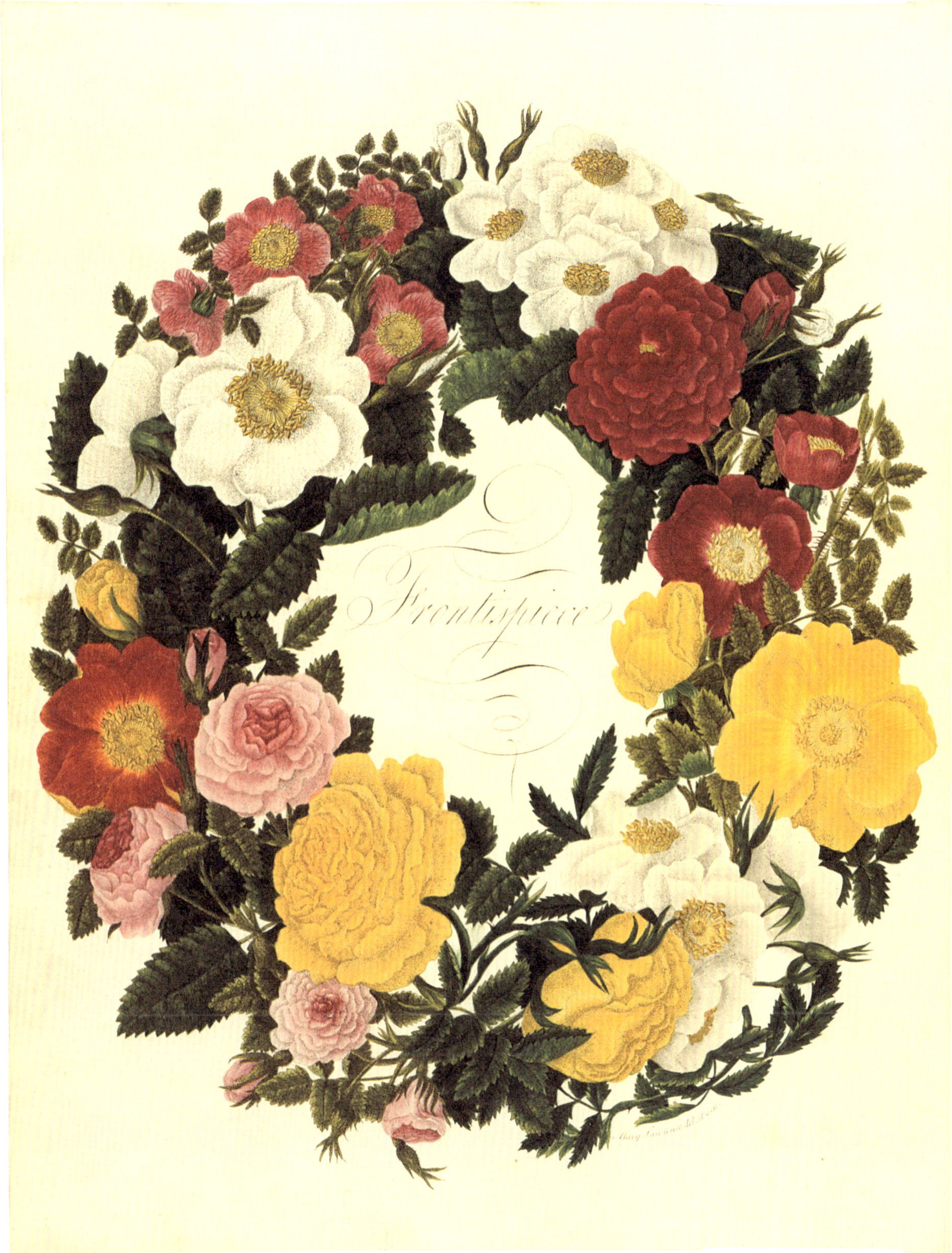

MARY LAWRANCE,
A Collection of Roses.
Frontispiece

encouraged to produce another volume, this time *A Collection of Passion Flowers*, the plates of which appeared between 1799 and 1802, two copies of which are at Oak Spring.

Lawrance's *Collection of Roses* is of considerable interest from a horticultural perspective, for it shows the varieties of the rose in existence at the end of the eighteenth century. Each plate presents a stem laden with buds and flowers; of particular note, because of its brilliant, yellowish-brown petals, is the 'Austrian Copper' (pl. 6), which the artist called 'Rosa lutea ß bicolor'. As an artist, Lawrance (who first exhibited her work at the Royal Academy in 1794) is generally not counted among the first rank. Her plates are somewhat summarily worked, and critics have faulted her less than sure hand with the new technique of stipple engraving, which, when skilfully used, can convey the most delicate modulations in hue of living flowers. Dunthorne (see his no. 215), for example, noted that she was not entirely successful in reproducing the complex modelling of the rose petal, so that while the plates are pleasingly decorative, they are not botanically accurate.

Lawrance's work, however, deserves a fresh evaluation; it cannot have been a simple task to present this flower for the first time in these ninety large folio plates. Furthermore, from an aesthetic standpoint, the sheer beauty of the frontispiece provides sufficient inducement to forgive some of the faults and inaccuracies to be found in the plates. It is dominated by a magnificent garland of roses, etched, stippled and hand-painted by the artist. The garland springs vividly from the page—its colours, in particular the rich range of greens and the silky red, yellow and white of the roses, rendered more brilliant by a coat of gum arabic. This page can certainly be counted among the most charming in the history of floral illustration.

79. Charles Malo (1790–1871)

Histoire des Roses par Charles Malo. Ornée de 12 Planches, en couleur, dessinées par P. Bessa. [*title vignette* 4 × 3.5 *cm.*] A Paris Chez Louis Janet, Libraire, Successeur de son Père, Rue S. Jaques N° 59. [*on the verso of the half-title*] Imprimerie de P. Didot, L'Ainé, Chevalier De L'Ordre Royal De Saint-Michel, Imprimeur du Roi. [1818]

12° 13 × 8 cm. *1–2* i–iv 1–240 p.

BINDING: Green morocco with gilt frame and blind-stamped central lozenge. Spine with 'Histoire des Roses' in gilt.

PLATES: 12 unnumbered plates, stipple engraved, printed in colour and finished by hand. Title vignette. Guard leaves throughout.

REFERENCES: Dunthorne 36; Nissen 1266; Plesch 324.

THIS BOOK is a typical example of the sentimental flower book, elegant little volumes that were enormously popular in Europe during the Romantic era. The writer Charles Malo excelled in this fashionable genre, producing numerous works, from *Guirlande de flore* (1815) and *La Corbeille des fruits* (1816) to the *Parterre de flore* (1821) and a *Histoire des tulipes* (1821).

This exquisite little volume, dedicated to the most popular flower of the first decades of the

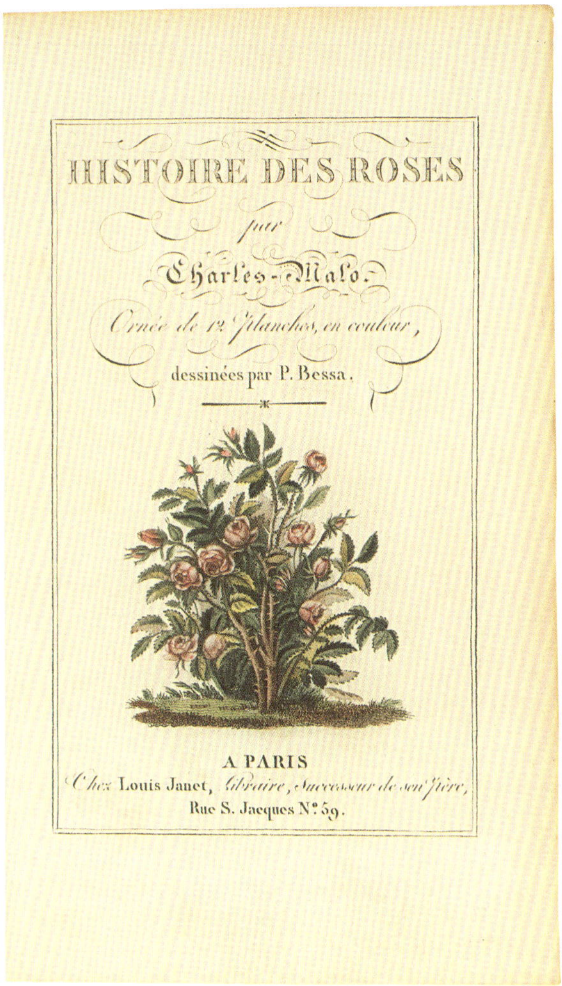

CHARLES MALO,
Histoire des roses.
Title-page

nineteenth century, provides, as the preface states, an 'histoire complète *des Roses* qui réunît, sous un format portatif, le luxe à l'agrément'. Malo traces the history of what he describes as 'le modèle le plus séduisant des graces et de la beauté, le symbole de la galanterie, de la mollesse et de la volupté' (p. 3), beginning with the rose in classical mythology, then describing its role in Christian iconography and historical episodes, and finally its appearance in contemporary tales and popular lore. Descriptions of both well-known and rarer species of the rose are followed by practical advice regarding their cultivation, and some notes on their medicinal, alimentary and cosmetic properties. The indispensable role played by the rose in the art of perfumery is not neglected: 'On en tire des pommades, des huiles et des essences; ainsi, la rose entre dans les eaux de senteur. . . . On fait avec la rose, des savonettes agréables, des pastilles suaves, d'excellents pots-pourris et des sachets odoriférants' (pp. 160–61). After suggesting various procedures for the drying of petals, Malo concludes with an anthology of verse on the theme of the rose.

It is, however, the twelve plates designed by Pancrace Bessa (1772–*c.* 1835), a renowned French

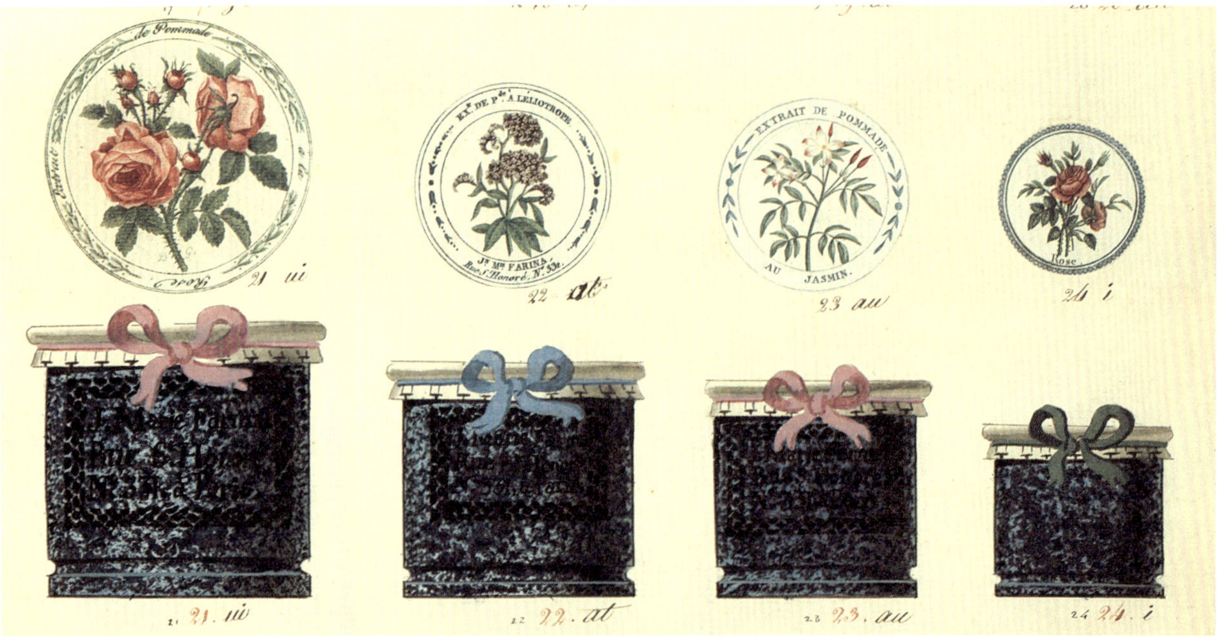

A manuscript trade catalogue of perfumes and other toiletries, produced and distributed by the firm Jean-Marie Farina, Paris, *c.* 1815–20, detail, numbers 21–24

painter of botanical subjects, that set apart this modest example of early nineteenth-century French printing as a true work of art. Bessa had studied with Spaendonck and with the famous Redouté; he was drawing-master to the duchesse de Berry and, beginning in 1816, completed many flower paintings for his patroness. One year before the publication of *Histoire des roses* Bessa was engaged in the engraving of some of the plates for Redouté's masterpiece, *Les Roses*; he thus acquired invaluable experience, which he would draw on frequently during the course of his career, as may be seen from his *Album* on the rose produced in 1830.

Although accustomed to working in far larger dimensions, Bessa demonstrates here a consummate skill in the design of miniaturized versions of the same subject. Each plate depicts two species, usually in contrasting colours, gathered in a graceful bouquet. His drawings for *Histoire des roses* were deftly engraved, using the new technique of stipple, by a certain Teillard, of whom, however, nothing is known.

80. James Bateman (1811–1897)

[*in a lithographed frame imitating Mexican handicraft, surmounted by an eagle with outspread wings holding a serpent in its beak; the eagle is perched on an Indian ficus-tree growing out of a rock*] The Orchidaceae of Mexico & Guatemala by Jas. Bateman, Esqre '—Like restless serpents, clothed | in rainbow and in fire, the parasites, | starred with ten thousand blossoms flow around. | the gray trunks. Shelley'. London, for the Author. Ackermann and Co exct. Jas. Ridgway & Sons. J. Brandard Del. et Lith. Printed by J. Graf LGt Castle SP.

75 × 57 cm. *i–viii 1* 2–12 *1–84* p., XL leaves of plates.

BINDING: Purple morocco binding with gilt borders of fillets and rolls.

PLATES: 40 hand-coloured lithographed plates numbered in arabic, titled with their Latin binomial at the bottom, signed by Ms Drake, Jane Edwards, Samuel Holden or Mrs Augusta Withers, lithographed by Maxime Gauci and printed by P. Gauci. Each plate is accompanied by two pages of descriptive letterpress. Uncoloured wood- or steel-engraved vignettes at the end of each letterpress by Lady Jane Walsh, Lady Grey of Groby, Mrs R. Wilbaham, George Cruikshank, J. Brandard, R. Branston, T. P. Woox, E. Landells, C. V. Skinner and G. Ackermann. Engraved by R. Branston and Ebenezer Landells.

REFERENCES: Savage 1925, p. 41; Pritzel (2nd edn), 470; Nissen, 89; RHS, pp. 45–6; Sitwell and Blunt, p. 48; Desmond 1987, pp. 147–8; P. Hayden, *Biddulph Grange, Staffordshire: A Victorian Garden Rediscovered*, London, 1989; Blunt and Stearn, pp. 249–51.

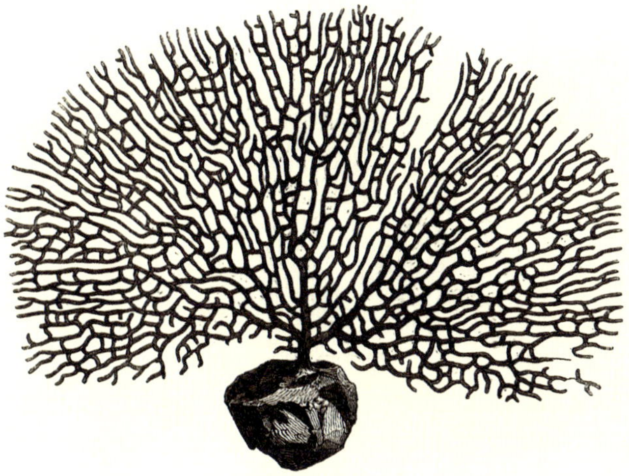

JAMES BATEMAN, *The Orchidaceae of Mexico and Guatemala*. Sea fan (*Gorgonia flabellum*), a vignette for the entry on 'Epidendrum Aloifolium'. This marine organism was formerly thought to be a botanical specimen

JAMES BATEMAN was a keen botanist and horticulturist, and certainly one of the most famous orchidologists of the nineteenth century. Born at Redivals near Bury in Lancashire, at a very early age he began to show an interest in plants, especially the tropical fruits and orchids that were to become his particular passion and to which he devoted a lifetime of study. With the encouragement and financial support of his wealthy father he was able to pursue his studies seriously. Thus, in 1833, when he was just twenty-two years of age, Bateman sent Thomas Colley, the foreman of an Oxford nursery garden, to British Guiana on an orchid-gathering expedition. Later he made the acquaintance of a merchant residing in Guatemala, George Ure Skinner, who shared his interest in tropical flowers. A long and fruitful collaboration grew out of this friendship, for Skinner collected and sent to Bateman many new species of orchid, and together they contributed significantly to the dissemination of these species in Europe.

In recognition of his signal contributions to the science of botany, Bateman was elected a fellow of the Linnean Society in 1833, and a member of the Royal Society in 1838. In the latter year he married Maria Egerton-Warburton, herself an ardent amateur botanist. Together they purchased a house at Biddulph Grange, Staffordshire, a forbidding and inhospitable site on which they nevertheless managed to construct one of the most beautiful landscapes in England, including an oriental garden in which they cultivated many plants native to China.

Bateman published three works on the genus Orchidaceae. The first, and without a doubt the most important, was *The Orchidaceae of Mexico and Guatemala*, published in London between 1837 and 1843. This was followed by *A Monograph of Odontoglossum* produced in 1864–74 with thirty hand-coloured plates lithographed by Walter Hood Fitch (see nos. 104 & 107), and, finally, in 1867 by *A Second Century of Orchidaceous Plants*, which contained a selection of orchids already presented in William Curtis's *Botanical Magazine*, again with plates by Fitch.

The Orchidaceae of Mexico and Guatemala created a considerable stir when it first appeared, at least in part due to its enormous size; indeed it remains today the largest work on the subject of botany ever printed and is familiarly referred to in botanical libraries as 'the librarian's nightmare'. Each of its magnificent plates is followed by two leaves of letterpress with taxonomic observations, advice on cultivation, and scientific, literary or historical notes (such as details about life in Central America), closing with apt literary citations and spirited vignettes. It was dedicated to Queen Adelaide, and had one hundred subscribers, a prestigious list that included members of some of the first families of the realm, as well as 'His Majesty the King of Belgium' and 'His Serene Highness the Grand Duke of Tuscany'. *The Orchidaceae* was printed in just 125 copies, which were sold for twenty guineas apiece, while a set of the lithographed and hand-coloured plates, issued one by one from 1837 to 1848, cost about £20 each (RHS, p. 46).

In the introduction to his work Bateman describes the current status of orchid cultivation in England, citing the Royal Horticultural Society as the most important centre for the study of this vast genus. He also expresses his grateful thanks to John Lindley, the greatest English expert on orchids of the period, who personally encouraged the author in his studies and assisted in the drafting of his book. Lindley himself wrote several important works on the subject, including *Sertum orchidaceum* (1837–41), with forty-nine illustrations painted by Ms Drake and lithographed by Maxime Gauci.

The Orchidaceae is a conspicuous work in part because of its unusual size and subject-matter but also because of the forty extraordinary plates with which it is illustrated. Each depicts an orchid plant in full bloom and in its natural dimensions. On the page the clusters of blossoms form an elegant cascade, their stupendous colours skilfully reproduced by hand-colouring the lithographed plates. Almost all represent newly discovered species which the author named after various English noble families or the estates on which the plants were first successfully brought to flower. Twenty-one of the plates are by Augusta Withers (*fl.* 1827–64) and sixteen by Ms S. A. Drake (*fl.* 1818–47) of Turnham Green, while plate 26 is signed by Jane Edwards (*fl.* 1840) and plate 18 by Samuel Holden (*fl.* 1840). Some of the original paintings are conserved in the Lindley Library in London. Very little is known about the lithographer, Maxime Gauci (1810–46), who signed each of the plates, although Blunt has declared him to have been a true 'master of the process' (Blunt and Stearn, p. 252). Plate

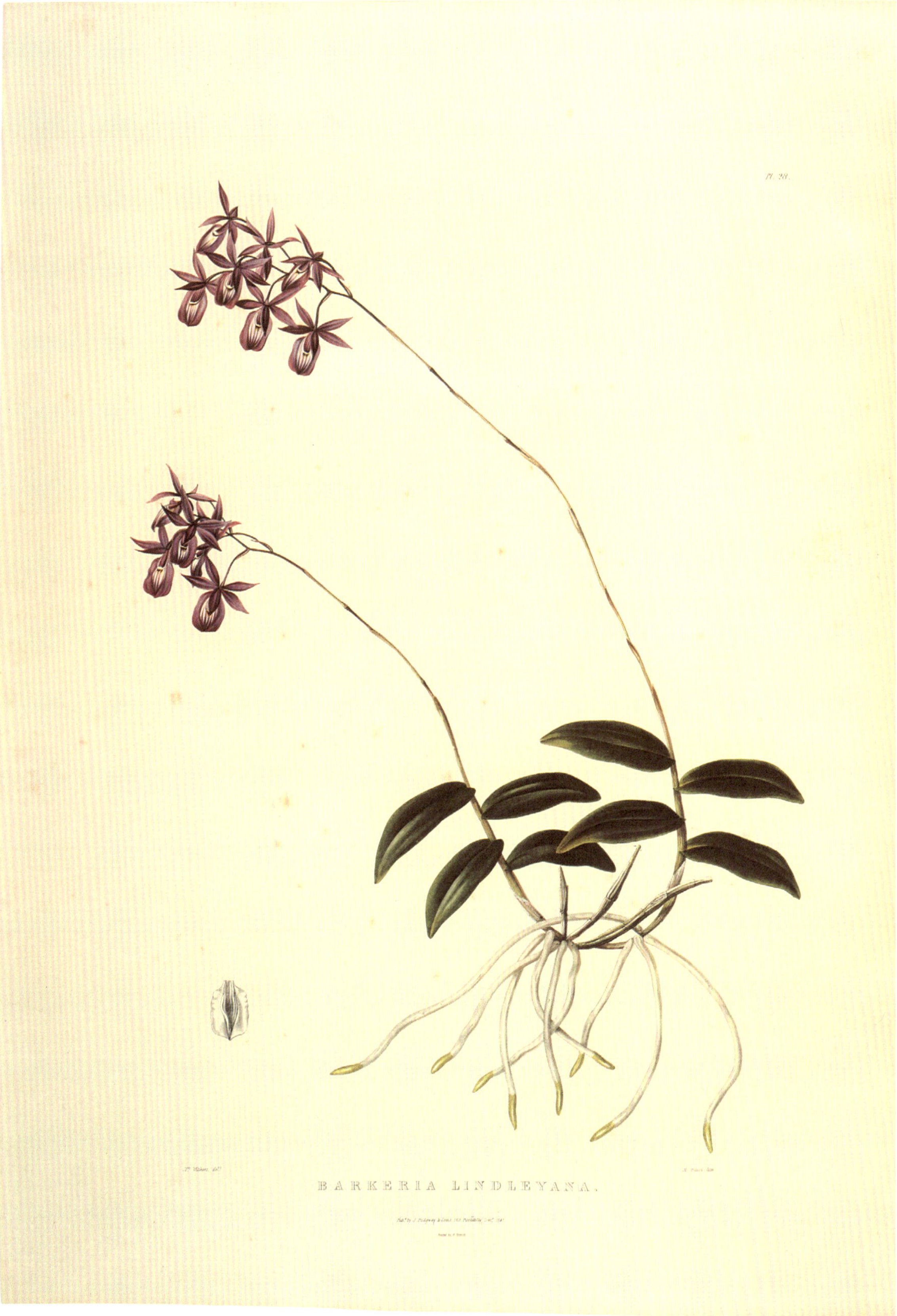

JAMES BATEMAN,
The Orchidaceae of Mexico and Guatemala. Epidendrum lindleyanum,
plate 28

XIII is unusually beautiful; it depicts the *Cattleya skinneri*, thus named by Lindley in honour of William Cattley of Barnet, one of the first botanists to attempt the cultivation of orchids in England, and George Skinner. But perhaps the most splendid illustration of all is plate XXVIII, the 'Barkeria lindleyana' (*Epidendrum lindleyanum*), named after Lindley himself and after George Barker of Springfield, 'one of the most ardent and successful collectors of Orchidaceous Epiphytes'.

A charming series of vignettes serves as a lively coda to the sections of letterpress in *The Orchidaceae*. They are frequently coupled with lines of verse from the great classical or English poets (Virgil, Horace, Juvenal, Shakespeare, Milton) and reflect the eclectic interests of the author. Some of the vignettes depict the handicrafts, architecture or scenes of daily life in Mexico. Others depict naturalistic specimens, such as the frond of algae at the end of the letterpress to plate XXV, which is captioned 'A kind of sea-weed, found with the shells figured under *Odontoglossum grande*'. Many of the vignettes are quite humorous, such as the tiny scene that appears at the end of the Introductory Remarks, in which a confused crowd of tiny figures struggles valiantly, with the aid of tall poles and a pulley, to lift a copy of *The Orchidaceae*. This vignette is signed by the celebrated English etcher and caricaturist George Cruikshank (1792–1878), who also drew the vignette for plate IX, in which various figures seek in vain to squash the enormous insects leaping from a box full of exotic flowers.

VIII · EXTRAORDINARY WOMEN PAINTERS

IT IS KNOWN that Christophe Plantin had three women employed at his celebrated printing-house, the Officina Plantiniana in Antwerp; their task was to colour by hand, under the supervision of expert botanical artists, copies of the important botanical texts that Plantin specialized in publishing towards the end of the sixteenth century (Arber, p. 308). Plantin clearly appreciated the fact that, with their qualities of patience and accuracy, women seemed to be uniquely adapted to this painstaking type of work.

While the annals of history have recorded the names of but a few women artists before 1800—and these mostly the wives or daughters of celebrated male artists—we can be certain that many women gifted with natural talent must have exercised their art in private. Due to social convention they were no doubt encouraged to avoid the more 'virile' themes typical of the monumental genres of historical and religious painting in favour of more 'feminine' subjects, which were generally decorative in nature. Flower painting was in this sense eminently suitable and was indeed sanctioned by long tradition for, as we have seen, floral designs constituted a recurrent theme in the decorative tradition of all the arts, and particularly in many of the feminine arts *par excellence*, such as embroidery and weaving.

Thus, in the first decades of the seventeenth century a handful of women began to establish reputations for themselves as professional artists, particularly in the genres of botanical illustration and flower painting. One of the first was Anna Maria Vaiana (active in Rome *c.* 1630), a contemporary and correspondent of Galileo Galilei, who specialized in garden and botanical subjects. She probably designed and engraved many of the illustrations for G. B. Ferrari's celebrated *Flora* (no. 29), although only one plate (p. 421) is actually signed by her. It depicts an elaborate vase of flowers, mostly ornamental bulbous species, beautifully composed and engraved. The artist was clearly fully acquainted with the latest innovations in the coeval genre of the floral still-life. Another talented, but almost entirely unknown, botanical artist who merits our attention is Giovanna Garzoni (1600–1670) of Ascoli, in Le Marche. At the age of sixteen she produced a charming painted herbal that is now in the collection at Dumbarton Oaks. Garzoni was quite famous in her day for her meticulous bodycolour paintings of flowers and bouquets, and received commissions from many of the most powerful families in Italy, including the Savoia in Turin and the Medici in Florence (Ciardi and Tongiorgi Tomasi, pp. 63–6; Tongiorgi Tomasi and Tosi, pp. 20–21).

The works of other flower painters, such as Clara Peters (?1589–after 1657) and Helena Rouers (*fl.* 1663) in the Low Countries and the French painter Louise Moillon (1609/10–90), show that many of these women artists possessed a thoroughly up-to-date knowledge of botany as well. For example, the meticulous accuracy that characterizes the paintings of Rachel Ruysch (1664–1750) no doubt in part reflects the training she had received from her father, Frederik Ruysch, an eminent professor of anatomy at the universities of Leiden and Amsterdam and an enthusiastic collector of anatomical and zoological specimens.

Many works by women botanical artists are to be found at Oak Spring, and a study of them, together with the fascinating personal histories of the artists who created them, allow us to appreciate the unique contribution they made to the genre of floral painting.

Without question, one of the most interesting and remarkable artists of the seventeenth century was Maria Sibylla Merian (1647–1717; see nos. 81, 82, 101). When she began her career a sumptuously decorative style of flower painting predominated in the Low Countries, but instead of allowing herself to be influenced by this style Merian remained true to her own artistic vision, which was the unique product of her character, education and profound, if somewhat unconventional, religious faith. She based her art on a meticulous study of the natural world, which in her imagination unfolded like a miraculous microcosm ruled over by a beneficent God, in which insects, plants and flowers coexisted in perfect harmony, and whose innermost workings only the most reverently attentive observer could begin to penetrate.

Merian's interest in botany was stimulated by the education she had received at home, for her father (Matthaeus the elder) was a skilled engraver who had collaborated on important scientific texts, among them Johann Theodor de Bry's *Florilegium novum* (no. 10), and after his death she benefited from the teaching of her stepfather, Jacob Marrel, a noted painter of *Tulpenboeken* (nos. 72–75). Merian's work reflected her mastery of a variety of media, for she more or less contemporaneously learned and then practiced the arts of painting, engraving and embroidery. Indeed, the art of embroidery, which required an acute sensitivity to the juxtapositioning of colours, provided excellent training, which she could capitalize on when engaged in flower painting or hand-colouring engravings.

Merian dedicated her career to the art of scientific illustration, and by her genius raised this ostensibly 'minor' genre to the level of a great art. Her indomitable character also led her to make a series of courageous and—certainly for a woman—unusual decisions regarding her personal life, decisions that perhaps were only possible for a woman in the seventeenth century who lived in the relatively free and open society maintained in the United Provinces. For example, at the age of fifty-two she realized one of her lifelong dreams and set off on a two-year expedition to Surinam, a Dutch colony in South America. Her interest in tropical flora and fauna prefigured the vogue for the exotic that would sweep across Europe a century later.

In the eighteenth century an increasing number of women artists began to win recognition for themselves as flower painters, as more intimate forms of painting, such as the floral still-life, began to replace the monumental genres in the public's favour. This shift in taste was encouraged by the exaltation of the Arcadian ideal of a 'return to nature' by the philosophers and writers of the period, while new species arriving from distant lands provided a constant source of fresh inspiration for artists. The demand for flower painters also grew apace as more and more books and treatises were written on these new plants and their cultivation, all of which required skilled botanical painters to prepare their illustrations. Antoine-Joseph Dézallier d'Argenville (1680–1765), the author of various treatises on art, gardening and natural history, affirmed that an artist 'qui a parfaitement imité la nature, n'eût-il peint qu'un insect ou des fruits, est aussi parfait dans son genre que Raphaël dans le sien' (Faré, p. 387), while Diderot declared that 'un peintre de fleurs ou de fruits peut être froid ou chaud comme un peintre d'Histoire' (Faré, p. 388).

In 1742 a woman artist, Madeleine Françoise Basseporte (1701–80), was chosen to succeed Claude Aubriet (no. 47) as 'Peintre du Roi pour la miniature', taking charge of the project begun by Daniel Rabel (no. 15) and Nicolas Robert (nos. 41 to 44)—documenting on precious *vélin* the botanical collection of Louis XV. So great was her skill that the marquise de Pompadour, who, as we know, made some attempts to paint flowers (see no. 63), more than once invited Basseporte to her château, Bellevue, to paint the flowers in her garden.

A somewhat unusual human story, and one that also underlines the growing commercial demand for botanical illustrations in the mid-eighteenth century, lies behind the inception of *A Curious Herbal* by Elizabeth Blackwell (no. 83). Although not an artist by profession, Blackwell compiled this illustrated herbal of medicinal plants, for which she had been told there would be a ready market among botanists and physicians, in order to earn sufficient money to pay her husband's debts and free him from prison. Happily, she succeeded in doing so with this commendably accurate, if artistically modest, little work.

The increasingly prominent role that women were beginning to play in the genre is also reflected in the large number of female pupils who clamoured to study under the most celebrated flower painters of the day. During the eighteenth century, for example, flower painting was considered a suitable pastime for well-born ladies, and the botanical painter Georg Dionysius Ehret counted many members of the aristocracy among his students (Calmann, pp. 81–4). Indeed a few of them, such as Frances Howard and Ann Hamilton (examples of whose work can be found at Oak Spring: see no. 56), became very fine painters in their own right.

With the explosion in popularity of the genre of flower painting at the end of the eighteenth century, middle-class women began to interest themselves in the art as well, and throngs of aspiring amateurs attended the lessons of such famous painters as Pierre Joseph Redouté (no. 61). Some of them went on to become established flower painters themselves (Hardouin-Fugier), among them Ernestine Panckoucke (1784–1860), Rosine Delaporte (1807–76) and Augustine-Hélène-Amélie Birat (1812–67). At Oak Spring, for example, there is a beautiful bodycolour painting of a *Paulownia imperialis* by one of Redouté's protégées, Adèle Riché (1791–1887; see *An Oak Spring Sylva*, no. 24). The painter Lise Cloquet also very probably benefited from Redouté's teaching, for she came from a family of artists who worked within his circle and who collaborated with him on a number of projects (Hardouin-Fugier, pp. 38–9). At Oak Spring is an album of thirty-two paintings in body-colour by Cloquet dating from 1820, including a particularly beautiful one of a white chrysanthemum shown against a grey background, as well as another four paintings completed between 1836 and 1837.

A painting of two hellebores and a butterfly by Barbara Regina Dietzsch (no. 84) provides an example of the union between the decorative tradition and the artistic tradition that distinguished the best of eighteenth-century floral still-life painting, a genre practiced by many of the members of her family.

In the nineteenth century, both the public's interest in botany and the vogue for flower painting grew rapidly, particularly in England. Teachers of flower painting sprang up everywhere, and manuals explaining the rudiments of the art found a ready market. At the same time, many women flower painters—some of real talent—appeared, including Mary Lawrance (no. 78), Clara Maria Pope (*c.* 1768–1838) and Priscilla Bury. Clara Pope designed five plates for the *Beauties of Flora* (1806–20) by the botanist Samuel Curtis (1779–1860), and all the plates for Curtis's *Monograph on the genus Camellia* (1819); copies of both works can be found at Oak Spring. Priscilla Bury, the daughter of a wealthy Liverpool merchant, made a special study of liliaceous plants, capturing their complex form and colouring in a series of paintings she called 'portraits'. Since her husband, Edward Bury, was a man of considerable means, she was able to publish a number of her works in a splendid edition, *A Selection of Hexandrian Plants*, prepared by the master printer and engraver Robert Havell (no. 86).

Another important figure of the time was Jane Loudon, who collaborated with her husband, John Claudius Loudon, the celebrated garden designer and writer, on many of his horticultural books and articles. She thus acquired an excellent knowledge of botany, which she very ably disseminated through her own works—gardening books addressed to a public of (mostly female) amateur gardeners, illustrated with charming lithographs drawn by Loudon herself (no. 85).

Destiny reserved a more difficult fate for Berthe Hoola Van Nooten (no. 87). Unlike Maria Sibylla Merian, who was inspired by her own interest in tropical flora to sail as far as Surinam, Van Nooten was constrained to follow her husband from Belgium to the faraway island of Java. When he died unexpectedly, she found herself alone with a family to maintain. Aware of the vogue in Europe for exotic flora, she decided to take advantage of her enforced exile and put to use the skill at flower painting that she had no doubt acquired as a girl. Thus she prepared forty magnificent plates for *Fleurs, fruits et feuillages choisis de l'Ile de Java peints d'après nature*. In her introduction to this work, Van Nooten describes her unhappy situation, but also admits that labouring on this project brought her 'a real and solid enjoyment'. Indeed, she reflected, no one who embarked on a study of the

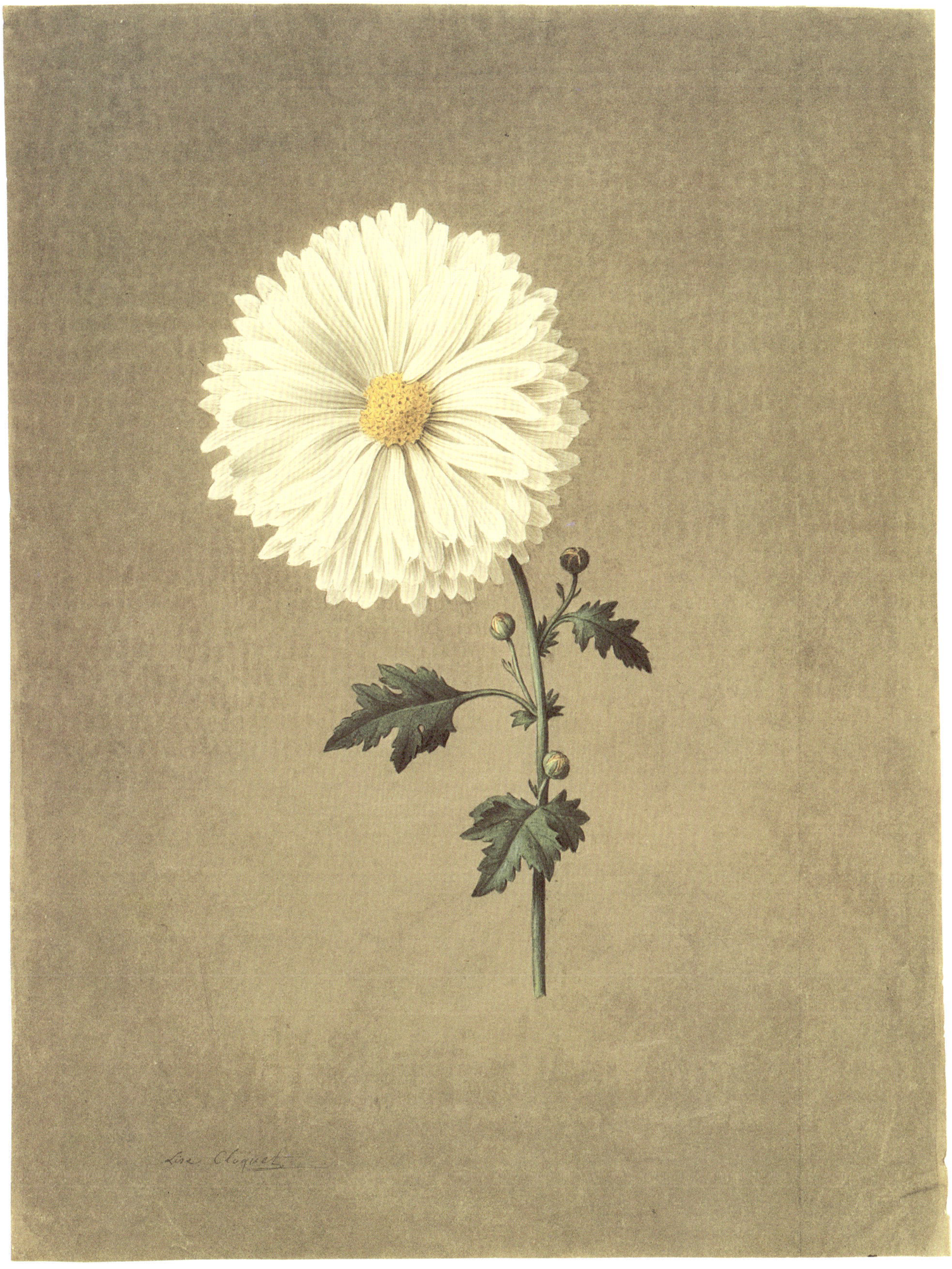

LISE CLOQUET,
Album of flowers.
Chrysanthemum sp.,
bodycolour, 1820,
folio 18

natural sciences could help but fall under its 'soft and balmy influences'. With becoming modesty she professed to leave to her male colleagues the more exacting rigours of serious scientific research, contenting herself with humble task of reproducing just a few of the pages in God's miraculous 'book of Nature'.

As we will see in this chapter, however, many of the women who succeeded Van Nooten did not accept the limits of this circumscribed role, and by their accomplishments proved themselves to be quite the equal of their most brilliant male counterparts in the history of botanical illustration.

81. Maria Sibylla Merian (1647–1717)

Manuscript florilegium.

39.4 × 25.5 cm. Dedication leaf and 90 vellum leaves.

BINDING: Red morocco with double gilt fillets. Two metal cockleshell-shaped clasps are attached to leather straps.

PLATES: Vellum dedication bearing an empty cartouche, surrounded by a garland of flowers coloured by hand in bodycolour, and a coat of arms. 90 plates of flowers, hand coloured in bodycolour and edged in gold, each with a guard leaf of laid paper; with 1 blank vellum leaf following dedication, 3 blank vellum leaves following plate no. [18], and 1 blank each following plate [58], [67], and [71], and 2 vellum blank leaves at end.

PROVENANCE: Coat of arms of Vincent Moller (1656–1737 or 1743) on the dedication leaf.

One of the greatest painters in the history of botanical illustration was a woman, Maria Sibylla Merian. She was born into a family of artists, her father being Matthaeus Merian the elder (1593–1650), a well-known engraver (see no. 16). Her natural talent was brought out by early training and further enriched by a lifetime of study and travel. She became a proficient painter, engraver and embroideress, with an unmistakable style and an unusually single-minded dedication to her art. After the death of her father, Merian learned the art of flower painting from her step-father, Jacob Marrel (see nos. 73, 74, 75). She also studied with two other artists from Marrel's workshop—Abraham Mignon, a painter specializing in still-lifes with insects, and the German Johann Andreas Graff, whom she married in 1665.

Even as a child Merian was fascinated by caterpillars and their various metamorphoses. One book had already appeared on the subject, *Metamorphosis naturalis* (1662–9) by the Dutch artist Johannes Goedaert, and between the years 1679 and 1683 she published her own two-volume work on the insects of Europe, *Der Rupsen Begin, Voedzel en Wonderbaare Verandering* (see no. 82). In 1705 she published her famous *opus magnum* on the insects of Surinam, *Dissertatio de generatione et metamorphosibus insectorum surinamensium* (see no. 101).

Although entomology was her main interest, plants and flowers also figured prominently in

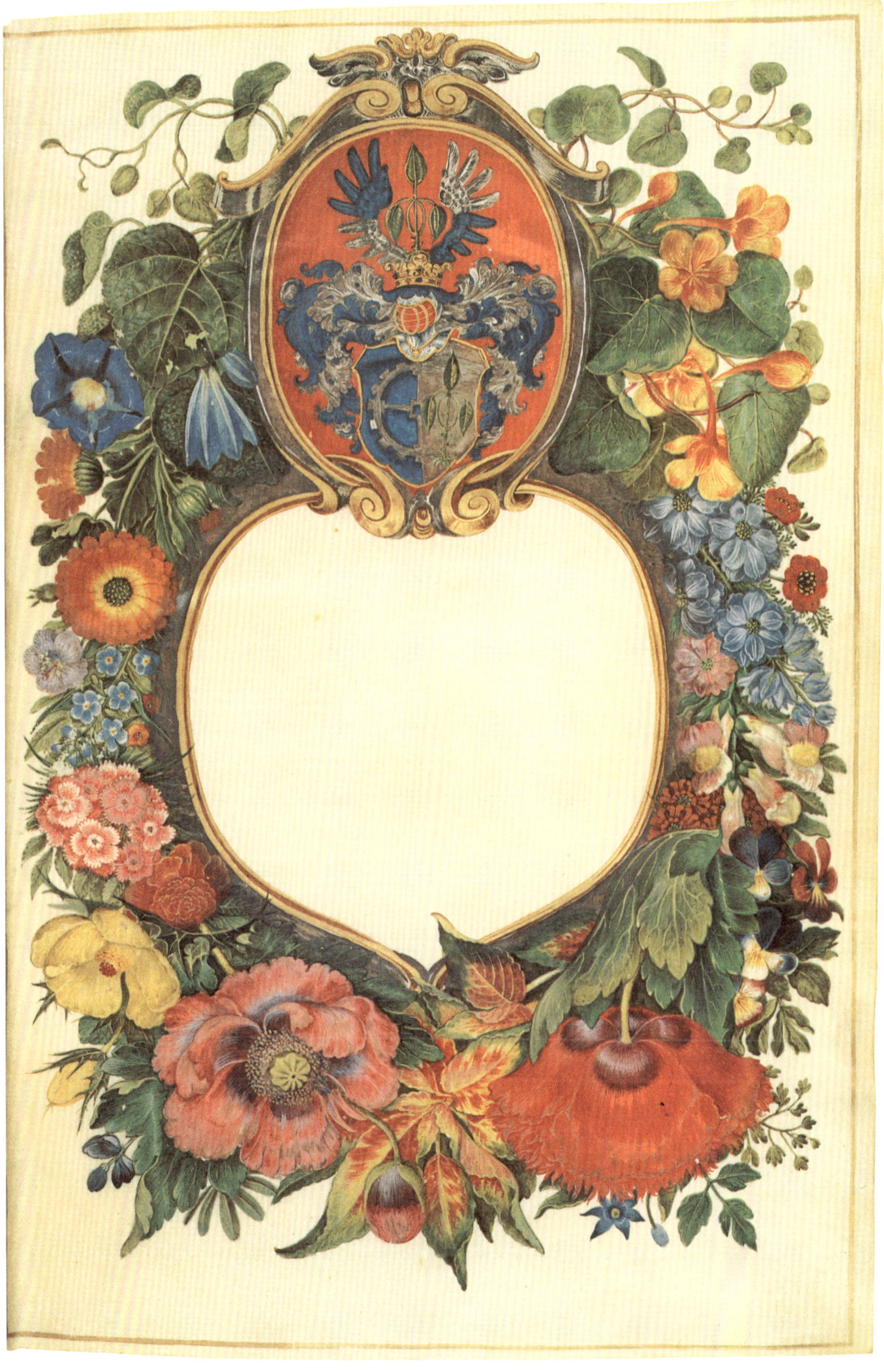

MARIA SIBYLLA MERIAN,
Manuscript florilegium.
Floral garland with the
coat of arms of
Vincent Moller

Merian's work, for she generally portrayed her insects together with a branch (often in flower) of the plant on which they habitually fed. Indeed, the great Swedish botanist Linnaeus was acquainted with her work, and cited her name more than once in his *Species plantarum* (1753). In addition to her highly original insect studies, Merian produced more traditional works in the popular genre of botanical painting. There is a fine picture by her of some irises in London's Natural History Museum (Mitchell 1973, pl. 234), and a group of paintings in bodycolour in the Royal Library at Windsor Castle (Morton, pp. 30–37). She also published an engraved volume of floral motifs for embroiderers and other artisans, the *Neues Blumenbuch* (1680), in which the influence of the French flower painter Nicolas Robert can be seen.

Much less well known is the fact that, perhaps after leaving the religious community of the Labadists in 1691, Merian painted a magnificent florilegium in three volumes for Vincent Moller, a nobleman who played a prominent role in the political, economic, social and religious affairs of the city of Hamburg (see U. P. Moller, *Die Hamburgische Familie Moller*, Hamburg, 1856, pp. 83–5). Moller's ornate coat of arms appears on the first leaf of the manuscript. (Fine as this work certainly is, it was thrown into shadow by the appearance, shortly after, of her book on the insects of Surinam.) Two volumes of Moller's florilegium were destroyed in Hamburg during the Second World War; the one that survives is now at Oak Spring.

The dedication in this volume is decorated with an elegant floral garland and the coat of arms of the author's distinguished patron. The ninety plates present a wide variety of flowers, each painted with the scientific precision that characterizes Merian's entomological studies. The specimens are beautifully arranged on the page, while the colours have been applied with great skill in infinitely subtle gradations of hue. In some of the more complex compositions, such as those of the *Ricinus* (fol. 66) and the *Datura* (fol. 78), the artist demonstrates her familiarity with the Dutch tradition of the floral still-life, while other paintings, such as those of the *Scabiosa* (fol. 68) or three varieties of *Verbascum* (fol. 7), have a spontaneity and naturalness that could only have come from a lifetime dedicated to the study of nature *dal vivo*.

Most of the flowers depicted in this florilegium are wild species, although the final folios present some popular cultivated flowers, mostly varieties of marigold. There is also a very realistic plate depicting some tomatoes (fol. 86) and three others showing different varieties of chili plants (fols. 87–9).

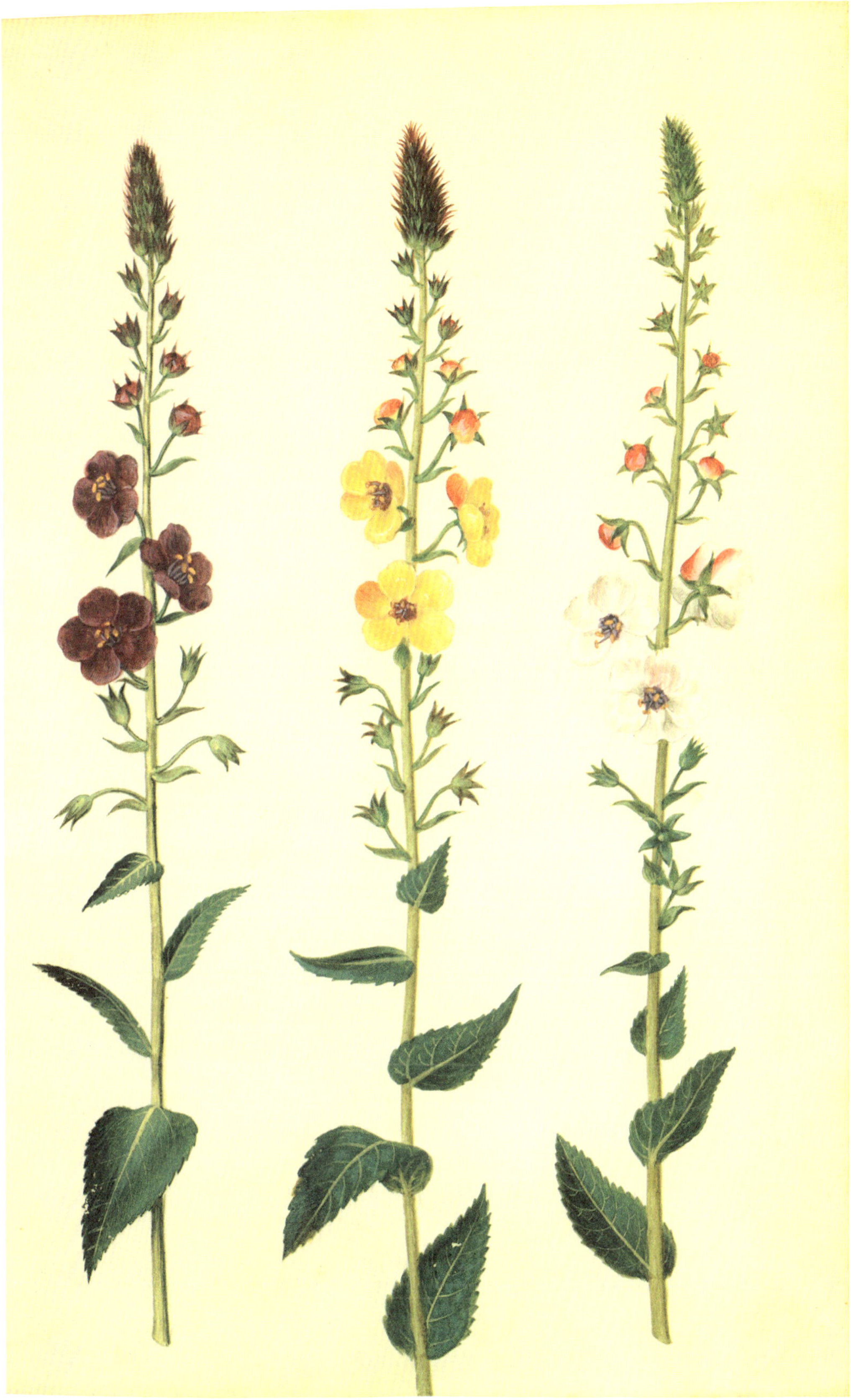

MARIA SIBYLLA MERIAN,
Manuscript florilegium.
Three moth mulleins
(*Verbascum blattaria*),
folio 7

82. Maria Sibylla Merian (1647–1717)

Der Rupsen Begin, Voedzel En Wonderbaare Verandering. Waar in De Oorspronk, Spys en Gestaltverwisseling: als ook de Tyd, Plaats en Eigenschappen der Rupsen, Wormen, Kapellen, Uiltjes, Vliegen, en andere diergelyke bloedelooze Beesjes vertoond word; Ten dienst van alle Liefhebbers der Insecten, Kruiden, Bloemen en Gewassen: ook Schilders, Borduurders &c. Naauwkeurig onderzogt, na 't leven geschildert, in Print gebragt, eu in 't kort beschreven Door Maria Sibilla Merian [*typographical ornament*] t'Amsterdam, Gedrukt voor den Auteur. Woont in de Kerkstraat, tusschen de Leidsche—en nieuwe Spiegelstraat over de Parssery de Swaan, alwaar dezelve Gedrukt, als ook afgezet te bekomen zyn. Als mede by Gerard Valk, op den Dam in de wakkere Hond [*3 volumes bound in one*]. [1713]–[1717].

4º 21.2 × 16.5 cm. A–C⁴ D⁴(–D4) *1–6* 7–30 p., 51 plates; Tweede Deel: Der Rupsen Begin, Voedsel en wonderbaare Verandering: –★A–C⁴ D⁴(–D4) *i–iv* 1–30 p., 51 plates; Derde Deel: Der Rupsen Begin Voedsel en Wonderbaare Verandering. A–C⁴ *1–4* 5–24 p., and 51 plates.

BINDING: Calf binding with panels of gilt tooling and a lozenge-shaped central ornament. Gilt lettering on spine: 'M.S. Merian Van der Rupsen Verandering.'

PLATES: 3 illustrated half-title pages and 3 sections, each of engraved, hand-coloured plates, numbered 1–50 (the second section using roman numerals).

PROVENANCE: Bookplate of Philippe Lévêque de Vilmorin (1872–1917). Vilmorin was a French botanist, geneticist, horticulturist, and author of *Manuel de floriculture* (Paris, 1908).

REFERENCES: Dunthorne, p. 205; Nissen 1342; Sitwell and Blunt, p. 67; Hunt 467; Landwehr 131; Stearn 1978; Gregorio and Celli.

This is the first complete edition of Merian's work on caterpillars and their transformation into butterflies, *Der Rupsen Begin, Voedzel en Wonderbaare Verandering* (Wonderful metamorphoses of caterpillars and their singular diet of flowers), originally published in three separate volumes. The first had appeared in 1679 and the second in 1683, both prepared by Merian's husband, the German printer and engraver Johann Andreas Graff. They appeared in three cities simultaneously—Frankfurt, Nuremberg and Leipzig. The fifty plates in each volume were accompanied by a simple, yet elegant and scientifically rigorous text written by the artist on the basis of her own observations. The third volume was published posthumously in Amsterdam, together with a Dutch translation of volumes I and II, under the supervision of the artist's daughter and close collaborator, Dorothea Maria Henriette. An edition in Latin entitled *Erucarum ortus*, prepared by Dorothea with illustrations printed from the original copperplates, also appeared in Amsterdam in 1717. In 1730 a French edition with eighteen additional plates, perhaps the work of Merian's other daughter, Johanna Helena, was published as *Histoire des insectes de l'Europe*. Another French edition was published together with Merian's famous work on the insects of Surinam in 1717 in Paris as *Histoire générale des insectes de Surinam et de toute l'Europe*.

Merian's interest in caterpillars and butterflies began when she was a child: it is known that she used to collect silkworms and grasshoppers and feed them leaves, scrupulously observing and recording their subsequent transformations (Stearn 1978, p. 10). After her marriage she continued her studies with the moths and butterflies to be found around her new home in Nuremberg, depicting them in the various stages of their life cycle, together with the plants on which they usually fed.

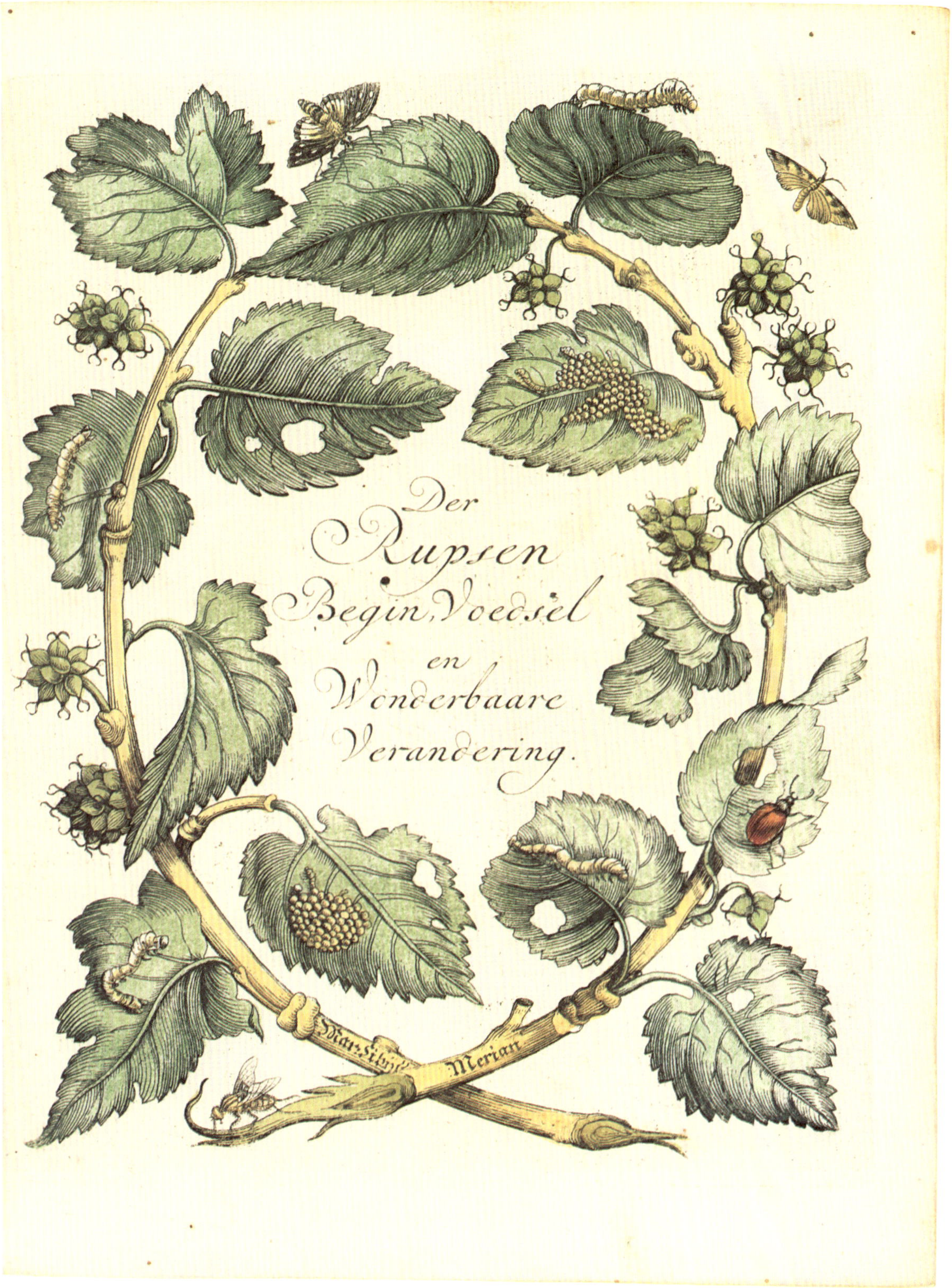

MARIA SIBYLLA MERIAN, *Der Rupsen Begin*. Title-page

Some of the preparatory drawings for *Der Rupsen Begin* are in the British Museum in London; others can be found in the Archives of the Academy of Sciences in St Petersburg. She was a fine draughtsman, but she had also beeen trained thoroughly in various printmaking techniques—engraving, etching, drypoint and monotype.

Merian's plates are wonderfully life-like. Among the most impressive are one showing the caterpillar, cocoon and butterfly of an *Aporia crataegi* together with a branch of flowering convolvulus (II, pl. 25) and another of a delicate sprig of sky-blue delphinium with the caterpillar and butterfly of a *Periphanes delphinii* (I, pl. 40), the original watercolour for which is preserved at St Petersburg (Ullmann, pl. 37). The description accompanying the latter plate explains: 'I have often seen on the sky-blue flowers of the *Consolida regalis* the enchanting little moth that I depict here: so well known is it for its beauty and unusual colouring that I found myself wondering more than once from what caterpillar it might spring. I therefore pursued my researches with zeal until I found the caterpillars I was looking for on the flowers of this very plant, to which they cause great damage, since they not only like to feed on them, but often devour the leaves and flowers with such voracity that they leave the stem completely bare.' The text continues with a description of the magnificent colours the insect assumes in its various mutations, and concludes with the following observations: 'I have portrayed one of these little moths in the centre of the picture, poised on two green leaves, the more to delight the eye of the nature lover the more attentive and acute that eye is, and to lend lustre to this tiny work of art of indefatigable nature.'

A fine title-page precedes each of the work's three volumes. The first is decorated with a simple garland of mulberry branches (*Morus alba*), whose leaves show the ravages of the silkworm (*Bombyx mori*), one of the first insects, we may recall, to be depicted by the artist as a child. The title-pages to the second and third volumes are decorated with more traditional floral garlands, examples of which may be found in an earlier work by the artist, *Neues Blumenbuch* (1680). She generally signed her works with her maiden surname, 'Maria Sibylla Merian', in keeping with the honoured tradition of artistic families, but the second title-page (dated 1683) bears the uncharacteristic signature 'Maria Sibylla Gräffin Sculpsit', in which she uses that of her husband.

In order to ensure that her illustrations were as true to life as possible, Merian usually coloured them herself, with the help of her two daughters. Since the plates in the Oak Spring copy have been painted with unusual skill, it is likely that they were done by one or more members of the family itself.

The Oak Spring collection also includes a copy of the first edition of volumes I and II (1679, 1683), with hand-coloured plates but no text, and one of the Latin edition of 1718.

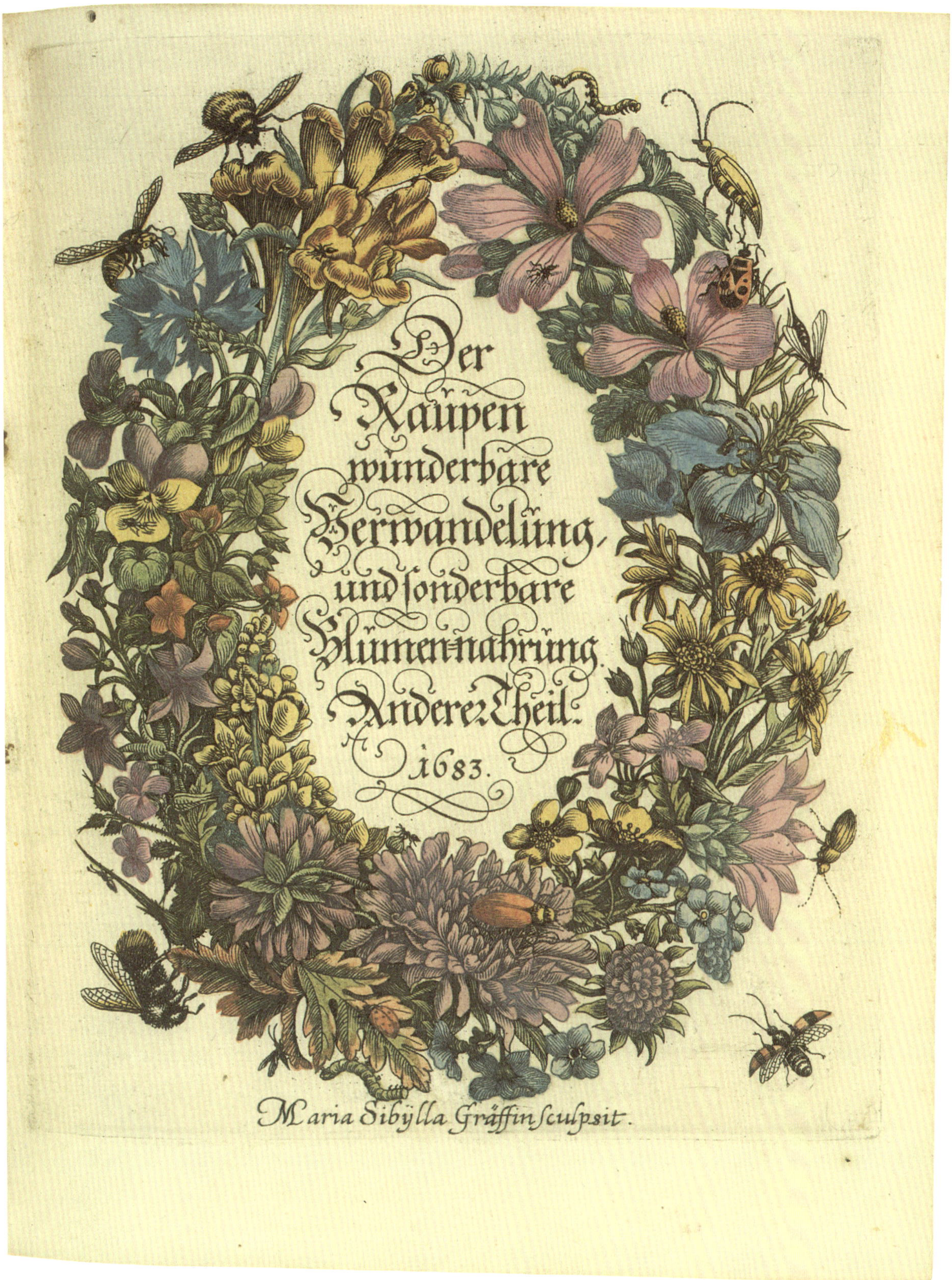

MARIA SIBYLLA MERIAN, *Der Rupsen Begin.* Title-page to Part Two

83. Elizabeth Blackwell (*c.* 1700–1758)

72 watercolour drawings for *A Curious Herbal.*

BINDING: Preserved in a drop-back oasis box.

PLATES: 72 watercolours (26 × 18.5 cm.) on their original lined mounts. They are the preparatory drawings for the following engraved plates of *A Curious Herbal* (London, 1737–9): 51, 53–55, 59, 64–66, 69, 72, 77, 81, 83, 124, 156, 165, 171, 173, 182, 215, 217, 219, 223, 225–228, 230–231, 233, 239, 243–244, 249–250, 252–254, 266, 268–269, 271, 273–275, 279–280, 283–284, 286–288, 290–292, 294, 307, 326, 345–348, 404, 410. The engraved plates are the reverse of the watercolours. There are also watercolours of the Wild Clary, Savin, Verbascum, Water Pepper, Wormwood, Roman Wormwood and Pitch Tree, which were apparently not published.

REFERENCES: Henrey, II.230–36; *An Oak Spring Garland*, no. 14.

THE FAME OF Elizabeth Blackwell (née Blachrie) is due more to the unusual personal circumstances that led to the publication of *A Curious Herbal* than to her inherent talents as a botanical illustrator. Her husband, Alexander Blackwell (1709–47), was an accomplished classicist who abandoned the idea of pursuing a professional career and set up a printing-house in London. His great success in this undertaking aroused the jealousy of certain rival printers who banded together to ruin him. They accused him of never having been apprenticed to the trade, which led to him being forced to pay a series of exorbitant fines and to close his shop. Finally, he was arrested for debt in 1734 and sent to prison. That the family's fortunes were in time restored was due entirely to the courageous efforts of his wife.

Having learned from Sir Hans Sloane, the President of the Royal Society, that botanists still lacked an up-to-date illustrated text of medicinal plants, Elizabeth Blackwell decided to remedy this lacuna if she could. She rented a house close to the Chelsea Physic Garden, the only botanical garden in London at the time, and began drawing its collection of medicinal herbs. She received considerable help in this project from various members of the Royal College of Physicians, including the President himself, Thomas Pellet, and Richard Mead, physician to the royal family. Other, lesser known, scientists, such as the pharmacist Robert Nicholls, also gave her help and encouragement, and all are mentioned with gratitude in the introduction to the published work.

Blackwell began her project in 1735. While she was sketching the plants, her incarcerated husband prepared the text that describes their use and properties, basing himself in large part on *Botanicum officinale*, published in 1722 by Joseph Miller (*fl.* 1722–48). *A Curious Herbal*, containing the illustrations of 500 flowering plants engraved and carefully hand-coloured by the artist herself, was published in parts between 1737 and 1739, and then reissued in two volumes by the London publisher Samuel Harding. The work enjoyed a modest success and Elizabeth was able to free her husband and regulate his affairs so that he could once more take up his profession. Unfortunately, Alexander Blackwell's travails did not end there. In 1742 he emigrated to Sweden, where he later became involved in a plot to alter the royal succession. Charged with treason, he was executed there without trial in 1747.

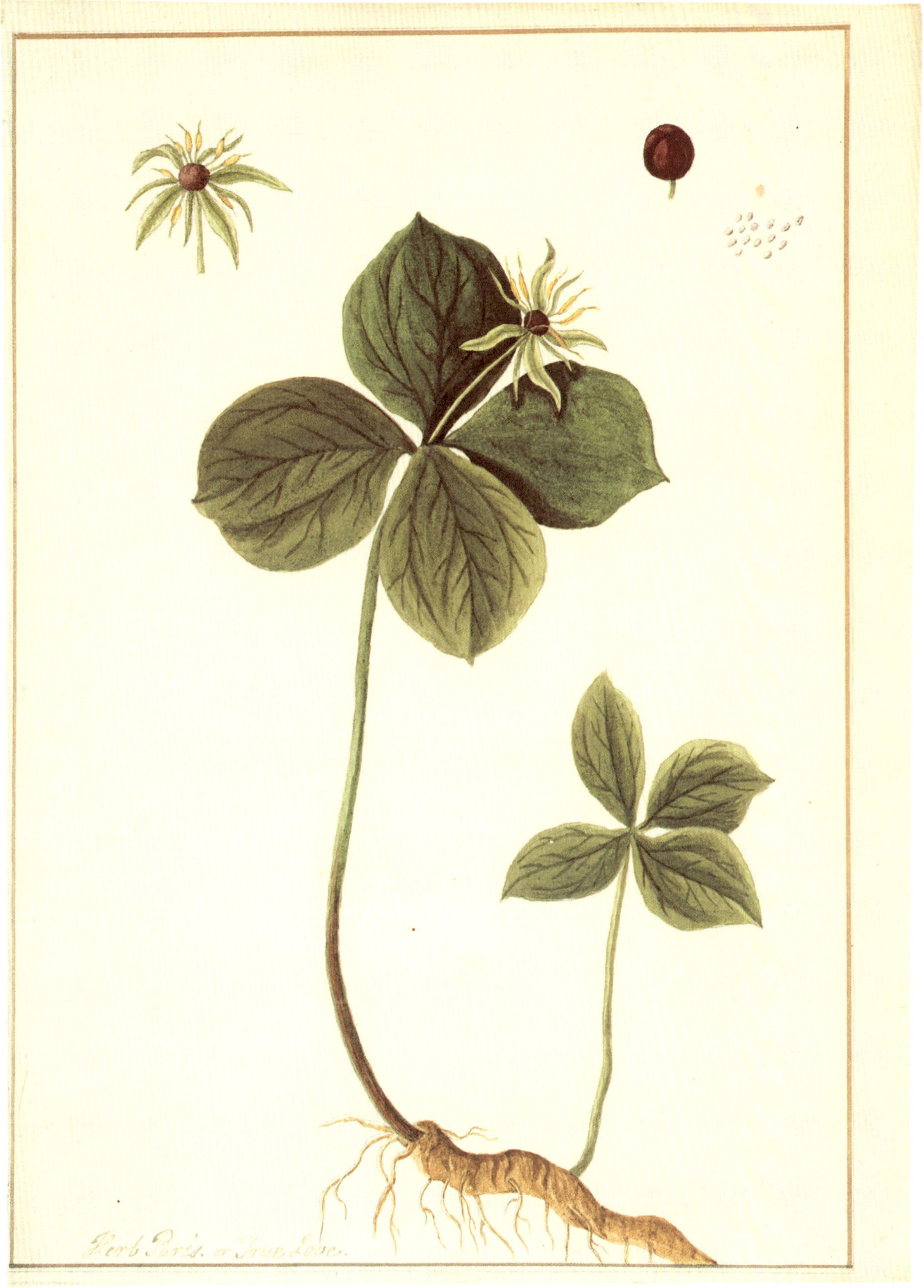

ELIZABETH BLACKWELL, drawings for *A Curious Herbal*. Herb Paris (*Paris quadrifolia*)

Elizabeth Blackwell's *Herbal* eventually came to the notice of the wealthy Nuremberg physician and botanist Christoph Jacob Trew (1695–1769), who decided to republish the work after having the plates re-engraved by N. F. Eisenberger (adding another one hundred plates of his own). This expanded version of the work, with texts in both German and Latin (by Trew and others), appeared in five volumes as *Herbarium Blackwellianum* in Nuremberg between 1750 and 1760, with a supplement (*Herbarii Blackwelliani auctarium*) in 1773, and in Leipzig in 1794.

The simplicity of Blackwell's drawings betray the fact that she was not a professional artist, but they none the less possess an immense charm and legibility. Each plant is rendered in simple outline, soberly coloured and carefully arranged on the page together with details of its flower and seeds. Particularly charming are the hollyhocks in plate 54 and the sprig of broom in plate 244 of the first volume, and the simple yet elegant 'Paris herb' (*Paris quadrifolia*) in plate 286 of the second. This forest plant flowers in April or May, while its shiny black berry, full of medicinal properties, appears in July. In pen and ink in her own graceful hand, Blackwell has added the popular name of this plant, 'True love'.

84. Barbara Regina Dietzsch (1706–1783)

Helleborus niger and butterfly.
28.9 × 20.9 cm. Bodycolour on black prepared paper.

REFERENCE: *European Drawings: Recent Acquisitions*, sale cat., Hazlitt, Gooden & Fox, London, 1988, no. 51.

Barbara Dietzsch was a celebrated painter of botanical and zoological subjects whose family counted many painters, engravers and musicians among its members. Although in the second half of the seventeenth century a handful of woman artists could be found working in the genre of naturalistic illustration, it was only in the eighteenth that a large number of them became quite well known and successful as painters of floral compositions. Barbara Dietzsch, the elder daughter of the landscape painter Johann Israel Dietzsch of Nuremberg, was one of these. As a painter she concentrated exclusively on flowers, birds and insects, a genre much in demand, for Nuremberg was an important centre of scientific and cultural activity in this period. Among the city's residents was the famous botanist and bibliophile Christoph Jacob Trew, author of the encyclopaedic *Plantae selectae* and a splendid florilegium, *Hortus . . . amoenissimorum florum*, which he produced in collaboration with Georg Dionysius Ehret (see nos. 50, 51).

Dietzsch's sister Margaretha Barbara was also a botanical painter (she provided Trew with some illustrations for his works), while her brother Johann Cristoph was a landscape painter. Although Dietzsch herself was without a doubt the most talented artist in the family, it is not always clear to which of the three siblings certain works should be attributed, for together with her brother and

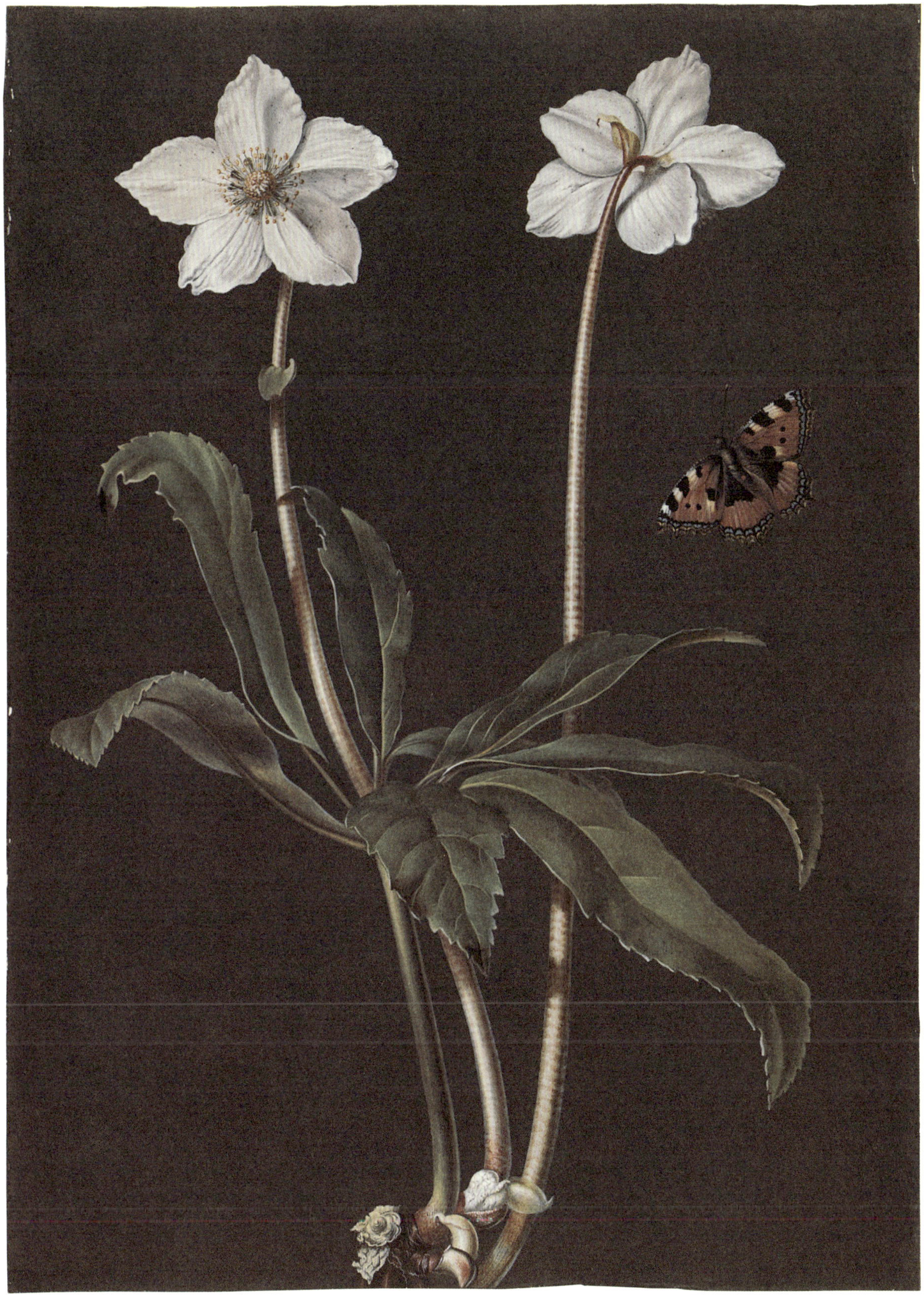

BARBARA REGINA DIETZSCH, Christmas rose (*Helleborus niger*) and butterfly sister she developed a highly personal style of painting on vellum using dense layers of bodycolour; she also frequently used a dark background that served as a striking foil for her botanical specimens.

Another hallmark of her work was the insertion of insects, lizards and other lively little creatures into her floral compositions, in keeping with an iconographic tradition initiated by Jan van Kessel (see no. 26) and subsequently adopted by numerous Dutch painters, including Herman Henstenburgh (see no. 27). While her works reflect the painstaking attention to detail of the German–Dutch school, they are immediately distinguishable by their great decorativeness, and by an intense chromaticism created through the use of well-charged brushstrokes brightened with touches of light-drenched colour reminiscent of certain 'herb pieces' by Otto Marseus van Schrieck (1619/20–78) and Matthias Withoos (1627–1703).

This exquisite painting depicts a hellebore (*Helleborus niger*) with two blooms balancing gracefully on long slender stems, one shown frontally and the other from the rear. The flowers with their five delicate white petals are arrestingly beautiful in their simplicity, while the long, undulating leaves with their serrated edges seem to thrust themselves out from the background, adding depth and movement to the composition. The subtly orchestrated chromaticism is unusually pleasing, with the snow white of the flowers and silvery grey-green of the leaves effectively set off by the velvety-black background. As in all her works, Dietzsch has added a small insect to the painting, in this case a charming tortoiseshell butterfly (*Aglais urticae*).

Oak Spring also has a painting in bodycolour by Dietzsch of *Two Quinces on a Branch* (see *An Oak Spring Pomona*, no. 83).

85. JANE LOUDON (1807–1858)

British Wild Flowers. By Mrs. Loudon. [*in the centre of the title-page*] 'Ye Field Flowers! the gardens eclipse you, 'tis true, | Yet, wildings of Nature, I doat upon you, | For ye waft me to summers of old, | When the earth teemed around me with fairy delight, | And when daisies and buttercups gladdened my sight, | Like treasures of silver and gold.' Campbell. Second Edition. London: William S. Orr & Co., Amen Corner, Paternoster Row.

31 × 25.5 cm. *i–v* vi–xvi *1* 2–311 *312* p., 60 leaves of plates.

BINDING: Green cloth binding with gilt lily of the valley and daisy in the centre of the front cover. Gilt floriated spine with the title 'British Wild Flowers'.

PLATES: 60 hand-coloured lithographs.

PROVENANCE: On the front free endpaper is inscribed: 'John Kelk'.

REFERENCES: Pritzel 5636 (1st edn); Nissen, 1233.

THIS IS A COPY of the second edition, printed *c.* 1855, of Jane Loudon's *British Wild Flowers*, first published in 1846 in London by the printer William Smith. Jane Loudon was the wife of John Claudius Loudon (1783–1843), the leading landscape gardener and horticultural writer of the first half of the nineteenth century and editor of *The Gardener's Magazine* (founded 1826), the first

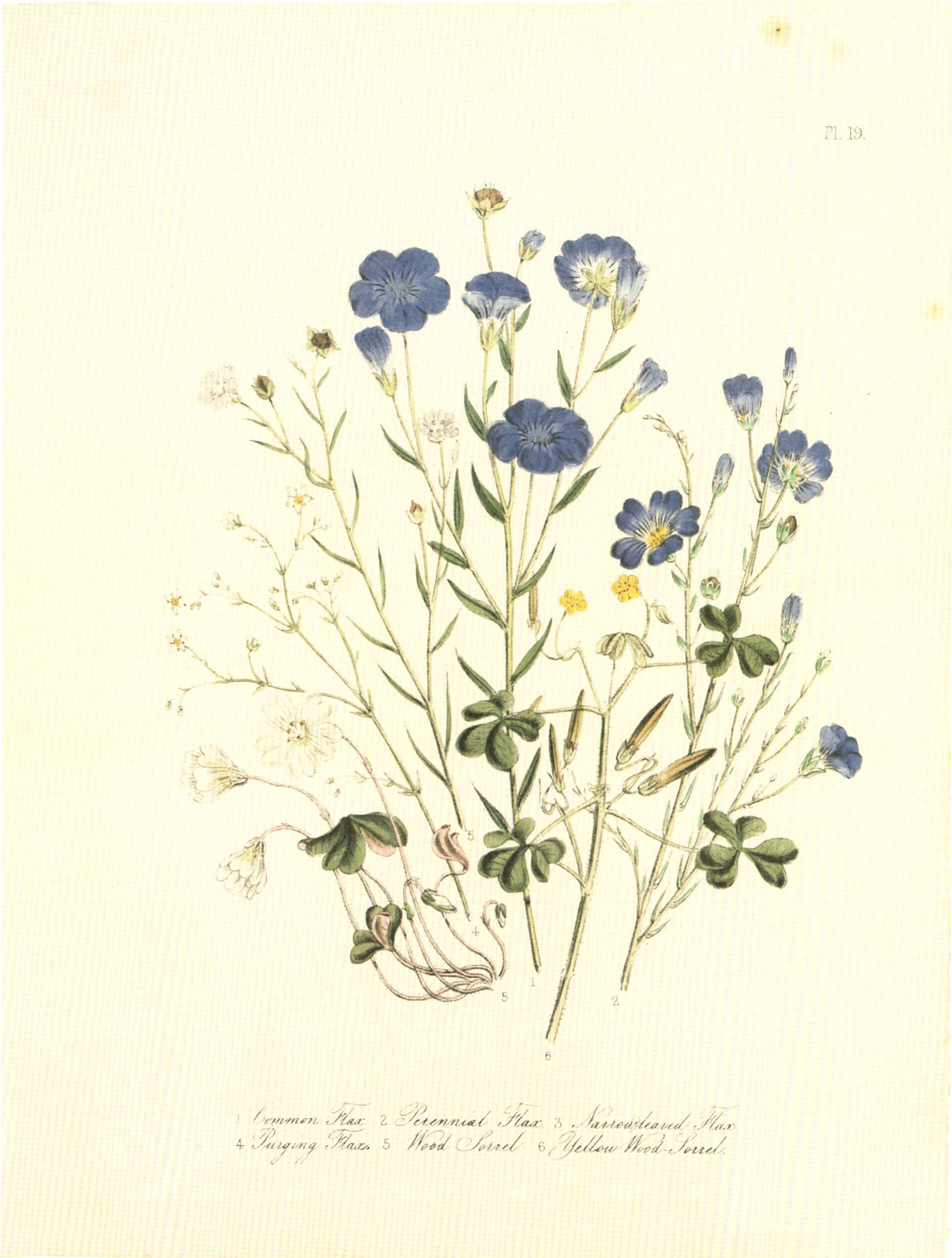

JANE LOUDON, *British Wild Flowers.* Plate 19: 1. Common flax (*Linum usitatissimum*); 2. Perennial flax (*Linum perenne*); 3. Narrow leaved flax (*Linum album*); 4. Purging or fairy flax (*Linum catharticum*); 5. Wood-sorrel (*Oxalis acetosella*); 6. Yellow wood-sorrel (*Oxalis stricta*)

such periodical addressed to the general public. In 1828 he met Jane Webb, the young author of a long historical novel that was to be reviewed in *The Gardener's Magazine*. Shortly afterwards they married, and in 1830 they began collaborating on numerous works devoted to the subject of gardening. After the death of her husband, Jane Loudon continued to write, specializing in practical gardening books.

In the 1840s Jane Loudon produced a highly successful series of works on ornamental plants, publishing in 1840 *The Ladies' Flower-Garden of Ornamental Annuals* and one year later *The Ladies' Flower-Garden of Ornamental Bulbous Plants*; 1843 saw the publication of *The Ladies' Flower-Garden of Ornamental Perennials*, and finally, in 1848, the series concluded with *The Ladies' Flower-Garden of Ornamental Green House Plants*. In 1842 she also produced a *Botany for Ladies*, and her *British Wild Flowers* in 1846. Loudon developed a simple formula for these works, combining a brief text with charming illustrations, an arrangement that appealed to her audience. For each plant a brief botanical description was provided, including its common name in English, its order and genus according to the system of Linnaeus, some historical and geographical notes, and instructions regarding its cultivation. Her principal source for this information was John Lindley's *A Synopsis of the British Flora* (1829).

The illustrations to *British Wild Flowers* are enormously attractive, Loudon using as her model the plates designed by James Sowerby (1757–1822) for his encyclopaedic *English Botany*, published in London between 1790 and 1814 (see no. 60). Employing the new and extremely rapid technique of lithography, which in the 1840s was replacing line engraving as the chief means for supplying illustrations, the author presents one or more species of flowers, so charmingly arranged on the page that they seem almost to take the form of bouquets. The illustrations were prepared for the author by Henry Noel Humphreys (1810–79), an artist, naturalist and numismatist who had studied for an extended period of time in Italy before returning to England around 1843 and establishing himself as a well-known book illustrator.

Although Blunt declares that the quality of the lithographs is considerably inferior in later editions of *British Wild Flowers* (Blunt and Stearn, p. 276), the plates in the Oak Spring copy of the second edition are technically of the highest standard. Those depicting the wild rose, poppies of various colours and the common celandine (*Chelidonium majus*) appear as fresh and spontaneous as the original drawings, and were impeccably printed. Plate 19 presents a variety of flowers, charmingly arranged on the page: (1) common flax (*Linum usitatissimum*); (2) perennial flax (*Linum perenne*); (3) narrow-leaved flax (*Linum album*); (4) purging flax (*Linum catharticum*); (5) wood-sorrel (*Oxalis acetosella*); and (6) yellow wood-sorrel (*Oxalis stricta*).

86. Priscilla Susan Bury (*fl.* 1831–1837)

A Selection of Hexandrian Plants, Belonging to the Natural Orders Amaryllidae and Liliacae, from Drawings by Mrs. Edward Bury, Liverpool. Engraved by R. Havell. This Work is most respectfully dedicated to Her Royal Highness Princess Victoria by her Obedient humble Servant. Rober.[t] Havell. London Pub[d]. 1831, End[d]. Dec[r]. 1, 1834, By Rob[t]. Havell, Zoological Gallery 77, Oxford S[t].

65 × 50 cm. 56 leaves, 51 leaves of plates.

BINDING: Blue leather with gilt rolls forming a double border. Gilt spine with the title 'Hexandrian Plants'.

PLATES: 51 hand-coloured aquatint plates by Robert Havell after Mrs Bury. Each plate is accompanied by a leaf of descriptive letterpress. The frontispiece bears an aquatint of an ornate vase containing a 'Lilium Tigrinum'. On the vase is a cartouche inscribed with the following lines of verse: 'Of all Hues, Celestial, Roseate, and gold | And glittering in elegant Splendour, behold | The Lilies, a race, to whom Nature has lent | All her Loveliest charms, of Form, Colour, and Scent. | With so many pleasing allurements endued | And by so many light-winged Votaries wooed. | That through all the wide circle of Flora's domain | Where the Loves, & the Graces so constantly reign | What Tribe can be found so varied, so fair, | Whose forms are so Noble, whose Painting so rare'. The Latin name of the plant and the signatures of the artist and the engraver-printer appear in the lower margin of each plate.

PROVENANCE: Inscribed on the front free endpaper: 'Philip B. Davies-Cooke Gwysaney Mold'. This name is also penciled in next to 'Mrs. Cook' on the printed leaf of subscribers.

REFERENCES: Pritzel (2nd edn) 1403; Dunthorne, p. 71; Nissen 306; Sitwell and Blunt, p. 53; McMillan; *A Magnificent Collection*, nos. 53, 54; Blunt and Stearn, pp. 248–9.

PRISCILLA SUSAN BURY (née Falkner) prepared the illustrations for one of the most splendid botanical works to be published in England in the nineteenth century, *A Selection of Hexandrian Plants*. The artist was the daughter of a wealthy Liverpool merchant whose family estate, Fairfields, just two miles from the city, boasted a garden with many rare and exotic plants (McMillan). At a very young age Priscilla Falkner began to devote herself to the study of botany. She was particularly interested in the lily and related species, and by 1829 had produced a sufficient number of paintings, which she called 'portraits', to begin to consider publishing them. In this project she was encouraged by a friend, the zoologist William Swainson, as well as by the botanist William Roscoe of Liverpool, a wealthy patron of the arts and sciences.

As originally conceived, the work was to have been published by Swainson in ten parts, each consisting of five plates prepared by the lithographer Charles Joseph Hullmandel of London. Subscribers were to pay one guinea per number for 'Drawings of Liliaceous Plants', and twenty-seven shillings for others. Around 1830, however, Priscilla married Edward Bury, a wealthy railway engineer, and with greater financial resources now at her disposal she was able to contemplate the publication of a more elaborate work. Thus, between 1831 and 1834 *A Selection of Hexandrian Plants* appeared, containing fifty illustrations of various six-stamen flowers. It was printed by Robert Havell (1793–1878), who was also an accomplished engraver of aquatints, justly celebrated for his etchings of theatre scenes and river views.

The seventy-nine subscribers to this work are listed on a page preceding the illustrations, and they include many of Liverpool's most prominent citizens. The name John James Audubon also appears, for it was Havell who engraved the plates for Audubon's magnificent *Birds of America*. As stated in the elegantly written preface, *A Selection of Hexandrian Plants* was conceived by Bury as 'an endeavour to preserve some memorial of the brilliant and fugitive beauties, for a particular splendid and elegant tribe of plants'. She admits with disarming candour that she was not in any way an expert botanist but rather an 'amateur' who, 'flattered by the opinion of her friends', exercised her modest talent in the portrayal of these flowering plants. She probably used as her model *Monandrian Plants*, a work by William Roscoe published between 1823 and 1829.

Bury was extremely fortunate to have Robert Havell as her collaborator, and she acknowledges her debt to him in the Introduction. A brilliant printmaker, Havell managed to translate the artist's fine watercolours into aquatints of even more striking beauty. The original watercolours are today preserved in the Garden Library at Dumbarton Oaks; they bear dates ranging from 1824 to 1832, thus the earlier ones are signed with the artist's maiden name, Falkner.

An analysis of the painting for the frontispiece sheds useful light on the creative process underlying Bury's work. The painting is dated 'Fairfield, June 1825', and depicts a modest vase containing four varieties of Turk's-cap lilies (*Lilium martagon*); the imposing cartouche that appears in the engraved frontispiece is missing. On the first page of the volume that contains the original watercolours for *A Selection of Hexandrian Plants*, which was compiled in 1848 by Bury for her son Edward James, this painting is described as 'Plate 1. Vase of Martagons. Original sketch for the frontispiece to "Hexandrian Plants"'. Thus, we may hypothesize that the painting was modified (no doubt with the permission of the artist) by the engraver, who in this way played an essential role in the creation of the work.

Bury's magnificent 'flower portraits' are reproduced in Havell's fine-grained aquatints with all their subtle variations in tone intact. In many of the illustrations exotic butterflies (not always realistically depicted) appear. After they were printed, the plates were retouched by hand by Havell's son, Robert. The printed frontispiece is unusually fine; it features a single stem of *Lilium tigrinum* in an ornate Baroque vase decorated with a cartouche. On the cartouche a poem dedicated to the lily with 'All her Loveliest charms, of Form, Colour, and Scent' has been inscribed. The composition is further embellished by a number of butterflies from India and Africa, somewhat transformed by the artist's imagination. Among the species depicted are two Lycaenidae, one shown in flight and the other resting on a flower stem, two Hesperiidae, a Papilionidae of the genus *Graphium*, and what appears to be a Casinidae. Among the most beautiful plates are those depicting a *Crinum augustum* (pl. 4), which flowered for the first time in Liverpool's botanical garden in May 1829, and a *Crinum pedunculatum* (pl. 11), whose snow-white flower and dark-green fleshy leaves form a striking composition.

Priscilla Bury also produced illustrations for two botanical magazines, *Botanic Garden* edited by Benjamin Maund and *The Botanist* edited by J. S. Henslow and John Stevens. Later she ventured even further afield as a scientific illustrator, collaborating on a work devoted to the radiolaria—wonderfully complex, microscopic animals belonging to the Rhizopod family.

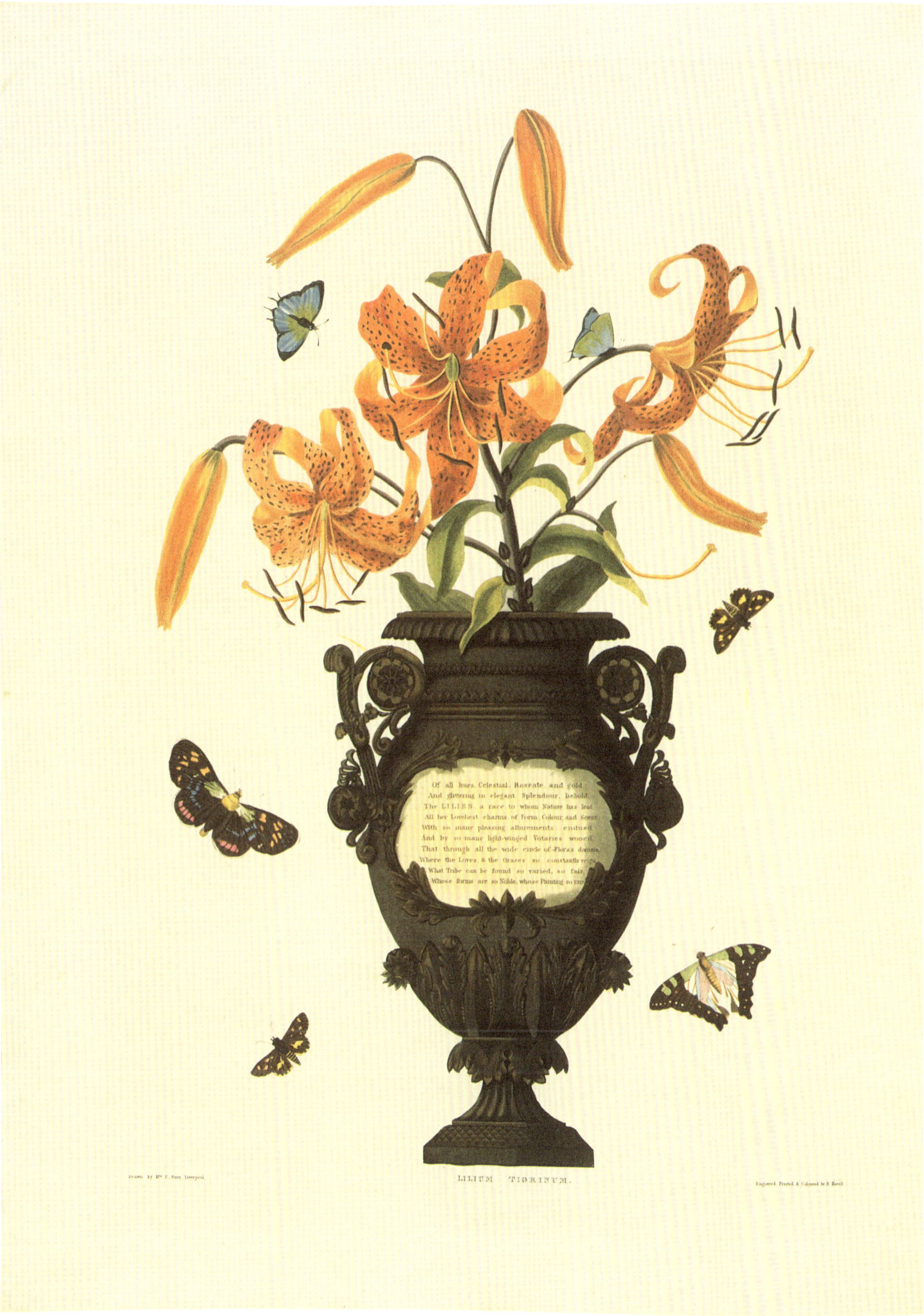

PRISCILLA SUSAN BURY, *A Selection of Hexandrian Plants.* Frontispiece: tiger lilies (*Lilium tigrinum*) and butterflies

87. Berthe Hoola Van Nooten (*fl. second half of the 19th century*)

Fleurs Fruits et Feuillages choisis de l'Ile de Java Peints d'après Nature par Madame Berthe Hoola Van Nooten. Troisième Édition publiée par Merzbach & Falk, Éditeurs Libraires de la Cour et de S.A.R. Le Compte de Flandre.—'Cependant, je vous dis que Salomon même dans toute sa gloire n'a pas été vêtu comme l'un d'eux.' Évangile, Matt. VI, 29.'—Bruxelles Librairie Européenne C. Muquardt Même Maison A Leipzig.

57 × 41.5 cm. *1–90* p., 40 leaves of plates.

BINDING: Half red buckram and printed paper boards, title in gilt on the spine.

PLATES: 40 chromolithographed plates by P. Depannemaeker from drawings by the author. At the foot of every plate is the Latin name of the plant and 'Peint d'après nature par Mme Berthe Hoola van Nooten, à Batavia. Chromolith: par P. Depannemaeker, à Ledeberg-lez-Gand. (Belgique)'.

PROVENANCE: On the front free endpaper: 'J. van Hupt'.

REFERENCES: Nissen 931; Stafleu and Cowan, II.3025.

This is the third edition, published in 1880, of a work that the Belgian artist Bertha Van Nooten produced around 1862 while she was living in Batavia (now Jakarta) on the Indonesian island of Java. The forty large plates in *Fleurs, fruits et feuillages choisis de l'Ile de Java* were printed in Belgium from Van Nooten's original sketches by P. Depannemaeker, using the new technique of chromolithography. Each plate is accompanied by a text in French that describes the plant and its history and enumerates its medicinal properties. Often, enlarged details of the flower and fruit are included. Van Nooten was clearly a more than competent artist, for the splendid tropical plants, with their lush foliage, vividly coloured flowers and exotic fruit, have been depicted with great skill. She managed to accentuate the splendour of each species by adopting a style that combined great precision and clarity with a touch of neo-Baroque exuberance, revelling in the rich forms and colours of the tropics. The reader's eye is immediately captured by the dark leaves, shown furled or crumpled or partly nibbled away by insects, the delicately rendered details of the follicles and seeds, and the heavy clusters of flowers that cascade down the page. The excellent reproduction of the artist's drawings in the form of chromolithographs lends an added tactility to these striking images.

The work opens with an illustration of the *Codiaeum variegatum*, with its tiny, delicate flowers and enormous lanceolate leaves splashed with yellow and red. Also quite effective are the plates of the *Poinsettia pulcherrima*, with its characteristic lanceolate brachts of bright vermilion, the *Saraca declinata* with its dense clusters of tube-shaped calyxes, which could be found in either bright yellow (as in this plate) or red, and the *Poinciana regia*, appropriately named the 'flame of the forest' for its brilliant scarlet flowers.

Europeans who beheld these plants for the first time, whether in the form of magnificent colour illustrations such as those of Van Nooten or as rare specimens cultivated in botanical gardens, could not but have been moved to wonder and amazement. Gradually, however, these plants were acclimatized in greenhouses, and today many are taken for granted as the most ordinary of ornamental house-plants.

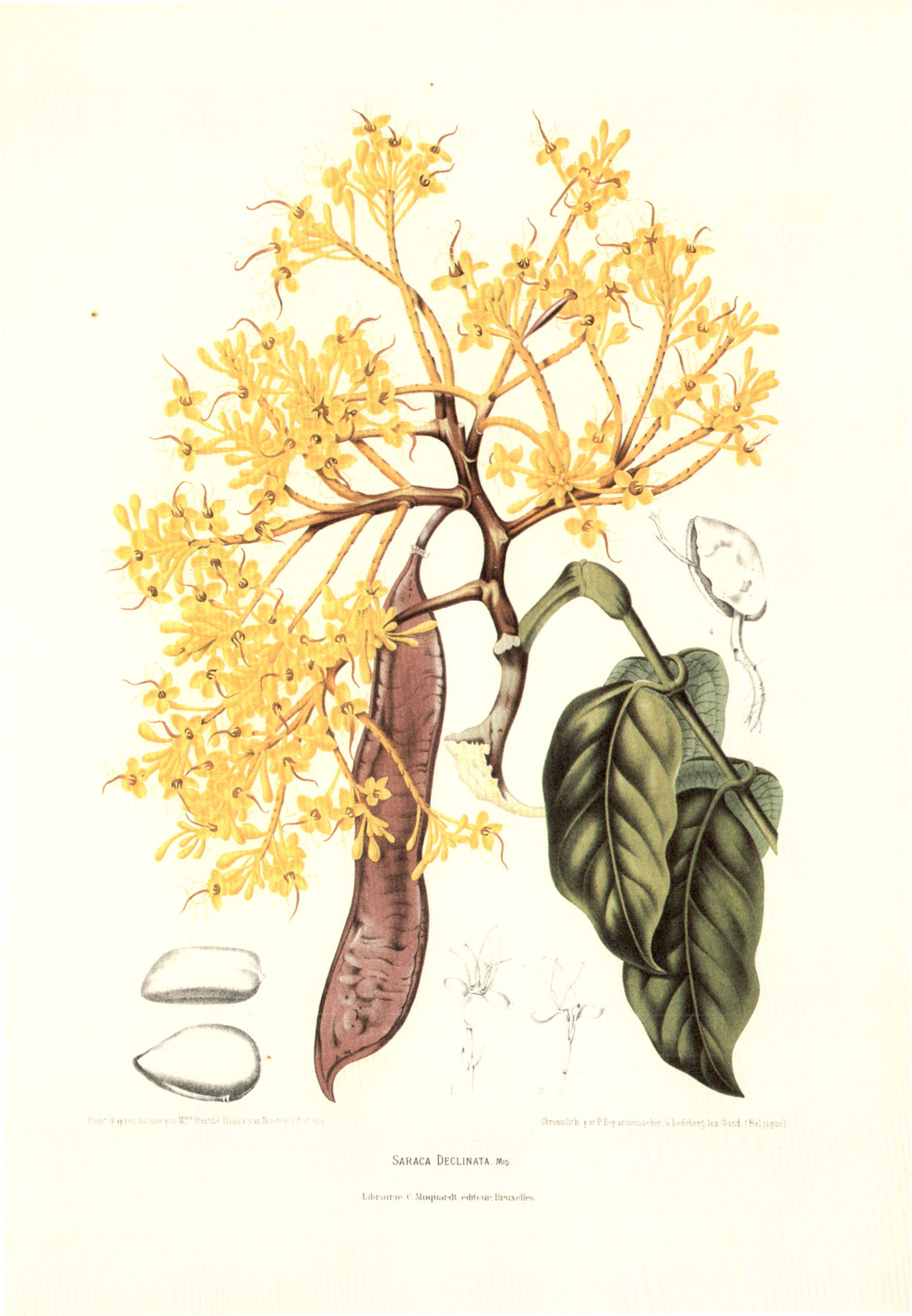

BERTHE HOOLA VAN NOOTEN, *Fleurs, fruits et feuillages*. Red Saraca or Sorrowless Tree (*Saraca declinata*)

IX · A GENTEEL PASTIME: FLOWER PAINTING FOR DILETTANTES

IN THE 'Epistre' to *Les Leçons royales*, which is dedicated to 'Madame la Dauphine', Catherine Perrot expressed her hope that her work would be well received by 'toutes les personnes de qualité, qui ont l'inclination pour la Mignature: Les personnes vulgaires ne seront pas moins obligées d'avoir agreables ces Leçons, puisqu'elles les aideront à faire des Couronnes de Fleurs pour couronner le Portrait de Vostre Personne Royale, et lui rendre les honneurs immortels qui luy sont dûs'. Perrot, who had studied painting under Charles Le Brun and Nicolas Robert (see no. 42), was accepted into the Académie Royale de Peinture in 1682, one of the few women ever to have received this honour. In 1686, one year after the death of her teacher, Robert, she wrote a treatise on bird and flower painting which represented the fruit of her studies under the greatest flower painter of the day, as well as of her own subsequent experience as a teacher. In 1693 she published a more general work on miniature painting, *Traité de miniature* (Bouchot, pp. 393–4).

In *Les Leçons royales* Perrot gives a detailed description of her teaching methods, which primarily consisted in having her students copy the engravings of her master, Robert, which they could afterwards colour in. 'Pour peindre les fleurs', she asserted, 'et pour bien réussir, vous ne pouvez prendre de meilleurs Exemples que les estampes de Nicolas Robert . . . le plus excellent fleuriste qui ait paru jusq'à present'. She also provided a list of the colours necessary for flower painting, together with their prices, and stated that these materials might be acquired at the shop 'A la Cornemuse', which was kept by the widow Foubert in the parish of Saint-Nicolas Deschamps. Perrot first had her students trace an engraving by Robert, either in silverpoint or, following the lines of the engraving, by pricking with a needle a series of tiny holes through onto a piece of high-quality vellum securely fixed to a wooden board. The colours, diluted with gum arabic, could then be spread within the outlines—taking care, Perrot admonished, to use only the most delicate tints where there were no shadows. For each one of the thirty illustrations by Robert, drawn from *Variae ac multiformes florum species* (no. 42), which she describes in her manual, Perrot specifies the colours to be used and their appropriate gradations. *Les Leçons royales*, which could be bought at the printing-house of François Poilly, was not written exclusively for 'personnes de qualité', even though Perrot counted among her pupils various members of the royal family, including Marie Louise d'Orléans. As the introduction declared, it could also be utilized, without the help of a teacher, as a 'do-it-yourself' manual by anyone interested in learning the art of flower painting. This typology would subsequently be adopted by many other authors of treatises on painting.

Robert and his student Perrot were the first botanical artists to specialize in the teaching of flower painting, but this genre soon became so popular that afterwards many other artists (including some of the very greatest exponents of the genre, such as Georg Ehret and Gerard van Spaendonck) were to follow suit. Their pupils became ever more numerous and included many ladies from aristocratic or wealthy families. Despite the great fame and success that these artists achieved as painting masters, however, their own treatises and memoirs at first only rarely reflected the didactic side of their activities, and they did not seem to feel it necessary to explain the purely technical aspects of their art.

One of the most celebrated flower painters who also engaged in teaching was undoubtedly Ehret (nos. 50, 51). After settling in England, his fame as a teacher was added to the renown he already enjoyed as a botanical painter. In his 'Memoir' he records with pride the names of some of the titled ladies who diligently followed his courses (no. 56). His teaching method was firmly based on the assumption that a sound knowledge of botany, following the Linnaean system of classification, was necessary for anyone who wished to study and paint flowers (Calmann, pp. 80–82).

Van Spaendonck, another eminent artist and one who had been made both 'Peintre en miniature du Roy' and 'Professeur de Fleurs' at the Museum du Jardin Royal des Plantes (formerly Jardin du Roi) in Paris, recalls that he did not in the least begrudge the time he dedicated to teaching, for he found 'mille occasions de propager son talent dans toutes les classes, même les plus hautes de la societé' (*Souvenirs*, p. xvii; see no. 92). As a contemporary wrote: 'The fame that M. van Spaendonck enjoys attracts to his studio a great number of many young ladies who come there with their mothers. A few of them have acquired considerable skill, and some of these, applying it to botany, have given us some excellent drawings of plants—as can be seen by looking through those that have been engraved for the *Annales du Museum*' (Blunt 1957, p. 6). The work of some of these students also appears in *Souvenirs*, a tribute in the form of series of floral engravings collected and published by van Spaendonck's pupils shortly after their teacher's death. However, in *Fleurs dessinées d'après nature*, a work compiled by van Spaendonck for his students, he merely provides examples of floral arrangements in the form of large plates, without entering into a discussion of the technical problems of flower painting.

A great number of students, mostly female, also crowded the studio of Pierre Joseph Redouté (who had himself studied under van Spaendonck at the Jardin des Plantes) in the rue de Seine, or frequented his lessons in 'Iconographie Végétale' at the Muséum d'Histoire Naturelle, where the artist taught painting in watercolours for thirty hours every week. Accuracy, composition and draughtsmanship constituted the fundamental precepts of Redouté's teaching. At first he insisted his pupils should always practice drawing from live botanical specimens, but eventually he agreed that they might exercise their skills during the colder months by drawing from copies and enlargements of floral engravings, while the flowers in the Muséum's gardens could serve as their motifs in spring and summer.

The maxim 'the art of flower painting is not a luxury', which Redouté liked to repeat, no doubt reflects the difficulties that botanical artists often encountered when they tried to make a living from their art. Indeed, Redouté encouraged his more talented pupils to dedicate themselves to less precarious, more lucrative activities, such as supplying designs for the applied arts. He himself on occasion provided floral motifs for manufacturers, including a paper-maker at Mulhouse and the Sèvres porcelain works (one pattern inspired by his work *Les Liliacées* was used to decorate a dessert service designed for the Empress Joséphine) (Hardouin-Fugier, pp. 14–16).

With the growing popularity of flower painting, swelled by the great numbers of ladies from the middle classes who now wished to learn this art, from the end of the eighteenth century to the mid-nineteenth, many enterprising publishers, particularly in England, issued treatises on the subject aimed at a more general audience. One of the first of these was *The Florist*, an elegant volume compiled *c.* 1760 by three London publishers—Robert Sayer, Thomas Bowles and John Bowles (no. 88). It offered not only 'the Rules . . . for executing the pleasing Branch of Painting' but twelve magnificent plates too, each depicting a bouquet made up of the flowers that could be found in bloom in one or other of the twelve months of the year.

Also interesting, and quite innovative in its modest way, was an album by another London printer, Carington Bowles (no. 89), which presented its female readers with motifs for their needlework as well as designs for flower painting. Bowles very intelligently provided directions on how to draw various flowers by first breaking them down into simple geometric forms, instructions that must have been of great help to those attempting flower painting for the first time. Other treatises were written by artists themselves, and are full of practical instructions for the beginner. Two typical examples are *The Seasons, or Flower-Garden* by Peter Henderson (no. 90), and *Groups of Flowers* by George Brookshaw (no. 91). The latter, which was dedicated by the author to his pupils, provides by means of stipple engravings both plain line drawings of various flowers and coloured versions of the same.

Oak Spring also possesses a copy of the treatise *A Botanical Drawing-Book; or An Easy Introduction to Drawing Flowers according to Nature* (1807 edition) by James Sowerby, and a manual by the artist Patrick Syme of Edinburgh (1774–1845), *Practical Directions for Learning Flower-Drawing* (Edinburgh, 1810). In the first, young beginners are led step by step through the process of drawing the individual parts of a flower to, finally, a sketch of the entire plant. The second provides simple, but very clear and complete, instructions on flower painting, one of the author's aims being to assist 'ladies in the country, who wish to learn flower-drawing, and particularly to those who have not an opportunity to receive instruction from a master'.

One of the most interesting and detailed treatises on flower painting to appear in the nineteenth century was written by a student of Redouté, Mme Augustine Dufur (1797–1831). In her *L'Art de peindre les fleurs à l'aquarelle*, published in Paris in 1834 (a copy of which is at Oak Spring), she describes in meticulous detail her technique for watercolour painting, which, she claimed, could create delicate and transparent effects far superior to those achievable by any other means.

After enumerating the tools necessary for the artist, Dufur passes to a discussion of colours. These, she explains, should be prepared by the artist from various basic pigments. For each hue it is best to select three tones—one for the highlights, another for the shadows, and a third for the middle gradations. Painting in watercolours requires great skill and quickness, the author reminds us, and one should always begin with the parts in shadow, working gradually toward the lighter sections. Every tint should also be lightened as it approaches the border of the painting, by dipping one's brush in water. The parts of the flower themselves should be painted in a certain sequence, beginning with the corolla, then passing to the calyx, the leaves, the peduncles and, finally, the stem. In painting the leaves one must be careful to follow the natural line of the veins. Since a wide range of tones is required to depict the many subtle hues of a flower, it is advisable for the student to begin with lighter tones rather than those that are too accentuated. The various colours must be retouched many times in order to achieve the most realistic effects possible, however, before a painting can be considered to be finished. The artist is free to utilize a variety of brushstrokes—from the finest of lines to stippling—but should take care to apply the strokes very closely to one another in order to create a perfectly homogeneous effect. The final step, which required a sure and careful hand, was to add the darkest shadows and to retouch the highlights in order to create the effect of glancing flashes of light.

These rules, so clearly and simply explained by Dufur, are just as valid today as they were in her own lifetime, and could be studied with profit by any modern painter in watercolours interested in the art of botanical illustration.

88. 'The Florist' [1760]

[*within a frame composed of a garland of flowers and a basket*] The Florist Containing Sixty Plates of the most beautiful Flowers regularly dispos'd in their Succession of Blowing. To which is added an Accurate description of their Colours, with Instructions for Drawing & Painting them according to Nature: Being a New Work intended for the use & amusement of Gentlemen and Ladies Delighting in that Art. London Printed for Robt. Sayer in Fleet Street, T. Bowles in St Pauls Church Y^d^ & John Bowles & Son, in Cornhil[l]. Price 6^s^ & Colour'd £1 1^s^ O^d^. [1760]

4° 22.5 × 15.5 cm. A–B⁴. Illustrated title-page with 60 numbered, hand-coloured plates and a head-piece. 61–76 p.

BINDING: Calf binding with gilt frame, floriated spine, and lettering on a red leather label: 'Florist'.

PLATES: Decorated title-page and 60 hand-coloured etched plates beginning with 'January'. The English name of the flower appears at the foot of the page, the name of the month in the upper left-hand corner, and the plate number in the upper right corner. Plates 6, 23, 34, 39, 43, 45, 48, 57 and 60 are signed either Tijou, M. Davies, A Smith or C. Phillips.

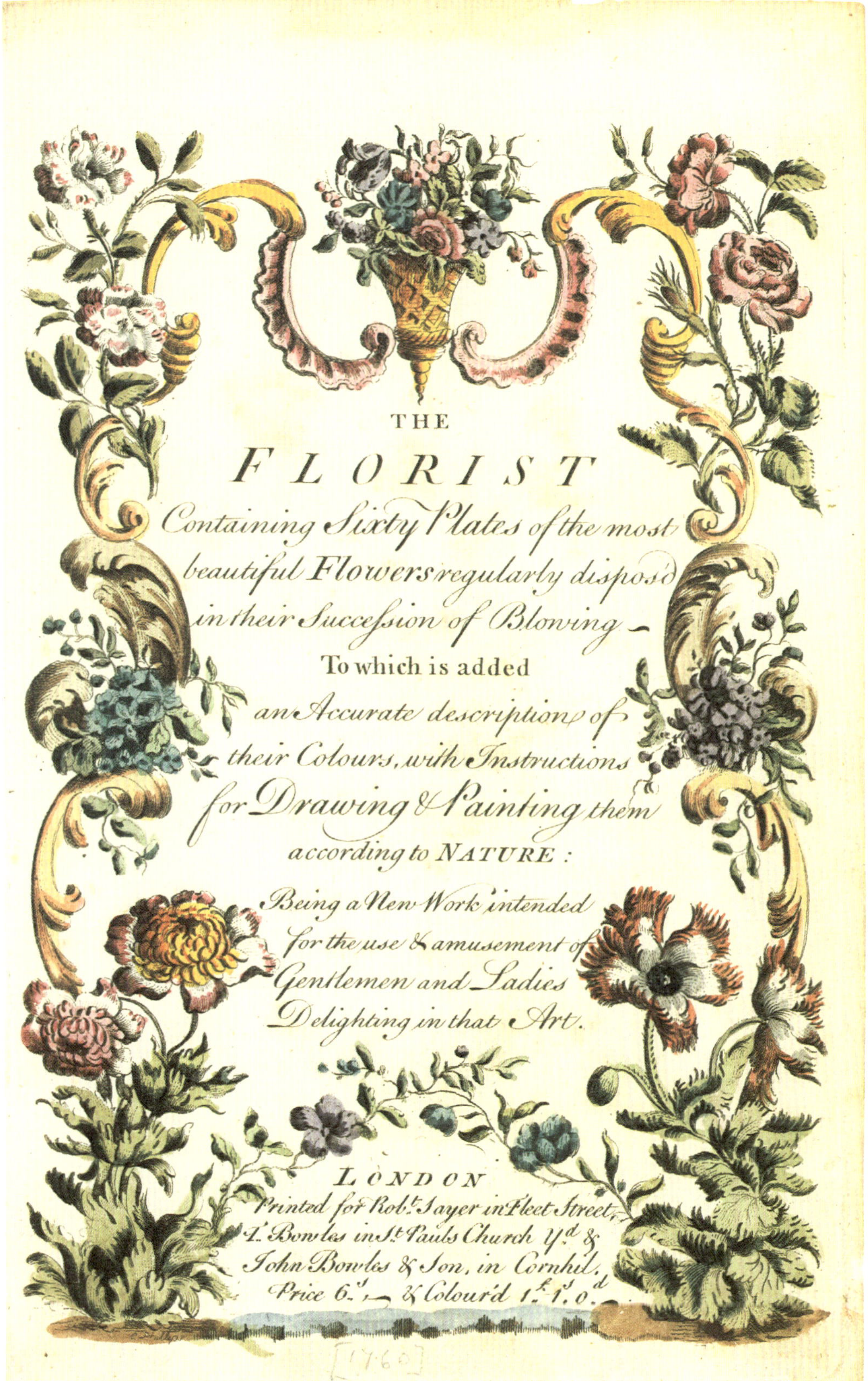

Title-page to *The Florist*, c. 1760

PROVENANCE: Armorial bookplate of John Hatt Noble. Inscribed on the first front free endpaper: 'John Caley I:L.s 1815'.

REFERENCES: Dunthorne 272; Nissen 1734; Henrey, II.590 and III.707.

THIS WORK, printed in London by Thomas Bowles (d. 1767), his son, John Bowles (1701–79), and his grandson, Carington Bowles (1724–93), and published by Robert Sayer (1725–94), represents an early example of one of the pattern-books for drawing and painting flowers (see Henrey, II.585–6). Sayer also published new editions of *Eight Beautiful Groups of Natural Flowers* (*c.* 1770) and *Small Bouquet of Flowers* (1787).

The introduction to *The Florist* contains some simple guidelines for the beginner, rules that, the anonymous author assures us, 'are the result of real Practice'. To produce a lifelike drawing, for example, the most important step is 'chusing [*sic*] the Flower in its proper State for Copying'. By drawing directly from nature, the author explains, one can avoid the stiff formality of the copied image. Regarding the use of colour, he recommends that the flowers be placed with the light falling from the left, casting the right half of the composition into shadow. He then adds a list of the paints necessary, with the reminder that 'the Painter, who may perhaps smile at the Plainness and intended Simplicity of these Instructions, should consider that this is a Work not address'd to those who are already Artists, but an Invitation to the young uninstructed Admirers of Painting to the Practice of this delightful Branch of it'. In a footnote he informs the reader that the requisite colours can be bought at a reasonable price from the printers of *The Florist*. Further instructions follow, regarding the best way to paint specific flowers: the hyacinth, cyclamen, crocus, snowdrop, anemone, almond blossom, auricula, daffodil, iris, rose, tulip, etc., many of which are depicted in the book.

The sixty plates in *The Florist* (engraved by Tijou, Adam Smith, M. Davies and Caspar Phillips) were intended not only to serve as models for the amateur artist; they could also be admired by flower lovers at the modest price of six shillings for an uncoloured copy, or one guinea for a hand-coloured one. The collection at Oak Spring includes both a hand-coloured copy and an uncoloured one of a different issue, also undated (octavo format, with pages numbered 1 through 16). Another edition containing the same text and plates was published in 1774 with the title *Bowles's Florist*.

89. CARINGTON BOWLES (1724–1793)

Bowles's Drawing Book for Ladies; or Complete Florist: Being An Extensive and Curious Collection of the most Beautiful Flowers, All drawn after Nature by A. Heckle. With A short Introduction to Drawing, and Directions for Mixing and Using of Colours. Also Several Proper and Easy Examples. The Whole adapted for the Improvement of Ladies in Needle-Work. London: Printed for the Proprietor Carington Bowles, at his Map and Print Warehouse, No. 69, St. Paul's Church Yard.

28 × 33.5 cm. *1–4* p., 23 leaves of plates.

BINDING: Three-quarter in green leather and cloth over boards.

PLATES: 23 engraved plates; plates 2–23 have been coloured by hand.

PROVENANCE: Inscribed on the front free endpaper: 'Bailey Decr 1st 1836'.

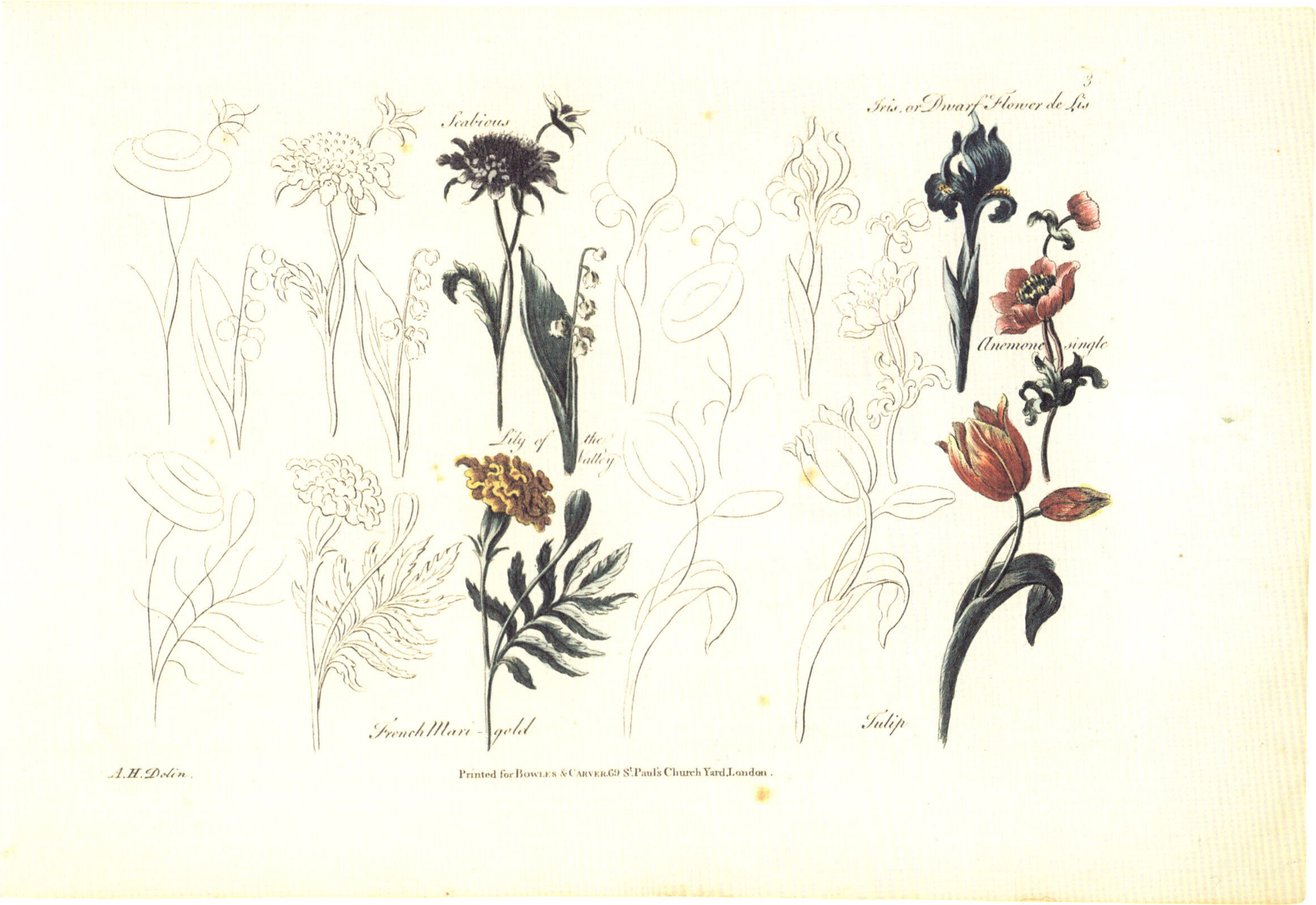

CARINGTON BOWLES, *Bowles's Drawing Book for Ladies*. Plate 3: Scabious (*Scabiosa* sp.), lily-of-the-valley (*Convallaria majalis*), French marigold (*Tagetes patula*), dwarf *Iris*, windflower (*Anemone blanda*), *Tulipa* sp.

THE ENGLISH engraver and publisher Carington Bowles is best known for his coloured engravings on allegorical and genre subjects and for the publication of numerous little treatises on artistic techniques, as well as guides and maps to London. In 1774 he published *Bowles's Florist*, with sixty engraved and hand-coloured plates of flowers arranged according to the months of the year.

Bowles's Drawing Book for Ladies was probably published in 1790 or thereabouts. Its aim was, as the title states, to provide floral motifs appropriate for use by women in their flower painting or embroidery. The introduction offers some simple rules and advice to be followed by those wishing to learn the art of flower drawing. 'Drawing is an art of some length and time', it observes, 'and to be perfect in it, it is necessary that the hand should be improved in practice, and the mind in judgment every day'. The starting-point of any drawing, according to Bowles, is the outline; in fact, he recommends that one should always begin in the upper right-hand corner, and the rest of the figure will then follow naturally, especially if one takes care frequently to compare the drawing with the original. These instructions are followed by a description of the 'proper materials for Drawing . . . Black Lead pencils, . . . Charcoal, red, black or white Chalk, . . . Pens, . . . Indian Ink' etc., and various observations on the use of colour, for it is 'the artful management of lights and shades that gives the appearance of substance, roundness, and distance'.

Much more useful than the text, however, are the plates, which—engraved from drawings executed *dal vivo* by the artist Abraham Heckle—were expressly designed to teach the rudiments of flower painting to non-artists. For example, the first illustration (not coloured) shows how many natural forms—leaves, branches, simple flowers—can be derived from basic geometric shapes and linear forms, such as the triangle, oval and spiral. The second plate demonstrates how to draw a flower in three steps, using two examples. First, a simple outline of each flower (in this case, a *Polyanthus* and a common marigold) is presented; then a drawing of each with many of the details added; and finally, two carefully coloured and very realistic finished drawings. The same exercise is repeated with other flowers in the following two plates. The remaining illustrations present different flowers suitable for copying, each one coloured by hand and labelled with its common name.

This little volume was a product of the popular culture of the period, when the vogue for flower painting and floral motifs was widespread and many people, not just the wealthy, took up the pastime of flower painting. Simple and unpretentious in its approach, *Bowles's Drawing Book* provided a useful beginner's manual for amateurs in the art of flower painting.

90. PETER HENDERSON (*fl.* 1791–1829)

The Seasons, or Flower-Garden: Being A Selection Of The Most Beuatiful [*sic*] Flowers That blossom at the Four Seasons of the Year, Spring, Summer, Autumn, Winter. The Whole Carefully Drawn From Nature. With a Treatise, on A Description Of, And General Instructions for Drawing And Painting Flowers. By P. Henderson. | 'Survey the whole, nor seek slight faults to find'. | London: Published as the Act Directs, By R. Ackermann, At his Repository of Arts, Strand. Hayden, Printer, Brydges Street, Covent Garden.

[*second part*] A Treatise: or, Instructions for Drawing and Painting Flowers, with General Observations on the Art. By P. Henderson. | 'Delightful art! how great thy friendly power, | That knows to cheer the melancholy hour; | To teach at once the troubled mind to bear | Oppressive ills, and soften fell despair'. *Guide through Wales*. | London: Published as the Act directs, by R. Ackermann, at the Repository of Arts, Strand. Hayden, Printer, Brydges Street, Covent Garden.

35 × 29 cm. *1–3* 4 *1–28* *1–5* 6–28 p., 23 leaves of plates.

BINDING: Mottled calf; on the spine in gilt: 'The Seasons, Henderson'.

PLATES: 23 engraved plates, hand-coloured with aquatint, some of them numbered in arabic numerals, presenting 6

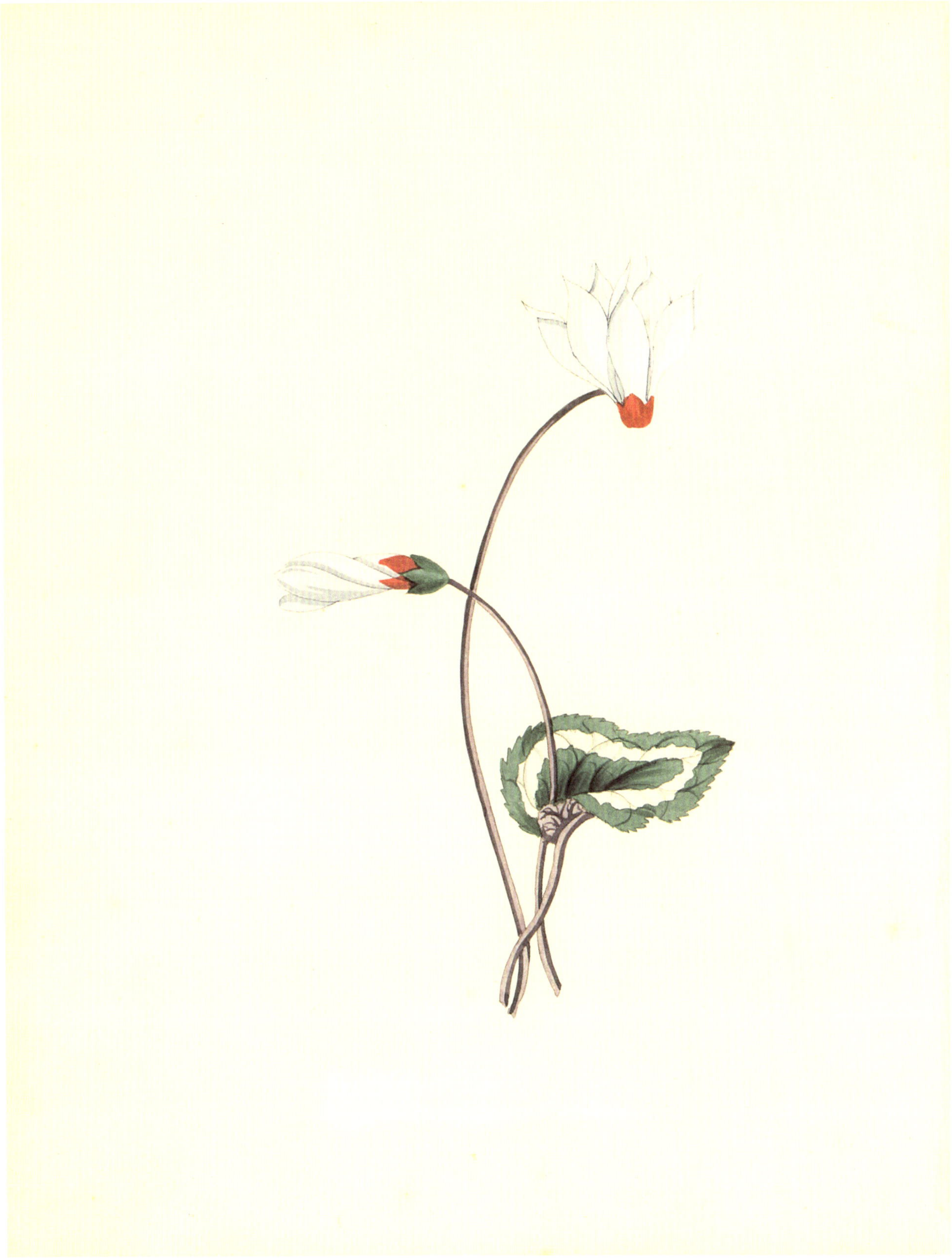

PETER HENDERSON, *The Seasons*. Two cyclamens with leaf, 'The Spring Cyclamen', plate 2

flowers for each season (lacking 3 plates: the Damask rose and the *Convolvulus minor*; and a frontispiece in line, aquatint and stipple engraving).

PROVENANCE: Ex-libris Paul Mellon.

REFERENCE: Dunthorne 126.

THE AUTHOR of this elegant volume of engravings, published in 1806–7, was a botanical artist of great accomplishment. Henderson, in fact, was one of the most important collaborators with Robert Thornton on the latter's megalomaniacal work *The Temple of Flora*, contributing the original paintings for two of its most impressive illustrations—the 'American cowslip' and the 'Dragon Arum' (see no. 93). In 1808 or 1809 Henderson also produced a *Pomona* on soft fruits, dedicating it to the artists who had assisted him with the work (see *An Oak Spring Pomona*, no. 39).

The first part of *The Seasons* is divided by the author into four sections, each one devoted to one season of the year and illustrated with six flowering plants. The plates were aquatinted by G. Hayden and retouched by hand, perhaps by Henderson himself. A frontispiece decorated with an allegorical scene opens the work; it shows Cupid using the point of his arrow, carving the title of the work into a stone. Various flowers can be seen growing at his side, including a convolvulus and some hollyhocks, while in the background is a densely forested landscape. This frontispiece was prepared by the combined techniques of stipple engraving (for the figure and foreground) and etching in line and aquatint (for the background). It seems more than possible that the scene was inspired by the frontispiece 'Cupid Inspiring the Plants with Love' (a reference to the sexual reproductive processes of plants) designed by Reinagle in 1800 for *The Temple of Flora*.

Unlike Thornton's *Temple of Flora*, however, this work by Henderson had no great scientific pretentions. The author's modest aim was to provide some advice on the art of flower painting, accompanied by plates of 'those flowers which appear to me the most beautiful and useful . . . to illustrate the Seasons'. He chose as his examples garden flowers whose simple charms could match the beauty of more exotic blooms. Among the many captivating illustrations in this work, the plates depicting the snowdrop, the spring cyclamen, the purple auricula and the scarlet poppy are particularly beguiling. The brief text accompanying each plate was also intended for the artist rather than the botanist. For example, the cyclamen is described in the following words: 'The leaves of this beautiful Spring flower are heart-shaped, variegated with white and green; the flowers are sweet scented, of a delicate white, with a border of carmine about the bottom of its pendulous cup.'

The second part of *The Seasons* offers some instruction, aimed primarily at the beginner, on the drawing and painting of flowers. Henderson recommends that one should learn the rudiments of drawing before attempting to use colours, for the painting of flowers requires 'delicacy, thinness, and transparency', which can only be learned through constant application and by a careful study of the great masters as well as of Nature herself. Appropriately, then, Henderson closes his work with this couplet from Alexander Pope's *Essay on Criticism* of 1711: 'Learn hence from ancient rules a just esteem; / To copy Nature is to copy them.'

91. George Brookshaw (*fl.* 1804–1822)

Groups of Flowers, Drawn And Accurately Coloured After Nature, With Full Directions For The Young Artist; Designed As A Companion To The Treatise On Flower Painting. By George Brookshaw, Esq. Author of The Pomona Britannica, Treatise On Flower Painting, &c. London: Printed for Longman, Hurst, Rees, Orme, and Brown, Paternoster-Row; and John Lepard, 108, Strand. By Augustus Applegath and Henry Mitton, 24, Nelson-Square, Great Surrey-Street. 1817.

38 × 26.5 cm. 1–18 p., 12 leaves of plates.

BINDING: Half-blue leather and marbled boards.

PLATES: 6 hand-coloured, stipple engraved plates and 6 uncoloured stipple engraved plates.

PROVENANCE: Stamped on front paste-down: A. W. Bahr.

REFERENCES: Dunthorne 53 (for 1819 edn); Nissen 246.

GEORGE BROOKSHAW was an artist and teacher of floral painting who also produced a number of highly popular treatises on the subject. In 1812 he published his most important work, *Pomona Britannica* (see *An Oak Spring Pomona*, no. 40), which he describes in the subtitle as 'A Collection of the Most Esteemed Fruits at present cultivated in the country; together with the Blossoms and Leaves of such as are necessary to distinguish the various sorts from each other. Selected principally from the Royal Gardens at Hampton Court, and the remainder from the most celebrated gardens round London.' This pomona was illustrated with ninety plates, depicting 256 varieties of fifteen fruits, often shown with their flowers.

Brookshaw then produced in rapid succession a series of shorter works intended for ladies and gentlemen wishing to learn the art of flower painting. In 1816 he published *A New Treatise on Flower Painting, or, Every Lady her own Drawing Master*, which was followed one year later by *Groups of Flowers* and then *Groups of Fruit, Accurately Drawn and Coloured after Nature* (see *An Oak Spring Pomona*, no. 41). The same year saw the publication of *Six Birds, Accurately Drawn and Coloured after Nature with Full Instructions for the Young Artist: Intended as a Companion to the Treatise on Flower Painting*. At Oak Spring is a copy of the first edition of *Six Birds* as well as the *Supplement to the Treatise on Flower Painting*, also published in 1817.

In the Advertisement that opens *Groups of Flowers*, Brookshaw states that the work was written for young, aspiring artists and for amateur painters. He offers various reflections on the art of floral painting, suggesting that it is an art particularly suitable for ladies, and furthermore provides 'an easy introduction to general painting, and does not require a previous knowledge of perspective'. In fact, 'When the learner has made a little progress, nature itself presents a boundless variety of copies'. He insists that floral painting does not in any way constitute a lesser genre, citing the great Sir Joshua Reynolds, who 'strongly recommends historical painters to study and paint from groups of Flowers, as objects the best calculated to form a free and graceful manner of composing'.

Each of Brookshaw's treatises contains a series of six plates designed to serve as models for the aspiring painter. *Groups of Flowers* includes plates depicting the moss rose, the anemone, the china aster, the ranunculus, the major convolvulus, and the dog rose. To provide the reader with a clearer

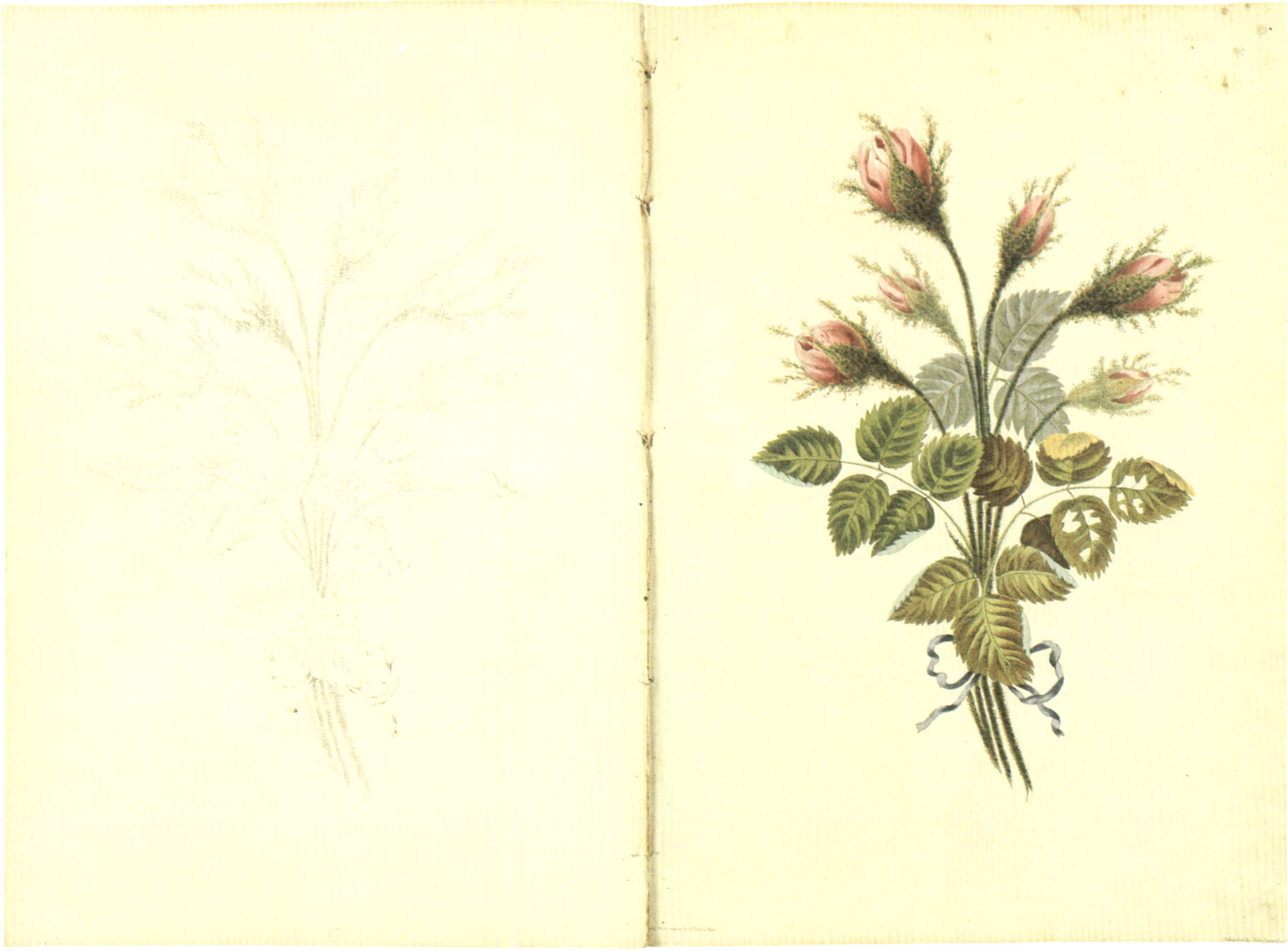

GEORGE BROOKSHAW, *Groups of Flowers*. 'Moss Rose' (*Rosa centifolia muscosa*), in both coloured and uncoloured states

understanding of the separate processes of drawing and painting, on the left-hand page the flower appears in a simple stipple engraving in black and white, while on the facing page the same printed image is presented in colour.

This is a much more modest work than the author's *Pomona Britannica*, in which many of the beautifully stipple engraved fruits appear against a dark aquatint background. *Groups of Flowers* none the less constitutes a pleasing example of that genre of works devoted to the teaching of flower painting which was so extraordinarily popular and widespread in the latter part of the eighteenth century (see no. 89) and for much of the nineteenth.

92. Gerard van Spaendonck (1746–1822)

Souvenirs de Van-Spaendonck, ou Recueil de Fleurs, Lithographiées D'après les Dessins de ce Célèbre Professeur, accompagné d'un Texte rédigé par plusieurs de ses Elèves. Paris, à la Librairie de Castel de Courval, Rue de Richelieu, N° 87. 1826. Imprimerie de E. Pochard. Rue du Pot-de-Feu, N° 14.

8° 21.5 × 29 cm. $\pi^2 \star^4$–$\star\star^4 \star\star\star^2$ *1–12*^{4} 13^2 *i–iv 1* ii–xix *xx* 1–98 *i–ii* p.

BINDING: Half red morocco and marbled boards. On the spine in gilt lettering 'Souvenirs de van-Spaendonck'. The binding is signed and dated 'Mercher 1961'. Original pink paper wrappers bound-in.

PLATES: 19 coloured plates of flowers lithographed by Godefroy Engelmann (1788–1839).

PROVENANCE: Bookplate of Arpad Plesch (1890–1974).

REFERENCES: Dunthorne 315; Pritzel 8810; Nissen 1880; Sitwell and Blunt, p. 77; Plesch, p. 414.

THIS VOLUME was edited and printed by Gerard van Spaendonck's students four years after his death, and testifies to the importance of this celebrated artist's role as Professeur de Peinture de Fleurs at the Jardin (Royal) des Plantes in Paris. It is presented as an affectionate visual antiphon to a volume of twenty-one stipple engravings of floral motifs (a copy is at Oak Spring), *Fleurs dessinées d'après Nature . . . recueil utile aux amateurs, aux jeuns artistes, aux élèves des écoles centrales et aux dessinateurs des manufactures*, published *c.* 1800 by van Spaendonck for his students, who included Adèle Riché, Maréchal, Oudard and Blanchard. *Souvenirs* provides much useful information on the life and personality of this artist, even if the admiration of the authors for their teacher led them to express themselves in perhaps excessively eulogistic terms.

We learn from this work that one of the artist's first patrons was the collector and connoisseur Claude-Henri Watelet. This important figure in artistic circles in Paris was the friend of François Boucher, Hubert Robert and Jean-Jacques Rousseau, author of a famous treatise, *Essai sur les jardins* (1774), and designer of a famous garden, Moulin-Joli (see *An Oak Spring Hortus*, no. 34). Watelet was instrumental in helping van Spaendonck to obtain his first official post, that of Peintre en Miniature du Roi.

When van Spaendonck arrived in Paris in 1770, flower painting was still considered a minor art, the 'nobler' genres of history and landscape painting being much more admired. Nevertheless, with his works van Spaendonck succeeded in expressing, in the words of his students, 'une langue à la fois philosophique et poétique', so that after having captivated the beholder with their magnificent colours his paintings 'font appelle à l'intelligence et à la science'. The text affirms that the artist singlehandedly transformed the genre of flower painting in France, introducing new elements such as elaborate vases and rich fabrics to embellish his sophisticated compositions. The authors were aware that the revived success of flower painting in the late eighteenth century owed much to the development of manufacturing and the industrial arts, which made it possible to produce cloth, rugs, and upholstering materials in vast quantities at low cost. They note that van Spaendonck's designs were admirably suited to such processes and found immediate and wide application.

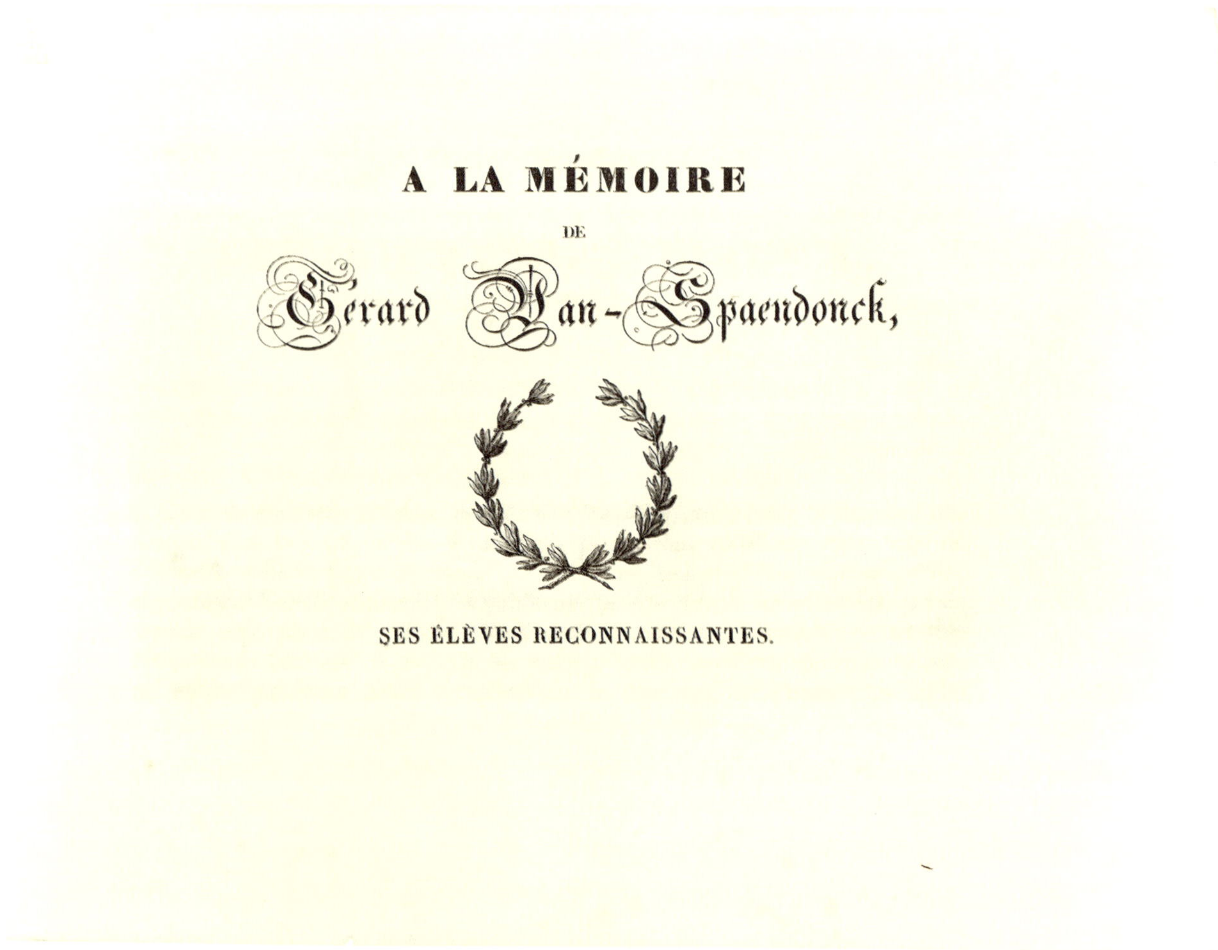
A LA MÉMOIRE

DE

Gérard Van-Spaendonck,

SES ÉLÈVES RECONNAISSANTES.

GERARD VAN SPAENDONCK, *Souvenirs de Van-Spaendonck.* Dedication page

The text then follows with an explanation of the important role played by the Jardin des Plantes in Paris, which was not only a centre for botanical and horticultural study, for it was also an 'espèce d'abrégé de l'univers', where artists could find creative stimulus as well as subjects for their canvases. In fact, it was perhaps in his role as teacher that van Spaendonck gave the best of himself, sharing his talent and communicating his enthusiasm to students and colleagues alike.

The biographical notes in this work are followed by the 'souvenirs', nineteen plates based on original drawings by van Spaendonck of the rose, ranunculus, poppy, lily, carnation and other favourite flowers of his. Each plate is accompanied by a brief text describing the flower's origins and its most popular varieties. They were lithographed by Godefroy Engelmann (1788–1839), which was a more direct, less cumbersome procedure for preparing book illustrations than traditional engraving techniques. Engelmann himself contributed to improvements to the process, developing new inks and lithographic pencils, even writing a manual that encouraged the spread of the tech-

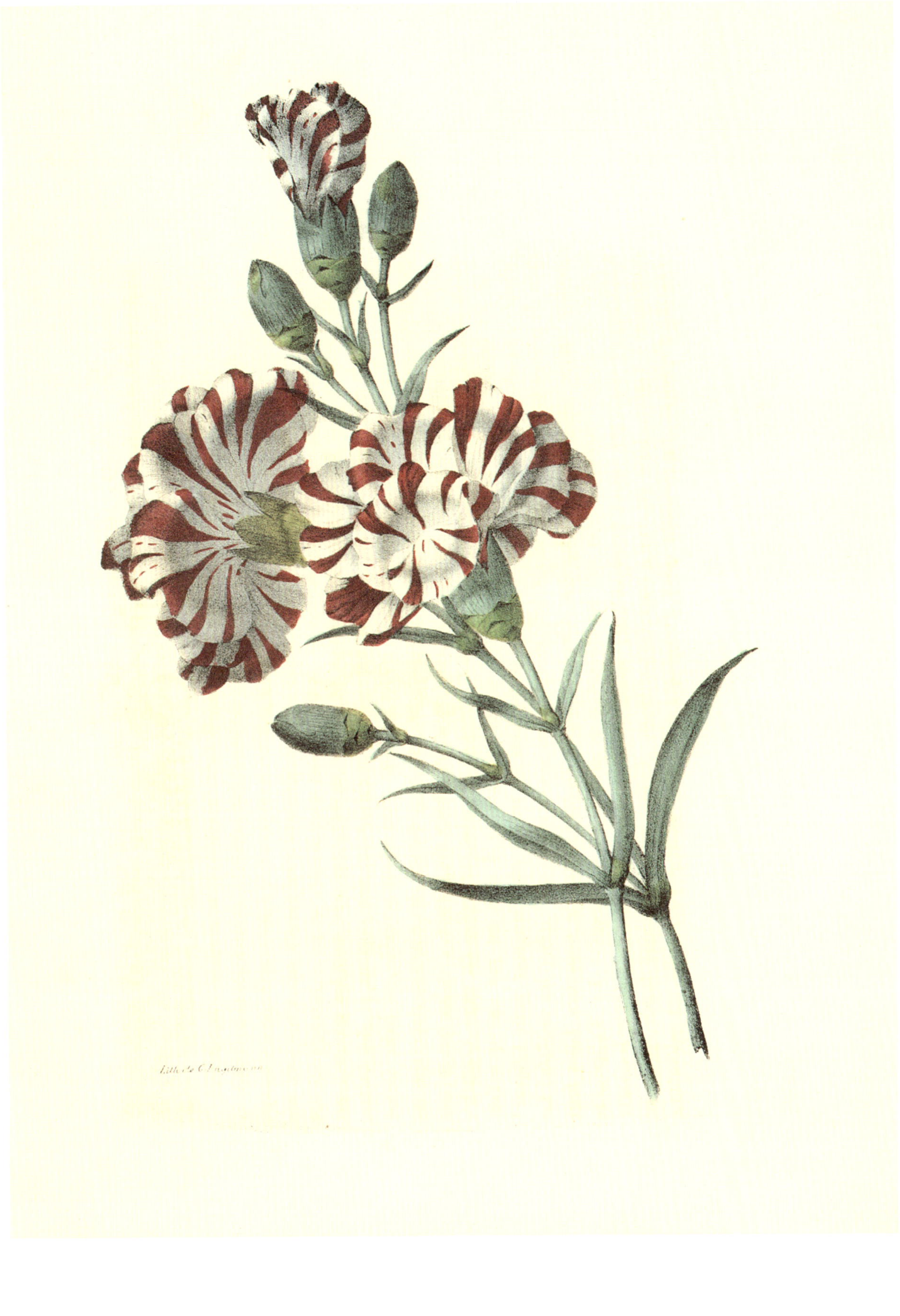

GERARD VAN SPAENDONCK, *Souvenirs de Van-Spaendonck.* Carnations (*Dianthus caryophyllus*)

nique, *Traité de lithographie* (1840). We can see from the plates in *Souvenirs* why this revolutionary printing process was such an immediate success, for here the precise outlines, fine half-tones and subtle colours that characterized van Spaendonck's original paintings in bodycolour are faithfully reproduced in all their original delicacy.

X • THE ROMANTIC PERIOD: SENTIMENT AND EXTRAVAGANCE

ONE OF THE most significant aspects of the Romantic movement was an interest in nature that expressed itself in many forms, ranging from the scientific to the purely emotional, in philosophy, science and the fine arts. It served to swell the already pronounced interest in botany and gardening to a genuine vogue that crossed all social barriers and manifested itself with particular enthusiasm in England and France (both of which are known to this day as countries replete with devoted gardeners). Every house, however modest, whether in the middle of the city or out in the country, had its own garden stocked with plants and flowers acquired at market or from floriculturists, whost flourishing businesses recalled the period of the great horticulturists and collectors of the seventeenth century.

The nineteenth century may without exaggeration be described as a 'century of flowers'. Among the middle classes and the *nouveaux riches* in particular, flowers were a cherished motif; not only did they place fresh flowers on their tables and paintings of flowers on their walls, floral designs could be found everywhere in their homes—in florid patterns on their wallpaper, woven into their carpets and upholstery, embroidered on their curtains, and printed on their dress material and decorated paper (Scourse, p. 15). The flower, associated since the dawn of history with the fair sex, was indeed utilized by the ladies with what might be considered excessive enthusiasm—woven into their hair or their hats, tucked coyly into their décolletage, printed on their gowns, praised in sentimental songs and poetry, painted in diaphanous watercolours, and used to construct charming 'tussie-mussies' (nosegays) and collages in coloured paper.

Thus, in the nineteenth century, flowers formed a ubiquitous theme in the decorative and industrial arts, although they continued to provide inspiration in the fine arts too. John Constable (no. 96), Gustave Courbet and Eugène Delacroix all painted floral still-lifes. In the private collection of Mr and Mrs Paul Mellon in Washington, D.C., for example, there are two striking watercolours of flowers by Delacroix, *Narcissus* (23 × 16.5 cm.) and *Flowers* (20.22 × 20.22 cm.).

The nineteenth century was also marked by significant advances in the science of botany, particularly in the study of exotic tropical species, and many important and lavishly illustrated works were published. At the same time, more and more laypersons were beginning to take an interest in botany, encouraged by the wider availability of small, easy-to-use microscopes. Book publishers were not slow to identify this new audience and began to produce works written especially for it. Thus, next to the ponderous scientific texts, specialized studies and the first great monographs, various sub-genres of less weighty literature made their appearance.

Following the success of Jean-Jacques Rousseau's *Lettres élémentaires sur la botanique* (1782), in the nineteenth century works of 'light Linnaean botany' and 'botany for ladies' became extremely popular. Some were written by eminent scientists, such as John Lindley, others by popular authors, such as Jane Loudon (no. 85). Botany offered an occasion for the dissemination of history or literature in a simple format designed to appeal to the widest possible audience. A characteristic work of this genre was *La Botanique historique et littéraire* by the marquise de Sillery, Mme Stéphanie Félicité Brulart de Genlis. Published in Paris in 1810, it presents a series of anecdotes, legends and episodes culled from various classical and contemporary sources, all linked by a floral motif.

At the beginning of the nineteenth century the flower came to be used more widely and more overtly than ever before as a metaphor in the realm of the sentiments. Flowery missives were used, in particular by women, to convey the state of mind (or heart) of the sender—joy, sadness, desire and, of course, love. Poetry on floral themes became a distinct, and not always minor, genre (Carter, pp. 151–2). One archetypal product of the Romantic period was the sentimental flower-book. These little volumes of poems or tales couched in the 'language of flowers', and elegantly bound and illustrated by means of charming lithographs, were presented as gifts to well-bred young ladies by their admirers. The illustrations to these works were sometimes prepared by noted artists: Charles Malo's *Histoire des roses*, for example, contains twelve plates by the botanical artist Pancrace Bessa (no. 79), who had collaborated with Redouté on his great work, *Les Roses*. No less captivating are other, more modest works, for example *Le Jardinier Fleuriste dédié aux dames* (no. 94), which contains a 'calendar of Flora' listing those flowers traditionally associated with each month of the year.

The 'language of flowers' adopted in these works was first introduced to Europe by Lady Mary Wortley Montagu, whose husband was Britain's ambassador to Constantinople from 1716 to 1718. In her letters home (first published in 1763), Lady Montagu described the customs of the people among whom she found herself, including the very beguiling one of assigning meanings, through verses and mottoes, to various objects, including flowers, fruits and plants as well as colours. Thus, she explained, a codex could be compiled that would allow one to express oneself 'without inking your fingers' (Coats, p. 15). In the first decades of the nineteenth century the public's fancy was captured by the charm of this unspoken language, and a series of 'dictionaries' were published, such as Delachénaye's *Abécédaire de flore, ou langage des fleurs, méthode nouvelle de figurer avec des fleurs les lettres, les syllables, et les mots* (Paris, 1811). The *Abécédaire* was quickly followed by many other 'grammar books', the most successful of all being Charlotte de Latour's *Le Langage des fleurs* (no. 97), which was translated, copied and imitated all over Europe.

Oak Spring possesses an unusually fine collection of Romantic flower books. One example is

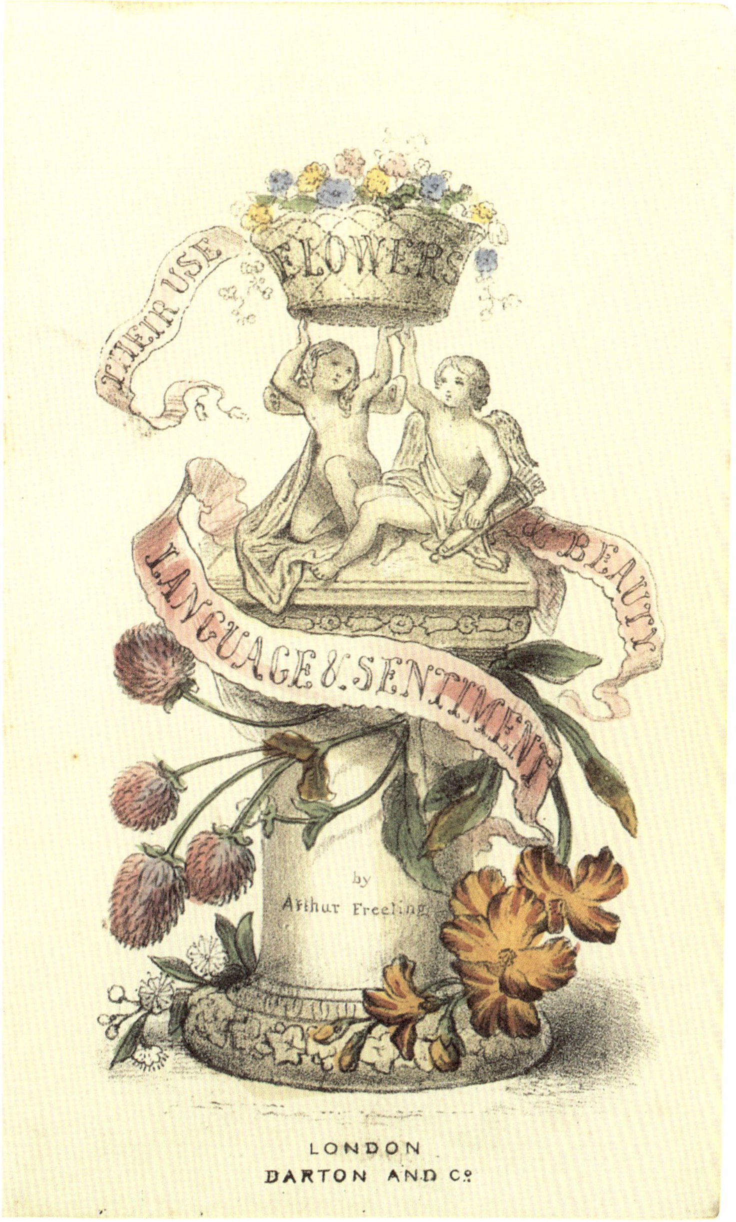

ARTHUR FREELING,
Flowers: Their Use and Beauty, in Language and Sentiment,
London, 1851. Lithograph title-page

the anonymous *Flowers: Their Use and Beauty, in Language and Sentiment*, published by Arthur Freeling in London in 1851. It presents a series of flowers, each with a short text explaining the sentiment associated with the flower, and lines of verse by contemporary poets expanding on the theme. The work is preceded by a somewhat naïve but pleasing lithographed title-page depicting a pair of young lovers standing on a pedestal and holding up a basket of flowers. Other works of the same genre combined poems and floral lithographs, such as *Flora and Thalia, or, Gems of Flowers and Poetry* (London, 1835) or *Flowers, Earth's Silent Voices*, published in Philadelphia in 1865 and illustrated with plates by the American artist Sophina Gordon (Oak Spring has a copy of each).

The vogue for the language of flowers became so widespread that even serious scientists were on occasion tempted to join in the game. Henry Phillips, for example, a banker and a member of the prestigious Royal Horticultural Society, briefly set aside his botanical studies to delve into poetry and literature in search of the symbolic meaning of various flowers. In *Floral Emblems* (no. 95), following the complex and highly artificial set of rules governing the seventeenth-century use of emblems and devices, he associated different flowers with different sentiments, utilizing short mottoes and verses to explain the symbolism of each. In his learned *divertissement* we find the same pleasing formula—combining prose, poetry and visual elements—typical of other works on floral subjects in the Romantic period. An almost contemporary game, *Das Reich der Blumenkönigin / The Realm of the Queen of Flowers / L'Empire de la Reine des Fleurs*, also found at Oak Spring, composed of colour plates of tree vases and a garland with cut-out colour flowers to insert, offered the possibility of forming bouquets and garlands not only 'in endless variation from the most beautiful flowers', but also to give particular significance according to the grouping of blooms of specific symbolism, precisely described in a list accompanying this amusing entertainment.

Another interesting work, this one in an Italian vein, is *L'Antotrofia ossia la coltivazione de' fiori* (no. 98), written and illustrated by Antonio Piccioli, a botanist, gardener, and talented botanical illustrator. In *L'Antotrofia* Piccioli presents, month by month, the flowers that could be found blooming in the garden of the Museum of Physics and Natural History in Florence, of which he was curator. In addition to his text and plates the author included verses by contemporary poets praising the beauties of nature and the charm of the world of flowers.

To the same cultural context belongs a remarkable album of flower collages fashioned from coloured paper by a hitherto unknown artist, a genuine 'master of paper mosaicks', Mary Wise (no. 100). This work, which dates from the mid-nineteenth century, represents the continuation of a (once again entirely female) artistic tradition begun by the brilliant English craftswoman Mary Delany (1700–88), who moved in the same circles as Handel, Hogarth and Jonathan Swift.

Many artists also found inspiration in the theme of metamorphosis and of the hybrid being, half man and half plant, which made their first appearances in the myths of antiquity, in the stories of Narcissus, Smilax and Crocus, as well as that of the nymph Daphne, who was transformed into a laurel tree so as to escape the pursuing Apollo. Metamorphosis formed a recurrent motif in the decorative arts, both Western and Byzantine, in the medieval period. One need only recall the *droleries* inserted in the borders of many illuminated manuscripts or the 'green men', disquieting leafy figures with human faces carved on the capitals of many Romanesque and Gothic churches (see Baltrusaitis).

This theme captured the imagination of the quintessentially Romantic artist J. J. Grandville, who, like many others of his generation, rejected the materialism of modern, industrialized society and sought refuge in a dream world populated by plants and animals in human guise. His works are filled with figures conceived and drawn with captivating wit. It seems that the greatest cartoonist of the twentieth century, Walt Disney, was inspired by the plates that Grandville designed for *Scènes de la vie privée et publique des animaux* (1842–4). Another work illustrated by Grandville, *Les Fleurs animées* (no. 99), narrates the fable of a revolt of the denizens of Flora's kingdom, who decide to assume human form in order to spy on men and discover what they are saying about them. A series of stories (containing satirical allusions to politicians and other figures) are recounted by different flowers, all of whom come to the same discouraging conclusion that it would be far better for them to resume their original physiognomy and return to their own world. Grandville used this occasion to create many enchanting female figures in floral attire (the word 'flower' in French carries the feminine article), in another graceful homage to the fair sex.

We must not forget, however, that the Romantic era was charged with contradiction and with emotional extremes. Next to the celebration of the gentler sentiments and of nature as an idyllic garden, the Romantics indulged themselves in reflections on the vanity of worldly things and cultivated a vision of the world as a place of mystery and brooding tragedy. Thus, at the very beginning of the Romantic period, the wealthy and eccentric physician Robert Thornton produced *The Temple of Flora* (no. 93)—probably the most unconventional flower book ever published. Despite the avowed scientific intent of this work, the flowers in the twenty-six colour plates are presented in no particular order, and each stands out—gigantic, disquieting, almost unreal—against its own mysterious and incongruous background. Drawn and engraved by various artists, but all under the direct supervision of the author, they represent the fantastical fruit of Thornton's fevered imagination, as may be deduced from the introductory 'Explanation of the Picturesque Plates'. In it Thornton writes:

> In the *night-blowing* CEREUS you have the moon playing on the dimpled water, and the turret-clock points XII, the hour at night when this flower is in its full expanse. . . . In the *Canada* LILY there is expressed the shade it delights in, with a sky whose clouds yet contain snow within their bosom. . . . In the DODECATHEON or *American* COWSLIP, a sea view is given, and a vessel bearing a flag of that country. . . . In the *maggot-bearing* STAPELIA you will find represented a green African snake, and a blow-fly in the act of depositing her eggs in the flower, with the maggots produced from this cause. The clouds are disturbed, and every thing looks wild and sombre about the *dragon* ARUM, a plant equally poisonous as foetid.

Every plate in this extraordinary work breathes drama and emotion, with lowering skies, stormy seas or restless dawns forming the backdrop to flowering plants, all of which, whether they are native or exotic species, looking anomalous and bizarre. Even the familiar white lily is depicted, sinuous and fleshy, against a dark background that breaks only in the right-hand corner to offer the glimpse of an anachronistic classical temple, here strange and somehow menacing.

Thornton's magnificent work inspired other elephant books on the subject of flowers, such as *The Beauties of Flora* (1820) by the floriculturist Samuel Curtis; illustrated with plates after Clara Maria Pope, a copy of this work is also to be found at Oak Spring.

93. Robert John Thornton (?1768–1837)

New Illustration of the sexual system of Carolus von Linnaeus: comprehending an elucidation of the several parts of the fructification; a prize dissertation on the sexes of plants; a full explanation of the classes, and orders, of the sexual system; and the Temple of Flora, or Garden of Nature, being picturesque, botanical, coloured plates, of select plants, illustrative of the same with descriptions. | '...SHALL BRITONS, IN THE FIELD | UNCONQUER'D STILL, THE BETTER LAUREL LOSE?– | IN FINER ARTS AND PUBLIC WORKS SHALL THEY TO GALLIA YIELD?... Thomson | By Robert John Thornton, M.D. Member of Trinity College, Cambridge; ...One of the Council of the London Medical Society; ... Honorary Member of the Medical and Phisical Societies of Guy's Hospital, of Bartolomew's Hospital, and of Lyceum Medicum Londinense; ... Member of Several Learned Academies and Societies Abroad; ... Lecturer on Medical Botany to the United Hospitals of Guy and St. Thomas; ... Late Physician to the St. Mary-Le-Bonne General Dispensary; ... Author of the Philosophy of Botany; ... The Philosophy of Medicine; ... and the Philosophy of Politics, &c. London: Printed, for the Publisher, by T. Bensley, Bolt Court, Fleet Street, [1799–] 1807.

60 × 48 cm. 3 parts bound in 1 volume. Copies, in terms of the number and colouring of the plates, vary considerably; no two appear to be alike.

BINDING: Contemporary panelled calf, gilt and blind-stamped fillets and rolls. All edges gilt. On the spine in gilt 'Thornton's New Illustration of the Sexual System and Temple of Flora'.

PLATES: Nine calligraphic engraved half-titles; 12 engraved tables, dedication and content leaves; 7 engraved portraits including two portraits in colour of Linnaeus (one showing him in Lapland dress); 3 allegorical title-pages, and 26 plates of plants engraved using mixed intaglio processes (aquatint, mezzotint, stipple and line engraving), printed in colour and finished by hand. Most of the plates are of the first state.

REFERENCES: Dunthorne 301; Grigson and Buchanan; Nissen 1955; Stafleu and Cowan (2nd edn), VI.14.283; King; Blunt and Stearn, pp. 236–43.

ROBERT THORNTON'S *New Illustration of the Sexual System of Linnaeus* is one of the most celebrated flower books ever published, and constitutes a fascinating legacy of the English Romantic period. Its publishing history was long and tortuous, however, and its author, a wealthy and eccentric scientist, dedicated a great part of his life and most of his personal fortune to its realization.

Destined for the Church, Robert Thornton instead set his heart on the study of medicine. Two developments in this period had a particular influence on his scientific formation—the publication of Linnaeus's plant classification system and, coinciding with this, the arrival in Europe of thousands of exotic new species from distant lands, all of them needing to be studied and correctly classified. The art of botanical illustration formed an important adjunct to these studies, and publishers specializing in the engraving of such illustrations emerged. In this area the British printing houses had reigned supreme until the middle of the eighteenth century, when, according to Thornton, they ceded their position to the detested French. The Revolutionary War with France had broken out in 1793 and the nationalistic sentiments that spurred Thornton to embark on his grandiose project are clearly reflected in the following lines from *Liberty* (1735–6) by the poet James Thomson that appear (slightly adjusted) on the title-page of *New Illustration*: 'Shall Britons, in the field unconquer'd still, the better laurel lose? In finer Arts and Public works shall they to Gallia yield?'

Falling heir to a considerable fortune, Thornton spent several years travelling in Europe, and then decided to dedicate his energy and material resources to the creation of a work that would, he vowed, become a genuine landmark in the history of science and an imperishable monument to the glory of Britain. After publishing a *Philosophy of Medicine*, in 1797 he began this ambitious new project. Conceived on a grand scale, *New Illustration* was to consist of three parts: a discussion of the sexual reproductive cycle of plants; an explanation of Linnaeus's plant classification system lavishly illustrated with botanical drawings and the portraits of famous botanists; and 'The Temple of Flora' comprising no less than seventy large plates of exotic plant species.

Publishing *New Illustration* was an affair plagued with difficulties. The plates were produced using a variety of processes that required the participation of a large number of artists. Deadlines were rarely kept, and many of the plates eventually went through more than one printing. As a consequence, no two extant copies are exactly the same. The work was advertised in 1797 and issued in parts, beginning in 1799 and continuing until 1807. At this point Thornton was forced to abandon his project, the single-minded pursuit of which had brought him to the brink of financial ruin. Only twenty-six of the originally projected seventy plates were actually completed.

The first two sections of *New Illustration* consist of a lofty rhetorical text of little scientific value, liberally sprinkled with elevating citations from various more or less celebrated poets, and illustrated with the portraits of eminent personages. Thornton took as his model for the text a poem by Erasmus Darwin (grandfather of Charles Darwin) entitled *The Botanic Garden* (1789–91), which happened to be one of the most popular works of the moment. Erasmus Darwin's portrait, together with one of Queen Charlotte—the consort of George III, to whom the work was dedicated—and one of the author himself, were the first to be commissioned for *New Illustration*. Thornton's portrait was undertaken by the renowned Francesco Bartolozzi, one of the greatest practitioners of the art of stipple engraving, after a painting by John Russell. Thornton appears in an oval framed by a garland of exotic flowers, with a hive of industrious bees at its apex. The author's hand rests on a volume of Linnaeus's revolutionary *Genera plantarum*, published in 1754. Two other engraved portraits of Thornton are to be found in this work, which gives one a fair idea of the author's sense of his own importance.

While the first two parts of *New Illustration* are of little interest save perhaps as a cultural document, the last section of Thornton's work represents a unique accomplishment. Obsessed by his personal vision of the perfect union between science and the fine arts, Thornton spared no effort or expense in the creation of *The Temple of Flora*, and the result was a series of plates of truly extraordinary beauty couched in the Romantic style (Grigson and Buchanan, p. 3).

Thornton engaged a host of artists and engravers to work on *The Temple of Flora*, but managed by vigilant supervision to maintain a remarkable homogeneity of style throughout. Four separate techniques were employed in the production of the plates—aquatint, mezzotint, stipple engraving and stippling with line engraving. Among the artists involved in the project were the painters Philip

Reinagle, who executed most of the preparatory drawings under Thornton's attentive eye, Abraham Pether and Andrew Henderson, and the engravers Richard Earlom, James Caldwall and Thomas Burke. Only the plate of the rose was drawn by Thornton himself, to be subsequently engraved by Earlom.

The title-pages to the three sections are imbued with the classicizing symbolism that was so popular during the eighteenth century. They depict *Flora Dispensing her Favours on the Earth*, engraved by Thomas Woolnoth from a painting by Richard Cosway; *Asculapius, Flora, Ceres and Cupid Honouring the Bust of Linnaeus* from a work painted by Reinagle and John Opie and engraved by Caldwall; and *Cupid Inspiring Plants with Love*, engraved by Burke from a painting by Reinagle. This third work, dominated by Cupid clutching his bow and arrow, clearly alludes to the 'loves' of the plants and thus to the sexual nature of their reproductive cycle, the principal theme of *New Illustration.*

As originally planned, *The Temple of Flora* was to consist of seventy plates of exotic plants arranged in sequence according to the classification system of Linnaeus. Each species was to appear against a landscape purporting to depict its native environment. Thornton reserved for himself the privilege of writing the text to the plates, a rhetorical exercise closing each time with some edifying lines of verse. Work began on 1 May 1798; the first flowers to be depicted were the 'American aloe' (*Agave americana*), engraved by Thomas Medland after a painting by Reinagle, and some 'Tulips' painted by Reinagle, engraved by Earlom, and accompanied by a short dissertation by Thornton on the interesting phenomenon of tulipomania (see no. 72). In May 1799 Reinagle completed a painting of the 'Superb lily' (*Lilium superbum*), a plant native to the eastern region of North America that was brought to Europe in 1738. From this painting, now in the Fitzwilliam Museum, Cambridge (Blunt and Stearn, pl. 85), two quite similar mezzotint engravings were prepared, one by Earlom and one by Ward.

In the following year one of the most celebrated plates in *The Temple of Flora* appeared—'The Night Blowing Cereus' (*Selenicereus grandiflorus*), called by Thornton 'Cactus grandiflorus'. This nocturnal flower from the Caribbean had already been depicted by Georg Ehret for two engravings dating from 1752 (Calmann, pls. 73, 74). In Thornton's plate the extravagant flower surrounded by its double corolla of blade-like sepals nearly fills the page, while in the background there appears an incongruous nocturnal landscape with a full moon and a ruined tower, a romantic motif that served to accentuate the exotic nature of the plant. The flower was very realistically painted by Reinagle, while the background was the work of Abraham Pether. So impressed were his contemporaries by this mysterious landscape that the artist came to be known as 'Moonlight Pether'.

In 1801 the illustration of the 'American cowslip' appeared. This plant was first described by the botanist Mark Catesby, who called it 'Meadia'; later Linnaeus gave it the name '*Dodecatheon meadia*'. Henderson succeeded in painting a quite striking picture of this modest plant, which dominates the foreground of the engraving by Warner with its crown of delicate, arrow-like flowers.

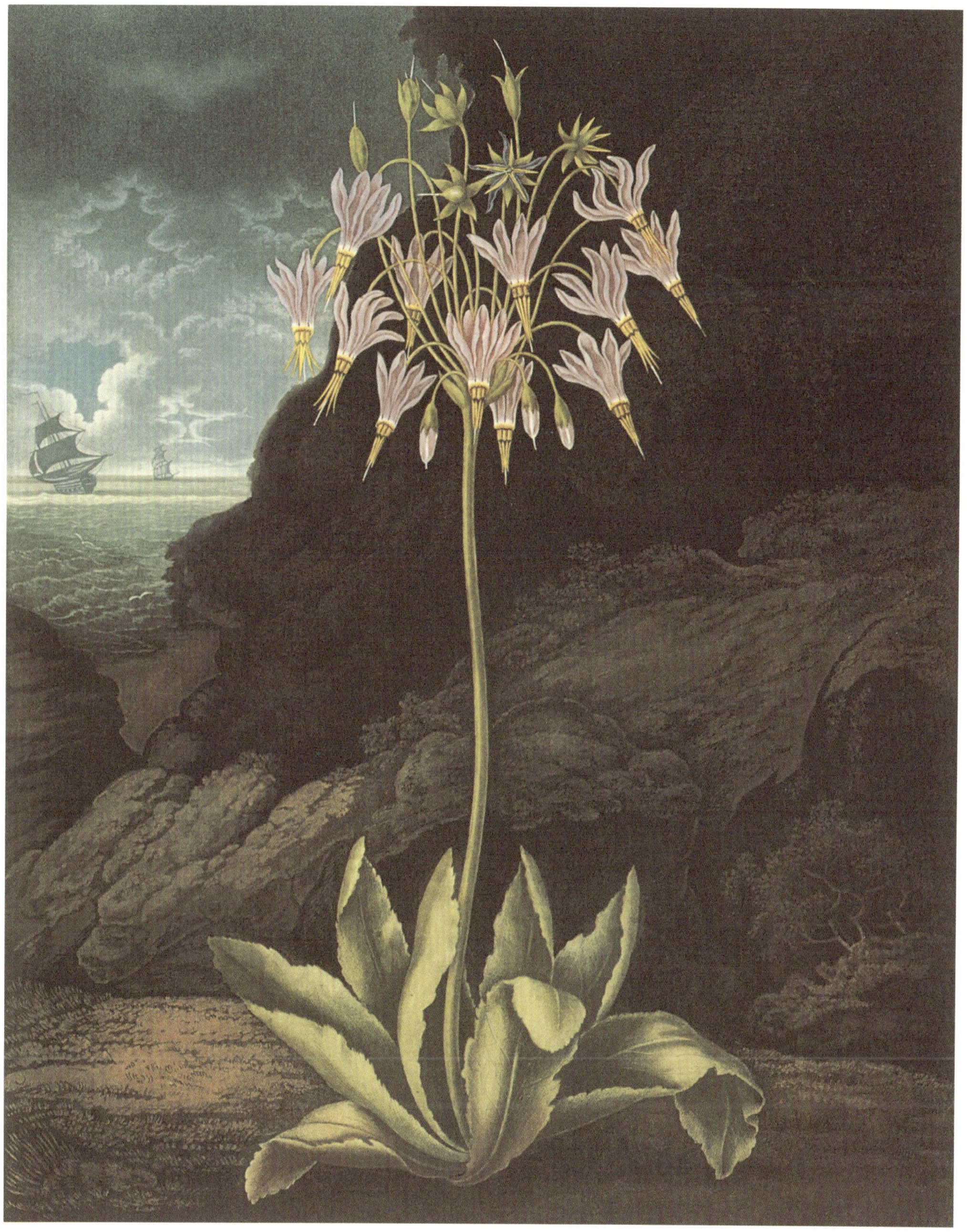

ROBERT JOHN THORNTON, *New Illustration of the Sexual System of Carolus von Linnaeus*. Shooting-star or 'American cowslip' (*Dodecatheon meadia*)

The rocky cliffs in the background part to reveal the glimpse of a stormy sea with two ships silhouetted against a horizon filled with billowing clouds. The description of the flower closes with the following dramatic, if not entirely apposite, lines by Erasmus Darwin: 'Alike to all, she bows with wanton air, / Rolls her dark eye, and waves her golden hair'.

All the plates in *The Temple of Flora* are impregnated with this disquieting atmosphere, so highly appealing to the Romantic temperament; even the shy auricula and the cheerful carnation are set against scenes of unrelieved gloom. The eerie mood and the almost frightening giganticism of the plants as they are portrayed in this work are particularly evident in the 'Dragon Arum' (*Dracunculus vulgaris*), described by Thornton as an 'extremely foetid poisonous plant'; its inflorescence emerges spear-like out of the violet depths of the spathes, to stab a sky filled with menacing clouds.

For several years Thornton continued his work on *The Temple of Flora*, hoping to emulate the resounding success of John Boydell's series of engravings illustrating the works of Shakespeare. To stimulate interest in his endeavour, in 1803 he organized an exhibition of the twenty engravings thus far completed, together with their original paintings (today dispersed) and twenty portraits, in a show in Bond Street entitled the 'Linneian Gallery'. Public response to this initiative was disappointingly lukewarm, however. At the same time, the enormous expense of his enterprise had considerably undermined the personal fortune of the author. In a desperate attempt to check the looming spectre of bankruptcy, Thornton launched a 'Botanical Lottery' for the 'promotion and encouragement of The Fine Arts, and Science' in Britain. The Grand Prize being offered was the entire 'Linneian Gallery' collection from the Bond Street exhibit—'paintings of the choicest Flowers, Allegorical Subjects, and Heads of Botanists, executed by the most eminent Painters', as the notice states, together with one of each of the lesser prizes (all copies of works by Thornton), the announced value of this prize being £5,080. This effort failed, as did the many other moneymaking schemes thought up by Thornton, such as that of publishing a smaller version of *The Temple of Flora*, and even the possibility of a collaboration with William Blake.

When Thornton died he left his family in poverty. Nevertheless, as Ronald King affirms: 'from the ashes of Thornton's supposed failure has risen the phoenix of success' (p. 43). In his fanatical pursuit of a highly personal vision of artistic perfection, Thornton succeeded in creating a genuine masterpiece, a series of plates that have fascinated generations of flower lovers and continue to cast their spell today. Certainly his work, in which word and image vie with one another to amaze and impress the reader, is out of step with contemporary aesthetic sensibilities. Nevertheless, *The Temple of Flora* represents a unique accomplishment in the history of botanical illustration.

94. 'Le Jardinier fleuriste' [1819]

Le Jardinier Fleuriste Dédié aux Dames Par un Amateur. [*vignette of a gardener with a spade standing between two flowering plants*]. à Paris. Chez Mareilly, Rue St Jacques, No 21.

12.5 × 8.7 cm. *i–vi 1* 2–199 *200 1–16* p., 12 leaves of plates.

BINDING: White leather binding with gilt floriated borders, housed in a matching box.

PLATES: 12 unnumbered plates and a colour-printed stipple vignette. Guard tissues.

PROVENANCE: On the front pastedown is a bookseller's label: 'Chez Magnier—Papetier—Rue de Richelieu, No 28.

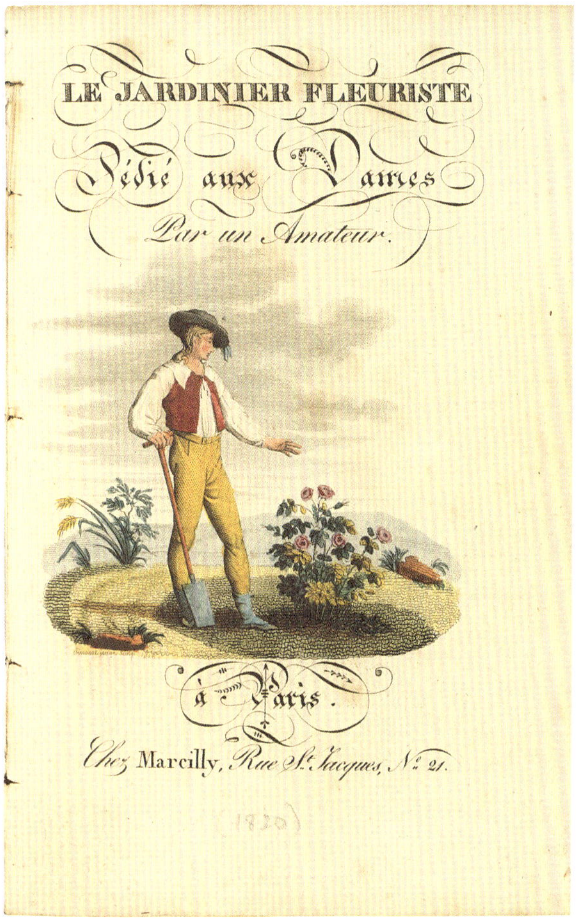

Le Jardinier Fleuriste. Title-page

THIS SMALL BOOK, bound in white leather with gilt decoration and illustrated with delicately tinted plates, is a charming example of the sentimental gift-book (see no. 79). The author, a flower lover who has temporarily assumed the guise of a *jardinier fleuriste*, gracefully addresses his lady readers with a short homage in verse printed on the second page. In it he declares that while he could only coax the flowers in his garden to grow at the expense of much time and labour, it merely sufficed for his charming readers to venture outdoors, and where they chose to tread, flowers would spring up immediately at their feet.

Alternating passages of prose with short poems, the author discusses the origins of the garden (going back to the time of Homer), and explains how to choose a suitable site for one's own garden. He then presents the history and morphology of various flowers and provides a 'calendar of Flora', listing the flowers that could be found blooming at each month of the year, beginning with the narcissus in January and concluding with heather in December.

Le Jardinier Fleuriste closes with verses extolling the beauty of flowers and the pleasures of botany: 'Parmi l'ombre des bois, sur des tapis de mousse, / C'est ainsi que ma vie échappe, obscure et douce, / Et que s'oublie, au sein de ces amusemens, / Les désirs inquiets, l'amour et ses tourmens.' These verses are signed by 'Bérenger (de Lyon)'—i.e., Laurent-Pierre Bérenger (1749–1822)—who may perhaps be the author of this work.

A calendar for the year 1820 is included, which tells us when this volume must have been published (late 1819). The text is accompanied by pleasing stipple engraved and hand-coloured plates of bouquets, made up of two or three flowers each.

95. Henry Phillips (1775–1838)

Floral Emblems. By Henry Phillips, F.I. & F.H.S. Author of Pomarium Britannicum, Etc. Etc. London: Printed for Saunders And Otley, British And Foreign Public Library, Conduit Street, Hanover Square. 1825.

22 × 14 cm. *i–iii* vi–xvi *xvii–xviii* *1* 2–352 p., 19 leaves of plates.

BINDING: Dark-blue calf binding, gilt spine with red leather label: 'Phillips's Floral Emblems'.

PLATES: 18 unsigned plates, added illustrated title-page, and head-piece, all lithographed and hand-coloured.

PROVENANCE: Armorial bookplate; ex-libris Paul Mellon.

REFERENCES: Pritzel 7129; Plesch 361.

HENRY PHILLIPS was a banker originally from Worthing who passed his life in London and Brighton. He was also a famous amateur botanist and horticulturist whose vast expertise led to his election as one of the first members of the Royal Horticultural Society, founded in London in 1804. He wrote a large number of works on his favourite subject, including *Pomarium Britannicum* (1820), *History of Cultivated Vegetables* (1822), *Sylva florifera* (see *An Oak Spring Sylva*, no. 9), *Flora historica* (1824) and, in 1825, *Floral Emblems*.

In contrast to the strictly scientific nature of Phillips's other works, *Floral Emblems* was intended as a 'divertissement' by the author, who delved into the past in order to rediscover the symbolic attributes associated with various flowers. In this work he describes a series of floral emblems and

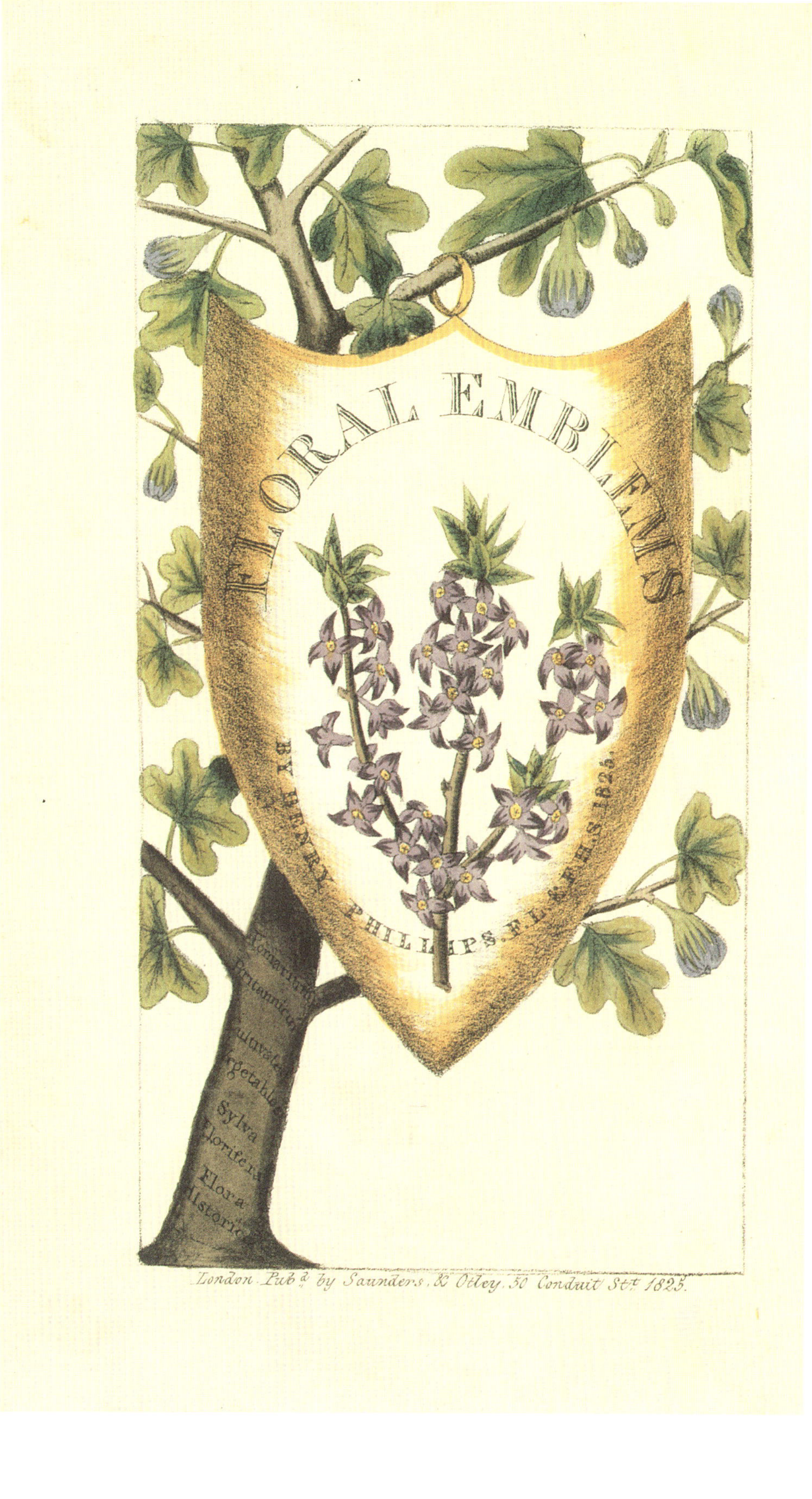

HENRY PHILLIPS,
Floral Emblems.
Frontispiece

their significance, presenting them in alphabetical order under such headings as 'Early Youth' and 'Injustice'. Some of the emblems are illustrated with hand-coloured, lithographed plates.

Classical mythology and biblical texts offered Phillips a rich source of botanical symbols and images. In classical times the flower was, of course, considered a symbol of feminine beauty, while the early Christians adopted it as a symbol of the life of the spirit. In the sixteenth and seventeenth centuries botanical images frequently appeared as emblems and devices, since convention forbade the use of the human figure. A motto or set of verses often appeared beneath the emblem, explaining its significance. In 1590 the German naturalist Joachim Camerarius the younger published in Nuremberg *Symbolorum et emblematum ex re herbaria desumtorum centuria una* (a copy is in the collection at Oak Spring), which describes a large number of emblems containing botanical elements. At the same time, many of the intellectual academies that were being founded in Europe, such as the Accademia della Crusca in Florence (established in 1594) and the German Fruchtbringende Gesellschaft (established in 1617), adopted flowers and other botanical images in their devices.

With the widespread popularity of flowers at all levels of society in the Romantic period, interest in the symbolism of flowers revived, and indeed a whole new set of associations—with the private world of the sentiments and most particularly with the language of lovers—was created around them. An immensely popular minor genre sprang up in which series of poems or vignettes on the theme of flowers were united with charming illustrations to form gift books suitable for presentation to young ladies and lovers.

Like all vogues, the passion for flowers led to an interest in every subject that could be linked with it, however tenuously. Thus authors such as Phillips were inspired to produce works that were intended for a much wider audience than their more serious efforts. The frontispiece to *Floral Emblems* contains an indirect reference to the popular nature of the work: it depicts a large shield hanging from a fig-tree on whose trunk the titles of Phillips's scholarly books have been carved, while the shield itself bears a flowering branch of *Daphne mezereum* (Mezereon), whose symbolic meaning is 'the desire to please'.

'The language of flowers', Phillips explains in his Introduction, 'is said to be as old as the world, and the antiquity of floral emblems as great as that of love itself, and by whom it is supposed to have been invented, since it is a kind of parable which speaks to the eye, and through that medium is transmitted to the heart'. The author also provides examples of 'numerical emblems', such as branches whose leaves represent the days of the week, or garlands each of whose flowers symbolizes a different month of the year. A wide range of sentiments, spiritual states and human attitudes are associated with specific plants or flowers, the author explaining the symbolism of each in a brief text accompanied by lines of verse from classical or contemporary poets. The daisy (*Bellis perennis*) is a symbol of innocence, for what small child playing in a meadow has not gathered these flowers to form his first daisy chain? The plum-tree (*Prunus domestica*) represents independence because it will flourish even without the care of a gardener, while the Spanish jasmine (*Jasminum grandiflorum*) is associated with sensuality because 'its highly odorous flower represents an immoderate desire for sexual gratification'.

There are whimsical lithographed plates illustrating some of the emblems, with short explanatory legends provided beneath. The illustration for 'Innocence', for example, is in the form of a cup with a handle made up of daisies and containing some sprigs of sweet pea; according to the caption, there are 'Delicate & lasting Pleasures arising from the Cup of Innocence'.

96. JOHN CONSTABLE (1776–1837)

Study of three 'parrot tulips'.

21.2 × 22 cm. Oil on board backed with panel. Inscribed on the back: 'Emma'.

PROVENANCE: Said to have been given by the artist to Emma Fisher; J. Rickett; Sabin Galleries 1961; Paul Mellon Collection.

REFERENCES: *Painting in England*, no. 98; Hoozee, p. 199; Reynolds, p. 243, no. 32.50.

THIS PAINTING of three tulips was sold together with an autograph copy of David Lucas's mezzotints after Constable, *English Landscape* (1830–32), which bears the inscription 'Miss Fisher—with Mr Constable's best wishes'. This Emma was probably the daughter of the Archdeacon John Fisher—a close friend of the artist—who died in August 1832, although another, less well supported, hypothesis is that the Emma meant here was the daughter of a certain Thomas Churchyard (Reynolds, p. 243).

Hoozee has dated the picture to *c.* 1814, the year in which Constable completed a painting of a vase of flowers today in the Victoria and Albert Museum in London (Mitchell, 1973, pl. 113). Stylistically, however, the *Study of Three Parrot Tulips* is much closer to another painting in the Paul Mellon Collection, *Study of Poppies*, which bears an inscription that has been deciphered as 'John Constable 24 July 1832'. The tulip study was most probably painted at the same time, that is, late in the artist's life. Another study of some poppy flowers executed in oils on paper, also in the Victoria and Albert Museum, has been dated by experts to around the same period (Walker, no. 46).

In keeping with the great Dutch tradition of the seventeenth century, Constable found constant inspiration in the study of nature, which can be traced not only in his remarkable landscapes but also in various unfinished paintings and sketches of flowers, which are imbued with the same immediacy and light-filled shimmer that mark his larger works. It is also interesting to observe that Constable was influenced by Henry Phillips (see no. 95), the prolific author of botanical works with whom he became acquainted around 1820, to introduce wild flowers into the foreground of his landscapes, a development that can be noted in the celebrated oil painting *The Cornfield* of 1826 in the National Gallery, London (Mitchell 1973, p. 91).

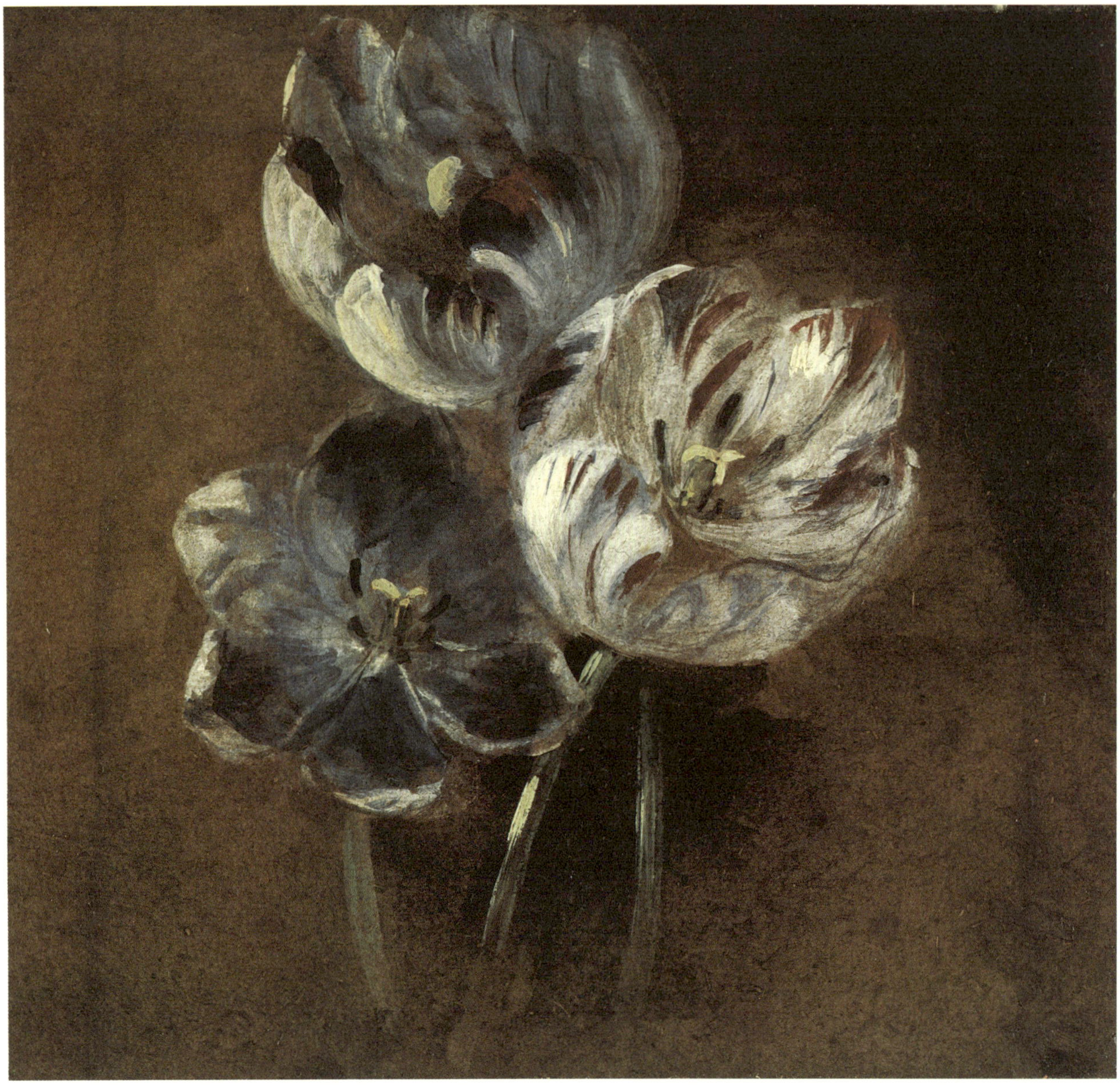

JOHN CONSTABLE, Three Parrot Tulips. Collection of Mr and Mrs Paul Mellon

The oil painting in the Paul Mellon Collection at Oak Spring depicts three variegated tulips whose corollas seem to be opening in succession before one's very eyes. The highly decorative style of the period has been replaced here by a more intuitive approach admirably suited to the naturalistic theme of the painting. The clotted brushstrokes of brilliant colour that Constable used to define the petals and stamens create an almost palpable feeling of volume, while the flowers themselves seem to be bathed in a light made up of a thousand reflections. His innovative technique provided an invaluable lesson for the French Impressionist painters who were soon to follow.

97. CHARLOTTE DE LATOUR (*fl. first half of the 19th century*)

Le Langage des Fleurs, par Mme. Charlotte De La Tour. Quatrième Édition. Paris, Audot, Libraire-Éditeur. Rue du Paon, 8, École de Médecine. 1833.

14.5 × 9.5 cm. *i–v* vi–xvi *1* 2–352 *353–354* *1* 2–18 p., 15 leaves of plates.

BINDING: Cloth binding with gold metallic threads. Original printed paper wrappers bound-in.

PLATES: 15 plates printed in colours with some colour highlights added by hand; engraved by Victor after drawings by Pancrace Bessa.

REFERENCE: Dunthorne 34.

THE SUBJECT of this charming gift-book is the Language of Flowers, an extremely popular conceit in the Romantic period. Books on the language of flowers first appeared in France and were soon being translated or copied all over Europe. Printing-houses already accustomed to publishing flower books in elegant editions, and artists specializing in the painting of flowers and bouquets, were quite prepared to turn their attention to the production of these lighter works, whose scope was decorative rather than scientific.

It is not surprising to find that a woman is the author of this delightful work, which in 1833 was already in its fourth printing. Charlotte de Latour was the pseudonym of Louise Cortambert, the wife of François Eugène (1805–81), a celebrated geographer and the librarian of the map department at the Bibliothèque Nationale in Paris.

In her introduction Latour addresses herself to those young ladies untouched by 'les folles joies du monde' who devoted themselves instead to the less worldly pastime of 'l'étude des plantes'. She points out to these students of nature that their gardens offered an inexhaustible font of pleasure and instruction, whether they wished to fabricate 'liqueurs parfumés' or 'conserves binfaisantes' from the essence of flowers, or were seeking instead to capture on canvas—in the tradition of the van Spaendonck brothers (see no. 92)—'les nuances trop fugitives de la plus belle des fleurs'.

The author goes on to explain the proper use of the language of flowers: in order to embellish a love both pure and chaste, 'on à imaginé le langage ingénieux des fleurs' that more eloquently than the written word 'se prête à toutes les illusions d'un coeur tendre et d'une imagination vive et brillante'. In fact, the language of flowers was governed by a set of rules as strict as that of any grammar: every colour had its significance, as did every means of presentation. When facing outward, a flower bore a certain meaning; if facing inward the opposite meaning was conveyed. A rosebud with leaves and thorns signified 'I am afraid, although I dare hope', but if the bloom was turned inwards it bore the discouraging message 'There is no need to fear, nor hope.' The pronoun 'I' was expressed by placing the appropriate flower on the right, the pronoun 'you' when the flower was placed on the left.

Charlotte de Latour then presents an entire series of flowers, arranged by season, discussing the 'meaning' of each in a short, informal text that combines historical notes with legend and literary

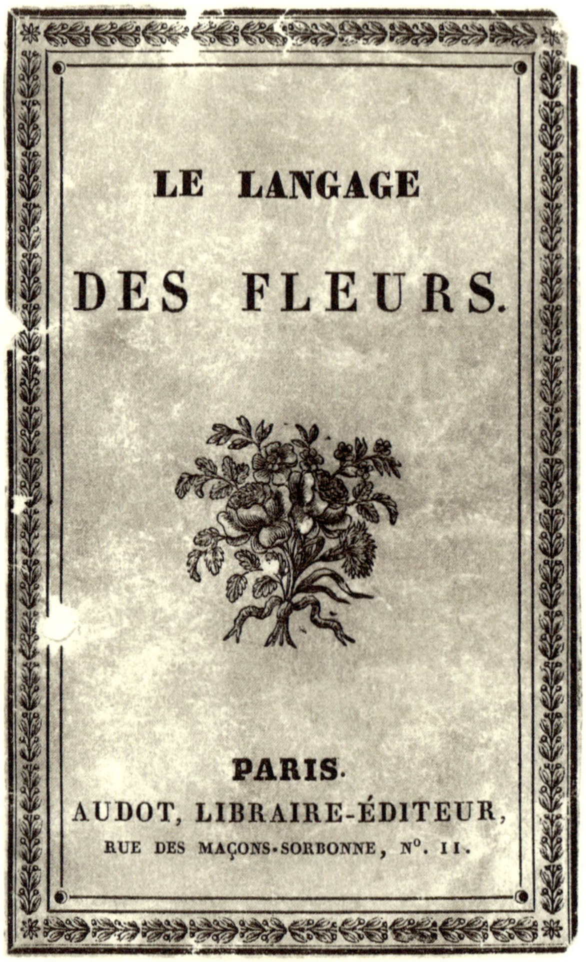

CHARLOTTE DE LATOUR, *Le Langage des fleurs.* Title-page

citations. The lilac, for example, represents the first stirrings of love so irresistibly associated with the return of spring, while grain symbolized wealth, and the late autumn amaranth immortality, because in ancient times wreaths of this flower were used to decorate the statues of the gods. The snowdrop instead expressed 'consolation' because as it miraculously pushed its way through the snow it seemed to be saying 'I am here to console you for the harshness of the lingering winter'.

The illustrations to *Le Langage des fleurs* constitute one of the chief sources of its charm. The original drawings were prepared by a floral artist of considerable renown, Pancrace Bessa (see no. 79); they were then engraved in colour by Victor, who had a workshop in Paris where he also produced lithographs. The book's frontispiece is decorated with a pretty bouquet composed of a rose (the symbol of beauty), a spray of ivy (constancy in friendship), and myrtle (love), while the title-page bears a vignette depicting a young girl being crowned with a floral wreath by a female figure. The motto underneath reads 'Te voila préservée de touts malheurs'. *Le Langage des fleurs* could be purchased in Paris from the bookshop and printing-house of Audot in rue du Paon at prices ranging from six-and-a-half francs for the most simple bound edition to ten francs for the edition bound in Russian leather.

98. Antonio Piccioli (1794–1848)

L'Antotrofia Ossia La Coltivazione De'Fiori Di Antonio Piccioli Giardiniere Botanico Nell'I. E R. M. Di Storia Naturale Di Firenze, Socio Corrispondente Dell'I. E R. Accademia Economico-Agraria dei Georgofili, Membro Della Societa' Di Orticoltura E Della Societa' Medico-Botanica Di Londra, E Membro Onorario Della Societa' Reale Di Orticoltura Di Berlino. Firenza, Per V. Batelli E Figli; M.DCCC.XXXIV.

22.5 × 13.5 cm. 1–798 *1–2* p. Two parts in one volume.

BINDING: Cathedral binding in half marbled boards and calf spine. Gilt lettering: 'Piccioli L'Antotrofia'.

PLATES: 72 plates printed in colour with some colour highlights added by hand; designed by Antonio Piccioli and engraved by Antonio Corsi.

PROVENANCE: Ex-libris Arpad Plesch (1890–1974).

REFERENCES: Sitwell and Blunt, p. 70; Plesch 361; Tongiorgi Tomasi and Tosi, p. 107; Pozzi, p. 345.

THE SON OF Giuseppe Piccioli, *giardiniere botanico* of the garden attached to the Museum of Physics and Natural History in Florence, Piccioli inherited his father's interest in the horticultural sciences, and eventually took over his position at the museum. In 1817 he presented to the prestigious Accademia dei Georgofili, founded in Florence in 1753, the first volume of what was destined to become a massive collection of watercolours of plants and flowers drawn from life, entitled *Plantarum pulcherrimarum horti botanici musei imperialis et regalis florentini icones* and dedicated to Ferdinand II, Grand-Duke of Tuscany. This work, which took Piccioli a lifetime to complete, is today conserved in the Biblioteca Nazionale Centrale, Florence.

Icones consists of 968 plates of flowering plants, divided into thirty-two volumes. Each of the plants, beautifully and accurately depicted, is accompanied by a brief description and notes regarding its cultivation. The fame that this immense undertaking procured Piccioli led to various marks of recognition: he became a corresponding member of the Accademia dei Georgofili, and was accepted by the Royal Horticultural Society, the Medico-Botanical Society of London and the Verein zur Förderung des Gartenbaus in den Königlichen preußischen Staaten.

In 1834 Piccioli published *L'Antotrofia* (derived from a Greek term referring to the 'cultivation of flowers'). This work in two volumes, each of which corresponds to a six-month period, describes the flowers, both native and exotic, to be found blooming in each month of the year in the garden that was overseen by him. There is a useful accompanying text; for example, a detailed treatise on grafting precedes the section for the month of March, while 'A new method for the propagation of plants by layerage' introduces the section on the flowers of April.

ANTONIO PICCIOLI,
L'Antotrofia. Snowdrop (*Galanthus nivalis*), plate 5

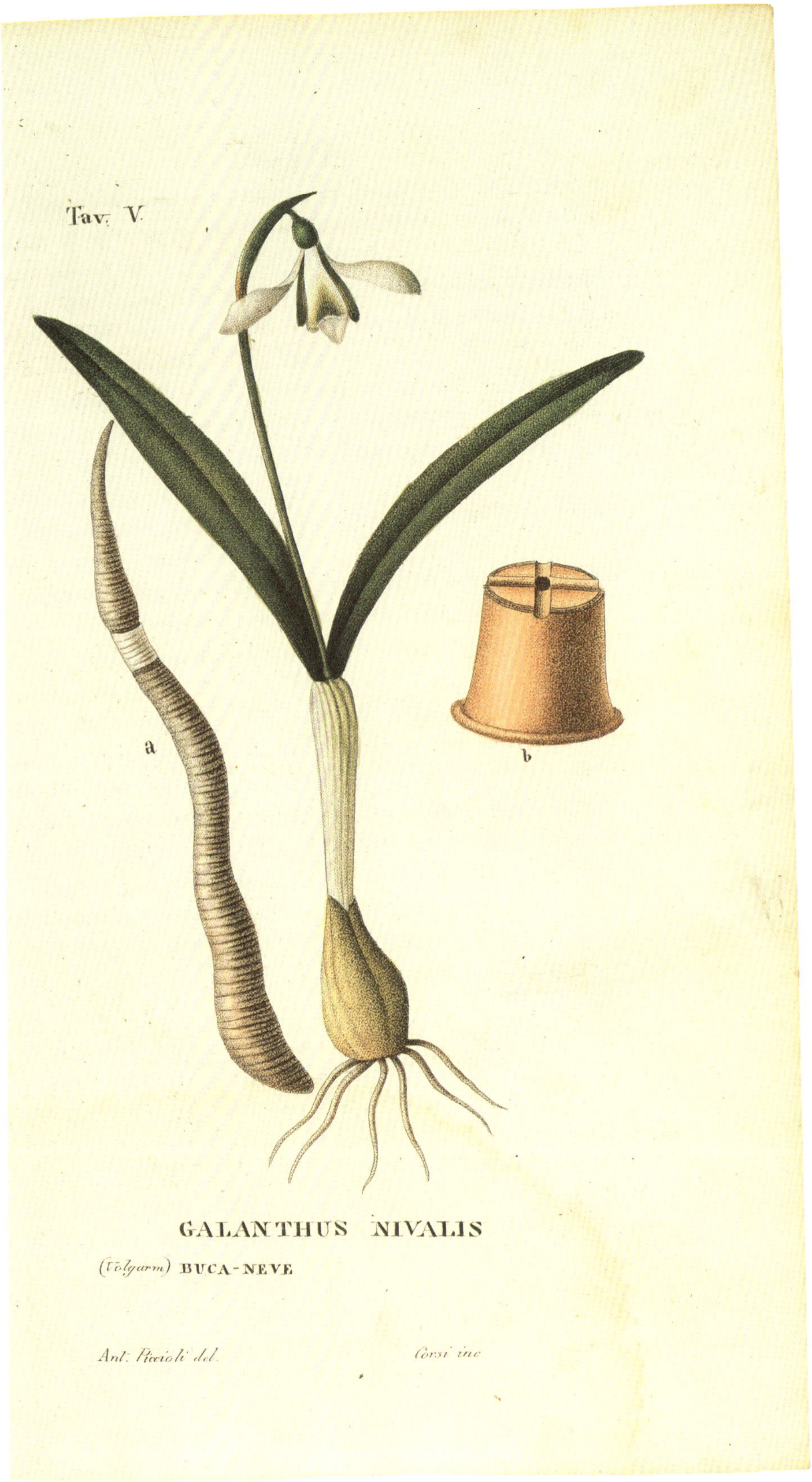

L'Antotrofia reveals Piccioli to be both a scientist and a man of letters. The work opens with an Italian translation of an idyll written by the Swiss poet Salomon Gessner (1730–88) and dedicated to the Greek myth of Lica, the first gardener. To his detailed descriptions of cultivation and propagation techniques, plant diseases and insects pests, Piccioli adds verses by classical and contemporary poets, ranging from Sappho and Metastasio to Lamartine and Gessner. Poems by various minor Italian authors, including Domenico Gazzadi of Sassuolo, are also included. Piccioli's *L'Antotrofia* provides an interesting example of the symbolism of flowers in an Italian vein. Next to each plate he presents verses in the form of an 'emblem' expressing the symbolic meaning of the flower depicted. For example, an image of the *Camellia japonica*, emblematic of 'Anxiety', appears next to a melancholy lyric by Sappho, while the *Anemone coronaria* is linked to a poem on the theme of 'Humility', and the *Lilium tigrinum* to one on 'Volubility'.

The charming plates, based on drawings by Piccioli, were colour-printed by the Florentine Antonio Corsi, one of the many artists who worked on the splendid *Pomona italiana* by Giorgio Gallesio, published in Pisa between 1817 and 1839. In the introduction to *L'Antotrofia* Piccioli expresses his hope that 'il dilettante di disegno, la ricamatrice, l'artista, il tappezziere' (the amateur draughtsman, the embroideress, the artist, the upholsterer) would find in these plates 'altrettanti modelli esatti e veridici di questa bella produzione della natura' (many exact and truthful models of these beautiful products of nature).

99. J. J. Grandville (1803–1847)

[*title-page of the first part*] Les Fleurs Animées par J. J. Grandville, Introductions Par Alph. Karr, Texte Par Taxile Delord. Premi re Partie. Paris Gabriel De Gonet, Editeur, Rue Des Beaux-Arts, 6. 1847. [*followed by a frontispiece wood-engraved half-title, showing a gracefully curved stem of flowers (honeysuckle, bignonaceae, and sage) which together form the words 'Les Fleurs', below which a damoiselle is poised in a dancer's attitude with a ribbon between her hands forming the word 'Animées'. Written on a leaf beneath her feet*] Dessins par J. I. Grandville Gravés sur acier par C. Geoffroi. Introductions par Alph. Karr. Texte par Taxile Delord. Botanique et Horticulture des Dames par le C.te Foelix. [*on a ribbon tied to the base of the stem*] G. de Gonet Edit. Rue des Beaux-Arts G. [*beneath the illustration*] Porret-Blanadet. [*beneath this*] Typ. Plon frères, à Paris.

[*title-page of the second part*] Les Fleurs animées par J. J. Grandville, Introductions Par Alph. Karr, Texte Par Taxile Delord. Deuxième Partie. Paris Gabriel De Gonet, Editeur, Rue Des Beaux-Arts, 6. 1847. [*followed by a frontispiece illustrated with a wood engraving of a branch divided into three stems which bear flowers and leaves forming the words 'Les Fleurs Animées'. Written on the four leaves at the base of the branch*] Seconde partie. G. De Gonet Editeur. [*below*] Quichon. Typ. Plon, à Paris. [*followed by*] Botanique des Dames, Introduction par Alph. Karr. [*and*] Horticulture des Dames, Introduction par Alph. Karr.

Both title-pages signed Cte. Foelix, the pseudonym of L. F. Raban (1795–1870).

25.5 × 18 cm. *1–5* 6–364 I–IV 1–62 I–IV 65–132 p.

BINDING: Red leather over boards. The spine is of red leather with gilt decoration and the words 'Grandville Les Fleurs Animées'.

PLATES: First part: Frontispiece of a wood engraving by A. D. Porret and Blandet, coloured by hand, and printed in Paris by Plons *frères*; 26 steel engravings by Charles Geoffroy from drawings by J. J. Grandville, coloured by hand, and depicting scenes with flowers in human (sometimes symbolic) form.

Second part: Frontispiece of a wood engraving by A. Quichon, hand coloured; 22 steel etchings by Charles Geoffroy from drawings by J. J. Grandville, coloured by hand, and depicting scenes with anthropomorphized flowers. Two steel engravings accompany the 'Botanique des Dames'.

REFERENCES: Wick; Renonciat, pp. 270–79; Pallottino; *Flowers into Art*, pp. 110–12.

PUBLISHED IN PARIS in eighty-three instalments between April 1846 and January 1847, *Les Fleurs animées* was the final work of J. J. Grandville, pseudonym of Jean Ignace Isidore Gérard. Born in Nancy, he received his training as an artist from his father, a painter of miniatures. Under his pseudonym he became extremely well known as a satirical cartoonist working for various periodicals, including *La Caricature, Le Charivari* and *La Silhouette*, and as an illustrator of lithographic albums and works of literature, among them the *Fables* of La Fontaine, *Les Voyages de Gulliver* by Swift and *Scènes de la vie privée et publique des animaux* by Hetzel.

In *Les Fleurs animées* Grandville abandoned his usual tone of caustic wit for a gentler, more poetic style. Aimed at an audience of female readers (the collection concludes with a 'Botanique' and a 'Horticulture des Dames'), his charming, yet subtly ironic 'femmes-fleurs' were an instant success, so much so that a second edition of the work went to press in 1847 immediately after the first, and many other editions followed, not only in France but in Germany, Belgium, Spain and the United States. For this work Grandville produced a series of fifty drawings in pen and ink, watercolour and bodycolour, depicting various flowers in human form (the greater part of these drawing are today conserved in the Bibliothèque Municipale, Nancy). Etched on steel by Charles Geoffroy (1819–82), they form an elegant visual accompaniment to a pleasing, if somewhat vacuous, text consisting of a series of moralizing tales by Taxile Delord (1815–77). The entire work was published in two volumes in Paris by the printer Gabriel de Gonet. The poetic grace of the illustrations was much appreciated by the author, Delord, who closed his tales with the epigram 'La plume c'est la bavarde du livre, le poète c'est le crayon' (p. 364).

Each of the fifty plates depicts a single flower (such as the 'Marguerite', 'Tulip', or 'Belle de Nuit') or more than one (such as the 'Bluet et le Coquelicot' and the 'Scabieuse et le Souci'), and was designed to accompany a short tale inspired by the nature of the flower, its name, or its historical or allegorical associations. In the plate illustrating the 'Scabieuse et le Souci', a 'mother scabiosa', holding by the hand two 'little girl scabiosae', seeks to comfort a melancholy maiden in the form of a calendula (or marigold), the popular French name for the calendula being *Souci* (worry). Only one flower appears in masculine guise, the *Soleil* or sunflower, while the *Giroflée* is accompanied by the only human figure in the entire series, an elderly, bespectacled herbalist who is savagely uprooting a terrified little gillyflower. This is one of the few illustrations in which the artist allowed his veiled sense of irony to verge on caricature.

In these drawings Grandville exhibits a profound knowledge not only of literature and the visual arts but of natural history as well. Despite its anthropomorphic aspect, the plant that inspired each

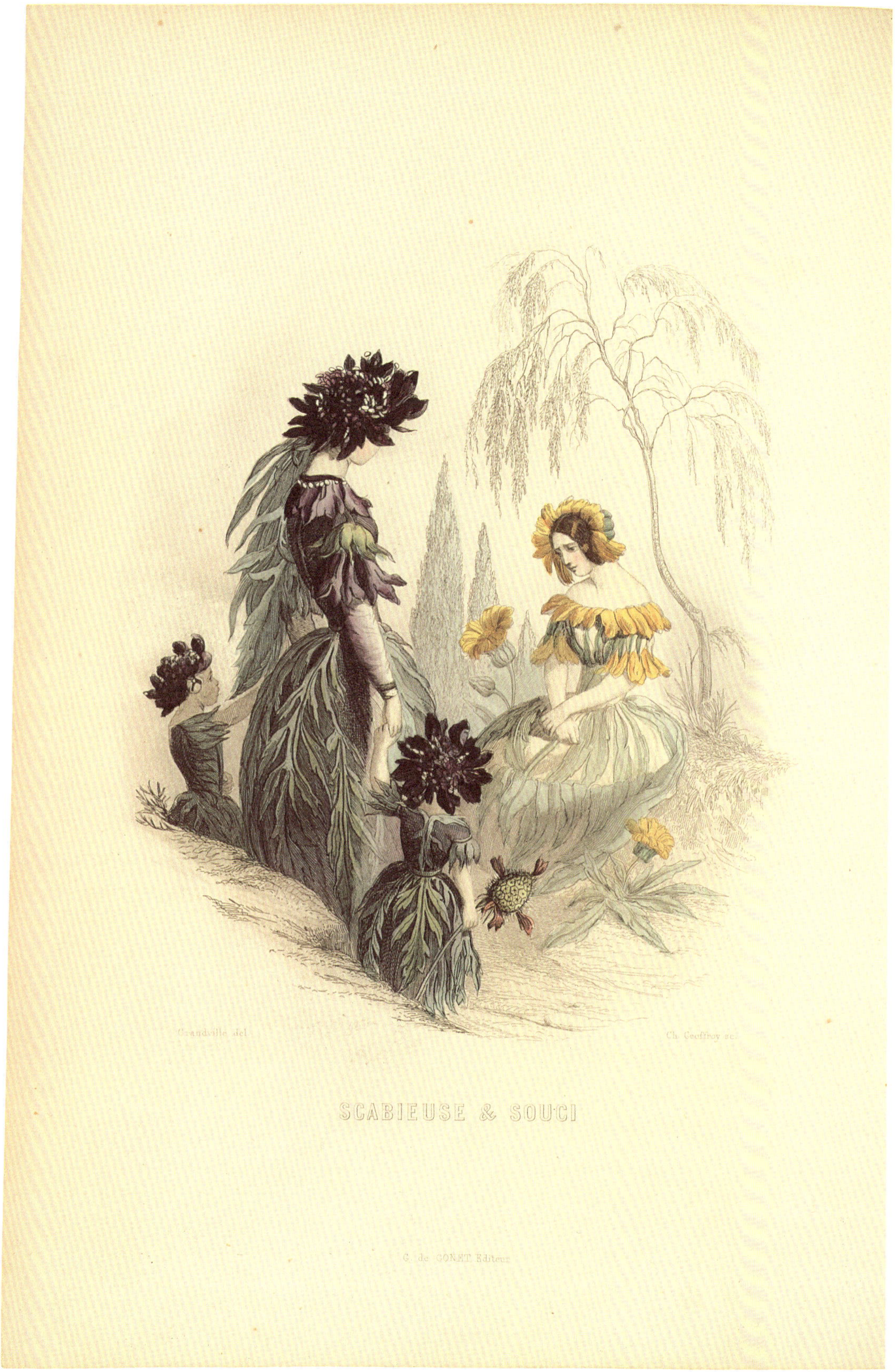

J. J. GRANDVILLE,
Les Fleurs animées.
'Scabieuse et Souci',
Part Two, between
pages 296 and
297

J. J. GRANDVILLE,
Les Fleurs animées.
Title-page to
Part Two

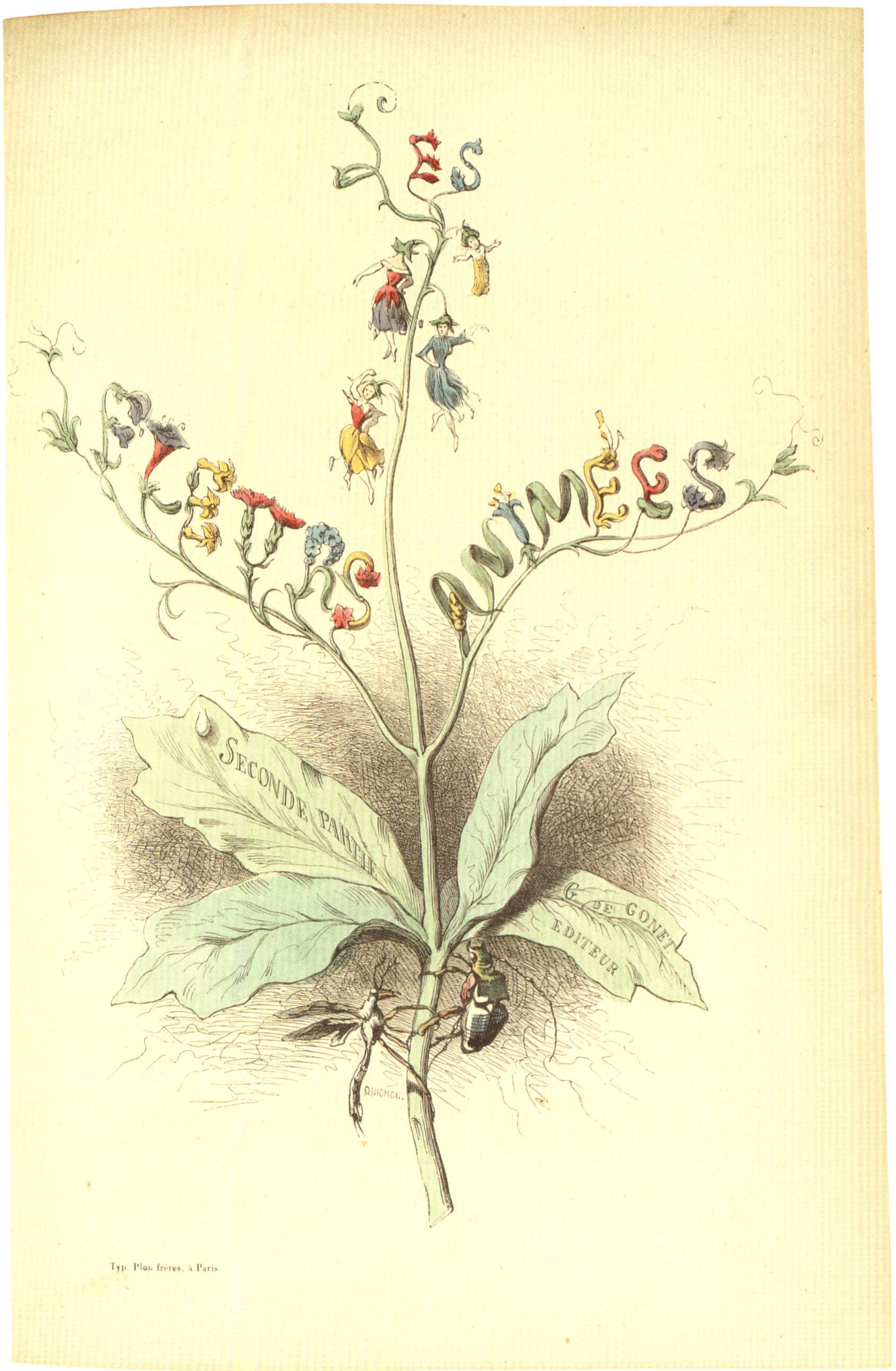

figure is in fact instantly recognizable. The artist draws the reader into a delightful game in which the fantastic hats, fluttering ribbons, silky skirts and delicate ornamentation that make up the corolla, petals, leaves and stems all provide distinct clues as to the species being portrayed. The landscape background to each illustration usually reflects the natural habitat of the plant, while any insect that may appear is always one closely associated with the species, such as the procession of chafers rendering homage to the regal *Rose*.

In addition to his work as a satirist, Grandville was in fact also a prolific botanical illustrator. Eighty illustrations of flowers drawn *d'après nature* are today in Nancy's Musée des Beaux-Arts. These watercolours, which date from 1846 and in which appear no less than forty of the species illustrated in *Les Fleurs animées* (*Grandville*, pp. 362–79), demonstrate Grandville's exceptional powers of observation, and confirm that wherever possible he used as his starting-point the actual plant, transforming it through the artistic process into an embodiment of the ideal feminine form. They represent one of the most charming examples of illustration from the Romantic period in Europe.

100. MARY WISE (*fl.* 1856)

Flower collages.

24 × 19 cm. 8 paper collages of flower arrangements.

BINDING: Bound in pink paper wrappers. On the front gilt foil pastedown is the inscription: 'A Token of Remembrance To Jane Lane, fashioned by the Hand of her affectionate Protectress and faithful friend Mary Wise. Queen's Parade, Bath | September 1856'. One of the collages is attached to the back pastedown. The whole is laid-in a portfolio of half pink leather and grey paper on embossed boards, trimmed with gold foil cut-outs; two watercolours depicting feathers form the central panels to the front and back covers. The portfolio is housed in a pink glazed cotton envelope.

PLATES: 8 plates of flower bouquets, made of coloured paper glued to the page.

THIS UNUSUAL and delightful work reflects the wholly feminine obsession in the Victorian period with flowers, and with arts and crafts utilizing floral motifs. Nothing is known of its creator, Mary Wise, but it is pleasing to imagine her seated at a table with scissors, a pot of glue and brightly coloured sheets of paper, absorbed in fashioning this collection of paper flowers for her beloved protégée, Jane Lane.

This period saw the emergence of many important woman artists who either taught flower painting—such as Mary Lawrance (see no. 78) and Mrs Withers—or who themselves became well-known painters, particularly of exotic flowers, for example Priscilla Bury (see no. 86) or Clara Pope. Most Victorian ladies, however, were interested neither in the science of botany nor in the documentation of rare species; they preferred to dwell on the beauty of flowers and on their sentimental associations, a predilection clearly reflected in the popularity of works on the language of flowers and similar topics in this period (see no. 97).

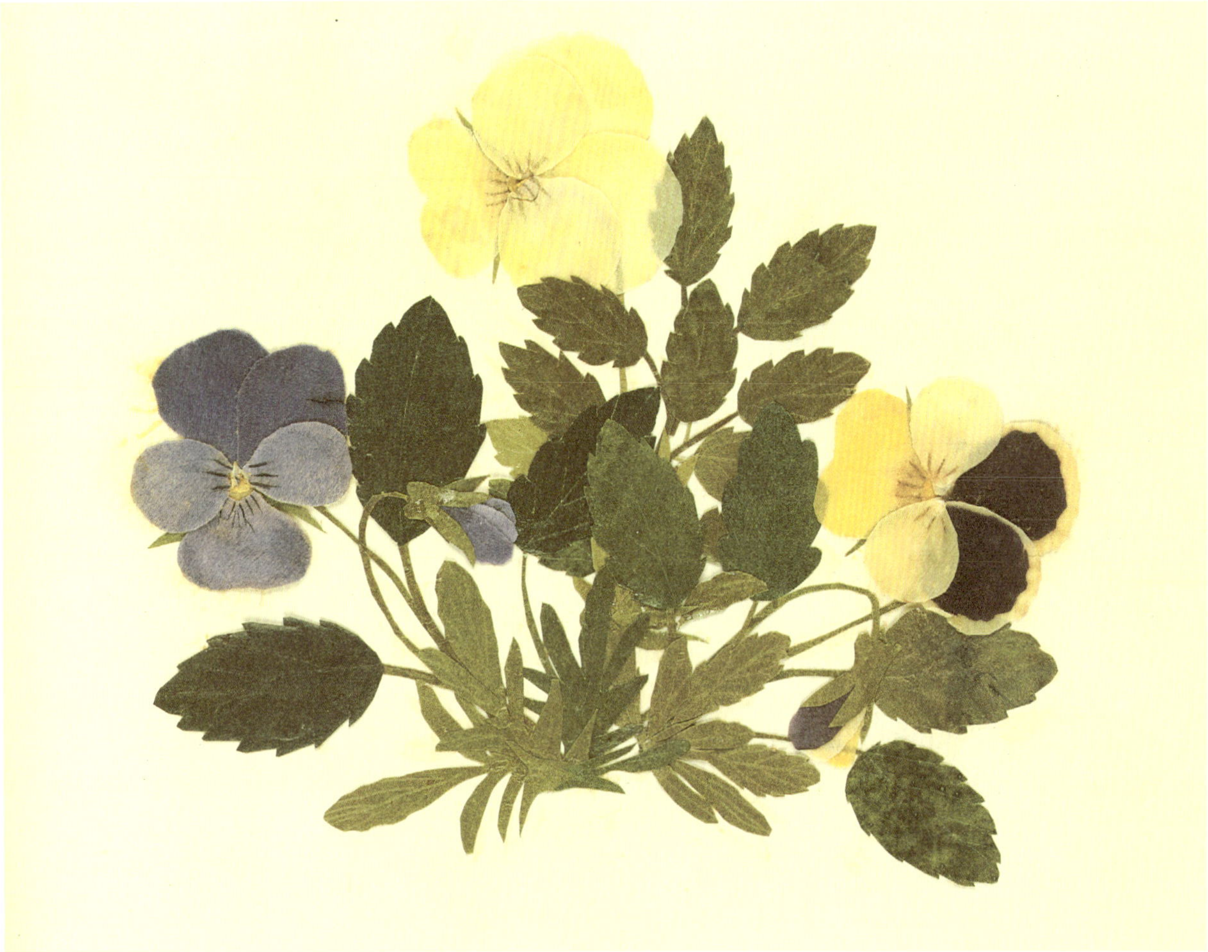

MARY WISE,
Flower collages.
Viola

The charming, but somewhat ingenuous, flower arrangements in this portfolio owe little to the august tradition of botanical illustration. They are instead reminiscent of the work of an earlier English artist, Mary Delany (1700–88), who produced, in addition to exquisite embroidery, decorative shellwork and landscape sketches, more than one thousand paper 'mosaicks' of flowers, which were greatly admired by George III and Queen Charlotte. Thus, the fashioning of flower collages was a sister art to the patient, eternally feminine accomplishments of embroidery, painting in watercolours and so on, in which Mary Wise must have been well versed.

The elaborate binding that was fashioned—perhaps by Wise herself—to hold these collages reflects the genteel tastes of the period and transforms this album into what an art critic of our own time might define as a 'multi-material object'. The pink, glazed cotton envelope was no doubt the product of Wise's own needle, and she perhaps also painted the two watercolours of feathers that appear on the beautifully embossed and decorated covers of the portfolio. As we peruse this work we can almost imagine a faint perfume of violets, reminiscent of some Victorian gentlewoman's private boudoir, rising from its folios.

XI · FLOWERS FROM DISTANT LANDS

EXOTIC PLANTS and their flowers have fascinated botanists and virtuosi ever since the arrival of the first specimens in Europe around the mid-sixteenth century. Explorers such as the Italian Giovanni Battista Ramusio (1485–1557) and naturalists, the Frenchmen Pierre Belon (?1517–64) and Jean Robin, for example, were among the first Europeans to mention (and in some cases to depict) the hitherto unknown species they encountered during the course of their travels. As we have already seen, Jacques Le Moyne de Morgues (nos. 4, 5, 6) was an explorer and cartographer as well as a botanical artist, while his English contemporary John White travelled in the wake of the famous expedition undertaken by Sir Walter Ralegh in 1585 to the land of 'Virginia', so named by Ralegh in honour of Elizabeth, the Virgin Queen.

Numerous colonizing expeditions to the Far East and the New World were organized by European powers during the course of the sixteenth and seventeenth centuries in order to defend their political or commercial interests. Many of the participants in these expeditions brought back with them interesting botanical and zoological samples, or notes and drawings documenting their observations *in situ*. During an expedition to Mexico organized by Philip II of Spain, which lasted seven years (1571–7), the physician Francesco Hernandez had artists complete more than 1,200 naturalistic works. Unfortunately this precious collection, which was carefully transported back to Spain, was subsequently destroyed in a great fire at the Escorial.

Jesuit priests, who spent long periods of time abroad on evangelizing missions, also collected examples of the local flora and fauna to bring back to Italy and France in the interests of science and for the benefit of wealthy collectors. Many of these learned priests were themselves amateur botanists, such as Michael Boym—the author of an important text on oriental flora, *Flora sinensis*, published in Vienna in 1656—and the German Athanasius Kircher, who assembled a famous *Wunderkammer* at the Collegio Romano and wrote many books on the natural sciences. The introduction of the camellia to Europe was originally attributed to the Jesuit George Camel (1661–1706), whose missionary zeal took him as far away as the Philippines, but we now know that the flower was in fact brought back from Japan by the German physician Engelbert Kaempfer (1651–1716). In his *Icones selectae plantarum* (London, 1791), edited and published by Sir Joseph Banks, Kaempfer describes no less than thirty species of this flower, which was particularly venerated in oriental countries for its beauty and symbolic associations (Fisher, pp. 123–8).

Thus, the seventeenth century saw the naturalist working side by side with the explorer. Indeed, it was often difficult to transport live specimens back to Europe, and many explorers took artists along with them to portray subjects of interest on the spot, so that at least a clear scientific record of these plants and animals could be brought back to Europe for study. When the Dutchman Prince Jan Maurits van Nassau-Siegen set off on his famous expedition to Brazil in October 1636, the naturalists Georg Marcgrav and Wilhelm Piso and the artists Albert Eckhout (*c.* 1610–65) and Frans Post (*c.* 1612–80)—Post was well known for his landscapes as well as for his naturalistic studies—formed part of the company.

In the eighteenth century, when travel was somewhat easier and the Grand Tour had become *de rigueur* for young gentlemen (who even dabbled in a little botanizing for their own pleasure), large-scale, highly prestigious scientific expeditions were organized regularly. Experts in every branch of the natural sciences were recruited, and special containers for transporting animals and plants were designed to be taken along with them (Desmond 1987, pp. 104–6). One of the most important scientific undertakings to be organized in this period was a French expedition to the Levant in 1700, in which the botanist Joseph Pitton de Tournefort and the painter Claude Aubriet participated (see no. 47).

While most artists had to be recruited for these scientific missions, Maria Sibylla Merian was motivated solely by her own thirst for knowledge to undertake a voyage to Dutch Surinam in South America, where she spent two years painting the native plants and insects (no. 101). Only when she caught malaria could this courageous woman be persuaded by her daughter, who had accompanied her on this journey, finally to return home.

John Bartram (1699–1777) a Quaker living in the American colonies and perhaps the founder of the first botanical garden in the New World, made significant contributions to the science of botany with the expeditions he conducted from Pennsylvania to Florida over a period of thirty years. He was one of the first botanists to describe the *Magnolia grandiflora* (see no. 51), which he saw at Bull Bay in South Carolina, and whose flower inspired enthusiastic admiration when the tree was first cultivated in Europe. Bartram also sent a specimen of the *Lilium philadelphicum* to the prestigious Chelsea Physic Garden in 1737. He documented his studies of the flora of the coast of North America with botanical illustrations he and his son William drew themselves (see Ewan).

In certain other countries where the colonial presence was particularly strong, institutions for botanical or zoological studies were established by the ruling powers or private interests. The great East India Company, for example, founded a botanical garden in Calcutta in 1786, from which specimens were regularly sent back to England.

Many Europeans who were sent abroad in various capacities were also amateur or semi-professional botanists who took advantage of their extended sojourns in exotic lands to conduct systematic studies of the native flora. At Oak Spring, for example, is a collection of four bound volumes of paintings of Chinese flora, accompanied by a manuscript index and a key to the Chinese names of the plants, compiled by John Bradby Blake, who had been sent to China by the East India Company in 1766. There are also five bound volumes of collected letters and notes sent by Blake from Canton to his father and to John Ellis of London. For eight years (Blake died in 1773 at the age of twenty-nine) he devoted all his spare time to the pursuit of natural history, seeking 'to procure seeds of all the vegetables found in China, which are used in medicine, manufacture of food, or in any shape serviceable to mankind'. He affirms that the paintings of these useful plants, almost all of them shown in flower, were executed by 'Mauk Sow-U . . . the most ingenious draughtsman of China'.

Significant contributions to the botanical sciences were also made by the botanist Sir Joseph Banks (1743–1820), who circumnavigated the world with Captain Cook on the *Endeavour*. On his return to England, Banks employed five artists (including John Frederick Miller and Frederick Polydore Nodder) to depict the plants he had collected during his travels, on the basis of his dried specimens and the sketches of Sidney Parkinson, a student of natural history who had participated in the expedition.

In the nineteenth century the imagination of Europeans was transfixed by the exoticism of these distant lands, news of which filtered back to them with the reports of explorers and colonizers. Their diaries describe in marvelling terms the vegetation to be found in strange climes, luxuriant and fertile on a scale unimaginable in Europe. The Amazon jungle, the Far East, the high plateau and mountains of the Himalayas, became the goals of more and more explorers and scientists, who gathered increasingly bizarre and unusual specimens to present to an amazed public on their return to civilization. Droves of professional 'plant hunters' were sent forth by private collectors and important botanical institutions to search for new species of particularly coveted plants. As these were introduced to Europe and gradually acclimatized, a rigorous study of their botany and native habitats followed.

At the same time, amateur and professional horticulturists vied with one another to grow the latest exotic species in their gardens and greenhouses. Indeed, the Victorian conservatory gave architects and builders an opportunity to display their virtuosity in the construction of ever higher and more transparent spaces, using new materials introduced by the Industrial Revolution. In these greenhouses lush tropical landscapes could flourish, with palm-trees of enormous girth and height and orchids of the most delicate form and colours, even in the depths of an English winter.

Documents show that even some army officers who were sent abroad, or members of their families, occupied themselves in their free time with botanical researches. For example, Mrs Cookson, the wife of George James Cookson of the Bengal Artillery, depicted some of the flowering plants and butterflies native to Bengal, a northeast region of India, in paintings that were subsequently reproduced in a magnificent series of large lithographed plates (see no. 103). Other botanizing officers asked local artists to document the results of their researches. During his period of service in India, Captain Frederick Parr had a group of five artists from Trichinopoly in East India depict the many botanical specimens he had collected (no. 105). In these paintings we can admire the talent of artists who, utilizing the style and techniques they had inherited from the great Moghul tradition of miniature painting, produced works at once beautiful and scientifically accurate.

At Oak Spring there is also an album of large watercolour paintings of flowering plants and fruit, birds and insects executed by a 'Miss Gough', daughter of Viscount Gough, in Madras between 1840 and 1843. In the same period Marianne North (1800–1860) produced a series of oil paintings of tropical plants that may be seen today in the gallery at Kew Gardens that bears her name.

It was the intrepid Joseph Banks, together with a handful of other eminent botanists and horticulturists, including John Wedgwood, William Forsyth, William Aiton, James Dickson, Richard Anthony Salisbury and Charles Greville, who, in 1804, founded the Royal Horticultural Society of London. One of the goals of the Society was to introduce and acclimatize new species of exotic plants. To this end, between 1820 and 1850 it organized numerous expeditions that resulted in the introduction to Europe of a large number of new plants and the documentation of countless others. Artists were also commissioned to prepare illustrations of some of these discoveries for publication in the Society's *Transactions*. John Lindley, who became the Society's secretary in 1858, was particularly active in this regard, identifying new plants of special interest and arranging for their depiction by talented artists (Elliott, p. 8). It was under the auspices of the Royal Horticultural Society that John Reeves and his son, both tea inspectors in Canton for the East India Company, undertook the project of documenting the local flora, utilizing the talent of Chinese artists practiced in the oriental tradition of flower painting (no. 102).

Kew Gardens too, under the guidance of Sir William Jackson Hooker and his son Sir Joseph Dalton Hooker, organized many important botanical expeditions to the least accessible regions of the world. The younger Hooker himself travelled to the Himalayas and conducted detailed researches on the flowering rhododendrons of the Sikkim, sending many specimens back to his father at Kew. Of the thirty-six species he succeeded in individuating, twenty-eight were entirely new and unknown in Europe. Hooker also completed many sketches that were later used by the botanical artist and lithographer Walter Hood Fitch as the basis for the illustrations to Hooker's *The Rhododendrons of Sikkim-Himalaya* (no. 104). At Oak Spring there is a replica, painted in watercolours by Walter Hood Fitch, of a celebrated painting of 1849 by William Tayler (1808–92) that shows Hooker in Sikkim receiving gifts of fruits and plants from the natives; in the painting Tayler has also depicted various rhododendrons (reproduced in Allan, facing p. 160).

Henry John Elwes, another intrepid voyager and naturalist, compiled a monumental monograph in seven parts on the genus *Lilium*, with lithographs prepared by Fitch, after an expedition to the Himalayas (no. 107).

One of the most celebrated plants from this period was the water-lily *Victoria amazonica*, originally called the *Victoria regia* by Lindley in honour of England's reigning monarch. Discovered at the beginning of the century by European explorers, it created a veritable sensation in England when the eclectic and versatile Joseph Paxton (1803–65) managed to coax the plant to flower while working at Chatsworth and Chiswick as head gardener for the Duke of Devonshire. With its astonishing floating leaves, which could exceed two metres in diameter, and its white flower, which

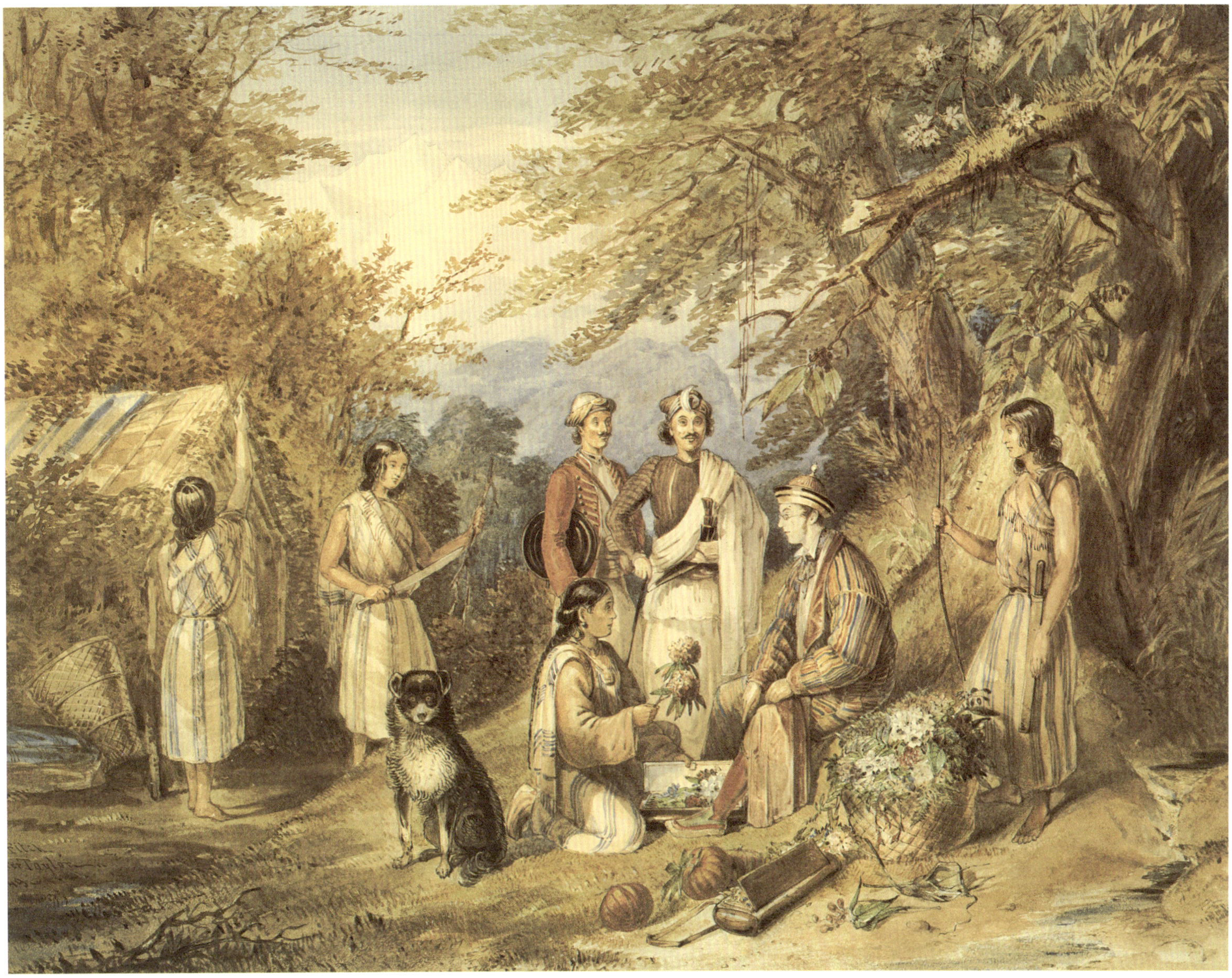

WALTER HOOD FITCH, *Sir Joseph Hooker Receiving Rhododendrons from Himalayan Natives*, watercolour on paper after a painting by William Tayler

opened at night and closed again at dawn, the *Victoria amazonica* attracted great crowds of visitors to the botanical gardens where it was cultivated. Paxton later said that he was inspired by the structure of the leaves of the *Victoria amazonica* when he designed his glass and iron Crystal Palace for the Great Exhibition in London of 1851. The flower was widely used as a decorative motif, and formed the subject of paintings by some of the greatest botanical artists of the nineteenth and twentieth centuries, including John Fisk Allen of Boston (no. 106), whose *Victoria Regia; or, The Great Water Lily of America* certainly represents one of the most beautiful books ever printed in the United States.

101. Maria Sibylla Merian (1647–1717)

[*title printed in black, retouched in gold*] Mariae Sibillae Merian Dissertatio de Generatione et Metamorphosibus Insectorum Surinamensium: in quâ, praeter vermes & erucas Surinamenses, Earumque Admirandam Metamorphosin, Plantae, flores & fructus, quibus vescuntur, & in quibus fuerunt inventae, exhibentur. His Adjunguntur Bufones, Lacerti, Serpentes, Araneae, Aliaque Admiranda Istius Regionis Animalcula; omnia manu ejusdem Matronae in America ad vivum accurate depicta, & nunc aeri incisa. Accedit Appendix Transformationum Piscium in Ranas, & Ranarum in Pisces. [*hand-coloured vignette with a lyre player*, 8.5 × 12 *cm.*] Amstelaedami, Apud Joannem Oosterwyk, 1719.

1° [*large broadsheets*] 55 × 38 cm. *i–viii 1* 2–67 62–66 p.

BINDING: Calf binding. Floriated gilt spine with lertteing: 'Merian Metam. Insetor Surinam'.

PLATES: Engraved and hand-coloured title-page from a drawing by F. Ottens, printed by J. Oosterwyk. Hand-coloured coat of arms of Balthazar Schott on the dedication page. 72 engraved plates of plants and insects.

PROVENANCE: On the title-page in manuscript: 'F.V.s'.

REFERENCES: Pritzel 6105 (2nd edn); Nissen 1341; Lendorff; Nebel; Ullmann; *An Oak Spring Garland*, no. 12.

This is the second edition of Maria Sibylla Merian's most important work. The first edition appeared in 1705 and was illustrated with sixty plates; this edition, published two years after Merian's death, contains an additional twelve plates, some her work and others perhaps by the hand of her elder daughter, Johanna Helena. Merian was an extraordinary woman, and *Dissertatio* was the crowning achievement of a highly singular life dedicated to art and to intellectual and spiritual enquiry.

In 1685 Merian was persuaded by her beloved half-brother, Caspar, to abandon husband and home and, together with her mother and two daughters, join the pious sect of the Labadists. This sect had been founded by the French ex-Jesuit Jean de Labadie, who preached a return to the virtues of the early Church, and to this end established a religious community at the remote castle of Waltha near Leeuwarden in Friesland. Taking advantage of the sect's belief in absolute equality between the sexes, the artist devoted all of her time to scientific study, having found at Waltha a magnificent collection of insects brought there from Surinam by a member of the Sommelsdijk family, proprietors of the castle. The colour, form and dimensions of these tropical insects must have been a real revelation to the artist, since nothing could have been more different from the European species she had been studying since childhood.

Merian left the Labadist community in 1691, after her half-brother and her mother had died, and moved to Amsterdam, where she made a living painting flowers and insects for wealthy art collectors (see no. 81). There she conceived and began preparation of the third volume of her work on caterpillars and butterflies, *Der Rupsen Begin, Voedzel en Wonderbaare Verandering* (see no. 82). She spent much of her time studying the insect collection of Nicolaas Witsen, burghermaster of the city and a member of the East India Society, and was also on good terms with two illustrious scientists, the anatomist Frederick Ruysch and the botanist Caspar Commelin, director of the city's botanical garden.

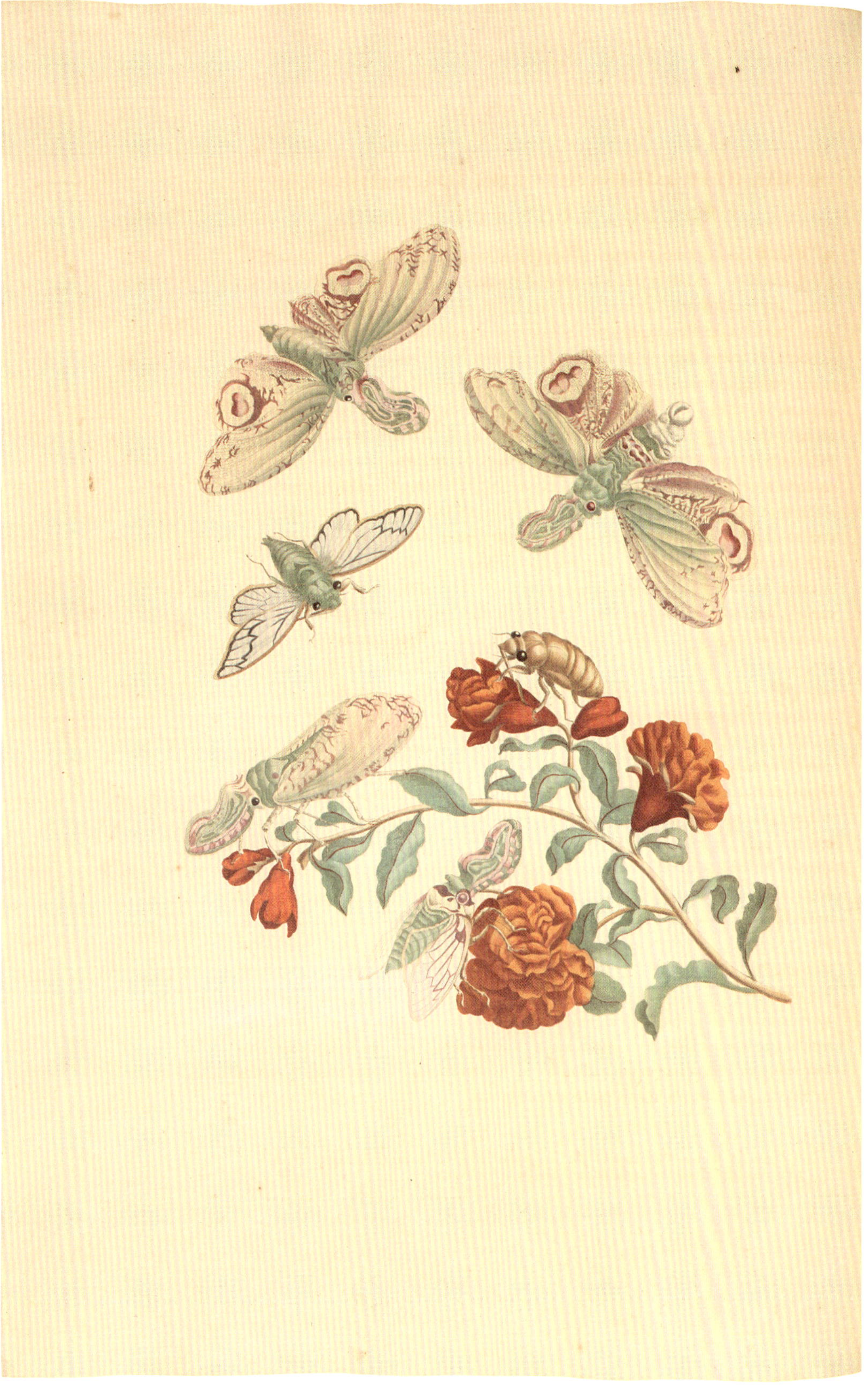

MARIA SIBYLLA MERIAN, *Dissertatio de generatione et metamorphosibus insectorum Surinamensium.* Pomegranate flowers (*Punica granatum*), insects and lantern flies

In 1699, with their encouragement and financial help, Merian—by now fifty-two years of age—set off for Surinam in the company of her daughter Dorothea Maria. As a result of this extraordinary two-year expedition she produced the *Dissertatio*, her most famous work. The two women suffered great hardships during their stay in the Dutch colony; as the artist remembered in 1702 on her return to Amsterdam: 'In that country there reigns a torrid heat, and every task becomes an enormous effort; I myself risked paying with my very life and this is the reason why I could not extend my stay any longer.' Indeed, she had contracted malarial fever, and thus in September 1701 she and her daughter set sail for home.

Merian spent the rest of her life completing this visual record of her entomological observations, arranging and retouching the paintings she had executed *in loco*; today these paintings can be seen in the British Museum, in the Royal Library at Windsor Castle, and at the St Petersburg Academy of Sciences, where are also some watercolours, executed in her later years, for the *D'Amboinsche Rariteitkamer* by Georg Eberhard Rumpf (Amsterdam, 1705). Her extraordinarily refined technique, which reflects her training as an engraver, can be appreciated in some of the works in the Windsor Castle collection: beneath the bodycolours one can trace the original marks of her pencil. These strokes are reminiscent of the fine lines incised by the engraver's burin, and in fact must have been of invaluable assistance to the artists charged with the task of transferring the drawings to the metal plates.

Merian was assisted in editing the scientific data by the botanist Commelin. She engraved perhaps twenty of the plates for her work, while the remainder were executed by Joseph Mulder and Jan Sluyer. The artist insisted, however, that only she and her daughter, who alone had studied the insects from life, should be entrusted with the task of colouring the plates (Stearn 1978, p. 14).

The first edition of *Dissertatio* was printed in 1705 in the artist's home city, Amsterdam; it appeared simultaneously in both Latin and Dutch, and consisted of sixty plates with as many pages of text. Its price was fifteen florins if ordered before publication, and eighteen if purchased afterwards, while a hand-coloured copy cost thirty florins. The magnificence of its engravings and its unusual scientific content make it one of the most beautiful and important books ever printed in Europe. The work opens with an imposing title-page, in this copy coloured by hand with gold highlights, which shows the artist studying specimens presented to her by six putti. In the background a spacious arch opens onto a tropical landscape. The foreword is full of fascinating information, the author describing in detail her venturesome and costly voyage ('longinquum et sumptuosum') and the methods she employed when painting. Each insect was carefully examined, often with the aid of a microscope, and depicted together with the plant, flower or fruit on which it normally fed. Each written entry begins with useful botanical information, thus providing us with indications as to how the artist composed her pictures.

Merian's sensibility to the minutest aspects of the natural world, and her rich visual vocabulary (the fruit of a lifetime of study and practice), is reflected in every detail of the *Dissertatio*. It contains a

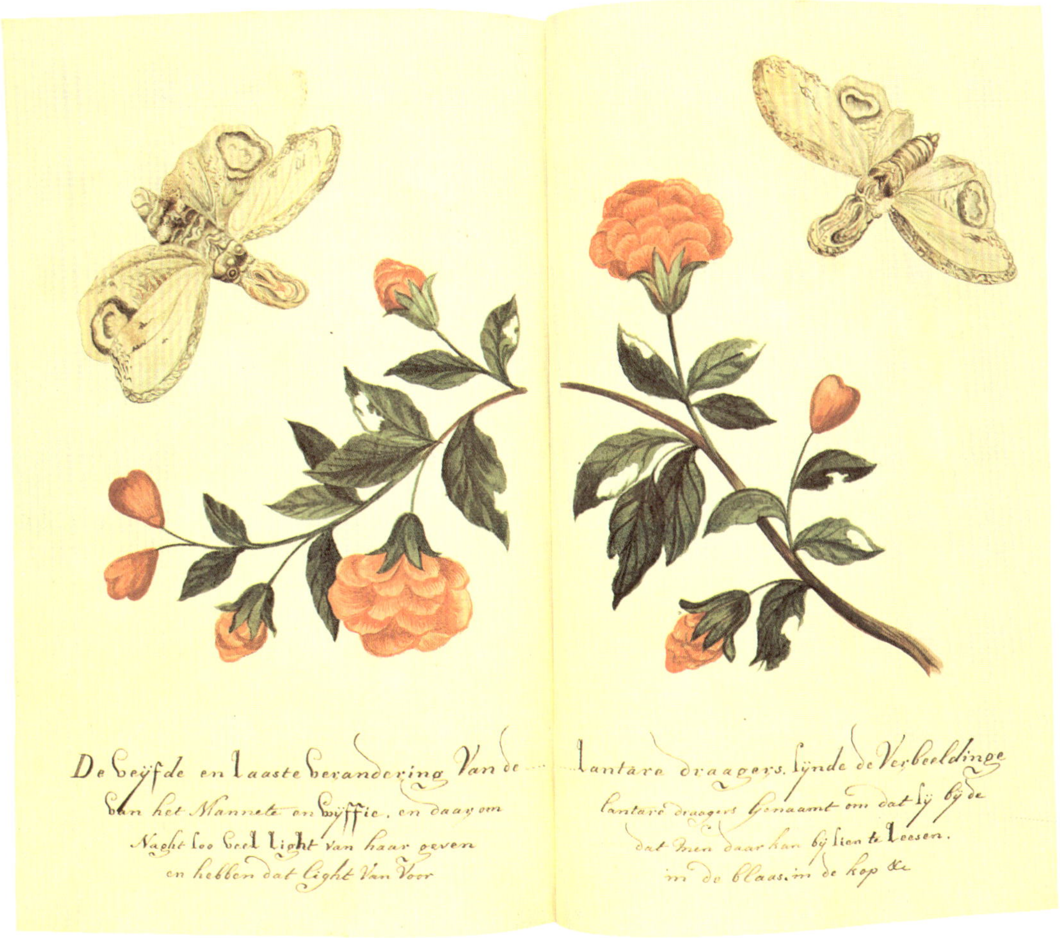

CORNELIUS MARKEE, *Naauwkeurige Verzameling van Europasche Insekten alles naauwkeurig gelchetst en gelchildert*, 1756. Volume V, Pomegranate flowers (*Punica granatum*) and lantern flies

myriad of exotic species, most of them shown in the various phases of their life-cycle. Their startling richness and variety of form make this work endlessly fascinating; one can imagine the amazement with which her contemporaries must have contemplated, for example, the monstrous lantern flies (*Fulgoria lanternaria*) that menacingly circle a branch of flowering pomegranate in plate XLIX.

Merian's vision was certainly not one of an idyllic tropical paradise: in not a few of her paintings she has depicted next to the insect its natural predator. As Luigi Figuier colourfully expressed it: 'Every one of her paintings depicts a drama in miniature. Next to the frightened and suspicious insect, one can see an avid lizard watching, or a ferocious spider weaving its sinister snare' (Figuier, p. 192). The implacable laws of nature do not spare the splendid tropical flowers depicted by the artist, who often saw the fresh green leaves and fleshy, vividly coloured petals as nourishment offered up to ravening insects.

Mention should here be made of Cornelius Markee, a Dutch artist from Middelburg who was to be deeply inspired by Merian's work. No less than 506 bodycolours of insects flitting around or resting on flowering plants are signed and dated (1756) by him. Collected as *Naauwkeurige*

Verzameling van Europasche Insekten, this five-volume work, in its original binding, is today conserved at Oak Spring. Markee, who also copied in eight volumes in watercolours the engraved plates of Pierre Hélyot's *Histoire des ordres monastiques, religieux et militaires* (Amsterdam, 1714–19), now at the Getty Center in California, devoted great attention to cocoons and caterpillars and to the butterflies that rest on or hover near flowering branches, which are often set in elegant vases. His butterflies in particular depend on those by Merian, as the plate that illustrates *Fulgoria lanternaria* clearly reveals.

102. CHINESE ARTISTS [*c.* 1820]

Album of Chinese flora painted for John Reeves the elder.

47.5 × 38 cm.

BINDING: Maroon morocco; sides panelled with broad and narrow gilt fillets, and an inner panel of a convoluted border of leaves with acorns. Gilt spine.

PLATES: 66 bodycolour drawings of flowers. The drawings, measuring 40 × 33 cm., were executed on special paper of English provenance and mounted on Whatman paper with a watermark dated 1832. Every drawing bears the name of the plant written in Chinese characters, and most of them a Latin transcription of the same.

PROVENANCE: On the front pastedown: 'Lady Winnington presented to Lesley Capt. Colgrane'.

THIS FINE ALBUM of Chinese flowering plants can be linked to the Reeves collection on the basis of its stylistic affinities with works now scattered in various other collections. A large number are to be found in the Natural History Museum and the Lindley Library of the Royal Horticultural Society, both in London.

John Reeves (1774–1856) was born in Essex and, along with his contemporaries the poets Coleridge and Charles Lamb, was educated in London at a bluecoat school, Christ's Hospital. He joined the counting-house of a tea-broker, where he gained so thorough a knowledge of teas as to recommend him to the office of the Inspector of Teas in England in the service of the East India Company. In 1812 Reeves was transferred to China, where he worked first as an assistant and then as Chief Inspector of Tea at the company's establishment in Canton. He lived in Canton and Macao until 1831, when he returned to England to pass the remainder of his days. In 1827 he was joined in Canton by his son, John Russell Reeves (1804–77), who was also employed as an Assistant Inspector of Tea and who continued the scientific studies begun by his father (Whitehead and Edwards, p. 17).

Despite the demands of his work, Reeves the elder managed to find time for his own particular passion—botany. He conducted systematic studies of the local flora, and in 1816 took back with him to England a large collection of plants and seeds. When he returned to China he continued to send specimens home regularly (RHS, p. 14; Elliott, pls. 37, 38). Thanks to his efforts a large number of entirely new species were introduced to England, and many natural history museums were able to update and replenish their oriental collections.

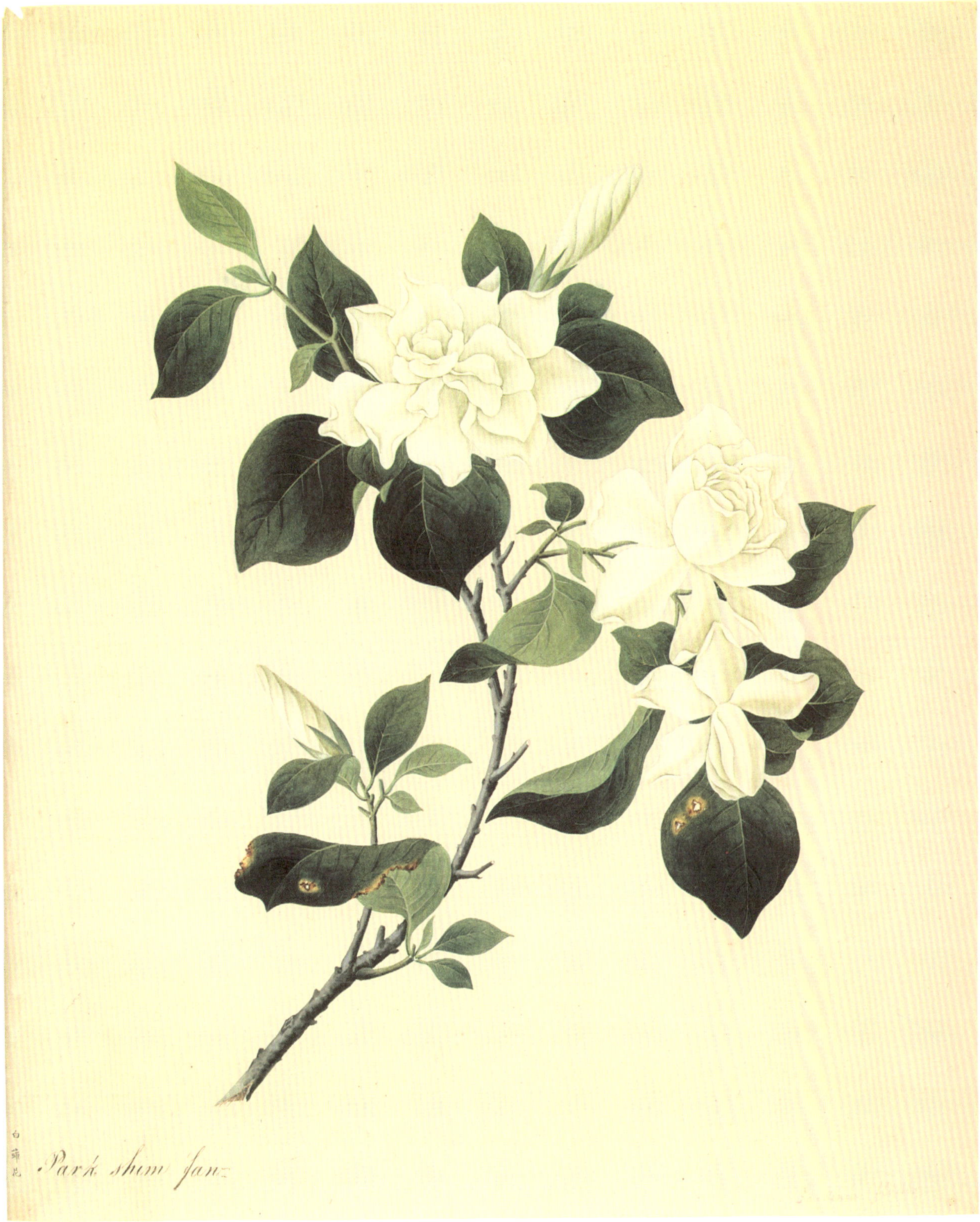

Chinese Artists,
Album of Chinese flora.
Cape jasmine (*Gardenia jasminoides*),
folio 7

Reeves became a member of the Royal Horticultural Society in 1817, and we find that in 1818 he sent the Society a specimen of *Wisteria sinensis*, just two years after Charles Hampden Turner and Thomas Carey Palmer had discovered the plant and introduced it to England (Coats, p. 90). Soon after the Royal Society asked Reeves to gather together a group of Chinese artists to paint the indigenous flora under his supervision. Reeves carried out his project conscientiously, sending many series of paintings back to England; in addition to those in the Natural History Museum and the Royal Horticultural Society, works produced for him can be found at the Royal Botanic Gardens, Kew, and in the Victoria and Albert Museum. At the Natural History Museum there is a noteworthy collection of paintings of plants and animals by the artist Chin Nung (1687–1764), while an interesting group of zoological paintings is today conserved in the Museum of Comparative Zoology at Harvard University.

Reeves himself only wrote a single work, *An Account of Some of the Articles of the Materia Medica employed by the Chinese*, published in 1828, but some specimens and drawings of fishes furnished by him were used in Sir John Richardson's *Report on the Ichthyology of the Seas of China and Japan*, published in 1846.

The sixty-six paintings in the Oak Spring album were painted by a number of artists, but all reflect the ancient and highly sophisticated Chinese tradition of floral painting. In the first plate a flowering branch of *Passiflora caerulea*, rendered with the precision of a miniature, meanders elegantly across the page. With perfectly controlled brushstrokes the artist has traced a sensitive and fluid outline of the branch, rendering the smallest details with simple strokes that belie years of disciplined training and practice. His superb technique permitted him to depict the fragile blooms with remarkable fidelity. The album presents different varieties of camellias, gardenias, peonies, magnolias and hibiscus, all rare flowers prized in Europe for their exotic beauty. The paintings themselves reflect an iconographic tradition dating back millennia, combining an exquisite aesthetic sense with a striving for the utmost realism. Each composition has a formal, almost abstract, beauty, while the details are rendered with great verisimilitude; the leaves cast tenuous shadows and the flowers with their slender stamens and pistils seem to tremble in the wind.

The painters working under Reeves also revived an ancient technique in which mineral-based pigments, frequently mixed with gum, were used to create paints with the consistency and lustre of polished lacquer. Such was their skill that they managed to create effects of shading that no Western artist of the period working in oils, body- or watercolours could emulate (Hulton and Smith, p. 62). No pains were spared in the execution of these botanical paintings, no doubt in part thanks to the careful supervision of both father and son. Particularly fine are two paintings of the *Gardenia jasminoïdes* (pls. 7 and 14) with its white flowers and lustrous dark-green leaves. The tender shoots, spiral-formed buds, and leaves—some of them just unfurling and others brown and drying—have been depicted with superb realism. Even the traces of damage left by insects have been meticulously added by the artist.

103. Catherina Teresa Cookson (*fl.* 1830–35)

Flowers Drawn and Painted after Nature in India by Mrs. James Cookson [?1835]

61 × 44 cm. *1–2* p., 31 leaves of plates.

BINDING: Half green leather and green silk moiré with gilt floriated borders.

PLATES: 31 lithographed and hand-coloured plates, 30 of which bear the names of the plants in Latin and English at the bottom of the page.

REFERENCES: Dunthorne 82; Nissen 399; *Stiftung für Botanik, Vaduz* I.135; *A Magnificent Collection*, no. 82; M. V. Matthew, *The History of the Royal Botanic Garden Library*, Edinburgh, 1987, pp. 123–4; Desmond 1994, p. 166.

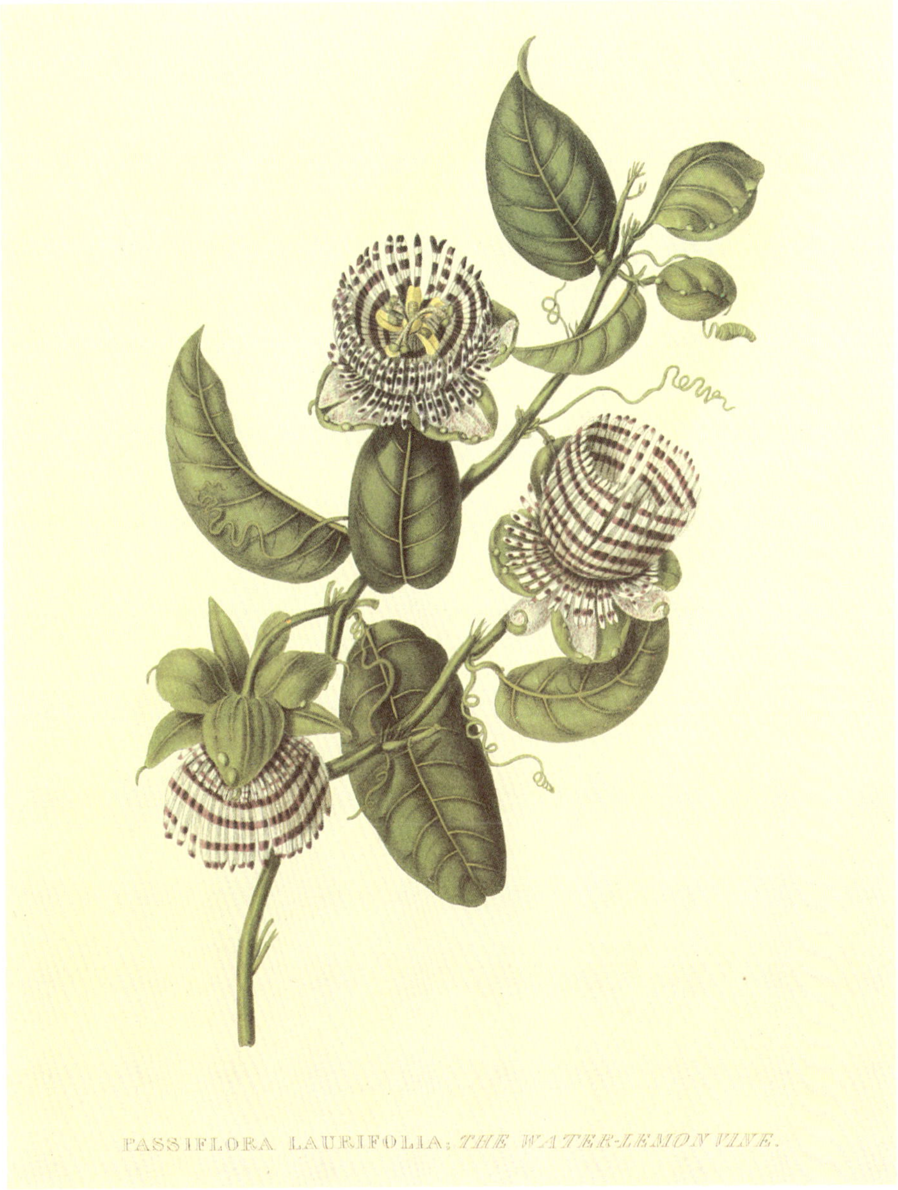

CATHERINA TERESA COOKSON, *Flowers Drawn and Painted after Nature in India.* Water lemon or Jamaica honeysuckle (*Passiflora laurifolia*)

In 1832 Catherina Teresa Murray married George James Cookson (1805–38), then serving in the Bengal Artillery. Unhappily, he died just six years later of smallpox at Karnal in the Punjab. While she was in India, however, Mrs Cookson produced this series of paintings of the indigenous flora, which was then published in the form of a volume of magnificent colour lithographs retouched by hand. Dunthorne and Nissen both assign the date 1830 to the work, but several copies recently on the rare book market bear a watermark dating from 1834, suggesting that it may in fact have been produced around the year 1835.

The volume opens with a landscape, 'A View taken in India of the Lotus growing in Water'. The thirty plates that follow depict various unusual flowering plants, including a *Datura metel* (Purple stramonium) with luxurient, chalice-shaped flowers, a magnificent deep purple *Impomoea quamoclit* (Quamoclit bindweed), a *Mirabilis jalapa* (Marvel of Peru), and a *Passiflora laurifolia* (the 'water-lemon vine') with extraordinary white flowers striped with violet.

This fascinating collection of flowers, many of them depicted with butterflies, reflects the interest of Victorian botanists and flower lovers in the exotic flora and fauna to be found in the most distant corners of the British Empire.

104. Joseph Dalton Hooker (1817–1911)

The Rhododendrons of Sikkim-Himalaya; Being an Account, Botanical and Geographical, of the Rhododendrons Recently Discovered in the Mountains of Eastern Himalaya, from Drawings and Descriptions Made on the Spot, During a Government Botanical Mission to that Country; By Joseph Dalton Hooker, R.N., M.D., F.R.S., F.L.S., &c., &c., &c., Edited by Sir W. J. Hooker, K.H., D.C.L., F.R.S., F.L.S., &c., Vice-President of the Linnean Society, and Director of the Royal Gardens of Kew. [*portrait vignette*] London: Reeve and Co., Henrietta Street, Covent Garden, 1849.

51 × 39 cm. *1–5* 6–14 *1* 2–7 *8* *1–58* p., XXX leaves of plates.

BINDING: Maroon cloth with title, 'Hooker's Rhododendrons', stamped in gilt on front cover.

PLATES: 30 hand-coloured lithographed plates, with Latin names, by Walter Hood Fitch (1817–1892) after the author and printed by Reeve, Benham & Reeve.

REFERENCES: Pritzel (for 2nd edn) 4200; Nissen 911; Sitwell and Blunt, p. 60; Allan; Coats, p. 126; Stafleu and Cowan (for 2nd edn), II.2969; Desmond 1987, pp. 131–43; Blunt and Stearn, pp. 261–4.

Hooker, an eminent botanist, naturalist and physician, grew up in a household where scientific studies and botanical illustration were considered to be a part of daily life. He was the son of William Jackson Hooker (1785–1865), professor of botany at the University of Glasgow and editor of *Curtis's Botanical Magazine*, for which he also provided many illustrations. In 1841, when the extensive botanical gardens at Kew passed out of private hands and were conveyed to the nation, becoming the Royal Botanic Gardens, William Hooker was named the first official Director; he occupied this position for twenty-four years (Allan, p. 15).

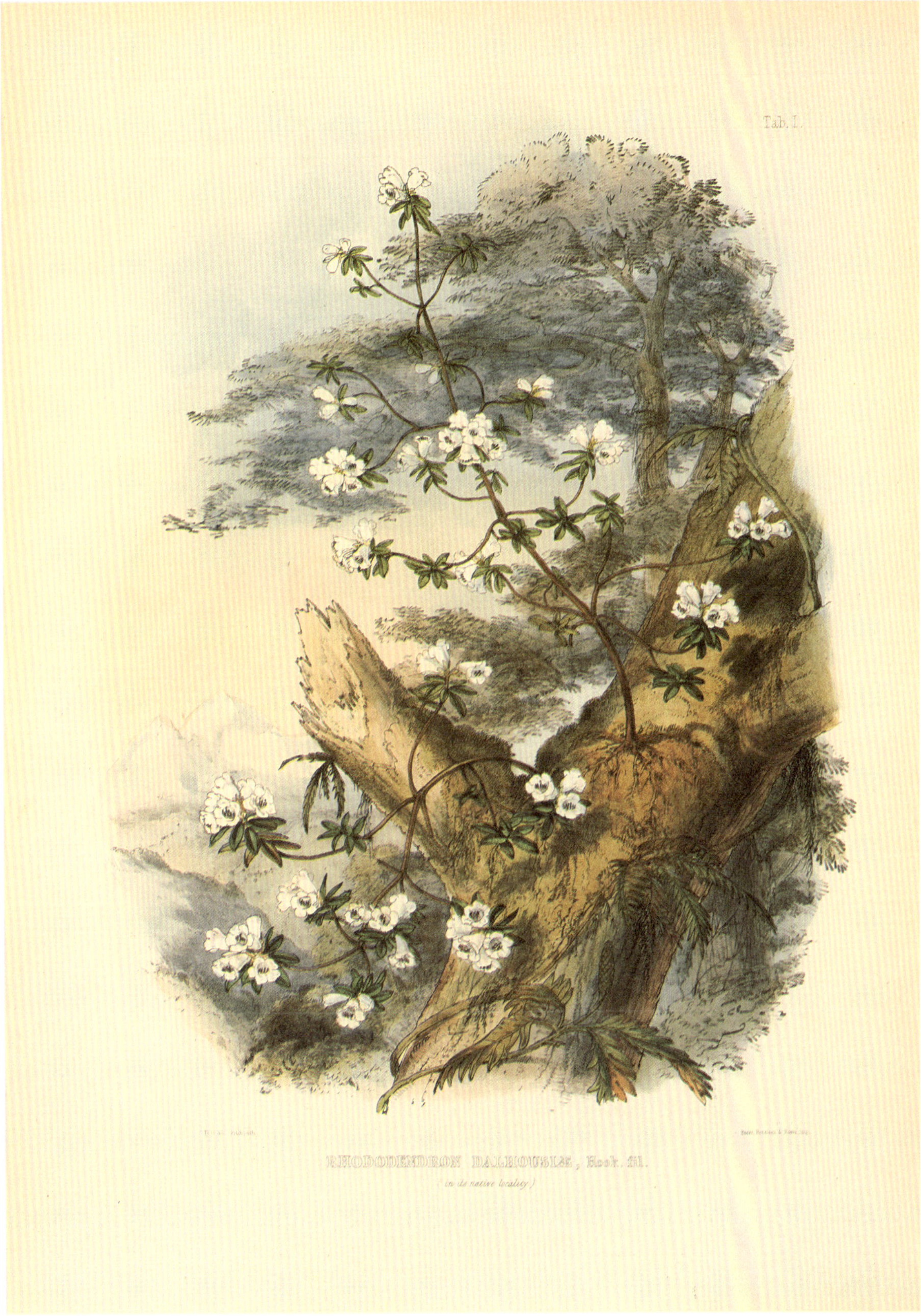

JOSEPH DALTON HOOKER, *The Rhododendrons of Sikkim-Himalaya. Rhododendron dalhousiae*, plate I

Hooker's scientific training began at an early age with research on the plants in his father's herbal garden (Daniels, p. 38), where it is known that he made a special study of mosses. In 1839, as assistant surgeon on HMS *Erebus*, he accompanied a scientific expedition to Antarctica. This was only the first of many voyages for Hooker, who in this way was able to make an exhaustive study of the flora of many distant countries. In 1855 he became Assistant Director at Kew Gardens, and in 1865 he succeeded his father as Director, filling this position with distinction for twenty years. The Hookers, father and son, oversaw the expansion of the Royal Botanic Gardens of Kew into the largest botanical garden in the world, and the unique scientific institution it remains to this day.

Hooker was the author of various important works on exotic flora, including *Flora Antarctica* (1844–7), *Flora Novae-Zelandiae* (1852–5), *Flora Tasmaniae* (1855–60) and *Flora Indica* (1855). He became a scientist of international renown, and in recognition of his achievements received numerous titles and appointments, including that of President of the prestigious Royal Society. (On his death he could have received the honour of being buried in Westminster Abbey, but by his own wishes his remains were placed in the family vault in a cemetery at Kew.) Like his father, Hooker was a highly skilled botanical illustrator, and he never abandoned the sound practice of sketching species of scientific interest from life whenever opportunity offered. Thus he utilized the large number of studies in watercolours and pen and ink drawn by him on the spot during a botanical expedition to the eastern Himalayas to illustrate *The Rhododendrons of Sikkim-Himalaya*.

Following his return to England, Hooker's drawings were lithographed by the artist Walter Hood Fitch (see no. 107). A native, like Hooker, of Glasgow, Fitch was a brilliant practitioner of the art of lithography, which well before the end of the nineteenth century had revolutionized the process of book illustration. After a period of apprenticeship to a firm of calico designers, he had the good fortune to meet William Jackson Hooker, who was then teaching at the University of Glasgow. Hooker recognized the young artist's talent and undertook to train him as a botanical illustrator. When Hooker was appointed Director of the Royal Botanic Gardens, he asked that Fitch accompany him to Kew. During his lifetime Fitch produced for Hooker, for his son Joseph, and for many other botanists a total of nearly 10,000 botanical plates. For *Curtis's Botanical Magazine* alone he prepared 2,900 illustrations. Fitch's intuitive grasp of plant structure and confident draughtsmanship permitted him to draw his illustrations directly on the lithographic stone, thus lending his printed plates a lifelike spontaneity. Although Fitch's last years were somewhat darkened by rancorous disagreements with Joseph Hooker, the long collaboration between the artist and the Hooker family was extremely profitable for both sides. While Fitch's reputation received added lustre from the fame of his two distinguished patrons, his work indubitably helped the Hookers to reach a wider audience (Lewis, p. 2).

This volume on the rhododendrons of the Himalayas is a fine example of Fitch's art. In his lithographs he has captured the exuberant form and colour of these flowering shrubs, many of which in this very period came to be acclimatized in European gardens. Each plate is accompanied by a meticulously written letterpress by Joseph Hooker. Sometimes at the base of the plate, magnified views of the pistils, stamens and sections of the ovaries are presented. The first plate is unusually attractive because the plant, a *Rhododendron dalhousieae* (dedicated to a certain Lady Dalhousie, in keeping with the custom of naming new species after eminent colleagues, patrons or friends), is shown in its native habitat, growing among the trunks of fallen trees against a hazy background of blue mountains.

In 1855 Fitch prepared the lithographs, this time after drawings by local artists, for another book by Joseph Hooker, *Illustrations of Himalayan Plants*. Among the subscribers to this work were the renowned scientists Charles Darwin and Alphonse de Candolle. At Oak Spring is Fitch's watercolour after Tayler, *Sir Joseph Hooker Receiving Rhododendrons from Himalayan Natives* (53 × 69 cm.), dated 1849. This famous incident had previously been painted by Frank Stone.

105. Hindu Artists (*mid-19th century*)

Specimens of Flowering Plants, collected by Captain Frederick Parr, During a Residence in India, and Drawn From Nature by Hindoo Artists under his Superintendence.

60.5 × 48 cm. On the verso of the front free endpaper is written in manuscript '*Parr* (*Capt: Frederick*) Specimens of East Indian Plants Drawn from Nature. Collection of 87, Original Water-Colour Drawings, mostly of Tropical Flowers but includes 6 of fish, 8 Birds, 9 Animals and 9 Butterflies & Reptiles etc.: a few of the plates are unfinished. M.S. nomenclature by Dr. R. Wight, loose but ready for insertion into large calf volume (*c.* 1850). *Added* are two original pencil Drawings of Elephants drinking at the River. *89 Drawings Altogether*.'

BINDING: Calf binding with octagonal panels of marbled paper and a red morocco label.

PLATES: 87 drawings in bodycolour of tropical plants and animals, some unfinished. They include 53 plants, 6 fishes, 9 birds, 9 mammals, 5 insects, 4 reptiles, and 1 crab. Most of the plants are inscribed with Latin names. Also, inserted are 2 drawings of elephants in landscapes; one printed leaf on the 'Convolvulus' and one on the 'Passion Flower'; one printed announcement for an 'Oriental Flower Show'; and two printed leaves with the title 'An Amateur's Collection of Water Colour Drawings'. On the verso of one of these last two sheets, in manuscript: 'A collection of drawings of Indian Plants etc. by native Artists, the nomenclature by Dr. Robert Wight M.D., F.L.S. etc. the oriental botanist. Lent by Mrs. Parr'.

PROVENANCE: From the descendants of Captain Frederick Parr with the bookplate: 'From the Library of [in manuscript] John C. W. Parr to his sister Catharine Harrietta Estelle Rees.' Inscribed along the edges 'Put in by Mrs Rees over Capt. Parr Bookplate.'

THE ANIMALS and flowering plants pictured in this album were collected by Captain Frederick Parr during his residence in India, and were depicted under his supervision by five artists from Trichinopoly in southern Madras. Very little is known of Parr apart from his service record, although in fact not a few officers in the British forces there (the Indian Army) were amateur naturalists who returned home with many new species from the Indian peninsula. Botanical and zoological studies were also sponsored by the wealthy East India Company, which had founded an important botanical garden in Calcutta in 1787 and a menagerie in Barrackpore. During the second half of the eighteenth century, Lady Impey—wife of the Chief Justice of Bengal—commissioned many artists to depict the birds, animals and insects of the region for a collection of paintings today unfortunately dispersed (S. Raphael, Introduction, *A Cabinet of Natural Curiosities: Drawings of Flora and Fauna, 1630–1830*, sale cat., London, Eyre & Hobhouse Ltd, 29 November–23 December 1983, n.p.). In addition to its fifty-three botanical illustrations, Parr's album contains six drawings of fishes, nine of birds, nine mammals, five butterflies, four reptiles and one crab, a number of which are somewhat eccentrically arranged on the sheet, perhaps in order to leave space for other sketches.

Hindu Artists, *Specimens of Flowering Plants*. Small lizard

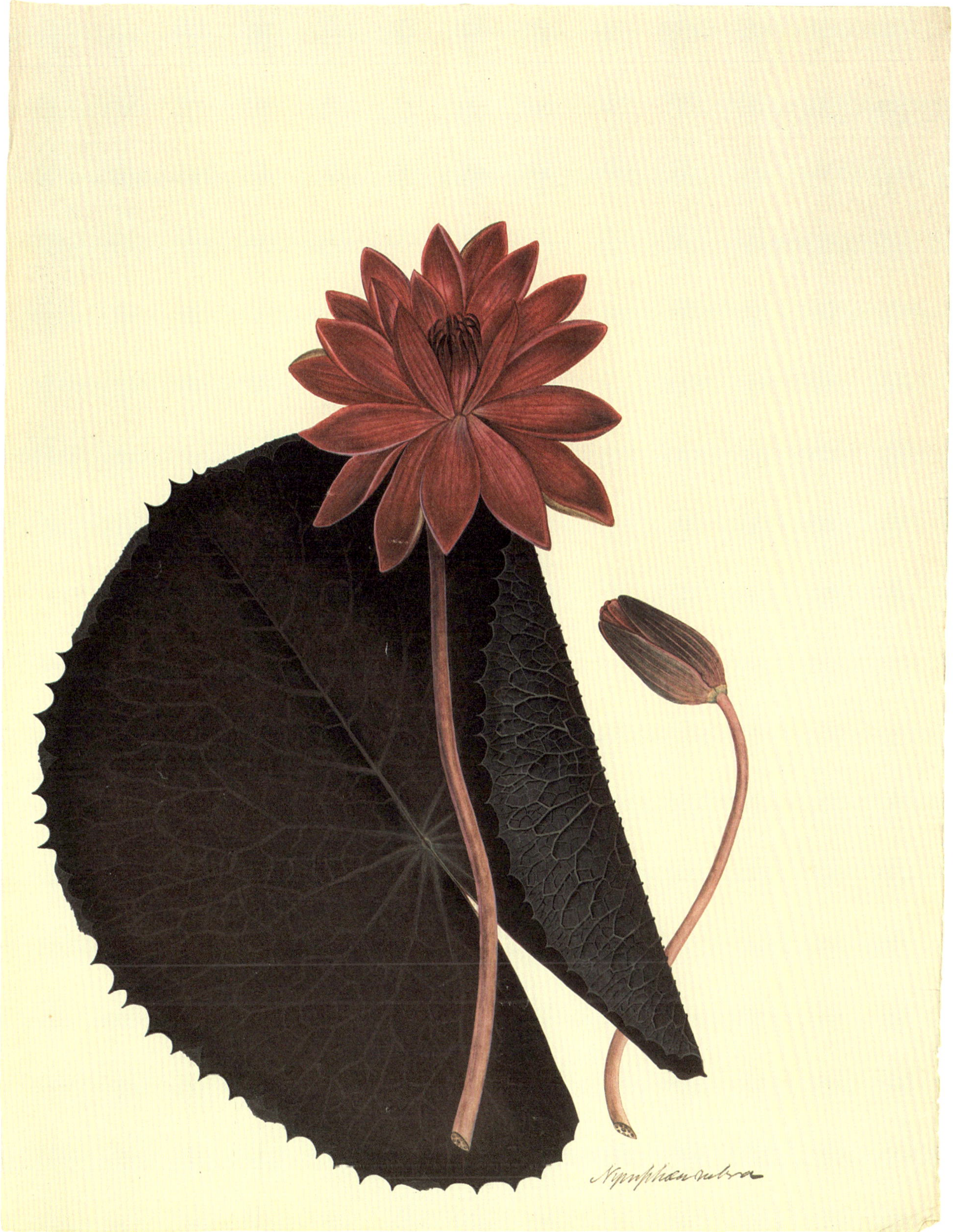

Hindu Artists, *Specimens of Flowering Plants.* India red water lily (*Nymphaea rubra*), folio 4

From a two-page printed leaflet inserted in the album, bearing the title *An Amateur's Collection of Water Colour Drawings of Oriental Plants (Chiefly Indigenous), Gathered and Depicted, during a Residence of many Years in the East Indies*, we learn that five artists accompanied Parr on a series of expeditions into the impenetrable forests of the 'Great Wynaud' and to the coast of Malabar. The botanical and zoological paintings produced by these 'Hindoo and Mahomedan' artists are remarkable for their scientific accuracy and artistic quality. Parr's artists were clearly masters in the great tradition of naturalistic miniature painting established by the Moghul emperors at the end of the sixteenth century, a tradition characterized by jewel-like precision and warm, brilliant colours (Hulton and Smith, pp. 55–60). The anonymous author of this printed sheet describes some of the hazards and difficulties encountered by Parr and his artists during their botanical expeditions, when they lived in 'tents pitched with difficulty . . . in the wildest scenes imaginable'. He concludes with a rhapsody on the natural beauties of this region, in which the richness of the flora and fauna were inseparable: 'It is indeed, the very garden of nature, affording the widest scope for botanical research. Here flourishes the graceful and tapering bamboo, clustered in impenetrable thickets; here the delicate moimosa [*sic*] of unnumbered hues . . . the pine-apple and the strawberry, amongst other fruits are likewise of spontaneous growth—and the stately Palm may also be met with in all its varieties. It is here the Zoologist may find subject for years of research, since the very decay of vegetation produces that wonderful connecting link between the vegetable and the animal kingdom.'

Since many of the plants which he had collected were entirely unknown in Europe, Parr asked a botanist specializing in Indian flora to examine them. Thus the nomenclature that appears at the foot of each plate is the work of Robert Wight (1796–1872), an assistant surgeon at the East India Company's Madras settlement and an amateur botanist who had served for one year as editor of the *Calcutta Journal of Natural History* (1846–7). Wight also wrote many works on the flora of India, including *Contributions to the Botany of India* (1834), *Illustrations of Indian Botany* (1840–50) and *Icones plantarum Indiae Orientalis* (1840–53). In the list of the subscribers to this last volume appears the name 'Parr F., Captain'.

Among the most beautiful plates in Captain Parr's album are a series of paintings of different water-lilies, a common aquatic plant of the region. One, the *Nymphaea rubra*, was introduced to Europe at the beginning of the nineteenth century from India, where the natives used to grind an edible flour from its seeds. Its deep-red flower and reddish-brown leaves with their delicate network of veins recall a plate of the same flower by Sydenham Edwards (1768–1819), which was printed in 1810 in *Curtis's Botanical Magazine* (XXXI, pl. 1280). Also quite impressive are the plates showing the 'Lagerstroemia regina' of Siam (*Lagerstroemia loudonii*) with its violet blooms, the 'Crossandra' (*Crossandra undulifolia*) with its lovely flowers shading from pale salmon to orange-yellow, and the *Clitoria ternatea*, an elegant leguminous plant with deep-blue flowers.

106. John Fisk Allen (1807–1876)

Victoria Regia; Or The Great Water Lily of America. With A Brief Account Of Its Discovery And Introduction Into Cultication: With Illustrations By William Sharp, From Specimens Grown At Salem, Massachusetts, U.S.A. By John Fisk Allen. Boston: Printed And Published For The Author, By Dutton And Wentworth, 37 Congress Street, 1854.

69 × 55 cm. *1–5* 6–16 *17–18* p., 6 leaves of plates.

BINDING: Half-red buckram and printed paper boards.

PLATES: 6 plates chromolithographed by Sharp & Son, Washington Village, Dorchester, Massachusetts, from drawings by John Fisk Allen.

REFERENCES: Hedrick, p. 488; Pritzel (2nd edn) 104; Nissen 16; Sitwell and Blunt, p. 47; *Flora Portrayed*, p. 77; *A Magnificent Collection*, no. 1; Hofer, p. 144.

THIS LARGE-SIZE folio monograph, published in Boston in 1854 by the amateur botanist John Fisk Allen, celebrates the *Victoria regia* or 'Great Water Lily of America', a flower that had been discovered along the banks of the Amazon river and taken to England, where it was first cultivated at the Royal Botanic Gardens, Kew. Three works on this flower had already been published in England: *Victoria regia* (1837) by John Lindley, *Description of the Victoria regia* (1847) by William Jackson Hooker, and *Victoria regia, or, Illustrations of the Royal Water Lily* (1851) by Walter Hood Fitch (see nos. 104 and 107). Allen's monograph was closely modelled after this third work.

The giant water-lily is a spectacular flower; nineteenth-century commentators describe with amazement the vast dimensions of its floating leaves, which could exceed two metres in diameter, and its great white flower, which opened in the evening and closed again at dawn in a truly lovely spectacle.

Allen was also a keen horticulturist and the author of a well-known work on viticulture, *A Practical Treatise on the Culture and Treatment of the Grape Vine*. The second edition of this work was published in Boston in 1848 by Dutton & Wentworth, the same printing-house that was to produce, six years later, his monograph on the water-lily.

Victoria regia was dedicated to Caleb Cope, then president of the Pennsylvania Horticultural Society, 'to whom we are indebted for the introduction to this country of many rare and beautiful plants'. After briefly describing the native habitat of this exotic flower, and recounting the story of its discovery at the beginning of the century, Allen explains how the plant should be cultivated. It was, in fact, first brought to bloom in America on 21 August 1851 at Springbrook, Cope's country seat near Philadelphia. This botanical event was described by Cope in a letter to Andrew Jackson Downing: 'I am sorry you were not here to witness the excitement which prevailed when the Victoria bloomed for the first time in this country, and when my grounds seemed to be in complete possession of the public. . . . The natural conditions of the plant in our country are, undoubtedly, more favorable than they can possibly be in England. . . . The Victoria is one of the few things that has not been exaggerated. It is truly a wonderful plant' (Allen, p. 10). And indeed he could boast

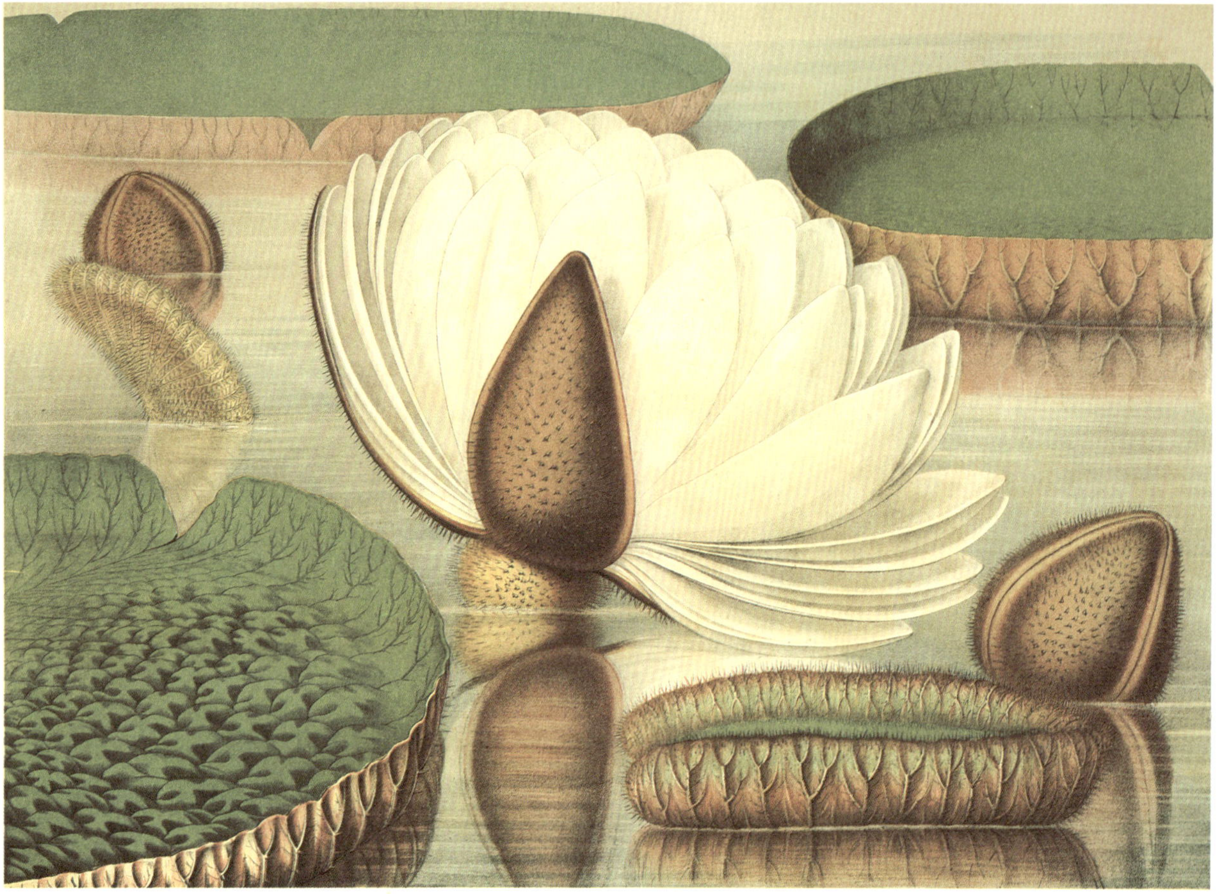

JOHN FISK ALLEN, *Victoria Regia*. Amazon water lily (*Victoria amazonica*), showing an intermediate stage of the bloom

that his flower was a full three inches larger, with leaves six inches wider, than any that English horticulturists had yet managed to obtain.

In his garden at Salem, Massachusetts, Allen succeeded in bringing forth, from a seed of Cope's flower, a plant whose growth he followed lovingly from January to July of 1853. It finally flowered on 21 July, in the impressive spectacle he describes in *Victoria regia*: beginning at four o'clock in the afternoon 'the petals, a pure white, began to unfold, and from 5 to 6 they rapidly opened, showing the flower in its first form. It remained in this condition until well after six, the next morning.'

Allen also provides a text to accompany the plates, six magnificent chromolithographs prepared by the English artist William Sharp (*c.* 1802–after 1862), a pioneer in the technique of colour lithography. Sharp had emigrated to Boston in 1838, taking with him his knowledge of this new technique. One of the most important commissions he received in America was for 110 colour plates, produced between 1847 and 1856, to illustrate Charles M. Hovey's *The Fruits of America*. Indeed, Sharp was still engaged on this project when he was called on to illustrate Allen's monograph.

The first plate in *Victoria regia* is based on a drawing by Allen himself, and depicts the young plant, a quarter life-size. The other five lithographs were prepared by Sharp after his own drawings and show, respectively, the opening flower, the underside of its unusually shaped leaf, the flower in an intermediate stage, a completely opened bloom, and a view showing the full form of the flower. Thus, in this series of plates the gigantic flower seems to open cinematographically before our very eyes. Mirrored in the water, we can admire its curious leaves, vast pale-green discs with turned up edges, tinted a purple crimson underneath, the coriaceous bud bristling with little spines, and the flower itself formed of numerous petals in three distinct sets, growing progressively smaller towards the stamens, and varying in colour from the pure white of the outer petals to the intense pink of the closely gathered smaller petals at the heart of the bloom.

107. Henry John Elwes (1846–1922)

[*title within floriated border, signed 'W.H.F. del' and 'W.G.S. sc'*] A Monograph of the Genus Lilium by Henry John Elwes, F.L.S. F.Z.S. Illustrated by W. H. Fitch. F.L.S. 1880.

56.5 × 40 cm. $\pi 1$ a^2 b–e^2 49 leaves of descriptive letterpress accompanying the lithographic plates. *i–vi i* ii–xv *xvi* 1–98 p.

BINDING: Three-quarter green morocco and cloth over boards.

PLATES: 1 map, 1 photograph, and 48 hand-coloured, lithographed plates of plants labelled with their Latin names.

REFERENCES: Nissen 594; Sitwell and Blunt, p. 56; Stafleu and Cowan, pp. 744–5; *Flora Portrayed*, pp. 77–81; *A Magnificent Collection*, no. 118.

ELWES's *Monograph of the Genus Lilium* was published in seven parts between 1877 and 1880 by the printers Taylor & Francis of London; it was so successful that nine supplements were produced, the last in 1962. The author was a man of versatile and eclectic interests—gardener, botanist, arboriculturist, ornithologist and lepidopterist. He was also an intelligent and tireless traveller, and a member of the Royal Society and the Linnean Society.

Born in Colesborne, Gloucestershire, and educated at Eton, Elwes first embarked upon a career in the army, serving for four years in the Scots Guards. Realizing, however, that his true bent was for exploration and study, he resigned his commission to become a naturalist. He immediately began travelling, first to Scandinavia, then the Near East, and then ever farther afield. During an expedition to the Himalayas he became interested in the genus *Lilium*, and took back a number of species to England for cultivation and study.

A Monograph of the Genus Lilium represents the fruit of these studies; it contains a long introduction and a scientific letterpress written by Elwes to accompany the book's forty-eight hand-coloured lithographed plates. In the introduction the author explains that he was not a professional botanist but a simple horticulturist who wished to present the various new forms of lilies he had discovered in the Himalayas and on the Korean peninsula. He provides a brief historical excursus,

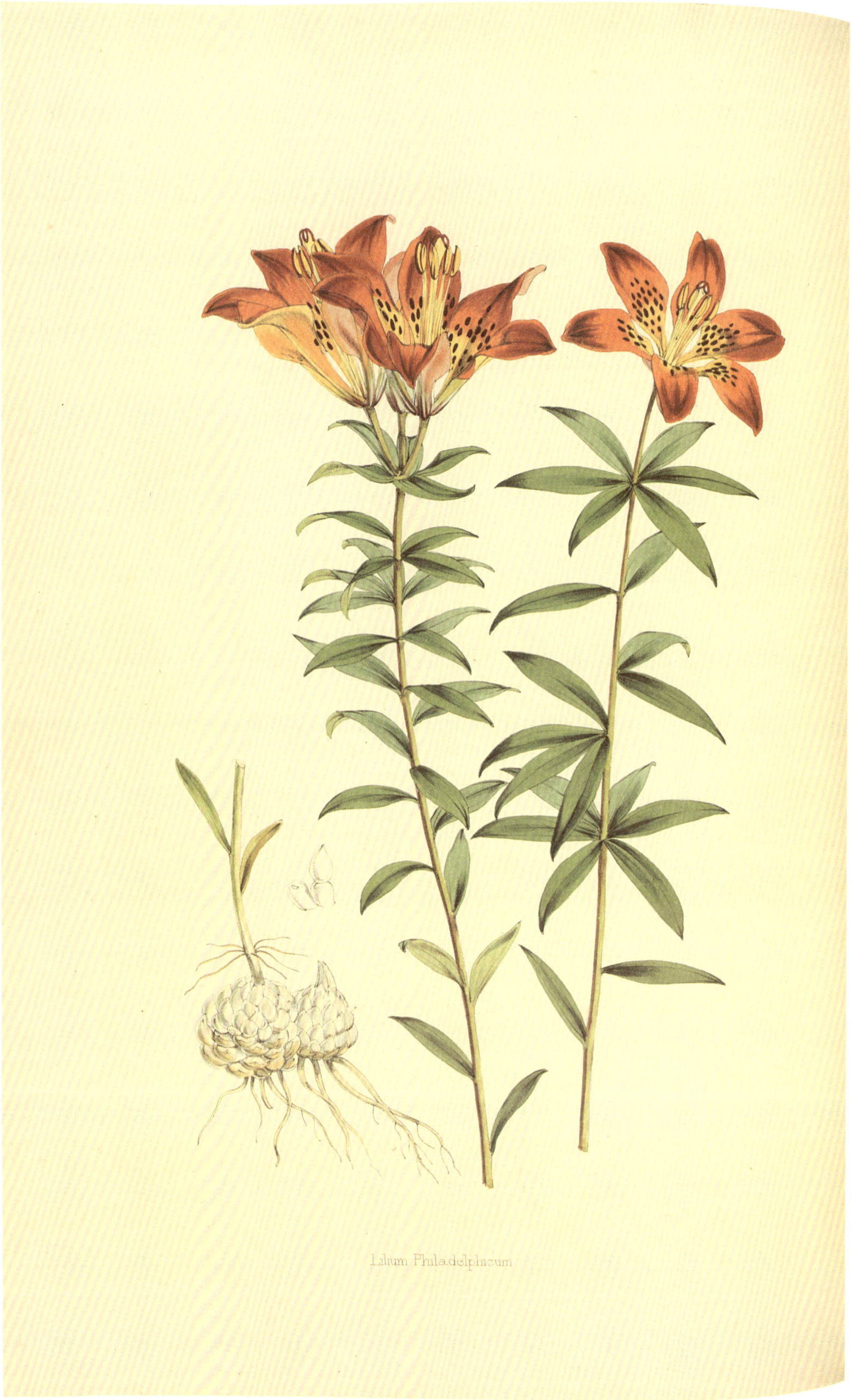

HENRY JOHN ELWES,
A Monograph of the Genus Lilium. Wood lily (*Lilium philadelphicum*)

citing such works as John Parkinson's *Paradisi in sole* of 1629 (see no. 40), which mentions five varieties of martagon lily, and Dieudonné Spae's 'Mémoire sur les espèces du genre Lis', published in 1847 in the *Mémoires de l'Académie Royale de Belgique*, which, Elwes explains, was probably the very first monograph written on the subject of lilies. The most recent work mentioned by Elwes is *Monographie historique et littéraire des Lis* by Frédéric Cannart d'Hamale, published in Malines in 1870, in which not only the botany of the flower, but also its use as an image in literature, is discussed.

Elwes provides a learned disquisition on the classification, the geographical distribution, and the cultivation of the lily, paying particular attention to the collecting and storing of bulbs. He also ascribes great importance to the illustrations for his work, and suggests that flora always be drawn from life. Elwes knew that using dried specimens as models, as was all too often the custom, could lead to errors in the portrayal of a plant and he therefore sought, as far as possible, to provide his artist with live specimens. This artist was Walter Hood Fitch (1817–92), perhaps the most prolific botanical illustrator of the nineteenth century. Fitch executed the plates for Joseph Dalton Hooker's *Rhododendrons of Sikkim-Himalaya* (see no. 104) and completed thousands of illustrations for *Curtis's Botanical Magazine*. Fitch worked almost exclusively in the medium of lithography, and the forty-eight plates in Elwes's *Monograph*, deftly drawn directly on the stone, confirm him to be an unsurpassed master of the technique.

The infinite variability of the lily from specimen to specimen made its portrayal difficult, but both Elwes and Fitch were determined that their illustrations should be scientifically accurate as well as aesthetically pleasing. The artist therefore made a careful study of live specimens before arranging his flowers in harmonious compositions. The flower itself was often accompanied by enlarged details of the bulbs and leaves (generally left uncoloured). Among the most remarkable plates are the 'Turk's Cap lily' (*Lilium superbum*), the 'Lance-leaved lily' (*Lilium speciosum*) with its long stem crowned by magnificent pink flowers, and the 'Philadelphia lily' (*Lilium philadelphicum*), a bright orange flower with brown papillae that could be found growing wild in many parts of the United States, including Nantucket Island, and as far north as Canada. Elwes explains that the plant was sent by John Bartram (1699–1777), the pioneering American botanist, to 'that prince of 18th-century gardeners, Philip Miller, then Curator of the Chelsea Botanic Gardens, in whose "Illustrations" a figure of it will be found'.

At Oak Spring can also be found all the subsequent supplements to the original monograph: *A Supplement to Elwes' Monograph of the Genus Lilium* by Arthur Grove and Arthur Disbrowe Cotton, published in seven parts between 1933 and 1940 with plates by Lilian Snelling, illustrating many new species discovered in China and Tibet; and parts eight and nine, published between 1960 and 1962 by William Bertram Turrill with plates by Margaret Stones (see no. 110).

XII · FLOWER PAINTERS OF OUR TIME

IN 1979 Paul Hulton and Lawrence Smith organized an exhibition, 'Flowers in Art from East and West', held at the British Museum in London, which dealt with themes already touched on in another show, 'A World of Flowers', organized by Henry Clifford for the Museum of Art in Philadelphia in 1963. The London exhibition marked a genuine turning-point, for it completely revolutionized our way of looking at, analysing and appreciating the botanical illustration as art. Hulton and Smith included three contemporary artists in their show—Lilian Snelling (1879–1972), Stella Ross-Craig (b. 1906), and Margaret Stones (no. 110)—whose works were displayed next to those of the greatest botanical illustrators of the past, including Le Moyne de Morgues, Merian, Ehret and Redouté, and masters of flower painting from India, Japan and China. The works were chosen to demonstrate the organizers' thesis that an unbroken continuity existed between the past and the present, between the great tradition of flower painting and contemporary botanical illustration, and that great art could still emerge from a symbiosis with modern science.

Two fundamental questions must be raised at this point, however. Is there really any need for the art of botanical illustration in an age where not only photography and cinematography, but also such revolutionary techniques as virtual reality can be used to capture—or even re-create—nature itself? And second, just where does the dividing line between illustration and art lie? The answer to the first question is a clear and unequivocal 'yes', at least as far as photography is concerned. The most sophisticated camera, even if used with great artistry, cannot match the ability of the human eye to receive and select information, including details that the camera and other instruments are entirely blind to. Indeed, many photographs may be required to convey the same amount of information that can be contained in a single drawing or painting. However, like the microscope or the herbal of dried specimens, the camera may often serve as an invaluable tool for the artist himself, a means of obtaining 'sketches' or visual memoranda that he can refer to while at work painting.

Naturally, certain techniques that once were fundamental to botanical illustration are no longer in use. For example, when lithography was invented by Alois Senefelder in 1798 it brought about a veritable revolution in the art of scientific illustration, and almost forty years later, when the chromolithograph was introduced, the technique resolved one problem that had plagued publishers since the invention of the printing press—how to run off colour images in large numbers. The chromolithograph represented only a partial solution, however, for it still began with an image drawn directly by the artist, which was then prepared for printing following a cumbersome procedure that had changed little since the early sixteenth century, when Hans Weiditz made the preparatory watercolours for the very first printed botanical illustrations—the woodcuts in Otto Brunfels's *Herbarum vivae eicones*. It was the rare artist who succeeded, like Walter Hood Fitch, when drawing directly on the lithographer's stone. With today's technology, however, it is possible to print accurate colour plates with great rapidity in an almost unlimited number of copies.

As in the past, most botanical illustrators today can be found working directly, and almost exclusively, for scientists, under the auspices of research institutions or botanical gardens, with their art destined mainly for publication in specialized journals. This represents the continuation of a long tradition, for botanical artists have been essential to botanical journals ever since William Curtis founded the *Botanical Magazine* in February 1787. Its first issue appeared with a number of fine engraved illustrations; chromolithography was introduced in 1845, but even up until 1949 images were sometimes painstakingly retouched by hand (one can imagine at what cost!). Curtis's *Botanical Magazine*, which eventually became *Curtis's Botanical Magazine*, during its long history availed itself of the talent of some of the greatest botanical artists in England including, in our own century, Lilian Snelling, Margaret Stones (no. 110) and Mary Grierson (b. 1912). This distinguished journal ceased publication in October 1983, having issued 184 volumes with a total of 10,570 plates. Its substitute is *The Kew Magazine*, the first issue of which appeared in April 1984 with a quite different format that includes a drastically reduced number of plates. Most of its illustrations are the work of talented artists on the permanent staff at Kew, including Stones, Pandora Sellars (b. 1936), Christabel King (b. 1950) and Mark Fothergill (b. 1959). From this list, and from the names to be found in the catalogues of national and international exhibitions, one can confirm the continuing, indeed growing, prominence of women artists in this genre.

The answer to the second question is more complex, in part because for much of this century realism in the figurative arts has been held in low esteem and the botanical painting has been dismissed as 'merely' illustration with little or no artistic content. It is a grave mistake to perpetuate this misconception, for in the history of art there always have been and always will be 'illustrators' and 'artists', and they can be told apart. Sometimes, of course, the naturalistic illustrator and the naturalistic artist may coexist and flourish within a single personality. Therefore, although collaborating with scientists may at first appear to limit an artist's freedom, it can in the end provide the stimulus for the creation of genuine works of art if the artist succeeds in finding a personal style within the formal limits imposed by the genre.

History has demonstrated many, many times that if an artist who is gifted with an innate sensibility to nature makes a study of past masters, and if he manages to add to this a thorough knowledge of botany, he may very well find—entirely on his own—that delicate balance between free expression and fidelity to nature that is the mark of the great botanical artist. One artist who succeeded in doing so was Rory McEwen (no. 108). Not constrained to make a living by his art,

McEwen was able to dedicate himself to innovative experimentation, and he developed a style that was at once realistic and deeply evocative, as may be seen by the works he presented at the London exhibition 'True Facts of Nature' in 1974.

The work of Margaret Mee is also distinguished by an unmistakable personal style: her paintings of flowering plants executed *in situ* in the Amazon forest (no. 109) are at once emphatic and grandiloquent. Margaret Stones's paintings, on the other hand, at first glance reflect the scrupulously analytical approach of the illustrator rather than the creative intuition of the artist, perhaps due to her experience as an illustrator at Kew. Nevertheless, their apparent simplicity belies a highly sophisticated artistic vision, and she has created many works of great purity and surprising visual impact (no. 110).

Very few recent artists outside of the genre of scientific illustration, even landscape painters, seem to have dedicated any of their time to the illustration of flowers. John Nash (1893–1977) represents one of the rare exceptions, and some striking woodcuts of poisonous plants by him are in the collection at Oak Spring. Another artist who has frequently utilized floral motifs in her work is Sophie Grandval-Justice (no. 111); through her fantastical re-elaborations of the patterns and colours to be found in the natural world she has created works of surprising poetry, apparently naïve but at the same time hypnotic and fascinating.

Oak Spring also has works by the botanical artist Mary Anderson Grierson and paintings on botanical subjects by Tony Forster (b. 1946), Mossie Fuller, Luis Marsans, Eliza Lloyd Moore (b. 1943), Mary Tift and Almina Dovati Fusi (1908–92).

ELIZA LLOYD MOORE, a box top with flower drawings in pencil and watercolour

108. Rory McEwen (1932–1982)

Group of Allwood's carnations. 71 × 51 cm. Watercolour on vellum. Inscribed in pencil on the verso: 'Started Nov. 14th 1961 / Finished March 6th 1962'. Signed lower right.

REFERENCES: *Rory McEwen*, no. 9; *An Oak Spring Garland*, no. 32.

RORY MC EWEN, descended from an aristocratic Scottish family, began to paint at the age of eight under the tutelage of his French governess, as we learn from 'An Autobiographical Fragment' published in the exhibition catalogue of 1988, *Rory McEwen* (n.p.). Indeed, an interest in the botanical sciences could be said to run in the family, for McEwen's mother was a direct descendant of the famous English botanist John Lindley. As a boy McEwen particularly admired the French artist Paul Cezanne, whose influence can be seen in many of McEwen's mature works, such as the landscape background to his painting of an *Allium ostrowskianum* (*Rory McEwen*, no. 59).

Between 1945 and 1950 McEwen had the good fortune to study under the artist and art historian Wilfrid Blunt at Eton College's drawing school. In this very period Blunt was at work on his magisterial *The Art of Botanical Illustration* (first published in 1950). Recognizing McEwen's unique talent—Blunt later pronounced him to be 'perhaps the most gifted artist to pass through my hands'—McEwen was encouraged to study the great botanical and floral painters of the past, including Albrecht Dürer, Jacques Le Moyne de Morgues and Nicolas Robert. McEwen also admired the masterpieces of Indian miniature painting to be found at the British Museum in London. The paintings of old florists' flowers that he produced in this period reflect his innate sensibility to the beauties of the natural world, as well as a striving for scientific accuracy generally to be seen only in more mature artists.

McEwen was furthermore able to study at close hand the work of Pierre Joseph Redouté (see no. 61), for eighty of the original watercolours for Redouté's most famous work, *Roses*, came into the possession of McEwen's sister on her marriage. Having a marked artistic identity of his own, McEwen was not unduly influenced by the exquisite style of the French master, although he did write a thesis on Redouté while he was at the University of Cambridge. There he also became a close friend of the aesthete Sacheverell Sitwell, author of *Old Fashioned Flowers* (1939).

It is interesting to study the technique underlying McEwen's remarkably realistic paintings. He generally began by tracing the image of the plant or flower on vellum with fine, parallel brushstrokes of dry watercolour. He then photographed his subject from various angles, and drew sketches of significant details from life, before finally completing the painting in his studio. As Martyn Rix had observed, in each of his works 'the feeling of texture was obtained by detailed modelling and by almost over-emphasising any highlights or shadow cast by veins or other uneven structures on the surface of petal or leaf' ('Rory McEwen and the Tradition of Botanical Art', in

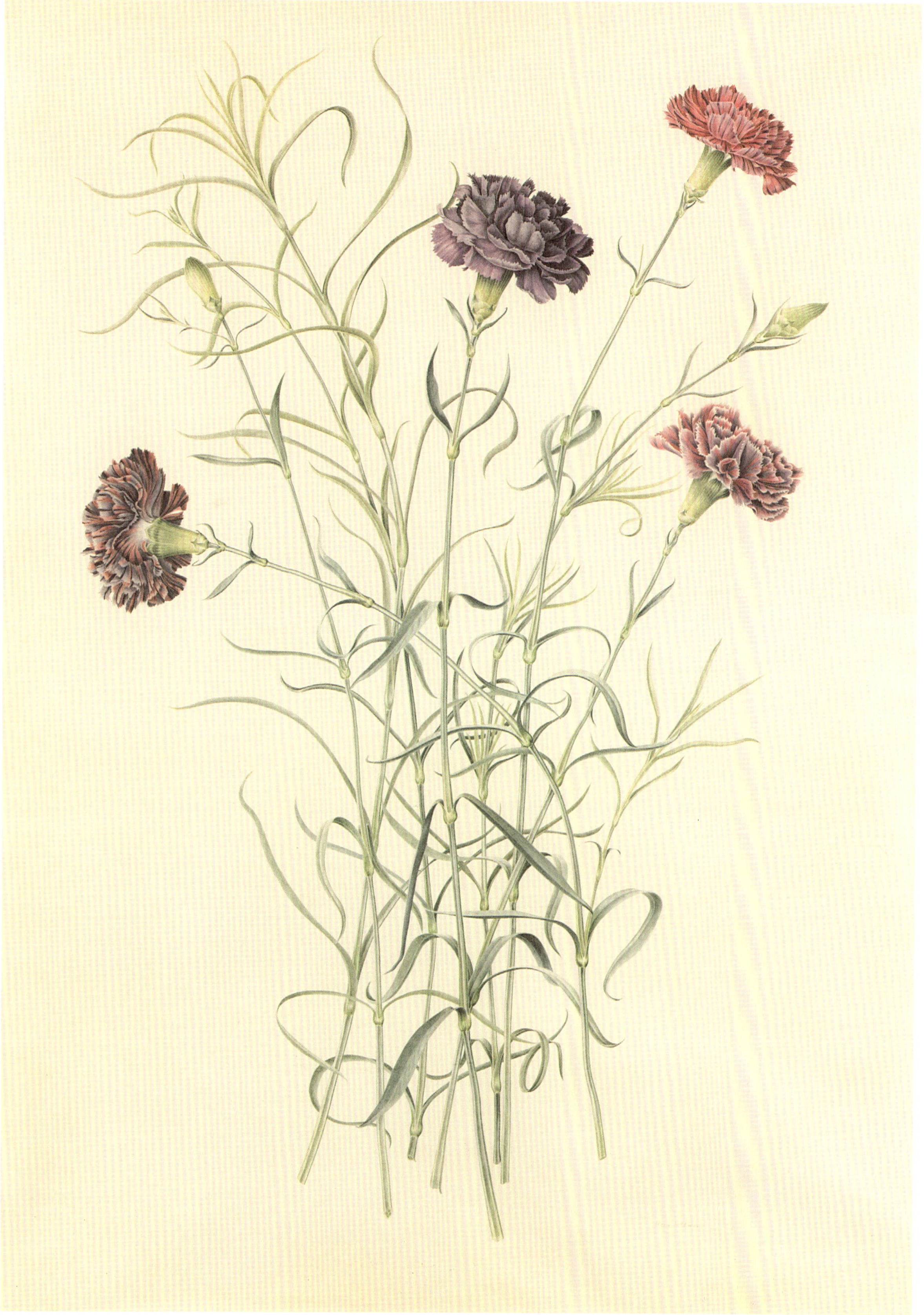

RORY MC EWEN,
Group of Allwood's
carnations

Rory McEwen, n.p.). McEwen's paintings attracted great attention at the exhibition 'True Facts of Nature' organized by the Redfern Gallery, London, in 1974, and he subsequently began to experiment with liquid watercolour on paper, a much more challenging medium for the botanical illustrator.

Although best known for his large floral paintings, in his later years McEwen also explored other themes, producing for example a haunting series of smaller paintings, each focusing on a single weathered leaf, curled and dried by the sun or damaged by insects. The artist's hyper-realistic approach served most effectively to heighten the impact of these almost surreal images of decay. He also completed many paintings for a monograph on the fritillary, which he began in collaboration with Rix, although this project was never completed.

McEwen's large painting at Oak Spring, which depicts several of Allwood's carnations in colours ranging from pale violet to deep pink, testifies to the extraordinary technique of this artist, whose death at the relatively early age of fifty has perhaps robbed him—at least for the present—of the recognition he certainly deserves as one of the most gifted botanical painters of the twentieth century. The flowers' long stems have been woven into an intricate background pattern that draws the eye irresistibly into its mysterious depths, before one's glance finally comes to rest on the exquisite blossoms with their delicate, saw-toothed petals.

At Oak Spring there is also a watercolour (dated 1962) by McEwen of two 'historic' tulips, as well as a copy of the fully revised 1977 edition of Wilfrid Blunt's *Tulips and Tulipomania*, with six lithographed plates (59 × 51 cm.) signed and dated (1976) by the artist. These plates depict tulips from the Wakefield Tulip Society (Yorkshire), and here—in an interesting departure from the traditional iconography of the Dutch *Tulpenboeken*—McEwen not only shows them with relatively short stems and therefore denuded of their lanceolate leaves, but also provides somewhat more unusual views, showing, for example, the inside of the corolla.

109. Margaret Ursula Mee (1909–1988)

Begonia egregia.

66 × 48.5 cm. Bodycolour on paper, inscribed in the lower centre of the page: 'Begonia egregia N. E. Br. Instituto de Botanica, San Paulo. flowered August 1957 Margaret Mee.'

BINDING: In a three-quarter white vellum and decorative paper dropback case with three other Mee drawings of Epidendrum vesicatum, Tillandsia pohliana and Tibouchina.

REFERENCE: M. Mee, *In Search of Flowers of the Amazon Forests*, Woodbridge, 1988.

MARGARET MEE (née BROWN) was a talented botanical painter and intrepid traveller whose life recalls that of the seventeenth-century German artist Maria Sibylla Merian: like Merian, Mee was irresistibly drawn to the natural beauties of the tropics. Born in Chesham near London, in

1952 at the age of forty-three she moved with her husband, Greville Mee, to Brazil, where she taught art at the British School in São Paolo. More significantly, she organized no less than fifteen expeditions into the Amazon jungle, the last of which took place just a few months before her death. She dedicated herself to documenting the flora of this tropical paradise, and it is a sobering thought that many of the flowers she portrayed are now extinct. Her autobiography, *In Search of Flowers of the Amazon Forests*, published early in the year 1988, is both a fascinating adventure story and a useful work on the natural history of the region; it is full of detailed ethnological observations and serves as an invaluable record of her work as an artist.

Mee was an exceptional personality. She was adventurous and resourceful by nature, and gifted with an inquiring mind that would not remain satisfied with information acquired at second hand, whether in her scientific studies or in the investigation of social and human conditions. Thus, back in 1932 she had travelled to Germany to make a close study of the political situation under Hitler. She made no secret of her democratic sympathies and during her extended sojourns openly frequented with persons of Jewish origin. Inevitably she fell under suspicion and, narrowly escaping arrest by Hitler's police, returned to England in the mid-1930s. At the end of the Second World War she decided to concentrate on developing her artistic talent and studied at various art institutes in London, where she came under the influence of the painter Victor Pasmore.

In 1957, five years after Mee and her husband had settled in Brazil, she set off on her first expedition—to explore the region of the Gurupi, a tributary of the Amazon river. 'But Margaret Mee was well prepared for her first expedition by a long background of challenges', notes the editor in his Introduction to *In Search of Flowers* (p. 27): 'Then forty-seven, she packed her artist's kit into a canvas rucksack and padded it with spare clothes. She also took a revolver.' The incredible landscapes and astonishing plants and wildlife that Mee found in the Amazon gripped her imagination and would henceforth never let go: 'On the first night', she recalls, 'I could not sleep for listening to the magic sound of the forest. Only the trees were silent, while the lake was alive with sparkling, splashing fishes, and the frogs' chorus mingled with the plaintive cries of night birds' (p. 34). She immediately began to make vivid landscape sketches, as well as careful botanical studies. When she returned to São Paulo, Mee showed her work to the botanists at the city's Botanic Institute, who were quite impressed and organized an exhibition of her paintings. This was followed by a larger exhibition in Rio de Janeiro, where she met Roberto Burle Marx, a garden designer of international repute. They became close friends, and Burle Marx was able to assist Mee considerably at various points in her career.

Mee also decided to undertake a systematic study of the botany of tropical plants and therefore began working at the São Paulo Botanic Institute, where she met many eminent botanists, among them Lyman Smith of Washington's Smithsonian Institute. Smith was then engaged in a study of the bromeliads, a family of epiphytic herbaceous plants of which many unusual species could be found in the Amazon forest.

Mee's second expedition, in 1962, took her to the almost impenetrable forests of the Mato Grosso, where the legendary Colonel Fawcett had disappeared without trace in 1925. During this trip she contracted a virulent form of malaria, but the inconvenience of the illness paled next to her excitement at being able to paint a stupendous variety of tropical plants *in situ*. She also gathered and brought back to São Paolo many interesting botanical specimens, among which two entirely new species were identified.

In her tireless search for the native habitats of rare plants, other expeditions followed in rapid succession—one in 1964, two in 1967, and then others in 1970, 1971, 1972, 1974, 1975, 1977, 1982 and 1984. While exploring the Rio Negro in 1967, she recorded with some frustration: 'I have found the petals of a qualea, a glorious gentian blue, but cannot locate the tree from which they have fallen. The forest canopy and lianas form a luxuriant mass of vegetation, with every tree and creeper struggling to reach the sun' (p. 136). In 1972, while voyaging along the Mamori river, she discovered 'On one of the enormous trees on the other side of the river . . . a mass of dark leaves hung in a heavy wreath around the trunk. Severino turned the boat in its direction, and on reaching the bank I could see the pendant white flowers of a beautiful white orchid (*Stanhopea grandiflora*) which I had not seen before' (p. 188).

In May 1988, at the age of seventy-nine, Mee was able to realize one of her most cherished dreams, that of painting from life a strophocactus or 'moonflower' (*Selenicereus wittii*), whose fabulous bloom remained open for only one night (Stearn 1990, p. 110). She first saw a specimen of the plant in 1964, and more than twenty years later finally came across a plant complete with buds along the banks of the Rio Negro. By the light of a battery powered torch she managed to watch the flower, and to sketch it as it opened: 'In the early stages an extraordinary perfume wafted from the flower, and we were all transfixed by the beauty of the delicate and unexpectedly large bloom, fully open in an hour' (p. 292). When she returned to São Paolo, Mee transformed her sketches into a large painting depicting the *Selenicereus wittii* in its native habitat. The plant is shown clinging to the trunk of an enormous tree and is laden with five blossoms, each in a different stage of flowering.

At least three other great artists in the history of botanical painting had succeeded in painting this flower, although from cultivated plants and not *in situ*. Ehret executed two plates of the *Selenicereus grandiflorus* for Trew's *Plantae selectae*, while legend has it that Redouté depicted a 'Queen of the Night' for Marie-Antoinette just before her execution. Finally, the 'Night Blowing Cereus' formed one of the most sensational plates in Robert Thornton's unfinished *Temple of Flora* (see no. 93).

Mee herself produced two folio volumes—*Flowers of the Brazilian Forests* (1968) and *Flowers of the Amazon* (1980). The first includes an affectionate preface written by Burle Marx.

At Oak Spring, which Mee visited in 1967, there are four large bodycolour paintings by her. The painting of the *Begonia egregia* was completed just after she had returned from her first expedition, and represents an excellent example of her early work. As Wilfrid Blunt affirms, in terms of

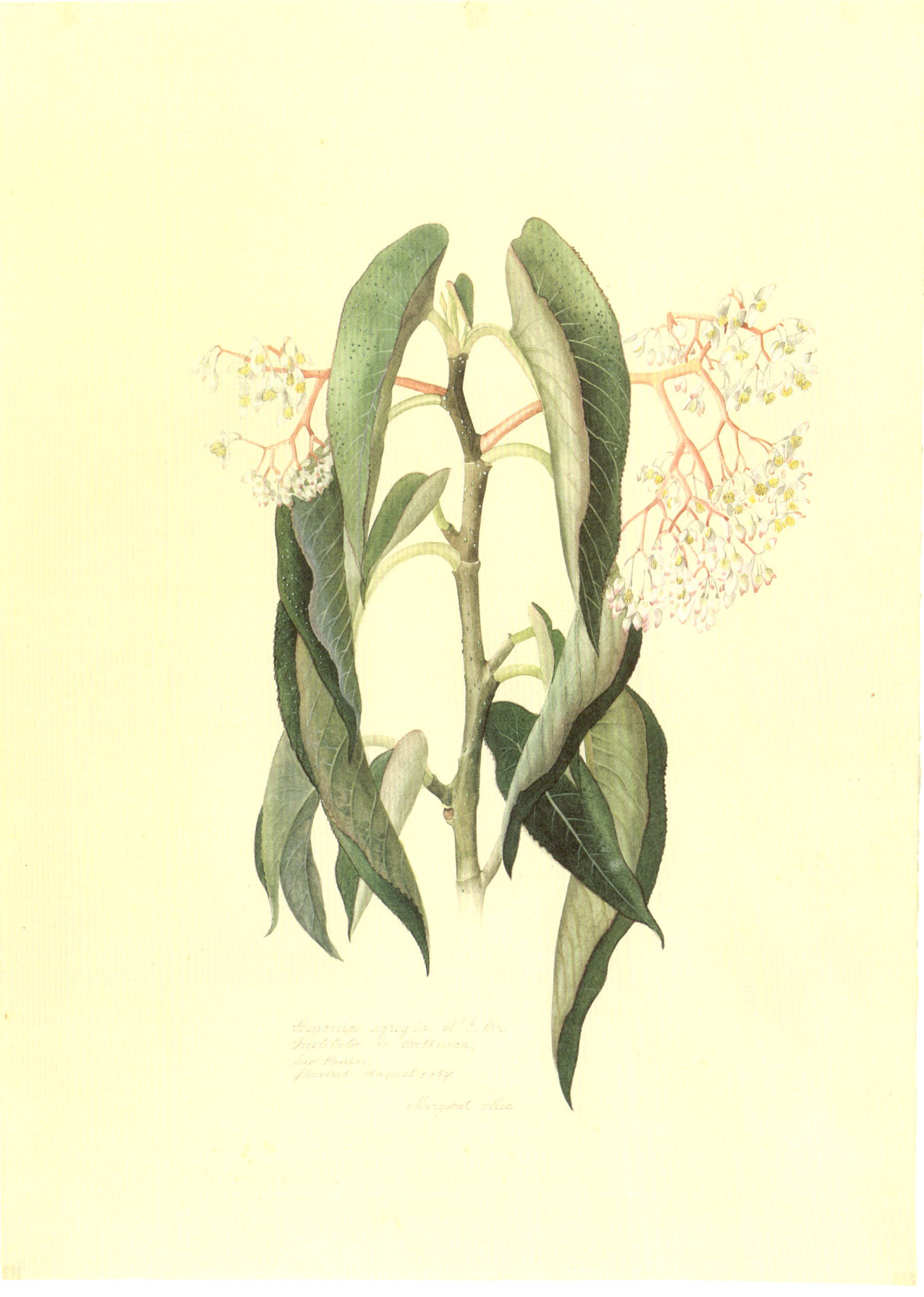

MARGARET URSULA MEE, *Begonia egregia*

artistic skill and scientific accuracy this British artist of our own times can bear comparison with the very greatest artists of the genre, including Ehret and Redouté.

In an ironic twist of fate, Mee, who for decades had courageously faced untold dangers in the trackless forests of the Amazon, died in a road accident in England just six months after returning from her latest expedition to celebrate the publication of her autobiography.

110. Margaret Stones (b. 1921)

Cyclamen cilicium var. intaminatum.

Cyclamen cilicium var. cilicium.

30.5 × 25.2 cm. Two paintings in watercolour over pencil on paper. Signed at the foot of the plant. Titles inscribed lower left and lower right. Also inscribed lower left: 'M. Koenen 197.79 01915, RBG Kew / Sept 12th 1979. / Very slightly flushed with pink. Pinkish-mauve veining on petals'; and lower right: '197–79–01908 / s.w. Turkey (Cilicia).'

Dampiera spicigera

38 × 29 cm. Watercolour over pencil on paper. Title inscribed lower left. Inscribed lower right: 'Coll: Coorow Sand plain. Oct. 26th 1983 Norman Brittan'. Signature running along the lower stem of the plant. The plant sections are identified in pencil at the lower left. Inscribed in the lower right corner: 'D.'

These two watercolours are the work of the Australian painter Margaret Stones, who is universally recognized as one of the greatest botanical artists of the twentieth century. She was born in Victoria, and although she studied art at Swinburne Technical College and at the National Gallery Art School in Melbourne, she first chose to become a nurse. After contracting a serious illness while working in Australia during the Second World War, however, Stones decided to abandon this strenuous profession and become a botanical artist. She embarked on an intensive course of study in botany and drawing, and in 1951 left her native land for England, where she completed her training at the Royal Botanic Gardens, Kew. In England she was also finally able to study original works by some of the greatest botanical illustrators of the past, Ehret among them.

Since first achieving recognition in the late 1950s, Stones has produced a continuous stream of works, including more than 400 drawings for *Curtis's Botanical Magazine*, a two-volume *Supplement to Elwes' Monograph of the Genus Lilium* (1960, 1962; see no. 107), a monumental six-volume work entitled *The Endemic Flora of Tasmania* (1967–78) and, more recently, 220 paintings for *Flora of Louisiana* (1991). She presently lives and works in Kew.

When the exhibition 'Flowers in Art from East and West' was organized by Paul Hulton and Lawrence Smith for the British Library in 1979, Stones was one of the three living artists they invited to participate. This show rekindled the public's appreciation of a somewhat neglected genre, and stimulated scholars and critics to study old masterpieces from a fresh perspective. Included in the exhibition was an extraordinary painting by Stones of a sunflower, with a series of enlarged details

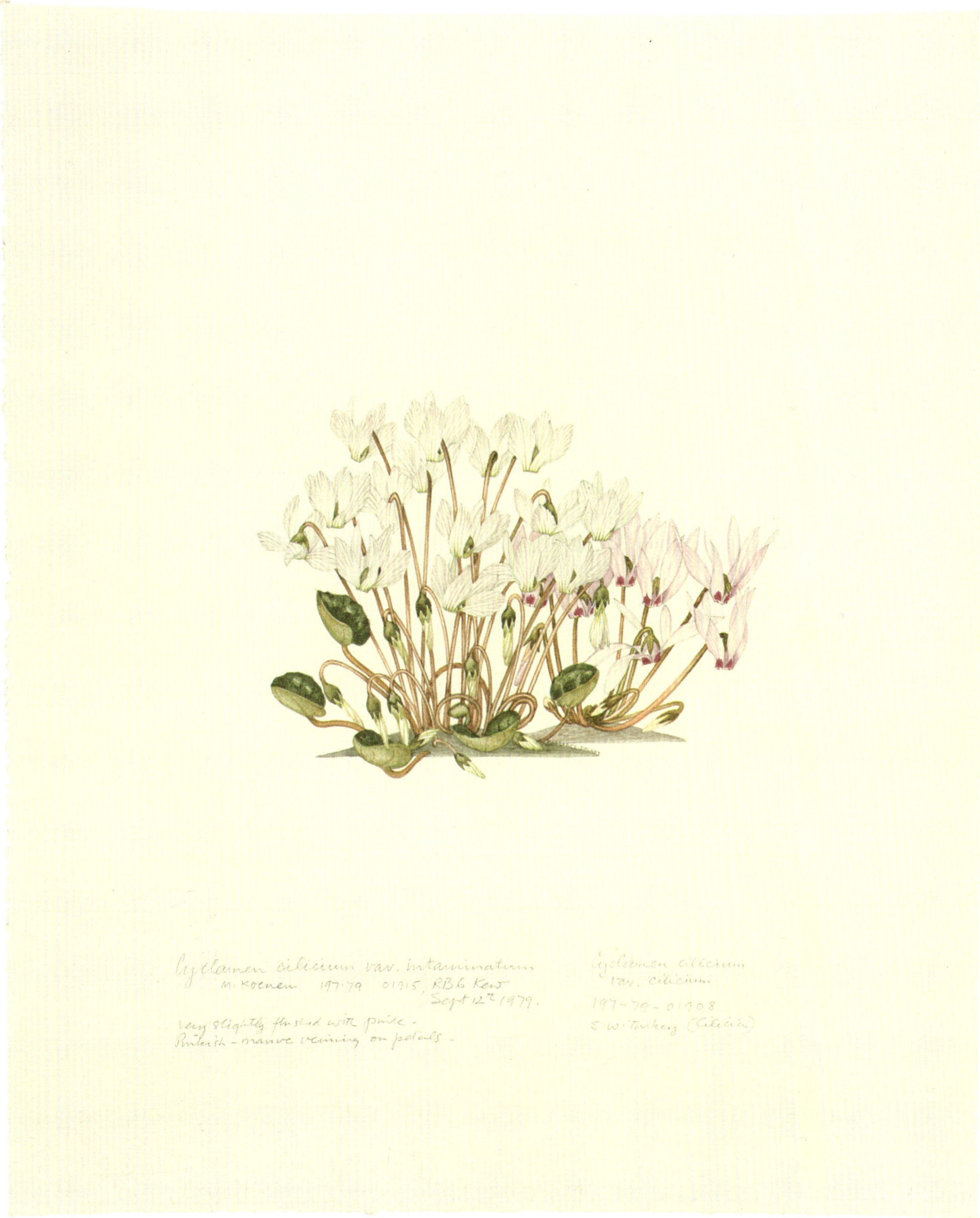

MARGARET STONES,
Two varieties of *Cyclamen cilicium*

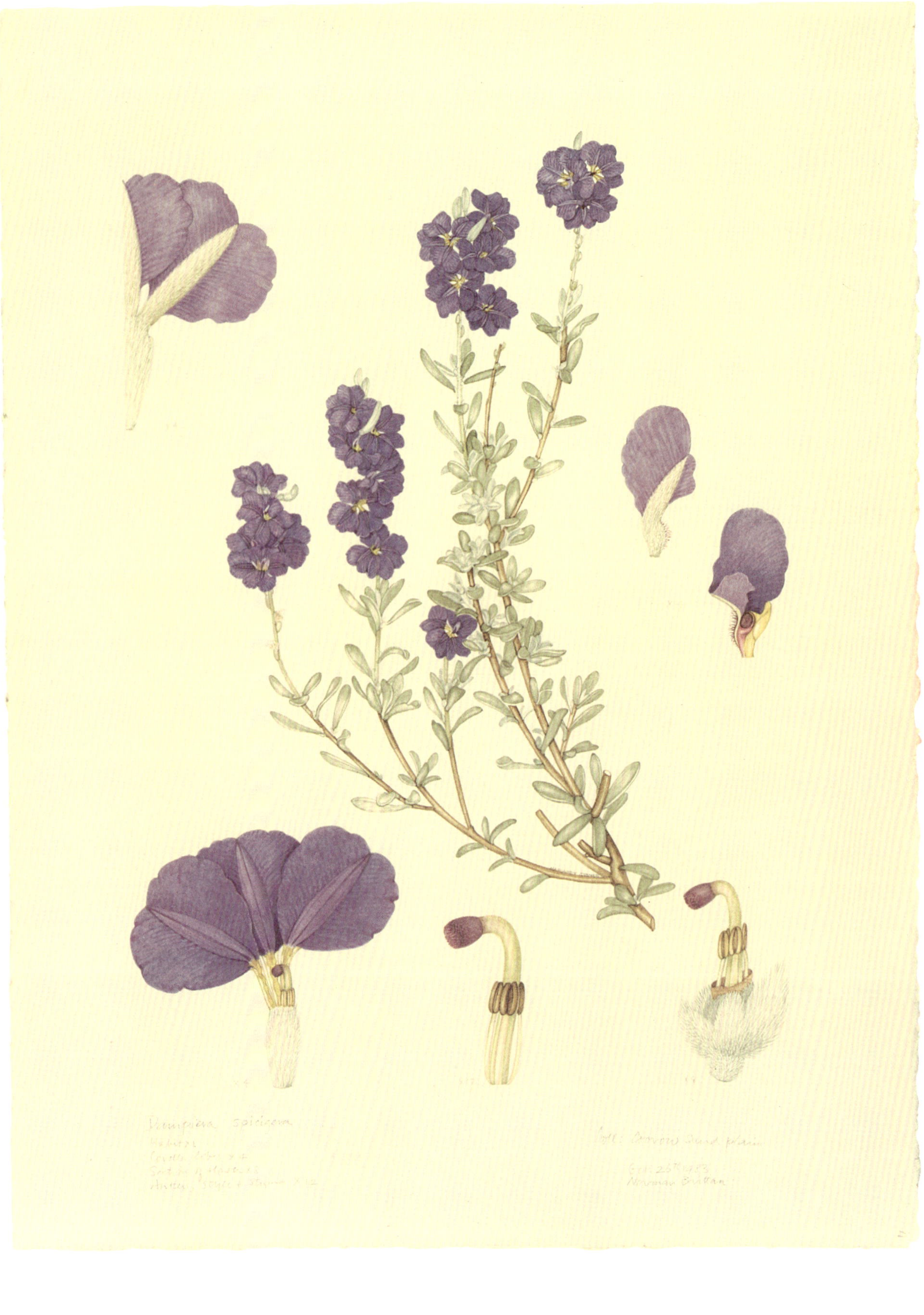

MARGARET STONES,
Dampiera spicigera

depicted in the lower portion of the canvas. According to Hulton, 'Her handling of the minutiae of botanical detail is indicative of unusual powers of observation and a highly effective technique. She has a strong sense of design and shows clearly in her work that she is aware of her place in a long European tradition' (Hulton and Smith, p. 145).

Stones's special gifts can be appreciated in these two paintings at Oak Spring. The first shows two cyclamens, one white and one pink, which spring from the earth in a rich profusion of stems, buds and leaves; the flowers with their delicately veined petals, some of them still half closed, seem to nod before our very eyes. In this painting Stones has created a tiny microcosm that is reminiscent, though in a modern key, of Dürer's famous work in water- and bodycolour on paper—*The Large Piece of Turf* (Albertina, Vienna). Her signature, half hidden in the soil, echoes a playful practice adopted by botanical illustrators in the sixteenth and seventeenth centuries.

In Stones's second painting, showing a *Dampiera spicigera*, Stones presents a highly realistic, and at the same time very personal, interpretation of her subject-matter. Here the *mise-en-page* gives way gracefully to the requirements of scientific illustration. The parts of the flower, such as the terminal sections of the pistils and anthers, are depicted in exquisite detail, while the velvety soft texture of the petals and their luminous violet colour have been reproduced with great skill.

Important works by Stones can be found at Kew's Royal Botanic Gardens, the Natural History Museum, the Royal Horticultural Society, the National Library in Canberra, Melbourne University, the Ashmolean Museum (Oxford), the Fitzwilliam Museum (Cambridge), and in many private collections. At Oak Spring there are twenty further watercolours by her.

111. Sophie Grandval-Justice (b. 1936)

Dandelion. 61 × 50 cm. Oil on canvas. Signed 'Sophie Grandval-Justice', and dated '1900' on the verso.

Sophie grandval-justice was born in Paris and studied engraving with the artist Johnny Friedlaender. In 1957 her striking designs on silk attracted the attention of Hubert de Givenchy, who introduced her to many of the great French fashion designers of the time. Soon she was working under commission for Balanciaga, Dior, Balmain and Dessès. In 1958 an exhibition of her designs for textile patterns was held at the Galerie des Beaux-Arts in Paris. One year later, encouraged by the collector and art dealer Lucien Durand, she decided to abandon textile designing and dedicate herself full-time to painting. Durand held an exhibition of her first oil paintings and watercolours in 1961 in Paris, and Delpire organized another in 1970. Grandval-Justice has become ex-

SOPHIE GRANDVAL-JUSTICE, Dandelion (*Taraxacum officinale*)

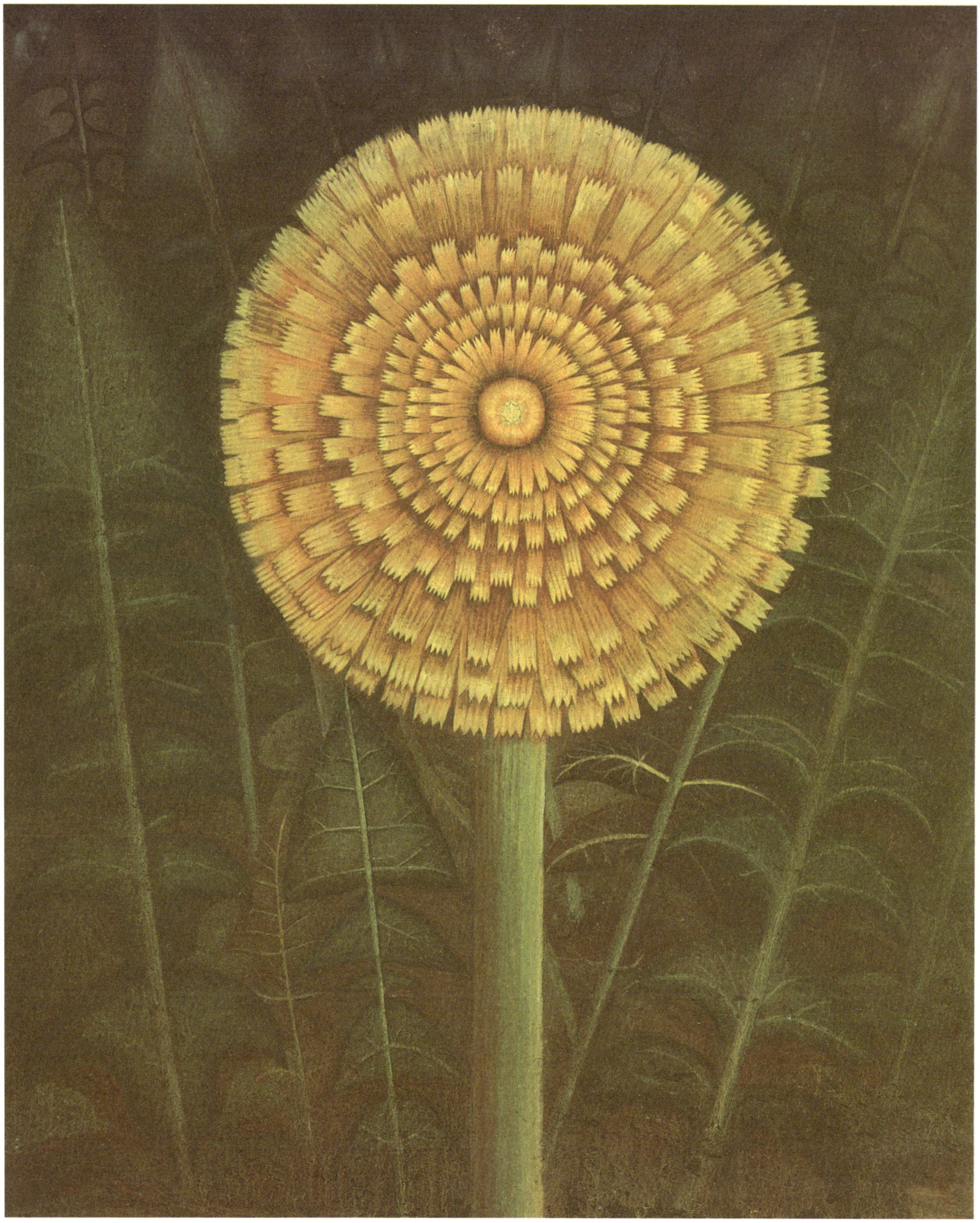

tremely well known, and many of her paintings are now in important private collections. She has travelled widely and worked both in London and Morocco; she presently lives in Auvergne, France.

The most intimate and hidden aspects of the natural world form the inspiration for Grandval-Justice's paintings. Details from Nature's vast tapestry, with its warp and weft of leaves spangled with bright flowers, fragile butterflies and vividly coloured birds, are translated by her fantastical imagination into works of striking originality, whose reiterated motifs reflect the tradition of textile decoration in which she began her training as an artist. This painting of the common dandelion (*Taraxacum officinale*), a subject that she had already painted back in 1966, offers a striking example of her unique mode of interpreting the natural world. The simple wild flower takes on an almost hierarchical grandeur, poised on its robust stem with its stylized bloom turned uncompromisingly towards the observer. The saw-toothed ligules radiate in concentric circles of orange and yellow from the central disc like brilliant rays of sunlight, while the dandelion's dark, runcinate leaves have been flattened into a highly decorative, but slightly disquieting, background pattern resembling a forest of spears.

At Oak Spring are two further oil paintings by Grandval-Justice, *A Dandelion, Buttercups, and other Plants, with Eight Butterflies and other Insects* (1983) and *Carrot Flowers* (1988), as well as forty-eight watercolours of plants and flowers undertaken between 1980 and 1984. They all reflect the highly sophisticated vision underlying her apparently naïve approach, reminiscent of early manuscript and printed herbals.

INDEX

INDEX

H

COMPOSED IN MONOTYPE BEMBO BY

MICHAEL AND WINIFRED BIXLER, SKANEATELES, NEW YORK

BLACK & WHITE AND COLOUR PHOTOGRAPHY BY

GREG HEINS, BOSTON, MASSACHUSETTS

PRINTED IN DUOTONE AND PROCESS COLOUR BY

THE HULL PRINTING COMPANY, MERIDEN, CONNECTICUT

BOUND IN SCHOLCO ALMOLINE CLOTH

BY ACME BOOKBINDERY

CHARLESTOWN, MASSACHUSETTS

DESIGNED BY

MARK ARGETSINGER

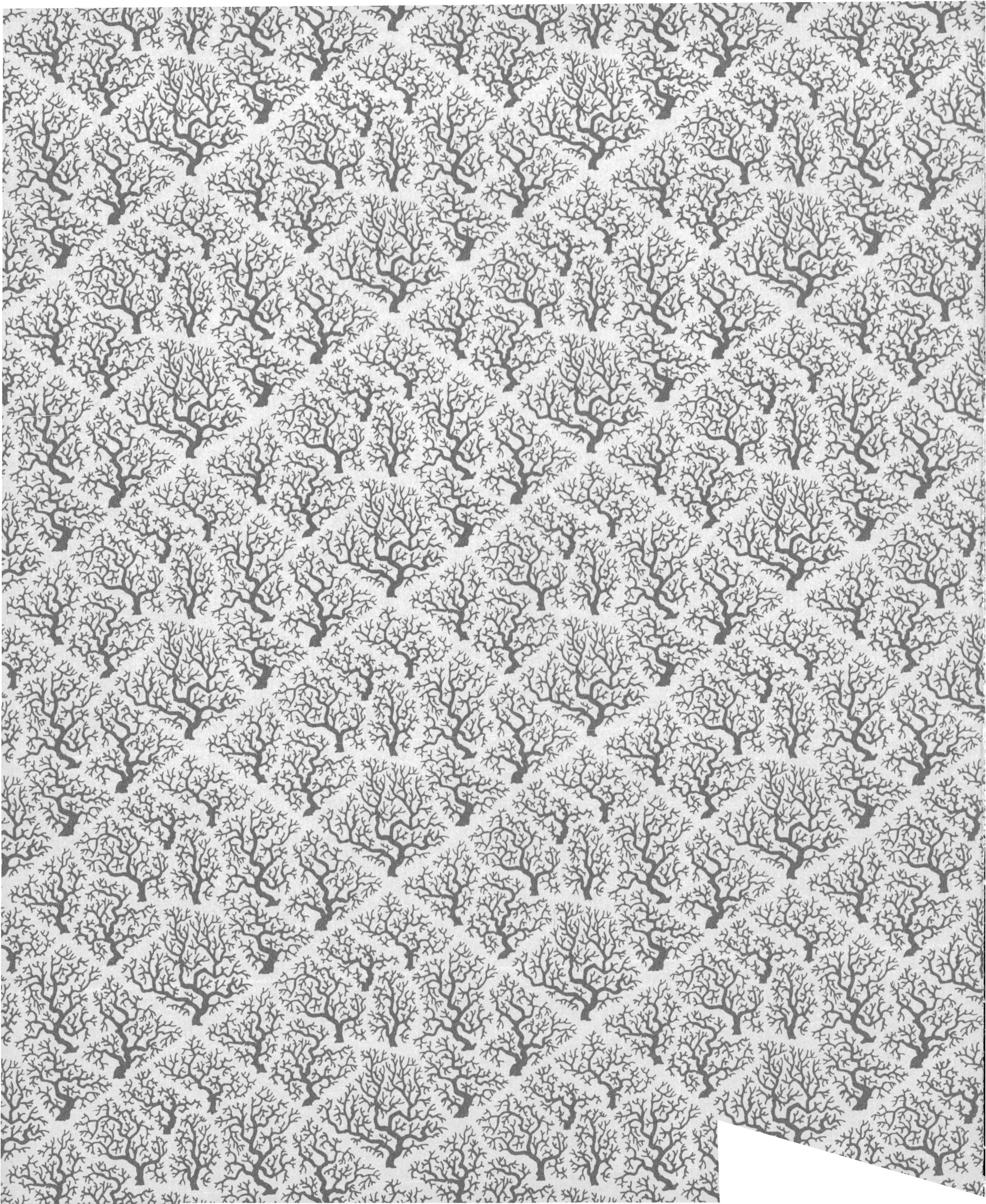